The Flora of Renfrewshire

The Flora of Renfrewshire

Keith Watson

With contributions by Simon Cuthbert and Michael Jarvis

Glasgow museums

piscespublications

First published in 2013 by Glasgow Museums and Pisces Publications.
www.glasgowmuseums.com
www.naturebureau.co.uk

ISBN 978-1874357-54-4

Designed by Pisces Publications
Edited by Kim Teo
Figures 1 and 4 by John Westwell

Colour photography:
© Simon Cuthbert: plate 1, lower left and right.
© David Fotheringham, Blue Leaf Nature: plate 3, upper; plate 4, upper; plate 10, upper right and lower left; plate 18, lower; plate 21, upper left; plate 24, upper right.
© David Palmar www.photoscot.co.uk: plate 18, upper left.
© Science Photo Library: plate 1, upper (with border overlay drawn by Keith Watson).
© T Norman Tait: front cover, lower left, centre and right; plate 4, lower; plate 7, all; plate 8, upper right and lower; plate 9, upper left and lower; plate 10, upper left; plate 11, all; plate 12, lower and inset; plate 13, upper right and lower left; plate 15, upper left (both); plate 17, lower; plate 21, upper right and lower; plate 22, lower; plate 23, lower left.
© Keith Watson: plate 3, lower; plate 5, all; plate 6, all; plate 8, upper left; plate 9, upper right; plate 10, lower right; plate 12, upper; plate 13, upper left and lower right; plate 14, all; plate 15, upper right and lower; plate 16, all; plate 17, upper left and right; plate 18, upper right; plate 19, all; plate 20, all; plate 22 upper; plate 23, upper left and right and lower right; plate 24, upper left and lower left and right.
© Gavin Young for Clyde Muirshiel Regional Park: plate 2, both.

Printed and bound in Britain by Gomer Press Ltd.

Supported by the Glasgow Natural History Society, the Botanical Society of the British Isles, East Renfrewshire Council, Renfrewshire Council, the Paisley Natural History Society and the University of the West of Scotland.

Contents

Forewords

This excellent publication is the first comprehensive flora devoted to the Vice-county of Renfrewshire (VC76). It is the result of almost 20 years of research and recording at the 1km-square level, and as such it is much more detailed than most floras. The background information, including chapters relating to geology, soils, climate, landscape and habitats, is comprehensive and will add to the reader's appreciation of the local area. The historical aspects of plant recording in the vice-county are also covered very fully, with due acknowledgement. From the maps within the Catalogue of Species one can see at a glance a plant's distribution – so much more informative than using only descriptive adjectives regarding frequency, or relying solely on the use of place names which may not be immediately meaningful to all readers.

The flora is the work of a professional botanist and the systematic, scientific and analytical input will make it an invaluable reference work for policy makers and those involved in environmental conservation. However, I am sure that amateur field botanists and others with an interest in the countryside will also find the book engrossing and derive much enjoyment from it. *The Flora of Renfrewshire* is a tribute to the skill and dedication of the author, who is to be congratulated on a significant achievement.

Peter Macpherson FRCP, FRCR, FLS
Past-presidencies: The Botanical Society of the British Isles and the Glasgow Natural History Society

As Glasgow Museums' Curator of Botany, Keith Watson has had regular access to a unique combination of resources housed at Glasgow Museums Resource Centre: the three principal herbaria for the local area – those of Glasgow Museums, the University of Strathclyde and the University of Glasgow, which all contain specimens of Renfrewshire plants, including many early records – together with Glasgow Museums' library and archives, which provide access to local publications and other relevant material. These resources, which by appointment are freely available to everyone, have been much consulted for *The Flora of Renfrewshire*; the verifiable plant records of the herbaria and literature sources have complemented the extensive fieldwork the book has drawn on – including 81,000 field records personally noted by the author and more than 60,000 records contributed by members of the Glasgow Natural History Society, Paisley Natural History Society and many other past and present local botanists.

The flora will be a standard reference book for many decades to come and represents a substantial body of data that will underpin future floristic studies of west central Scotland. Glasgow Museums is therefore delighted to have contributed to *The Flora of Renfrewshire* and to be co-publishing it with Pisces Publications; we are very grateful for the generous support that has made this publication possible, from the Glasgow Natural History Society, the Botanical Society of the British Isles, East Renfrewshire Council, Renfrewshire Council, the Paisley Natural History Society and the University of the West of Scotland.

Richard Sutcliffe, Research Manager (Natural Sciences), Glasgow Museums

List of Plates

List of Figures

List of Tables

Acknowledgements

The Flora of Renfrewshire would not represent such a comprehensive account of the local flora without the invaluable contributions of a number of past and present botanists. Many of these are highlighted in Chapter 1.8. Special thanks are due to several local botanists who have been active in the last 30 years or so and have made their records freely available: notably the late Allan Stirling, Dr Peter Macpherson and Dr Alan Silverside, and also Ian Green, who contributed a number of new local first records, and John Day and Clive Dixon, who provided useful modern records generated from, respectively, loch and grassland surveys in the 1990s. The many records compiled by previous Botanical Society of the British Isles vice-county recorders and members have greatly enhanced the depth and coverage of the data generated during the 1950s–80s period: of particular note are the late Robert Mackechnie, Elizabeth Conacher and Basil Ribbons, and their many helpers (see p. 49). The accounts of several critical groups and species have been made far more reliable and comprehensive thanks to the comments and records from several specialists, particularly David McCosh (hawkweeds), George Ballantyne (brambles), Dr Alan Silverside (eyebrights, monkeyflowers and brideworts), Jeannette Fryer (cotoneasters) and Bert Reid (dandelions).

The accounts of geology and soils (Chapter 1.3) have been generously provided by, respectively, Dr Simon J Cuthbert of the University of the West Scotland and Dr Michael C Jarvis of the University of Glasgow. The geology text and maps are based on the account previously used in the Renfrewshire Local Biodiversity Action Plan and reproduced by permission from Renfrewshire Council. The local climate data provided by John Pressly of the Coats Observatory at Paisley Museum has added greater local depth to the climate statistics reproduced from the Meteorological Office. The colour illustrations have been greatly enhanced by kind donations from local photographers: T Norman Tait, David Fotheringham, David Palmar and Gavin Young (see p. iv for details). Norman Tait also provided additional help and advice with the preparation of the images for the publication. Richard Weddle of Glasgow Museums' Biological Record Centre provided local botanical records and extracted the vice-county records from the database of *The Changing Flora of Glasgow*.

The author would like to thank the many other people who have provided assistance or advice, including: Nicola McIntyre of Paisley Museum and Valerie Boa of the McLean Museum, Greenock, for allowing access to their herbarium collections; Alan Brown of Clyde Muirshiel Regional Park for local plant records and arranging permission to reproduce aerial photographs; Susan Ramsay of the University of Glasgow, for assistance on the archaeobotany chapter; Francesca Pandolfi of Glasgow City Council for helping with GIS mapping of the vice-county boundary; and Alan Turner, also of Glasgow City Council, for providing population statistics.

David Pearman and Jim McIntosh of the Botanical Society of the British Isles have provided valuable support and encouragement throughout the development of the flora, and the author is also grateful for the support from Jenny Gough, formerly of Renfrewshire Council, Julie Nichol of East Renfrewshire Council and Paul Tatner of the University of the West of Scotland. A number of stimulating discussions on the natural heritage of Renfrewshire have been held with Iain Gibson and Dave Mellor, and the latter provided enjoyable company on a number of excursions into the Renfrewshire countryside.

The BioBase database developed by Alan Thurner has proved to be a user-friendly database for the botanical records, and Paul Griffiths of Adit Ltd has provided more recent BioBase support. The distribution maps have been produced using Alan Morton's computerized mapping system DMAP and the base map was created by the author using Alan's digitizing software.

The publication would not have been possible without the generous financial support provided by several organizations. The Glasgow Natural History Society has once again helped to ensure a local botanical publication has been published through essential sponsorship from the Professor Blodwen Lloyd Binns Bequest Fund, incorporating a very recent and kind donation from Professor Jim Dickson. Further sponsorship was provided by the Botanical Society of the British Isles, East Renfrewshire Council, Renfrewshire Council, the Paisley Natural History Society and the University of the West of Scotland.

The author is grateful to Dr Peter Macpherson for kindly reading through the manuscript and making a number of useful comments, and for providing a foreword to the flora.

And finally, special thanks to Sue and Peri for support and patience throughout the compilation of this flora.

PART 1

Background

1.1 Introduction

This flora covers a region equivalent to the historical county of Renfrewshire – a region that comprises the current local authority areas of Renfrewshire, East Renfrewshire and Inverclyde, plus some land from the eastern edge of the old county that is now part of the City of Glasgow – and throughout the publication 'Renfrewshire' should be understood to refer to this area.

Renfrewshire is located on the western fringe of the Glasgow conurbation, between the River Clyde and the upland borders of Lanarkshire and Ayrshire, with a maritime coastline at the Firth of Clyde. It is a small county but has a varied topography and is strongly influenced by the north-westerly Atlantic climate (plate 1).

The flora follows on from the surveying of the Glasgow area that was undertaken mainly in the mid 1980s and led to the publication *The Changing Flora of Glasgow* (Dickson, Macpherson and Watson 2000). This surveying included the study of the parts of north- and south-west Glasgow which were once within the old county of Renfrewshire – work which *The Flora of Renfrewshire* has continued, whilst primarily focussing on the rest of the county.

The aim of the flora has been to list all the plants – native or alien, past or present – recorded as growing wild in Renfrewshire. The records have been derived principally from field record cards, but literature sources, herbarium specimens and various file notes have all made important contributions. The plant records presented in the Catalogue of Species (Part 2) have been generated from the author's database, which consists primarily of field card records dating from 1993 onwards and a number of slightly older records generated between the late 1950s and 1986 by previous Botanical Society of the British Isles recorders for Renfrewshire. The post-1993 distribution data is supplemented by a number of relevant records, mainly dating from 1983 to 1993, generated during *The Changing Flora of Glasgow* survey work, held by Glasgow Museums' Biological Records Centre (see Chapter 1.8, p. 47).

More than 143,000 records have been used in the compilation of the individual species accounts and the accompanying maps. In total 1533 species – including microspecies, infraspecific taxa and hybrids – are documented in the flora.

Historically, Renfrewshire has never been considered a remarkable county on the strength of its floristic diversity, but it is hoped that this publication will help to redress this perception. Writing back in the nineteenth century, the local naturalist John Wood (1893) was aware that the county did contain botanical treasures: 'Can anything rare or beautiful find a home in dirty, smoky, wet and muddy Renfrewshire? Come and see. Let me take you with me in imagination out from your mighty city's noise and dust and bustle, out into the pleasant fields and along the green burn-braes (for such things are) of my adopted county.'

1.2 The Vice-county of Renfrewshire

The study area of this flora is described as 'Renfrewshire', referring to the old county boundary (see Chapter 1.1), but it is also described as a botanical recording unit – the Vice-county of Renfrewshire, abbreviated to VC76 in the Catalogue of Species. The Vice-county of Renfrewshire differs slightly from the old political county of Renfrewshire in that it includes a small section which is now part of the local authority of Ayrshire, near Lugton, which has not been surveyed for this flora, but for convenience in this flora the terms 'Renfrewshire' and 'Vice-county of Renfrewshire' are used synonymously.

Botanical vice-counties

The vice-county system dates from 1852 when HC Watson (1804–1881) divided Great Britain into 112 divisions; based on existing political counties, or parts of these where large, these areas eventually became known as vice-counties and were fully utilized in Watson's *Topographical Botany*, published in 1873. The vice-county system represents a stable basis for botanical recording as it has remained unchanging since the nineteenth century, unlike political boundaries, which are subject to frequent reorganization.

Renfrewshire is one of the smaller vice-counties, covering some 655 square kilometres (*c.* 253 square miles) of land. The whole or parts of 17 hectads (10km squares), or 211 tetrads (2km squares), or 753 monads (1km squares) comprise the vice-county recording area (Biological Record Centre – www.brc.ac.uk). Hall (1915) described the old political county of Renfrewshire as an area of 244 square miles, out of a total of 30,408 square miles for Scotland as a whole, and as being the 27th in size of the then 33 counties in Scotland.

Renfrewshire has an elongated elliptical shape, lying roughly north-west to south-east, with its longest axis being some 50km (31 miles) and greatest width about 21km (13 miles). It is situated at 55° 51' N and 4° 21' W, and occurs entirely within the Ordnance Survey 100km National Grid Square NS (26). The boundary is shown in figure 1, which also includes several key place names; further information on place names and their locations is provided in the Gazetteer.

The River Clyde forms most of Renfrewshire's long northern border, and becomes increasingly estuarine until it merges with the broad Firth of Clyde at Renfrewshire's western coastline opposite Argyll (VC98). The southern boundary is formed mostly of upland ground shared with the Vice-county of Ayrshire (VC75). From the coast at Wemyss Bay the boundary follows the Kelly Burn up to the Renfrewshire Heights, from where it follows the North Rotten Burn to the highest ground of the Misty Law watershed. The boundary then descends, via the Maich Water, to cross a broad low-lying area, sometimes known as 'the Paisley Ruck' (a geological fault line), at the southern end of Barr Loch, before climbing up the Roebank Burn just north of Beith (in VC75). From here the route roughly follows the Ayrshire/Renfrewshire watershed and historical county boundary towards the high ground of the moorland above Eaglesham.

The eastern boundary is shared with the Vice-county of Lanarkshire (VC77) and follows the White Cart Water and its upland tributaries into the lowland Glasgow conurbation, where the boundary then follows a rather obscure route that eventually reaches the Clyde near Renfrew: this route leaves the White Cart Water at Linn Park and heads north towards Polmadie, and then west through Langside and Pollok Country Park, before heading north to Braehead, where an even more curious diversion takes it north of the Clyde to Scotstoun and Jordanhill, adjacent to the Vice-county of Dunbartonshire (VC99).

Figure 1. The old county of Renfrewshire, on which the botanical Vice-county of Renfrewshire (VC76) was based. The boundary encompasses the current local authority areas of Inverclyde, Renfrewshire, East Renfrewshire and also parts of Glasgow. The conurbations of Glasgow, Paisley and Inverclyde and other built-up areas are indicated with dark shading; larger waterbodies are represented with bordered light shading and main watercourses are shown.

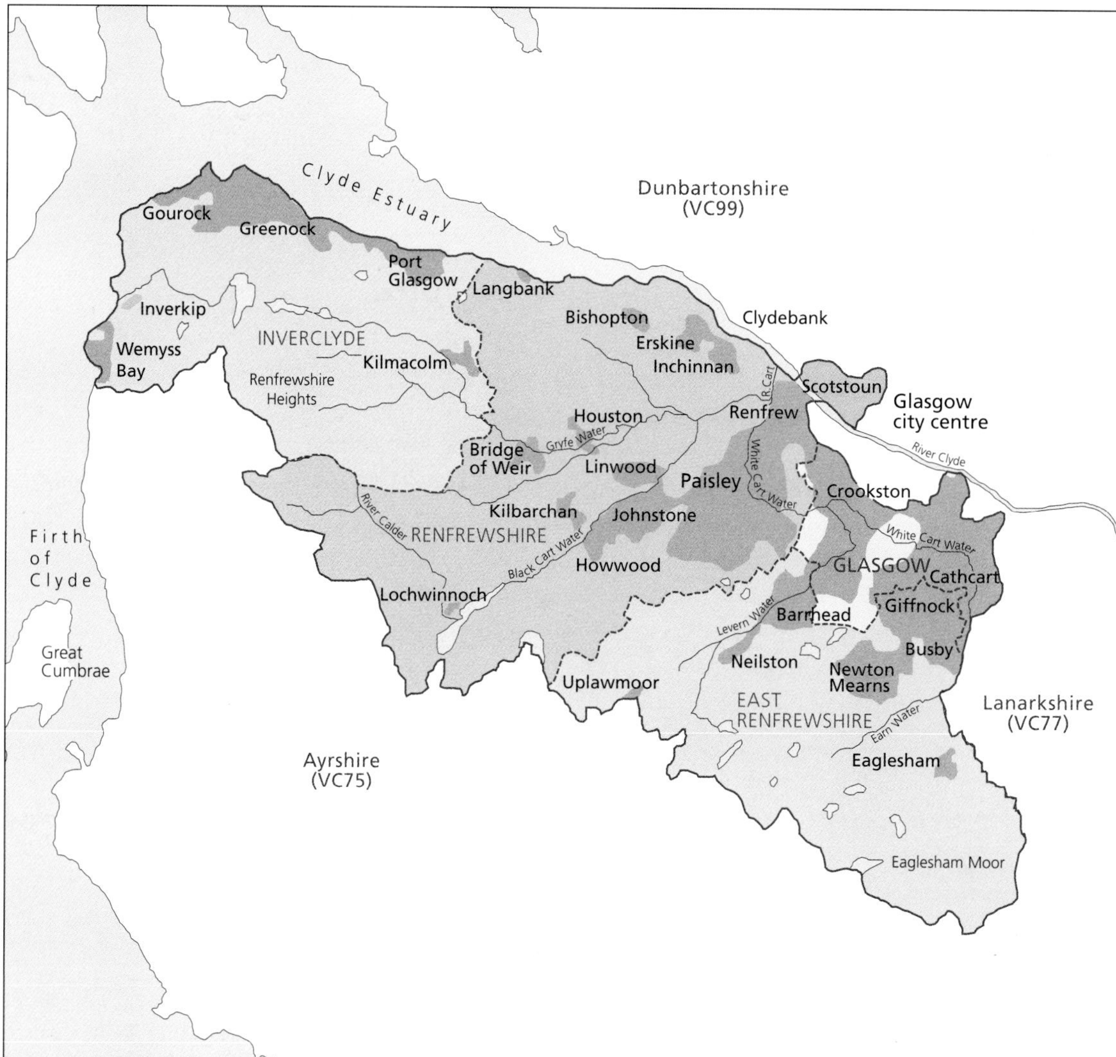

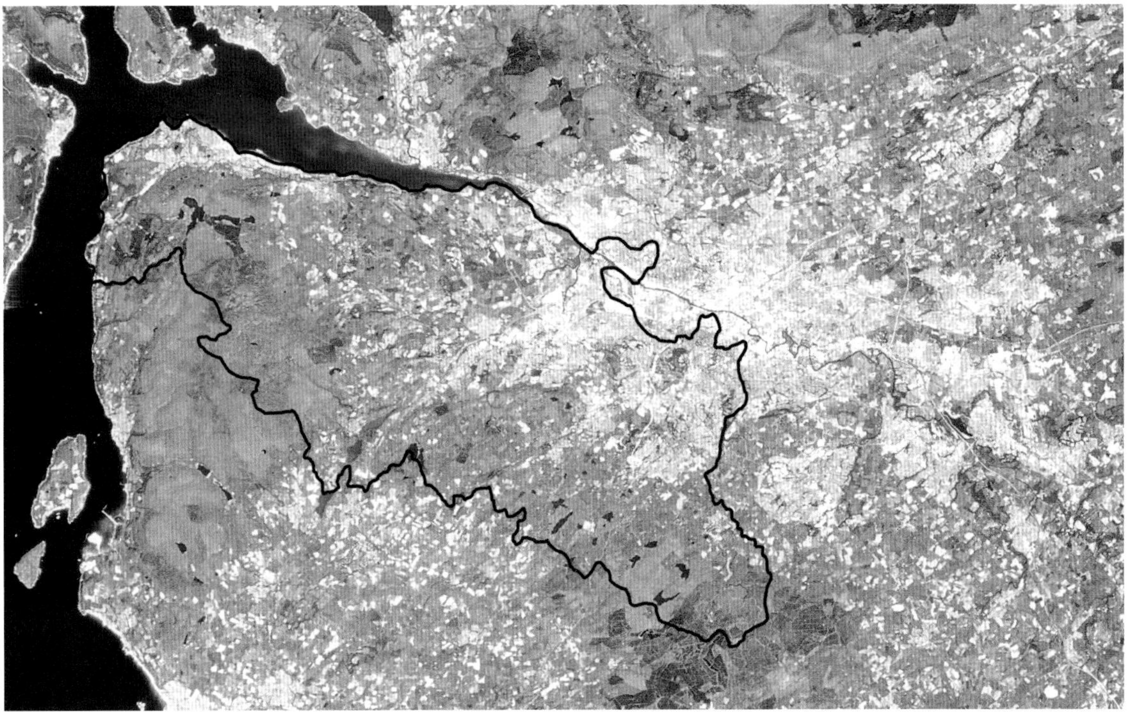

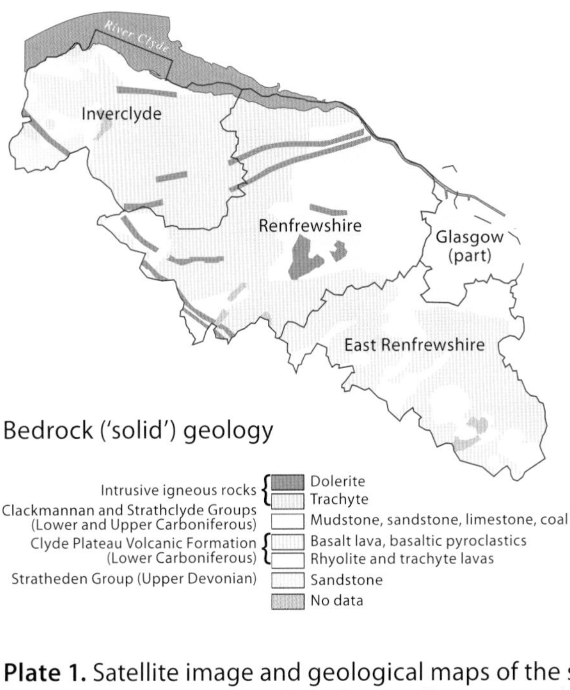

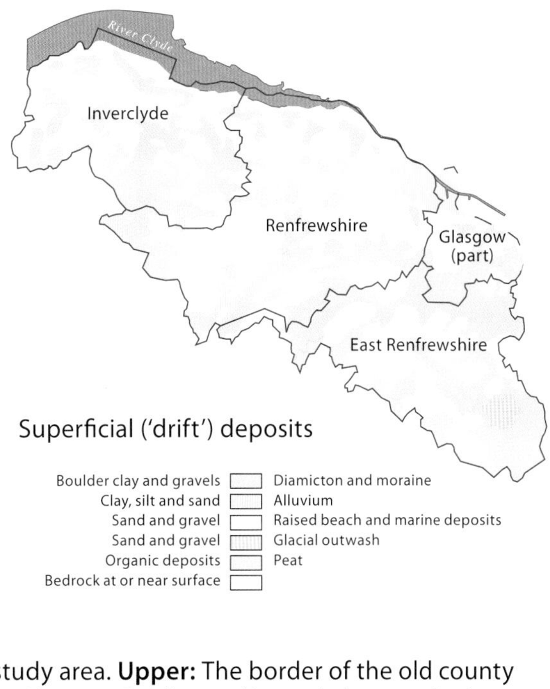

Plate 1. Satellite image and geological maps of the study area. **Upper:** The border of the old county of Renfrewshire is overlaid on this satellite image; the Glasgow, Paisley and Inverclyde conurbations show as grey, brown indicates the Renfrewshire Heights in the west and Eaglesham Moor in the south-east, and the Firth of Clyde is visible to the west. **Lower left:** Bedrock ('solid') geology of Renfrewshire. **Lower right:** Superficial ('drift') deposits of Renfrewshire.

Plate 2. Aerial photographs. **Upper:** Looking north to the Clyde Estuary, with Loch Thom in the foreground and Shielhill Glen in the centre left. **Lower:** Looking east across Barr and Castle Semple lochs, with Lochwinnoch to the centre left.

Plate 3. Landscape views. **Upper:** The view eastwards towards Glasgow from Barscube Hill, near Bishopton, with the Clyde Estuary in the north.
Lower: The Paisley urban fringe to the north of the Gleniffer Braes.

Plate 4. Upland scenes in the Renfrewshire Heights. **Upper:** Upland vegetation at Craig Minnan. **Lower:** The River Calder with the summit of Misty Law in the distance.

Plate 5. Raith Burn and upland plants in the Renfrewshire Heights. **Upper left:** Alpine Clubmoss at Hannah Law. **Upper right:** Parsley Fern near the Raith Burn. **Lower:** The Raith Burn, with Queenside Hill behind.

Plate 6. Moorland landscape and vegetation. **Upper left:** Lesser Twayblade on Duchal Moor. **Upper right:** Few-flowered Sedge at Muirshiel. **Lower:** Duchal Moor – blanket mire with eroding peat hags.

Plate 7. Glen Moss plants and landscape. **Upper left:** Coralroot Orchid. **Upper right:** Tufted Loosestrife. **Lower:** Glen Moss SSSI, Kilmacolm.

Plate 8. Dargavel Burn plants and aerial view. **Upper left:** Narrow Buckler-fern. **Upper right:** Bog-sedge. **Lower:** Species-rich mire and carr at Dargavel Burn SSSI, near Kilmacolm.

Plate 9. Bog and mire plants. **Upper left:** Cranberry at Shovelboard SSSI. **Upper right:** Tall Bog-sedge at a moss at Black Hill. **Lower:** Common Cottongrass in profusion at Glen Moss SSSI.

Plate 10. Wetland plants. **Upper left:** Bogbean at Glen Moss. **Upper right:** Round-leaved Crowfoot at Barscube Hill. **Lower left:** Nodding Bur-marigold at Barscube Hill. **Lower right:** Greater Spearwort at Caplaw Dam.

Plate 11. Plants by the River Cart at Blythswood. **Upper:** Hoary Cress on the east bank of River Cart with the confluence of the White Cart and Black Cart waters in the background. **Lower and inset:** Holy Grass on the east bank of the River Cart.

Plate 12. Coastal and estuarine habitats. **Upper:** The rocky coastline of the Firth of Clyde at Cloch lighthouse with Thrift in the foreground. **Lower and inset:** Common Reed by the estuarine Clyde, with the Erskine Bridge above.

Plate 13. Grassland plants. **Upper left:** Heath Milkwort and Mouse-ear Hawkweed on a rocky bank, Barmore Hill. **Upper right:** Mountain Pansy above Knapps Loch. **Lower left:** Spignel growing outside a field wall near Bridge of Weir. **Lower right:** Common Stork's-bill, with Little Mouse-ear in the foreground, at Erskine.

Plate 14. Local rarities of species-rich grassland. **Upper left:** Field Gentian at Commore Dam. **Upper right:** Frog Orchid at Commore Dam. **Lower left:** Mountain Everlasting at Commore Dam. **Lower right:** Moonwort near Knockindon Burn.

Plate 15. Flushed grassland at Cauldside, Strathgryfe. **Upper left:** Whorled Caraway inflorescence with (image below) a section of stem and leaf. **Upper right:** Heath Spotted-orchid. **Lower:** Flushed grassland with abundant Whorled Caraway and orchids.

Plate 16. Orchids. **Upper left:** Heath Fragrant-orchid × Heath Spotted-orchid at Cauldside. **Upper right:** Heath Fragrant-orchid × Northern Marsh-orchid at Greenside. **Lower:** Heath Fragrant-orchid in species-rich grassland at Whinnerston.

Plate 17. Plants of rocky outcrops. **Upper left:** Rock outcrops at Walls Hill with Bell Heather and Wood Sage prominent. **Upper right:** Slender St John's-wort by the Green Water, near Hillside. **Lower:** English Stonecrop, a feature of exposed rock outcrops in the west of the vice-county.

Plate 18. Woodland plants. **Upper left:** Bird's-nest Orchid at Castle Semple Loch. **Upper right:** Toothwort at Linn Park. **Lower:** Oak woodland at Craigmarloch with Bracken and Bluebell.

Plate 19. Upland woodland species. **Upper left:** Aspen by the North Rotten Burn. **Upper right:** Wood Crane's-bill by the Raith Burn. **Lower left:** Northern Bedstraw by the Gryfe Water. **Lower right:** Globeflower at Harelaw Reservoir.

Plate 20. Plants of open places. **Upper left:** Purple Ramping-fumitory at Malletsheugh. **Upper right:** Common Broomrape at Port Glasgow. **Lower left:** Barberry in a hedgerow at Bishopton. **Lower right:** Pale × Common Toadflax hybrid, Greenock.

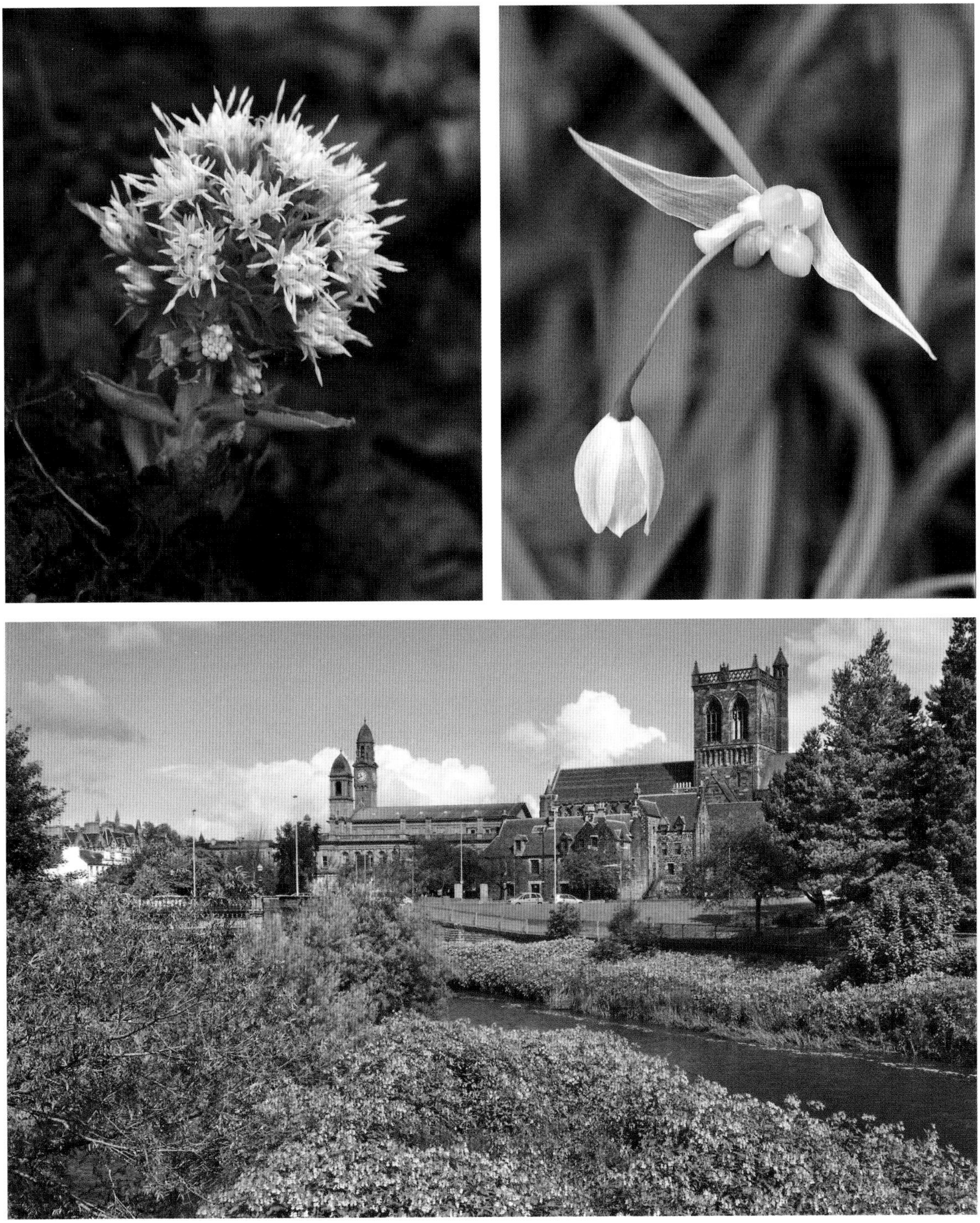

Plate 21. Alien plants. **Upper left:** White Butterbur at Finlaystone. **Upper right:** Few-flowered Garlic. **Lower:** The White Cart Water near Paisley Abbey, with banks dominated by Indian Balsam.

Plate 22. Alien plants growing by the Gryfe Water. **Upper:** Hybrid Monkeyflower at Cauldside. **Lower:** Purple Toothwort at Kilmacolm.

Plate 23. Alien plants. **Upper left:** American Blue-eyed-grass at Black Hill. **Upper right:** Lamb's-tail at Lochwinnoch. **Lower left:** Photo montage of a specimen of Grass Vetchling from Inchinnan. **Lower right:** Greater Cuckooflower at Formakin Estate.

Plate 24. Changing land use and conservation. **Upper left:** Developing wetland at the drained Midton Loch, Whittliemuir, viewed from Walls Hill. **Upper right:** Former variously improved acid grassland now with new plantation woodland at Knockmountain. **Lower left:** Bog-myrtle population devastated during pipeline work at Gall Moss, near Loch Thom. **Lower right:** Roadside verge with orchids near Woodhall, following cessation of amenity mowing.

1.3 Geology and Soils

Geology

Dr Simon J Cuthbert

The geology of the Vice-County of Renfrewshire is described in the publications of the British Geological Survey (BGS), especially British Geological Survey (1985, 1989, 1990, 1993, 1994), Hall (1998) and Patterson (1990); up-to-date geological data may be found at the BGS website www.bgs.ac.uk. The bedrock can be divided into two main types, closely related to the topography (plate 1). Low ground in the eastern part of the vice-county is underlain by mainly sedimentary rocks of the Strathclyde and Clackmannan Groups of early to mid Carboniferous age (354–316 million years BP). These relatively soft, easily eroded rocks are mainly claystones (shales), sandstones, coal seams and thin but regionally persistent limestones. They will tend to produce relatively fertile soils, which are likely to become quite alkaline on the limestones. These are locally intruded by sills of dolerite, a base-rich igneous rock similar to basalt. These sills produce distinctive crags and waterfalls where exposed. Dolerite, limestone, sandstone, coal and associated ironstones have all been economically exploited at various times.

A narrow strip of red sandstones, siltstones and limestones of earliest Carboniferous or latest Devonian age (*c.* 360 million years BP) extends along the coastal strip from Langbank through Port Glasgow and Greenock to Gourock. The characteristic limey 'cementstones' of this formation tend to give a higher pH to the soils, making them alkaline in content. These rocks reappear in another coastal strip from Lunderston Bay to Wemyss Bay, along with upper old red sandstones and conglomerates of the Stratheden Group. The latter extend eastwards from Inverkip up into the high ground around Loch Thom.

High ground, including Muirshiel between Houston and Loch Thom, the Gleniffer and Brownside Braes, and eastwards to Eaglesham, is underlain by more resistant volcanic igneous rocks of the Clyde Plateau Volcanic Formation of the earliest Carboniferous age (*c.* 350 million years BP). These mainly basaltic rocks dominate the moorland and steeper slopes of the western and southernmost parts of the county. While potentially rich in nutrients, basalt is somewhat impermeable and thus poorly drained, and when combined with the high rainfall this tends to produce rather waterlogged soils with large areas of blanket peat cover. The most acid (silicic) and resistant rocks are trachytes, rhyolites and volcanic agglomerates (volcanic vent deposits) and these form the highest ground along the watershed from Ballageich Hill to Corse Hill, south of Eaglesham.

A linear belt of shattered rock known as 'the Paisley Ruck' extends from Renfrew to Lochwinnoch. This reflects an area of ancient activity along a major fault, and the consequent weakening and preferential erosion of the rock has resulted in the formation of the Black Cart Water valley and Castle Semple Loch.

Superficial deposits

The cover of superficial deposits or 'drift' over the bedrock is the result of the action of glaciers, seas and rivers from ancient times (plate 1). The lower, flatter ground from the northern parts of Paisley and Johnstone, plus Renfrew, Glasgow Airport and the area up to Bishopton, are underlain by glaciomarine silts and clays, with local patches of sand and gravel around Erskine and Bishopton. These were deposited in the shallow seas as the glaciers retreated.

Further south the ground becomes increasingly hummocky, with low rounded hills known as 'drumlins' characteristic of south Paisley and Johnstone and north of the Gleniffer Braes. These are tills or 'boulder clays' – stiff silts and clays with abundant embedded cobbles and boulders deposited and then moulded by glacial ice. Moving south on to the braes, and westwards on to Muirshiel and the Loch Thom area, these isolated hummocks of till merge into a more continuous cover, but are thin or absent on the highest ground, where the bedrock is close to the surface. Tills are overlain by blanket peat deposits extensively in the uplands but also in broad hollows such as at Caplaw Dam. Extensive areas of peat are also found on the poorly drained lowland glaciomarine clays at Linwoood Moss and Barochan Moss.

In the eastern part of the vice-county the River Clyde, White Cart Water and Black Cart Water have deposited linear belts of sand and gravel. West of Renfrew, extending to Gourock and around Lunderston Bay, a narrow strip of the coast is marked by one or more flat terraces of sand and gravel that are former beaches, now raised above sea level by post-glacial uplift and known as raised beaches. These provide level ground suitable for transport links such as the railway and M8 motorway. The superficial deposits tend to dominate the soil and drainage characteristics, so that the influence of bedrock will only be seen where it is close to the surface, or has been uncovered by mining or quarrying activities.

Soils

Dr Michael C Jarvis

There is a close relationship between soil types and geology. Indeed the soil is the link between geology and vegetation. Scottish soils have been classified and mapped by the Soil Survey of Scotland, but they published soil maps covering the old county of Renfrewshire at only the relatively small scale of 1:250,000 (Bown et al. 1982). For neighbouring areas of Ayrshire and North Lanarkshire there are much more detailed soil maps at 1:63,360 scale, showing the distribution of the most detailed soil classification units, soil series (Mitchell and Jarvis 1956). This level of detail is necessary before the correspondence between soil type and vegetation becomes evident. *The Changing Flora of Glasgow* (Dickson et al. 2000) includes a map of the soil series within that survey area ('the Glasgow Rectangle'), including part of Renfrewshire, and some discussion of their relationship to the distribution of plant species. The soil series represented there include most of the series found elsewhere in the Vice-county of Renfrewshire, with the exception of those at higher altitude and the red soils of the upper old red sandstone area towards the Clyde Coast.

A group of soil series formed on one geological parent material, but differing in drainage and other factors, is called a soil association. Soil associations are shown on the published 1:250,000 map covering Renfrewshire and their relationship to the underlying geology is clearly visible at this scale (Bown et al. 1982).

The majority of Renfrewshire's agricultural soils have been formed on rather fine-grained glacial till formed from Carboniferous shales, sandstones and limestones of the Strathclyde and Clackmannan Groups. The contribution from the sandstones is too slight to permit free drainage. These soils belong to the Rowanhill Association, except where there is a significant contribution of basaltic till and the soils are classified within the Kilmarnock and Darleith Associations. Together the Rowanhill and Kilmarnock Associations make up the best of the dairy-farming land in Renfrewshire, as in Ayrshire. They are too wet in the autumn and early spring for the well-timed cultivation needed for crop farming, but their water-holding capacity is a great advantage for grass yields and sustains the growth of grass in dry summers.

Permanent grass on dairy farms is not noted for its biodiversity. Although the Carboniferous sedimentary rocks are quite diverse, the glacial till overlying them is generally well mixed and rather uniform. It is only in places where burns have cut through the till that rock is exposed or thin soils have developed directly upon it and reflect its local variations. Small glens of this kind are in many cases wooded, and the older woods of this type show interesting local variation in their ground flora according to the nature of the rock involved, with very wet gley soils overlying shales and drier brown earths overlying sandstone. It might be expected that distinctive alkaline soils would form on limestone in these situations, but good examples are hard to find.

In the north of the vice-county, close to the Clyde and in the small area of the old county of Renfrewshire over the river, the soil types depend less on the parent material and more on the history of disturbance by water. Fierce meltwater rivers at the end of the last glaciation followed the courses of the Kelvin and Clyde, leaving sandy deposits scattered by their banks and coating the surface of the till on the nearby drumlins and terraces. This complex topography is described in *The Changing Flora of Glasgow*. Later the flat land south of the Clyde was covered by a shallow sea that left behind the heavy soils known as carse clays.

The carse clays of the flat land under Glasgow Airport and other areas from Mosspark, a southern suburb of Glasgow, to Bishopton belong to the Stirling Association. These carse lands were probably once entirely covered by peat mosses, raised bogs like the surviving Flanders Moss in the Carse of Stirling. Only fragments of these mosses remain, such as Linwood Moss, overgrown with birch and degraded by draining. The heavy clay soils revealed by the removal of the peat, probably for fuel in the medieval period or later, have been farmed wherever the Black Cart or Gryfe waters flow close enough for successful drainage, but much of the area is now covered by industry or housing. It is still vulnerable to flooding, as some of the occupants have found.

There were once similar flat peat mosses occupying flat valley bottoms at higher altitude at either end of Loch Libo near Uplawmoor, and between the lochs of Kilbirnie and Castle Semple. Again, almost all the peat has been removed at an unrecorded early date, with evidence surviving only in place names such as Mossend, near Lochwinnoch Railway Station. Here the soil exposed by removal of the peat is alluvial and locally variable in composition and clay content. Its propensity to flooding has prevented parts of this area from being farmed and the Lochwinnoch marshes have been a centre of biodiversity throughout recorded history.

Early Carboniferous basalt is the predominant underlying rock on the higher ground in the south and west of the vice-county, including the hills from Muirshiel northward to Greenock, the Gleniffer Braes and, in East Renfrewshire, the ground rising towards the Ayrshire border. Less fertile and stonier than the Rowanhill Association soils, and sometimes better drained but mainly on higher and steeper ground, they support a mix of improved and unimproved grassland, patches of woodland, and bogs where silt and clay have accumulated in hollows. Where the underlying till is finer in texture, as on the top of the Gleniffer Braes or on the high ground south of the Gryfe Valley, the grassland is particularly poor and rushy. This part of the county has a long history as cattle- and sheep-rearing land, evident today in its intricate networks of drystone walls.

The highest ground, in the Muirshiel Hills and along the East Renfrewshire–Ayrshire border, is mainly covered in blanket peat. It has a close affinity with similar hill land in Argyll. The mineral soils under the peat are derived from basalt and might be relatively free-draining, at least in the neighbourhood of Muirshiel, if it were not for the presence of a fragipan – a densely compacted layer about 50cm below the surface formed originally under periglacial conditions. Otherwise these soils resemble the dry Darleith brown earths of the lower basaltic ground, but the fragipan impedes drainage, making them much wetter, and they tend to carry a mixture of *Juncus* and *Molinia* grassland and heather, with peat cover on the higher land. In many places below about 400m birchwood is visible growing in the mineral soil at the base of the peat, but it does not extend to the tops of the hills. Remains of old peat slides can be discerned on many of the higher ridges and saddles. In the south-east part of the vice-county, on either side of the M77, poorly drained soils with similar vegetation grade into peat as the ground rises towards the south. However, here their wetness is often due to an increasing percentage of clay rather than fragipan formation.

The hills around Misty Law and the Hill of Stake differ in being formed from rhyolites and trachytes, more acidic volcanic rocks than are found in most of Renfrewshire's basaltic uplands. Similar rocks make up the high ground above Greenock and occur in patches south of Neilston and Newton Mearns. The mineral soils found there are, therefore, more acidic than the Darleith and Baidland series of the Darleith Association, although otherwise similar. These soils and the associated peatlands support much of the heather moor that is found in Renfrewshire. The dark heather-clad areas are readily visible from the air. These, with associated acidic screes and occasional small rocky outcrops, are of central importance for land use and biodiversity. But some of the less widespread soil types of lowland Renfrewshire provide more unusual, and sometimes diverse, habitats.

Further west, on the west-facing slopes above the Clyde Estuary, the till is developed from Old Red Sandstone and gives rise to red soils of the Largs Association. On the high ground these soils appear to show some evidence of fragipan formation, are rather poorly drained with peat cover in places, and support heather moor with *Juncus* and *Molinia* grassland. However, at lower altitude and where burns have cut steep-sided glens, the red soils are more freely drained and their calcareous nature becomes more evident, with glen woods of high biodiversity and fertile pastures. In the neighbourhood of the Sheilhill Glen above Inverkip, basalt and calcareous rocks of the Upper Old Red Sandstone provide the parent material for richer and finer-textured soils of the Sorn Association, which would be poorly drained were it not for the steep slopes. The Sheilhill Glen owes its exceptional biodiversity to the combination of these mixed but fertile soils and a maritime climate. At the coast, raised beach soils formed from Old Red Sandstone or basalt are very freely drained and, where they are too steep for agricultural use, they support rich mixed woodland facing the sea.

Although there appear to be no good examples of natural limestone soils in Renfrewshire, there are shallow, immature patches of alkaline soil in a number of old limestone quarries that carry fragments of calcicolous vegetation. Also, as discussed in *The Changing Flora of Glasgow*, in the city of Glasgow a variety of crumbling buildings, derelict building sites, railway lines and other man-made habitats have become vegetated over time and have accumulated deposits that may be called 'urban soils' in the sense of Gilbert (1989). Similar deposits are common in Paisley, Greenock and the smaller post-industrial area around them. The former car plant at Linwood has returned to grassland and the large expanse of the old ordnance factory at Bishopton is now largely birchwood, although still not accessible for detailed investigation. In these unpromising places biodiversity is recovering and so, presumably, are the soils.

1.4 Climate

The climate of Renfrewshire is strongly influenced by its northern latitudinal position and its western exposure to the influence of the Atlantic. It is characteristically an oceanic climate, with generally mild temperatures, frequent cloud cover, high rainfall and strong winds, resulting in a low potential water deficit. The Atlantic climate is shared with much of western Britain and Ireland and other European countries extending from western France, across the Low Countries to western Scandinavia. The western weather pattern was noted by Ferguson (1915) in the introduction to his Renfrewshire plant list, where he wrote that with 'the prevailing westerly and south-westerly winds, the county enjoys an equable though rather moist climate'. Writing even earlier, in relation to agriculture, J Wilson (1812) remarked that 'the quantity of rain is not so much to be dreaded as its frequency'.

Weather data relevant to the area are gathered at the Meteorological Office's weather station at Paisley in the lowland, central-east part of the vice-county. Table 1 shows climate averages for Paisley for 1981–2010 (www.metoffice.gov.uk), with figures 2 and 3 providing graphic representation of, respectively, the average maximum and minimum temperatures and amounts of sunshine and rainfall.

Table 1. Climate data for Paisley 1981–2010. Source: www.metoffice.gov.uk

Month	Maximum temperature (°C)	Minimum temperature (°C)	Air frost (days)	Sunshine (hours)	Rainfall (mm)	Rainfall ≥ 1mm (days)	Wind at 10m (knots)
January	6.9	1.8	8.8	37.6	148.2	17.3	6.6
February	7.4	1.8	7.9	66.9	104.6	13.2	6.8
March	9.6	3.0	4.5	98.6	112.3	14.9	6.3
April	12.6	4.8	1.3	134.5	63.6	11.6	5.5
May	15.9	7.3	0.1	180.1	67.5	11.9	5.2
June	18.1	10.1	0.0	158.9	66.4	11.1	5.0
July	19.7	12.0	0.0	154.3	73.0	12.0	4.9
August	19.2	11.7	0.0	146.8	92.5	12.8	4.9
September	16.4	9.7	0.0	114.9	112.5	13.8	5.5
October	12.7	6.7	1.1	85.2	143.1	16.8	5.6
November	9.4	4.0	4.5	54.0	126.4	16.0	5.9
December	6.9	1.7	9.3	33.1	135.2	15.5	5.4
Year	12.9	6.2	37.3	1265.0	1245.1	167.0	5.6

The Meteorological Office data for Glasgow Airport (Anon. 1981) showed an average monthly mean temperature of 8.6°C, which is less than the averages for Ayrshire, Edinburgh and many other Scottish coastal locations. April through to October is the seven-month period when the average exceeds 6°C, commonly accepted as the threshold for plant growth (Dickson et al. 2000). However, frosts can occur in April and October, and indeed occasionally in May or even September. For the 1981–2010 period the average maximum

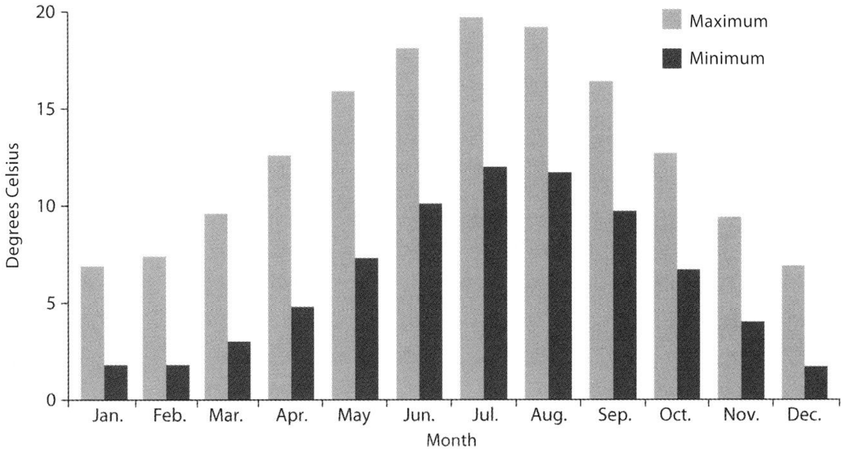

Figure 2. Average maximum and minimum temperatures in degrees Celsius for Paisley 1981–2010.
Source: www.metoffice.gov.uk

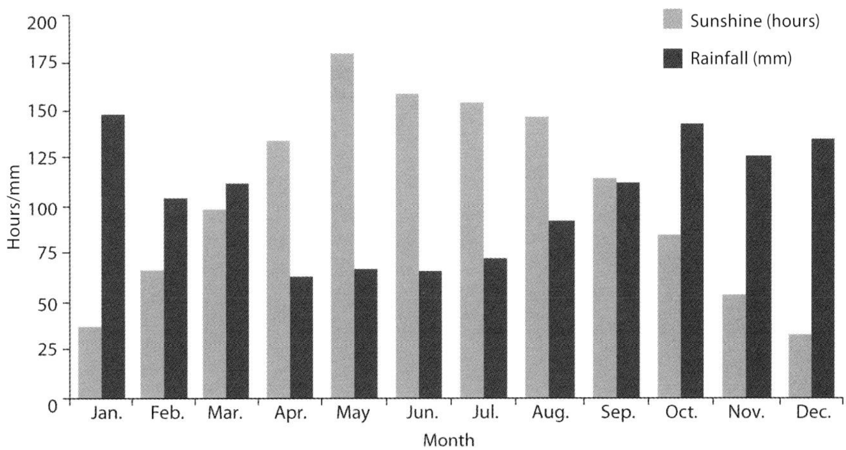

Figure 3. Average monthly sunshine in hours and rainfall in millimetres for Paisley 1981–2010.
Source: www.metoffice.gov.uk

temperature was 12.9°C and the minimum 6.2°C. In common with many parts of western Scotland, May and June tend to have the highest sunshine averages.

The yearly average rainfall from 1981 to 2010 stands at 1265mm (49in.), and there were between 11 and 17 days every month with precipitation greater than 1mm. April through to July are the driest months (63mm–70mm of precipitation a month), with August showing a marked increase. October through to January have been the wettest months (more than 125mm of precipitation a month). This persistent rainfall, combined with the low or mild temperatures, results in low chances of drought conditions inhibiting plant growth and encourages soil formations such as podsols, gleys and organic peats (see Chapter 1.3). Also of note is the east–west trend of increasing rainfall across the Clyde Valley, as shown in figure 4, which, although somewhat dated, shows the collated data from all local rain gauges for the period 1941–70 (Anon. 1981). Greenock had some 1750mm (69in.) of rain each year, whereas the highest ground of the Renfrewshire Heights received over 2400mm (94in.). Even in the relatively drier south-east, the high ground above Eaglesham received up to 1600mm (63in.) a year. This range in rainfall is likely to be a key factor in plant distribution patterns, such as those for English Stonecrop or Whorled Caraway, which show abrupt boundaries even though the underlying geology is similar throughout (see fig. 15).

Additional topical data has been kindly provided by John Pressly of the Coats Observatory at Paisley Museum, where detailed records have been kept since 1885. These data show a general trend of rising temperatures and increased rainfall (figs 5 and 6). The

Figure 4. Average rainfall across Renfrewshire in millimetres for the period 1941–70.
Adapted from *The Climate of Great Britain Climatological Memorandum 124 Glasgow and the Clyde Valley*, published by the Meteorological Office, Edinburgh, 1981.

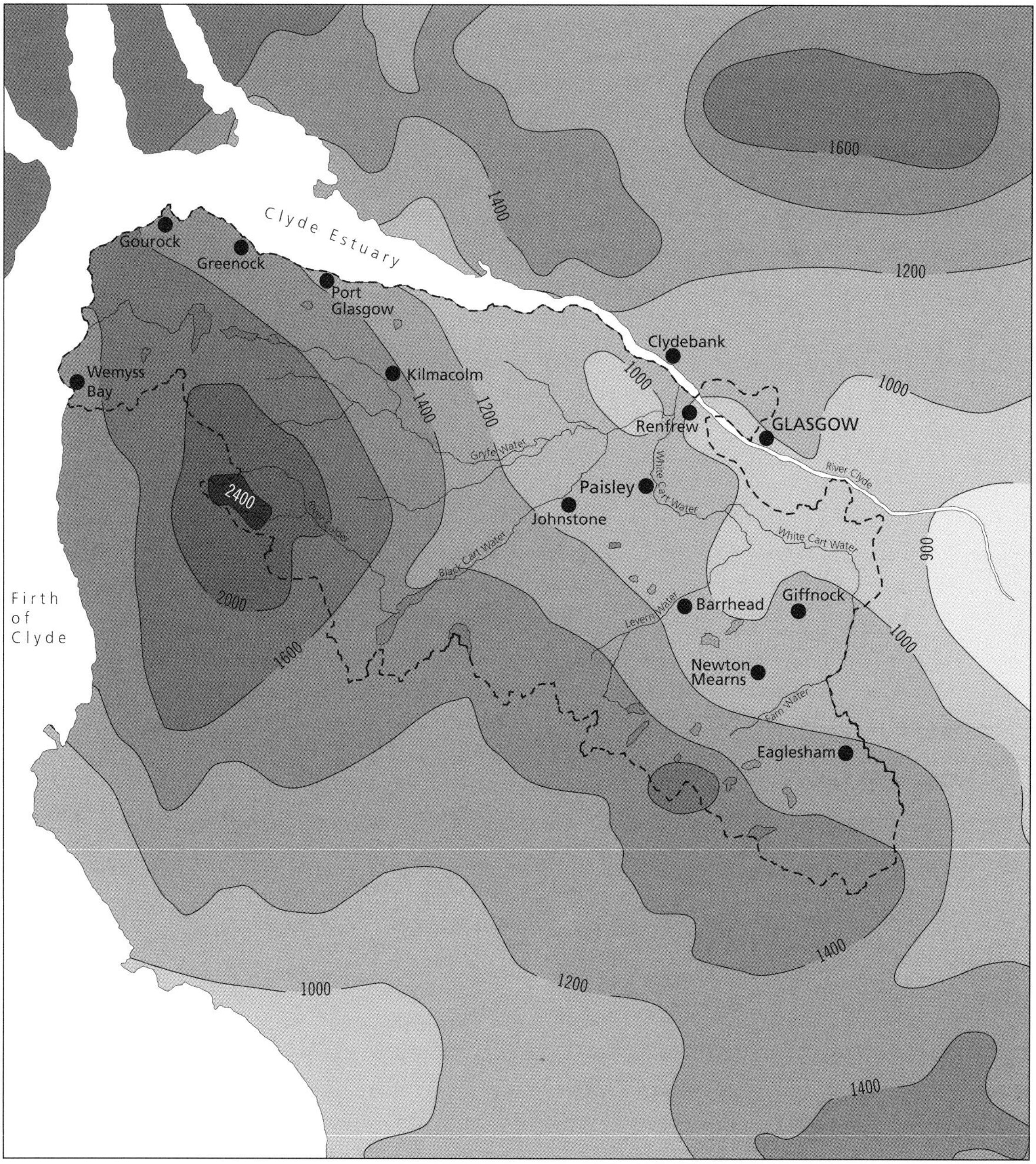

mean annual temperature rose steadily from the 1990s but since 2008 there has been a notable cooling and the rainfall pattern has been remarkably erratic. There is little evidence of any correlation between climate patterns and local floristic changes over this period, but, as suggested by much of the current thinking on climate change, such continued trends – from whatever cause – will likely affect future plant distribution. However, it is interesting to note that writing in 1901 the botanist Peter Ewing described arable crops as declining due to climate deterioration.

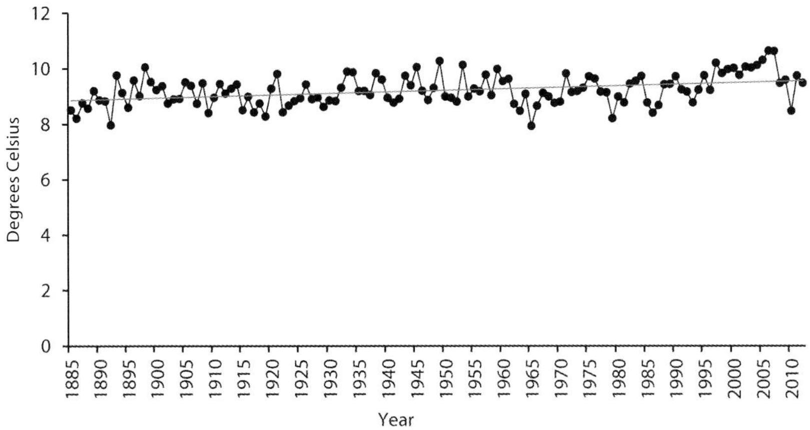

Figure 5. Mean annual temperature for Paisley in degrees Celsius 1885–2012.
Courtesy of Coats Observatory at Paisley Museum.

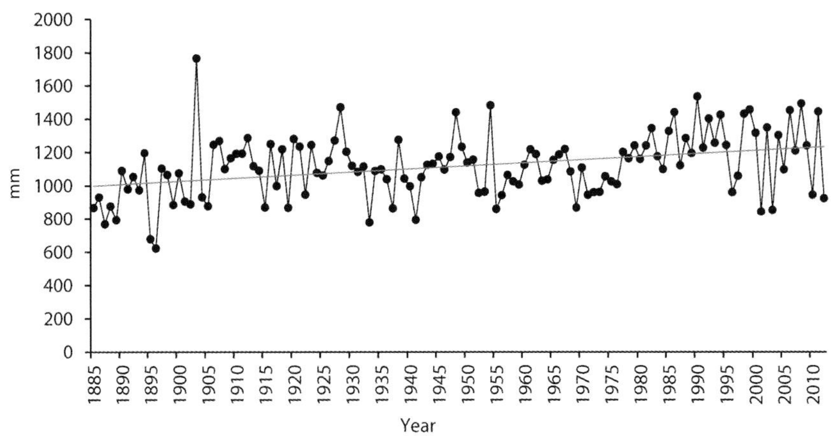

Figure 6. Annual rainfall in Paisley in millimetres 1885–2012.
Courtesy of Coats Observatory at Paisley Museum.

1.5 Landscape

The geology, soils, topography and climate of Renfrewshire influence the range of plants that grow naturally in the vice-county, and together these form the basis of the natural environment. After the end of the last Ice Age (*c.* 10,000 years BP), vegetation will have naturally recolonized the area, creating a landscape that will have been dominated by wild woodland, with open ground restricted to wet depressions and flood plains, including developing peatlands, coastal fringes and perhaps some of the more exposed, shallow soils of the uplands.

The impact of humans has greatly changed the natural heritage of Renfrewshire. Neolithic and later settlers dramatically reduced the woodland cover, creating a more pastoral environment with cultivated farmland. However, over the last 150 years or so, even more dramatic changes have occurred, notably because of increased industrialization, residential development, agricultural intensification and associated pollution and disturbance.

Physical features

Although Renfrewshire can be considered a 'lowland' county, lacking alpine heights, it is not low-lying and has quite a varied topography. Historically, the area has been described (Wilson 1812) as comprising three zones: the 'hilly', the 'gently rising' and the 'flat', terms that are repeated in Groome's gazetteer account (1882–5). The hills rarely exceed 500m and generally lack distinctive peaks, having more flat-topped ridges or plateaux separated by several broad, but often steep-sided, valleys. However, Wilson (1812) greatly admired the 'gently rising land' when he wrote, 'In no part of Britain, perhaps, has nature formed a more beautiful surface of the ground, than in the district of Renfrewshire'. The uplands afford long views eastwards up the Clyde Valley, with contrasting upland rough grazing, enclosed pastures and the lowland conurbations of Paisley and Glasgow; more dramatic views north and west across the Firth of Clyde towards the mountainous Argyll can be attained from upland vantage points, notably from the hills above Greenock and, on a clear day, from the moors above Eaglesham.

The main upland feature is the Renfrewshire Heights, nowadays more likely to be known as Clyde Muirshiel Regional Park (CMRP), which dominates the western third of the county and extends down to the narrow coastal fringe between Port Glasgow and Wemyss Bay (plate 1). The highest ground in the county can be found here, with the summits of Hill of Stake (522m) and Misty Law (507m) on the Ayrshire border, and a little to the north Creuch Hill (441m), Hyndal Hill (379m) and Dunrod Hill (298m). In the extreme north-west the steep-sided valley about Spango, between Inverkip and Greenock, separates the geologically distinct Earn Hill ridge (202m) from the main upland area to the south.

In the south-east of the vice-county the moorlands above Eaglesham, bounded by Ayrshire and Lanarkshire, form a hilly backdrop to the White Cart Water catchment, with the main summits being Corse Hill (376m), Drumduff Hill (364m), Melowther Hill (303m) and Ballageich Hill (333m). A third but lower-lying, hilly region is formed between what Hall (1915) describes as the two 'rift' valleys of the Black Cart Water east of Lochwinnoch (sometimes known as 'the Paisley Ruck') and the narrower valley associated with Cowdon Burn, near Neilston, isolating a plateau rising to 259m at Corkindale Law and 236m at Broadfield Hill, with steep marginal hillsides such as the Gleniffer and Gavin Braes and the Liboside Hills forming strong landscape features.

To the north-east is an extensive low-lying former alluvial flood plain – the 'laich lands' (low-lying lands) mentioned by Groome (1882–5) – marking out the arable lands about Houston and the urbanized ground of Paisley and south-west Glasgow. In between the flood plain and the higher ground are extensive areas of undulating upland fringe, a landscape of enclosed farmland of variously improved pastures bounded by stone walls or hedgerows, with local blocks or belts of old plantation woodland.

Water and drainage

As Renfrewshire receives a moderately high rainfall it is not surprising that waterbodies and watercourses are important landscape features. The main watercourses are the Gryfe, White Cart and Black Cart waters; the latter two merge about Inchinnan to form the very short River Cart, which meets the estuarine Clyde along the northern boundary of the vice-county.

The White Cart Water rises in the uplands in the extreme south-east of the vice-county and is fed by tributaries such as Ardoch Burn, Polnoon Water and the Earn Water and, later, in Glasgow, the Auldhouse Burn and Levern Water; the last arises above Commore Dam near Neilston and is fed by the Cowdon Burn, Killoch Water and finally the Brock Burn before joining the White Cart Water at Crookston Castle in Glasgow.

Another major drainage feature starts with the River Calder, which drains a large part of the highest ground of Misty Law and Duchal Moor before entering Castle Semple Loch at Lochwinnoch. The Black Cart Water flows east out of the loch towards Johnstone, where it is fed by the Old Patrick Water and other smaller burns, and then merges with the Gryfe Water at Blackstone before meeting the White Cart Water at Inchinnan; its name is derived from the peaty coloured waters which developed as it flowed through the mosses west of Paisley before their reclamation some 200 years ago.

The Gryfe Water arises on the eastern slopes of the Renfrewshire Heights and is fed by a number of burns, notably Green, Burnbank and Locher waters and Mill, Houston and Dargavel burns, before joining the Black Cart Water near Glasgow Airport. The rest of the western upland area is characterized by the numerous, but often short-lived, fast-flowing burns cascading down from the plateau towards the coast, the larger ones being the Kip Water, Kelly Water and Devol Burn.

The open waterbodies include a number of small lochs and lochans, dams and reservoirs – of which few, if any, have been unaltered by human activities. In the western hills peaty waterbodies are found on the upland plateau, for example Daff Reservoir, Loch Thom, the Gryfe reservoirs (including several small bodies of water associated with the Greenock Cut aqueduct), Harelaw Reservoir, Queenside Loch below Misty Law and, isolated to the east side, Kaim Dam. In the central area waterbodies are not so frequent but are represented by Auchendore and Leperstone reservoirs (the latter recently breached), Knapps Loch, Whitemoss Dam and Houstonhead Dam.

In the south-east there are several waterbodies near the Paisley and Glasgow conurbations, including Harelaw Reservoir, Glanderston Dam, Commore Dam, Stanely Reservoir, Balgray Reservoir and Waulkmill Reservoir. In the more rural parts of the south-east there are many waterbodies, including Harelaw Dam, Long Loch, White Loch, Brother Loch, Black Loch, Bennan Loch, Lochcraig Reservoir, Lochgoin Reservoir and Dunwan Dam. The main central valley ('the Paisley Ruck') is marked by the largest lowland waterbodies of Barr and Castle Semple lochs, with the smaller Loch Libo situated in the other parallel 'rift' valley around Cowdon Burn to the south, and between the two is the large Barcraigs Reservoir.

Woodlands

Although woodlands account for less than ten per cent (5532ha) of the total land cover of the vice-county (Forestry Commission 2001), they can be an important visual feature, helping to soften the otherwise open pastoral landscape of farmland and small villages or dwellings.

Only a small percentage of the total woodland has been defined as Ancient, Semi-natural or Long Established in the Forestry Commission inventory (see Chapter 1.7, p. 29). Many of the more natural (or semi-natural) woodlands survive on the steeper, less accessible ground of small hills and braes, such as about the Gleniffer Braes, Bridge of Weir and Uplawmoor, and on the steep coastal ground between Inverkip and Langbank. The best examples of semi-natural woodland are the 'valley woods' associated with watercourses, where the sheltered environments and often rich soils support equally rich floras; notable examples include Kelly Glen, Shielhill Glen, Devol Glen, Calder Glen and several less accessible, relict stretches along the larger watercourses. Many of the smaller watercourses throughout the area support scattered woodland blocks, some extending into urban areas; the numerous small, steep-sided valleys draining from the Renfrewshire Heights are fine examples.

Plantation woodland, often integrated with old woodland sites, form part of the designed landscape and have important cultural significance, in addition to being valuable habitats. The mature woodlands of policy estates, supporting a number of now familiar exotic species, are of greatest interest; examples include estates at Ardgowan, Barochan, Caldwell, Cowdon Hall, Erskine, Formakin and Finlaystone. Other plantations, shelterbelts or farm woodlands were widely planted in the eighteenth and nineteenth centuries. During the twentieth century large-scale, monotonous commercial conifer plantations were created, such as at Leapmoor, Garvock, Thornlybank, Hartfield Moss and Whitelee.

Agriculture

The impact of farming, from the time of the early pastoralists to today's highly mechanized, modern farming, has been one of the key shapers of the landscape of modern Renfrewshire.

Today farming represents the major rural land use, from the intensively managed lowlands to the more extensively grazed uplands. Descriptions of the nineteenth-century agricultural landscape are provided in Chapter 1.6.

An attractive rural pastoral scene can still be found in Renfrewshire, particularly in the upland margins. The fields are divided by hedgerows or by walls in the uplands or where the soils are shallow or poorly draining. However, much of the rural lowland landscape, and increasingly that of the uplands, is now represented by extensive areas of open, intensively managed pasture, many with removed or defunct boundary features; the impact of such intensification is discussed further in Chapter 3.3.

1.6 The Changing Environment

Prehistoric Renfrewshire

There is little direct data about what the flora of Renfrewshire would have been like at the end of the last Ice Age. However, the post-glacial vegetation of central Scotland has been studied and summarized by Ramsay and Dickson (1997) and much of their work is relevant to the vice-county. Additionally, there is direct local evidence from pollen-sample study sites at Linwood Moss, to the north-west of Paisley (Boyd 1986) and Walls Hill (Whittliemuir), *c.* 5km west of Paisley (Ramsay 1996); these both provide interesting sketches of the changes in different local landscapes over several thousand years.

The late-glacial information from Garscadden Mains (Dunbartonshire – to the north of the vice-county), as discussed in *The Changing Flora of Glasgow* (Dickson et al. 2000), and the inter-glacial data from nearby Sourlie (Ayrshire, VC75) to the west (Bos et al. 2004), provide some evidence of the type of flora present at, or before, the end of the current interstadial; at the Sourlie site, some 30,000 years BP, the landscape was thought to be 'a grassy, sedgy, herb- and moss-rich tundra with a low growing shrubby community comprised mainly of *Salix* and *Betula nana*'.

Analysis of the evidence of the late-glacial vegetation is hampered by a lack of dating, but some samples indicate a late Devensian tundra-type flora, followed by a dwarf shrub community (Crowberry, Juniper, willows, Dwarf Birch) and some tree birches. The Lomond Re-advance (*c.* 11,000 BP) reversed the trend but from 10,000 BP onwards rapid climatic warming was underway, with a spread of birch woodland which soon covered much of central Scotland. Hazel quickly followed and there is evidence for elm from around 9000 BP. Oak also appeared about this time, followed later by Alder. Pine is considered never to have had a significant presence in central Scotland, but pine stumps which date to 4270 BP were found at Walls Hill bog (Midton Loch, Whittliemuir); it is speculated that pine colonized drier areas of peat bogs, but pollen production appears to have been low.

It is considered that Scotland 6000 years ago would have been drier and warmer than at present, and with little evidence of high levels of human impact a wooded landscape would be expected. Trees grew wherever it was physically possible for them to do so and the woodland cover was very extensive (Dickson and Dickson 2000). However, as with much of Britain, there is strong evidence of Elm decline over the following thousand years or so.

In central Scotland, most evidence of human impact on the vegetation is associated with the Neolithic age (*c.* 4000–2500 BC) – about the time of the Elm decline, although not necessarily the cause. There is more evidence of woodland clearance in the following Bronze Age, and by the pre-Roman Iron Age (beginning *c.* 750 BC) the clearance had increased dramatically. At the time of the Roman occupations (in the first to early third century AD) the landscape of central Scotland is considered to have been pasture and meadows with associated ruderal plant communities, with only little pollen evidence for arable agriculture (Ramsay and Dickson 1997). However, local evidence indicates that following the Romans there was a return of woodland vegetation at the expense of agriculture. Climatic deterioration may have contributed to this, as woodland cover remained for several hundred years.

Sample site: Linwood Moss (summarized from Boyd 1986)

Two sample spots were studied in a broad basin-like area of low-lying terrain, *c.* 8km across, surrounded by higher ground to three sides: Linwood Moss (NS4360) and Moss Cottage (NS4466). The area, situated on the flood plain between the Gryfe and Black Cart waters, was formerly extensively covered by raised bogs overlaying raised marine clay-rich sediments. The sites were inundated at least twice by the sea during the deglaciation period; the first inundation more than 9000 years ago, and a second, but lower, inundation some 5000 years ago.

The landscape at Linwood Moss (*c.* 10,000 BP) is considered to have been very open, and in places damp with a moderately rich herb flora locally dominated by rushes and Meadowsweet; grass pollen was varied but included Cock's-foot and Common Reed. Tree pollen was sparse, perhaps restricted to a few damp woodlands, but later Birch–Hazel woodland followed, in a pattern that was the same across Scotland. Willow was locally quite common, indicating local minerotrophic wetland conditions. By 7000 BP the landscape was thought to comprise bogs dominated by Heather and cottongrasses, with nearby wet woodland and regional oak rich woodland.

Sampling from nearby Moss Cottage (some 5000 BP) revealed an influx of sedimentary material under marine conditions; this was followed by a hydroseral development from saline saltmarsh and reed swamp to freshwater swamp and Alder fen. Pollen samples included much grass pollen (possibly Common Reed) with rush, sedges and herbs, and also some coastal indicators such as Sea-milkwort and Sea Rush. Oak and Elm pollen occurred in the local region on higher ground, with birch, Alder and Hazel growing close to the parts of the site fringing the coastal reed swamp, saltmarsh and non-saline swamps and marshes. Subsequent changes to more open vegetation, including local docks, plantain and nettle, were indicative of human agricultural clearance, although it is considered that although this was happening on the surrounding high ground, the wetter areas, for which there is a lack of evidence of charcoal burning, were probably not so heavily affected. Following the decrease in woody vegetation, open Heather-dominated wet heath vegetation appears to have become established on the peaty soils, and in turn, with increasingly wet conditions, Bog-moss dominated raised bog. Later drying events, indicated by increased sedges and Heather, are linked to drainage and peat removal, although the evidence from upper layers is compromised by disturbance at the sample sites.

Sample site: Walls Hill (summarized from Ramsay 1996)

The sample was taken from a raised bog adjacent to the now drained Midton Loch (Whittliemuir) just to the east of Walls Hill (*c.* NS4158), an Iron Age fort. Older samples from this site, dating to some 5000 years ago, show a native woodland comprised of birch, oak and Alder with, to a lesser extent, Scots Pine, Ash and Elm; Heather was well represented, indicating some human impact but this is not considered to have been widespread. This was followed, particularly in the Iron Age, by extensive woodland clearance, indicating a change to pastoral (shown by increased grass-, buttercup- and daisy-types and Ribwort Plantain pollen) rather than arable agriculture (although some cereal and weed pollen was found), and there was also an increase in Heather. At this time clearances and agriculture were thought to be temporary, perhaps due to exhaustion of the acidic soils and movement to new sites.

There is strong evidence from after the Roman occupation for woodland regeneration at Walls Hill and it appears that there was little cultivation in the area about the bog throughout much of the following millennium. At the end of the first millennium AD, a return of human influence is thought to account for declining tree pollen and increased grass and weedy species at Walls Hill, but the clearance was temporary. However, woodland declined dramatically in the second millennium and there is evidence that by the medieval period there was an open landscape, which has remained to the present day. Pastoral farming appears to be the main agriculture but at the boggy sample sites, Heather pollen shows a strong rise, reflecting tree clearance and the drying of the bog surface.

Medieval Renfrewshire

By the start of the second millennium AD there is strong evidence from sites across central Scotland for increased woodland clearance and a rise in pastoral agriculture. Dickson and Dickson (2000), wrote that at the time of the Medieval Warm Period, from the late tenth to the twelfth century, 'Scotland experienced hot, dry summers and rather cold winters', which they noted as 0.8°C warmer than in the first half of the twentieth century, and which was 'considered a golden age in Scotland where farming established at higher levels on hillsides than at present. These temperatures encouraged population increase, monastic sheep farms and feudal estates.'

There is very little recorded evidence of local vegetation from the thirteenth to the end of the eighteenth century. One exception, though, comes from Paisley Abbey. The opportunity to study old plant remains in the heart of Paisley provided a fascinating glimpse into human and plant relationships in the medieval period, as described by Camilla Dickson (1996). Paisley Abbey, situated in the centre of Paisley (NS4863), was founded in the late twelfth century by monks of the Cluniac order. The drains that were constructed in the late fourteenth century were studied in 1990. Extensive silt deposits had accumulated from river water (the adjacent White Cart Water) used to flush out the drains. Analysis of these deposits revealed various remains, including those of plants, dating from the first half of the fifteenth century. In total 140 plants were represented in two samples, comprising species from semi-natural habitats and those from the abbey's kitchens and latrines. The remains from the latter two sources included plants that would have been used for medicinal purposes (such as Greater Celandine, Hemlock, Opium Poppy, Monk's Rhubarb, Flax and Caper Spurge), as sources of food (such as cereals, including Bristle Oat, and fruits including Domestic Apple and Damson) and for dyeing (for example Weld). Some of these plants were imported, including Fig, Mace and presumably Walnut, but a number of native plants in this category could have been grown or collected locally from the wild. Camilla Dickson divided the native plant finds by the various semi-natural habitats they would have grown in (table 2).

Table 2. Native and archaeophyte plant fragments found in medieval remains at Paisley Abbey (details from Dickson 1996 and Dickson and Dickson 2000).

Woodlands
Trees: Alder, Ash, Blackthorn, Downy Birch, Elder, Hazel, oaks, Rowan, Wild Cherry.
Ground flora: Remote Sedge, Wood Avens, Wood Dock, Wood Sedge.

Grasslands
Autumn Hawkbit, Broad-leaved Dock, Carnation Sedge, Common Sedge, Compact/Soft Rush, Cow Parsley, Creeping Buttercup, Glaucous Sedge, Heath Bedstraw, Heath Wood-rush, Lady's-mantle, Mouse-ear, Oval Sedge, Ox-eye Daisy, Ribwort Plantain, Self-heal, Sharp-flowered/Jointed Rush, Sheep's Sorrel, Tawny Sedge, White Clover, Yarrow.

Wetlands
Blinks, Bogbean, Bristle Scirpus, Common Duckweed, Common Spike-rush, Common Water-starwort, Floating Sweet-grass, Marsh Ragwort, Marsh Violet, Marsh Woundwort, Meadowsweet, Ragged-Robin, Spearwort, Toad Rush, Trifid Bur-marigold, Water-crowfoot, Water-pepper, Water-plantain, Winter-cress, Wood Club-rush, Yellow-sedge.

Heath and Bog
Blaeberry, Bracken, Broom, Cranberry, Cross-leaved Heath, Gorse, Hare's-tail Cottongrass, Heather, Heath Rush, Marsh Lousewort, Tormentil.

Arable and rural weeds
Annual Knawel, Annual Meadow-grass, Burdock, Chickweed, Cleavers, Common Goosefoot, Common Nettle, Corn Marigold, Corn Spurrey, Dead-nettle, Equal-leaved Knotgrass, Fool's Parsley, Greater Plantain, Hedge Mustard, Hemp-nettle, Long-headed Poppy, Meadow-grass (Rough?), Nipplewort, Orache, Pale Persicaria, Prickly Sowthistle, Radish, Redshank, Scentless Mayweed, Small Nettle, Smooth Sowthistle, Sun Spurge, Wild Turnip.

The nineteenth-century rural landscape

There are very few early descriptive accounts of the local flora or vegetation but some literature sources from the early nineteenth century and map-based evidence together help to illustrate the state of the environment at the time of the earliest botanical recording. Two valuable publications are Crawford and Robertson's (1818) *A General Description of the Shire of Renfrew* and J Wilson's (1812) *General View of the Agriculture of Renfrewshire*.

There are quite a few map sources covering Renfrewshire, as discussed by John Moore of Glasgow University Library in his *Early Cartography of Renfrewshire to 1864* (Moore 2000); many of the maps are accessible via the web pages of the National Library of Scotland. The seventeenth-century maps of Timothy Pont (the basis for the first atlas of Scotland, published by Blaeu) and the eighteenth-century maps produced for General Roy's military survey provide an impression of the open landscape before the botanical recording period. Later maps such as the John Thomson map (1826) and then the Ordnance Survey maps (1860 onwards) chart the changing urban infrastructure of the county.

One of the most fascinating maps, and relevant to the end of the eighteenth century, is the *Map of the County of Renfrew* by John Ainslie (1745–1828), surveyed in 1796 and published in 1800 (fig. 7). Moore wrote, 'John Ainslie was without doubt, the outstanding cartographer of his generation' and he describes the map as 'superbly detailed … gives a comprehensive picture of the county's roads, tolls, woodlands, plantations and policies, mosses, settlements,

Figure 7. *Map of the County of Renfrew* by John Ainslie (1796). Courtesy of the National Library of Scotland.

The inset shows the area between Paisley (to the south-east) and Houston (to the north-west), clearly showing four large mosses – areas of peatland – associated with the low-lying flood plains of the Black Cart and Gryfe waters.

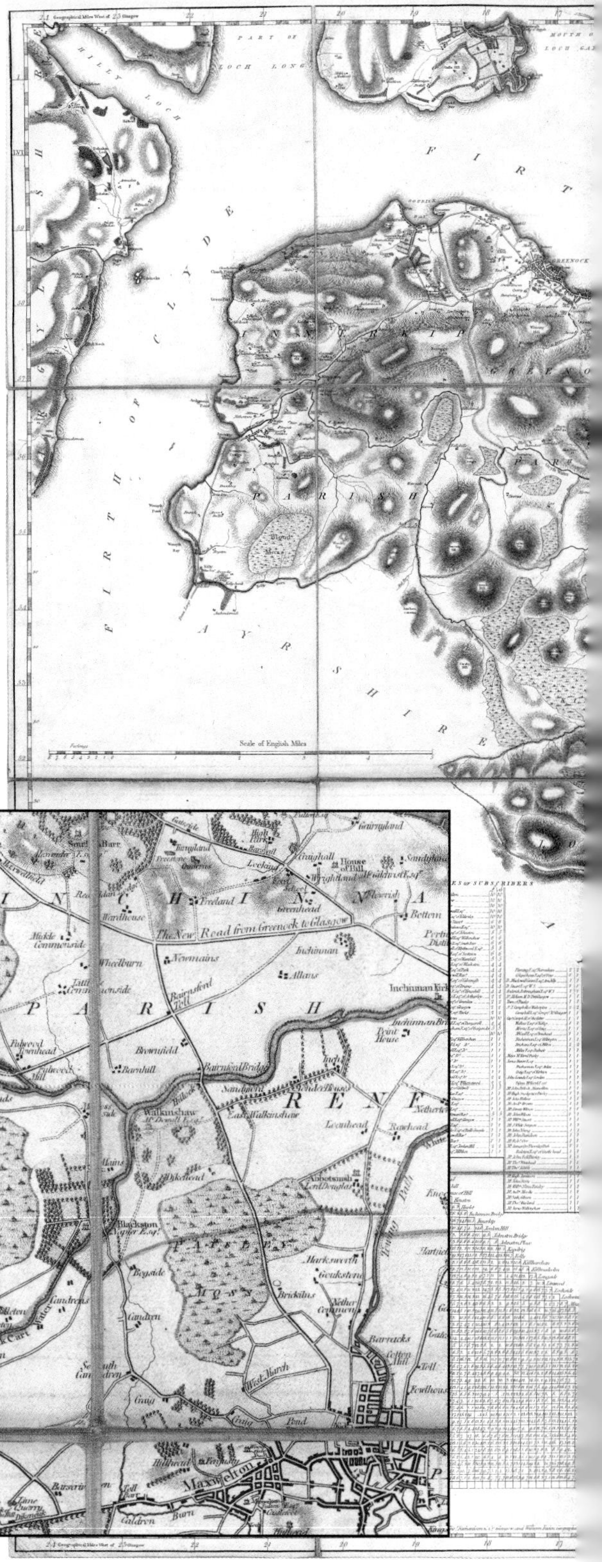

MAP
of the
COUNTY
of
RENFREW
Surveyed by JOHN AINSLIE in 1796
To William Macdowal Esqr.
OF GARTHLAND
Lord Lieutenant
OF THE
COUNTY OF RENFREW
This Map is most humbly Inscribed
by his much obliged and
most obedient Servt.
John Ainslie
PAISLEY
RENFREW
GLASGOW
CATHCART
Scale of English Miles

industries and drainage pattern, which stands comparison with the later work of the Ordnance Survey'. Hills are shaded and named and parish boundaries were shown for the first time. The local estates are marked with names of the proprietors and details about local industries, including mills, quarries and bleachfields. The map also shows tables of parish populations (table 3) and acreages, and a distance table to 91 identified places.

Crawford and Robertson (1818) provide valuable details about the land use of the then 16 parishes within Renfrewshire. They show the surface coverage of arable land, 'sound pasture', moss or moors, urban features and woodlands (see table 4). These statistics complement the graphic representation of Ainslie's map from a few years earlier.

Table 3. Population totals for Renfrewshire parishes, from John Ainslie's map of 1796 (plus, in italic, two small sections of Ayrshire included on the Ainslie map sheet).

Cathcart	777	Lochwinnoch	2613
Eaglesham	1020	Mearns	1430
Eastwood	2672	Neilston	2510
Erskine	800	Paisley/Abbey	24,592
Greenock	14,397	Port Glasgow	4378
Houston	1266	Renfrew	1638
Inchinnan	300	*Part of Beith*	*49*
Inverkip	1200	*Part of Dunlop*	*50*
Kilbarchan	2506	**Total**	**63,128**
Kilmacolm	930		

Table 4. Extract from the 'Table of Surface in English Acres' in the *History of the Shire of Renfrew* by G Crawford and G Robertson (1818, p. 431).

Old Parish	Arable	Sound pasture	Moss or moors	Houses, roads or waters	Woods, natural or planted	Total (acres)
Cathcart	2450	10		25	120	**2605**
Eaglesham	6100	3960	*5100	230	60	**15,450**
Eastwood	5750	25		145	425	**6345**
Erskine	3000	2500		65	800	**6365**
Greenock	2315	930	2780	300	40	**6365**
Houston	4000	2500	350	150	500	**7500**
Inchinnan	2600	100		60	300	**3060**
Inverkip	4500	1500	5860	140	540	**12,540**
Kilbarchan	5400	2000	1200	140	460	**9200**
Kilmacolm	8500	8000	2500	100	700	**19,800**
Lochwinnoch	8000	5750	4000	500	1000	**19,250**
Mearns	7520	2400	950	250	100	**10,950**
Neilston	10,500	1200	1000	256	1000	**13,956**
Paisley/Abbey	12,500	1500	260	900	1000	**16,160**
Port Glasgow	440	350		18	36	**844**
Renfrew	3600	36		60	80	**3776**
Total (acres)	**86,905**	**32,761**	**24,000**	**3339**	**7161**	**154,166**

* Note: The value of the Eaglesham 'Moss or Moors' cell has been changed from the printed '5200' to '5100', so that the totals of both its row and column tally.

From Crawford and Robertson: 'Explanation of terms:
1. By arable is meant all lands in cultivation, whether in tillage or in grass.
2. Sound pasture, natural grass distinct from heath, whether on hills or plains.
3. Moss or moors, dark heath and peat mosses in which little or no green pasture is found.
4. Woods, not merely clumps and belts of planting, but hedge-rows, and even single trees where-ever growing.'

From the above sources it is evident that at the start of the nineteenth century Renfrewshire was a very open, rural area with few large towns or associated infrastructure. However, there were plenty of small farms and other hamlets and agriculture was widespread, but intensification and other improvements were not well developed. It is also evident that woodlands – natural or plantation – were a scarce resource and unevenly distributed. The account of Kilmacolm Parish by Crawford and Robertson is perhaps a useful description which could be applied more widely: 'The quarter next to the sea is very steep and rough, but is much beautified by woodlands both natural and planted. A great part of the parish particularly on the southern and on the western extremities is moorish land, rising to a considerable height, and very bleak and barren. But the greatest expanse of country, of uniform feature, is a hollow plain, shelving both from south and north towards the Gryfe, and its tributary streamlets in the centre. This is thickly scattered over with farm hamlets, whilst the soil, which is incumbent on rotten rock, is naturally fine pasture land. Much of it, indeed, is in cultivation and produces good crops of grain and potatoes, also some clover and turnips. More than 6000 acres of this description of soil are situated in one unbroken expanse, in the heart of the vale of Strathgryfe in this Parish.'

Crawford and Robertson described the lands before the mid eighteenth century as being 'entirely open with the beginnings of plantations, and general spirit of improvement', indicating a lack of enclosed fields and a more basic level of farming; but they go on to state that by the early nineteenth century there was 'hardly a single acre susceptible of improvement, that has not been rendered greatly more productive' and further that 'The whole arable part has been enclosed and fields divided by hedgerows or stone walls.' Wilson (1812) noted that prior to 1780 'uninclosed sown grasses were not attempted' but subsequently described enclosures of fields of between five and 12 acres, and noted that 'A thorn hedge planted on the surface, with a stone dike, produces a strong and beautiful fence', adding that 'Sweet-briar has been found to be a good addition to the plants usually employed for hedges on account of its vigorous and long shoots'.

Wilson described the main crops at this time as oat and barley, plus wheat, beans and, rarely, peas, and with potatoes introduced in about 1750 'with great success'; he wrote that there were few turnips and added that clover was 'never sown as a green crop'. Flax is mentioned for the area about Lochwinnoch and Kilbarchan. He also described the four main cornfield weeds as being thistles, Wild mustard (Skillocks), Couch (Felt), and Corn Marigold (Gule). The meadow species mentioned are Meadow Buttercup (Crowfoot/Crow-toe), Ox-eye (Gowan) and Ragwort, and Wilson added, 'but no care is taken to exterminate weeds, or to prevent them from spreading. It is lamentable to notice thistles and other pernicious weeds accumulated on both sides of public and private roads and on banks of ditches.'

Of grazing pasture Wilson noted the improvement through lime and dung, along with draining and, rarely, burning. Hay was cut in July but 'boghay' of higher ground, and bogs or marshy places where not cultivated, was not cut until the end of July and difficult to manage: 'The quantities of soft-grass, sprits, rushes and aquatic plants, which abound in our meadows, rendering the process for saving this kind of hay, in many instances, extremely troublesome and precarious.'

The more fertile lowlands were heavily cultivated with arable crops, or as Crawford and Robertson put it, 'the lower part of the county abounds with corn, and the higher with pasturage'. In table 4 arable is reckoned to be easily the largest land-use category at 86,905 acres, 56 per cent of the total land, with 'sound pasture' accounting for 26 per cent.

Towards the end of the nineteenth century Groome (1882–5) calculated that 'nearly two thirds' of the land area (156,785 acres) was cultivated ('crop, bare fallow and grass'), with nearly 3.5 per cent 'under wood' and the remaining land 'occupied by buildings and roads etc., or by rough hill grazings and waste ground'. Groome also provided tables of different crops and livestock. It is interesting to note that he remarked on the gradual reduction of arable crops but the increase in permanent pasture, which he thought 'entirely due to the large towns in the neighbourhood, which afford a ready market for stock and for dairy produce'. This trend has continued into modern times.

There is little detail about the upland areas but Crawford and Robertson did note that 'A range of hills forms the southern boundary. The most easterly of these, in the parish of Eaglesham, are covered with heath. There are some of the same hue, towards the west, in the parishes of Lochwinnoch, Kilmacolm and Innerkip [Inverkip]. The intermediate hills are in generally fine pasture; some of them naked rock.' In table 4 moss or moors accounted for 24,000 acres, which is 15.5 per cent of the area. It is likely that the upland moors escaped excessive attention of farm improvement, and Ferguson writing in 1915 considered the Renfrewshire plateaux flora of moorland and natural or semi-natural grass as 'largely a natural one, little influenced by the operations of man…'.

The lowland mosses or mires though did not escape attention. Wilson considered reclamation as 'an attempt of great importance' and noted that in the lower and fertile districts there were four main mosses named as 'Blackstoun, Paisley, Dargavel and Barrochan', estimated at some 1970 acres – these are shown in detail on the insert of Ainslie's map (fig. 7). These lowland mosses were targeted for improvement through digging, grazing and liming, rather than just by burning and/or manuring. He also described how great care and pains were taken over drainage 'to carry off the stagnant water'. He reported that nearly the whole of 'Paisley Moss' which in 1793 covered 130 acres 'is now improved and well cultivated', and that of 674 acres of Hartfield Moss, some 450 had been reclaimed.

There is little evidence of the condition of former 'natural' open waterbodies, but it is likely that none has escaped some degree of drainage or manipulation. In 1812 Wilson wrote that the level of Castle Semple Loch (Lochwinnoch) was reduced and land acquired along the edge of the lake. The created Barr and Peel meadows were described as '250 acres of spongy land abounding with aquatic plants and producing hay of little value'. The numerous watercourses have seen many changes from bank stabilization or canalizing to changes brought about by enrichment or pollution of the water. Wilson (1812) described how in 1774 the Black Cart Water near Johnstone was deepened to produce a two-mile canalized stretch; such actions along with similar extensive canalizing of the urban White Cart Water and the Clyde shore, and the drainage of wetlands and mires, has resulted in dramatic changes to the lowlands about Paisley.

It has already been noted that woodland was a scarce resource 200 years ago, although Wilson noted that 'it appears in former time to have abounded with woods', quoting place names, such as Fullwood and Linwood, and other ancient records. He considered that most natural or copse woods were in Paisley, Houston or Eastwood, notably the latter, and that these were cut every 30 years. Crawford and Robertson described parts of Eastwood as 'still under its original growth of dwarf wood (oak, hazel, alder &c)' and noted 'considerable stools of natural wood, as well as plantation, on the steep banks of the Cart' in Cathcart. However, both parishes were well under cultivation at this time.

Woodland planting was widely practised in the nineteenth century and Crawford and Robertson noted that 'Every gentleman's seat and many farm hamlets stand now encircled by plantations' and added that 'Even many of the hill sides and hill tops, originally bleak and barren, are now waving with forest trees'. In table 4, however, only 7161 acres (4.5 per cent) are noted and this should have included the then recent plantations. Later, in Kilbarchan Parish (NSA 1845), it was noted that 'The parish contains no natural woods; but almost every considerable property has plantations'. In Paisley Parish (NSA 1845) 'several extensive woods and plantations occur throughout the parish, beside numerous ornamental clumps and belts affording shelter'. Trees recorded here on the high ground included Birch, Larch, Spruce, Fir and Scot's fir (pine) with, to the lowlands, Oak, Elm, Plane (Sycamore), Horse-chestnut and Ash, plus 'other showy, less valuable species'. The plantations were nursed by fast-growing deciduous trees, such as various poplars and also conifers, which were 'removed from time to time to allow more valuable ones to enlarge'.

The growing population of Renfrewshire

Renfrewshire has a distinguished early history, being associated with the first High Steward of Scotland (Walter Fitzalan, *c.* 1157) who came to reside in Renfrew, where the influential House of Stewart was established. Much of the area was known in early records as Strath-Gryfe and became a distinct Sheriffdom in 1404 – the Barony of Renfrew – as the area's status and importance grew. The actual name Renfrew is thought to derive from 'Rhyn' (point of land) and 'Fraw/Frew' (flow of water). By the time of *The Statistical Account of Scotland (1791–1799)*, the county had been divided into 16 parishes (see table 3), many of these relating back to historical estates or religious diocese boundaries.

Over the last 200 years the impact of human development other than agriculture has had an increasing impact on the landscape of the vice-county. Today more than 500,000 people reside in the old county area. This contrasts with a population of under 70,000 at the time of Ainslie's map, rising to well over a quarter of million by the end of the nineteenth century according to Groome (1882–5) (see table 5).

Table 5. Population census of the county of Renfrewshire (Groome 1882–5).

1801	1811	1821	1831	1841	1851	1861	1871	1881
78,501	93,172	112,175	133,443	155,042	161,091	177,561	216,947	263,372

At the start of the nineteenth century the Ainslie map (fig. 7) shows a largely rural landscape, with the largest population centres being Paisley and Greenock parishes. The western fringes of Glasgow were relatively rural but contained larger settlements such as Pollokshaws, Cathcart and Eastwood. Elsewhere across the county there were scattered small towns or villages such as Johnstone, Kilbarchan, Lochwinnoch, Bridge of Weir, Houston, Kilmacolm, Renfrew, Barrhead, Neilston, Eaglesham, Mearns and Inverkip. Dotted between these small towns were numerous farmsteads.

In the late nineteenth century and through the twentieth century increasing industrialization resulted in a growing population and a greater associated infrastructure of roads and railways, and Paisley Canal linking to Glasgow. Renfrewshire was an important industrial centre, second only to Lanarkshire in Scotland; in addition to the working of minerals and general manufacturing and engineering, shipbuilding, leather tanning, rope making, sugar refining, distilling and printing were important industries, and notable products included cotton, linen, silk and other textiles.

Today there are large areas of conurbation stretching westward from Glasgow, and along much of the Clyde coast. Many once isolated settlements are now part of large urban areas. Data about the size and populations of the modern political units that are within the old county of Renfrewshire are shown in table 6. The total population within the boundary exceeds half a million, *c.* 511,320.

Table 6. Surface land area and population sizes for the local authority areas encompassed within the old county of Renfrewshire boundary.

Modern local authority	Area (sq. km)	Population*
East Renfrewshire	160	89,850
Inverclyde	174	79,220
Renfrewshire	261	170,650
Glasgow (part)	46	171,600

* Population data obtained from the National Records of Scotland (pre-Census mid-2011 population estimates). The figure for the Glasgow portion is based on estimates by data zone, kindly provided by Glasgow City Council, Development and Regeneration Services, Development Plan Team.

1.7 Habitats and Vegetation

The following section discusses the floristic composition of the main habitat types that are found in Renfrewshire. Where possible a reference is made to the corresponding National Vegetation Classification (NVC) communities (Rodwell 1991–2000).

Woodlands

Table 7. Woodland areas by type from the Renfrewshire Local Biodiversity Action Plan – original source *Consensus of Woodlands* (Forestry Commission 2001).

Semi-natural woodlands	358ha
Mixed woodlands	641ha
Plantation of Broad-leaved trees	2660ha
Plantation of Conifers (non-natives)	1873ha
Total	5532ha

Woodlands include the familiar mature broad-leaved types but also plantations of conifers or mixed species; large patches of scrub, on dry or wet ground, and hedgerows are also included. Inventory figures produced by the Forestry Commission indicate that less than ten per cent of the vice-county supports woodland and of this only a very small amount, less than one per cent, is defined as ancient or semi-natural (see table 7). The latter low percentage reflects the amount of cover shown on nineteenth-century maps and indicated by accounts of that time, which depict or describe a paucity of woodland habitat in Renfrewshire, as discussed in Chapter 1.6.

Much of the woodland seen today is made up of plantations: long-established, policy-estate types, farm shelterbelts, commercial conifers or the recent phenomenon of community woodlands – many of uncertain 'native' provenance. Most of these have little floristic interest except for the old policy woodlands. The latter often originated in what were urban-fringe areas, but many are now absorbed into built developments such as housing estates, hospital grounds and public parks. The plantation influence can be readily seen by the range of canopy species, which typically includes Beech (*Fagus sylvatica*), Sycamore (*Acer pseudoplatanus*), Common Lime (*Tilia* x *europaea*), Horse-chestnut (*Aesculus hippocastanum*) and conifers such as larches (*Larix* spp.), pines (*Pinus* spp.) and spruces (*Picea* spp.), plus a range of rarer trees. A troublesome feature can be the understorey scrub – often evergreen and introduced as game cover – which is now widespread and typically includes Rhododendron (*Rhododendron ponticum*), Snowberry (*Symphoricarpos rivularis*), Cherry Laurel (*Prunus laurocerasus*), dogwoods (*Cornus* spp.), Yew (*Taxus baccata*), hollies (*Ilex* spp.), privets (*Ligustrum* spp.) and several currants (*Ribes* spp.). The more exotic ground-flora species can include various 'knotweeds' (*Persicaria* and *Fallopia* spp.), bamboos (for example *Sasa palmata*), Irish Ivy (*Hedera* 'Hibernica'), Pink Purslane (*Claytonia sibirica*), Wood Forget-me-not (*Myosotis sylvatica*), Welsh Poppy (*Meconopisis cambrica*), Honesty (*Lunaria annua*), Yellow Archangel (*Lamiastrum galeobdolon* ssp. *argenteum*) Garden Solomon's-seal (*Polygonatum* x *hybridum*), Fringe-cups (*Tellima grandiflora*), Lesser Periwinkle (*Vinca minor*), Ground-elder (*Aegopodium podagraria*), and bulbs such as daffodils (*Narcissus* spp.), Snowdrop (*Galanthus nivalis*) and Spanish and hybrid Bluebell (*Hyacinthoides* spp.).

Semi-natural woodlands are now mainly confined to steeper ground, and are particularly well represented along watercourses, where they are known as 'valley woods'. The valley woods tend to be the richest in old woodland species and reflect the varying soil conditions of the slopes, from acidic to more base-rich soils, and also some riparian influence, resulting in complex vegetation zoning. However, other semi-natural woodlands can also be diverse, reflecting the range of ground conditions, past management actions and local species introductions. The following summarizes the main semi-natural woodland types, annotated with NVC affinities.

Acidic Birch woodlands

In the upland areas a number of woods occur on acidic soils (NVC W11/17), with domination usually by birch (mainly *Betula pubescens*), although there can be some Pedunculate Oak (*Quercus robur*) and Rowan (*Sorbus aucuparia*), and often also planted conifers and naturalized Beech (*Fagus sylvatica*). The ground flora shows heathland affinities but usually it has an acidic grassland appearance, encouraged by grazing pressure. The ground flora typically includes examples of the following: Wavy Hair-grass (*Deschampsia flexuosa*), bent grasses (*Agrostis* spp.), Sweet Vernal-grass (*Anthoxanthum odoratum*), Creeping Soft-grass (*Holcus mollis*), Blaeberry (*Vaccinium myrtillus*), Broad Buckler-fern (*Dryopteris dilatata*), Hard-fern (*Blechnum spicant*), Bracken (*Pteridium aquilinum*), Hairy Wood-rush (*Luzula pilosa*), Wood Sorrel (*Oxalis acetosella*), Foxglove (*Digitalis purpurea*) and mosses (which can include *Polytrichum* spp. *Dicranum scoparium, Pleurozium schreberii, Plagiothecium undulatum* and *Hypnum cupressiforme*).

Oak–Birch woodlands

The less markedly acidic soils away from the uplands or drained mire areas support a more varied lowland woodland flora (NVC W10/11); however, these woodlands have invariably suffered from land-use pressures and are now confined to steeper ground or exist as relicts, usually modified, in other woodland mosaics. Pedunculate Oak (*Quercus robur*) is the dominant species, but is not as common as might be expected, and native Sessile Oak (*Q. petraea*) populations seem rare; associates include birch (some *Betula pendula*) and usually also Sycamore (*Acer pseudoplatanus*), Ash (*Fraxinus excelsior*) and Beech (*Fagus sylvatica*). The ground flora typically includes Creeping Soft-grass (*Holcus mollis*), Common Bent (*Agrostis capillaris*), Bluebell (*Hyacinthoides non-scripta*), Broad Buckler-fern (*Dryopteris dilatata*), Male-fern (*Dryopteris filix-mas*), Wood Sorrel (*Oxalis acetosella*), Red Campion (*Silene dioica*), Pignut (*Conopodium majus*), Greater Stitchwort (*Stellaria holostea*), brambles (*Rubus fruticosus* agg.), Honeysuckle (*Lonicera periclymenum*) and Bracken (*Pteridium aquilinum*); common bryophytes include *Rhytidiadelphus squarrosus, Kindbergia praelonga, Mnium hornum* and *Lophocolea bidentata.*

Ash–Elm woodlands

Woodlands on neutral or less acidic soils – often shallow or clayey soils with less humus development – are usually associated with watercourse valleys (NVC W8/9). In addition to Ash (*Fraxinus excelsior*) and Elm (*Ulmus glabra* – many of which are diseased or have died),

Sycamore (*Acer pseudoplatanus*) can be common, due in part to its rapid recent spread; other associates can include Hazel (*Corylus avellana*), Hawthorn (*Crataegus monogyna*) or, more locally, Blackthorn (*Prunus spinosa*); a number of other trees often related to past planting invariably occur, such as oaks, Beech, Cherry, Common Lime, Horse-chestnut and the occasional conifer. The ground flora tends to be variable but is usually diverse, and supports many of the rarer woodland species; typical species include Dog's Mercury (*Mercurialis perennis*), Wood Avens (*Geum urbanum*), Lesser Celandine (*Ficaria verna*), Tufted Hair-grass (*Deschampsia cespitosa*), Tuberous Comfrey (*Symphytum tuberosum*), Ramsons (*Allium ursinum*), Common Nettle (*Urtica dioica*), Red Campion (*Silene dioica*), Primrose (*Primula vulgaris*), Dog-violet (*Viola riviniana*), Enchanter's Nightshade (*Circaea* spp.), Wood Brome (*Bromopsis ramosa*), Wood False-brome (*Brachypodium sylvaticum*), Woodruff (*Galium odoratum*), Sanicle (*Sanicula europaea*), Male-fern (*Dryopteris filix-mas*), brambles (*Rubus fruticosus* agg.) and larger mosses (including *Kindbergia praelonga, Eurhynchium striatum, Atrichum undulatum, Thuidium tamariscinum* and *Plagiomnium undulatum*). The woodlands in steep, shaded valleys are, in addition, notable for diverse ferns (including *Polypodium vulgare, Polystichum aculeatum, Phegopteris connectilis, Gymnocarpium dryopteris* and *Asplenium* spp.) and bryophytes; note should also be made of the locally abundant Greater Wood-rush (*Luzula sylvatica*), which is a common feature of steeper, often north facing, valley side slopes.

Wet woodlands

The wetter conditions along watercourses, flushed valley sides and some low-lying marshy areas encourage the growth of wet woodlands (NVC W6/7), where Ash–Elm associate species are reduced and the canopy is marked by more Alder (*Alnus glutinosa*), Grey Willow (*Salix cinerea*) and occasionally Bird Cherry (*Prunus padus*). Ground flora elements typically include Tufted Hair-grass (*Deschampsia cespitosa*), Soft-rush (*Juncus effusus*), Remote Sedge (*Carex remota*), Reed Canary-grass (*Phalaris arundinacea*), Lesser Celandine (*Ficaria verna*), Opposite-leaved Golden-saxifrage (*Chrysosplenium oppositifolium*), Yellow Pimpernel (*Lysimachia nemorum*), Bugle (*Ajuga reptans*), Northern Bitter-cress (*Cardamine amara*), Rough Meadow-grass (*Poa trivialis*), Marsh Hawk's-beard (*Crepis paludosa*), Wood Horsetail (*Equisetum sylvaticum*), Meadowsweet (*Filipendula ulmaria*), Yellow Iris (*Iris pseudacorus*), Creeping Buttercup (*Ranunculus repens*), Lady-fern (*Athyrium filix-femina*) and Common Nettle (*Urtica dioica*).

Carr woodlands

Wet, usually scrubby willow woodlands ('carr') also occur at many fen or marshy sites (NVC W1/3). Grey Willow (*Salix cinerea*) is the commonest tree, with, at more species-rich sites (NVC W3), Bay Willow (*S. pentandra*), found locally, and, rarely, Tea-leaved Willow (*S. phylicifolia*). The ground vegetation is not so distinctive and usually consists of shaded elements of the adjacent swamp or fen vegetation. Typical species include Soft-rush (*Juncus effusus*), Creeping Bent (*Agrostis stolonifera*), Reed Canary-grass (*Phalaris arundinacea*), Meadowsweet (*Filipendula ulmaria*), Creeping Buttercup (*Ranunculus repens*), Water Avens (*Geum rivale*), various sedges (*Carex* spp.), Marsh Cinquefoil (*Comarum palustre*), Northern Bitter-cress (*Cardamine amara*), Marsh Hawk's-beard (*Crepis paludosa*), water forget-me-nots (*Myosotis* spp.), Hemlock Water-dropwort (*Oenanthe crocata*), Marsh-marigold

(*Caltha palustris*) and Common Marsh-bedstraw (*Galium palustre*); the various bryophytes can include *Calliergoniella cuspidata, Calliergon cordifolium*, *Plagiomnium* spp. and the occasional bog-moss (*Sphagnum* spp.).

Bog woodlands

Acidic bog-type woodland (NVC W4) is related to the drier, acidic birch woodlands, but differs quite dramatically due to water-logged ground conditions. The distinctly wet peaty soils tend to be dominated by Downy Birch (*Betula pubescens*), sometimes, though rarely, with Grey Willow (*Salix cinerea*); the ground flora reflects mire affinities, usually with much Bog-moss (*Sphagnum fallax*, *S. palustre* and *S. fimbriatum*) and Hair-moss (*Polytrichum commune*) with Brown Bent (*Agrostis canina* s.s.), Wavy Hair-grass (*Deschampsia flexuosa*), Purple Moor-grass (*Molinia caerulea*), Soft-rush (*Juncus effusus*), Common Sedge (*Carex nigra*), Heath Bedstraw (*Galium saxatile*), Tormentil (*Potentilla erecta*), Marsh Violet (*Viola palustris*) and occasionally, where not too shaded, cottongrasses (*Eriophorum* spp.) or Heather (*Calluna vulgaris*).

Thorn scrub and hedgerows

Thorn scrub is quite widespread but local (NVC W21/22). It is usually dominated by dense Hawthorn (*Crataegus monogyna*) with more local Blackthorn (*Prunus spinosa*), plus roses (*Rosa* spp.) and brambles (*Rubus* spp.). Scrub is often found along steeper banks by watercourses or in woodland margins and, of course, as hedgerows; the associate vegetation may indicate colonization of adjacent grasslands but can also reflect former mature woodlands modified through grazing pressure, where persistent woodland relicts can be seen.

Hawthorn is the main species of the extensive managed hedgerows, although various other woody elements are occasionally to frequently found, for example Elder (*Sambucus nigra*), Ash (*Fraxinus excelsior*), Elm (*Ulmus glabra*), Hazel (*Corylus avellana*), roses (*Rosa* spp.), Blackthorn (*Prunus spinosa*) and, though only locally, plums (*P. domestica* or hybrids) and Crab Apple (*Malus sylvestris*). Beech (*Fagus sylvatica*) or Privet (*Ligustrum ovalifolium*) hedges are frequent in urban-fringe areas, usually associated with old estates or urban gardens. Many of the agricultural hedgerows are gappy and neglected and heavy grazing has obliterated any diverse hedge-bank flora; elements of the latter are best observed along roadside verges, notably in upland fringes.

Gorse scrub

Gorse (*Ulex europaeus*) and, to a much lesser extent, Broom (*Cytisus scoparius*) (NVC W23) can be frequent components of marginal, low-fertility (acidic) grasslands in less heavily managed fringes, such as the edges of fields, slopes and so on, and ridges where not heavily grazed. The Gorse typically occurs with relict or invasive Rowan and, on less acidic soils, Hawthorn, and if longer established with occasional Ash or Oak. There are often dense patches of bramble (*Rubus fruticosus*) and Bracken (*Pteridium aquilinum*) may also be present. The associated flora includes bent-fescue grassland relicts (see p. 34), but species that are prominent include Foxglove (*Digitalis purpurea*), Wood Sage (*Teucrium scorodonia*), Honeysuckle (*Lonicera periclymenum*), Slender St-John's-wort (*Hypericum pulchrum*),

Cat's-ear (*Hypochoeris radicata* and various ferns (such as *Dryopteris* spp. and *Oreopteris limbosperma*) and bryophytes.

Bracken scrub

Bracken (*Pteridium aquilinum*) dominates many of the steep brae slopes, notably at the Gleniffer Braes and the hills above Gourock and Greenock, but also at many other local ridges and slopes. At some sites this reflects invasion of bent-fescue-type pastures (NVC U20), where grassland species may persist, but other areas support relict woodland species (NVC W25), often indicated in spring by displays of profuse Bluebell (*Hyacinthoides non-scripta*) and other woodland elements such as Foxglove (*Digitalis purpurea*), Creeping Soft-grass (*Holcus mollis*), Pignut (*Conopodium majus*) and ferns (*Dryopteris* spp.).

Urban scrub

Scrub can be a prominent feature of urban-fringe areas, typically representing, in the absence of grazing or repeated disturbance, late waste-ground succession stages. Goat Willow (*Salix caprea*) can be very common, often with birches (*Betula* spp.) and various other woody species including Elder (*Sambucus nigra*) and Hawthorn (*Crataegus monogyna*), and occasionally aliens such as Butterfly-bush (*Buddleia davidii*), Apple (*Malus pumila*) and whitebeams (*Sorbus* spp.), plus various young trees, notably Sycamore (*Acer pseudoplatanus*) and Ash (*Fraxinus excelsior*). The ground flora is influenced by the mineral and nutrient conditions of the existing open ground substrate, for example rubble, boulders, fine ash or ballast, or compacted ground, and can develop varied species complements; over time more shade-tolerant elements are encouraged.

Grasslands

Renfrewshire is well endowed with grasslands as a result of its long history of agricultural development and historical woodland clearance. The predominant agricultural practice is pasture grazing, mainly for sheep in the uplands with dairy and, less frequently, beef cattle elsewhere. Traditional hay meadows have been replaced by early silage cuts, although areas in the upland fringes still receive only single cuts late in the year. However, the spread of grassland 'improvement' techniques has resulted in diverse areas of grassland being lost or modified, even during the course of modern surveying. Apart from areas within the Renfrewshire Heights, and very locally in the south-east uplands, the vast majority of the grassland encountered can be classed as improved (or 'species poor' semi-improved). Open, bright-green fields can now be seen stretching from the lowlands to the upland slopes. Today areas of species-rich or unimproved grassland are scarce and tend to be restricted to upland areas or steeper ridges and valley sides. Locally, roadside verges, stretches of disused railways and areas or urban waste ground can support diverse grasslands, although these seldom receive sympathetic management.

Unimproved grasslands are classified in the NVC as neutral (mesotrophic), acidic (calcifuge) or calcareous (calcicole) reflecting the soil pH, which is in turn dependent on the underlying geology, soil depth and degree of flushing, but such categories are often obscured by past agricultural treatment.

Acid grasslands

Acid grasslands develop where the underlying rock is base-poor or on surface deposits such as sand, gravel or drained peat. They are a common feature of the uplands (moorland) where they can be somewhat monotonous and more acidic (NVC U5/U6), marked by Mat-grass (*Nardus stricta*), Velvet Bent (*Agrostis vinealis*), Sheep's-fescue (*Festuca ovina*) and Wavy Hair-grass (*Deschampsia flexuousa*), with Heath Bedstraw (*Galium saxatile*) and Tormentil (*Potentilla erecta*); the presence of ericoids (Blaeberry and Heather) indicates affinities or gradations to heath vegetation; other associates include Heath Rush (*Juncus squarrosus*) and Lousewort (*Pedicularis sylvatica*), plus several mosses (notably *Pleurozium schreberii*, *Dicranum scoparium*, *Hypnum cupressiforme*, *Sphagnum capillifolium* and *Polytrichum commune*). The highest ground can support a few local rarities including Stiff Sedge (*Carex bigelowii*) and Viviparous Sheep's-fescue (*Festuca vivipara*), a few clubmosses and, by burn sides, Mossy Saxifrage (*Saxifaga hypnoides*).

Also classified under acidic grasslands are the extensive areas of bent-fescue pasture (NVC U4) which is the typical pasture grassland of the undulating foothills on freer draining soils and can be quite variable in composition. Common grass species include Common Bent (*Agrostis capillaris*), Sheep's-fescue (*Festuca ovina*), Sweet Vernal-grass (*Anthoxanthum odoratum*), Wavy Hair-grass (*Deschampsia flexuosa*), Creeping Soft-grass (*Holcus mollis*), Yorkshire-fog (*Holcus lanatus*) and Spreading Meadow-grass (*Poa humilis*), with herbs such as Heath Bedstraw (*Galium saxatile*), Tormentil (*Potentilla erecta*), Green-ribbed Sedge (*Carex binervis*), Field Wood-rush (*Luzula campestris*), Common Sorrel (*Rumex acetosa*), Cat's-ear (*Hypochoeris radicata*), Pill Sedge (*Carex pilulifera*), Pignut (*Conopodium majus*), Heath Milkwort (*Polygala serpyllifolia*) and occasionally short-grazed Blaeberry (*Vaccinium myrtillus*); bryophytes include *Pleurozium schreberii*, *Rhytidiadelphus squarrosus*, *Hypnum cupressiforme* and *Dicranum scoparium*.

The grassland associated with rock outcrops can be distinctive, due to the shallow skeletal soils of low-fertility and fast drainage (NVC U1 affinity); characteristic species include English Stonecrop (*Sedum anglicum*), Sheep's-sorrel (*Rumex acetosella*), Cat's-ear (*Hypochoeris radicata*), Early Hair-grass (*Aira praecox*) and occasionally Fine-leaved Sheep's-fescue (*Festuca filiformis*), Common Whitlowgrass (*Erophila verna*), Wall Speedwell (*Veronica arvensis*) and Blinks (*Montia fontana* ssp. *chondrosperma*); mosses include short *Polytrichum* spp. and *Racomitrium* spp., usually with various shrubby, foliose and crustose lichens.

Neutral grasslands

Neutral grasslands (NVC MG5/6a) occur on more fertile, often lowland soils, and have tended to suffer the most from agricultural improvement. They typically have as major constituents grasses such as Common Bent (*Agrostis capillaris*), Red Fescue (*Festuca rubra*), Smooth Meadow-grasses (*Poa pratensis* agg.), Crested Dog's-tail (*Cynosurus cristatus*) and Yorkshire-fog (*Holcus lanatus*); associate species include Ribwort Plantain (*Plantago lanceolata*), Meadow Buttercup (*Ranunculus* acris), Common Sorrel (*Rumex acetosa*),

Yarrow (*Achillea millefolium*), Knapweed (*Centaurea nigra*), Ox-eye Daisy (*Leucanthemum vulgare*), Lady's Bedstraw (*Galium verum*), Devil's-bit Scabious (*Succisa pratensis*), Greater Butterfly-orchid (*Platanthera chlorantha*), Bird's-foot Trefoil (*Lotus corniculatus*), Autumnal Hawkbit (*Scorzoneroides autumnalis*), Meadow Vetchling (*Lathyrus pratensis*), Yellow Rattle (*Rhinanthus minor*), Ragwort (*Senecio jacobaea*), Smooth Lady's-mantle (*Alchemilla glabra*), clovers (*Trifolium* spp.), Self-heal (*Prunella vulgaris*), Oval Sedge (*Carex leporina*) and bryophytes such as *Rhytidiadelphus squarrosus*, *Kindbergia praelonga* and *Lophocolea bidentata*.

Locally short-grazed grasslands can be notably diverse and several species present indicate relatively more calcicolous conditions (NVC CG10, but also U4c and grading to MG5c) and occur at various places, usually on slopes or ridges. They can be formed at local and relatively base-rich igneous rock outcrops, or in places with under-lying superficial deposits, but any distinct calcareous grasslands of lowland carboniferous limestone rock exposures have long succumbed to urbanization. The various indicator species typically include Bird's-foot Trefoil (*Lotus corniculatus*), Lady's Bedstraw (*Galium verum*), Mouse-ear (*Pilosella officinarum*), eyebrights (*Euphrasia* spp.), Common Milkwort (*Polygala vulgaris*), Fairy-flax (*Linum catharticum*), Harebell (*Campanula rotundifolia*), Mountain Pansy (*Viola lutea*), Bitter-vetch (*Lathyrus linifolius*), Hairy Lady's-mantle (*Alchemilla filicaulis* ssp. *vestita*), Spignel (*Meum athamanticum*), Burnet-saxifrage (*Pimpinella saxifraga*), Spring-sedge (*Carex caryophyllea*), Heath Speedwell (*Veronica officinalis*) and Wild Thyme (*Thymus polytrichus*), and can include several rarities such as Field Gentian (*Gentianella campestris*), Moonwort (*Botrychium lunaria*), Frog Orchid (*Coeloglossum viride*) and Heath Fragrant-orchid (*Gymnadenia borealis*).

Wet grasslands

Where grassland is poorly draining but not heavily grazed it can be marked by an abundance of tussocky Tufted Hair-grass (*Deschampsia cespitosa*) – NVC MG9, usually with limited associates such as Common Sorrel (*Rumex acetosa*), buttercups (*Ranunculus* spp.), Marsh Thistle (*Cirsium palustre*), Angelica (*Angelica sylvestris*), willowherbs (*Epilobium* spp.), Sneezewort (*Achillea ptarmica*) and various other marshy grassland elements. Where the water table remains fairly high, rushes and sedges tend to dominate, accompanied by an increase in species more typical of true wetlands (see 'Mires' on p. 37).

Wet pastures (NVC MG10) are a very common feature of agricultural land, and are marked by increased Yorkshire-fog (*Holcus lanatus*), Creeping Buttercup (*Ranunculus repens*) and Soft-rush (*Juncus effusus*) and various other marshy elements. These wetter pastures generally occur with, and grade into, rush-dominated pastures or marshy grasslands (see 'Mires' on p. 37).

On a small scale, but quite common, are swampy mats of Floating Sweet-grass (*Glyceria fluitans*), often with Creeping Bent (*Agrostis stolonifera*), which occur about shallow ditches, pond margins and marshy depressions, usually reflecting enriched water input (NVC MG11/S22). Associates of swamp zones and ditch sides include Brooklime (*Veronica beccabunga*), Bog Stitchwort (*Stellaria alsine*), water forget-me-nots (*Myosotis* spp.) and water-cresses (*Nasturtium* spp.). Related flush-vegetation types (NVC M35), including local springs and seepage areas, can support additional species such as Small Sweet-grass (*Glyceria declinata*), Blinks (*Montia fontana* ssp. *fontana*) and often the water crowfoot species (*Ranunculus omiophyllus* and *R. hederaceus*) and various bryophytes.

Poorly draining pasture hollows or margins of open waterbodies are subject to inundation and can develop a distinctive flora (NVC MG13). Of the grasses, Creeping Bent (*Agrostis stolonifera*) is common, as is Marsh Foxtail (*Alopecurus geniculatus*), and frequent low-growing herbs – several annuals – include Toad Rush (*Juncus bufonius*), Marsh Cudweed (*Gnaphalium uliginosum*), Redshank and Water-pepper (*Persicaria* spp.), Curled Dock (*Rumex crispus*) and Marsh Yellow-cress (*Rorippa palustris*).

Dwarf shrub heaths

Heaths are characterized by vegetation dominated by ericoid shrubs, although graminoid vegetation can be prevalent. Typically they occur on acidic soils with a low nutrient status, including peat and sand. Away from the uplands they are frequently small scattered sites, often on steep slopes, and in mosaics with acidic grassland or mires. There are two main divisions, which can intergrade: dry and wet – the latter implies poor drainage and often occurs on leveller ground or in high rainfall areas, where it is usually classed as mire.

Dry heath

Dry heath (NVC H12) is typified by Heather (*Calluna vulgaris*), usually with subordinate Blaeberry (*Vaccinium myrtillus*), and various acid grassland elements such as Wavy Hair-grass (*Deschampsia flexuosa*), Sheep's-fescue (*Festuca ovina*) and Velvet Bent (*Agrostis vinealis*), plus Heath Bedstraw (*Galium saxatile*) and Tormentil (*Potentilla erecta*) and bryophytes (such as *Hypnum cupressiforme*, *Dicranum scoparium*, *Pleurozium schreberii*, *Polytrichum* spp.) and various *Cladonia* spp. lichens. Blaeberry (*Vaccinium myrtillus*) tends to be especially common on steep, north-facing slopes about rock outcrops, usually with much *Pleurozium schreberii* and other bryophytes (NVC H18). The NVC Survey (1992) of Clyde Muirshiel Regional Park recognized large areas of the more oceanic Calluna–Erica heathland (NVC H10), characterized by the increase of Bell Heather (*Erica cinerea*) in the west.

Wet heath

Wet heath (NVC M15) is distinguished by the presence of Purple Moor grass (*Molinia caerulea*), Deergrass (*Trichophorum germanicum*) and Cross-leaved Heath (*Erica tetralix*), in addition to Heather and Blaeberry, and also by the increased presence of bog-mosses such as *Sphagnum capillifolium* and more locally *S. compactum* and other bryophytes; other species include Heath Rush (*Juncus squarrosus*), Heath Wood-rush (*Luzula multiflora*), Green-ribbed Sedge (*Carex binervis*), Tormentil (*Potentilla erecta*), Heath Lousewort (*Pedicularis sylvatica*), Heath Spotted-orchid (*Dactylorhiza maculata*) and various calcifuge grasses. Wet heath can be well represented at the margins of the Renfrewshire Heights where grazing suppresses the shrubby heather 'canopy' more typical of within the CMRP boundary.

Mires

Mires are classed as peat-forming vegetation occurring on peat deposits of a depth greater than 0.5m, with the water at or just below the surface for most of the year. They comprise flushes, fens, bogs and wet heaths (see Dwarf shrub heath above); a major division occurs when the peat is flushed with mineral groundwater (minerotrophic – fens or flushes) or fed purely by rainwater (ombrotrophic – bogs).

Peat bogs

Vegetation occurring on peat with a depth of greater than 0.5m, fed solely by rainwater, is commonly referred to as 'bog' or 'moss'. Typically bogs are described as being of two main types: blanket bog (NVC M19), an upland type, occurring in areas of high rainfall, where the peat carpets the rolling hillsides, and raised bog (NVC M18), discrete units of often very deep peat, usually in lowlands, occurring in water-logged depressions or on flood plains. In Renfrewshire the latter are now quite rare with relict examples – degraded or mostly destroyed – on the flood plain to the north-west of Paisley, but elsewhere blanket bogs occur extensively on the higher ground of the Renfrewshire Heights and the south-east uplands, notably Eaglesham Moor.

The typical wet raised bog flora consists of bog-mosses (*Sphagnum* spp., particularly *S. papillosum* and *S. magellanicum*), usually in thick carpets or characteristic hummocks, with cottongrasses (*Eriophorum* spp.), occasionally some Purple Moor-grass (*Molinia caerulea*) or Deergrass (*Trichophorum germanicum*) and the ericoid shrubs Heather (*Calluna vulgaris*) and Cross-leaved Heath (*Erica tetralix*); other species include Bog Asphodel (*Narthecium ossifragum*), Cranberry (*Vaccinium oxycoccos*), Crowberry (*Empetrum nigrum*) and Round-leaved Sundew (*Drosera rotundifolia*), and in addition to bog-mosses (*Sphagnum* spp.) there can be a range of other bryophytes, notably leafy liverworts, and shrubby lichens. On the highest ground of the Renfrewshire Heights the blanket bog 'moorland' additionally supports Cloudberry (*Rubus chamaemorus*), Lesser Twayblade (*Neottia cordata*) and Cowberry (*Vaccinium vitis-idaea*).

Bog pools

These are acidic mires where the groundwater only picks up limited minerals. They are associated with bogs as internal pools or marginal zones and typically support carpets of luxurious bog-mosses (*Sphagnum* spp.). Common Cottongrass (*Eriophorum angustifolium*) often marks out bog pool communities (NVC M2/3) with or without *Sphagnum* spp.; associates are limited but may include Common Sedge (*Carex nigra*), White Sedge (*Carex canescens*), Purple Moor-grass (*Molinia caerulea*) and other bog species.

Poor fens

With increasing mineral enrichment, for example in bog-margins 'laggs' or other peaty areas where there is lateral water movement, the vegetation is characterized by increased sedges (NVC M4/5) such as Bottle Sedge (*Carex rostrata*), White Sedge (*Carex canescens*) and Common Sedge (*Carex nigra*), and also a number of herbs, reflecting mineral enrichment, and characteristic bryophytes.

The commonest poor fen is sedge-and-rush-dominated (NVC M6), where typically Soft-rush (*Juncus effusus),* Sharp-flowered rush *(J. acutiflorus)* or Common Sedge (*Carex nigra*) dominates and associates include Star Sedge (*Carex echinata*), Marsh Cinquefoil (*Comarum palustre*), Brown Bent (*Agrostis canina*), Marsh Violet (*Viola palustris*), Marsh Willowherb (*Epilobium palustre*), Heath Bedstraw (*Galium saxatile*), Tormentil (*Potentilla erecta*), occasionally Narrow Buckler-fern (*Dryopteris carthusiana*) and often luxurious bog-mosses (notably *Sphagnum palustre, S. fimbriatum* and *S. fallax*) and Hair-moss (*Polytrichum commune*).

Rich fens

These fens have distinct mineral-enriched seepage and usually support the highest species diversity, often including several species of local rarity. They are marked by sedge domination over a carpet of bryophytes (NVC M9). Bottle Sedge (*Carex rostrata*) often dominates, but not with the monotony of swamp stands; associates include Marsh Cinquefoil (*Comarum palustre*), Marsh-marigold (*Caltha palustris*), Bogbean (*Menyanthes trifoliata*), Common Marsh-bedstraw (*Galium palustre*), Creeping Forget-me-not (*Myosotis secunda*), Spearwort (*Ranunculus flammula*), Marsh Willowherb (*Epilobium palustre*), Marsh Pennywort (*Hydrocotyle vulgaris*) and other sedges (including *Carex nigra* and *C. panicea*).

Of high interest are the local flush areas (NVC M9b), often channels, supporting distinctive species including various sedges (some locally rare such as *Carex paniculata, C. limosa, C. diandra, C. lepidocarpa*), Marsh Arrow-grass (*Triglochin palustris*), Marsh Lousewort (*Pedicularis palustris*), Bog Pondweed (*Potamogeton polygonifolius*), Bulbous Rush (*Juncus bulbosus*) and locally rare bog-mosses and other 'brown-mosses' (such as *Calliergon* spp., *Drepanocladus* spp., *Plagiomnium* spp.). Other short-sedge and moss-dominated, often species-rich flushes, occur locally on sloping ground (NVC M10); sedges can often include Dioecious Sedge (*Carex dioica*), Flea Sedge (*C. pulicaris*) and Tawny Sedge (*C. hostiana*) and Few-flowered Spike-rush (*Eleocharis quinqueflora*).

Rush pastures

Rush-dominated pastures are a common feature of the Renfrewshire agricultural landscape, occurring on waterlogged mineral or peaty soils; they range from large flushes on upland hill slopes to poorly draining, low-lying flood plains, where maintained by grazing. Sharp-flowered Rush (*Juncus acutiflorus*) and Soft-rush (*Juncus effusus*) are the major dominants of the marshy vegetation (NVC M23a/b), the latter being more characteristic of low-lying wet areas. Associates vary, and may often be limited, but often can be quite diverse; species typically include Cuckooflower (*Cardamine pratensis*), Common Marsh-bedstraw (*Galium palustre*), willowherbs (*Epilobium* spp.), Spearwort (*Ranunculus flammula*), Marsh Ragwort (*Senecio aquaticus*), Marsh-marigold (*Caltha palustris*), Ragged-Robin (*Silene flos-cuculi*), Meadowsweet (*Filipendula ulmaria*), Angelica (*Angelica sylvestris*), Greater Bird's-foot Trefoil (*Lotus pedunculatus*), Marsh Thistle (*Cirsium palustre*), Yorkshire-fog (*Holcus lanatus*), Creeping Bent (*Agrostis stolonifera*) and various bryophytes.

Purple Moor-grass mire

Purple Moor-grass (*Molinia caerulea*) is a widespread and common species of degraded bogs, wet heaths and some rush pastures, and can dominate distinct mire areas (NVC M25), but usually only locally. Tormentil (*Potentilla erecta*) is a strong associate, although the vegetation is often species poor; other species recorded include Wavy Hair-grass (*Deschampsia flexuosa*), Brown Bent (*Agrostis canina*), Carnation Sedge (*Carex panicea*), Whorled Caraway (*Carum verticillatum*), Devil's-bit Scabious (*Succisa pratensis*), Angelica (*Angelica sylvestris*) and Sharp-flowered Rush (*Juncus acutiflorus*), while the presence of bog-mosses (*Sphagnum* spp.), Bog Asphodel (*Narthecium ossifragum*), Cross-leaved Heath (*Erica tetralix*) and Heather (*Calluna vulgaris*) show affinities to more acidic, often modified, bog vegetation (NVC M25a).

Tall herb meadows

Marshy areas dominated by tall herbs (NVC M27) occur infrequently along watercourse margins and at the margins of lochs or ponds, and are dependent on reduced grazing pressure – and hence are now quite rare. They are an attractive and colourful sight with indicator species such as Meadowsweet (*Filipendula ulmaria*), Angelica (*Angelica sylvestris*), Valerian (*Valeriana officinalis*) and Yellow Iris (*Iris pseudacorus*), with various associates such as Ragged-Robin (*Silene flos-cuculi*), Marsh-marigold (*Caltha palustris*), Water Avens (*Geum rivale*) and other species noted in this section about rush pastures (p. 38).

Swamps

Swamp vegetation occurs on mineral soils when the water table is above ground level for most of the year. It is formed from emergent vegetation in standing water, dominated by reeds or tall sedges, either in single-species or mixed stands, usually with few associates, and typically where mesotrophic or eutrophic waters occur.

Swamp communities frequently seen, often at a large scale, are those dominated by Bottle Sedge (*Carex rostrata*) (NVC S9/S27) and Water Horsetail (*Equisetum fluviatile*) (NVC S10). Other common but more local swamp dominants of wetlands or open-water margins include Reedmace (*Typha latifolia*) (NVC S12), Bur-reed (*Sparganium erectum*) (NVC S14), Common Spike-rush (*Eleocharis palustris*) (NVC S19), Reed Sweet-grass (*Glyceria maxima*) (NVC S5), Common Reed (*Phragmites australis*) (NVC S4) and Floating Sweet-grass (*Glyceria fluitans*) (NVC S22). Reed Canary-grass (*Phalaris arundinacea*) (NVC S28) is a common dominant of seasonally dry swamps at a number of wetland sites, and is frequent along flood-plains and on the margins of lowland watercourses.

Open water

Open water can support a range of free-floating and submerged aquatic plants such as various pondweeds, duckweeds and other 'waterweeds'. Systematic sampling of the open water has not been carried out in Renfrewshire and although several lochs were surveyed in detail by the SNH Loch Survey (1996) NVC codes were not applied.

Broad-leaved Pondweed (*Potamogeton natans*) is the commonest floating macrophyte found in the area, occurring at most open waterbodies and usually tolerant of some enrichment; in similar places but less frequently seen are Amphibious Bistort (*Persicaria amphibia*) and Yellow Water-lily (*Nuphar lutea*). Many of the rural open waterbodies (still or slow flowing) are minerotrophic but low in bases and the submerged macrophytes are seldom luxurious but frequently include Perfoliate Pondweed (*Potamogeton perfoliatus*), Alternate Water-milfoil (*Myriophyllum alterniflorum*), Small Pondweed (*P. berchtoldii*), Blunt-leaved Pondweed (*P. obtusifolius*), Canadian Pondweed (*Elodea canadensis*), Water Starworts (*Callitriche* spp.) and, on stony loch beds, Shoreweed (*Littorella uniflora*).

Coastal vegetation

Although Renfrewshire has a long coastline, coastal vegetation is not extensive but there is a complex range of communities still to be found. The conurbation between Gourock and Port Glasgow effectively separates the western Inverclyde coast from the more estuarine Clyde further to the east.

The extensive tidal mudflats about Langbank support Dwarf Eel-grass (*Zostera noltei*) and Beaked Tasselweed (*Ruppia maritima*) (NVC SM1/2), but glasswort (*Salicornia* sp.) (NVC SM8) has not been seen for a very long time. Sea Aster (*Aster tripolium*) and scurvy-grasses (*Cochlearia* spp.) colonize muddy and stony margins to the fringes of the saltmarsh zones (NVC SM12).

Local areas of saltmarsh (NVC SM13/16) occur scattered in pockets along the Inverclyde coast but are better developed, but still small, east of Langbank; species seen include Common Saltmarsh-grass (*Puccinellia maritima*), Red Fescue (*Festuca rubra* sspp.), Creeping Bent (*Agrostis stolonifera*), Sea Arrow-grass (*Triglochin maritima*), Sea Plantain (*Plantago maritima*), Sea Milkwort (*Glaux maritima*), Sea-spurrey (*Spergularia maritima*), Salt Rush (*Juncus gerardii*) and goosefoots (*Atriplex* spp.).

Brackish-influenced swamp communities occur by the Clyde at Erskine and by the estuarine River Cart, with large populations of Sea Club-rush (*Bolboschoenus maritimus*) (NVC S21) and Common Reed (*Phragmites australis*) (NVC S4d); scurvy-grasses (*Cochlearia* spp.) can be marginal associates.

Strandline and sandy grasslands occur to the landward side of saltmarshes but, due to land reclamation and disturbance, are scarcely well developed. Strandline-vegetation species (NVC SD1-4) are quite variable but indicators include Lyme-grass (*Leymus arenarius*), goosefoots (*Atriplex* spp.), Sea Radish (*Raphanus raphanistrum* ssp. *maritimus*), Sea Mayweed (*Tripleurospermum maritimum*), Sea Purslane (*Honckenya peploides*), Perennial Sowthistle (*Sonchus arvensis*), Curled Dock (*Rumex crispus* ssp. *littoreus*), Couch-grass (*Elymus repens*) and Tall Fescue (*Festuca arundinacea*). A small area of coastal influenced, sandy grassland (NVC SD8?) at Erskine is of note for supporting populations of Spreading Meadow-grass (*Poa humilis*), Little Mouse-ear (*Cerastium semidecandrum*),

Squirrel-tail Fescue (*Vulpia bromoides*), Bird's-foot (*Ornithopus perpusillus*), Stork's-bill (*Erodium cicutarium*) and *Taraxacum* spp. (Section *Erythrosperma*).

On the Inverclyde west coast there are small areas of beach and coastal rock receiving more typical maritime waves; here Thrift (*Armeria maritima*), Sea Campion (*Silene uniflora*) and colourful lichens have colonized the rock (NVC MC2), and there is a narrow strandline in places, but little saltmarsh.

Tall herb stands and rank grasslands

This category covers open areas, often neglected and nutrient enriched, where tall herbs and grasses predominate. Pure stands of tall herbs are rarely extensive; they tend to occur particularly on disturbed alluvial ground beside watercourses and on open waste ground or neglected grasslands, often where there are nutrient-enriched soils and generally a lack of any cropping apart from occasional fires.

Tall grasses are invariably present to some degree and therefore intergrade with neutral grasslands. The commonest tall grasses (NVC MG1) are False Oat-grass (*Arrhenatherum elatius*) and Cock's-foot (*Dactylis glomerata*), with Couch-grass (*Elytrigia repens*) where conditions are damper; at many low-growing former pasture sites a lack of cutting encourages dense Creeping Soft-grass (*Holcus mollis*) or even Common Bent (*Agrostis capillaris*) and other short grasses forming dense suppressive turf. Bracken (*Pteridium aquilinum*) is also a common invader of neglected pastures in the upland fringe (NVC U20).

Commonly seen tall herb stands or populations include Rosebay Willowherb (*Chamerion angustifolium*), Common Nettle (*Urtica dioica*), Creeping Thistle (*Cirsium arvense*), Cow Parsley (*Anthriscus sylvestris*), Hogweed (*Heracleum sphondylium*), Raspberry (*Rubus idaeus*), brambles (*R. fruticosus*) and more local stands of neophyte aliens such as goldenrods (*Solidago* spp.), Lupin (*Lupinus polyphyllus*), Russian Comfrey (*Symphytum* x *uplandicum*), Confused Michaelmas-daisy (*Aster novi-belgii*) where damp and, notably along watercourses, the three notorious species Indian Balsam (*Impatiens glandulifera*), Japanese Knotweed (*Fallopia japonica*) and Giant Hogweed (*Heracelum mantegazzianum*).

Short herb and ephemeral communities

Recently disturbed sites are rapidly colonized by low-growing plants that are short-lived and often ephemeral. Such casual species do not usually persist at these sites unless the area is re-disturbed or the soil conditions are particularly stressful, for example nutrient poor, toxic, unstable or arid. Normally such sites in urban situations, if left alone, soon develop a denser cover of taller rank grassland or scrub.

Typical open-ground species include Colt's-foot (*Tussilago farfara*), groundsels and ragworts (*Senecio* spp.), hawkweeds (*Hieracium* spp.), dandelions (*Taraxacum* spp.), Perforate St John's-wort (*Hypericum perforatum*), Ox-eye Daisy (*Leucanthemum vulgare*), clovers (*Trifolium* spp.) and several other legumes, and usually also a broad range of other herbs, grasses and mosses, which can include a number of aliens, some rare casuals (in total, too numerous to list).

A distinctive feature of the local urban waste ground (though not really ephemeral) is disturbed, often clayey, ground that is poorly drained due to compaction, which supports some marshy or flush elements; this vegetation is widespread at urban sites and can be remarkably consistent in its species complement: typical indicators are Jointed Rush (*Juncus articulatus*) and Compact Rush (*J. conglomeratus*), occasionally Hard Rush (*J. inflexus*), Selfheal (*Prunella vulgaris*), Crested Dog's-tail (*Cynosurus cristatus*), White Clover (*Trifolium repens*), Oval Sedge (*Carex leporina*), Glaucous Sedge (*Carex flacca*), Red Bartsia (*Odontites vernus*) and orchids (*Dactylorhiza* spp.), and a notable moss is *Calliergoniella cuspidata*.

A traditional place to see open vegetation is at weedy agricultural fields, but these are a rare sight today, and the few arable fields – mainly barley – surveyed tend to be species poor. Species found are usually restricted to marginal zones and mostly common species; typical species are Common Chickweed (*Stellaria media*), Annual Meadow-grass (*Poa annua*), hemp-nettles (*Galeopsis* spp.), Knotgrass (*Polygonum aviculare*), Redshank (*Persicaria maculosa*), Shepherd's-purse (*Capsella bursa-pastoris*), Corn Spurrey (*Spergula arvensis*), Fat-hen (*Chenopodium album*), Red Dead-nettle (*Lamium purpureum*), Pineappleweed (*Matricaria discoidea*), Prickly Sowthistle (*Sonchus asper*), Scentless Mayweed (*Tripleurospermum inodorum*), speedwells (*Veronica* spp.), Groundsel (*Senecio vulgaris*), Common Orache (*Atriplex patula*) and Field Pansy (*Viola arvensis*).

1.8 Recorders and Recording History

The earliest written records about plants in Renfrewshire date from the late eighteenth century, but it was not until the early nineteenth century that the first substantial species lists and descriptions were written. Information about the flora of the vice-county gradually increased through the nineteenth century, and by studying the various sources produced by the end of the 1860s it is possible to obtain a picture of the flora of this time.

Details about the flora and vegetation from long before the eighteenth-century have been obtained from modern research, notably from the archaeobotanical work of Professor JH Dickson, CA Dickson and Dr Susan Ramsay, all at the University of Glasgow. This research has provided records of the flora from after the last Ice Age to the medieval period (see Chapter 1.6).

The following summarizes the various sources and recording periods which provide the historical basis of this flora. It also serves as an acknowledgement of the many people who have contributed records during the last 200 years of local botanical recording. Abbreviation codes for individual recorders, authors or publications – as used in Part 2, The Catalogue of Species, and listed on p. 54–5 – are provided within parentheses, as are bibliographical references.

The earliest known plant lists for Renfrewshire are to be found in the descriptions of some of the parishes in *The Statistical Accounts of Scotland* and *The New Statistical Accounts of Scotland*, which were written by local church ministers and date from 1791–9 and 1834–45 respectively. The accounts provide little detail and vary by author, but in general furnish a few interesting records and comments on the landscape of the time, and some include short species lists. Additional useful sources are the local accounts of farming by Wilson (1812) and general history by Crawford and Robertson (1818); both are primarily about landscape but contain a few references to certain species – for example trees, farm weeds and moorland plants – representing a number of first written records for the area (see Chapter 1.6).

The remarkable publication in 1813 of *Flora Glottiana* by Thomas Hopkirk (TH) represented the first substantial account of plants found in the Glasgow area (670 flowering plants and 32 ferns). Most named localities are only relevant to the eastern parts of the vice-county, but statements such as 'around Glasgow' or even just 'common' or similar would indicate a plant was very likely to be present in Renfrewshire at the time.

Flora Scotica (1821) by Professor Sir William Jackson Hooker (FS) listed quite a few plants from Renfrewshire locations, plus again a few comments about plants 'around Glasgow'. Another valuable early nineteenth-century source is 'A catalogue of plants observed in Renfrewshire' by J Montgomery (MC; dated 1834) reproduced in *The New Botanist's Guide* (Watson 1837). The catalogue lists 70 species, some from more than one location within the vice-county.

The specimens held in the various local herbaria provide an important source of verifiable plant records dating from the 1830s onwards. The oldest and most numerous herbarium records belong to the collection of the University of Glasgow Botany Department

(GL), but the Herbarium of Glasgow Museums (GLAM) also contains many records dating from the 1840s. The herbarium of the Anderson's Institution, later Anderson's University and then the University of Strathclyde, (GGO), is of note for the specimens collected by Roger Hennedy (Chair of Botany in 1863); Hennedy was very active in the Paisley area, taking his botany class to local sites. Unfortunately, a number of his specimens were trimmed at a later date, deleting early locality details: the act of an unidentified 'vandal', according to Professor Blodwen Lloyd Binns, who resurrected and cared for the collection in the 1950s (Macpherson and Watson 1996). These three herbaria (GL, GLAM and GGO) are now all part of Glasgow Museums' Collection and housed at Glasgow Museums Resource Centre, Nitshill. For further details on these collections see Patton (1954) regarding GL, Jones (1980) regarding GLAM and Lloyd (1964) regarding GGO.

The many Renfrewshire records in the Herbarium of Paisley Museum (PSY) tend to date from the late nineteenth to early twentieth century, but have provided very useful vouchers, particularly due to the efforts of Thomas Henry and later Daniel Ferguson. The smaller collection at the McLean Museum in Greenock (GRK) also contains a number of interesting local records.

The publication of *The Clydesdale Flora* (1865) by Roger Hennedy (RH) marked the first detailed and thorough census of local plants, providing evidence for the first records for many species: 'Paisley Canal Bank' and the 'hills above Gourock' appear frequently in the text. In all there were five editions of his flora produced (two published after Hennedy's death in 1876); the final version appeared in 1891 and was revised by Thomas King, Professor of Botany at Anderson's University. Richard McKay's plant lists in *Notes on the Fauna and Flora of West Scotland* (FFWS; 1876) reproduce many of Hennedy's observations, supplemented by additional records from the growing body of local naturalists.

Two other related sources of early plant records for the area are two unpublished lists: Morris Young's 'Flora of Renfrewshire' (MY) and the Paisley Philosophical Institution's 'The Flora of Renfrewshire (Paisley and Neighbourhood)' (PPI). The latter, dated *c.* 1869, was produced by the Botanical Section and appears to be a list compiled by Thomas Henry, who collected many specimens now in the Herbarium of Paisley Museum. The Morris Young catalogue also dates from around 1869 and represents a list of plants with some locality names annotated in pencil. Some of the species annotated are not from the county so there is some ambiguity about its title and content (Weddle 2008). However, together both lists, if treated with some caution, appear to represent a useful source of plants and complement Hennedy's text, which often lacks a definite indication that an otherwise common plant was actually recorded from the vice-county.

From the early 1850s to the beginning of the twentieth century, various natural history society journal articles and excursion reports have generated many interesting finds and a number of species lists for favoured botanical haunts. The main journals are the *Reports and Transactions of the Glasgow Society of Field Naturalists* (TGSFN; 1873–7), *Proceedings and Transactions of the Natural History Society of Glasgow* (NHSG; 1858–1908), *Annals of the Andersonian Naturalists' Society* (AANS; 1893–1914), and later *The Glasgow Naturalist* (GN; 1909 onwards); additionally there are the *Transactions of the Paisley Naturalists' Society* (TPNS; 1912 onwards). However, apart from the odd exception, the flow of botanical records in natural history society journals appears to dry up in the first half of the twentieth century.

A few other occasional publications also provide local plant records, often with descriptive text on the local environment; these include *Rambles Round Glasgow* by Hugh MacDonald (HMcD; 1854, 1856), *Busby and its Neighbourhood* by the Rev. W Ross (1883) and *Kilmacolm: A Parish History* by James Murray (1898). 'Rambles with the Paisley Botanists' (RPB), an article in *The Paisley & Renfrew Gazette* (Anon. 1869), provides a month-by-month account of findings by an anonymous 'occasional companion' of local botanists of the Paisley Philosophical Society, adding that they 'tend to work and somewhat secretively make their forays … before decent folk are out of bed'.

The *Topographical Botany* publications (TB; Watson 1873) gradually pulled together the various plant records and established the vice-county system. The second volume (Baker and Newbold 1883) contains records for Scotland; the main source for Renfrewshire records was a manuscript catalogue provided by Roger Hennedy entitled the 'Hennedy Catalogue' and dated 1872. Peter Ewing published 'A Contribution to the Topographical Botany of the West of Scotland' and following editions to this (Ewing 1890, 1892, 1892, 1897), which contain several first named sources for Renfrewshire. At the start of the twentieth century the British Association's *Fauna, Flora and Geology of the Clyde Area* (FFCA; Elliot et al. 1901) was published. The vascular plant section was compiled by Ewing and listed all the plants of the Clyde area with reference to their geographical zones; unfortunately, many of the records are difficult to localize due to the grid system employed, but most can be related back to the earlier publications.

The plant list 'Renfrewshire Plants' (TPNS 1915) represents the first published checklist of the Renfrewshire flora. The work was based on the efforts of the Botanical Committee of the Paisley Naturalists' Society – Daniel Ferguson (DF) was the convenor and the others D Crilley, J Finnie, Rev. C Hall, R Houston and A Stewart). The list recorded '740 species' but it does not include any locality details, a deliberate act: 'one of the main objects of the Paisley Naturalists' Society is to guard against the extermination of rare plants'. The list included 'truly indigenous plants' but also those described as being 'introduced through human agency and now thoroughly established'; casuals were included if their appearance had been of a recent date. It also lists 24 'Excluded Species' – species previously listed in 1901 as occurring in Renfrewshire but presumed extinct by the time of the 1915 publication. Species considered rare are also annotated (see Chapter 3.2 p. 360).

The work of Robert Grierson (RG), the celebrated recorder of 'waste' places, as highlighted in *The Changing Flora of Glasgow* (Dickson et al. 2000), produced a number of interesting records of casuals from the eastern edge of the vice-county; most of his work is summarized in his *Clyde Casuals* publications (1930). Another useful source of casual rarities from literature and herbarium sources is the work of Ferguson in Paisley around 1910 (Ferguson 1911 and PSY). Additionally, the collecting by R and T Wilkie (*c*. 1892; specimens now in GL) also provided a number of old, interesting and verifiable casual records.

JR Lee's 1933 *Flora of the Clyde Area* is the last flora of the wider Clyde region, collating the work of many authors over the preceding years, and is a key reference source for interpreting old literature records. It indicates whether a species has been recorded in Renfrewshire, and often names a particular station; a drawback is that the records are undated and may relate back to older nineteenth-century finds.

Although the general flow of records declines during the middle part of the twentieth century, this lack is in some measure countered by the efforts of Robert Mackechnie (RM). He was a very keen, knowledgeable and thorough botanist who kept detailed field notebooks, which list many plants from named and dated locations, and amassed a very large herbarium; the latter is mainly housed at the Royal Botanical Gardens Edinburgh (E), but some 10,000 specimens are deposited in Glasgow (GLAM), of which a good proportion are from Renfrewshire. His collection, literature and notebook records are important sources for recording the appearance and spread of alien species, in addition to locating many scarce or rare natives present in the twentieth century.

In the second half of the twentieth century an active band of Botanical Society of the British Isles (BSBI) fieldworkers amassed a wealth of data, mostly between the late 1950s and 1986, systematically recording the vice-county on a 1km-square basis. Robert Mackechnie (RM) was an early leader of the group, in his capacity as BSBI recorder; later this mantle was passed to Elizabeth Conacher (ERTC). Basil Ribbons (BWR) was a key player at this time and was very active in organizing field visits and botany classes and encouraged a large number of keen helpers. Many field cards were completed during this period, representing more than 33,000 individual records, the majority of which are now entered into the database compiled for this flora, and are displayed on the distribution maps. Basil Ribbons provided the author with a list of the many helpers and students on these field forays, and these are listed in table 8. One of the helpers, JD Morton (JDM), was an avid researcher and extracted many records from old literature sources, as well as providing field records. A further recorder at this time was JH Penson (JP), whose notes refer to a number of rare and unexpected finds, but unfortunately most lack verification or corroboration and so several records have been omitted.

Table 8. People who assisted Basil Ribbons with recording between 1960 and 1986.

Mary Allan	J McKee
John M Blackwood	Jean Mackinlay
Eleanor Byers	M Mathieson
ME Byres	Amelie R Miller
Dr Kathleen Calver	Jean Millar
Arthur Copping	Kathleen Niven
Nancy CT Conacher	J Parkinson
Agnes R Drysdale	CB Peach
Flora M Elder	Alfred A Percy
MB Kerr	Grant Pollit
Annie Laird	Alex Shaw
JM Lennie	AD Smith
May Little	Peter J Wanstall
AD Macfarlane	John Watt
Esther LP Mackechnie	

In the 1980s Dave Mellor (DM), later joined by Shona Allen (SA), working in the Natural History section of Paisley Museum, established a biological records centre for Renfrewshire – the records are now part of Glasgow Museums' Biological Records Centre (LRC) – and encouraged the cataloguing of existing data and also developed an active fieldwork programme, linked to growing interest in plant and habitat conservation. In 1981 Paisley Museum produced an annotated list, entitled 'Uncommon Plants of Renfrewshire', and in 1989 produced a provisional checklist called 'The Plants of Renfrewshire'.

Other valuable records from around the 1980s came from the work of two very knowledgeable and expert local botanists. Allan McGregor Stirling (AMcGS) botanized extensively in the Clyde area, and although he did not amass many field cards for the vice-county, he was still responsible for many interesting finds – in particular he was an important source for rose, bramble and dandelion records. Dr Alan Silverside (AJS) came to the now University of the West of Scotland in the mid 1970s and produced species lists for several localities, and in 1976 collated a list of notable species, many of which were unrecorded locally in the national plant atlas; he has a keen eye for the unusual rarity and expertise in critical genera, notably monkeyflowers, eyebrights, brideworts and dandelions. Additionally Dr Peter Macpherson (PM), although concentrating on neighbouring Lanarkshire (VC77), has been the source of many locally rare, mostly alien, records.

Other contemporary recorders include Ian Green (IPG), a very sharp-eyed botanist who in a short period contributed many valuable records, notably from the Lochwinnoch area; his brother Paul (PRG) has also contributed some notable finds. Modern records were also generated during local field surveys, including those from the 1992 vegetation survey of the Clyde Muirshiel Regional Park (CMRP-NVC), those made by Clive Dixon (CD) during a survey of Renfrewshire grasslands for the Nature Conservancy Council Scotland (1995), and those of John Day who worked on the 1996 Scottish Natural Heritage Loch Survey (SNH-LS); the last are particularly valuable records from a number of local waterbodies. Several interesting records from the extensive Clyde Muirshiel Regional Park (CMRP) were provided by Alan Brown (AB) and the CMRP ranger service.

Work for *The Changing Flora of Glasgow* (Dickson et al. 2000) throughout the 1980s and into the early 1990s generated many records, mainly at the tetrad level, relevant to the part of Renfrewshire within the 'Glasgow Rectangle' study area. Some 6000 tetrad records, representing 15,000 field records, are reproduced in the species distribution maps within this flora, unless covered by more recent site visits.

Additionally, Richard Weddle has extracted some 5000 relevant modern records from the Glasgow Museums' Biological Records Centre database; most of these records originate from the CMRP ranger service or those compiled by Paisley Museum staff.

The author was invited to takeover as the BSBI Recorder for Renfrewshire in 1993 and spent much of the following years recording throughout the vice-county, tending to target areas outside the recently recorded CFOG area and concentrating on more rural areas. In total more than 81,000 field records were made by the author during the period 1993–2012.

More than 123,00 records are held in the author's database for *The Flora of Renfrewshire*; table 9 lists the main contributors. In total the number of records used in this flora exceeds 143,000. Figure 8 shows the distribution and abundance of records across the vice-county.

Table 9. Main contributors to the database for *The Flora of Renfrewshire.*

Keith Watson	81,198
Basil Ribbons and associates (BWR)	33,137
Robert Mackechnie (RM)	1,751
Scottish Natural Heritage Loch Survey (SNH-LS)	908
Clyde Muirshiel Regional Park (CMRP-NVC survey)	787
Ian and Paul Green (IPG and PRG)	618
Dr Alan J Silverside (AJS)	529
Allan McGregor Stirling (AMcGS)	449
Dr Peter Macpherson (PM)	326

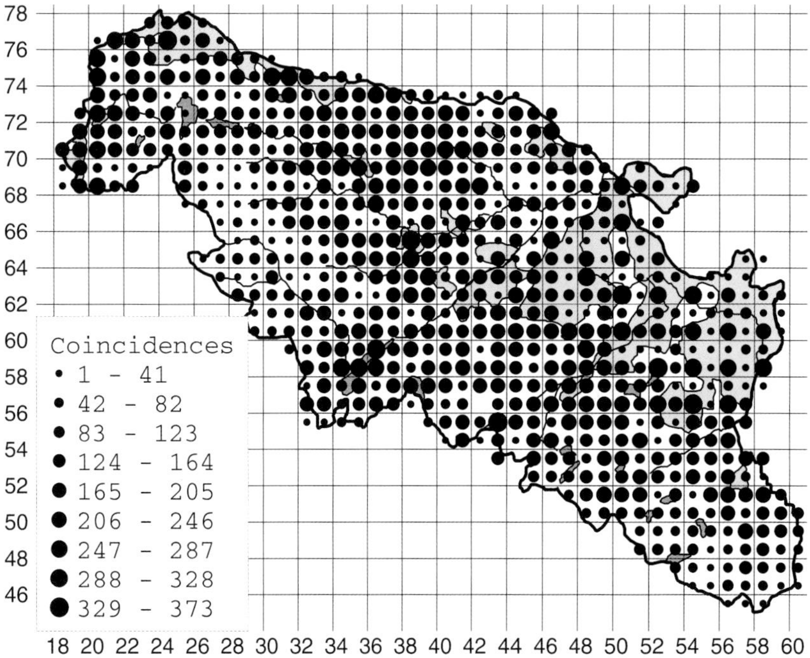

Figure 8. Coincidence map from the database for *The Flora of Renfrewshire*, showing all records and the total number of records from each 1km square.

PART 2

Catalogue of Species

Introduction to the Catalogue of Species

The following catalogue lists all the flowering plants and ferns that have been recorded as growing wild in the vice-county of Renfrewshire (VC76). The taxonomic arrangement follows the layout and nomenclature of the third edition of the *New Flora of the British Isles* (Stace 2010). This edition contains some rather radical changes, based chiefly on modern molecular data, to the systematic layout and a number of changes to scientific names. In the following catalogue, where there are new species names or new family groupings, previously used names are provided – from the second edition of Stace (1997) – to facilitate cross-referencing with other recent publications, in particular *The Changing Flora of Glasgow* (Dickson et al. 2000).

How the species accounts are organized

Each account starts with the plant's names: the scientific name and standard authority abbreviation – with the synonym below where applicable – and the standard common name. The names are taken from Stace (2010) with some minor additions from other literature, notably *Alien Plants of the British Isles* (Clement and Foster 1994). This is followed by summary information and then descriptive text. A few of the old, now long extinct or casual records have just summary information. This is also the case for a number of the critical taxa, notably brambles, dandelions and hawkweeds.

Throughout the species accounts, abbreviations are used to save space: the abbreviations are listed at the end of this introduction.

Distribution maps are shown for most extant native species and many of the aliens if present in ten or more 1km squares, and especially if it helps to demonstrate a distribution pattern.

Summary information

For each species account the lines of text immediately following the names contain summary information about the national and local distribution and status, and also information about the first record of a species within the vice-county.

- **The first element of the summary information** within each species account presents details of the species' British status, distribution and frequency, with additional information about Scotland where appropriate; where it is relevant to the wider phytogeographical perspective occasional reference is made to plants in Ireland. Unless specified otherwise, a species should be taken to be native within Britain; for aliens the place of origin is also provided. The information is chiefly interpreted from the maps in the *New Atlas of the British and Irish Flora* (Preston et al. 2002) or the species accounts in Stace (2010).

- **The second element of the summary information** notes the species' status within Renfrewshire (whether native or alien, and if alien how brought to the area), the number of records for it and an indication of the plant's current distribution status within the vice-county.

  The number of records shows the number of modern 1km-square records followed by the total number of 1km-square records for a species made since recording began, for example: 16/25. The number of 1km squares from which a plant has been recorded during modern surveying is used as a measure of frequency:

0 = extinct or considered to be extinct	21–35 = occasional
1–3 = very rare	36–100 = frequent
4–10 = rare	101–250 = common
11–20 = scarce	251 or more = very common

Such usage may be qualified if a plant has a particular geographical affinity, for example 'locally common in the uplands'.

Natives and Aliens

The definition of natives and aliens (non-natives or introductions) and methods of arrival used in this flora follow those adopted by the BSBI (Macpherson et al. 1996) and used in the *New Atlas of the British and Irish Flora* (Preston et al. 2002) and *The Changing Flora of Glasgow* (Dickson et al. 2000).

Native: a plant that is present in the study area without human intervention, either intentional or unintentional, and having arrived from an area in which it is native.
Alien: a plant brought to the study area by people – intentionally or unintentionally – even if native in the source area, or one that has come to the study area without human intervention but from an area in which it is present as an alien.

Aliens occurring in Britain are defined as one of two types:
Archaeophyte: a plant which became naturalized before AD 1500.
Neophyte: a plant first introduced, or which became naturalized, after AD 1500.

For aliens the method of introduction is defined as:
Hortal: intentionally brought to the study area.
Accidental: unintentionally brought to the study area.

When referring to Renfrewshire, the definitions 'hortal' and 'accidental' have been applied to plants otherwise native in Britain if it appears clear that they have only arrived in the vice-county through the action of humans.

Extinct is used where there is strong evidence that a native or alien species is no longer present within the vice-county. This is also used for casual species which have not been reported recently.

Endangered or **threatened** are occasionally used to indicate that a native plant's survival in the vice-county appears to be precarious. A comment may also be provided if there are doubts about the reliability of records or if a species is thought likely to have been overlooked.

- **The third element of the summary information** provides the location, year and recorder abbreviation for the first recorded occurrence of the species in the vice-county. The first known record is taken as a dated literature reference or a label from a herbarium sheet indicating the presence of a species in the vice-county, but does not include records from earlier archaeological finds (see Chapter 1.6). The first dates are more reliable for rare or unusual plants, as many species presumed common were not given site details in the earliest literature, often being just noted as 'common' and so on in the local area, but not specifically in Renfrewshire. Recorder abbreviations are included in the list of abbreviations at the end of this introduction, with further details in Chapter 1.8. The absence of a recorder's abbreviation implies the author's field record.

Main descriptions in the species accounts

The main descriptive part of each species account provides information about past records and also modern finds, with reference to the species' ecology and habitat preferences in the local area.

The records date from three main periods (see also Chapter 1.8):

- **Old records** are generally considered as being those dating from the earliest recording up until 1950.
- **Records from 1950 to 1987** – a separate note is made of the records generated during these years, for the most part by the previous BSBI recorders for the vice-county; these records are mainly annotated with 'BWR' as a convenient abbreviation and also in honour of the leading role played by Basil Ribbons during this period, but the records may not be his personal ones. The vast majority of the records were generated during the 1960s to early 1980s.
- **Modern records** date from 1987, although most field records date from 1993 onwards. Records from *The Changing Flora of Glasgow* have also been included as modern, even though a few may date from slightly earlier than 1987, but they are distinguished on the distribution maps.

Locality names are usually followed by a four-figure grid reference, the date and the recorder initials or with a lack of initials indicating the author's record; further details on place names can be found in the gazetteer (p. 386).

Occasionally reference is made to hybrids, infraspecific taxa or related cultivars or casual species which do not have separate individual species accounts.

Distribution maps

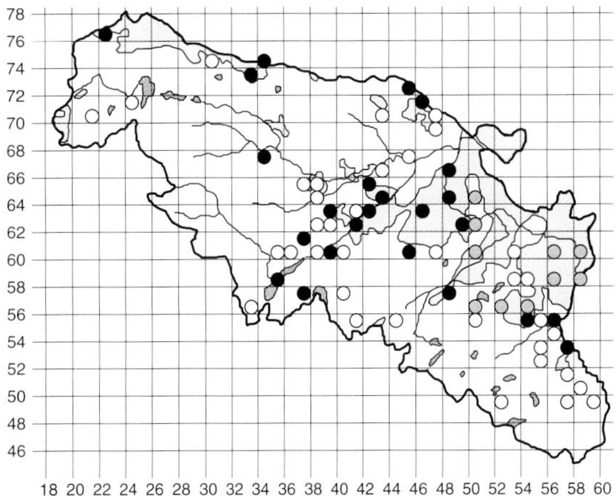

Figure 9. Circles representing records from different recording periods.

The distribution maps have been produced using Alan Morton's DMAP mapping programme. The study area boundary map was digitized by the author using DMAP digitizing software; this shows as a background the main urban zones (light grey), watercourses and larger waterbodies (darker grey) (see fig. 9).

The distribution maps show the National Grid Reference System grid lines spaced at 2km intervals, creating 2km × 2km grid squares (tetrads – equivalent to four 1km squares). The grid lines are numbered with the easting numbers on the horizontal axis and the northing numbers on the vertical axis; all grid lines occur within the National Grid Reference System 100km-square NS (26).

The recording unit has been primarily site based, but with an attempt to record on a 1km-square basis, as mapping at the tetrad level obscures some of the finer detail of local patterns. However, a few record cards cover more than 1km square – in such cases the record has been assigned to the site's central grid reference, and the circle will appear in the relevant 1km square.

There are three types of circles used:

- **The solid black circles** represent modern records (other than those taken from *The Changing Flora of Glasgow* database) from 1987, though most date from 1993 onwards.
- **The open circles** indicate older records (pre-1987). A few pre-1987 records, such as from herbarium sheets and old literature, may not appear on the distribution maps due to lack of detailed locality information.
- **The grey circles** are records transferred from *The Changing Flora of Glasgow* database (via Glasgow Museums' Biological Records Centre). These are at the tetrad scale and it is not possible to convert most to 1km-square precision. It is not possible to position these circles centrally within each tetrad square so they are placed in the lower-left 1km square of each tetrad; this requires the reader to remember that whereas solid black and open circles refer to 1km squares, grey circles represent the whole of the 2km × 2km squares in which they are placed. However, their inclusion helps to graphically demonstrate a species' total distribution and provides a link to the distribution maps in *The Changing Flora of Glasgow*; a few subsequent corrections to the Glasgow flora's dataset may result in slight discrepancies between the published maps in the Glasgow flora and those in this flora.

Abbreviations used in the catalogue of species

The standard herbarium codes used follow those in *British and Irish Herbaria* (Kent and Allen 1984). Further information about recorders is provided in Chapter 1.8 and further bibliographic details can be found on p. 388.

AANS	*Annals of the Andersonian Naturalists' Society* (1893 onwards)
AAP	Alfred A Percy
AB	Alan Brown
AGK	Archie G Kenneth
AHJ	Andrew Halcro-Johnston
AJS	Dr Alan J Silverside
AM	Alison Moss
AMcGS	Allan McGregor Stirling
AN	Alan Newton
ANG	Andersonian Naturalists of Glasgow excursions
AR	Alison Rutherford
ASNH	*The Annals of Scottish Natural History* (1892–1939)
BIRM	Herbarium of the University of Birmingham
BM	Herbarium of the British Museum
BS	Brian Simpson
BWR	Basil W Ribbons and collaborators
CB	Chris Balling
CD	Clive Dixon
CFOG	*The Changing Flora of Glasgow* (Dickson et al. 2000)
CGE	Herbarium of the University of Cambridge
CMRP	Clyde Muirshiel Regional Park
CMRP-NVC	Clyde Muirshiel Regional Park National Vegetation Classification survey (1992)
DF	Daniel Ferguson
DH	David Hawker
DM	Dave Mellor
E	Herbarium of the Royal Botanic Gardens Edinburgh
EL	Dr Elspeth Lindsay
ERTC	Elizabeth RT Conacher
FB	Frances Black
FFCA	*Fauna, Flora and Geology of the Clyde Area* (Elliot et al. 1901)
FFWS	'Vegetabilia', in *Notes on the Fauna and Flora of the West of Scotland* (McKay 1876)
FS	Sir William Jackson Hooker; *Flora Scotica* (Hooker 1821)
GC	*The Glasgow Catalogue of Native and Established Plants*, 2nd edn (Ewing 1899)
GGO	Herbarium of the University of Strathclyde (now in Glasgow Museums' Collection)
GHB	George Ballantyne
GL	Herbarium of the University of Glasgow (now in Glasgow Museums' Collection)
GL	Gary Linstead
GLAM	The Herbarium of Glasgow Museums (formerly called the Herbarium of Glasgow Art Gallery and Museum)
GMRC	Glasgow Museums Resource Centre
GMTC	GMT Conacher
GN	*The Glasgow Naturalist* (1909 to present)
GRK	Herbarium of McLean Museum and Art Gallery (Greenock)
GW	Graeme Walker
Hennedy	*see RH*
HG	Hector Galbraith
HLU	Herbarium of the University of Hull
HMcD	Hugh MacDonald; *Rambles Round Glasgow* (MacDonald 1854, 1856)
Hooker	*see FS*
Hopkirk	*see TH*
IG	Iain Gibson
IPG	Ian P Green
JB	John Blackwood
JC	Jo Birkin (née Ciuduski)
JD	John Douglass
JDM	James D Morton
JG	Joe Greenlees
JHD	Professor JH Dickson
JM	John Mitchell
JP	John Hubert Penson
JMY	James Murray; *Kilmacolm: A Parish History* (Murray 1898)
JW	John Wilson; *General View of the Agriculture of Renfrewshire* (Wilson 1812)
KC	Dr Kathleen M Calver
KF	Keith Futter
LB	Liz Buckle
Lee	John R Lee; *The Flora of the Clyde Area* (Lee 1933)
LP	Liz Parsons
LR	Laurie Ritchie

LRC	Local Records Centre (Glasgow Museums' Biological Records Centre)
MC	J Montgomery catalogue (Montgomery 1834), in *The New Botanist's Guide*, Vol. II (Watson 1837)
MG	Mark Gurney
MMGNS	*Manuscript Magazine of the Glasgow Naturalists Society*, IV (1863)
MY	Morris Young: 'Morris Young's "Flora of Renfrewshire" (VC76)', *Glasgow Naturalist* (Weddle 2008)
NHSG	*Proceedings of the Natural History Society of Glasgow* (1858–83), *Proceedings and Transactions of the Natural History Society of Glasgow* (1883–96) and *Transactions of the Natural History Society of Glasgow (including the Proceedings of the Society)* (1896–1908)
NPA	*New Atlas of the British and Irish Flora* (Preston et al. 2002)
NSA	*The New Statistical Account of Scotland* (1845)
NVC	National Vegetation Classification (Rodwell 1991–2000)
OS	Olga Stewart
OSA	*The Statistical Account of Scotland, 1791–1799*
P&TNT	Pearl and Norman Tait
PB	Petrina Brown
PM	Dr Peter Macpherson
PPI	*The Flora of Renfrewshire (Paisley & Neighbourhood)*, Paisley Philosophical Institution (Henry *c.* 1869)
PRG	Paul Green
PSY	Herbarium of Paisley Museum
REM	Rennie Mason (Whitelee rangers)
RG	Robert Grierson; *Clyde Casuals* (Grierson 1930)
RH	Roger Hennedy; *The Clydesdale Flora* (Hennedy 1865, 1869, 1874, 1878 and 1891)
RM	Robert Mackechnie
RPB	'Rambles with the Paisley Botanists', *Paisley & Renfrewshire Gazette* (Anon. 1869)
RW	Richard Weddle
SA	Shona Allan
SJL	Sarah J Longrigg
SNH	Scottish Natural Heritage
SNH-LS	Scottish Natural Heritage Loch Survey (1996)
SWT	Scottish Wildlife Trust fieldwork surveys *c.* 1990–2000
TB	*Topographical Botany* (1873 onwards) (Baker and Newbold 1883)
TGSFN	*Report and Transactions of the Glasgow Society of Field Naturalists* (1873–7)
TH	Thomas Hopkirk; *Flora Glottiana* (Hopkirk 1813)
TNT	Norman Tait
TPNS	*Transactions of the Paisley Naturalists' Society* (1912 onwards)
Wood	John Wood's 'Rarer flowers of East Renfrewshire' (Wood 1893)
WR	Rev. William Ross; *Busby and Its Neighbourhood* (Ross 1883)

Standard botanical terms used

agg.	aggregate – group of closely related species, often microspecies
auct.	(*auctorum*) of various authors but not the original one
f.	(*forma*) form
nothossp.	nothosubspecies – hybrids between subspecies
pro parte	partly, in part – used for previous family names now divided or amalgamated
s.l.	(*sensu lato*) in a broad sense – for closely related species group
sp., spp.	species, singular and plural
ssp., sspp.	subspecies, singular and plural
var., vars.	variety, varieties

Standard abbreviations used

c.	(*circa*) about, approximately
et al.	(*et alii*) and others
no.	(numero) number

Geographical abbreviations used

N, S, E, W	north, south, east and west
C	central, or central area

PTERIDOPHYTES (FERNS and FERN ALLIES)

LYCOPHYTES (CLUBMOSSES)

LYCOPODIACEAE

Huperzia selago (L.) Bernh. ex Schrank & Mart.
Fir Clubmoss

Britain: common in N Scotland and uplands elsewhere.
Native; 21/31; occasional on high ground.
1st Lochwinnoch, 1845, NSA.

This, the most frequent clubmoss, can be found on the high ground of the Hill of Stake/Mistylaw area and on scattered rock outcrops, scree or peaty banks elsewhere in the Renfrewshire Heights. There are no records from the south-east uplands, except for Brother Loch (NS5052, 2003, JD). Several sites are at or below 200m, such as Everton (NS2170, 2004), Gimblet Burn (NS2370, 2001), Harelaw Reservoir (NS3173, 1998) and Marshall Moor (NS3662, 1994, CD), although old records indicate a decline at other former lowland sites. There are no modern sightings from previous BWR localities at Barmufflock (NS3664, 1961), Knocknair (NS3074, 1969) and Paisley Glen (NS4760, 1974).

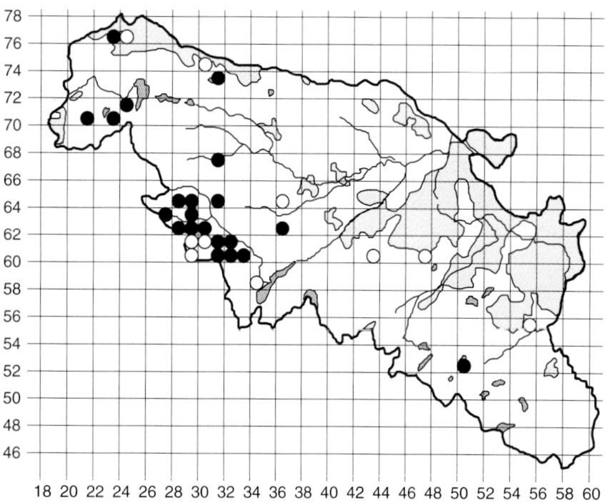

Huperzia selago

Lycopodium clavatum L.
Stag's-horn Clubmoss

Britain: common in N Scotland and uplands elsewhere.
Native; 9/22; rare, scattered.
1st Gourock, 1850, GGO.

There are only a few scattered modern records of this clubmoss of open heathy areas, and it appears to have undergone a marked decline in the last century. It is perhaps under-recorded from the Renfrewshire Heights, where it is only known from Queenside Loch (NS2964, 1998), Barnshake Moor (NS3266, 1999), near Daff (NS2270, 2001) and Darndaff (NS2673, 2001); a northern outlier is on Earn Hill (NS2375, 2006). There are also two records from the south-east uplands: Dickman's Glen (NS5847, 2001) and near Carrot (NS5848, 2003, JD). BWR lowland records include Ranfurly Golf Course (NS3664, 1969), Kaim Dam (NS3362, 1971) and Paisley Glen (NS4760, 1974), and JDM noted it at a railway cutting at Barrhead (NS4958, 1960). The presence and decline of this species at the two Glasgow locations was discussed in CFOG.

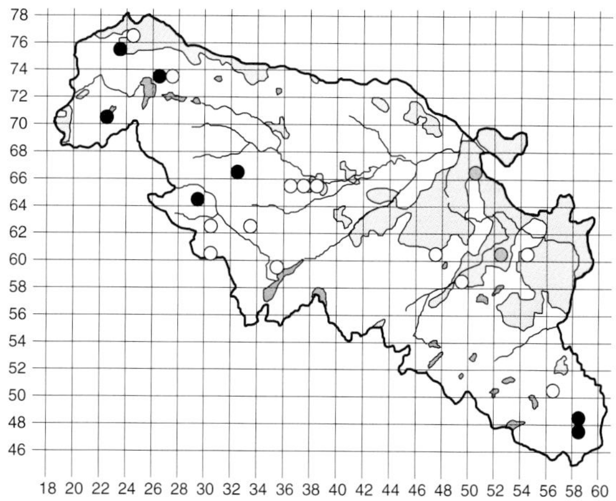

Lycopodium clavatum

Lycopodium annotinum L.
Interrupted Clubmoss

Britain: frequent in N Scotland.
Native; 0/1; presumed error.
1st Gryfe Farm, 1901, FFCA.

The first record, which was repeated by Lee (1933), is presumably an error as a labelled specimen in GRK from 'Near Gryfe Farm' by AO Leitch is Stag's-horn Clubmoss.

Diphasiastrum alpinum (L.) Holub
Alpine Clubmoss

Britain: common in N Scotland and uplands elsewhere.
Native; 5/10; rare, uplands.
1st Lochwinnoch, 1845, NSA.

Hennedy collected Alpine Clubmoss from Gourock (1848, GGO) and noted it at Inverkip and Paisley Canal bank in 1865, but considered the species extinct at this last, unexpected, lowland site in 1891. Three of the modern records are from the high ground of the Mistylaw area, where this rarest clubmoss grows on rocky outcrops at Queenside Loch (NS2964, 1998), Muirshiel Barytes Mine

(NS2864, 1998) and Hannah Law (NS3061, 1997); it was reported from Queenside Loch shore in 1972 (NS2964, HG). JD recorded it from the south-east uplands (NS5847, 2002) and also found a small plant on open scree near Carrot Farm (NS5848, 2003).

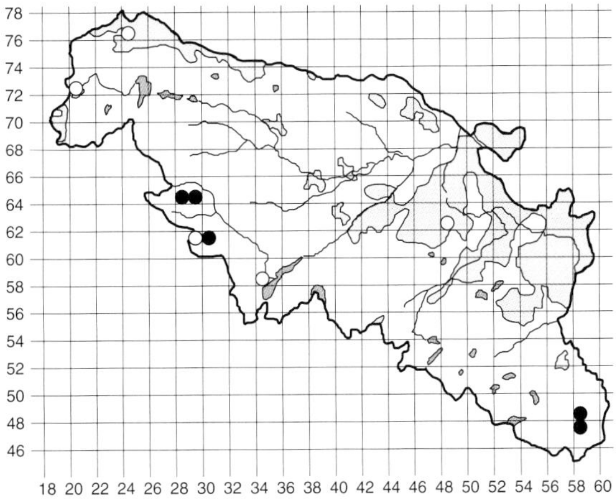

Diphasiastrum alpinum

SELAGINELLACEAE

Selaginella selaginoides (L.) P. Beauv.
Lesser Clubmoss

Britain: common in N Scotland and uplands elsewhere.
Native; 17/21; occasional in the W.
1st Gourock, 1865, RH.

Apparently rarely recorded by early authors, although described as common in 1876 in FFWS, this diminutive clubmoss of base-rich, burn-side or rocky flushes today has a very strong western distribution, with all records from high elevations (above 140m). Several sites occur below Mistylaw, e.g. Raith Burn (NS2862, 1996), with a further cluster to the north-west, e.g. North Rotten Burn (NS2667, 2000); two sites are at Earn Hill (NS2375, 2003) and an isolated one is at Jock's Craig (NS3369, 2007).

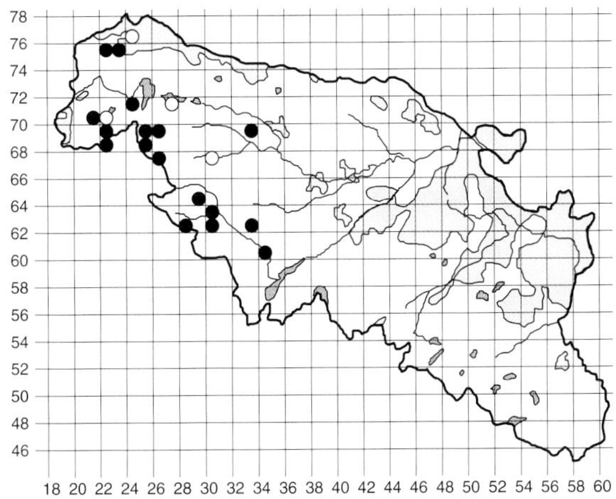

Selaginella selaginoides

EUSPORANGIATE FERNS (ADDER'S-TONGUE and MOONWORTS)

OPHIOGLOSSACEAE

Ophioglossum vulgatum L.
Adder's-tongue

Britain: frequent in England, scattered in Scotland, rare in the NE.
Native; 6/15; rare, declining.
1st Paisley Canal, 1863, MMGNS.

Adder's-tongue was recorded from Shielhill Glen (1881, NHSG), Gourock (1865, RH) and Bardrain Glen (1930, TPNS), and collected from Ferguslie (1896, PSY); Wood (1893) thought that, like Moonwort, it was common between Corkindale Law and Neilston Pad. In 1869 it was known (RPB) from one or two places about Paisley but the author was 'not ready to disclose for fear of extermination'. Today Adder's-tongue is found in short, species-rich grassland on Barscube Hill (NS3871, 2002) and a small population also occurs by the path at Paisley Moss (NS4665, 2003). A further find is from under Bracken near the seventh tee at Ranfurly Old Castle Golf Course (NS3864, 2007, AM); it may well be hidden in other Bracken stands elsewhere. AJS found it to the edge of Midtown Wood (NS4060) in 1978, but there are no more recent records from here. A site at Maxwelton, Paisley (NS4663) was lost to development in the 1990s (DM pers. comm.). It had also disappeared from the disused railway scrub at Hillington (NS5266, CFOG) by the 1990s and has not been reported again from Whitecraigs Golf Course (NS55N, 1987, CFOG).

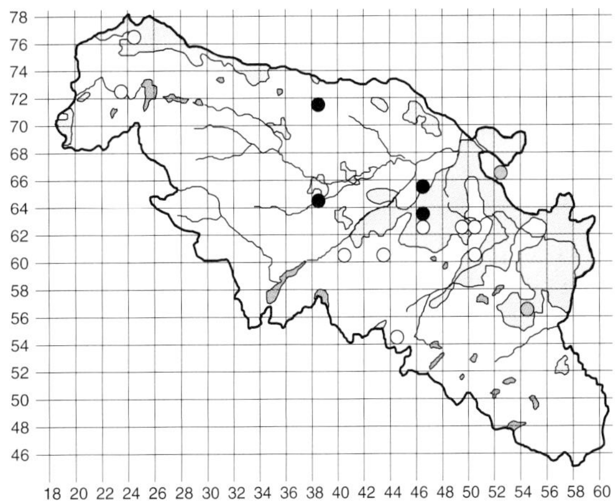

Ophioglossum vulgatum

Botrychium lunaria (L.) Sw.
Moonwort

Britain: frequent and widespread, mainly in the N and W.
Native; 4/26; rare and declining.
1st Greenock, 1845, NSA.

Moonwort was often considered rare in the past – old localities include Gourock and Neilston Pad (1876, FFWS) – but Wood (1893) thought it common between Corkindale Law and Neilston Pad. Today Moonwort is known from two relatively unimproved pastures in the Gleniffer Braes at Knockindon Burn (NS4659, 1995) and Wardlaw Braes (NS4660, 1995), and a large population occurs in the species-rich grassland of Barscube Hill (NS3871, 1997), where it had been observed for several years (DM pers. comm.). A single spike was also found by the Maich Water (NS3160, 2003). There are a number of 1970s' records (BWR), suggesting that since then there has been a sharp decline or modern under-recording: Greenock Cut (NS2574, 1970), west of Netherwood (NS3369, 1970), Craigmarloch (NS3471, 1971), Floak (NS4950, 1974) and pastures near Old Hall (NS3671, 1976); the local stations were discussed by BWR (Ribbons 1971).

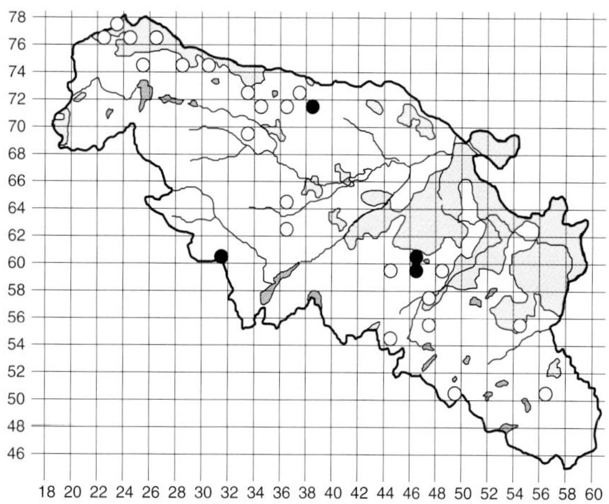

Botrychium lunaria

CALAMOPHYTES (HORSETAILS)

EQUISETACEAE

Equisetum fluviatile L.
Water Horsetail

Britain: common throughout.
Native; 221/264; common in wet places.
1st Renfrewshire, 1883, TB.

A common dominant plant of deeper water, this horsetail is found at many open waterbodies, where it can form large, often pure swamp stands. Its distribution shows a rural pattern; it is absent from many urban areas, but this likely reflects a lack of wet places there, as Water Horsetail is capable of colonizing disturbed flooded areas such as at old quarries or disused railways. It avoids the strongly peaty waters of upland areas, but occurs at over 400m at Queenside Loch (NS2964, 1998) and Creuch Hill (NS2658, 1997).

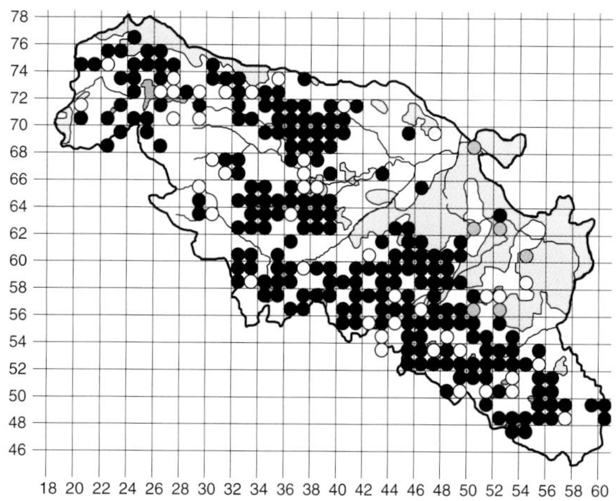

Equisetum fluviatile

Equisetum × litorale Kühlew. ex Rupr.
(*E. arvense × E. fluviatile*)
Shore Horsetail

Britain: widely scattered.
Native; 2/5; very rare but probably under-recorded.
1st Shielhill to Daff Reservoir, 1973, RM.

This hybrid was also recorded in 1976 by RM from Houston Estate (NS4167) and in 1980 AMcGS found it at the 'shore marsh, Cardwell, between Inverkip and Cloch' (NS2074). It was more recently recorded from roadside wet grassland at Ardgowan Estate (NS2073, 1998) and from the flooded disused railway (cycleway) near Birkmyre Park, Port Glasgow (NS3174, 2003).

Equisetum arvense L.
Field Horsetail

Britain: common throughout.
Native; 285/383; very common and widespread.
1st Renfrewshire, 1883, TB.

Field Horsetail is found throughout Renfrewshire in grasslands and moderately wet marshes, on waste ground and it tolerates shade in urban scrub and some woodlands. The map shows an absence from both high and peaty ground; it is seldom recorded above 250m.

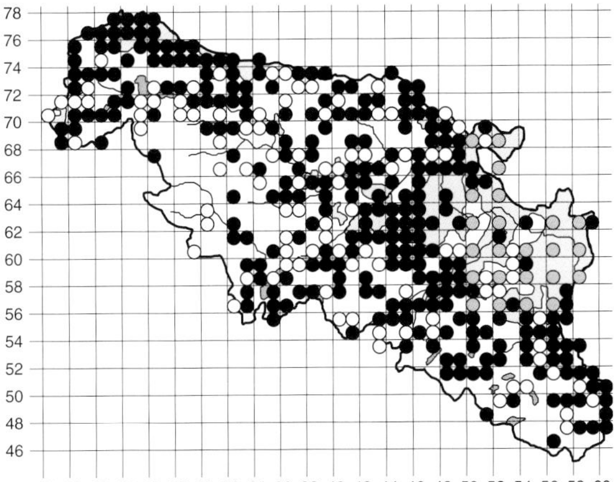

Equisetum arvense

Equisetum pratense Ehrh.
Shady Horsetail

Britain: frequent in upland Scotland.
Native; 0/2; doubtful records.
1st Castle Semple Loch, 1887, NHSG.

This nationally 'Scarce Plant' has few regional records, and has only ever been noted a couple of times in the VC, and both records may be questioned. The first was from an excursion and the second, more recently, (1942, TPNS) from Bardrain Glen, but the common name used is Corn Horsetail (i.e. *E. arvense*), so is very likely a transcription error. There are no other records from either of these well-botanized localities.

Equisetum sylvaticum L.
Wood Horsetail

Britain: common in the N and W.
Native; 103/142; common in rural fringes.
1st Renfrewshire, 1883, TB.

Wood Horsetail is found in wet woodlands and scrub areas, and also in open flushed grasslands that are not heavily grazed. It is widespread in the lowlands, often occurring in urban areas, but here is usually related to relict semi-natural habitats. It is very rare in the farmland of the north central area and becomes scarce in the uplands above 200m.

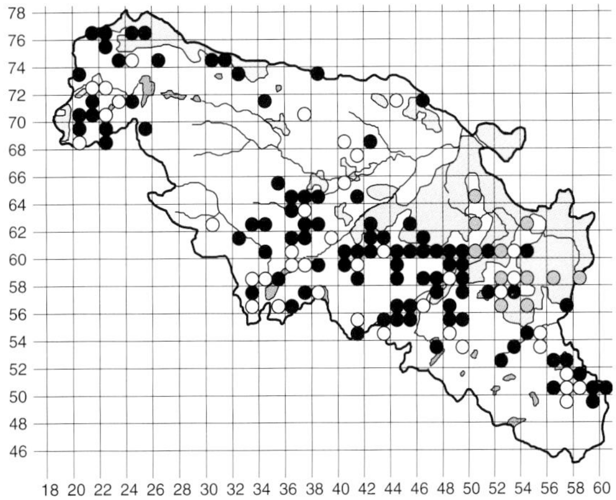

Equisetum sylvaticum

Equisetum palustre L.
Marsh Horsetail

Britain: frequent throughout.
Native; 57/68; frequent in upland-fringe flushes.
1st Gourock, 1847, GGO.

This horsetail was recorded in the 19th century, e.g. Foxbar Dam (1882, PSY) and Paisley (1855, GGO), but rarely. Today Marsh Horsetail is widespread in suitable habitat, which tends to be flushed, slightly base-enriched marshes or mires. It can occur associated with disturbed waste ground, as at Muirshiel Barytes Mine (NS2864, 1998). It avoids acidic peaty areas but occurs at the fringes of the Renfrewshire Heights, and there are a good number of records for the species about the Gleniffer Braes (NS4460 etc.) and especially from scattered sites across Eaglesham Moor. The hybrid with *E. fluviatile*

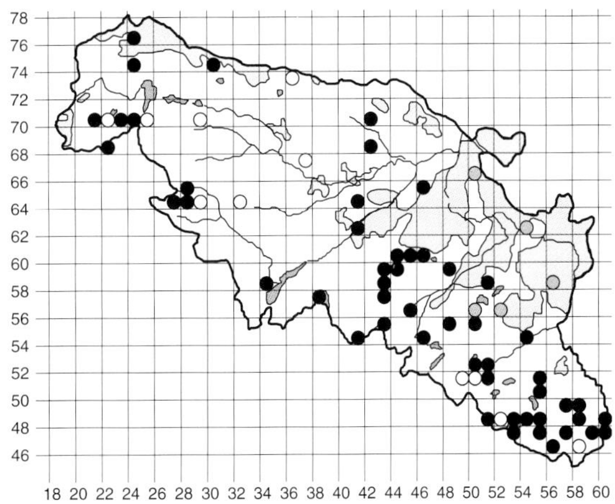

Equisetum palustre

(*E.* × *dyceii* C.N. Page) was found at the marshy quarry floor at Knocknairs (NS3074, 1998).

Equisetum telmateia Ehrh.
Great Horsetail

Britain: frequent in England, scattered in C Scotland.
Native; 1/4; very rare.
1st Wemyss Bay, 1874, TGSFN.

The first record may well relate to Kelly Glen, which runs into Wemyss Bay (although Hennedy, in 1891, named both localities). Today Great Horsetail in Kelly Glen is only known from the VC75 side, with further populations in adjacent Skelmorlie. However, it can still be found in the wet woodlands in the lower section of Shielhill Glen (NS2471, 2000), where it was first recorded in 1881 (NHSG); a couple of plants appear to be hybrid (with Field Horsetail) but none has been expertly determined.

LEPTOSPORANGIATE FERNS (TRUE FERNS)

OSMUNDACEAE

Osmunda regalis L.
Royal Fern

Britain: frequent in the W.
Native and hortal; 4/7; presumably extinct as a native.
1st McInroy's Point, 1854, HMcD.

Recorded as a native between Inverkip and Gourock (1865, RH), Royal Fern was also known from near Skelmorlie (presumably VC75), but it was excluded from the Renfrewshire list (TPNS 1915). Hennedy (1865) wrote, 'Owing to the rapacity of dealers who are more alive to regal impress than royal beauty, this fern is rapidly disappearing from localities to which it lent a charm'. The only extant locality is from the roadside edge of the willow carr at Sergeant Law Moss (NS4459, 1995), where it appears to be a garden outcast; it was first recorded from here, along with other garden material, in 1976 by AJS. It is listed on a record card from Kelly Glen (NS1968, 1959, BWR) but this appears to be the only mention of it, and it has not been reported since. There are three modern records from Glasgow parks in VC76 (Rosshall, Pollok and Queen's), but all as garden plants (CFOG).

HYMENOPHYLLACEAE

Hymenophyllum tunbrigense (L.) Sm.
Tunbridge Filmy-fern

Britain: local in the W.
Native; 0/3; probably extinct.
1st Gourock and Inverkip Glen, 1865, RH.

A further, less likely, locality was Elderslie in 1865 (MY). This rare plant of shady, wet rocks and woodlands may still lurk in Kelly Glen or one of the narrow, steep-sided valleys (cleughs) above Gourock, but it has not been seen during recent surveys and was excluded from the Renfrewshire list (TPNS 1915).

Hymenophyllum wilsonii Hook.
Wilson's Filmy-fern

Britain: frequent in the W.
Native; 9/14; rare, strongly western.
1st Gourock, 1863, MMGNS.

This filmy-fern shows a strong western bias, and can be found in the wet, shaded and rocky steep-sided valleys (cleughs) that cut down from the Renfrewshire Heights, e.g. Hole of Spango (NS2373, 1993), Raith Burn (NS2862, 2005), River Calder (NS3361, 1997) and Blackwater (NS3267, 1999). It was known at Shielhill Glen (1881, NHSG) and AMcGS recorded it from Devol Glen (NS3174) in 1981, but it has not been seen at either locality recently.

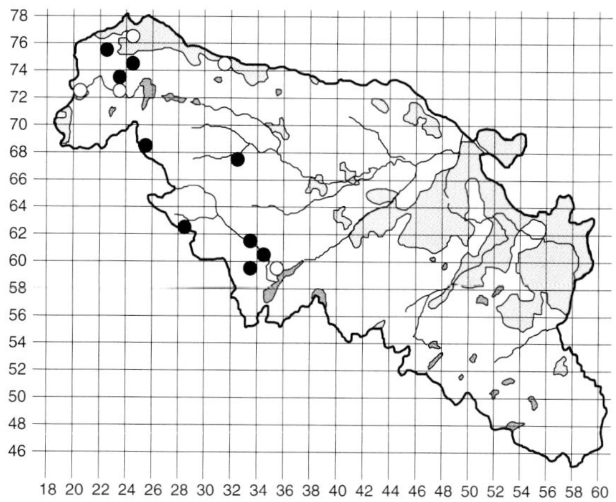

Hymenophyllum wilsonii

MARSILEACEAE

Pilularia globulifera L.
Pillwort

Britain: widely scattered, mainly western.
Former native; 0/2; extinct.
1st Renfrewshire, 1872, TB.

Lee (1933) also listed Pillwort for Renfrewshire but there are no historical site details, so it must be assumed long gone from the VC.

SALVINIACEAE
(AZOLLACEAE)

Azolla filiculoides Lam.
Water Fern

Britain: neophyte (W N America); frequent in the S, rare in C Scotland.
Hortal; 3/3; very rare.
1st Underheugh, 2004.

This aquatic fern has colonized a small pool in an old quarry at Underheugh (NS2075) near Lunderston Bay; the nearby garden centre may well be the source. Recently Water Fern has appeared in ponds in Glasgow, including Pollok (NS5562) and Mansewood (NS5660) (2011, Froglife surveys).

DENNSTAEDTIACEAE

Pteridium aquilinum (L.) Kuhn
Bracken

Britain: common throughout.
Native; 230/265; common.
1st Paisley, 1845, NSA.

Bracken is a common plant of woodlands, heaths and acid grasslands, and can be found in urban areas. It occurs on the lower reaches of the Raith (NS2862) and Cample (NS3062) burns, but most records are from well below 300m and there is a distinct absence from the highest ground. It can dominate the steep north-facing brae slopes above Gourock and Greenock and along the Gleniffer Braes, where grazing pressure has declined.

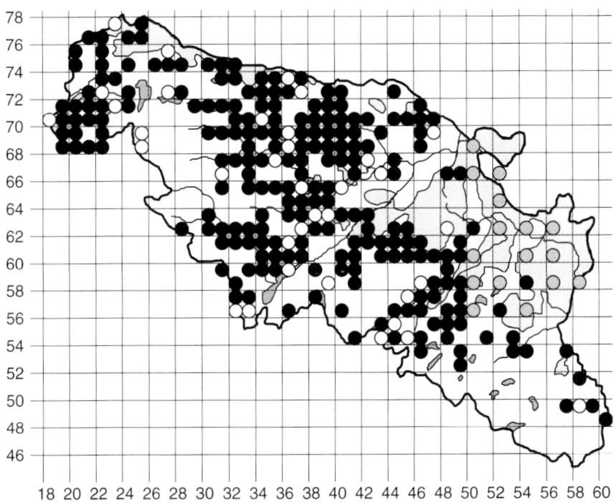

Pteridium aquilinum

PTERIDACEAE
(ADIANTACEAE, CRYPTOGRAMMACEAE)

Cryptogramma crispa (L.) R. Br. ex Hook.
Parsley Fern

Britain: frequent in uplands.
Native; 3/11; declining, now very rare.
1st Lochwinnoch, 1845, NSA.

Some 19th-century records include 'near Kilmacolm' (1880, NHSG), a burn side near Barrhead (1865, MY), Neilston Pad (1876, FFWS) and Milton Farm (1898, JMY). In more recent times Parsley Fern was recorded from Black Loch (NS4951, 1960, BWR) and Glenbrae (NS2872, 1978, ERTC), and a 1959 BSBI database record is from NS3361 (Turnave Hill?); despite modern searches it has not been seen at these sites. A couple of patches occur on rocky scree at the top of Rough Burn (NS3161, 2011) and the same habitat at one of the steep-sided feeders of the Raith Burn (NS2962, 2003). There is also a single plant on scree at Craigmuir (NS3770, 2008, AM), where it is seriously threatened by gravel extraction. Wood (1893) wrote of the demise of this fern at the Neilston Pad: 'dealers from the city found it out' and 'left not so much as one little plant on the hillside'.

Adiantum capillus-veneris L.
Maidenhair Fern

Britain: occasional in the SW, scattered alien elsewhere.
Former hortal; 0/1; extinct.
1st Milliken, 1904, PSY.

This is an unusual record, presumably of planted material, from the estate at Milliken; Maidenhair Fern is not mentioned elsewhere.

ASPLENIACEAE

Asplenium scolopendrium L.
Phyllitis scolopendrium (L.) Newman
Hart's-tongue

Britain: common in C Scotland, rare in N Scotland.
Native; 101/109; locally common.
1st Port Glasgow, 1840, GL.

This lime-loving fern can be found on both natural rock outcrops and on stone walls. It is well represented in the wooded fringes circling the Renfrewshire Heights, from the west and north coasts to the central valley, including urban-fringe areas, but it is absent from the south-eastern uplands. Lamenting its disappearance from Rouken Glen, due to collecting, Wood (1893) said that 'it stands small chance of survival in the neighbourhood of any large town'; however, at least this fern seems to have made a comeback, perhaps thanks, in part, to the basic mortar used in walls.

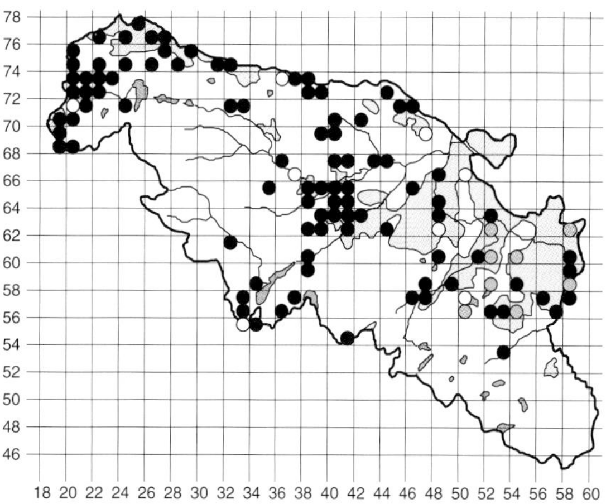

Asplenium scolopendrium

Asplenium adiantum-nigrum L.
Black Spleenwort

Britain: frequent, mainly in the W.
Native; 13/21; scarce.
1st Lochwinnoch and Renfrew, 1845, NSA.

Never present in large numbers, this fern tends to be tucked away on rock outcrops in shady glens, such as by the burn above Spango (NS2474, 1998) and at Parkglen (NS3970, 1998) and Busby (NS5756, 2001), and it still can be found in Kelly Glen (NS2068, 1995) where it was recorded in 1887 (NHSG). An isolated site is a shaded rock outcrop at Cleaves Wood (NS4068, 1998). Curiously, in 1898 JMY wrote that 50 years prior it 'flourished on every roadside', although Hennedy (1891) only described it as frequent; either way it is much rarer today.

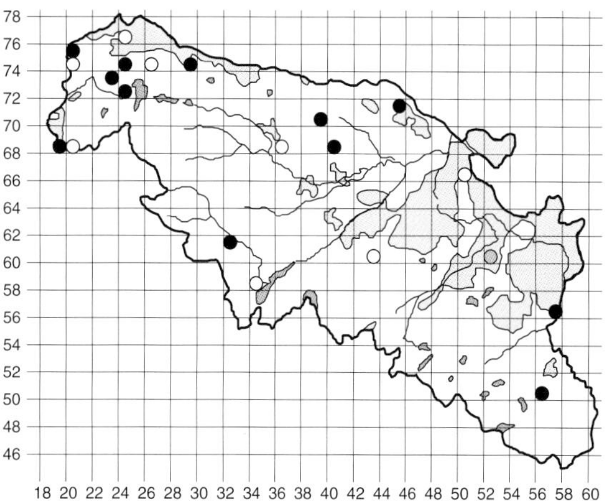

Asplenium adiantum-nigrum

Asplenium marinum L.
Sea Spleenwort

Britain: common on the W coast, rare in E England.
Native; 0/1; probably extinct.
1st Wemyss Bay, 1865, RH.

Sea Spleenwort was also recorded from Wemyss Bay in 1874 (TGSFN) and Lee (1933) also mentions the original location. It was noted on a card from Wemyss Point (NS1870, 1986, BWR), but there are no further details. It may still occur here but the numerous recent coastal developments are likely to have destroyed any relict sites.

Asplenium trichomanes L.
Maidenhair Spleenwort

Britain: frequent throughout, scarcer in the E.
Native; 115/130; common in the W.
1st Lochwinnoch, 1845, NSA.

More common than Wall-rue, this fern grows on similar stonework but is more likely to be found on natural rock outcrops. The map shows a chiefly lowland pattern, well represented to the north-west and along the coast and the central valley. It is assumed that most records refer to ssp. *quadrivalens* D.E. Mey., but there are two records for ssp. *trichomanes*: from NS47A (Finlaystone area) in the BSBI 1987 monitoring report and 'At fall on the Earn above Waterfoot' (near Eaglesham) by D Patton in

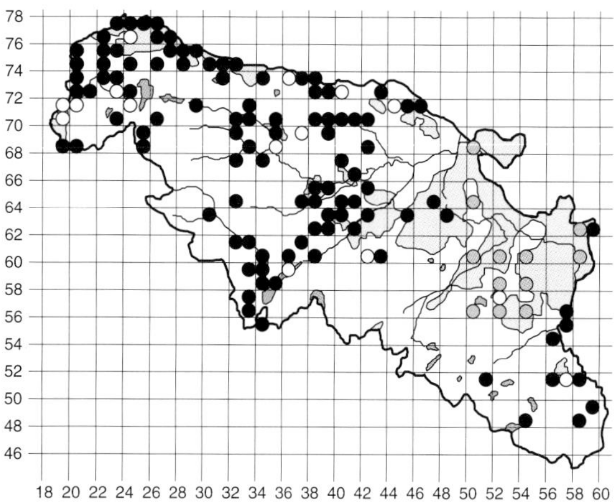

Asplenium trichomanes

1906 (GL, determined JD Lovis); from the latter Lovis also determined plants with abortive spores as the hybrid between the two subspecies (nothossp. *lusaticum* (D.E. Mey.) Lawalrée).

Asplenium viride Huds.
Green Spleenwort

Britain: frequent in uplands.
Native; 4/9; rare.
1st Near Bridge of Weir, 1860, GL.

Today this fern is found on shaded, base-rich rocks along larger watercourses such as the River Calder near Glenward Hill (NS3261, 1993 and NS3361, 1997), Shielhill Glen (NS2437, 2000) and the North Rotten Burn (NS2568, 1997); it was reported from the last site in 1890 (NHSG). It has been known from Devol Glen since 1884 (NHSG) and although last recorded there in 1981 (NS3174, AMcGS), it probably still occurs at this site.

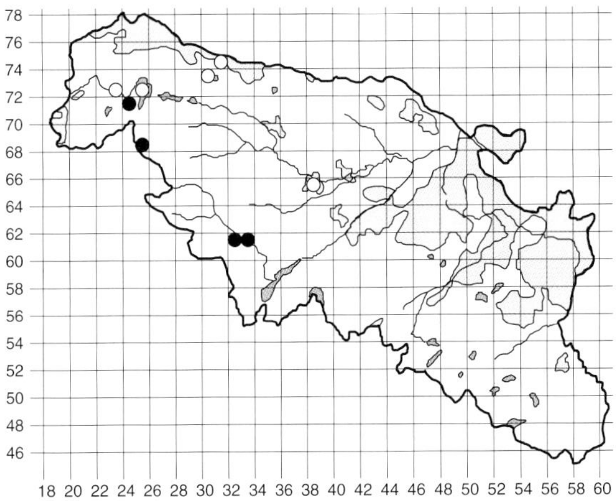

Asplenium viride

Asplenium ruta-muraria L.
Wall-rue

Britain: frequent throughout except in N Scotland.
Native; 33/47; scattered, locally frequent.
1st Lochwinnoch and Renfrew, 1845, NSA.

This diminutive fern is often seen on old stonework, particularly in milder western or coastal locations such as between Ardgowan (NS2073, 1998) and Greenock (NS2975, 2003). There is also a cluster of records from about Kilbarchan (*c.* NS4163), again reflecting the presence of old stonework. Hennedy (1891) thought Wall-rue frequent, but in 1869 (RPB) it was considered rare about Paisley and 'not plentiful enough to stand a raid'.

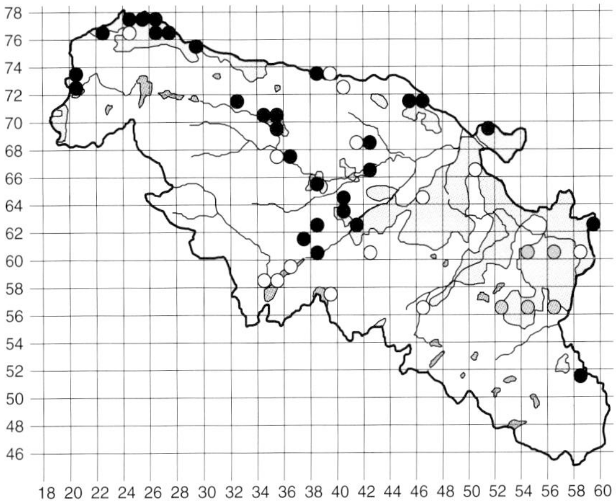

Asplenium ruta-muraria

Asplenium ceterach L.
Ceterach officinarum Willd.
Rustyback

Britain: common in the SW, scattered in S Scotland.
Native; 2/7; very rare.
1st Barrhead and Paisley, 1842, GL.

Stone walls are the exclusive habitat for this rare fern. There are several 19th-century records from walls in the Paisley area, including Dundonald (1845, NSA), and MY is annotated, 'near Paisley Moss' (1864). Commenting on Hennedy's 1865 Paisley location, Ferguson (1915) wrote, 'unfortunately, owing to the rapacity of plant-hunters anxious to acquire specimens for their herbaria, the fern was finally exterminated'. In 1973 (Conacher 1974) recorded five plants from a south-facing wall in Bridge of Weir and later (1985) recorded it from an east-facing wall near Clochodrick Stone (NS3761), where it has not been seen since; BWR recorded it from the Greenock Cut (NS2674, 1965). The only modern records are from a roadside wall by Barcraigs Reservoir, near Reivoch (NS3957, 1999), and a wall in Langbank (NS3873, 2006).

THELYPTERIDACEAE

Phegopteris connectilis (Michx.) Watt
Beech Fern

Britain: frequent in the N and W.
Native; 54/70; frequent in the W.
1st Lochwinnoch, 1845, NSA.

Beech Fern is a local fern of shaded woodlands, with a strong western pattern, usually found on steep banks, but it can occur on sheltered ledges on rock outcrops (often north-facing) at open, grazed upland areas; it seldom occurs in large numbers. It can still be found at 19th-century locations such as Carruth Bridge (NS3565, 1997), Bardrain Glen (NS4360,

2003), Duchall Wood (NS3368, 1994), Shielhill Glen (NS3471, 2000) and Kelly Glen (NS2069, 2011).

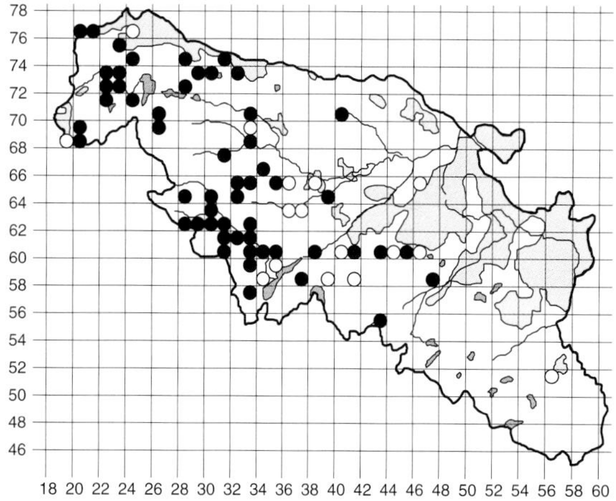

Phegopteris connectilis

Oreopteris limbosperma (All.) Holub
Lemon-scented Fern

Britain: common in the N and W.
Native; 103/129; common in upland fringes.
1st Gourock, 1855, GGO.

This is a locally frequent fern of upland fringe grasslands and rocky screes, particularly by burns and ditches. Considered very common by Hennedy (1891), there are few named locations. Today it is found in rural places, most frequently in the west, but can extend close to urban areas where they are adjacent to steep ground, e.g. along the Gleniffer Braes and the braes above Gourock and Greenock; it occurs by a ditch side at Cowglen Golf Course in Glasgow (NS5460, 2007).

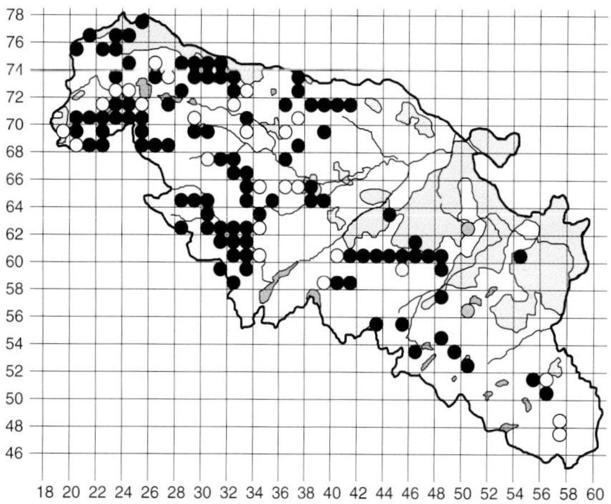

Oreopteris limbosperma

WOODSIACEAE

Athyrium filix-femina (L.) Roth
Lady-fern

Britain: common throughout.
Native; 379/442; very common except on the highest ground.
1st Paisley, 1869, RPB.

Lady-fern is a common fern of woodlands, usually preferring damper ground than other woodland ferns. It is common along burn sides, whether shaded or not, and can also be found on shaded rocky ground and walls. It becomes scarce on the highest peaty ground and on heavily grazed or improved farmland.

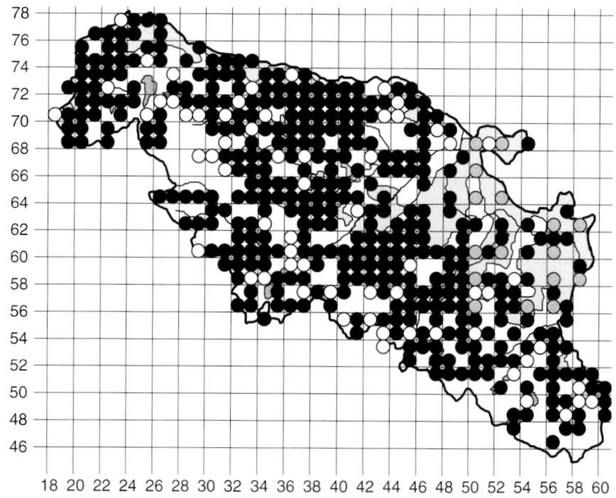

Athyrium filix-femina

Gymnocarpium dryopteris (L.) Newman
Oak Fern

Britain: common in the N and uplands elsewhere.
Native; 41/48; frequent, mainly western.
1st Lochwinnoch, 1845, NSA.

More frequent in the west, Oak Fern is most likely to

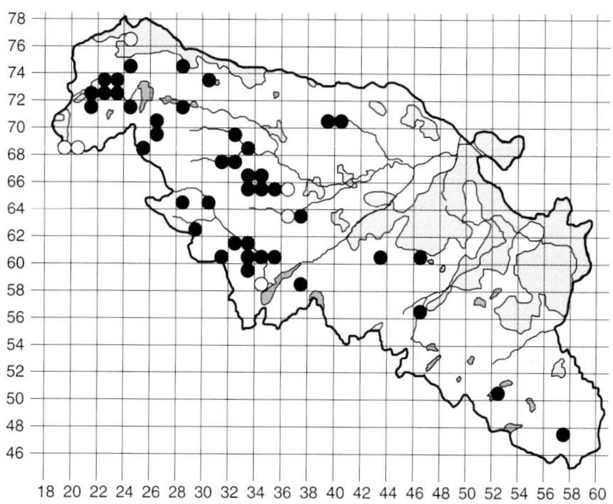

Gymnocarpium dryopteris

be encountered on heavily shaded, steep embankments of local wooded burn sides as they cut down from upland areas. Old records include Duchal Wood (1881, NHSG), Shielhill Glen (1881, NHSG), Devol Glen (1884, GL), Kelly Glen (1887, NHSG), Carruth Bridge (1890, AANS) and Bardrain Glen (1898, PSY). Today it has been refound at all these woodlands with the exception of Kelly Glen. The easternmost records are from shaded burn sides at Carrot (NS5747, 2000) and Bennan Loch (NS5250, 1998).

Cystopteris fragilis (L.) Bernh.
Brittle Bladder-fern

Britain: frequent in the N and uplands elsewhere.
Native; 29/38; occasional, scattered.
1st Gourock, 1865, RH.

Other old records include glens at Shielhill (1881, NHSG), Kelly (1881, NHSG), Devol (1883, GL) and Bardrain (1902, PSY). Today it can still be found in several woodlands, usually associated with watercourses, generally on base-rich rock outcrops or overhangs and often near the water's edge. AJS knew it from an urban wall near Paisley Canal Station (NS4863, 1975).

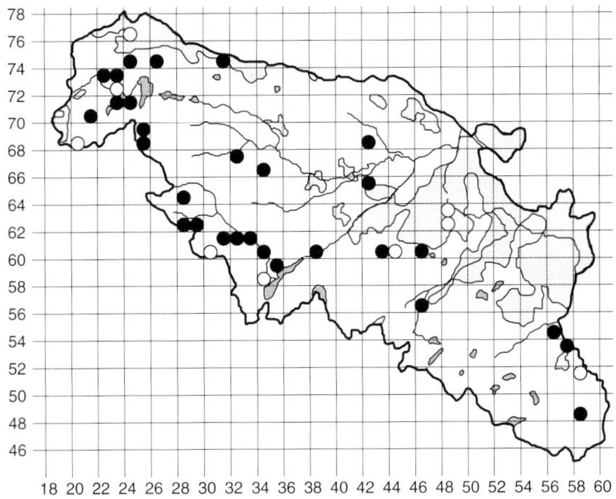

Cystopteris fragilis

BLECHNACEAE

Blechnum spicant (L.) Roth
Hard-fern

Britain: common in the N and W.
Native; 241/283; common.
1st Gourock, 1860, GL.

A common and widespread fern of acidic woodlands, heaths and grasslands, occurring in both lowland and upland locations, but scarce in urban areas. Hard-fern can be frequent on steep, grazed open hillside terraces mixed in with acidic grassland and heath species.

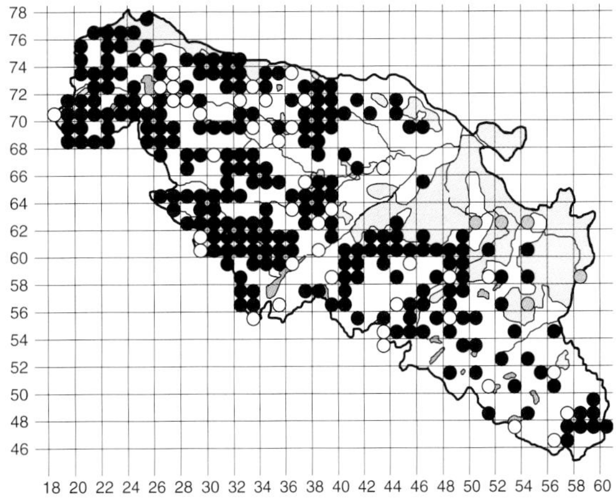

Blechnum spicant

DRYOPTERIDACEAE

Polystichum setiferum (Forssk.) T. Moore ex Woyn.
Soft Shield-fern

Britain: common in the SW, frequent elsewhere, absent from N Scotland.
Native and hortal; 3/8; found in the extreme W, very rare.
1st Inverkip, 1863, MMGNS.

Old records for Soft Shield-fern include Cloch and Gourock (1880, NHSG) and a field excursion record from Wemyss Bay (1887, NHSG), and it was collected from Ferguslie (1900, PSY). Commenting on its presence at Inverkip, Somerville (1907) wrote, 'used to be plentiful', indicating a loss by the end of the 19th century. RM collected it from Ferryhill Plantation in 1976 (E) and in 1978 it was recorded from Lunderston Bay (NS2074, BWR), where it was also found in 1996, growing below the wooded cliffs. AMcGS recorded it from 'nearby Lunderston' (NS2173) in 1983, not far from the modern find at Daff Burn, Inverkip (NS2172, 2002). It was reported from Pollok Castle woods in 1964 (Mackechnie 1971), but this may well refer to planted material or spread from nearby gardens (e.g. at Haggs Castle, NS5662, where it grows). A cultivated form was found under trees by a burn in Clarkston (NS5657, 2011).

Polystichum aculeatum (L.) Roth
Hard Shield-fern

Britain: locally common throughout.
Native; 66/77; frequent in valley woodlands.
1st Cathcart, 1813, TH.

A fairly widespread woodland fern occurring, though seldom abundantly, throughout the VC, but noticeably absent from the south-east and high ground; the south-east outlier is the upper Earn Water, Loganswell (NS5151, 2005). Hard Shield-fern is mainly found on more basic rock outcrops in shady woodland valleys. The hybrid with Soft Shield-fern, *P. × bicknellii* (H. Christ) Hahne, was recorded from Inverkip Glen (NS2072, 1987, AMcGS).

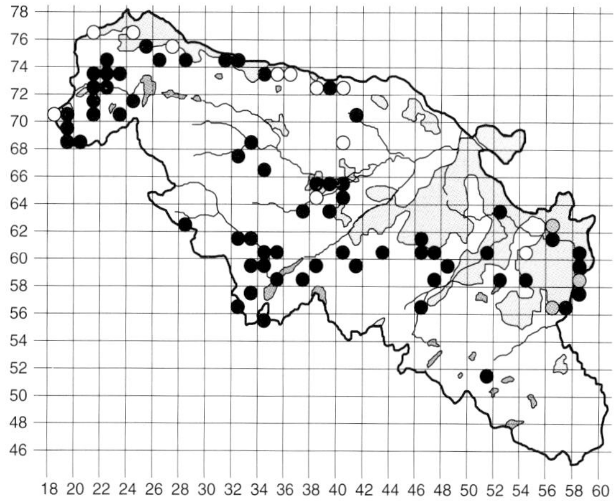

Polystichum aculeatum

Dryopteris filix-mas **(L.) Schott**
Male-fern

Britain: common throughout.
Native; 324/382; very common.
1st Renfrewshire, 1869, PPI.

This fern is found in most woodlands but prefers less acidic soils. It is also common on open or partially shaded ground including urban waste ground and walls, burn sides and rocky scree, becoming less frequent in upland areas. Some of the older records may refer to Scaly Male-ferns. There is one tentative record for the hybrid with *D. affinis* (*D. × complexa* Fraser-Jenk.) from Big Wood, Erskine (NS4472, 1996) and it is likely this hybrid is widespread.

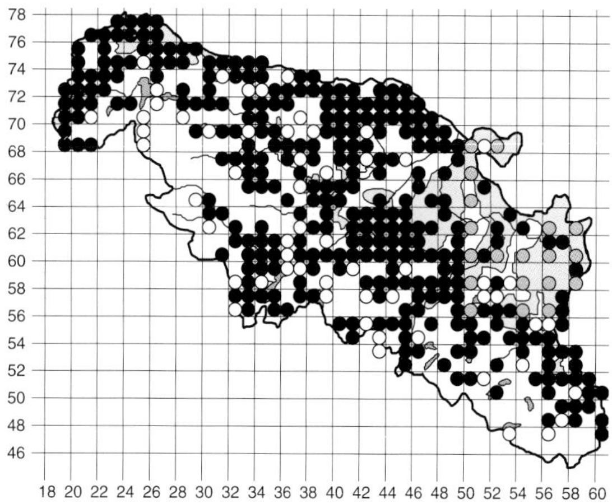

Dryopteris filix-mas

Dryopteris affinis **agg.**
Scaly Male-ferns

Britain: common in the N and W.
Native; 251/302; very common.
1st Cloch light, 1857, GL.

Under-recorded in the past and also confused with the previous, Scaly Male-ferns are widespread and common. They grow in similar places to Male Fern, from remote woodlands to urban waste ground, and can be notable on rocky scree in the uplands. Recording has been at the aggregate level. Many are assumed to be Golden-scaled Male-fern (*D. affinis* (Lowe) Fraser-Jenk.), but AMcGS collected *D. borreri* (Newman) Newman ex Oberh. & Tavel from Inverkip (NS2072, 1990, E) and there are also eight tentative field records (mainly from woods to the north-west) and one for *D. cambrensis* (Fraser-Jenk.) Beitel & W.R. Buck from the River Calder (NS3361, 1997).

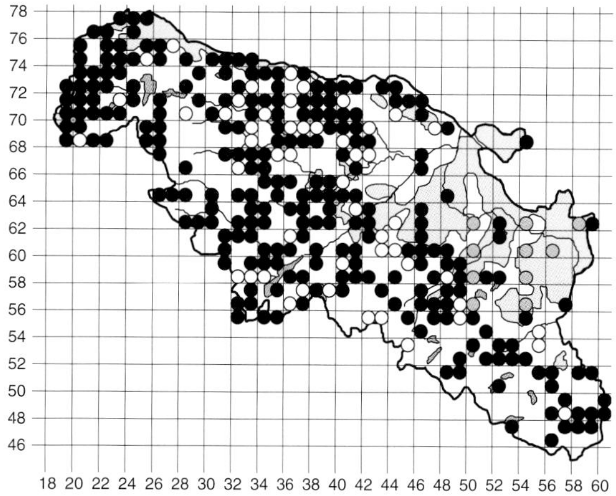

Dryopteris affinis agg.

Dryopteris aemula **(Aiton) Kuntze**
Hay-scented Buckler-fern

Britain: frequent in the W.
Native; 2/3; very rare.
1st Kelly Glen, 1891, RH.

Very likely overlooked, this western fern had not been reliably recorded until PRG observed it twice in 2010, in birch woods at Forbes Place (NS1969) and Leapmoor (NS2171).

Dryopteris cristata **(L.) A. Gray**
Crested Buckler-fern

Britain: very rare, now only in E England.
Native; 0/1; very rare, likely extinct.
1st Loch Libo, 1863, GL.

RM refound and collected this fern from the Loch Libo

site (NS4355) in 1945. It was last recorded in 1969 by Hugh McAllister, who then only saw one plant, on a rocky promontory. Older records for a 'bog beyond Crofthead' (1865, RH) presumably relate to the loch or the marshy carr wetlands to the east. The site has been well searched in recent years without success, so this Renfrewshire rarity may well be a recent extinction.

Dryopteris carthusiana (Vill.) H.P. Fuchs
Narrow Buckler-fern

Britain: scattered throughout.
Native; 49/51; frequent in boggy places.
1st Neilston, 1876, GL.

This paler relative of the Broad Buckler-fern occurs in acidic 'poor-fen' mires, usually associated with bog-mosses. It is well represented in upland peaty soils to the south-east but is notably rare in the Renfrewshire Heights, being restricted to the eastern fringes. In addition to the first record, the only other named 19th-century record comes from Loch Libo (1876, FFWS), where Narrow Buckler-fern still occurs. The hybrid with Broad Buckler-fern (*D.* × *deweveri* (J.T. Jansen) Jansen & Wacht.) has been tentatively recorded from a few sites including a pond, Barscube Hill (NS3871, 1999), Loch Libo (NS4355, 1996), mire at Dargavel Burn (NS3771, 2005) and Harelaw Reservoir (NS3073, 2007); it was first noted at boggy moorland at Gleniffer Braes Road (NS4459) in 1975 (BWR).

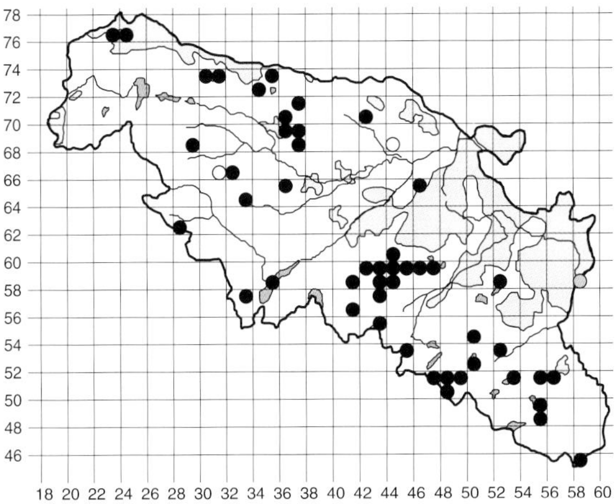

Dryopteris carthusiana

Dryopteris dilatata (Hoffm.) A. Gray
Broad Buckler-fern

Britain: common throughout.
Native; 438/497; very common and widespread.
1st Wemyss Bay, 1848, GL.

This is a very common fern of woodlands, tolerating heavily shaded plantations and acidic soils, and also occurring in boggy birch woodlands. It can be common in urban scrub and colonizes waste ground, heaths where not too heavily burned or grazed, rocky hillsides and burn sides; it is found from the lowlands to the peaks of the Renfrewshire Heights, where it can even occur at the drier margins of mires and along drains.

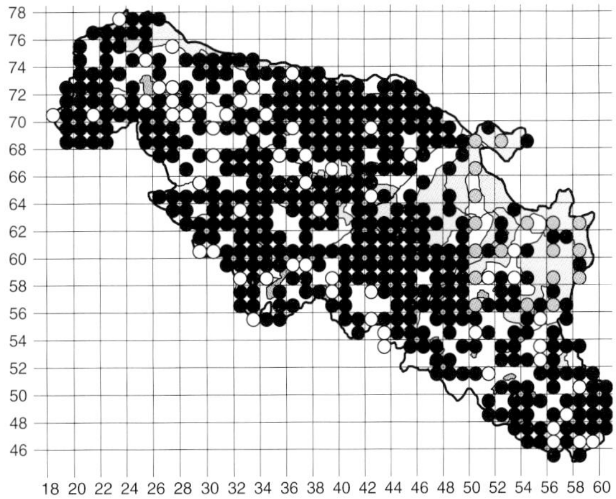

Dryopteris dilatata

POLYPODIACEAE

Polypodium vulgare L.
Polypody

Britain: common throughout except in E C England.
Native; 223/264; common.
1st Port Glasgow, 1840, GL.

Polypody was considered very common by Hennedy (1891) but named stations are few: Kelly Glen and Castle Semple Loch (1887, NHSG) and Howwood (1900, PSY). Polypody is frequently found on small ridges or rocky outcrops throughout the upland areas, where it is inaccessible to grazers. It also

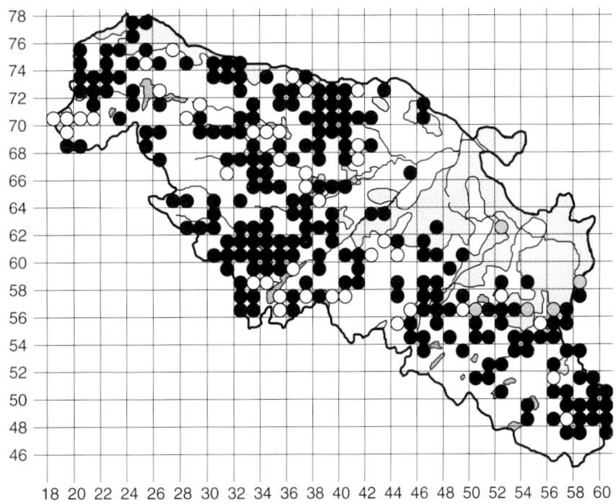

Polypodium vulgare

occurs on rocks and trees in woodlands and on old stone walls in urban areas. The map may include old, and some modern, records of Intermediate Polypody.

Polypodium interjectum Shivas
Intermediate Polypody

Britain: frequent in the W, local but widespread elsewhere.
Native; 11/12; scarce, but under-recorded.
1st Mearns, 1883, GL.

This fern was not distinguished by earlier recorders and today is probably still under-recorded. Most of the modern records are from woodlands in the north and west coastal areas, but inland records include Garpel Burn (NS3459, 1998), Ranfurly (NS3865, 1997) and Enoch Burn, Whitelee (NS5750, 2003, JD).

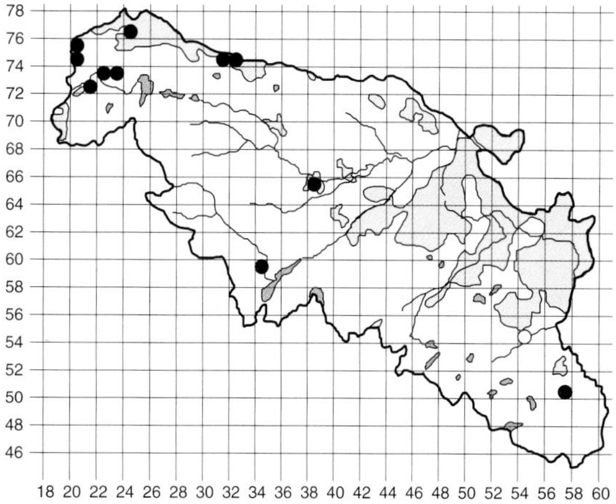

Polypodium interjectum

GYMNOSPERMS (CONIFERS)

PINACEAE

Tsuga heterophylla (Raf.) Sarg.
Western Hemlock-spruce

Britain: neophyte (W N America); scattered.
Hortal; 1/1; very rare.
1st Glentyan Estate, 2007.

There is only one record, of a young sapling which had colonized a shady wall at Glentyan Estate (NS3963), but Western Hemlock-spruce may well be spreading at other similar estates where it has been planted.

Picea sitchensis (Bong.) Carrière
Sitka Spruce

Britain: neophyte (W N America); widely planted.
Hortal; 37/39; frequent, scattered.
1st Finlaystone, 1976, AJS.

Sitka Spruce was widely planted in the 20th century and can be found in dense commercial plantations. Young plants have been noted from many of the widespread mapped sites, which include acidic heaths, moors, grasslands and screes.

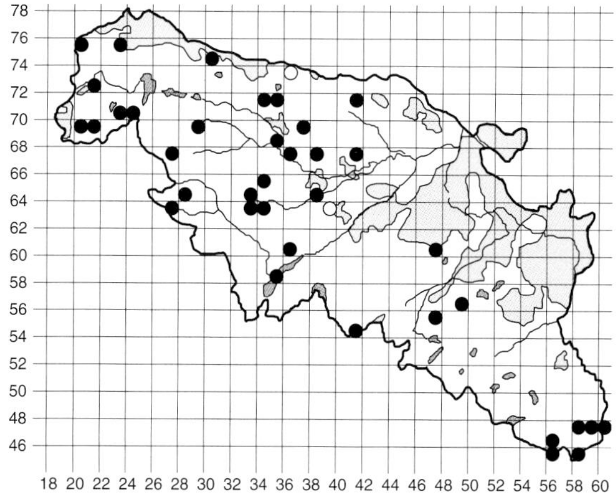

Picea sitchensis

Picea abies (L.) H. Karst.
Norway Spruce

Britain: neophyte (N Europe); widely planted.
Hortal; 31/35; occasional.
1st Neilston Pad, 1972, BWR.

Like Sitka, this spruce is widely planted but several records refer to young plants, as at West Gavin (NS3758, 1998), North Barlogon (NS3768, 1998), Linn Park (NS5859, 2005) and Glentyan (NS3963, 2007).

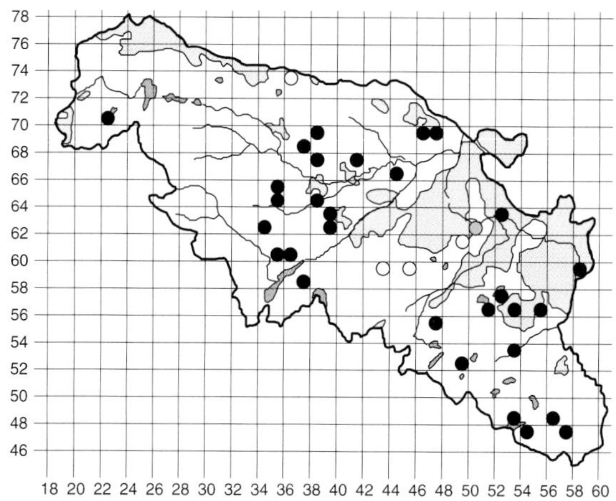

Picea abies

Larix decidua Mill.
European Larch

Britain: neophyte (C Europe); widespread.
Hortal; 47/67; frequently planted.
1st Renfrew, 1777, JW.

Larch is found throughout the area, being a feature of many estate woodlands and farm shelterbelts. Many records refer to mature trees and planted stands but there are several records for young plants, although it can hardly be classed as naturalized. Young saplings have been noted at Neilston Pad (NS4755, 1993), Bardrain Glen (NS4360, 1993), Broadfield Hill (NS4059, 1997), Gotter Water (NS3466, 1999), Cowdon Burn (NS4757, 1999), Dunrod Glen (NS2173, 2003) and Underheugh (NS2075, 2004); a few of these records may refer to the hybrid with Japanese Larch (*L.* × *marschlinsii* Coaz), which has been recorded at eight widespread sites. Japanese Larch (*L. kaempferi* (Lamb.) Carrière) has been noted as obviously planted trees, but there is one field record from Finlaystone (NS3673, 1976, AJS).

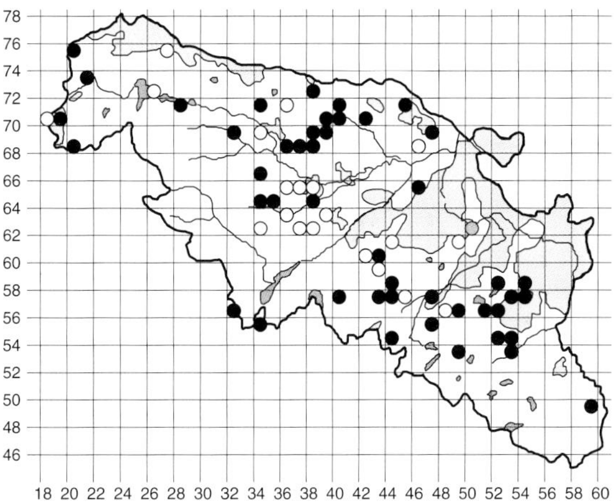

Larix decidua

Pinus sylvestris L.
Scots Pine

Britain: rare native in upland Scotland, common alien elsewhere.
Hortal; 108/159; commonly planted, occasionally seeding.
1st Renfrew, 1777, JW.

Scots Pine is a commonly planted tree of policy estates and farm shelterbelts, but it is not today native to this part of Scotland. However, the mature plantations on the deep peat of Barochan Moss (NS4268, 1994) and the Royal Ordnance Factory, Georgetown (NS4270, 1996) have developed a quite 'natural' pinewood atmosphere. Young saplings are not uncommon, being recorded at a dozen or so of the sites.

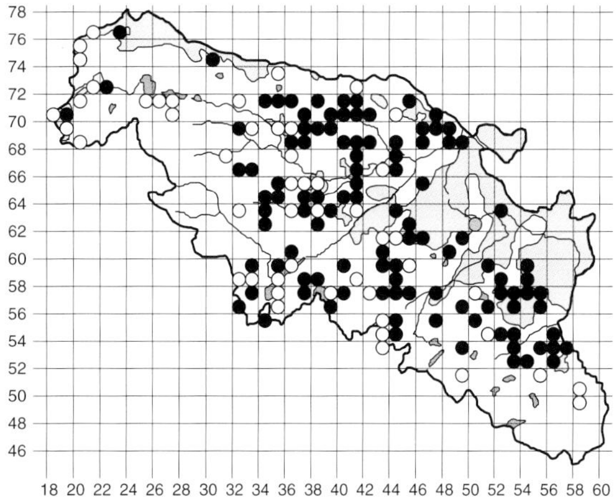

Pinus sylvestris

ARAUCARIACEAE

Araucaria araucana (Molina) K. Koch
Monkey-puzzle

Britain: neophyte (Chile and W Argentina); widely scattered.
Hortal; 1/1; very rare.
1st Wemyss Bay, 2002.

A young sapling of this locally rarely planted tree of estate woodlands and larger gardens was seen in a scrubby plantation near Inverkip Power Station (NS1970, 2002). Its origin is unclear but it appeared to be self-sown.

TAXACEAE

Taxus baccata L.
Yew

Britain: common in England, frequent in C Scotland.
Hortal; 50/58; frequent as planted trees, local as seedlings.
1st Dargavel House, Erskine Parish, 1845, NSA.

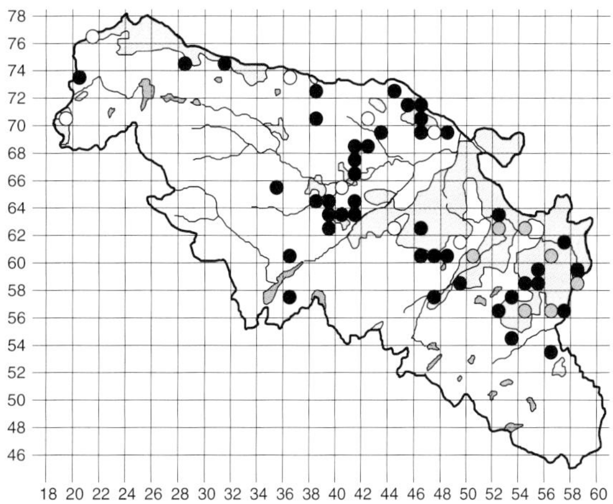

Taxus baccata

Yew is frequently planted in estate and plantation woodlands and can be found as seedlings, especially on the less acidic soils of watercourse valleys. The earliest record presumably refers to the massive specimen at Dargavel House (NS4369, 1999), although the tree at Craigend Estate, Crosslee (NS4065, 2000) is larger. In 1845 the Dargavel tree was described as 'one of the greatest ornaments and antiquities of the parish'. A specimen recorded from Elderslie (1859, PSY) is said to be from 'William Wallace's tree'. However, Yew does not appear to be a tree that can be described as established anywhere, except as long-lived individuals.

CUPRESSACEAE

Chamaecyparis lawsoniana (A. Murray bis) Parl.
Lawson's Cypress

Britain: neophyte (W N America); frequent in S England, local in Scotland.
Hortal; 8/8; occasional seedlings, but probably overlooked.
1st Crookston (NS56B), 1986, CFOG.

This conifer was recorded from urban-fringe waste ground at Greenock (NS2776, 2000), Neilston (NS4757, 2004) and Cowdon Burn (NS4657, 2007), but is likely to occur elsewhere, however doubtfully persisting. Other sapling records include Ardgowan (NS2073, 2000), Bankfoot (NS2173, 2003), Waterside (NS5160, 2005), Earn Hill (NS2473, 2006) and Bow Hill Cemetery (NS2676, 2007); it is possible that some records may refer to related Cypress taxa.

Juniperus communis L.
Common Juniper

Britain: frequent in uplands.
Native; 7/10; rare, threatened.
1st Renfrewshire, 1869, PPI.

Juniper was presumably more frequent in the 19th century, but there are no localized records, except for 'Gourock' (1876, FFWS). An old Juniper was noted in the gorge at Bardrain Glen in a report by the 'Stickleback Club' (Stewart, 1930), but this is the only record of this shrub from that site. Common Juniper is present as six mainly isolated bushes in the Renfrewshire Heights, two of which are known to be female: at Kilmacolm High Dam (NS3167, 2000) and along the Cample Burn (NS3062, 2000, AB). Large sprawling male plants are known from Lyles Hill (NS3164, 2000) and at Cample Burn (NS3062, AB); a smaller, indeterminate bush also occurs at the last (2000, AB) and recently a similar dwarf shrub was found on a ledge at the Forking of Raith (NS2862, 2004, AB). The local survival is threatened by the individual's isolation plus pressures of grazing and some burning, but efforts are underway to conserve the populations by CMRP staff. Several shrubs, including some young plants, occur to the edge of Paisley Golf Course, but presumably are of planted origin (NS4559, 2010).

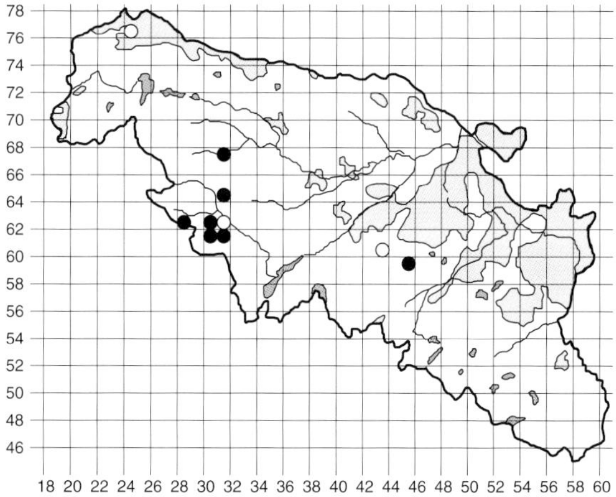

Juniperus communis

ANGIOSPERMS (FLOWERING PLANTS)

PRE-DICOTS (PRIMITIVE ANGIOSPERMS)

NYMPHAEACEAE

Nymphaea alba L.
White Water-lily

Britain: common in S England and NW Scotland, rare in NE.
Native (and hortal?); 7/11; rare.
1st Lochwinnoch and Paisley, 1845, NSA.

White Water-lily can still be found at Loch Libo

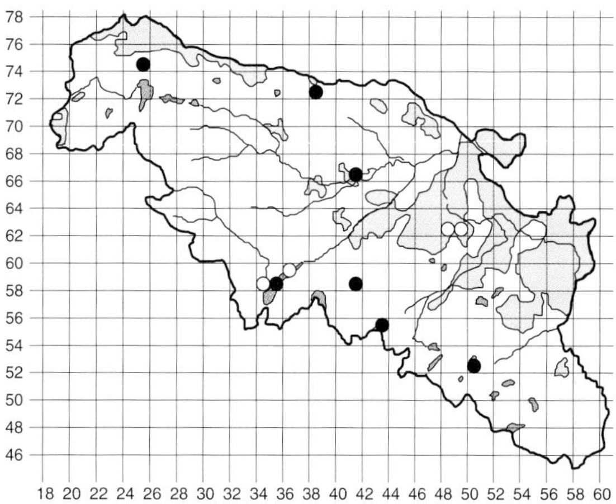

Nymphaea alba

(NS4355), where it was recorded in 1887 (NHSG), but it has not recently been seen at Jenny's Well pond, where it was recorded in 1942 (TPNS). It was recorded at Castle Semple Loch in 1934 (RM) but it is now only known from a small pond there, probably planted (NS3659, 2012). It is also found (hortals?) at two other ponds at Houston (NS4167, 1996, SNH-LS) and Gleddoch estates (NS3872, 1998). More natural settings are at Walls Loch (NS4158, 1993) and Little Loch (NS5052, 1997, SNH-LS); JDM recorded it at the latter in 1957. The origin of the isolated population at Gryfe Reservoir No. 4 by Greenock Cut is unknown, but possibly it was recently introduced (NS2974, 2007).

Nuphar lutea (L.) Sm.
Yellow Water-lily

Britain: common in S England and SW Scotland, rare alien in N Scotland.
Native; 19/21; rare away from the Black Cart Water.
1st Paisley and Lochwinnoch, 1845, NSA.

The map readily shows the linear distribution of this species along the Black Cart Water, from Barr and Castle Semple lochs (NS3557) downstream to near Inchinnan (NS4666); away from this river Yellow Water-lily is rare. It was found in the White Cart Water at Crookston in 1856 (HMcD) but is not known from this river today. It is found in open water at Loch Libo (NS4355, 1999), Caplaw Dam (NS4358, 1993), Little Loch (NS5052, 1996) and Walls Loch (NS4153, 2002); it is also found in the Roebank Burn, Knowes (NS3756, 1999) and Old Patrick Water, Hartfield (NS4358, 1999).

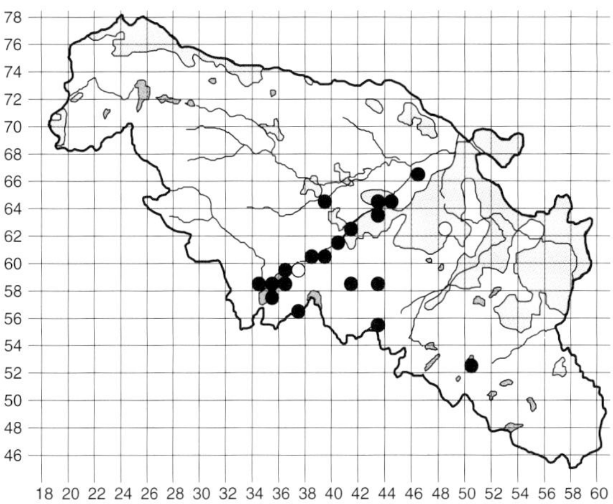

Nuphar lutea

ARISTOLOCHIACEAE

Asarum europaeum L.
Asarabacca

Britain: neophyte (Europe); scarce but widely scattered.
Former hortal; 0/1; extinct.
1st Levernholme Wood (*c.* NS5062), 1890, AANS.

There is only one record for Asarabacca (presumably a garden plant or escape), which was not mentioned by Hennedy or Lee, and was excluded from the Renfrewshire list (TPNS 1915).

EU-DICOTS (TRUE DICOTYLEDONS)

CERATOPHYLLACEAE

Ceratophyllum demersum L.
Rigid Hornwort

Britain: scattered in the S, rare in Scotland.
Native; 0/1; extinct or error.
1st Stanely Castle, 1865, MY.

This is a dubious early record for a plant excluded from the Renfrewshire list (TPNS 1915) and otherwise first recorded in the Glasgow area in 1938 (CFOG).

PAPAVERACEAE

Papaver pseudoorientale (Fedde) Medw.
Oriental Poppy

Britain: neophyte (SW Asia); scattered in England and lowland Scotland.
Hortal; 3/3; very rare outcast.
1st Teucheen Wood (NS4869), 1996.

This tall garden outcast has also been recorded from Kaimhill (NS4065, 1999) and East Langbank (NS3973, 1998).

Papaver somniferum L.
Opium Poppy

Britain: archaeophyte; common in S England, frequent in lowland Scotland.
Hortal; 7/8; rare casual.
1st Rubbish tip at Barr Loch, 1969, AMcGS.

There were several records made for CFOG from places such as Queen's Park (NS56R, 1983), Ralston Golf Course (NS56B, 1987) and the White Cart Water, Busby (NS55T, 1984), but there are only two more recent records for this garden outcast: Gryfe Wraes (NS3966, 1998) and Brodie Park (NS4762, 2003).

Papaver rhoeas L.
Common Poppy

Britain: archaeophyte; common in the S, extending to C and E Scotland.
Former accidental and hortal; 5/7; rare casual.
1st Paisley and Lochwinnoch, 1845, NSA.

This cornfield weed was perhaps never common in this part of the country; this is suggested by the paucity of old records, and there are only two relevant CFOG records (NS56K and NS55N). Modern field records may well be from wildflower mixes: roadsides, Erskine (NS4572, 1998, PM), Linwood Moss (NS4465, 1998) and Inchinnan (NS4767, 1999).

Papaver dubium L.
Long-headed Poppy

Britain: archaeophyte; common in England and lowland Scotland.
Accidental; 22/26; occasional.
1st Cathcart, 1865, RH.

The commonest poppy in the area, Long-headed Poppy is much more frequent in the drier east; many of the records are from urban-fringe waste ground.

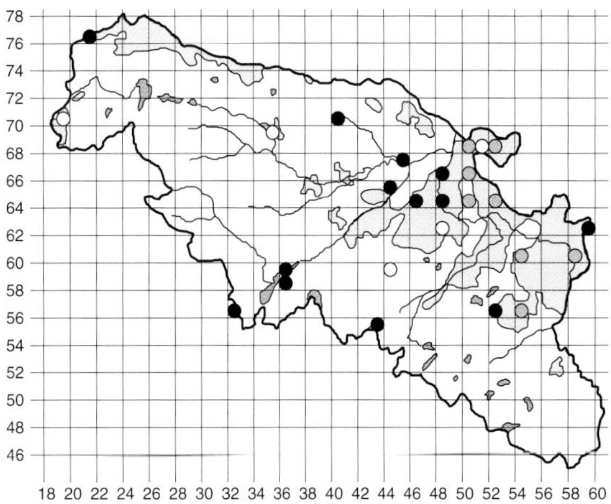

Papaver dubium

Papaver argemone L.
Prickly Poppy

Britain: archaeophyte; frequent in SE England, local in E Scotland.
Former accidental; 0/2; extinct.
1st Lochwinnoch, 1845, NSA.

The only other old record is from Elderslie (1896, NHSG), repeated by Lee in 1933.

Meconopsis cambrica (L.) Vig.
Welsh Poppy

Britain: local in Wales and SW England, common alien in the N and W.
Hortal; 101/112; common, lowland.
1st Carruth Bridge, 1890, AANS.

Welsh Poppy can still be found at its original location and at Glentyan Estate (NS3963), where it was first noted in 1921 (AANS), but it has not been refound at Finlaystone, where it was first noted in 1890 (AANS). It is now quite widespread, occurring in old estates, but is probably more frequent now as a garden outcast or escape. Its distribution reflects an urban pattern but it can occur at more isolated sites.

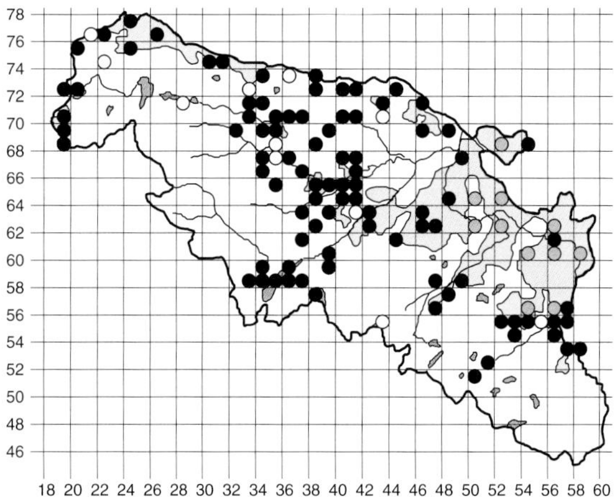

Meconopsis cambrica

Glaucium flavum Crantz
Yellow Horned-poppy

Britain: common on coasts in the S, rare in Scotland.
Former native; 0/1; extinct.
1st Renfrewshire, 1883, TB.

Hennedy (1891) noted Yellow Horned-poppy as frequent on sea shores but only named it from a few islands, while Ferguson (1915) and Lee (1933) made no mention of the species.

Glaucium corniculatum (L.) Rudolph
Red Horned-poppy

Britain: neophyte (SE Europe); scarce casual.
Former accidental; 0/1; extinct.
1st Paisley, 1910, PSY.

Chelidonium majus L.
Greater Celandine

Britain: archaeophyte; common in S England, scattered in lowland Scotland.
Hortal; 2/6; very rare.
1st Lochwinnoch, 1845, NSA.

Greater Celandine was recorded at Duchal in 1862 (NHSG) and Finlaystone (1898, JMY) and more recently at Bridge of Weir (1957, PSY). The only modern records are from Greenbank (NS56T, 1984, CFOG) and a small population in West Lady Wood (NS5256, 2000).

Eschscholzia californica Cham.
Californian Poppy

Britain: neophyte (SW N America); scattered, mainly in the S.
Hortal; 1/1; very rare casual.
1st Lyoncross, 1999.

The sole record is from dumped garden waste on a trackside at a farm in Barrhead (NS5157).

Dicentra formosa (Haw.) Walp.
Bleeding-heart

Britain: neophyte (W N America); scattered, mainly in the W.
Hortal; 7/7; rare.
1st Bridge of Weir, 1970, BWR.

The modern records (which may include Turkey-corn or hybrid with it) are all presumably recent garden outcasts: River Calder (NS3459, 1995), Ranfurly (NS3865, 1997), the Gryfe Water at Milton (NS3568, 1997) and Kilmacolm (NS3569, 1997), and Teucheen Wood (NS4868, 2007). A large population occurs in Darnley [Waulkmill] Glen (NS5258, 2006) and presumably refers to the old Turkey-corn record. In 1978 BWR recorded Eastern Bleeding-heart (*D. spectabilis* (L.) Lemaire) at Lunderston (NS2074) but there is no specimen or further details.

Dicentra eximia (Ker Gawl.) Torr.
Turkey-corn

Britain: neophyte (E N America); scattered garden escape.
Hortal; 0/2; extinct.
1st Darnley, 1920, TPNS.

The original record may refer to Bleeding-heart which now occurs in the glen. More recently AJS recorded this species as a garden outcast on a roadside at the Gleniffer Braes (NS4459, 1975).

Corydalis solida (L.) Clairv.
Bird-in-a-bush

Britain: neophyte (Europe); scattered, mainly in the SE.
Former hortal; 0/1; extinct.
1st Ruins at Montreal, 1898, JMY.

With no further information or records this garden plant is best considered extinct.

Pseudofumaria lutea (L.) Borkh.
Yellow Corydalis

Britain: neophyte (S Alps); common in the S, frequent in E Scotland.
Hortal; 5/7; rare.
1st Maxwell Park, 1903, GL.

Yellow Corydalis was collected in 1921 from a garden in Kilbarchan (GL). Modern records are from walls in Kilbarchan (NS4063, 1999), Jordanhill (NS5468, 2002, AMcGS), Langbank (NS3873, 2006), Glentyan Estate (NS3963, 2007) and Newlands (NS5760, 2010, PM).

Ceratocapnos claviculata (L.) Lidén
Climbing Corydalis

Britain: widespread but local, frequent in W C Scotland.
Native; 26/36; occasional, local.
1st Renfrewshire, 1834, MC.

The map shows a widespread scattering of isolated sites, but with a remarkable concentration in the Barochan area from Scart Wood (NS3867, 2002) across to Barochan Moss (NS4268, 1994). At most of these sites Climbing Corydalis occurs in humus-rich or peaty woodlands, usually associated with Bracken.

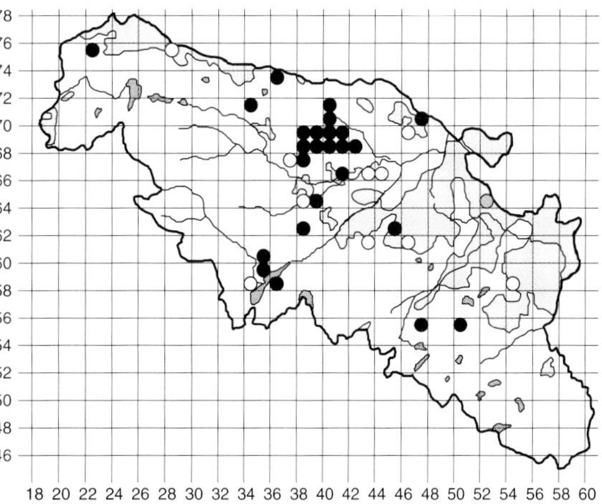

Ceratocapnos claviculata

Fumaria capreolata L.
White Ramping-fumitory

Britain: widely scattered, mainly coastal.
Former native; 0/2; extinct.
1st Stanely Castle, 1845, NSA.

There is doubt about the identity of the original report and although Hennedy (1865) noted White Ramping-fumitory at Cathcart, he made no mention of Common Ramping-fumitory, for which it could be mistaken. There is, however, a specimen from the banks of the White Cart Water at Paisley, dated 1883 (GL), confirming its former presence in the VC.

Fumaria bastardii Boreau
Tall Ramping-fumitory

Britain: locally frequent in the W, rare in Scotland.
Native; 2/3; very rare or overlooked.
1st Nether Auldhouse, 1948, RM.

Not distinguished in early literature, this fumitory was more recently recorded (CFOG) at Arkleston (NS5065, 1985) and Cowglen Golf Course (NS5561, 1991), but there are no modern field records.

Fumaria muralis Sond. ex W.D.J. Koch
Common Ramping-fumitory

Britain: common in W England and E Scotland, local elsewhere.
Native or accidental; 24/27; frequent.
1st Renfrewshire, 1883, TB.

This is the commonest fumitory in the VC; it has a scattered distribution strongly associated with disturbed ground such as soil heaps in lowland, urban areas. All records appear to be for ssp. *boroei* (Jord.) Pugsley.

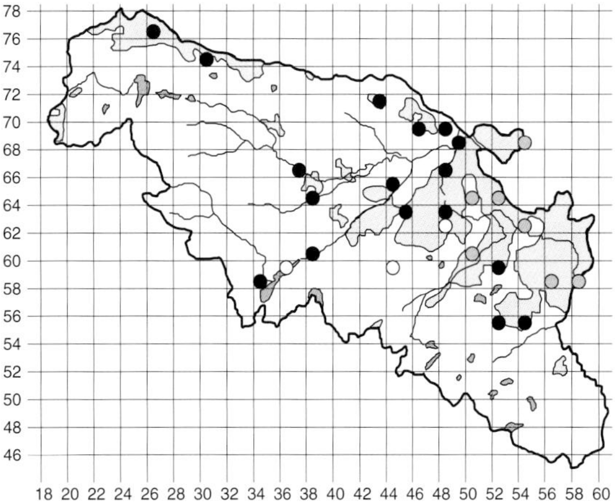

Fumaria muralis

Fumaria purpurea Pugsley
Purple Ramping-fumitory

Britain: scattered and local, mainly in the W, local in SE Scotland.
Native or accidental; 6/7; rare.
1st Bishopton, 1916, GL.

This British endemic has gone from its CFOG location at Cowglen (1991) but can be found on similar spoil heaps at Elderslie (NS4562, 2002) and Lochwinnoch (NS3558, 2003, IPG); it was also found on disturbed roadside soil by the M77, Malletsheugh (NS5255, 2001) and trackside scrub, near new housing at Stanely Reservoir (NS4661, 2005). AJS recorded it at the University of the West of Scotland Paisley campus in 1992 (NS4863). A 1942 RM specimen of Purple Ramping-fumitory from Netherlee was redetermined (by Tim Rich) as Common Ramping-fumitory.

Fumaria officinalis L.
Common Fumitory

Britain: archaeophyte; common in England and E Scotland.
Accidental; 13/22; scarce.
1st Kilbarchan, 1845, NSA.

Less frequently encountered than the Common Ramping-fumitory, this fumitory has a similar distribution pattern, and occurs in similar places and also rarely as an arable weed, e.g. South Mains (NS4266, 2003), and may well have declined at this latter habitat.

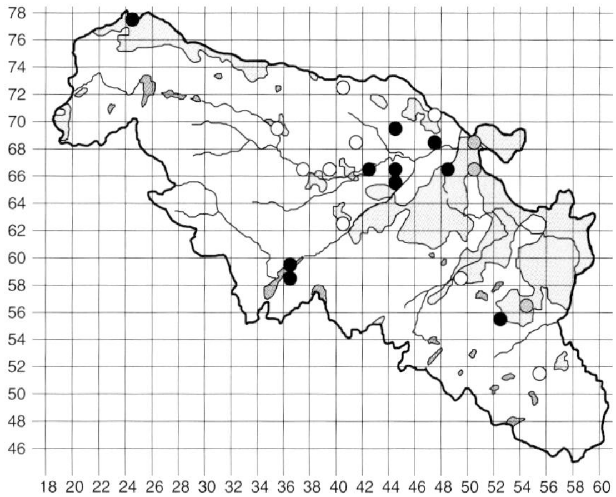

Fumaria officinalis

Fumaria densiflora DC.
Dense-flowered Fumitory

Britain: archaeophyte; local in SE England and E Scotland, rare elsewhere.
Accidental; 0/1; extinct or overlooked.
1st Allotments at Merrylee, 1942, GLAM.

BERBERIDACEAE

Berberis vulgaris L.
Barberry

Britain: frequent and scattered, extending to C and E Scotland.
Native or hortal; 7/10; rare.
1st Inverkip, 1847, GGO.

Old records include herbarium sheets (GL) from Hazeldean Dam, Newton Mearns (1909) and Kilbarchan (1912). More recently BWR recorded Barberry from a farm track, Finnockbog (NS2070, 1976) and it was recorded at a wood edge, west of Levernholm (NS56A, 1986, CFOG). Today there is large specimen in a hedgerow along Old Greenock Rd, Bishopton (NS4271, 2001), perhaps the same place as mentioned in a 1948 ANG excursion report. Other modern records are from Milliken (NS4162, 2001), Corsliehill House (NS3969, 1998), Gleddoch Estate (NS3872, 1998), a hedge by the Gryfe Water, Bridge of Weir (NS3865, 2005) and by Castle Semple Loch (NS3659, 2011).

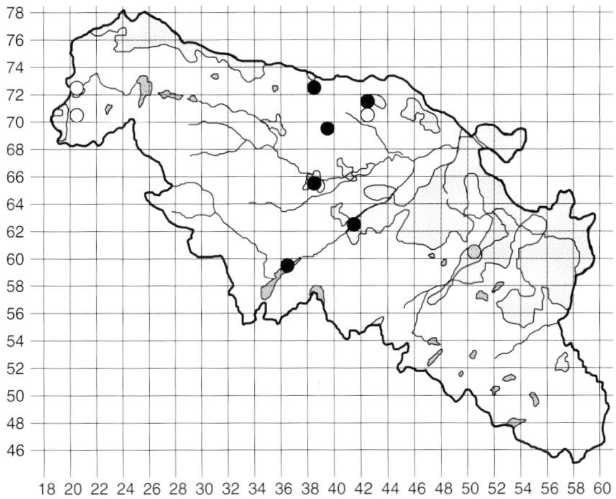

Berberis vulgaris

Berberis thunbergii DC.
Thungberg's Barberry

Britain: neophyte (Japan); scattered.
Hortal; 3/3; very rare.
1st Teucheen Wood, Erskine (NS4869), 1996.

Two other records are from Whitecroft cliff, Port Glasgow (NS3274, 1997) and Spango (NS2374, 1997).

Berberis glaucocarpa Stapf.
Great Barberry

Britain: neophyte (C Asia); scarce, mainly in S England.
Hortal; 1/1; very rare.
1st Inverkip (NS2060), 1999.

Berberis gagnepainii C.K. Schneid.
Gagnepain's Barberry

Britain: neophyte (W China); scattered, mainly in S England.
Hortal; 2/2; very rare.
1st Rouken Glen (NS5458), 1996.

A further record is from a path edge, Moorpark (NS4966, 2011).

Berberis darwinii Hook.
Darwin's Barberry

Britain: neophyte (Chile and Argentina); scattered in England and C Scotland.
Hortal; 7/7; rare.
1st Craigton (NS5464), 1994, CFOG.

Six other modern records are Formakin Estate (NS4170, 2003, IPG), the White Cart Water, Paisley (NS4863, 2004), Rouken Glen (NS5458, 2005), Bridge of Weir (NS3865, 2005), Battery Park (NS2527, 2012) and Larkfield (NS2375, 2012). This barberry was mentioned from Bishopton in an ANG excursion report (1948) but the plant's status is unclear, although the author described the plant as 'well-known'.

Berberis* × *stenophylla Lindl.
(*B. darwinii* × B. *empetrifolia* Lam.)
Hedge Barberry

Britain: neophyte (garden origin); scattered, mainly in the S.
Hortal; 1/1; very rare.
1st Waste ground, Mearns (NS5455), 2001.

Mahonia aquifolium (Pursh) Nutt.
Oregon-grape

Britain: neophyte (W N America); common in S England, frequent in C Scotland.
Hortal; 7/9; rare.
1st Duchal, 1891, NHSG.

The original record was for cultivated plants in fruit. There are seven scattered modern records, mainly from estate grounds: at Formakin (NS4070, 1994), Kilmacolm (NS3569, 1997), Ardgowan (NS2073, 1998), Durrockstock (NS4662, 1999), Barochan (NS4165, 1999), Dykebar Hospital (NS4961, 1999) and Auldhouse Road (NS5660, 2005). The records may include hybrids or cultivars.

Epimedium alpinum L.
Barren-wort

Britain: neophyte (S Europe); scarce, mainly in the N.
Former hortal; 0/2; extinct.
1st Carruth to Bridge of Weir, 1890, AANS.

When this excursion record was made the comment was 'not seen recently' (Paterson 1893). It was collected from the 'Cemetery' (Greenock area) by T Fisher in 1893 (GRK), where it was presumably planted, but it was excluded from the Renfrewshire list (TPNS 1915).

RANUNCULACEAE

Caltha palustris L.
Marsh-marigold

Britain: common throughout.
Native; 274/325; widespread and common in wet places.
1st Paisley, 1845, NSA.

A common plant of marshes, fens, swamps and along water margins, Marsh-marigold is well represented in rural areas but also quite frequent along urban rivers and relict marshy areas. It is rare in the peaty uplands and also in the urban north-eastern area or more intensively managed farmland.

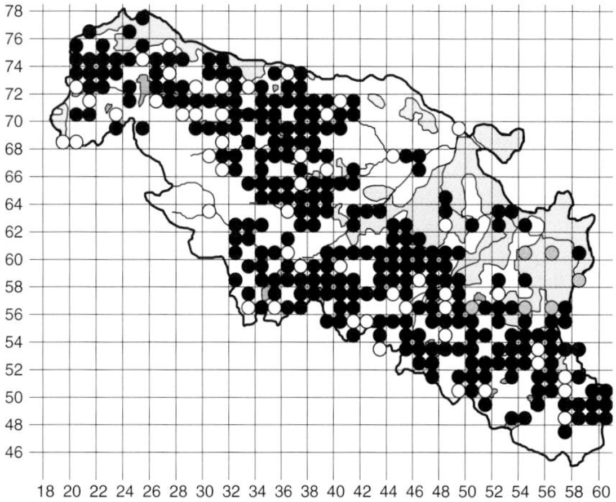

Caltha palustris

Trollius europaeus L.
Globeflower

Britain: frequent in Scotland and upland England.
Native; 9/22; rare.
1st Cloak, 1834, MC.

Globeflower was last recorded from Bardrain Glen in 1942 (TPNS) and seems to have been lost from other old localities such as Carruth Bridge (1891, AANS) and Eaglesham (1895, PSY). Today it is found along the Gryfe Water at Strathgryfe (NS3370, 1998), Kilmacolm (NS3469, 1999) and Cauldside (NS3270, 2005), along the Locher Water at Lawmarnock (NS3766, 2002) and Ranfurly (NS3864, 2002), and along the River Calder (NS3361, 1997). It occurs in some abundance in a marshy willow carr at Levan (NS2176, 1999) and a flush by Harelaw Reservoir (NS4859, 1999); it was recorded from the latter in 1960 (GL). It was recorded from Shielhill Glen in 1960 (BWR) and more recently in 1976 by DM (pers. comm.) and has been reported from Castle Semple (NS3658, 1992, CMRP).

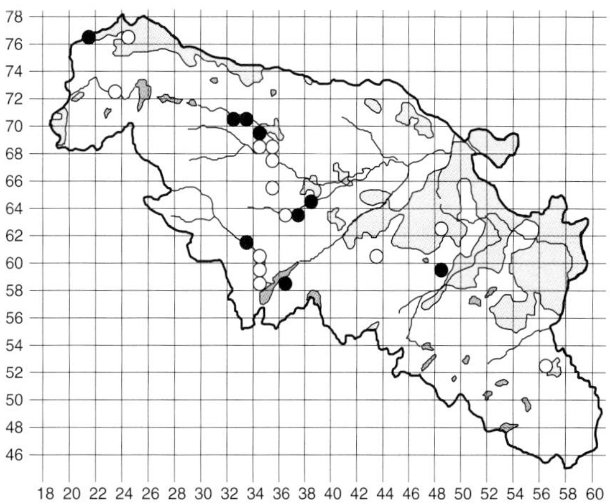

Trollius europaeus

Helleborus foetidus L.
Stinking Hellebore

Britain: scattered in S England, scattered alien elsewhere.
Former hortal; 0/1; extinct garden plant.
1st Carruth Policies, 1898, JMY.

Helleborus viridis L.
Green Hellebore

Britain: scattered in England, scattered alien in the N.
Hortal; 2/5; very rare.
1st Renfrewshire, 1887, NHSG.

The first local record for Green Hellebore was from Paradise, Finlaystone Estate (1890, AANS), which Lee (1933) repeated, but it has not been recorded during recent surveys there. It was reported in 1901 from Barrhead (FFCA) and more recently JHD collected it from woods at Lunderston Bay in 1958 (GL); the latter may well be the same place as PRG's Lunderston record (NS2074, 2010). The only other modern record is from the west bank of the White Cart Water at Busby (NS5756, 1991, GW), although it was not refound during a recent visit.

Eranthis hyemalis (L.) Salisb.
Winter Aconite

Britain: neophyte (S Europe); frequent in the SE, scattered in E Scotland.
Hortal; 1/1; very rare.
1st Darnley, 2000.

A small population of Winter Aconite occurs below an ancient lime tree at the site of the old house (NS5258) in the estate woodland at Darnley.

Aconitum napellus L.
Monk's-hood

Britain: rare in SW England, widespread alien elsewhere.
Hortal; 8/16; rare.
1st Roman Bridge, Inverkip, 1880, NHSG.

Monk's-hood was recorded at the Inverkip Road (1893, GRK), Castle Semple (1889, AANS – also collected from there in 1904, PSY), Darnley [Waulkmill] Glen (1920, RG) and Inverkip (1933, Lee). In the recent past (1966) it was found by JDM from 'High Wood' (NS3367) and from Mearns (NS5454) and BWR recorded it twice at Houston (NS4066, 1969) and by the Gryfe Water, Hatterick (NS3567, 1971); there were two relevant records in

CFOG. It is likely that some of the later records, at least, may be Hybrid Monk's-hood (*A.* × *stoerkianum* Rchb.) [*A.* × *cammarum* auct.], which was first recorded from Lochwinnoch (NS3658, 1973, PM), and is the only Monk's-hood reliably recorded in modern times: at Lochside Station (NS3557, 1993), Nether Johnstone (NS4153, 1998) and a roadside at Netherton (NS5749, 2000); there are two vegetative records from by the White Cart Water in Glasgow: Rosshall (NS5163, 2005) and Langside (NS5761, 2005), which are possibly also of the hybrid.

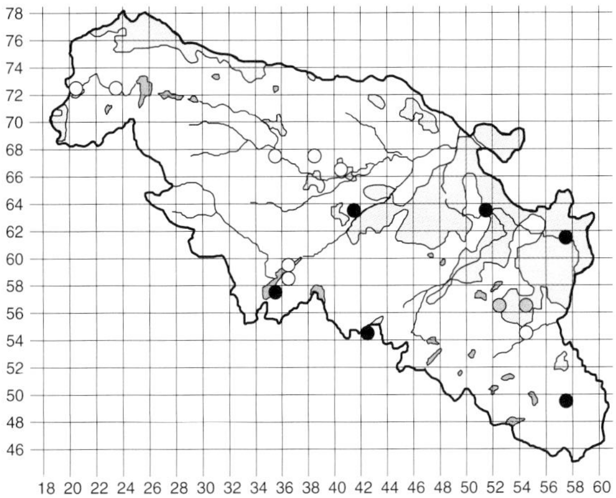

Aconitum napellus

Consolida ajacis (L.) Schur
Larkspur

Britain: neophyte (S Europe); frequent in SE England, very rare elsewhere.
Former hortal; 0/2; extinct casual.
1st Renfrewshire, 1895, NHSG.

Larkspur was also collected from wasteland near Paisley (1910, PSY) and RG (1923) knew it from 'near Glasgow' (CFOG).

Anemone nemorosa L.
Wood Anemone

Britain: common throughout except in extreme N Scotland.
Native; 134/156; common in old woodlands.
1st Neilston, 1792, OSA.

Wood Anemone can be readily encountered in semi-natural woodlands throughout the rural lowlands and upland and urban fringes. It is usually associated with woodland flushes and where soils are not strongly acidic. It is scarce in the uplands, except along some wooded upper-burn stretches, as at the Forkings of Raith (NS2862, 1997), but can be found in open moorland, e.g. Marshall Moor (NS3762, 1997), under Bracken along the Gleniffer Braes (*c.* NS4560, 1993) and on rocky outcrop grasslands as at Mearnskirk (NS5354, 1997) and Barnbrock (NS3564, 2007).

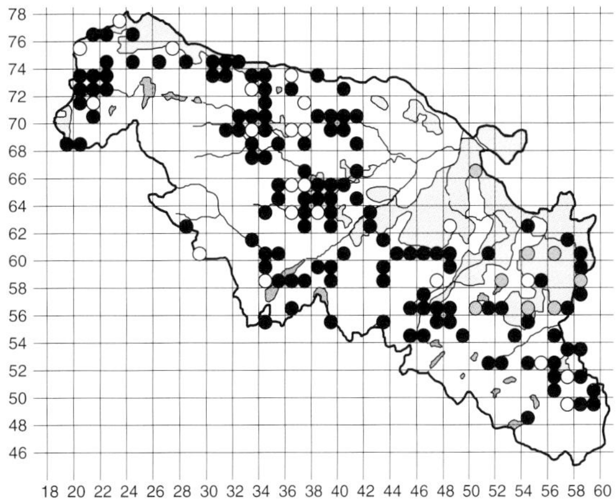

Anemone nemorosa

Ranunculus acris L.
Meadow Buttercup

Britain: common throughout.
Native; 398/469; very common.
1st 'Renfrewshire meadows', 1812, JW.

This is a very common and variable plant of slightly acidic to somewhat improved or amenity-managed grasslands, and it is also frequent at waste-ground sites; it extends to high altitudes but not where there is deep peat.

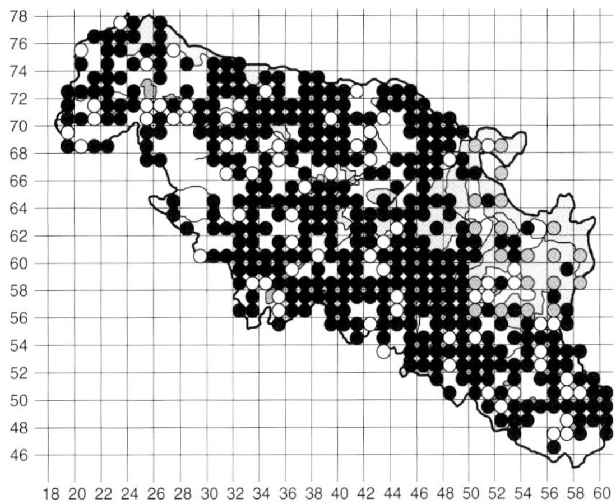

Ranunculus acris

Ranunculus repens L.
Creeping Buttercup

Britain: common throughout.
Native; 474/531; very common.
1st Paisley, 1869, RPB.

This buttercup is found just about everywhere, usually

in some abundance, except in the upland peaty moorlands. Most typically it is associated with damp grasslands and the margins of open waterbodies or watercourses; however, it can also be common on freely draining waste ground and as a garden weed.

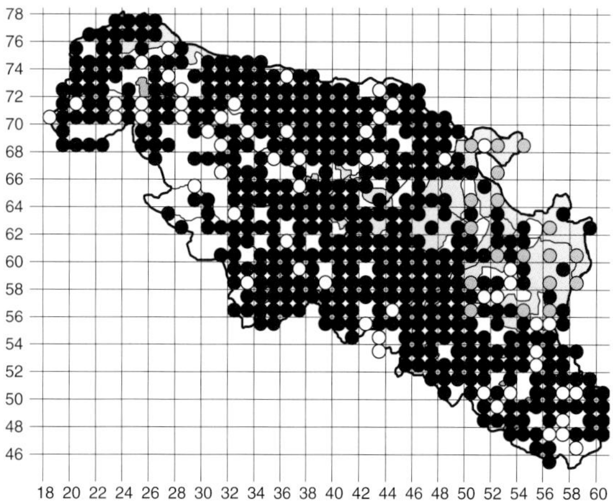

Ranunculus repens

Ranunculus bulbosus L.
Bulbous Buttercup

Britain: common in England, frequent in E Scotland but local and often coastal elsewhere.
Native; 6/10; rare.
1st Paisley, 1869, RPB.

There are very few records, old or modern, for this buttercup of old grasslands, indicating it may never have been common in the VC; it was noted from 'only one spot in the neighbourhood of the town [Paisley]' in 1869 (RPB) and Hennedy (1891) noted it at Gourock. Old herbarium records (GL) are from the White Cart Water area in Glasgow (1896 and 1916), where it has not been recorded recently. However, it is still likely to be under-recorded; modern locations include unimproved pastures by the White Cart Water, Millarston (NS5755, 2002), Earn Water, Moorhouse (NS5252, 2003) and Gryfe Water, Kilmacolm (NS3569, 2003). It has also been found on a coastal rocky outcrop at Inverkip (NS2071, 2006) and on the front lawn at Glentyan House (NS3963, 2007). A record from path-edge waste ground, by the White Cart Water at Renfrew (NS4967, 2001), may represent an 'alien' colonist.

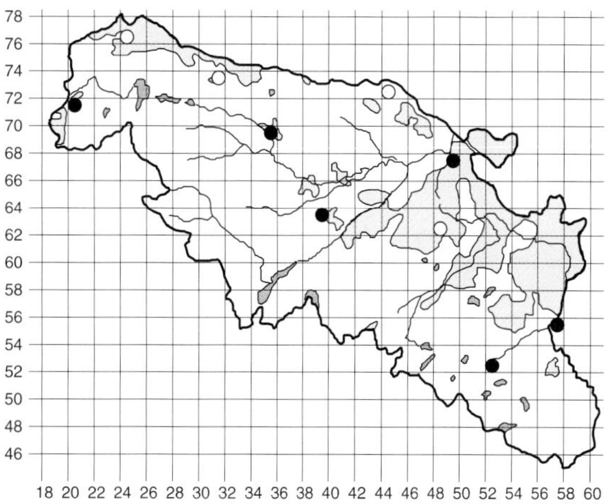

Ranunculus bulbosus

Ranunculus sardous Crantz
Hairy Buttercup

Britain: frequent and local on the coast in S and E England, scattered and rare inland.
Former accidental; 0/1; extinct.
1st Rosshall, Crookston, 1891, GL.

Hairy Buttercup was considered rare in Clydesdale by Hennedy (1891); there are no other Renfrewshire records.

Ranunculus arvensis L.
Corn Buttercup

Britain: archaeophyte; frequent in SE England, very rare in Scotland.
Former accidental; 0/3; extinct arable weed.
1st Paisley, 1910, PSY.

By the end of the 19th century this arable weed was very rarely found within the Glasgow area (CFOG). The last, and only recent, record was from the 'edge of an oat field, Georgetown' in 1957 (Conacher 1959). However, it is marked on a card from the Levern Water, Neilston (NS4656, 1960, BWR) but there is no further detail on this record.

Ranunculus auricomus L.
Goldilocks Buttercup

Britain: common in England (except in the W), frequent in C and S Scotland.
Native; 4/10; rare and endangered.
1st Renfrewshire, 1872, TB.

Goldilocks Buttercup is probably overlooked and likely was never a frequent local plant, but it seems that this species of wooded riverbanks is now very rare, possibly due to increased competition from aliens such as Pick-a-back-plant exploiting the enriched nutrient conditions along most watercourses, as at the Gryfe Water, Houstonhead (NS3965, 2002); the other modern field record is from Mill Burn, Pomillan (NS3470, 2002) and at both places populations appear precariously small. It has not been recorded recently from the two CFOG stations at Linn Park (NS55Z, 1981) and Capelrig Burn (NS55N, 1986), nor from the only two localities noted by BWR: Lochwinnoch (NS3558, 1956) and Locher Water (NS3864, 1969).

Ranunculus sceleratus L.
Celery-leaved Buttercup

Britain: common in S and E England, frequent in C Scotland.
Native; 15/17; scarce and local.
1st Banks of the Clyde, Renfrew, 1865, RH.

This buttercup appears to have a strongly eastern distribution, becoming commoner towards the Glasgow area (CFOG). Most of the modern records are associated with brackish influenced water, notably about the Newshot Island coast, but also extending up the Black Cart Water at Inchinnan (NS4667, 2002) and the Gryfe Water at Selvieland (NS4567, 2001). It has not been seen at the Langbank coast for more than 130 years (1880, NHSG). Inland records include a drainage pond at Priesthill (NS5261, 2005) and Paisley Moss (NS6465, 1997, SA), where it has been known since 1981 (AMcGS). There are also a couple of records from Castle Semple (CMRP), including bare soil near the car park (NS3658, 1992).

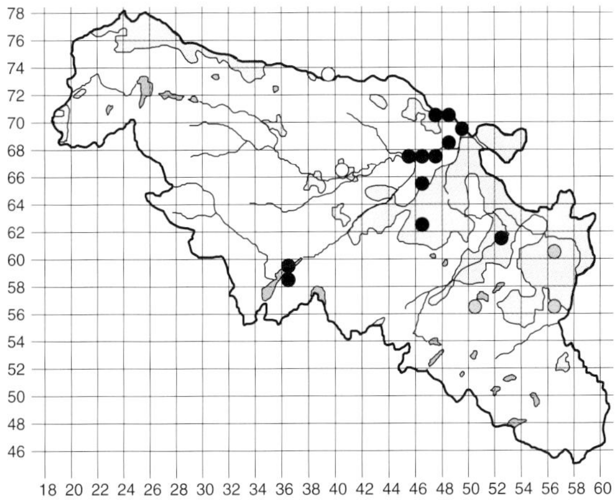

Ranunculus sceleratus

Ranunculus lingua L.
Greater Spearwort

Britain: widely scattered in England and E and C Scotland, scattered alien elsewhere.
Native; 2/5; very rare.
1st Paisley, 1845, NSA.

Greater Spearwort was recorded from Gourock in 1876 (FFWS), and in 1869 it was stated (RPB) that this showy spearwort did not grow in the locality of Paisley but that 'some transplant it'. It was collected from Castle Semple Loch in 1894 (GRK) and it was recorded on an ANG excursion there in 1955, and more recently (1992) it was noted at Barr Loch (NS3558, 1992, CMRP-NVC) and Aird Meadow (NS3558, 2010, MG). The other modern record is from the swampy fringe of Caplaw Dam (NS4368, 1993).

Ranunculus flammula L.
Lesser Spearwort

Britain: common throughout.
Native; 272/316; very common in wet places.
1st Gourock, 1856, GL.

Lesser Spearwort is commonly found in wetlands throughout the area. It favours marshy areas but is tolerant of deeper swampy waters, although it avoids more acidic mires; it can also be found along watercourse margins and is particularly common to the margins of open waterbodies.

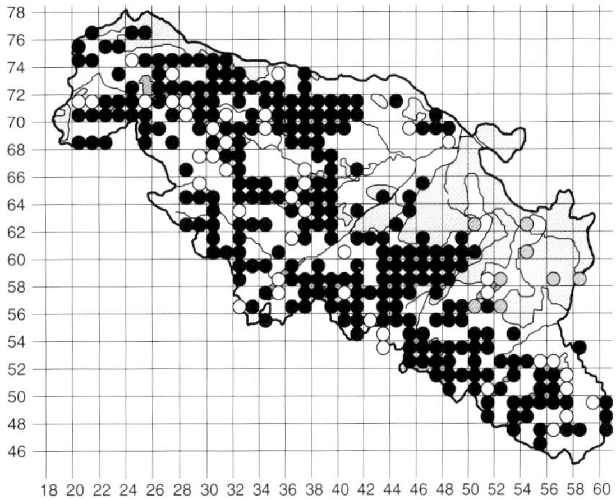

Ranunculus flammula

Ranunculus Subgenus *Batrachium*

Ranunculus hederaceus L.
Ivy-leaved Crowfoot

Britain: common in W England and lowland Scotland.
Native; 68/84; frequent.
1st Paisley, 1845, NSA.

Ivy-leaved Crowfoot may have undergone a decline

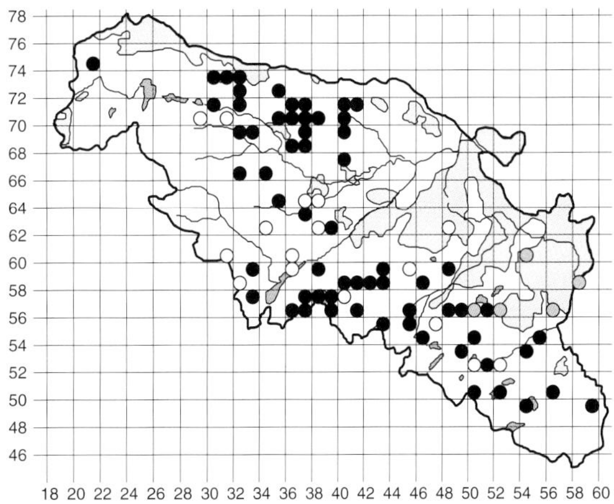

Ranunculus hederaceus

at some of its rural locations but it can still be found scattered throughout the VC, although it is notably rare in the west and in urban-fringe areas. It is usually found in slow-flowing burns or ditches, very wet flushes and towards the shallower margins of waterbodies, often where trampled by livestock.

Ranunculus omiophyllus Ten.
Round-leaved Crowfoot

Britain: locally common in the W, only in SW in Scotland.
Native; 67/91; frequent.
1st Gourock, 1865, RH.

Round-leaved Crowfoot shows a similar distribution pattern to Ivy-leaved Crowfoot. It populates two distinct zones, both between upland and urban ground. It occurs in similar places but tends to favour more acidic or less nutrient-rich waters; its absence from the west is unexpected.

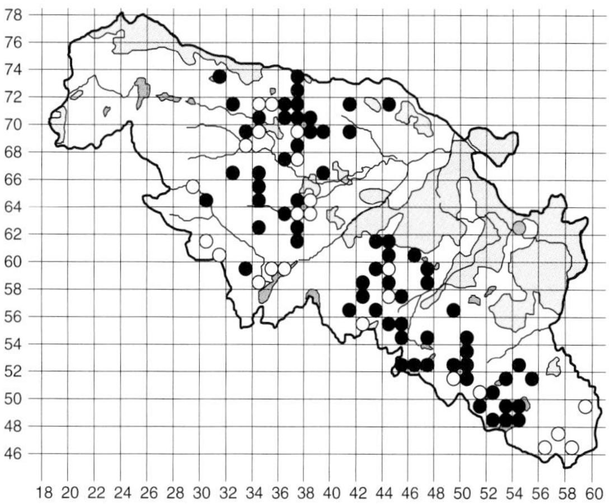

Ranunculus omiophyllus

Ranunculus baudotii Godr.
Brackish Water-crowfoot

Britain: coastal, mainly in England, but local in NW Scotland.
Native; 0/1; doubtful record.
1st Renfrewshire, 1883, TB.

Hennedy (1865) reported this water-crowfoot from the Clyde near Dumbarton Castle (VC99), but in a later edition of his flora (1891) a Gourock locality was questioned, and Brackish Water-crowfoot was not mentioned by Lee (1933); there are no other records or specimens to verify if this was ever a Renfrewshire plant.

Ranunculus trichophyllus Chaix.
Thread-leaved Water-crowfoot

Britain: frequent and scattered in England, local in E Scotland.
Former native; 0/1; extinct.
1st Loch Libo, 1883, GL.

Neither Hennedy (1891) nor Lee (1933) mention this water-crowfoot as being from Renfrewshire, but the specimen from Loch Libo appears to be correct so this species may still occur there, but no water-crowfoot has been recorded from there since.

Ranunculus aquatilis s.l.

Britain: common in the S, mainly lowland and eastern in Scotland.
Native; 25/31; occasional, local.
1st Paisley and Kilbarchan, 1845, NSA.

There are a number of records for this group comprising the following three species from several localities. It appears that Stream Water-crowfoot is the commonest, certainly along watercourses, but the two other species in this aggregate are known to be present. However, there has to be some caution over the identification of quite a few of the field records.

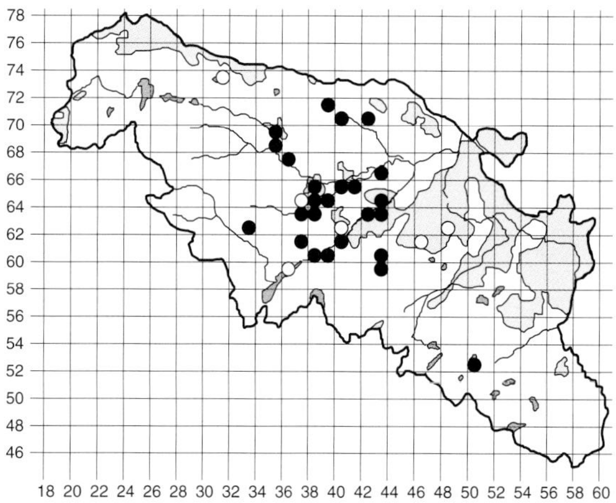

Ranunculus aquatilis s.l.

Ranunculus aquatilis L.
Common Water-crowfoot

Britain: frequent in England, local in Scotland, mainly in the E.
Native; 4/4; rare.
1st High Bardrain, 1993.

The four modern records for Common Water-crowfoot are from ditches at High Bardrain (NS4359) and Craigmuir (NS4360, 1994), a small inundation pool near Harelaw (NS3863, 1996) and the North Loch, Drum Estate (NS3971, 1996, SNH-LS); all occur in low-nutrient or peaty water.

Ranunculus peltatus Schrank
Pond Water-crowfoot

Britain: frequent in England and C Scotland.
Native; 1/4; very rare.
1st Renfrewshire, 1872, TB.

Pond Water-crowfoot was collected from Lochwinnoch in 1883 (GL) and it was also recorded by RM in 1934 from Castle Semple Loch but it has not been refound in modern surveys. Other old collections include Stanely Reservoir (1904, PSY) and Snipes Dam (1922, GL); Lee (1933) noted var. *floribundus* Bab. from 'lochs above Neilston'. The sole modern record is from Brother Loch (NS5052, 1996, SNH-LS), but it may well occur at other waterbodies.

Ranunculus penicillatus (Dumort.) Bab. ssp. *pseudofluitans* (Syme) S.D. Webster
Stream Water-crowfoot

Britain: frequent in England, rare in Scotland except in the SE.
Native; 7/8; local, in central watercourses.
1st Locher Water (NS3764), 1965, BWR.

This water-crowfoot, apparently the most frequent, can be locally common in the relatively clear waters of the Dargavel Burn, Locher Water and the Gryfe Water, and has been tentatively recorded from the Black Cart Water at Millikenpark (NS4061, 2001).

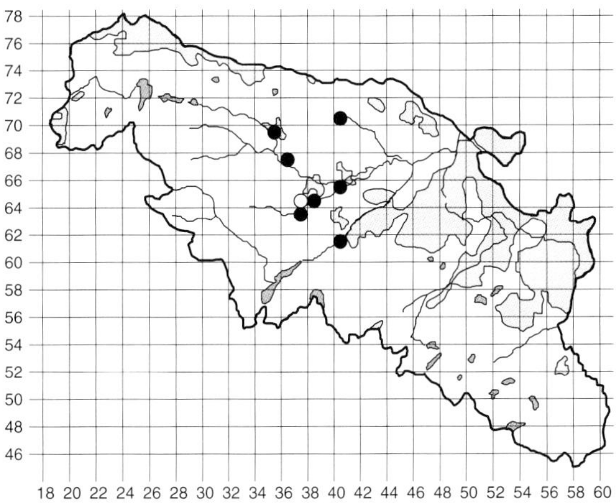

Ranunculus penicillatus

Ranunculus fluitans Lam.
River Water-crowfoot

Britain: scattered but local in England, scarce in Scotland, mainly in the E.
Native; 0/3; extinct.
1st Black Cart Water, Inchinnan, 1865, RH.

Lee (1933) mentions the presence of this species in the 'Cart' (as well as the Clyde), but no specimens exist to confirm the identity, so the records may refer to one of the preceding three species. In the Renfrewshire list (TPNS 1915) it was an 'excluded' species.

Ficaria verna Huds.
Ranunculus ficaria L.
Lesser Celandine

Britain: common throughout except in upland Scotland.
Native; 218/248; common.
1st Stanely Castle, 1864, PSY.

Except in the uplands, Lesser Celandine is a common sight in woodlands in early spring, but also can be found in open grasslands, often where seasonally damp, and occasionally on some disturbed waste-ground sites. Many records – old and modern – do not distinguish the two more widespread subspecies, but there are a number of modern records for both: ssp. *fertilis* (Lawalrée ex Llaegaard) Stace [*R. ficaria* ssp. *ficaria*], although this may include immature bulbiferous plants, is recorded from 101 1km squares, and ssp. *verna* [*R. ficaria* ssp. *bulbifer* Lambinon] from 81; the individual maps show a similar distribution pattern for both subspecies.

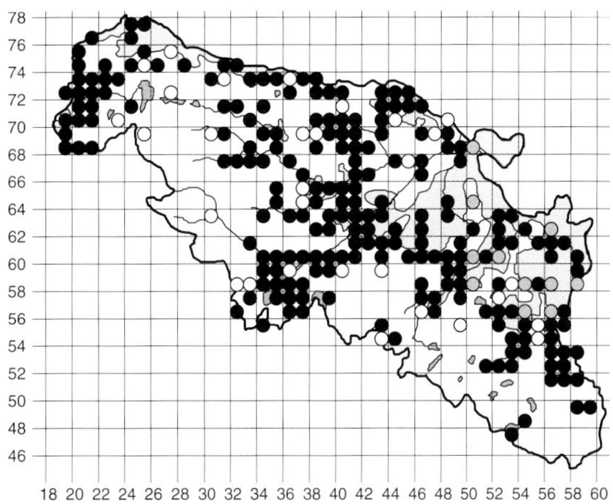

Ficaria verna

Aquilegia vulgaris L.
Columbine

Britain: common in SW England, widely scattered elsewhere.
Hortal; 25/29; occasional.
1st Renfrewshire, 1887, NHSG.

Old records include the railway, Meikleriggs (1901, PSY) and the Kilmacolm area (1924, RM). In more recent times BWR noted it near Port Glasgow (NS3173, 1963) and north of Kilmacolm (NS3671, 1962) and there were four relevant CFOG records. Today Columbine is known from several scattered locations, usually on waste ground near houses, but also along more remote roadsides. Many of the plants have variously coloured flowers indicating a range of cultivars.

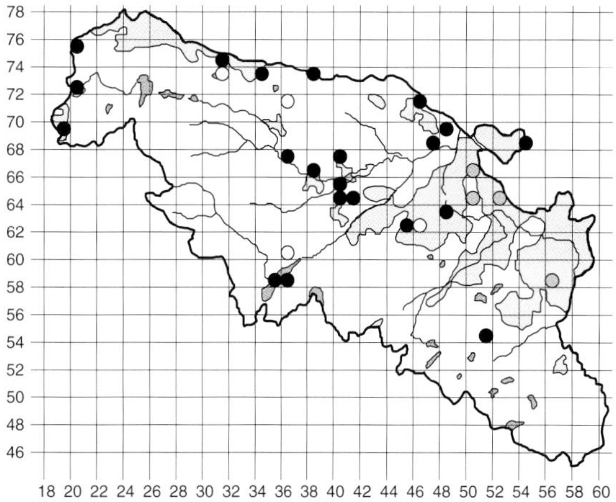

Aquilegia vulgaris

Thalictrum aquilegiifolium L.
French Meadow-rue

Britain: neophyte (S Europe); scarce but widely scattered.
Hortal; 0/1; probably extinct.
1st Cardonald (NS56H), 1983, CFOG.

This garden outcast has not been seen again although it was reported from Braehead just over the border with Lanarkshire (VC77) (1984, CFOG).

Thalictrum flavum L.
Common Meadow-rue

Britain: common in SE England, rare alien in Scotland.
Former native or hortal; 2/3; extinct as a native.
1st Old canal bank, Hawkhead, 1900, PSY.

Lee (1933) recognized Common Meadow-rue from the VC but also noted that it was 'probably extinct as a wild species; but it may be found occasionally as an escape'. The two modern records are for ssp. *glaucum* (Desf.) Battand., first found at Haugh Hill, Crookston (NS5261, 1991, CFOG), which was developed shortly after the find; a more recent find was from rough grassland, Cowglen (NS5360, 2005, LR).

Thalictrum minus L.
Lesser Meadow-rue

Britain: scattered and local in the N, often coastal, alien in much of the S.
Native; 2/5; very rare.
1st Wemyss Bay, 1887, NHSG.

Lesser Meadow-rue was collected from Craigton in 1916 (GL) and also found on the shore at Inverkip in 1960 (NS2072, BWR). The sole modern record is from waste-ground grassland at Ferguslie Park (NS4664, 2000). The CFOG site at Priesthill (NS5160, 1991) was destroyed subsequently by housing.

BUXACEAE

Buxus sempervirens L.
Box

Britain: rare native in the SE, but mostly alien elsewhere, frequent in E C Scotland.
Hortal; 8/9; rare.
1st Dargavel, 1895, PSY.

Box was recorded at Glentyan in 1961 (BWR) and can still be found there today (NS3963, 1999). Most other records are associated with old estates, all presumably planted, as at Formakin (NS4070, 1994), Ardgowan (NS2073, 1998), Gleddoch (NS3872, 1998), Houston (NS4167, 1999) and Craigends (NS4166, 2000); it was also recorded from Ranfurly Golf Course (NS3864, 1999) and an isolated young plant at Linn Park (NS5859, 2005) appears to be self-sown.

GUNNERACEAE

Gunnera tinctoria (Molina) Mirb.
Giant-rhubarb

Britain: neophyte (W S America); scattered, mainly western.
Hortal; 2/2; very rare.
1st Glentyan Estate, 2007.

A young plant has colonized a shaded wet hollow in the grounds of Glentyan Estate (NS3963). Young plants also occur in woodland near Ashton but are thought to be planted (NS2276, 2007).

PAEONIACEAE

Paeonia officinalis L.
Garden Peony

Britain: neophyte (S Europe); scattered, mainly in the S and W.
Hortal; 9/9; rare garden outcast.
1st Kennishead, 1986, CFOG.

There are eight other modern records of garden outcasts, with no evidence of local spread. It is possible that some records may refer to other species or cultivars. Plants occur on waste ground or roadsides at Boglestone (NS3373, 2001), Sergeantlaw (NS4459, 1995), Long Wood (NS5452, 1998) and Ingliston (NS4371, 1999); other records are from the edges of woods near houses: Teucheen (NS4869, 1996), Craigends (NS4166, 2000), West Lady Wood (NS5256, 2000) and Busby Glen (NS5756 2005).

GROSSULARIACEAE

Ribes rubrum L.
Red Currant

Britain: common throughout except in upland Scotland.
Hortal; 74/80; frequent.
1st Cardonald, 1912, GL.

There are also specimens from Cardonald (1921, GL) and Hawkhead (1931, PSY). As with Black Currant, there are a few records from the 1960s–80s period (BWR), but most records are modern and show a very similar pattern. Red Currant grows in woodlands and shaded habitats, especially along riverbanks, but seldom in abundance.

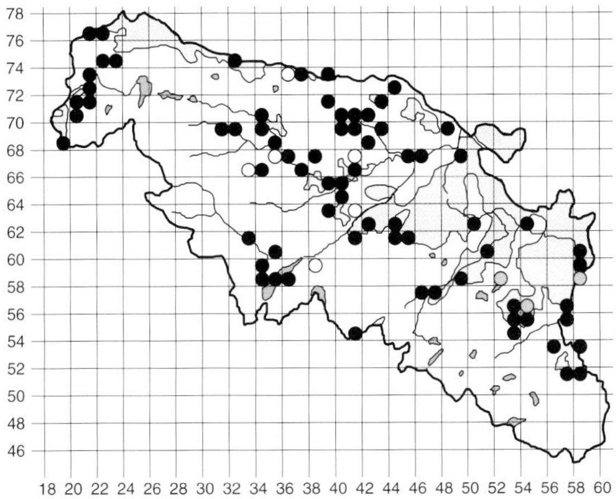

Ribes rubrum

Ribes spicatum E. Robson
Downy Currant

Britain: local in N England and NE Scotland, rare alien elsewhere.
Hortal; 2/2; very rare.
1st Capelrig Burn, 2001.

The original record is from a scrubby burn side near houses (NS5455). A more doubtful cultivar was noted in the hospital grounds at Mearnskirk (NS5354, 1997).

Ribes nigrum L.
Black Currant

Britain: neophyte (N Eurasia); frequent throughout except in upland Scotland.
Hortal; 54/63; frequent.
1st Renfrewshire, 1834, MC.

The only other 19th-century record for Black Currant is from Castle Semple Loch (1887, NHSG), but there are a few from the 1960s–80s period (BWR). Today it is scattered throughout the area; it is most often found in damp woodlands and along shaded riverbanks, and occasionally in more open marshes, but, like Red Currant, although appearing naturalized it is seldom found in any abundance.

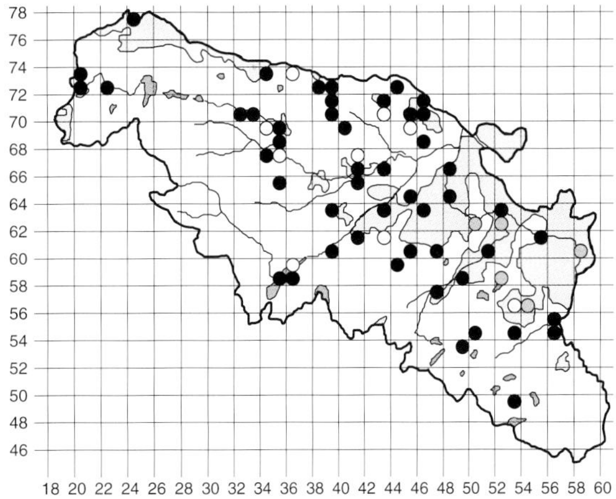

Ribes nigrum

Ribes sanguineum Pursh
Flowering Currant

Britain: neophyte (W N America); scattered throughout except in NW Scotland.
Hortal; 46/47; frequent.
1st White House, Kilbarchan (NS4163), 1976, BWR.

A more recent arrival than the other currants, the Flowering Currant is apparently spreading fast and is more likely to be seen as it tends to favour urban-fringe open waste ground, railway banks and waysides.

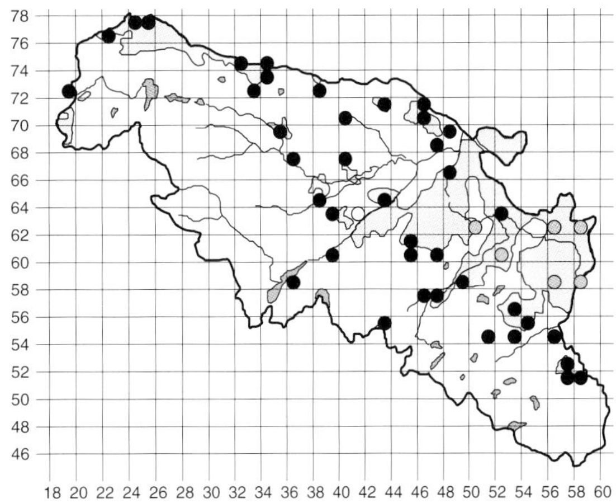

Ribes sanguineum

Ribes alpinum L.
Mountain Currant

Britain: local in N England, frequent alien in Scotland.
Hortal; 4/6; rare.
1st Castle Semple, 1834, MC.

A further old record was from Carruth Glen (1898,

JMY), and in CFOG the species is noted at two sites: Levern Water, Nitshill (NS5160, 1986) and Rouken Glen (NS5458, 1987, PM). A more recent record is from an abandoned garden behind a hotel in Erskine (NS4671, 1997) and Mountain Currant is also found along the Garpel Burn, Lochwinnoch (NS3459, 1998).

Ribes uva-crispa L.
Gooseberry

Britain: neophyte (Europe); common throughout except in upland Scotland.
Hortal; 98/117; frequent.
1st Gourock, 1865, RH.

Gooseberry is well established in the VC, but as with the other currants seldom found in any abundance. It grows in many shaded old woodlands, sometimes near to habitations but not always. It is absent from upland areas.

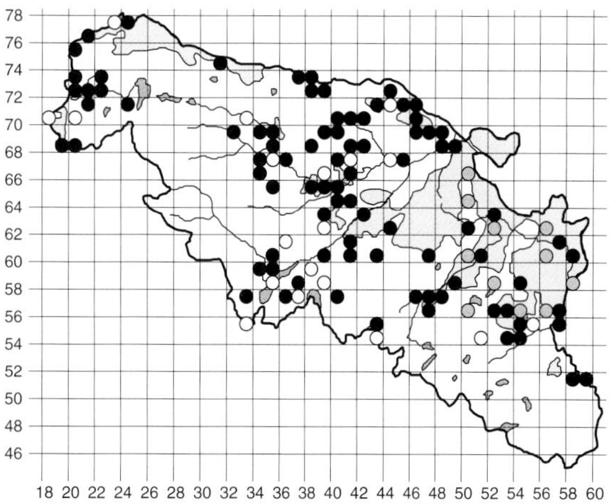

Ribes uva-crispa

SAXIFRAGACEAE

Astilbe × arendsii Arends
Red False-buck's-beard

Britain: neophyte (cultivated); scarce and scattered, local in W C Scotland.
Hortal; 8/9; rare.
1st Gleniffer Braes (NS4459), 1975, AJS.

There are eight widely scattered modern records for this garden outcast, although some may refer to different species or cultivars: River Calder, Lochwinnoch (NS3459, 1995), Cowdon Hall (NS4757, 1997), Ardgowan Estate (NS2073, 1998), Mearns (NS5455, 2001), Houston (NS4067, 2002), Busby Glen (NS5756, 2005), Overton (NS2674, 2007) and Darnley (NS5258, 2007).

Rodgersia podophylla A. Gray
Rodgersia

Britain: neophyte (E Asia); a few scattered records.
Hortal; 4/4; rare.
1st Skiff Wood (NS4059), 1993.

PM recorded Rodgersia from Cathcart (NS5860) in 1996 and it also occurs at Formakin Estate (NS4176, 2007). It has been planted in Linn Park, where it persists (NS5859, 2005).

Bergenia crassifolia (L.) Fritsch
Elephant-ears

Britain: neophyte (Siberia); widely scattered in the S.
Hortal; 1/1; very rare outcast.
1st Teucheen Wood (NS4868), 1996.

The only record for Elephant-ears is as a garden outcast, which does not appear to have persisted (2007).

Darmera peltata (Torr. ex Benth.) Voss ex Post & Kuntze
Indian-rhubarb

Britain: neophyte (W N America); widely scattered, mainly in England.
Hortal; 1/3; very rare.
1st Finlaystone Estate, 1975, AJS.

AJS thought Indian-rhubarb well naturalized along the burn at Finlaystone, but there are no modern records from there. It appears to have made a similar spread along the Dargavel Burn, Formakin (NS4070, 1994). A further record is from Rouken Glen (NS5458, 1978, BWR).

Saxifraga stellaris L.
Starry Saxifrage

Britain: frequent on mountains.
Native; 1/1; sole record.
1st North Rotten Burn, 2011.

Not mentioned for the VC by the earlier authors or more recent field botanists, the first record for Starry Saxifrage was a rather unexpected find of a single plant on a rocky ledge by the North Rotten Burn (NS2568). However, it is known from NS26 in VC75 but there are no site details. The site is at a relatively low altitude (270m) and not readily connected to the higher hills where this species may be expected to occur.

Saxifraga × urbium D.A. Webb
(*S. umbrosa* L. × *S. spathularis* Brot.)
Londonpride

Britain: neophyte (garden origin); frequent, mainly in W to NE Scotland.
Hortal; 57/66; frequent.
1st 'Banks of Cart, opposite the mill, Cathcart', 1813, TH.

It is assumed that old records for '*S. umbrosa*' refer to this garden escape or outcast. The next records are from glens at Inverkip (1886, GRK), Carruth (1890, AANS) and Darnley (1891, GL), and Londonpride can still be found today at these places; it was also collected from Castle Semple Loch in 1900 (PSY). It was encountered on several occasions during the 1960s–80s period (BWR) and today it is widespread, generally occurring in woodlands, disused railways and on waste ground.

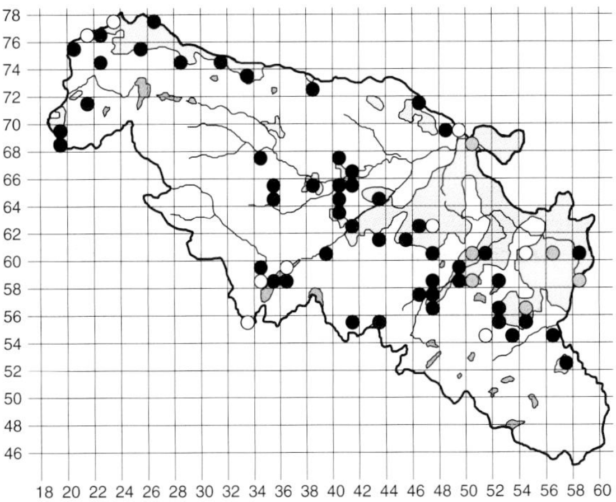

Saxifraga × urbium

Saxifraga × geum L.
(*S. umbrosa × S. hirsuta*)
Scarce Londonpride

Britain: neophyte (garden origin or Pyrenees?); rare and local.
Hortal; 5/5; rare outcast or escape.
1st Auchmountain Glen, 1996.

There is confusion over the identity of '*S. geum*' in the literature, so the only confident records are from the modern period. The first site is a shaded glen (NS2674) with other records being from Daff Burn (NS2172, 2002), Boylestone Quarry (NS4959, 1999), Locher Bridge (NS4064, 1999) and Shielhill Glen (NS2471, 2000). All the localities tend to be on shaded wet rocks or burn sides.

Saxifraga hirsuta L.
Kidney Saxifrage

Britain: neophyte; widely scattered in the N and W. (Native in SW Ireland.)
Hortal; 10/12; rare outcast or escape.
1st Renfrewshire, 1915, TPNS.

The earliest record, as '*S. geum*', may refer to other garden outcasts (Scarce Londonpride?); Lee (1933) reported '*S. geum* L.' at Carruth Glen and Inverkip. Modern records include Levern Water, Arthurlie (NS4958, 1997), Bank Brae, Glentyan (NS3962, 1997), Devol Glen (NS3174, 2005) and Ranfurly, Bridge of Weir (NS3865, 1997); it was collected from the latter in 1956 (PSY). There is a cluster of records from the Daff Burn, Inverkip (NS2072, NS2172 and NS2171) and Ardgowan Estate (NS2073, 1998). It is typically associated with sheltered, damp shaded conditions. An individual from a population at Ranfurly (NS3865, 1997), growing with Londonpride, appeared to be intermediate and could be a hybrid.

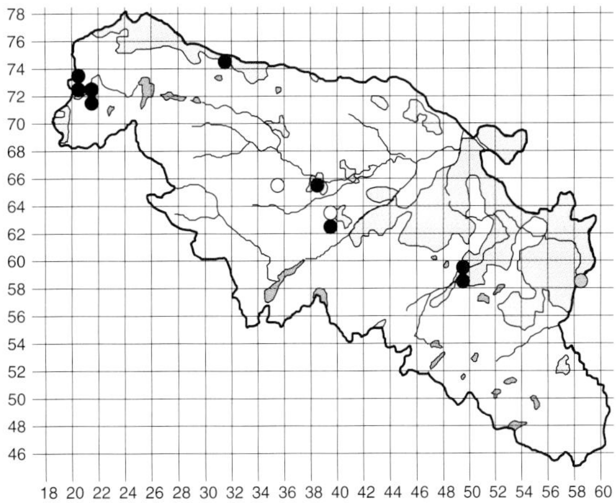

Saxifraga hirsuta

Saxifraga aizoides L.
Yellow Saxifrage

Britain: common in NW Scotland, elsewhere only in NW England.
Native; 2/4; very rare.
1st Gourock, 1865, RH.

Yellow Saxifrage was recorded from Shielhill Glen in 1888 (NHSG), but the only other record is from moors and flushes south of Daff Reservoir (NS2270, 1977, BWR). It is still there today in a series of small, base-rich flushes along the Glenshilloch Burn (NS2270 and NS2269, 2001).

Saxifraga granulata L.
Meadow Saxifrage

Britain: frequent, but rare in N Scotland.
Native; 1/3; very rare.
1st Renfrewshire, 1872, TB.

Meadow Saxifrage was reported from Busby (1891, RH) and was recorded from near the White Cart Water in 1985 (CFOG). The only modern record from this location is from beneath a planted Weeping Ash near the mansion house in Linn Park (NS5859, 2002). There is also an old record from Barrhead (1901, FFCA).

Saxifraga hypnoides L.
Mossy Saxifrage

Britain: frequent in upland Scotland, local in N England and Wales.
Native; 8/17; rare.
1st Renfrewshire, 1834, MC.

This upland saxifrage was known from Marshall Moor and Lochwinnoch in 1845 (NSA) and there are several 19th-century records from Shielhill and Calder glens, while Hennedy (1891) mentions Gourock. Today it can be found along the North Rotten Burn, where it was first reported in 1890 (NHSG) and more recently by BWR (NS2568, 1975 and NS2569, 1986); BWR also reported Murchan Spout (NS2960, 1986). Other modern records are mainly from upland burn sides in the Renfrewshire Heights: River Calder (NS3261, 1993 and NS3361, 1997), Forkings of Raith (NS2862, 1997 and NS2962, 2003), Queenside (NS2864, 2003) and Shepherd's Burn (NS2569, 1997). The sole south-eastern record is from the Carrot Burn (NS5747, 2003).

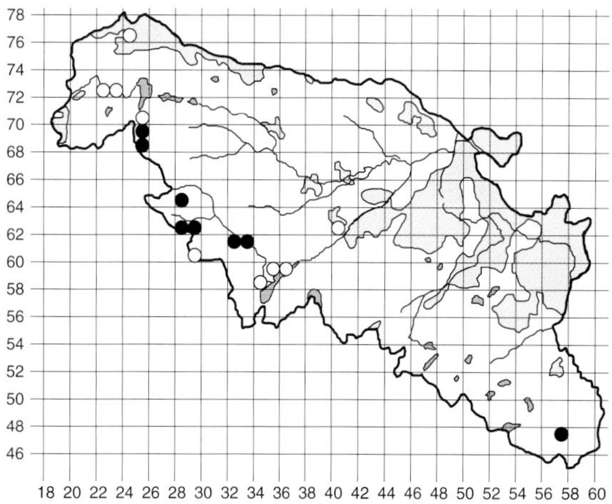

Saxifraga hypnoides

Saxifraga × *arendsii* group
Garden Mossy-saxifrages

Britain: neophyte (garden origin); rare casual.
Hortal; 2/3; very rare.
1st Gourock, 1976, AJS.

Garden outcasts, usually well established, have been recorded from a couple of sites: to the edge of Fulton Wood (NS4165, 1999) and by the Levern Water, Barrhead (NS5059, 2003).

Saxifraga tridactylites L.
Rue-leaved Saxifrage

Britain: frequent in England, rare and local in Scotland.
Accidental; 1/1; very rare.
1st Paisley, 2002, AJS.

The only record of Rue-leaved Saxifrage in the VC was a noteworthy find made by AJS from cobbles by shrubs at a new car park in Hunter St, Paisley (NS4864).

Chrysosplenium oppositifolium L.
Opposite-leaved Golden-saxifrage

Britain: common throughout except in E England.
Native; 258/282; common.
1st Paisley, 1845, NSA.

This golden-saxifrage is found in many semi-natural woodlands, usually marking out the wetter soils of flushes or depressions, or along burn sides. It can occur in the open at upland flushes and sometimes beneath marsh and swamp dominants (best seen early in the year), only becoming absent where conditions are peaty and acidic. It is generally rare in urban areas.

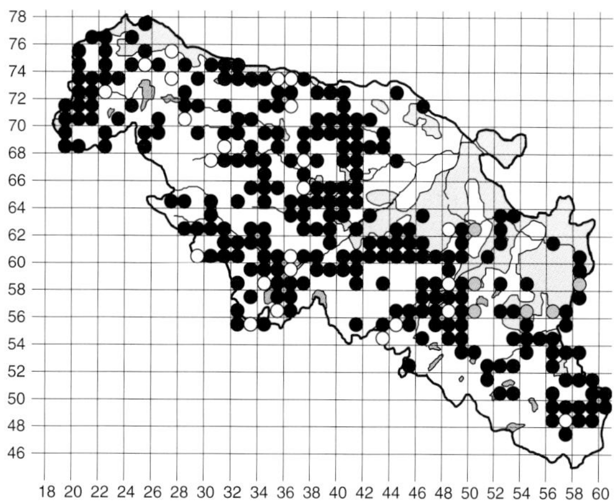

Chrysosplenium oppositifolium

Chrysosplenium alternifolium L.
Alternate-leaved Golden-saxifrage

Britain: frequent from SW England to NE Scotland, but rare elsewhere.
Native; 36/49; occasional.
1st Cloak, 1834, MC.

This species, the rarer of the two golden-saxifrages, is found to the sides of wooded burns and rivers at various locations, usually upstream where less polluted. Occasionally Alternate-leaved Golden-saxifrage can occur in open flushes, such as on the Gleniffer Braes (NS4660, 1995). There are plenty of old literature records and specimens of this species and it is likely to have declined, certainly at lowland sites.

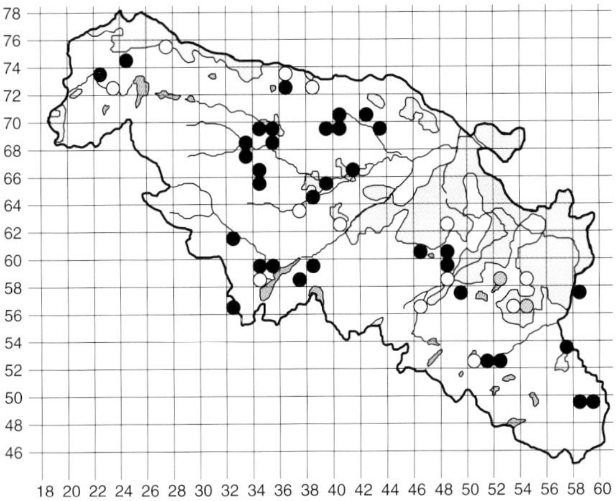

Chrysosplenium alternifolium

Tolmiea menziesii (Pursh) Torr. & A. Gray
Pick-a-back-plant

Britain: neophyte (W N America); scattered, mainly in the W, frequent in C Scotland.
Hortal; 77/85; frequent.
1st Craigends House, 1969, BWR.

Although only apparently present in the VC for some 40 years or so, this plant is now well established along many shaded riverbanks, usually where damp. Populations can be large and dense, and may well outcompete other low-growing native species.

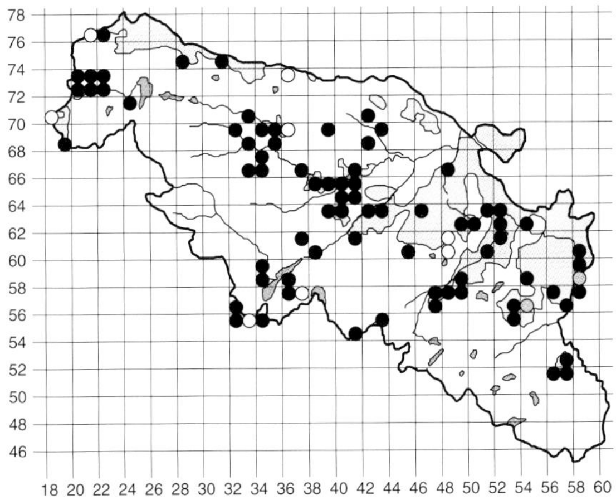

Tolmiea menziesii

Tellima grandiflora (Pursh) Douglas ex Lindl.
Fringecups

Britain: neophyte (W N America); widely scattered.
Hortal; 36/37; becoming frequent.
1st Waterfoot, 1973, BWR.

AJS recorded Fringecups at Finlaystone (NS3673) in 1975, but all other records are from the modern recording period, including several CFOG records. It is widely scattered and usually encountered in woodlands, often along riverbanks, but can occur on waste ground and along roadsides. It is notably well established and apparently spreading along the White Cart Water in Glasgow.

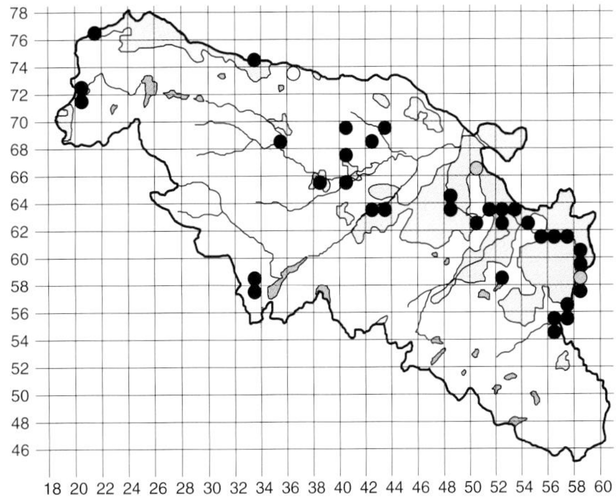

Tellima grandiflora

CRASSULACEAE

Crassula helmsii (Kirk) Cockayne
New Zealand Pygmyweed

Britain: neophyte (Australia and New Zealand); established in the S, scattered in the N and W.
Hortal; 3/3; very rare, eradicated?
1st Castle Semple Loch, 2010, CMRP.

This invasive aquatic weed was recently discovered during survey work in the Lochwinnoch area. Populations were found at three sites: a ditch in the RSPB reserve (NS3558), a pond at Barrs of Cloak (NS3460) and a school pond in Howwood (NS4060). Swift action was taken to remove the plants before any became established.

Umbilicus rupestris (Salisb.) Dandy
Navelwort

Britain: common in the SW, extending to W C Scotland.
Former native; 0/4; extinct.
1st Inverkip, 1821, FS.

Navelwort was also recorded from 'Renfrewshire coasts' in 1845 (NSA), from Wemyss Bay in 1873 (MY) – and collected from there in 1887 (GRK) – and listed in 1901 from Gourock (FFCA), but there have been no other records for more than 100 years.

Chiastophyllum oppositifolium (Ledeb.) Stapf ex A. Berger
Lamb's-tail

Britain: neophyte (Caucasus); very rare escape.
Hortal; 1/1; very rare.
1st Lochwinnoch, 1997.

This small garden escape was first recorded from walls and stonework at a bridge over the River Calder at Bridgend, Lochwinnoch (NS3559). It was still present in 2003 (IPG).

Sempervivum tectorum L.
House-leek

Britain: neophyte (S and C Europe); scattered in England, rare in E Scotland.
Former hortal; 0/4; extinct.
1st Lochwinnoch, 1845, NSA.

House-leek probably occurred on local roofs in the 19th century, as indicated by Hennedy in 1865 (and CFOG). Writing in 1898 JMY noted Mathernock and Peacemuir in the Kilmacolm area.

Sedum rosea (L.) Scop.
Roseroot

Britain: common in NW uplands, rare hortal elsewhere.
Hortal; 1/2; very rare garden outcast.
1st Renfrewshire, 1915, TPNS.

The original record, in the Renfrewshire List (TPNS 1915), is unexpected and presumably of a garden outcast. A small clump of Roseroot was recently seen rooting on maritime rocks below the sea wall, near housing, at Wemyss Bay (NS1869, 2012).

Sedum spectabile Boreau
Butterfly Stonecrop

Britain: neophyte (China and Korea); scattered in the S.
Hortal; 0/1; very rare, probably overlooked.
1st Flenders (NS5656), 1983, BWR.

This is the sole record for this widely grown garden plant. Some non-flowering records of Orpine may well refer to this species.

Sedum telephium L.
Orpine

Britain: widespread but local, frequent in lowland Scotland.
Native or hortal; 32/45; occasional.
1st Lochwinnoch, 1845, NSA.

Orpine can occasionally be found on waste ground, disused railways and roadside hedge banks, generally in urban-fringe areas. The majority of the modern records are from a central area to the immediate west of Glasgow, perhaps representing a local distribution but it is difficult to class any as native populations, although Orpine was considered frequent by both Hennedy (1891) and Lee (1933).

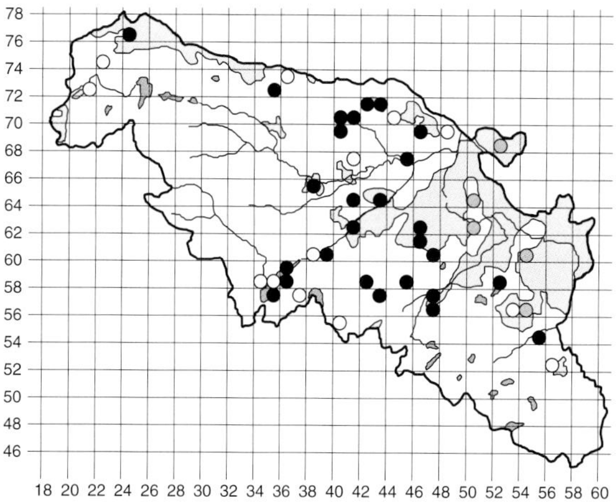

Sedum telephium

Sedum spurium M. Bieb.
Caucasian-stonecrop

Britain: neophyte (Caucasus); widely scattered.
Hortal; 6/6; rare garden outcast.
1st Crossmyloof (NS56R), 1988, CFOG.

This stonecrop has also been recorded from Woodhall (NS3473, 1997), Eaglesham (NS5851, 2000), the coast near Cloch (NS2075, 2006), a road verge, Brueacre (NS1970, 2010, PRG) and near the Kip Water, Inverkip (NS2172, 2011).

Sedum rupestre L.
Reflexed Stonecrop

Britain: neophyte (Europe); common in S England, scattered in lowland Scotland.
Hortal; 1/3; very rare.
1st Gourock, 1976, AJS.

AJS also recorded this stonecrop from Lochside Station in 1976 (NS3657), and more recently it has spread along a roadside wall at Malletsheugh (NS5255, 2001).

Sedum forsterianum Sm.
Rock Stonecrop

Britain: local in the SW, scattered alien elsewhere.
Hortal; 4/4; rare.
1st Waste ground, Ladyburn (NS3075), 1998.

There are three other records for this garden outcast: from disused railways at Kilbarchan (NS4162, 1998) and Auchenbothie (NS3371, 2005), and from waste ground, Millarston (NS4563, 2002).

Sedum acre L.
Biting Stonecrop

Britain: common in England to lowland and coastal Scotland.
Native; 24/31; occasional.
1st Near Paisley, 1858, PSY.

The majority of records for Biting Stonecrop occur to the edge of Glasgow and Paisley, and there are a couple of outliers in Greenock. It can be frequent on waste ground, as at the Royal Ordnance Factory, Georgetown (*c.* NS4467, 1996). Habitats include waste ground and old walls, but it does not appear to be part of natural rocky grasslands.

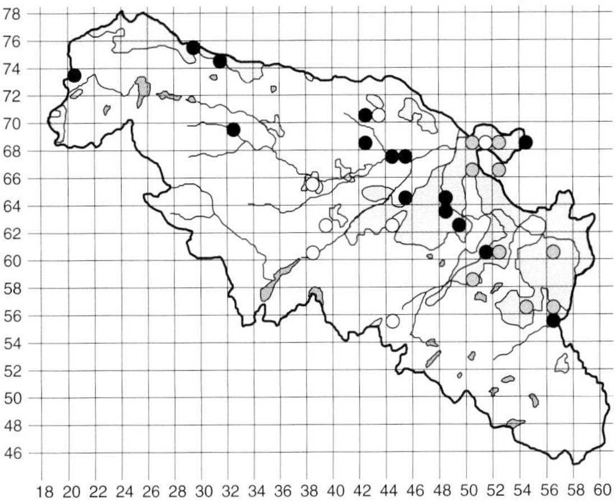

Sedum acre

Sedum album L.
White Stonecrop

Britain: archaeophyte; common in England, scattered in lowland Scotland.
Hortal; 9/9; rare.
1st Levern Water (NS55D), 1988, CFOG.

This garden outcast has been recorded from waste ground at eight widespread modern localities: Ladyburn (NS3075, 1998), Georgetown (NS4268, 1999), Eaglesham (NS5851, 2000), Lochwinnoch (NS3458, 2001), Paisley (NS46R, 1999, AJS), Murdieston Park (NS2675, 2002), Trumpethill (NS2176, 2012) and Gourock (NS2775, 2012).

Sedum anglicum Huds.
English Stonecrop

Britain: common in the extreme W, rare elsewhere.
Native; 106/121; common in the W.
1st Renfrewshire, 1834, MC.

English Stonecrop shows a remarkable western distribution, presumably reflecting the influence of climate (rainfall) rather than geology; the most easterly record is from Boylestone Quarry (NS4959, 1999). It is typically associated with shallow soils of rock outcrops and ledges in grazed upland areas.

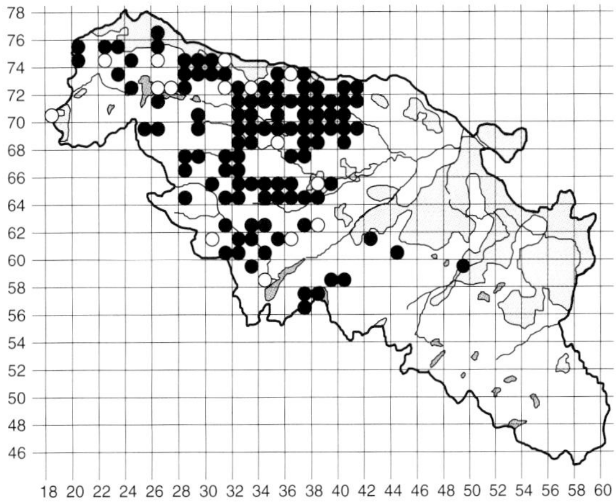

Sedum anglicum

Sedum dasyphyllum L.
Thick-leaved Stonecrop

Britain: neophyte (S Europe); local in S England.
Hortal; 0/2; extinct.
1st Caldwell Estate, 1856, TPNS.

This stonecrop, presumably planted, was also recorded from Finlaystone Estate (1858, PSY). There are no other records.

Sedum villosum L.
Hairy Stonecrop

Britain: local in N England and upland Scotland except in the NW.
Native; 6/22; rare, declining.
1st Lochwinnoch, 1834, MC.

There are many 19th-century literature references to Hairy Stonecrop and several herbarium specimens; it was also recorded on a few occasions during the

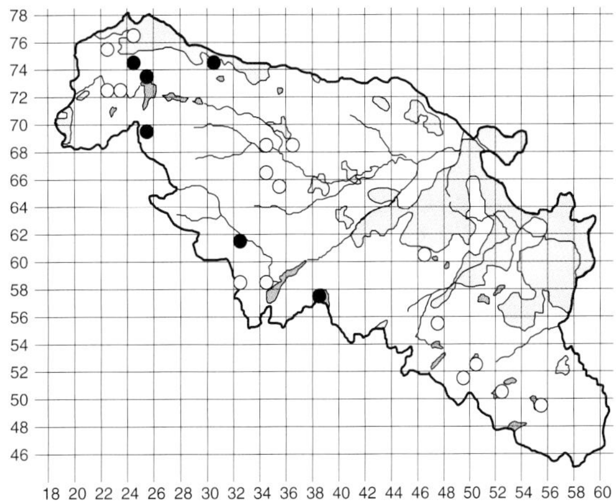

Sedum villosum

1960s–80s (BWR), but there are only a few modern records, indicating a sharp decline. There are only a few records from the upland Renfrewshire Heights, although several were made about Loch Thom: flushes on the steep slopes of Garvock Hill (NS2569, 1997), a burn entering a reservoir at Greenock Cut (NS2474, 2003, SL) and a ditch, north-west of Loch Thom (NS2573, 2006, SL); it can also be found in a flush on Mistylawmuir (NS3261, 2003). The other isolated sites are flushed rocky gravels at Burnthills (NS3857, 2001) and Knocknairs Hill SSSI (NS3074, 1994, SNH). There are no modern records for the south-east uplands, where RM noted several sites, and the last record was made by BWR (Neilston Pad, 1972).

HALORAGACEAE

Myriophyllum spicatum L.
Spiked Water-milfoil

Britain: common in the S and E extending to C Scotland, rare in the N and W.
Native; 7/12; rare.
1st Lochwinnoch, 1845, NSA.

Although Spiked Water-milfoil was considered the commonest water-milfoil by early authors, this is not the case now. Old records include the 'Cart' (1861, GGO), Castle Semple Loch (1887, NHSG) and Thornliebank (1909, GL); more recently RM reported it from the south-west corner of Long Loch (NS4752, 1966) and JDM noted it at Knapps Loch (1963) – although AJS (1976) noted Alternate Water-milfoil from the latter, casting some doubt on this record. The SNH Loch Survey only found it once, at North Loch, Drum Estate (NS3971, 1996) – in contrast with the seven sites where the survey found Alternate Water-milfoil. The other modern records are from South Loch, Drum Estate (NS3971, 1997), Picketlaw Reservoir (NS5651, 1998), Fereneze (NS4859, 1999), Reilly Quarry pool (NS4169, 1991, PSY), the Black Cart Water, Inchinnan (NS4667, 2002) and the White Cart Water, Paisley (NS4863, 2004); Spiked Water-milfoil was also recorded from the White Cart Water at Crookston (NS5362, 1991, CFOG).

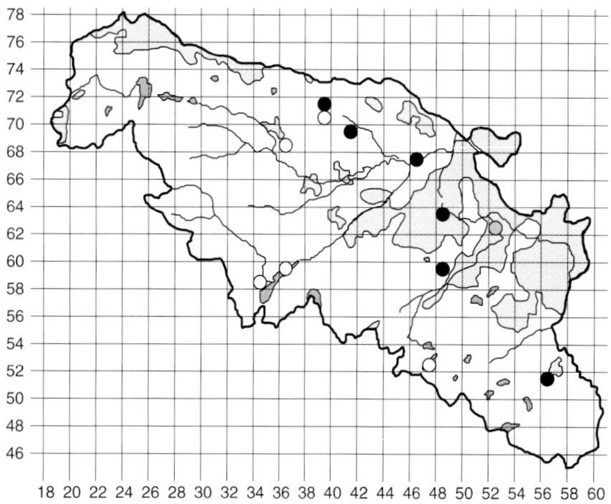

Myriophyllum spicatum

Myriophyllum alterniflorum DC.
Alternate Water-milfoil

Britain: common in the N and W.
Native; 42/47; frequent.
1st Castle Semple Loch, 1887, NHSG.

Earlier authors considered this water-milfoil rare. It was collected (GL) from Snypes Dam in 1922 by Lee, and by 1933 he had elevated its status to frequent and mentioned 'Small lochs in the Mearns district'. Today Alternate Water-milfoil is strongly represented in the open waterbodies to the south-west of Mearns but also scattered across the central and western parts of the VC. The sites are usually more nutrient poor than those preferred by its larger relative Spiked Water-milfoil and virtually all are standing open waterbodies, although it has been recorded from the Levern Water at Neilstonside (NS4655, 1999) and Carswell (NS4653, 1999), and a burn at Lochend (NS3664, 2004). It occurred in some abundance on the wet mud of the recently breached Leperstone Reservoir (NS3571, 2000).

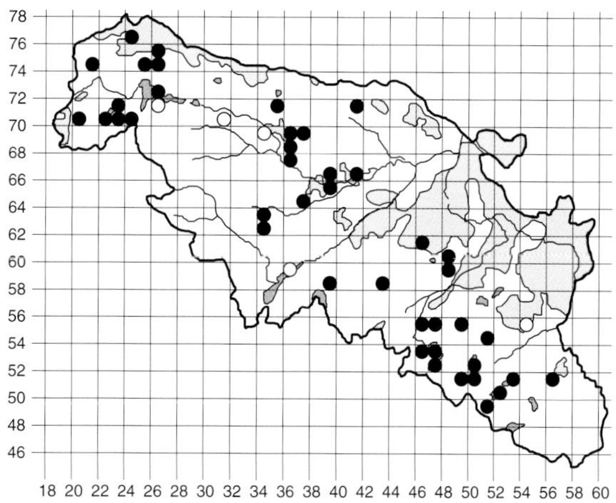

Myriophyllum alterniflorum

VITACEAE

Parthenocissus quinquefolia (L.) Planch.
Virgina-creeper

Britain: neophyte (N America); frequent in S England, very rare in Scotland.
Hortal; 1/1; very rare.
1st Glentyan Estate, 2007.

Virginia-creeper has spread (seeded?) along the outside of a high stone wall opposite the Glentyan Estate grounds in Kilbarchan (NS4063).

FABACEAE

Astragalus cicer L.
Chick-pea Milk-vetch

Britain: neophyte (Europe); rare casual.
Former accidental; 0/1; extinct casual.
1st Giffnock quarry, 1919, RG.

Onobrychis viciifolia Scop.
Sainfoin

Britain: local in C S England, scattered alien elsewhere.
Former accidental; 0/1; extinct.
1st Caldwell, 1893, PSY.

The location of the sole record lies on the boundary of VC75 (Ayrshire).

Anthyllis vulneraria L.
Kidney Vetch

Britain: throughout but local, often coastal.
Native or accidental; 9/12; rare.
1st Gourock, 1865, RH.

There are literature and herbarium records for Kidney Vetch from Giffnock (between 1891 and 1911) and it was recorded during CFOG surveys from Braehead (NS56D, 1985), but there are only a few modern field records and these are from waste places such as about Ferguslie Park (NS4664, 2000), Paisley (NS4665, 1992, DM) and a cycle path near Howwood (NS3961, 2001, DM); it is also known from Lochwinnoch (NS3559, 1986, DM). Other records include Barrangary Tip (NS4469, 2001) and, at a more remote location, a roadside at Harelaw (NS3072, 2011). In Hennedy's time it may have occurred as a native of base-rich pastures but there are no obvious native populations recorded today.

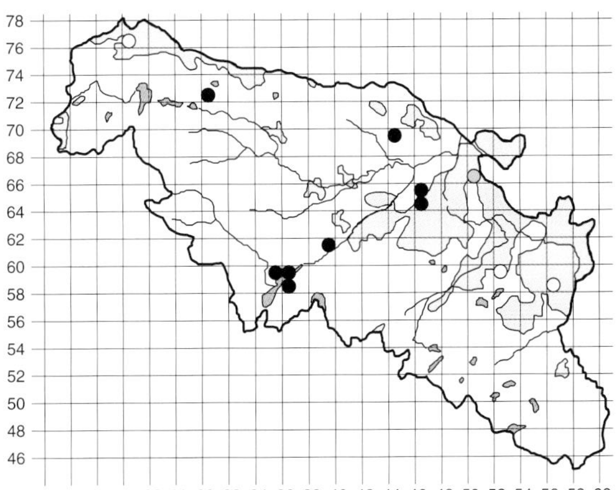

Anthyllis vulneraria

Lotus tenuis Waldst. & Kit. ex Willd.
Lotus glaber Mill.
Narrow-leaved Bird's-foot-trefoil

Britain: frequent in the SE, rare in Scotland.
Former accidental; 0/1; extinct.
1st Giffnock, 1928, GL.

Lotus corniculatus L.
Common Bird's-foot-trefoil

Britain: common throughout.
Native; 274/363; very common.
1st Renfrewshire, 1869, PPI.

This is a very common plant of short grasslands, which tolerates grazing but not heavy improvement treatments; it may have declined in recent years but is still a feature of relict unimproved grasslands, although such are now largely confined to steeper embankments or waysides. It extends from the uplands where not peaty to coastal rocks. It is also frequently found in waste-ground grasslands in urban areas, although often as sown cultivars (var. *sativus* Hyl.), which have been recorded from 13 widespread waste-ground sites, all in urban fringes.

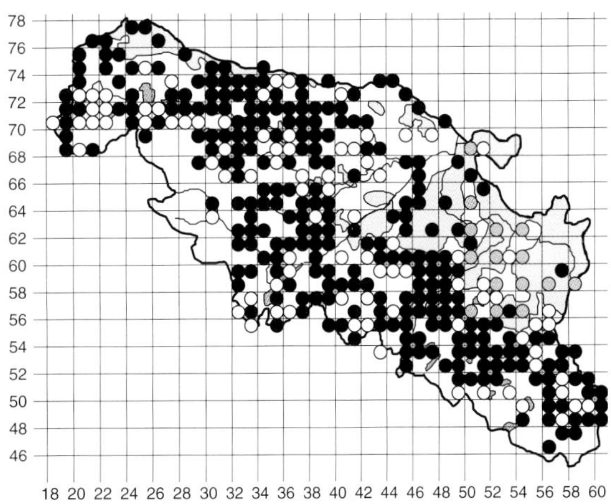

Lotus corniculatus

Lotus pedunculatus Cav.
Large Bird's-foot-trefoil

Britain: common throughout except in N Scotland.
Native; 284/351; very common.
1st Paisley Canal Bank, 1865, RH.

This bird's-foot trefoil shows a very similar distribution pattern to Common Bird's-foot-trefoil, but it reflects wetter ground, and is a frequent associate of rushes, notably Sharp-flowered Rush. It persists in urban areas but is absent from the peaty uplands.

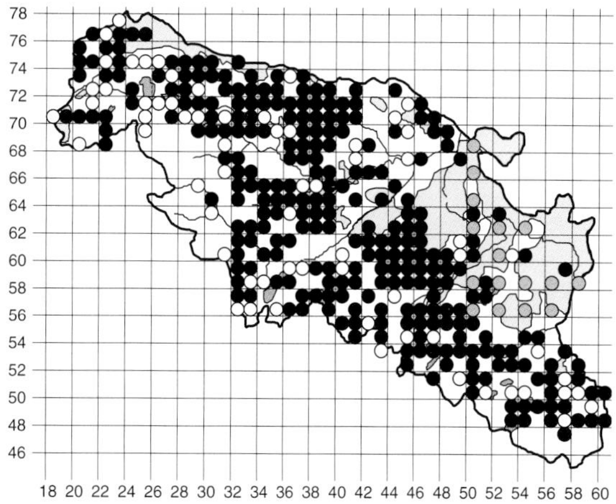

Lotus pedunculatus

Ornithopus perpusillus L.
Bird's-foot

Britain: frequent and scattered in the S, local in C and E Scotland.
Native and accidental; 1/4; very rare.
1st '1.5 miles west of Kilmacolm', 1880, NHSG.

Lee (1933) mentioned Bird's-foot at Greenock and it was known as a garden weed in Newlands (NS5760), from where it was first recorded in 1974 (CFOG). The only modern record is of small population at the coastal grassland just below the Erskine Bridge (NS4672, 2008). Although easily overlooked, this small plant of open soils may always have been a very rare plant.

Securigera varia (L.) Lassen
Crown Vetch

Britain: neophyte (S Europe); frequent in the S, extending to E Scotland.
Accidental; 1/1; very rare casual.
1st Paisley, 1999, AJS.

The sole record is from waste ground in central Paisley (NS46R).

Scorpiurus muricatus L.
Caterpillar-plant

Britain: neophyte (S Europe); scarce casual.
Former accidental; 0/1; extinct casual.
1st Giffnock, 1892, GL.

Vicia orobus DC.
Wood Bitter-vetch

Britain: frequent in Wales, scattered in Scotland.
Former native; 0/1; extinct.
1st Gourock, 1893, GL.

Collected by R Brown, Gourock appears to be the only place where this vetch has been recorded; it does not appear in any old lists for the VC.

Vicia cracca L.
Tufted Vetch

Britain: common throughout except in N Scotland.
Native; 234/327; common.
1st Barrhead, 1856, HMcD.

Tufted Vetch is commonly found throughout the lowlands, growing in various grasslands including tussocky damp pastures, along waysides and hedgerows, and it readily colonizes waste ground.

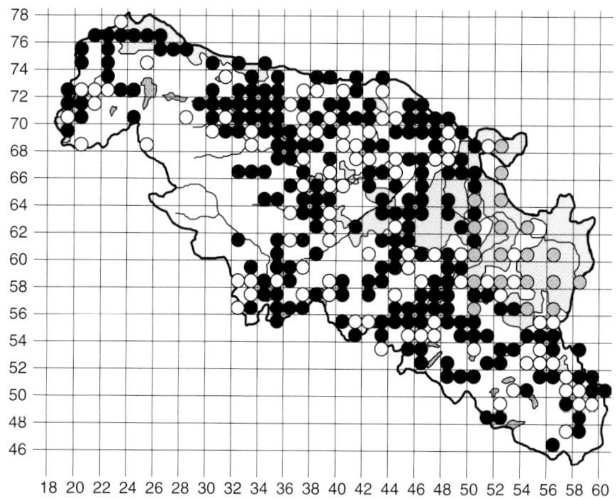

Vicia cracca

Vicia sylvatica L.
Wood Vetch

Britain: scattered throughout.
Former native; 0/1; extinct.
1st Shielhill, 1933, Lee.

There appears to be no earlier record for Wood Vetch than Lee's 1933 note of it at Shielhill, or subsequent sighting of it at this well-surveyed location.

Vicia hirsuta (L.) Gray
Hairy Tare

Britain: common in England to lowland Scotland.
Native; 44/51; occasional.
1st 'About Renfrew', 1813, TH.

Considered frequent in the 19th century (Hennedy 1891), Hairy Tare is most likely to be found in urban-fringe areas where it grows on waste-ground grassland and other freely draining, disturbed places. The few remote rural localities include waste ground at Fauldside (NS5253, 2003), the disused railway, Auchenbothie (NS3471, 2001; NS3271, 2005) and Shillford Tip, Uplawmoor (NS4455, 1999).

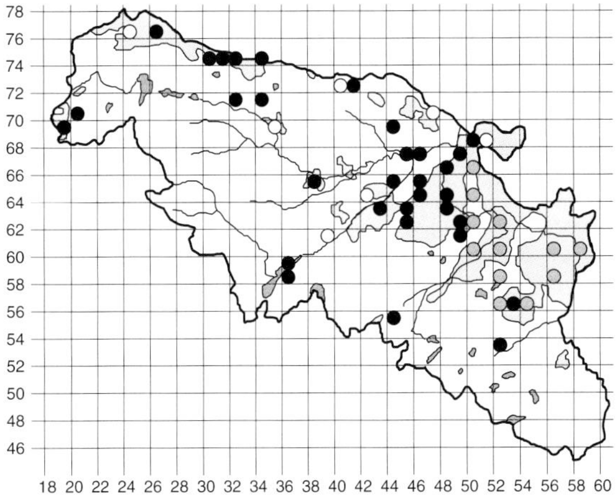

Vicia hirsuta

Vicia tetrasperma (L.) Schreb.
Smooth Tare

Britain: common in SE England, scarce alien in Scotland.
Accidental; 2/5; very rare.
1st Renfrewshire, 1869, PPI.

A specimen from Linwood Moss 1927 (PSY) is actually Hairy Tare, leaving only Giffnock (1892, GL) and Hawkhead Estate (1942, TPNS) as other early named locations. The only modern records are from the disused railway (cycleway) at Scart (NS3667, 2001) and waste ground at Port Glasgow (NS3174, 2012).

Vicia sepium L.
Bush Vetch

Britain: common throughout.
Native and accidental; 188/266; common.
1st Wood below High Dam, Eaglesham, 1841, GL.

There are many old and modern records for Bush Vetch and it is undoubtedly still widespread today, although there appears to have been some decline at lowland rural places. It has a scattered distribution occurring on urban waste ground, grassy places including waysides and hedgerows, and scrubby woodland edges. It can extend into the uplands where it is seen along the steeper embankments of burn sides.

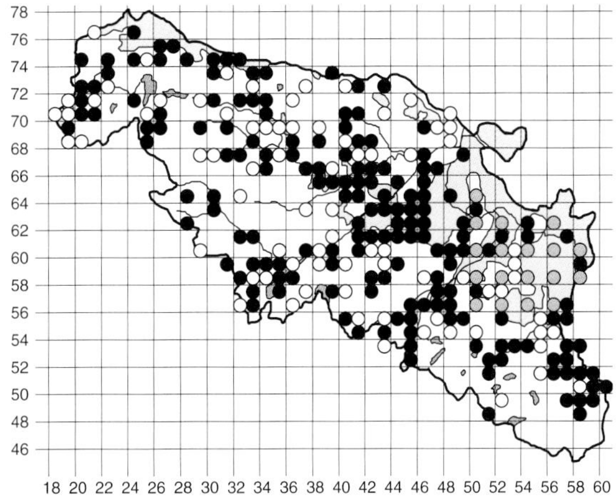

Vicia sepium

Vicia sativa L.
Common Vetch

Britain: common in the S and lowland Scotland.
Native or accidental; 66/80; frequent.
1st Paisley Canal and Gourock, 1865, RH.

This is a frequent vetch of short free-draining waste-ground grassland, most visible in early summer. The distribution is clearly lowland. Recording has been mostly at the species level, though many records are assumed to be ssp. *nigra* (L.) Ehrh., but several will refer to the two other subspecies (both of which were considered rare in CFOG). There are five records for ssp. *segetalis* (Thuill.) Gaudin: Formakin Estate (NS4170, 2003, IPG), roadsides at Locher Water (NS4064, 2011) and Ladymuir (NS3463, 2011), and waste ground at Yoker (NS5169, 2011) and

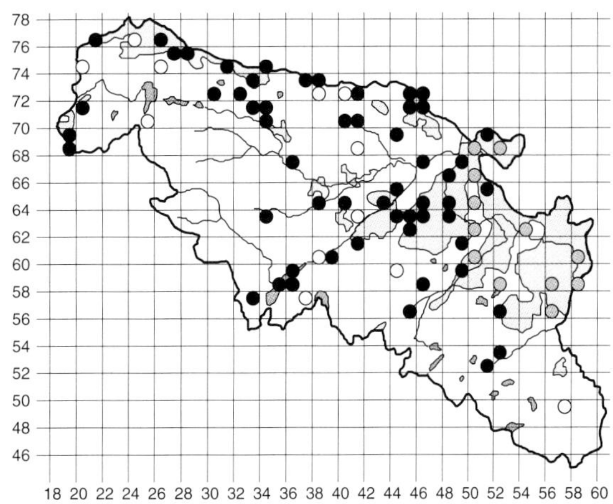

Vicia sativa

Trumpethill (NS2176, 2012). The ssp. *sativa* appears to be the rarest with only one record: waste ground, Gourock (NS2775, 2012).

Vicia lathyroides L.
Spring Vetch

Britain: frequent on coasts.
Accidental; 0/1; extinct casual.
1st Garden weed, Paisley College, 1980, AJS.

Spring Vetch is native to the coastal Clyde islands (VC100) and Ayrshire (VC75); there are no native records in the VC.

Vicia lutea L.
Yellow-vetch

Britain: rare in coastal S England and SW Scotland, rare alien elsewhere.
Former accidental; 0/2; extinct.
1st Near Renfrew, 1873, TGSFN.

The other old record is from wasteland near Paisley (1910, PSY).

Vicia bithynica (L.) L.
Bithynian Vetch

Britain: local in S England, rare elsewhere.
Former accidental; 0/2; extinct.
1st Renfrewshire, 1899, GC.

This vetch was collected from wasteland near Paisley (1910, PSY).

Lathyrus linifolius (Reichard) Bässler
Bitter-vetch

Britain: common in the N and W.
Native; 75/115; frequent in upland fringes.
1st Gourock, 1865, RH.

There are quite a few older records for Bitter-vetch, including many from the 1960s–80s (BWR) and there may well have been a decline in some of the west-central upland fringes. However, it can still be found on wooded banks that are not too heavily shaded and grassy slopes, usually near rocky outcrops such as along watercourses. It is scarce in the lowlands where soils are more enriched and it also avoids the grazed, wetter upland soils; it appears scarce in the extreme west.

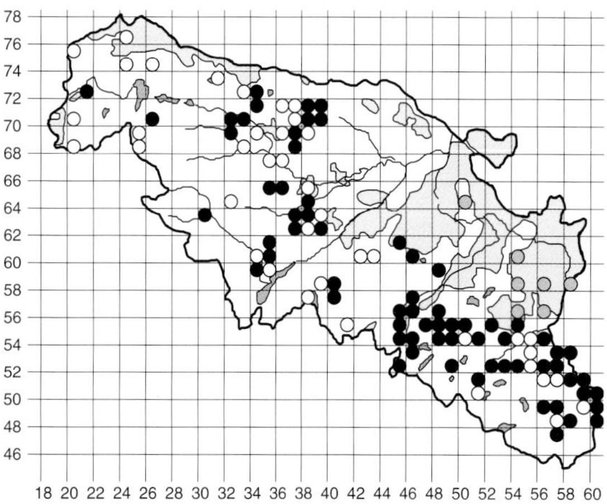

Lathyrus linifolius

Lathyrus pratensis L.
Meadow Vetchling

Britain: common throughout except in N Scotland.
Native; 273/359; very common.
1st Renfrewshire, 1869, PPI.

Meadow Vetchling is a common sight in neutral grasslands and along waysides, but it does not tolerate heavy improvement or grazing. It can be found in slightly marshy tall grasslands and can also be a common feature in rough grassland at waste-ground sites.

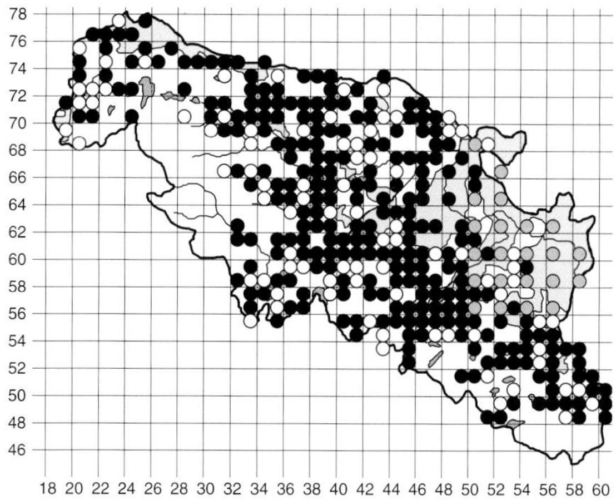

Lathyrus pratensis

Lathyrus grandiflorus Sm.
Two-flowered Everlasting-pea

Britain: neophyte (SE Europe); scattered in lowland Scotland.
Hortal; 4/4; rare outcast or escape.
1st Mosspark (NS56L), 1986, CFOG.

This large-flowered pea is also known from waste ground at Kirktonmill (NS4856, 1999), Paisley (NS4864, 2000) and Howwood (NS3960, 2001).

Lathyrus latifolius L.
Broad-leaved Everlasting-pea

Britain: neophyte (Europe); frequent in the SE, scarce in lowland Scotland.
Hortal; 4/4; rare outcast or escape.
1st Yoker Ferry (NS5168), 1985, AMcGS.

Other CFOG records include Maxwell Park (NS56R,

1986), Levern Water (NS55E, 1988) and Carnwadric (NS55P, 1989). Broad-leaved Everlasting-pea has perhaps been overlooked but there have been no records made during more recent recording.

Lathyrus nissolia L.
Grass Vetchling

Britain: frequent in the S, scattered alien in N England, rare alien in Scotland.
Accidental; 2/2; very rare casual, increasing?
1st Lochwinnoch, 2011.

This was a star find at a recent BSBI field meeting, discovered on grassy stonework by a drainage pipe (NS3558). Only a few days later P&TNT found it in Bishopton (NS4768). It appears to be increasing in the west-central Scotland area, perhaps as a contaminant of seed mixes.

Lathyrus aphaca L.
Yellow Vetchling

Britain: local in the S, scattered alien further N up to C Scotland.
Former accidental; 0/3; extinct casual.
1st Renfrewshire, 1895, NHSG.

This vetchling was reported from Giffnock (1901, FFCA) and also collected from wasteland near Paisley (1910, PSY).

Lathyrus cicera L.
Red Vetchling

Britain: neophyte (S Europe); rare casual.
Former accidental; 0/1; extinct casual.
1st Giffnock, 1892, GL.

Pisum sativum L.
Garden Pea

Britain: neophyte (S Europe); scattered in the S and up to C Scotland.
Hortal; 1/1; very rare casual.
1st East of Renfrew (NS56D), 1987, CFOG.

Cicer arietinum L.
Chick Pea

Britain: neophyte (Middle East); casual, mainly in the S.
Former accidental; 0/1; extinct casual.
1st Allotments, Queen's Park, 1983, GL.

Ononis repens L.
Common Restharrow

Britain: common in England, mainly in the E in Scotland.
Native; 3/7; very rare.
1st Gourock, 1837, GL.

There are several old records for Common Restharrow including Paisley (1845, NSA), Hawkhead Plantation (1890, AANS) and Paisley and Potterhill Railway (1904, PSY). It was also recorded, in some abundance, from Hurlet (1890, AANS), which is near to the location for a modern record from a grassy embankment at Hurlethill (NS5161, 2010); it was first recorded at this site in 1991 (CFOG). The other modern sites are waste-ground grassland at the Phoenix Business Park (NS4463, 1995) and by Spango Burn (NS2374, 2006).

Melilotus altissimus Thuill.
Tall Melilot

Britain: archaeophyte (Europe); frequent in S and E England, scarce in C Scotland.
Accidental; 5/8; rare casual.
1st Renfrewshire, 1869, PPI.

The PPI list includes both '*M. officinalis*' and '*M. arvensis*', indicating that both Tall and Ribbed Melilots were known. RG (1930) refers to Darnley but the next record is from Barr Loch (NS3557, 1963, JDM). There are five modern records, all from waste ground: Paisley (NS46R, 1999, AJS), Porterfield (NS4967, 2001), Trumpethill (NS2176, 2012), King's Inch (NS5168, 2012) and Port Glasgow (NS3174, 2012).

Melilotus albus Medik.
White Melilot

Britain: neophyte (Eurasia); frequent in S England and C Scotland.
Accidental; 3/8; very rare casual.
1st White Cart Linn, 1887, AANS.

White Melilot was collected from Giffnock (1890, GL) and The Cottage, Paisley (1902, PSY); Wood (1893) mentioned it being present near Barrhead and also stated that it was known there 'five or six years previous' and RG (1930) noted it at Darnley. The three modern records are from waste ground: it was found, in some abundance, at the entrance to Barrangary Tip (NS4469, 2001), at Port Glasgow (NS3174, 2012), and three plants were found to the east side of the road near Calder Bridge, Lochwinnoch (NS3558, 2003, IPG).

Melilotus officinalis (L.) Pall.
Ribbed Melilot

Britain: neophyte (Europe); common in S and E England, frequent in lowland Scotland.
Accidental; 6/12; rare casual.
1st Renfrewshire, 1869, PPI.

This melilot may have been confused with Tall Melilot in the literature ('*M. officinalis* auct.'), but an 1890 specimen from Giffnock is this species (GL). Other old records include Inchinnan (1873, TGSFN), Greenock (1880, NHSG), by Queen Mary's tree, Darnley (1888, AANS) and wasteland near Paisley (1910, PSY). In more recent times AMcGS knew it from Lonend, Paisley (NS4863, 1980). The modern records are from tips at Linwood (NS4465, 1998), Shillford (NS4455, 1999) and Barrangary (NS4469, 2001), and waste-ground sites at Clydeport (NS3174, 2001), Paisley (NS4860, 1988, DM) and Castle Semple (NS3558, 2011). For the 1888 record Wood (1893) wrote, 'It is not a plant likely to become fixed in our country', which appears to have been a fairly accurate prediction.

Melilotus indicus (L.) All.
Small Melilot

Britain: neophyte (S Europe); scattered in the S and in lowland Scotland.
Former accidental; 0/3; extinct casual.
1st Crossmyloof, 1892, NHSG.

Two herbarium sheets (GL) labelled as '*M. officinalis*' are Small Melilot: Nitshill brickworks (1909) and a coup near Paisley (1911), indicating that this species may have been formerly more widespread.

Melilotus messanensis (L.) All.
Sicilian Melilot

Britain: neophyte (S Europe); rare casual.
Former accidental; 0/1; extinct casual.
1st Crossmyloof, 1894, GL.

Trigonella foenum-graecum L.
Fenugreek

Britain: neophyte (SE Europe); rare casual.
Former accidental; 0/1; extinct casual.
1st Crossmyloof, 1894, GL.

Trigonella glabra Thunb.
T. hamosa Forssk. non L.
Egyptian Fenugreek

Britain: neophyte (Mediterranean); rare casual.
Former accidental; 0/1; extinct casual.
1st Strathbungo, 1865, RH.

RH (1865) wrote of this species, 'Found on road from Strathbungo to Paisley Canal; considered by Dr W Arnott to be introduced with seed from Egypt'.

Trigonella monspeliaca L.
Star-fruited Fenugreek

Britain: neophyte (S Europe); rare casual.
Former accidental; 0/1; extinct casual.
1st Crossmyloof, 1894, GL.

Trigonella polyceratia L.

Britain: neophyte (S Europe); rare casual.
Former accidental; 0/1; extinct casual.
1st Crossmyloof, 1894, GL.

Medicago lupulina L.
Black Medick

Britain: common in England and lowland Scotland.
Native or accidental; 62/68; occasional.
1st Paisley Canal, 1858, GGO.

Black Medick shows a strong urban-lowland pattern, with only a few records in the agricultural lowlands; most localities are short or open grasslands of waste-ground areas, disused railways or some waysides. There are few old records: it was recorded from Ferguslie in 1898 (PSY) and only recorded on a handful of occasions by BWR, with most other records for CFOG and later.

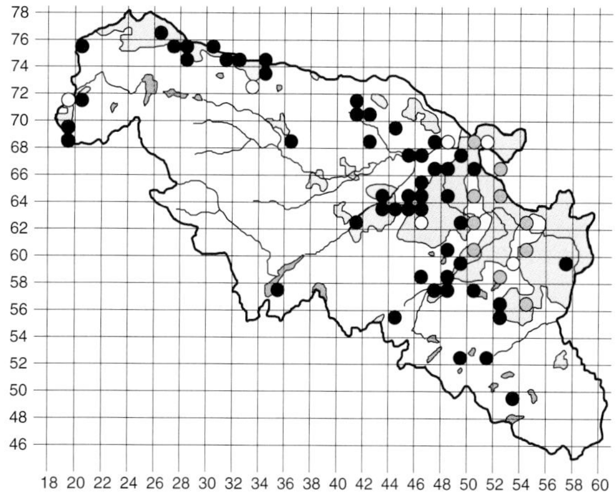

Medicago lupulina

Medicago sativa L.
Lucerne

Britain: rare native in E Anglia, scattered casual elsewhere.
Accidental or hortal; 1/6; very rare casual.
1st Racecourse, Paisley, 1900, PSY.

Other old records include a coup at Newlands (1892, GL) and Maxwell Park, Pollokshaws (1903, GL). The most recent record for Lucerne was from Barshaw Park in 1985 (NS56C, CFOG). Sickle Medick (ssp. *falcata* (L.) Arcang.) was noted from near Renfrew (1873, TGSFN) and Sand Lucerne (nothossp. *varia* (Martyn) Arcang.) was collected from a roadside between Strathbungo and Paisley Canal in 1863 (GGO).

Medicago truncatula Gaertn.
Strong-spined Medick

Britain: neophyte (S Europe); rare casual.
Former accidental; 0/1; extinct casual.
1st Giffnock, 1892, GL.

Medicago polymorpha L.
Toothed Medick

Britain: rare on S coast, scattered alien in NE Scotland.
Former accidental; 0/5; extinct casual.
1st Renfrewshire, 1869, PPI.

An undated specimen from Paisley Canal (GGO) probably dates from the time of the first record. Toothed Medick was also collected from wasteland near Paisley (1910, PSY), Crossmyloof (1892, NHSG) and Giffnock (1892, GL), but there are no modern records.

Medicago arabica (L.) Huds.
Spotted Medick

Britain: common in SE England, scattered alien in the N.
Former accidental; 0/2; extinct casual.
1st Renfrewshire, 1869, PPI.

RG (1930) considered Spotted Medick an 'occasional casual around Glasgow' but the only named location is Jenny Wood (1870, MY).

Medicago intertexta (L.) Miller

Britain: neophyte (S Europe); rare casual.
Former accidental; 0/1; extinct casual.
1st Giffnock, 1892, GL.

Medicago suffruticosa Ramond ex DC.

Britain: neophyte (S Europe); very rare casual.
Former accidental; 0/1; extinct casual.
1st Giffnock, 1892, GL.

Medicago tornata (L.) Miller

Britain: neophyte (S Europe); very rare casual.
Former accidental; 0/1; extinct casual.
1st Giffnock, 1892, GL.

Trifolium ornithopodioides L.
Bird's-foot Clover

Britain: occasional on southern coasts, rare casual in Scotland.
Accidental; 0/1; extinct or error.
1st Cartside, 1865, MY.

There is strong doubt about the identity of this record. It appears to be the sole record and was excluded from the Renfrewshire list (TPNS 1915), where it was referred to as 'Fenugreek'.

Trifolium repens L.
White Clover

Britain: common throughout.
Native and hortal; 408/501; very common.
1st Renfrewshire, 1869, PPI.

One of the commonest plants found in the VC, White Clover only really becomes scarce in the peaty uplands, but even here it can sometimes be found on grassy patches by rocks or burns. It is found in all types of neutral or improved grassland, often sown, tolerates some damp conditions and can be very common at open waste-ground sites.

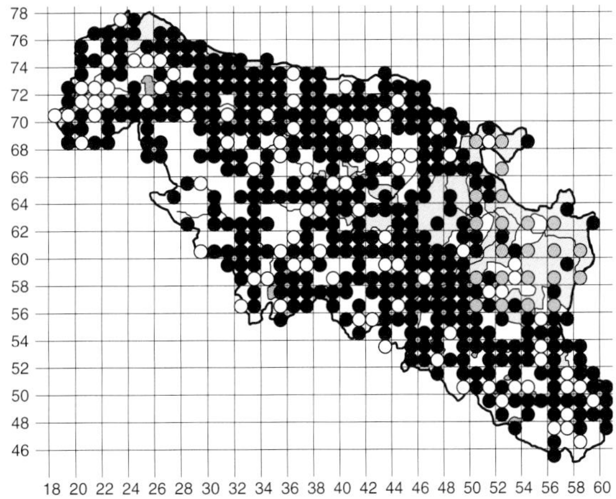

Trifolium repens

Trifolium hybridum L.
Alsike Clover

Britain: neophyte (S Europe); common in C and NE Scotland.
Accidental; 57/67; frequent.
1st Canal Bank, 1899, PSY.

The next record for this clover was not made until 1959 (NS5851, BWR). The distribution pattern shows a very strong urban preference, although it can be seen in more rural locations such as on a roadside at Dodside (NS5053, 1998); it may have been mistaken elsewhere for White Clover cultivars. Alsike Clover tends to be found, sometimes in abundance, on open waste ground, short grasslands or along waysides.

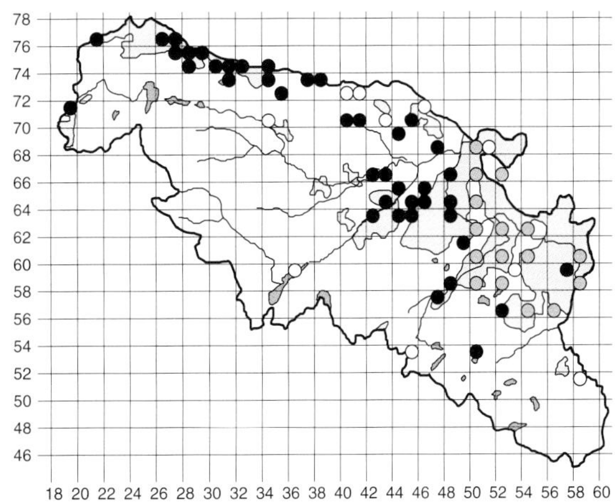

Trifolium hybridum

Trifolium resupinatum L.
Reversed Clover

Britain: neophyte (SE Europe); scarce in S England.
Former accidental; 0/3; extinct casual.
1st Roadside between Strathbungo and Paisley Canal, 1863, GGO.

This rare, casual clover was also collected from Giffnock (1892, GL) and from railway sidings, Thornliebank (1909, GL); the latter was named as '*T. fragiferum*'.

Trifolium aureum Pollich
Large Trefoil

Britain: neophyte (Europe); scarce casual in C Scotland.
Accidental; 1/1; very rare.
1st Paisley, 2003, AJS.

Although there are a few records in adjacent Glasgow (CFOG) the only VC76 record is from sown grass near Paisley Gilmour Street Railway Station (NS4864).

Trifolium campestre Schreb.
Hop Trefoil

Britain: common in lowland England to C and NE Scotland.
Native or accidental; 54/57; frequent in lowlands.
1st Paisley Canal Bank and Gourock, 1865, RH.

Most of the records for this trefoil are from after 1950, although it was collected from old quarries at Giffnock (1883, GL) and from a wayside in Kilbarchan in 1921 (GL), and Hennedy (1891) thought it common. The modern sites are usually open or short grassland on waste ground and waysides, notably disused railways or cycleways (as between NS3271 and NS3865), but mainly in lowland urban areas. It was only recorded a few times by BWR, so this distinctive trefoil may have become more frequent in recent years.

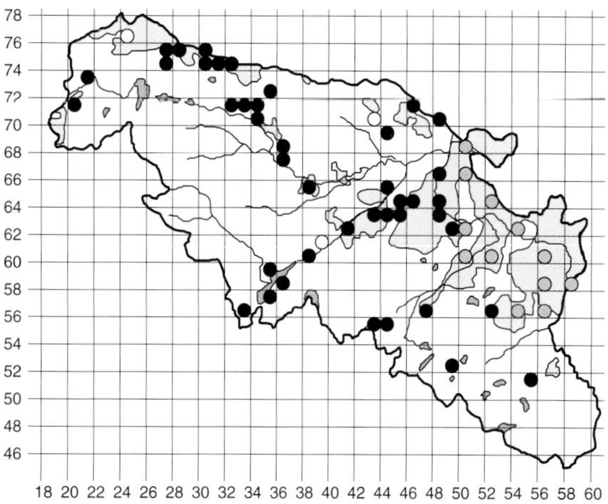

Trifolium campestre

Trifolium dubium Sibth.
Lesser Trefoil

Britain: common throughout except in N Scotland.
Native; 139/188; common.
1st Paisley Canal Bank, 1858, GGO.

Lesser Trefoil is the commonest of the yellow trefoils. It is found on urban waste ground and lawns, and it can also be found locally in short pasture grasslands. However, it becomes rare or absent in the uplands. There are only a few named old localities but it was well recorded during the 1960s–80s (BWR). Old records for '*T. filiforme*', which was excluded from the Renfrewshire list (TPNS 1915), presumably refer to this species.

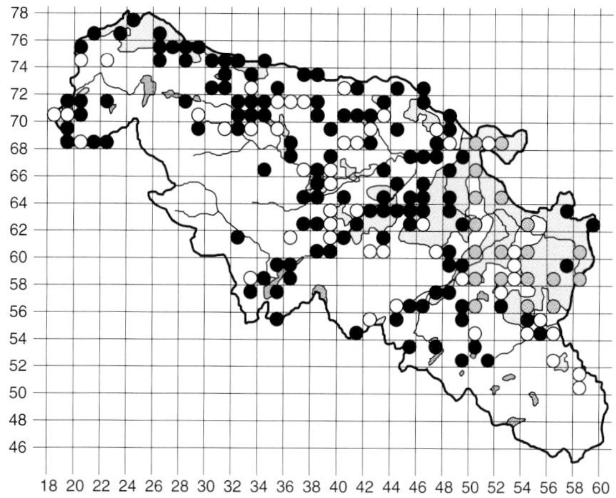

Trifolium dubium

Trifolium pratense L.
Red Clover

Britain: common throughout.
Native and hortal; 282/361; very common.
1st Renfrewshire, 1869, PPI.

Red Clover is common throughout the VC and

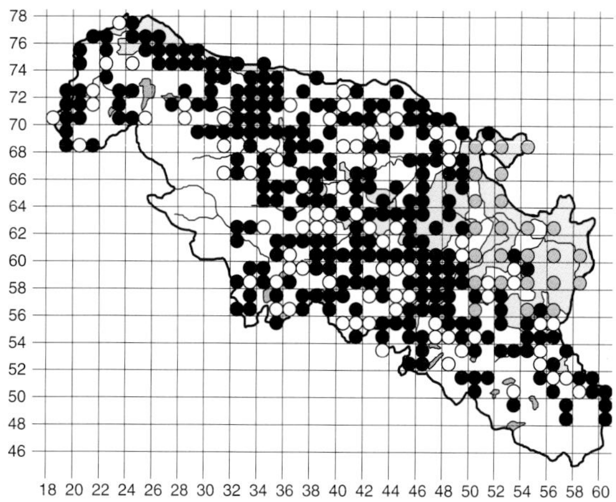

Trifolium pratense

only absent from the acidic uplands. It occurs in unimproved neutral grasslands, can be sown, and is well represented on all types of urban waste ground and wayside grasslands, except where marshy.

Trifolium medium L.
Zigzag Clover

Britain: common in C and NE Scotland.
Native; 46/74; frequent, eastern.
1st Lochwinnoch and Kilbarchan, 1845, NSA.

There are few old records but Hennedy (1891) described Zigzag Clover as common. The pattern shows an obvious preference for the east; this clover is absent from suitable habitats in the west, presumably for climatic reasons, and it may well have declined in the central area. It is often found in unimproved pastures but it can also be found at short waste-ground and wayside grasslands, including disused railways.

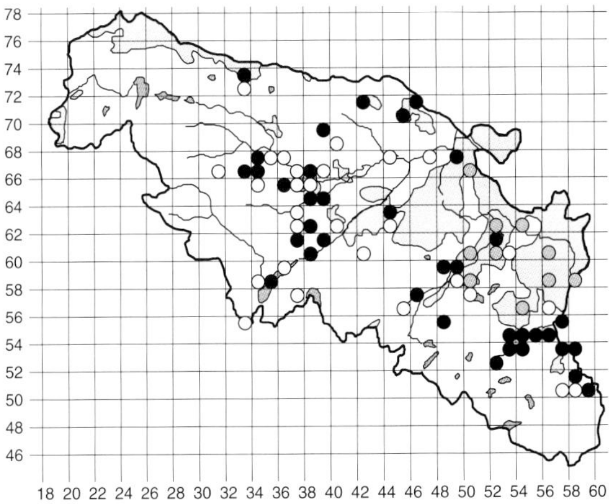

Trifolium medium

Trifolium striatum L.
Knotted Clover

Britain: frequent in the S, local in E Scotland.
Former accidental; 0/1; extinct casual.
1st Paisley, 1980, AJS.

A rare clover known from coastal Ayrshire but the only VC76 record is as a weed of lawn turf (NS4864).

Trifolium lappaceum L.
Bur Clover

Britain: neophyte (S Europe); rare casual.
Former accidental; 0/1; extinct casual.
1st Giffnock, 1892, GL.

Trifolium arvense L.
Hare's-foot Clover

Britain: frequent in the S, mainly in the E in Scotland.
Native; 3/12; very rare.
1st Lochwinnoch, 1845, NSA.

Other old records include Greenock west (1880, NHGS), Giffnock Quarry (1891, GL), wasteland, Paisley (1900, PSY), near Paisley (1915, TPNS) and more recently (BWR) Loch Libo (NS4355, 1965), Slatesmill, Kilmacolm (NS3469, 1964), Lunderston Bay (NS2074, 1983) and Barcraigs Reservoir (NS3957, 1984). The three modern records are all from waste ground at Spango (NS2374, 1997), Howwood (NS3860, 2005, IPG) and Cowglen (NS5560, 2005, LR).

Trifolium squamosum L.
Sea Clover

Britain: local in S England.
Former accidental; 0/1; extinct.
1st Giffnock, 1892, GL.

Lupinus polyphyllus Lindl.
Garden Lupin

Britain: neophyte (W N America); local in England, frequent in C Scotland.
Hortal; 33/35; occasional.
1st Kilmacolm, 1984, BWR.

Garden Lupin was commonly recorded in urban Glasgow (CFOG) but there are very few old records. The modern records are mostly from urban fringes, where it can form locally large stands on waste ground and along roadside verges or watercourse banks. It is likely that some of the records refer to cultivars or hybrids.

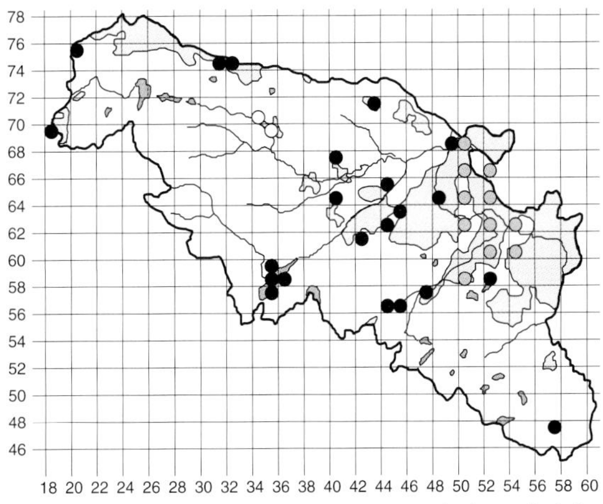

Lupinus polyphyllus

Laburnum anagyroides Medik.
Laburnum

Britain: neophyte (C Europe); frequent in England and lowland Scotland.
Hortal; 18/25; occasional.
1st Court Knowe, Cathcart, 1856, HMcD.

There is a 1907 specimen from a hedgerow attributed to VC76 (GL) but there is no named locality; Lee (1933) noted Laburnum at Gourock and a few scattered localities were recorded in the 1960s–80s (BWR). Today Laburnum is known from waste ground and old woodland estates, all in the east, reflecting relicts of past planting but also some spread. The latter, however, appears to be a rare event. Some records, particularly the older ones, may refer to the following two accounts.

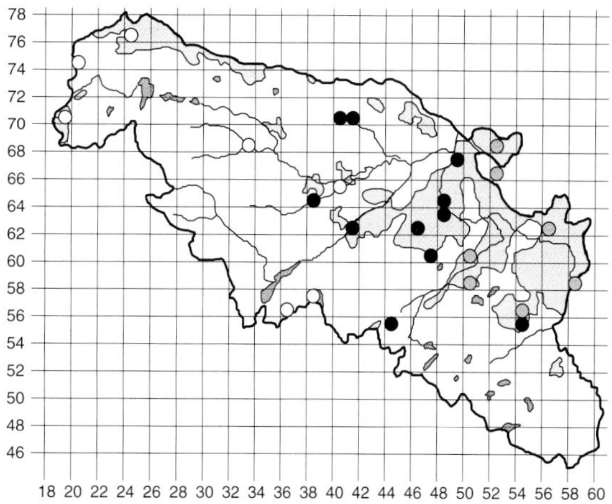

Laburnum anagyroides

Laburnum × watereri (Wettst.) Dippel
(*L. anagyroides × L. alpinum*)
Hybrid Laburnum

Britain: neophyte (garden origin); local, mainly in the W.
Hortal; 4/4; rare.
1st Darnley (NS5258), 1993.

Hybrid Laburnum was also recorded at Ardgowan Estate (NS2073, 2004) and Formakin Estate (NS4070, 1994) – both are likely to be relicts of past planting. A fourth recorded site is the edge of waste ground at Neilston Station (NS4757, 2004), but again this may have been planted in the past.

Laburnum alpinum (Mill.) J.S. Presl
Scottish Laburnum

Britain: neophyte (C Europe); scarce, mainly in the N.
Hortal; 4/5; rare.
1st Langside, 1856, GGO.

Although present in the area for more than 150 years, the first field record for Scottish Laburnum is from woodlands by the roadside at Lunderston Bay (NS2074, 1996), where it appears as if it could have been self-sown. Two other records, which may also be self-sown, are from Wemyss Castle wood (NS1970, 1999) and Porterfield (NS4967, 2001). It also grows at Dargavel House (NS4270, 2005) where it was presumably planted.

Cytisus scoparius (L.) Link
Broom

Britain: common throughout.
Native; 255/299; very common in lowlands.
1st Paisley, 1845, NSA.

Broom shows a very neat distribution pattern, avoiding the peaty or water-logged soils of the upland areas, though there is a trackside record at High Myres (NS5646, 2000). It is usually associated with free draining, sandy soils ranging from hillside pastures to urban waste ground.

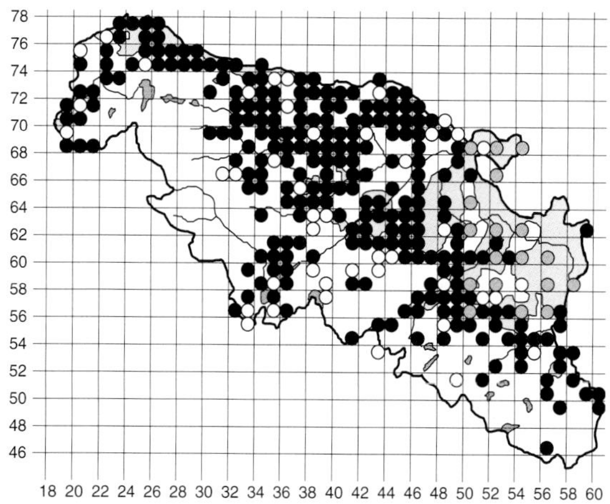

Cytisus scoparius

Ulex europaeus L.
Gorse

Britain: common throughout.
Native; 253/294; very common in lowlands.
1st Paisley, 1845, NSA.

Gorse is a common and often highly visible, colourful feature of hilly pastures such as the Liboside Hills, Gleniffer Braes and the north-facing braes at Inverclyde. It readily invades acidic (sandy or dry and peaty) grasslands if grazing is not too severe, but is rare or absent in wet peaty soils of the uplands or where the land is heavily agriculturally improved; it also readily colonizes waste ground in urban areas.

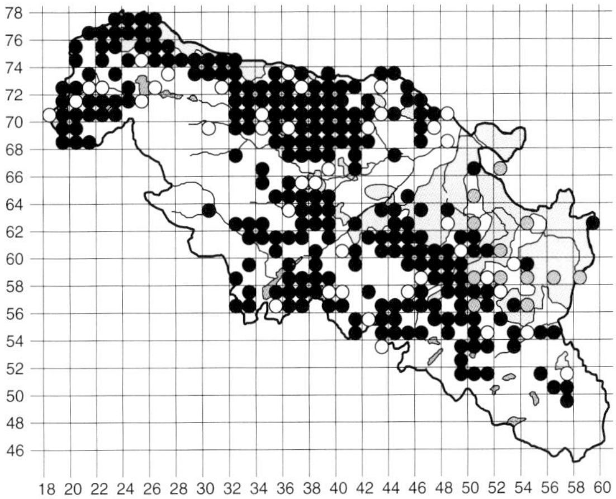

Ulex europaeus

Ulex minor Roth
Dwarf Gorse

Britain: mainly in SE England and rare elsewhere.
Hortal; 1/1; very rare.
1st Inverkip, 1999.

Presumably originally planted, this gorse is found in thick scrub in the grounds of Inverkip Power Station (NS1971).

POLYGALACEAE

Polygala vulgaris L.
Common Milkwort

Britain: scattered or locally common throughout.
Native; 17/20; scarce, possibly overlooked.
1st Paisley, 1845, NSA.

There is some doubt as to the former frequency of this milkwort of more base-rich soils; it was considered by Hennedy to be very common, but some old records may refer to Heath Milkwort. However, it is possible that given its preference for better soils it may well have suffered more from agricultural improvement. It may have been overlooked as the modern sites are widespread, but it is undoubtedly much the rarer of the two species. There appears to be a stronghold in the south about Muirhead (NS4855, 1999) and about Eaglesham (NS5650, 1991, LB); another cluster occurs above Port Glasgow, e.g. Cauldside (NS3270, 2005). Outlying sites include the slopes of Garvock Hill (NS2569, 1997) and one record from Earn Water (NS5252, 2003).

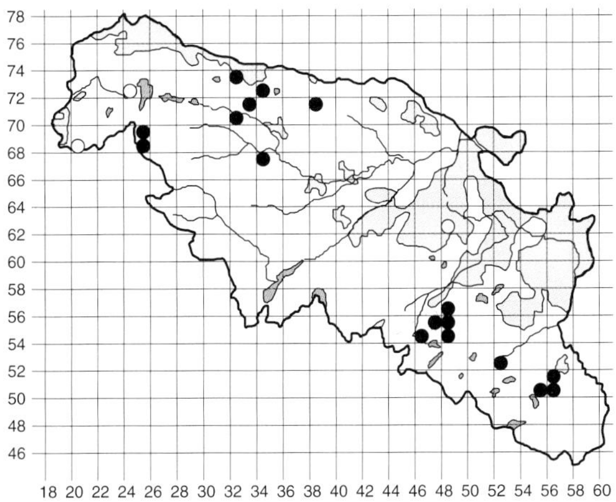

Polygala vulgaris

Polygala serpyllifolia Hosé
Heath Milkwort

Britain: common in the N and W.
Native; 125/169; locally common at upland fringes.
1st Port Glasgow, 1884, GL.

By far the commonest milkwort, this species is found in strongly acid grasslands including heath mosaics. It is a good indicator of such grasslands and its distribution pattern reflects this vegetation. Absent from the improved or urbanized lowlands, and also the boggy uplands, Heath Milkwort can be commonly found on upland-fringe slopes. It has probably declined on farmland with increasing agricultural improvement.

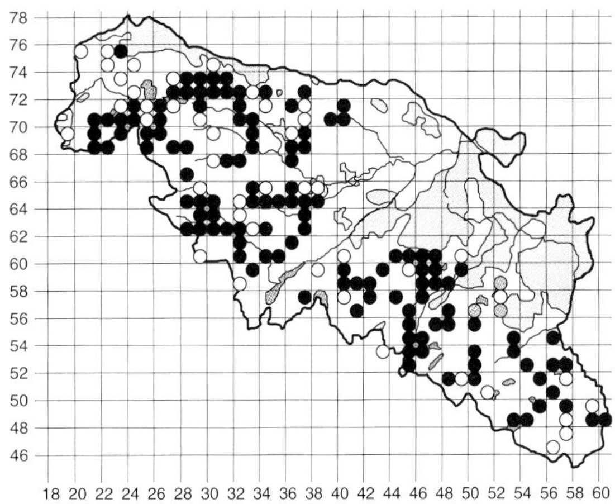

Polygala serpyllifolia

ROSACEAE

Physocarpus opulifolius (L.) Maxim.
Ninebark

Britain: neophyte (E N America); widely scattered.
Hortal; 2/2; very rare.
1st Formakin (NS4070), 1994.

Ninebark was also recorded from the wooded Rouken

Glen (NS5458, 1994, CFOG) and appears to have spread (2005).

Spiraea spp.
Brideworts

Britain: neophytes (N America and garden origins); widely scattered.
Hortals; 55/66; becoming frequent, widely scattered.
1st Banks of Loden, 1834 MC.

Brideworts were also recorded from Lochwinnoch and Kilbarchan (1845, NSA) and a little later from Castle Semple Loch (1887, NHSG), Kilmacolm (1888, PSY) and Levernholme Wood (1890, AANS). Lee (1933) refers to '*S. salicifolius*' as 'perfectly naturalised' but only names Kilmacolm. These records reflect a long history of Brideworts in the area, although specific identities are uncertain. The combined map shows vegetative or undetermined shrubs, and all records for the two commonest hybrids and Steeple-bush.

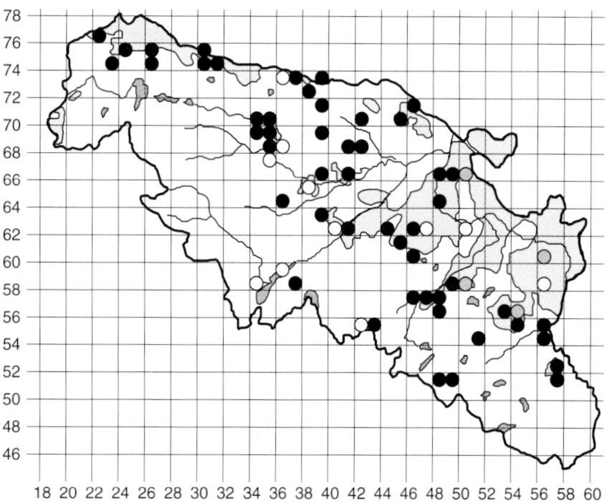

Spiraea spp.

Spiraea × *rosalba* Dippel
(*S. salicifolia* L. × *S. alba*)
Intermediate Bridewort

Britain: neophyte (garden origin); widely scattered, mainly in the N and W.
Hortal; 8/8; rare.
1st Barrhill Plantation (NS4671), 1995.

Difficult to distinguish from Confused Bridewort, this hybrid may well be under-recorded as there are only a few modern records, all from urban fringes: Eaglesham (NS5751, 1997), Kilbarchan (NS4162, 1998), Kirktonmill (NS4856, 1999), Mill Dam (NS3469, 1999), Duchal House (NS3568, 2000), a quarry at Kilmacolm (NS3570, 2011) and Mearns (NS5455, 2001).

Spiraea × *pseudosalicifolia* Silverside
(*S. salicifolia* × *S. douglasii*)
Confused Bridewort

Britain: neophyte (garden origin); scattered, mainly in the W, frequent in C Scotland.
Hortal; 23/28; occasional.
1st Paisley (NS4762), 1975, AJS.

This is the commonest Bridewort recorded, although a few records may refer to Intermediate Bridewort, particularly the older records. It is mostly found on waste ground, along waysides or in estate woodlands, usually, but not exclusively, in urban fringes.

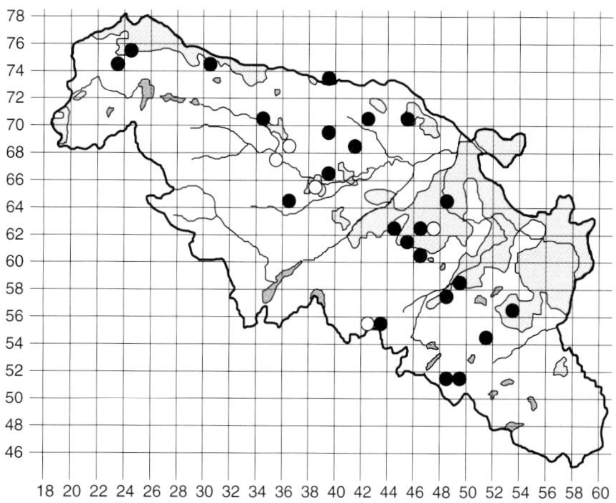

Spiraea × *pseudosalicifolia*

Spiraea alba Du Roi
Pale Bridewort

Britain: neophyte (N America); widely scattered, mainly in the W and N.
Hortal; 3/4; very rare.
1st Neilston (NS4856), 1986, AMcGS.

Near the first recorded location Pale Bridewort appears to have spread along the Levern Water to Holehouse (NS4756, 1997). It was recorded from the Aurs Burn, Barrhead, during CFOG surveying (NS55E, 1989). It is also known from Formakin Estate (NS4070, 1994), where it was presumably planted and has persisted or spread locally.

Spiraea douglasii Hook.
Steeple-bush

Britain: neophyte (W N America); widespread but local.
Hortal; 17/20; occasional.
1st Williamwood, 1941, RM.

Several records are from old estates where Steeple-bush may have been planted, but at most it appears to have spread. Other records, though, are from grassy waste places where it may have been dumped and can often form large clumps.

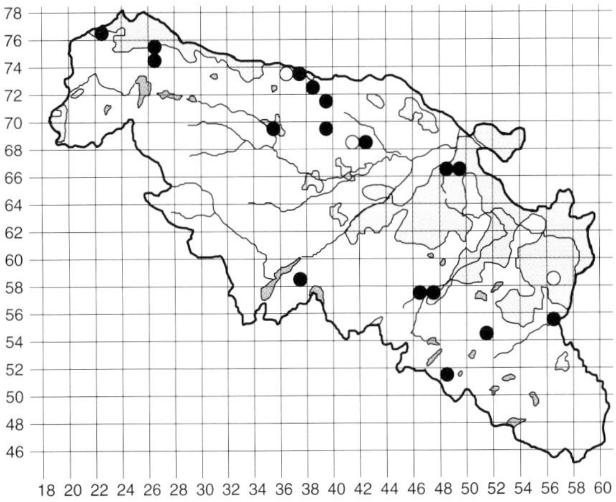

Spiraea douglasii

Spiraea japonica L. f.
Japanese Spiraea

Britain: neophyte (Japan); scattered.
Hortal; 1/1; very rare.
1st Port Glasgow, 2012.

This is a popular shrubbery plant but rarely, as along a roadside at Port Glasgow (NS3075), noted as spreading.

Spiraea × *vanhouttei* (Briot) Carrière
(*S. cantoniensis* Lour. × *S. trilobata* L.)
Van Houtte's Spiraea

Britain: neophyte (garden origin); rare, scattered, mainly in the S.
Hortal; 1/1; very rare.
1st Mearns (NS5455), 2001.

This cultivated shrub is occasionally planted and can persist but there are no definite records of it as self-sown.

Spiraea × *arguta* Zabel
(*S. multiflora* Sieb. ex Blume × *S. thunbergii* Zabel)
Bridal-spray

Britain: neophyte (garden origin); rare, a few Scottish sites.
Hortal; 2/2; very rare.
1st Corsliehill Road (NS3968), 1996.

The first record is for a large bush growing in a hedgerow along a remote country road but close to Corsliehill House. Another record is from the roadside at Fauldhouse (NS5253, 2003), presumably a garden outcast.

Aruncus dioicus (Walter) Fernald
Buck's-beard

Britain: neophyte (circumboreal); scattered, mainly in the NW.
Hortal; 7/8; rare.
1st Gleniffer (NS4459), 1975, JDM.

Buck's-beard was also recorded during CFOG surveying from Ralston Golf Course (NS56B, 1987) and Eastwood Park (NS55P, 1987). There are five more recent records: it is present in some abundance on path sides and walls to the edge of the woodland at Ranfurly (NS3865, 1997) and is also known from Cowdon Burn (NS4757, 1999), Locher Bridge (NS4064, 1999), Jenny's Well (NS4962, 2001) and Linn Park (NS5859, 2010).

Holodiscus discolor (Pursh) Maxim.
Oceanspray

Britain: neophyte (W N America); scarce but widely scattered.
Hortal; 0/2; very rare.
1st Kilmacolm, 1888, PSY.

AJS found Oceanspray on an old wall and rock face at Kilmacolm (NS3669, 1975), although this was unlikely to have been at the same location as the original record (which was named *Spiraea salicifolius* but re-determined by AJS).

Prunus spinosa L.
Blackthorn

Britain: common throughout except in N and upland Scotland.
Native; 142/165; common.
1st Cathcart, 1865, RH.

Prior to recent recording, there were few named localities. Today Blackthorn is commonly found at

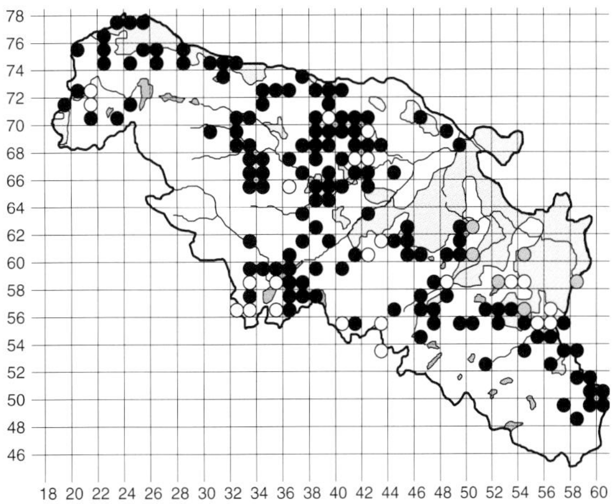

Prunus spinosa

woodland edges, forming local scrub patches and in hedgerows. It is absent from the wetter or more acidic soils. There are four records for the hybrid with Wild Plum (*P. × fruticans* Weihe): Humbie Burn (NS5454, 1997), Corsliehills Wood (NS3869, 1998), Gotter Water (NS3466, 1999) and at the Royal Ordnance Factory, Georgetown (NS4368, 2005), and the map may well include a few others.

Prunus domestica L.
Wild Plum

Britain: archaeophyte; common in the S, frequent in lowland Scotland.
Hortal; 30/37; occasional.
1st Inverkip, 1865, RH.

Plum trees have been recorded at a number of scattered localities, mainly from old hedgerows. Most of the records probably refer to Bullace (ssp. *insititia* (L.) Bonnier & Layens), which was listed in 1883 (TB), collected from Kilmacolm in 1904 (PSY) and named from an old walled garden at Nitshill (NS5160, 2005) and Capelrig Burn (NS55N, 1986, CFOG). Plum (ssp. *domestica*) was recorded by AJS from Lochwinnoch (NS3558, 1974) and Finlaystone Estate (NS3673, 1976).

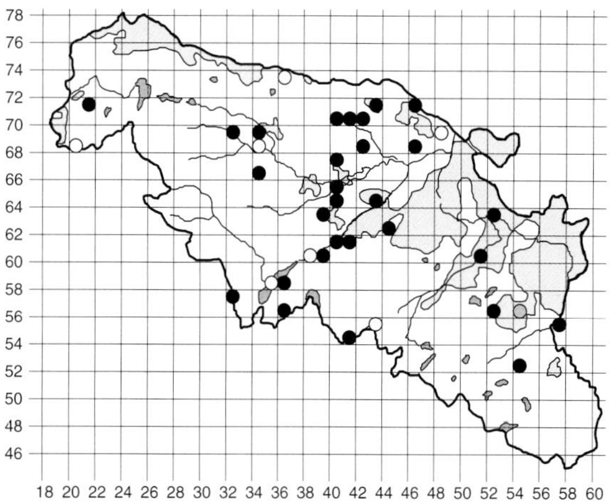

Prunus domestica

Prunus avium (L.) L.
Wild Cherry

Britain: throughout except in N Scotland.
Native and hortal; 69/86; frequent.
1st Lochwinnoch, 1845, NSA.

The records refer to presumed native trees in woodlands and also some newer colonists which may include cultivars or even other species. There are few old references: Cathcart (1891, RH), River Calder, Lochwinnoch (1899, NHSG) and RM noted it in the Gourock area (1923). Wild Cherry is found in several woodlands, including estate plantations, but normally as scattered individuals, e.g. River Calder (NS3459, 1995) and Devol Glen (NS3174, 1996).

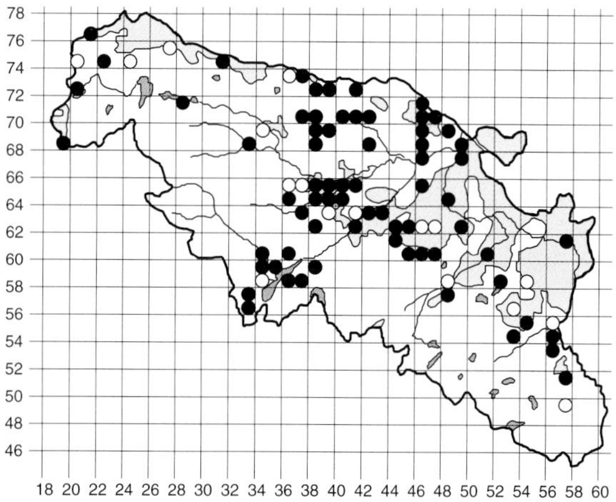

Prunus avium

Prunus cerasus L.
Dwarf Cherry

Britain: neophyte (SW Asia); scattered, mainly in the S.
Hortal; 1/1; very rare.
1st Nether Barfod, 2003, IPG.

The sole record is from a roadside hedgerow (NS3657). Old literature records for '*P. cerasus*' refer to Wild Cherry or in some cases Cherry Laurel.

Prunus padus L.
Bird Cherry

Britain: common in the N, scattered as alien in S England.
Native; 82/92; frequent.
1st Banks of Cart, 1813, TH.

Bird Cherry has a widespread, scattered distribution, generally in rural lowland locations, usually by watercourses. It is rare or absent from urban areas and

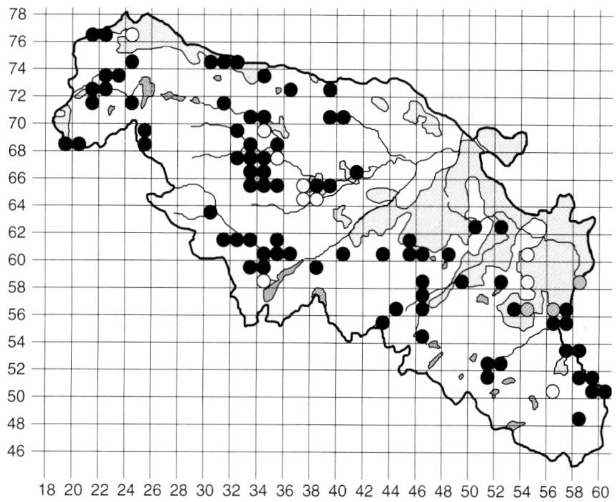

Prunus padus

intensively farmed land; it avoids the acidic uplands, but it can be found by the Raith Burn (NS3063) and North Rotten Burn (NS2568).

Prunus lusitanica L.
Portugal Laurel

Britain: neophyte (SW Europe); scattered in S, and in C Scotland.
Hortal; 12/13; rare.
1st Wemyss Bay, 1848, GL.

There are several scattered records for this laurel, mainly from estate woodlands, but no evidence of active spread as seen for the Cherry Laurel. Many, if not all, of the records refer to planted relicts.

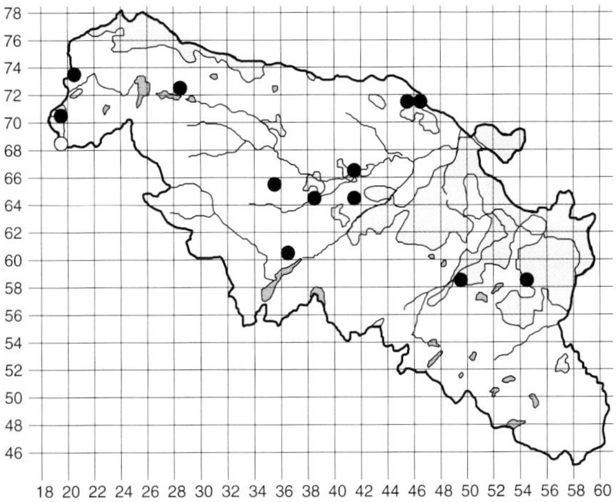

Prunus lusitanica

Prunus laurocerasus L.
Cherry Laurel

Britain: neophyte (SE Europe); common in the S, frequent in lowland Scotland.
Hortal; 48/51; occasional.
1st Kelly Glen, *c.* 1887, GRK.

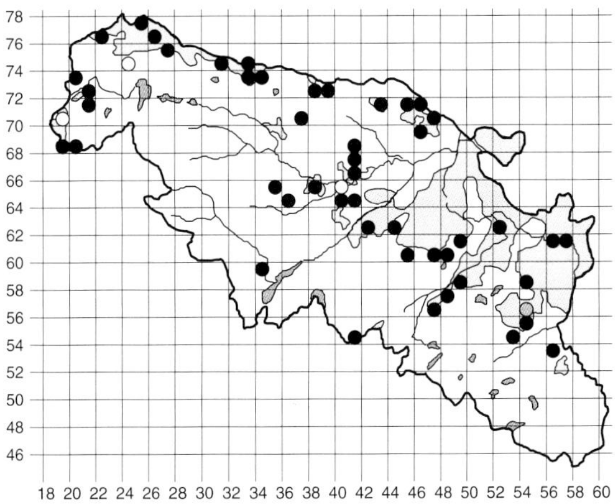

Prunus laurocerasus

The original specimen was labelled '*P. cerasus*'. The next records are from the Ravenscraig–Gourock area (1923, RM) and, more recently (BWR), from estates or parks such as at Crosslee (NS4065, 1972), Paisley Glen Park (NS4760, 1974) and Wemyss Bay (NS1970, 1974). Such old policy estates remain the major source of the modern records, where Cherry Laurel can occur in dense stands and it seems to spread freely. Its distribution is lowland, with many records from urban fringes, especially along the coast.

Pyrus communis L.
Pear

Britain: archaeophyte; scattered in England, local in C Scotland.
Hortal; 5/7; very rare.
1st Renfrewshire, 1915, TPNS.

RM knew Pear from Finlaystone (1962) and there are two CFOG records, both from 1991: an old orchard at Langside (NS5761) and a single tree by a burn at Thornliebank (NS5459). A single tree grows by the old stonework of the ruins at Walkinshaw (NS4666, 2003), and Pear was also seen in an old walled garden at Glentyan (NS3963, 2007) and at Cowdon Hall (NS4657, 2007). All are presumably relicts of past planting.

Malus sylvestris (L.) Mill.
Crab Apple

Britain: common in S and lowland Scotland.
Native or hortal; 29/35; occasional.
1st Renfrewshire, 1869, PPI.

Crab Apple can occasionally be seen in old hedgerows or woodland edges in agricultural areas, but it does not appear to be actively spreading and is likely to be in decline. Some of the older records may refer to Apple.

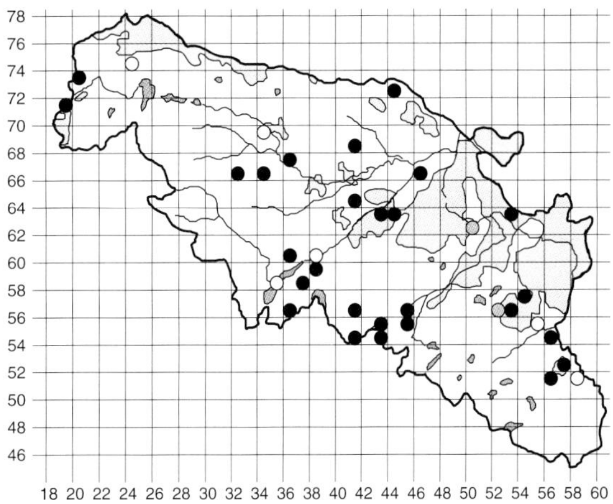

Malus sylvestris

Malus pumila Mill.
M. domestica Borkh. nom. illeg.
Apple

Britain: neophytes (cultivated); widespread, common in lowlands.
Hortal; 48/49; locally frequent.
1st Eaglesham, 1864, GL.

Apart from the first record, which will likely refer to a garden tree, all the records for Apple are from 1983 onwards (CFOG). Today Apple can be frequently seen in waste places near habitation which have been left undisturbed for a period. Discarded cores are the presumed source of most records.

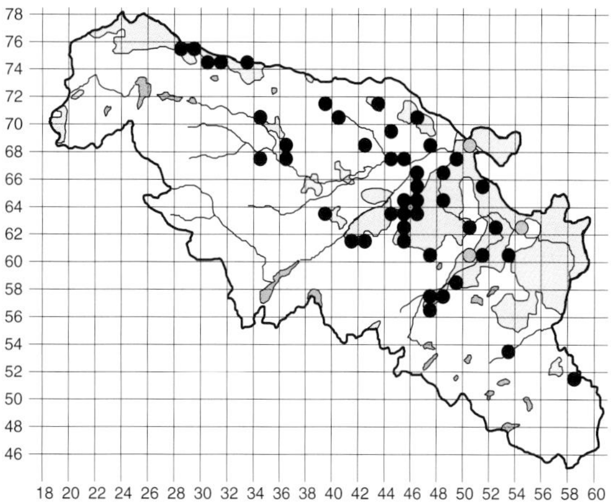

Malus pumila

Sorbus aucuparia L.
Rowan

Britain: common throughout.
Native; 352/404; very common.
1st Cart–Earn confluence (Waterfoot), 1856, HMcD.

Rowan is a very common small tree occurring throughout the VC, extending into the uplands except where the land is heavily grazed or very wet and peaty. It is seldom found in any abundance and is often present as singletons on rocky knolls or along boundary features. It prefers the less nutrient-enriched upper slopes of river valleys. Some of the lowland waste-ground records may refer to cultivated material.

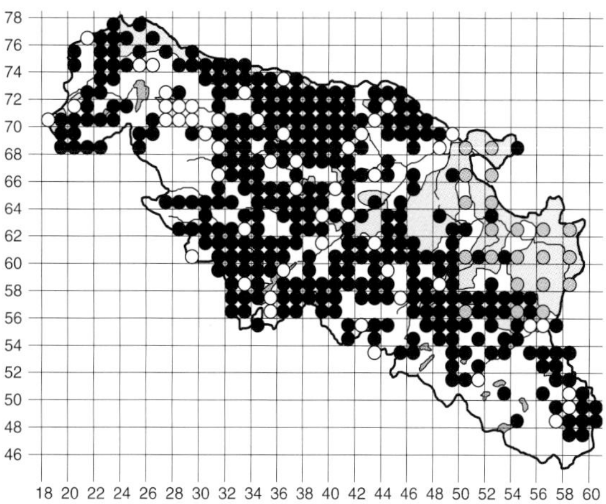

Sorbus aucuparia

Sorbus intermedia agg.
Swedish Whitebeam

Britain: neophyte (Baltic region); scattered in C and NE Scotland.
Hortal; 34/37; occasional.
1st Red Burn, *c.* 1966, JDM.

The original record (NS5762) for Swedish Whitebeam is from a file note with little detail. JP recorded it at Blythswood in 1968 and later AJS knew a single bush by an old railway in Paisley (NS4762, 1974). It has a similar pattern to that of Common Whitebeam, both being most likely to be seen on waste ground in urban fringes. It appears to have been better recorded during the surveys for CFOG.

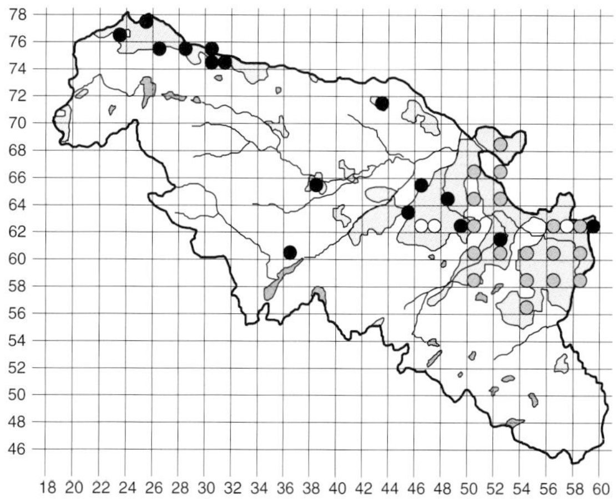

Sorbus intermedia agg.

Sorbus aria agg.
Common Whitebeam

Britain: local in S England, local alien elsewhere.
Hortal; 25/32; occasional.
1st Lochwinnoch, 1883, GL.

This whitebeam was collected from Nitshill in 1888 (GL) but, like the first, this record likely refers to planted specimens. It was noted a few times by BWR: Kelly Glen (NS1968, 1970), Dykebar (NS5062, 1973), Bridge of Weir (NS3667, 1982), Kilmacolm (NS3569, 1984) and Auchengrange (NS3757, 1985), and there are several records in CFOG. Not all of the records refer to self-sown plants, as some are relicts of past planting. Common Whitebeam is most likely to be encountered on waste ground in urban fringes.

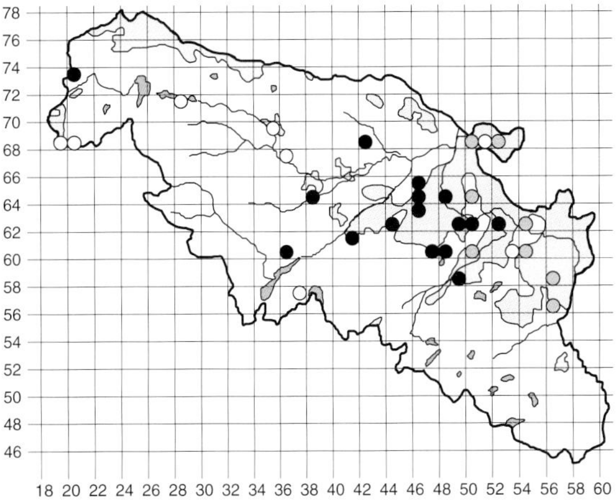

Sorbus aria agg.

Sorbus hybrids

German Service-tree *Sorbus × thuringiaca* (Ilse) Fritsch (*S. aucuparia × S. aria*) was collected (GL) from Cathcart Castle (1857) and Mill Hall, Eaglesham (1883). A mature specimen of *S. × liljeforsii* T.C.G. Rich (*S. aucuparia × S. intermedia*) occurs today in Cowdon Hall grounds (NS4757, 1997). There is no evidence of any being self-sown.

Sorbus rupicola (Syme) Hedl.
Rock Whitebeam

Britain: scattered, mainly in the W.
Native; 1/1; possibly now extinct.
1st Wemyss Bay, 1990, AR.

This is the sole record (NS1969) for Rock Whitebeam, and there is some doubt about its accuracy (AR pers. comm.). The locality has been extensively developed recently and the cliffs, although still a suitable habitat, are not readily accessible for surveying.

Amelanchier lamarckii F.G. Schroed.
Juneberry

Britain: neophyte (N America); occasional in the S.
Hortal; 1/1; very rare.
1st Erskine Home Farm Wood, 1996.

Juneberry is occasionally planted locally but the sole record is for a young sapling in a woodland (NS4472).

Cotoneaster spp.
Cotoneasters

This is a difficult group to identify as it contains many cultivars and hybrids, and although they have been occasionally observed as garden outcasts or bird-sown individuals in urban areas, critical field identification has not occurred. A few modern samples have been determined by Jeanette Fryer and older records from CFOG are mostly derived from PM's efforts (Macpherson & Lindsay 1993 and 1996) or field records for a few more distinctive species. A few species seem to be well established and spreading.

Cotoneaster apiculatus Rehder & E.H. Wilson
Apiculate Cotoneaster

Britain: neophyte (W China); rare.
Hortal; 1/1; very rare.
1st Base of a wall, Shawlands (NS5661), 1992, PM.

Cotoneaster bullatus Bois
Hollyberry Cotoneaster

Britain: neophyte (W China); frequent, but over-recorded.
Hortal; 18/18; scarce.
1st Side of a lane, Langside (NS5761), 1989, PM.

PM also recorded Hollyberry Cotoneaster from a lane in Newlands (NS5660, 1997). There have been several tentative field records for large, rugose-leaved plants but the following map may well include similar looking taxa; most records are from urban fringes.

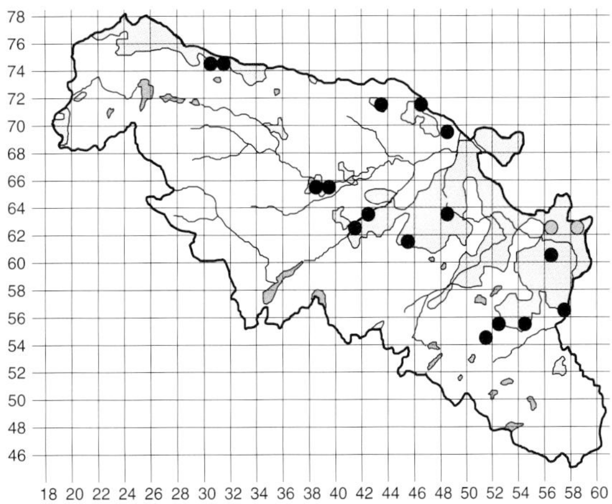

Cotoneaster bullatus

Cotoneaster dammeri C.K. Schneid.
Bearberry Cotoneaster

Britain: neophyte (C China); scattered.
Hortal; 1/1; very rare.
1st Waste ground by a quarry, Tower Hill, Gourock (NS2477), 2001.

Cotoneaster dielsianus E. Pritz. ex Diels
Diel's Cotoneaster

Britain: neophyte (China); scattered.
Hortal; 2/2; very rare.
1st Under a tree in a garden, Newlands (NS5760), 1990, PM.

PM also recorded Diel's Cotoneaster at Pollok (NS5360, 1996).

Cotoneaster hjelmqvistii Flinck & B. Hylmö
Hjelmqvist's Cotoneaster

Britain: neophyte (W China); scattered.
Hortal; 1/1; very rare.
1st By a wall, Shawlands (NS5561), 1992, PM.

Cotoneaster horizontalis Decne.
Wall Cotoneaster

Britain: neophyte (W China); frequent in the S, scattered in C Scotland.
Hortal; 21/22; occasional.
1st Disused railway, Kilmacolm, 1984, BWR.

This cotoneaster has been noted at several scattered locations, though some modern field records may include other small-leaved species. It appears to be persistent, growing on waste ground and waysides.

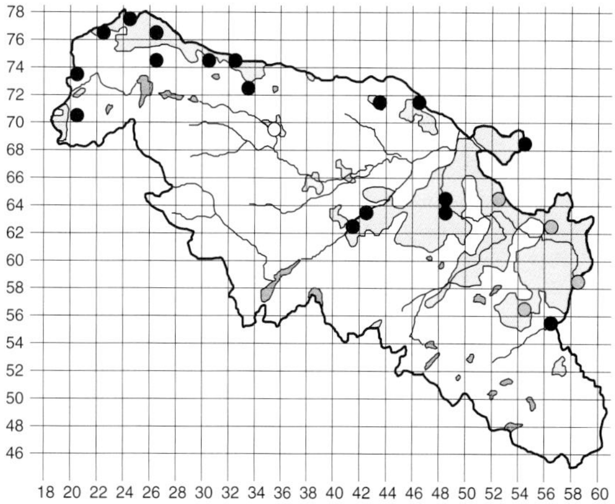

Cotoneaster horizontalis

Cotoneaster integrifolius (Roxb.) G. Klotz
Entire-leaved Cotoneaster

Britain: neophyte (W China); frequent, but confused with other species.
Hortal; 5/5; rare.
1st Pollok (NS5360), 1996, PM.

Jeanette Fryer confirmed the original record and two other collections: waste ground, Port Glasgow (NS3274, 1998) and below a wall by the White Cart Water, Paisley (NS4863, 2004). Unconfirmed records are from Bow Hill, Gourock (NS2577, 2004) and Langbank (NS3873, 2006).

Cotoneaster salicifolius Franch.
Willow-leaved Cotoneaster

Britain: neophyte (W China); frequent in the S, rare in C Scotland.
Hortal; 4/4; rare.
1st Eastwood Cemetery (NS5559), 1987, JHD.

Other confirmed records are from Linn Park (1992, PM), waste ground, Port Glasgow (NS3274, 1998) and by the White Cart Water, Paisley (NS4863, 2004).

Cotoneaster simonsii Baker
Himalayan Cotoneaster

Britain: neophyte (Himalayas); frequent throughout.
Hortal; 46/46; becoming frequent.
1st Crookston, 1986, CFOG.

This is the commonest Cotoneaster in the VC, occurring throughout urban-fringe areas on waste ground and along waysides, and extending into rural fringes. An unusual find was a well established population on a wet heath at North Barlogan (NS3768).

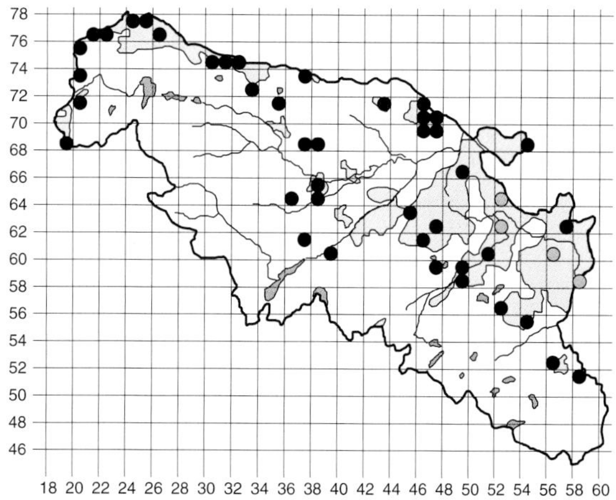

Cotoneaster simonsii

Cotoneaster × suecicus G. Klotz
(*C. dammeri × C. conspicuus* C. Marquand)
Swedish Cotoneaster

Britain: neophyte (garden origin); scattered throughout.
Hortal; 1/1; very rare.
1st Port Glasgow, 1998.

This hybrid is only known from waste ground by a fence (NS3274).

Pyracantha coccinea M. Roem.
Firethorn

Britain: neophyte (S Europe); scattered in S England, very rare in Scotland.
Hortal; 2/2; very rare.
1st Ladyburn, 1998.

The Ladyburn site (NS3076) is waste ground by

old railway tracks where the plant was presumably dumped. It has also been recorded from Durrockstock (NS4662, 1999), where it was probably planted.

Crataegus monogyna Jacq.
Hawthorn

Britain: common throughout except in N Scotland.
Native and hortal; 393/449; very common.
1st Barrhead, 1856, HMcD.

Hawthorn is found just about everywhere in the VC, only becoming rare or absent in the wet or peaty uplands. In addition to forming most hedgerows, it can be found in woodlands and scrub land, and also invades under-grazed grasslands and can colonize waste ground and waysides. Records will include some modern plantings, which may include cultivated or hybrid stock.

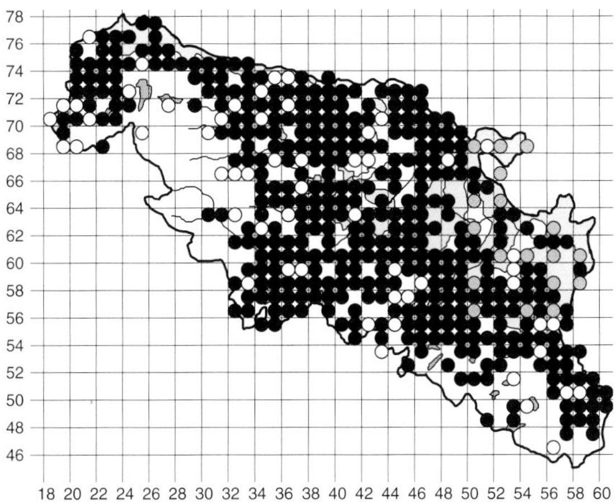

Crataegus monogyna

Filipendula vulgaris Moench
Dropwort

Britain: locally frequent in the S, extending to C Scotland.
Former hortal; 0/1; extinct.
1st Giffnock, RG, 1930.

Grierson, in his Clyde Casuals (1930), referred to this sole record as a hortal.

Filipendula ulmaria (L.) Maxim.
Meadowsweet

Britain: common throughout.
Native; 354/401; very common.
1st Paisley, 1845, NSA.

There are several old records in lists from Wemyss Bay, Castle Semple Loch and Loch Libo (all 1887, NHSG). Today Meadowsweet is a common plant of marshes, fens and wet woodlands, and can occur on poorly draining waste ground. It does not tolerate heavy grazing and is absent from the more acidic mires in the uplands.

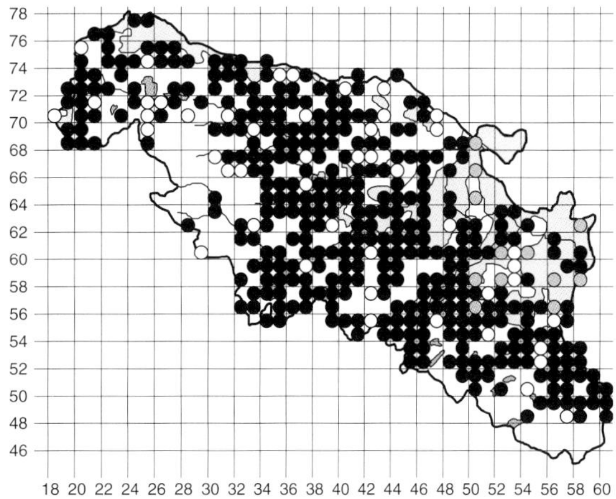

Filipendula ulmaria

Filipendula camtschatica (Pall.) Maxim.
Giant Meadowsweet

Britain: neophyte (E Asia); scarce, mainly in W C Scotland.
Hortal; 3/3; very rare.
1st Crookston, 1991, CFOG.

The site at Crookston has been planted with trees so the population of Giant Meadowsweet there is now very much reduced. A more recent record is from the estate woodlands at Formakin (NS4070, 1994), and it can be found in some number in the wet woodland at Rouken Glen (NS5458, 2005).
A smaller, pink-flowered garden cultivar or hybrid (*F.* × *purpurea* Maxim.) outcast grows by the Levan Burn (NS2176, 1999).

Rubus chamaemorus L.
Cloudberry

Britain: frequent in uplands in the N.
Native; 2/4; very rare.
1st South-west of Queenside Loch (NS2864), 1965, BWR.

Surprisingly, Cloudberry is not reported from the VC in the 19th-century literature and it is also omitted by Lee (1933). Hennedy (1891), without naming Renfrewshire, described Cloudberry as rare on peaty moors; this is descriptive for the VC today, as this species is only known from a few patches, though some are quite large, as on the boggy moor north of the Hill of Stake (NS2763), where it extends just into NS2863 (IG pers. comm.); it also grows on the south side of the summit in VC75. It was recorded from moorland north-west of Queenside Loch (NS2964, 1971, HG), but there are no recent sightings from here.

Rubus tricolor Focke
Chinese Bramble

Britain: neophyte (China); scattered.
Hortal; 2/2; very rare.
1st Port Glasgow, 2012.

This widely planted bramble was noted spreading in roadside scrub (NS3075) and to the edge of trees by a path at Glasgow Airport (NS4665, 2012, IPG).

Rubus saxatilis L.
Stone Bramble

Britain: frequent in the N and W.
Native; 9/17; rare.
1st Gourock, 1821, FS.

Old records include Lochwinnoch (1845, NSA), Kelly Glen (1887, NHSG) and Bardrain Glen (1904, PSY), and Hennedy (1865) noted Stone Bramble at Inverkip and Gourock. More recently (1978), BWR recorded it from Blacketty Water (NS3067), where today it is known slightly downstream (NS3167, 2000). The other modern records are from rocky outcrops, usually along upland watercourses as along the Raith and Cample burns, and down the River Calder (NS2862 to NS3261), along the upper reaches of the Gryfe Water at Little Creuch Hill (NS2669, 1997), at a north-facing rock outcrop at Earn Hill (NS2375, 1998) and by the North Rotten Burn (NS2568, 2011).

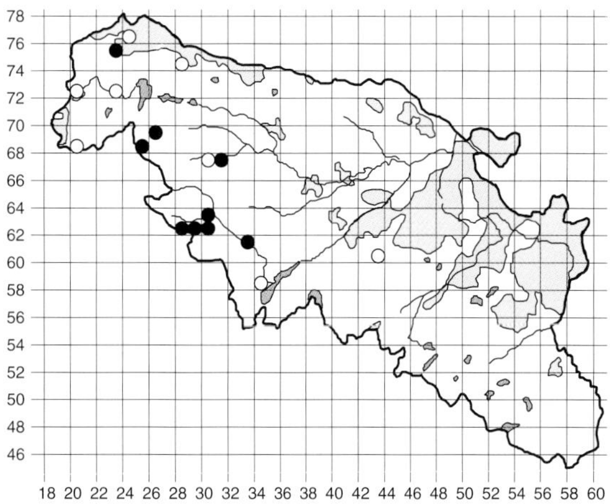

Rubus saxatilis

Rubus idaeus L.
Raspberry

Britain: common throughout.
Native or hortal; 291/342; very common.
1st Lochwinnoch, 1845, NSA.

Raspberry grows, often in abundance, throughout the lowland rural and urban-fringe area. It becomes rare in the uplands but has been recorded from the Raith Burn (NS2862, 1997) and North Rotten Burn (NS2569, 1997) in the Renfrewshire Heights. It is commonly found along roadside hedgerows and scrub, woodland edges and clearings, and also on long-neglected waste ground or in old gardens, generally reflecting a richer soil.

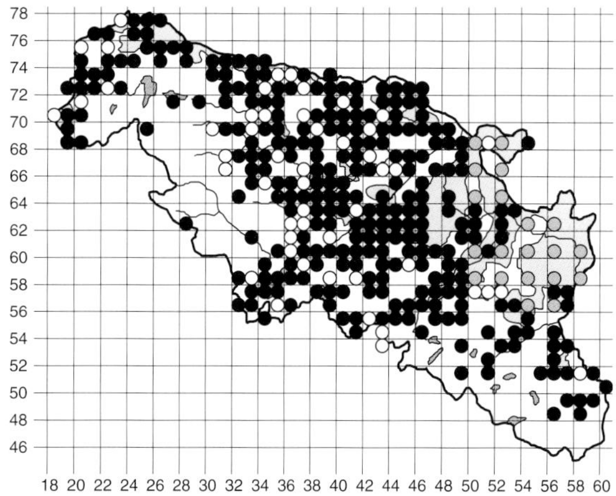

Rubus idaeus

Rubus spectabilis Pursh
Salmonberry

Britain: neophyte (W N America); frequent in C Scotland, local elsewhere. (Widespread in the N of Ireland).
Hortal; 11/13; scarce.
1st Glentyan, 1961, BWR.

This introduced shrub is found at a cluster of old estate woodlands in the north about Erskine: Boden Boo (NS4571, 1995), Barrhill (NS4671, 1995), Big Wood (NS4572 and NS4672, 1996) and Erskine Home Farm (NS4472, 1996). It is also known from Brodie Park (NS4762, 2003), Ardgowan Estate (NS2073, 1998) and by the Kip Water (NS2172, 2011). Salmonberry can still be found at Glentyan (NS3963, 2007), but

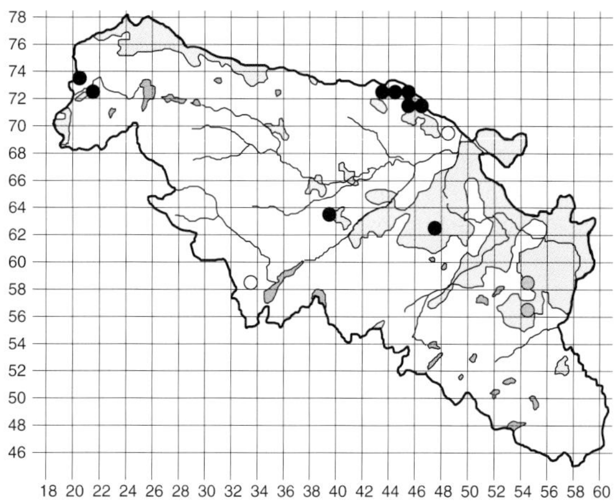

Rubus spectabilis

has not been seen at North Barr (NS4869, 1977) and Glenlora (NS3358, 1985) where recorded by BWR.

Rubus fruticosus agg.
Brambles

Britain: common throughout except in N Scotland.
Native or hortal; 344/397; very common.
1st Lochwinnoch, 1845, NSA.

Brambles are found throughout the VC, only becoming rare or absent from the wet, acidic uplands. Recording has been predominantly at the aggregate level but there are a few records for the various microspecies. AMcGS recorded a number of species in recent times (some are included in CFOG) and George Ballantyne (GHB) has produced VC lists of Scottish Brambles as well as making records in the VC following visits in 1983 and 1999 (at 10km-sq. level); GHB and Alan Newton (AN) have also checked herbarium records. GHB has provided comment on historical and modern records, helping the following accounts to be more comprehensive.

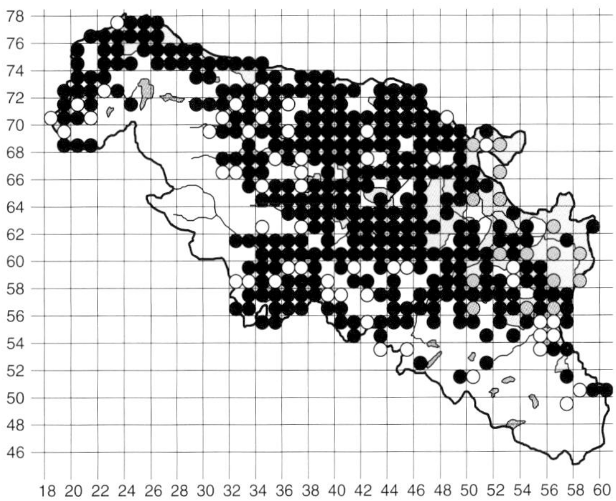

Rubus fruticosus agg.

Hennedy's *Clydesdale Flora* contains a reasonable account of brambles and the 19 taxa listed in the 1891 edition can mostly be equated with modern nomenclature. Further reliability followed the visit to the Clydesdale area by Rogers (1902). Trail (1902) was able to include Rogers' species in his list for the VC along with a few others and most of these may also be linked to modern nomenclature. In total, 24 species have definitely been recorded over the years, while a further four are considered possible by GHB. The species are arranged according to the *Atlas of British and Irish Brambles* (Newton & Randall 2004) and include all records except for some of the commoner species, where at least one record (usually 10km sq.) is provided. Other likely candidate species for the VC are *R. dumnoniensi* Bab., *R. furvicolor* Focke, *R. pyramidalis* Kaltenb. and *R. wirralensis* A. Newton (GHB pers. comm.).

Section *Rubus* (*Sub-erecti*)

The few Scottish members mostly favour more acidic ground and now appear to be locally rare or extinct. '*R. suberectus*' was recorded from a 'roadside between Govan and Renfrew' by Hooker in 1821 (FS); this is a collective name referring to either *R. fissus* or *R. scissus*.

R. fissus Lindl.: NS27, Gourock, 1845, Adamson in E (as '*R. scissus*' redetermined B Miles); NS37, Langbank, 1901, Rogers (confirmed AN); NS36, Kilmacolm, 1901, Rogers (confirmed AN).

R. scissus W.C.R. Watson: NS37, hill between Kilmacolm and Langbank, 1901, Rogers (as '*R. plicatus*', redetermined AN).

Section *Hiemales* (Series *Sylvatici* to *Hystrices*)

R. armeniacus Focke [*R. procerus* auct.]: an aggressive hortal which appears to be increasing. Lee first recorded it from Newlands in Glasgow in 1958 (GL). Modern records include Barshaw Golf Course (NS5064, 1986, CFOG), by the White Cart Water, near Moss Park (NS56, 1999, GHB), Rouken Glen (NS55, 1999, GHB) and roadside, Jordanhill (NS5468, 2002, AMcGS).

R. dasyphyllus (W.M. Rogers) E.S. Marshall: considered to be widespread locally, it was first recorded from Gourock and Cathcart (1891, RH) and a hill above Ashton (1901, Rogers). Later AJS noted Finlaystone (NS3673, 1976) and AMcGS knew it from Giffnock (NS5659, 1980). Modern records (all 1999, GHB) are from Pollok Country Park (NS56), near Cowglen Golf Course (NS56), Rouken Glen (NS5458) and Patterton Station (NS55).

R. elegantispinosus (A. Schumach.) H.E. Weber: this hortal is listed for the VC in Edees & Newton (1988) and although no precise record has been traced it is becoming common in E C Scotland, so may well be present locally (GHB pers. comm.).

R. errabundus W.C.R. Watson: first recorded from Langbank (1901, Rogers – as '*R. scheutzii*'), there are no modern records for this widespread species in lowland Scotland, but it was known from Finlaystone (NS3673, 1976, AJS) and Ranfurly (NS36, 1983, GHB).

R. infestus Weihe ex Boenn.: widespread and common across C Scotland and apparently so in the

VC. It was first recorded in 1883 (TB – as '*R. rudis*') and later from Ashton (1901, Rogers); it was collected from Bishopton (1949, Lee in GL). GHB recorded it from seven localities in 1983 (in NS35, NS36, NS46 and NS47) and there are two modern records (1999, GHB) from east of Neilston (NS45) and east of Barrhead (NS55).

R. laciniatus Willd.: a hortal first recorded at Bishopton Station (NS4370, 1975, BWR) and later from Rouken Glen (NS5458, 1981, AMcGS); there are two modern records – from the disused railway at Kilbarchan (NS4062, 1998) and the edge of a small clump of pines in Queen's Park (NS5762, 2007).

R. leptothyrsos G. Braun [*R. danicus*]: probably the commonest taxon in the VC (GHB), it was recorded from Ashton in 1901 (Rogers) and collected 'Between Erskine and Bishopton' by Lee in 1948 (GL). More recent records include localities such as Langbank (NS37), Bridge of Weir (NS36), Gleniffer Braes (NS46), Houston (NS46) and Castle Semple Loch (NS35). Modern records (GHB, 1999) are from Pollok Country Park (NS56), east of Neilston (NS45), Patterton Station (NS5357) and Barrhead (NS55).

R. lindleianus Lees: a western species, often near the coast in Scotland. Described as frequent by Hennedy in 1865 and by Rogers (1901); it was collected in 1944 by Lee near Loch Libo (GL) and other records include Finlaystone Estate (NS3673, 1976, AJS) and Langbank and Yetston (NS37, 1983, GHB). The most recent records (all NS27, 1987) were made by AN and AMcGS: south of Gourock, Lunderston Bay and south of Inverkip.

R. mucronulatus Boreau: this bramble is very common in east Scotland and through the Central Belt, so is likely to be present locally, but, although listed for the VC in Edees & Newton (1988), no detail has been traced.

R. nemoralis P.J. Müll.: widespread and common throughout much of the UK, it was considered 'frequent' by Hennedy (as '*R. rhamnifolius* var. *affinis*') and by Rogers (as '*R. villicaulis* var. *selmeri*'); Lee collected it from Erskine Ferry in 1937 (GL). There are more recent records (1970s–80s) from Finlaystone and Langbank (NS36), Howwood (NS46) and Erskine Ferry (NS47), and modern ones (1999, GHB) from Pollok Country Park (NS56), east of Neilston (NS45) and Patterton Station (NS5357).

R. polyanthemus Lindeb.: a common species most frequent in the west in Scotland, especially near the coast. Hennedy thought it 'frequent' (as '*R. rhamnifolius*') and Rogers described it as 'very common' (as '*R. pulcherrimus*'). Lee collected it from Inverkip (1938, GL) and more recently (NS27, 1987, AN and AMcGS) records have been made from south of Inverkip, Lunderston Bay and south of Gourock. GHB knew it from Langbank (NS37, 1983) and made the sole modern record from by the White Cart Water near Moss Park (NS56, 1999).

R. radula Weihe ex Boenn.: common in east Scotland but less so in the west, with only one record from a roadside at Howwood (NS3960, 1972, AMcGS).

R. raduloides (W.M. Rogers) Sudre: in Scotland most records for this bramble are concentrated about the Clyde and it appears to be common in the VC. The first record is from near Gourock (1845, Adamson in E – as '*R. rudis*') and it was also known to Hennedy (as '*R. glandulosus* var. *rudis*') and Rogers (as '*R. melanoxylon*'). More recent records include Finlaystone Estate (NS3673, 1976, AJS) and GHB (1983) noted it from the shore near the start of the M8, Ferryhill (NS47), Ranfurly, Bridge of Weir (NS36) and east of Howwood (NS36). There are three modern records (1999, GHB): Pollok Country Park (NS56), east of Neilston (NS45) and Rouken Glen (NS5458).

R. scoticus (W.M. Rogers & Ley) Edees: an endemic bramble almost confined to the Clyde area. Old records are from near Gourock (1845, Adamson in E – as '*R. rudis*' var.) and Kilmacolm and Ashton (1901, Rogers as '*R. radula* ssp. *sertiflorus*'). There are more recent records from Langbank (NS37, 1983, GHB) and from south of Gourock, Lunderston Bay and south of Inverkip (all NS27, 1987, AN and AMcGS).

R. septentrionalis W.C.R. Watson: scattered throughout Scotland, probably under-recorded in the VC. The sole records are from roadsides: by the B789 south of Langbank and by the B787 east of Howwood (both NS46, 1983, GHB).

R. ulmifolius Schott: this bramble is uncommon in Scotland away from the SW and often associated with railways. The sole record for the VC is from Gleniffer, near Paisley (NS46, 1999, GHB).

R. vestitus Weihe: rare in Scotland with only one VC record, from an industrial estate in Paisley (NS4865, 1987, AMcGS).

Section *Corylifolii*

R. eboracensis W.C.R. Watson: rare in Scotland, mainly in the east. This species was listed in Edees

& Newton (1988) from NS36 and dated 1975, but no other details are known.

R. hebridensis Edees: an uncommon species from several islands off the west coast of Scotland. It was first described from material that had been collected at Gourock in 1845 by F Adamson (CGE). Other old records include Ashton (1901, Rogers – as '*R. corylifolius* var. *cyclophyllus*') and from between Inverkip and Wemyss Bay (1939, Lee in GL – as '*R. incurvatus*'). More recently it was collected from West Bay (NS2377, 1980, R Pankhurst) and Ashton, Gourock (NS27, 1987, AG Kenneth) and south of Gourock (NS27, 1989, AN and AMcGS).

R. latifolius Bab.: this species (as '*R. corylifolius*') was mentioned by Hopkirk (1813), though not from the VC, and noted by Hennedy as 'very common in Clydesdale', which is still likely the case. More recent records include Castle Semple (NS35, 1983, GHB), Houston (NS46, 1983, GHB) and a roadside, Loch Libo (NS4355, 1984, AG Kenneth). Modern records are from south of Gourock and Lunderston Bay (NS27, 1989, AN and AMcGS), and GHB (1999) recorded it from Paisley (NS46), east of Neilston (NS45), Pollok Country Park, Crookston and Cowglen (NS56) and Barrhead (NS55); AMcGS also knew it from Jordanhill (NS5468, 2002).

R. pictorum Edees: scattered throughout the Central Belt, but probably rare in the VC. The sole record is from Cathcart (NS56, 1974, AMcGS).

R. tuberculatus Bab.: fairly frequent in C Scotland, often associated with railways and industrial sites, possibly increasing. Not listed by Trail (1902), it may be that Hennedy's '*R. carpinifolius* var. *tuberculatus*' is this species (Pollokshaws and Busby). There are three modern records (1999, GHB): Brownside, Paisley (NS46), by the White Cart Water, Moss Park (NS56) and Waulkmill, Barrhead (NS55).

Section *Caesii*

Rubus caesius L.
Dewberry

Britain: common in S and E England, rare possible alien in Scotland.
Native or accidental; 1/2; very rare.
1st Gourock, 1865, RH.

Many old records for Dewberry are errors for prostrate forms of *R. latifolius*; Hennedy thought it 'very rare' and Lee (1933) omitted it altogether. Apparently Rogers' 1901 record from Ashton is an error. The only modern record is from waste ground, Braehead (NS5167, 1992, GL, confirmed by AN) and its native status is questionable.

Potentilla fruticosa L.
Shrubby Cinquefoil

Britain: rare in N England, scattered alien elsewhere.
Hortal; 5/6; rare.
1st Bridge of Weir, 1980, BWR.

Also recorded from waste ground at Nitshill Station (NS56F, 1986, CFOG), this garden outcast has only a few other modern records: waste ground, Underheugh (NS2075, 1997) and disused railways at Elderslie (NS4463, 1999), Paisley (NS4864, 2000) and Langbank (NS3873, 2006).

Potentilla anserina L.
Silverweed

Britain: common throughout except in uplands.
Native; 204/292; common.
1st Gourock, 1860, GL.

Silverweed is commonly found on grassy waste ground and often along roadsides. Although found in dry places it seems to prefer inundated grassy margins of open waterbodies or marshes, and is also frequent at coastal grasslands. It is absent from the acidic uplands.

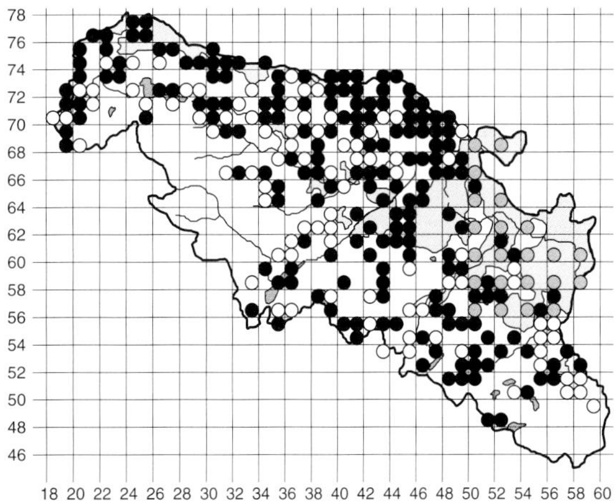

Potentilla anserina

Potentilla inclinata Vill.
Grey Cinquefoil

Britain: neophyte (C and SE Europe); rare in S England.
Former accidental; 0/1; extinct casual.
1st Giffnock Quarries, 1909, GL.

Potentilla norvegica L.
Ternate-leaved Cinquefoil

Britain: neophyte (N and C Europe); scattered in C Scotland.
Accidental; 4/10; rare.
1st Renfrewshire, 1895, NHSG.

Ternate-leaved Cinquefoil was collected from Stony Brae, Paisley (1899, PSY) and Langbank 1905 (GL), and Lee (1933) noted it at Elderslie and Ferguslie Park. More recently AMcGS found it at Lonend, Paisley (NS4863, 1980). It was found at Yoker (1986, CFOG), and other modern records are from waste ground by the White Cart Water near Glasgow Airport (N4866, 1995), waste ground near the Royal Ordnance Factory, Georgetown (NS4467, 2012, LP) and Woodhall (NS3473, 2012).

Potentilla erecta (L.) Raeusch.
Tormentil

Britain: common throughout.
Native; 380/444; very common.
1st Neilston, 1845, NSA.

Tormentil is one of the most common herbs found in unimproved acidic grasslands throughout the VC, extending from the lowlands to the highest hills, and only becoming rare in urban or intensive agricultural land. It can occasionally be found on open waste ground but also tolerates the wet acidic soils of mire margins and wet heaths.

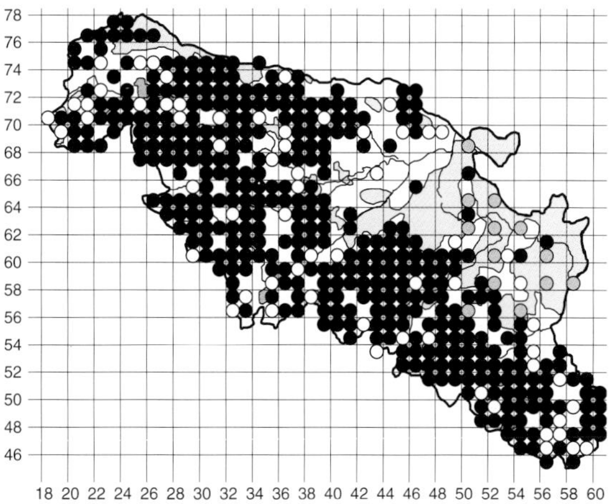

Potentilla erecta

Potentilla anglica Laichard.
Trailing Tormentil

Britain: frequent, mainly in the SW, very rare in N and E Scotland.
Native; 13/21; scarce, possibly overlooked.
1st Gourock, 1876, FFWS.

Trailing Tormentil was also recorded from the Loch Libo area and Castle Semple Loch (1887, NHSG); it was collected from Cathcart in 1887 (GL) and Hennedy (1891) noted it at Gourock. In more recent times it was noted from several scattered locations and it has probably been overlooked during modern recording. Records include the Royal Ordnance Factory, Georgetown (NS4268, 1999 and NS4270, 2005), Muirhead (NS4855, 1999), Balgray Reservoir (NS5057, 2000) and the disused railway (cycleway) at Auchenbothie (NS3371, 2005) and Bridge of Weir (NS3865, 2005).

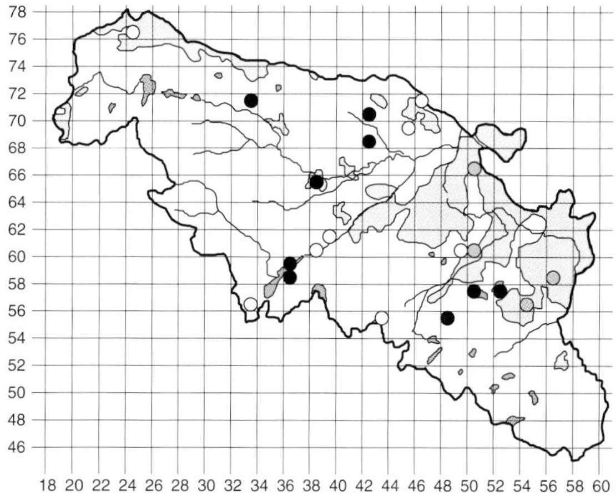

Potentilla anglica

Potentilla reptans L.
Creeping Cinquefoil

Britain: common in England and up to C Scotland, scattered in N Scotland.
Native or accidental; 20/21; occasional.
1st Lochwinnoch, 1845, NSA.

Lee (1933) repeats Lochwinnoch (where he collected Creeping Cinquefoil in 1916, GL), and was also collected from Thornliebank (1890, GL) and Pollok

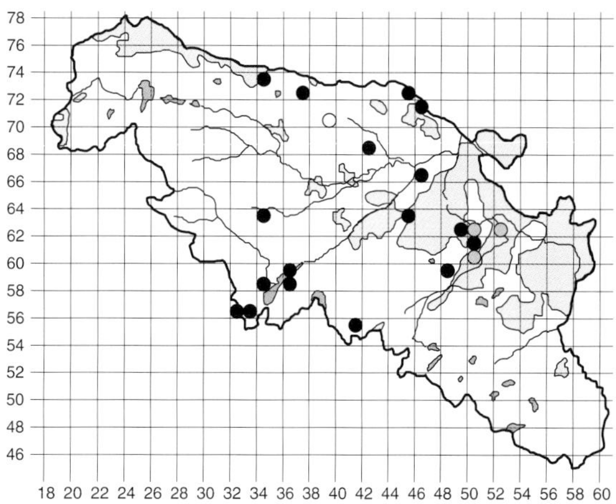

Potentilla reptans

(1907, GL). There were no other records until CFOG surveying. Localities tend to be waysides, waste ground and short grassland; it appears absent from the extreme east and west of the VC.

Potentilla sterilis (L.) Garcke
Barren Strawberry

Britain: common throughout except in N Scotland.
Native; 52/62; locally frequent.
1st Renfrewshire, 1869, PPI.

Barren Strawberry is locally frequent in less acidic woodlands, such as occur alongside many watercourses. It can also be found along hedge banks or on scrubby embankments, possibly reflecting old woodland locations. There are some notable gaps in the distribution pattern, e.g. from the upper reaches of the White Cart Water in the south-east and along the upper Gryfe Water, indicating a lowland preference. It occurs by the Gimblet Burn (NS2370, 2006), one of its few occurrences in the Renfrewshire Heights.

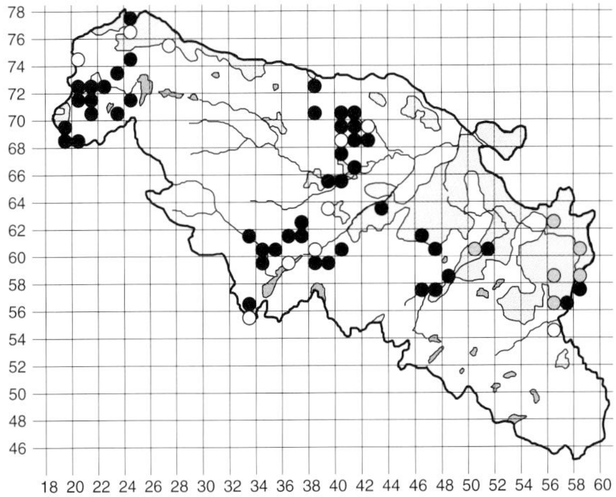

Potentilla sterilis

Comarum palustre L.
Potentilla palustris (L.) Scop
Marsh Cinquefoil

Britain: common in the N and W.
Native; 180/213; common.
1st Renfrewshire, 1834, MC.

Marsh Cinquefoil is a common plant of wet places, mainly in the rural areas. It is usually found in the deeper water of marshes, fens or swamps, often with Bottle Sedge, but can also grow with bog-moss in flushed bogs and mires.

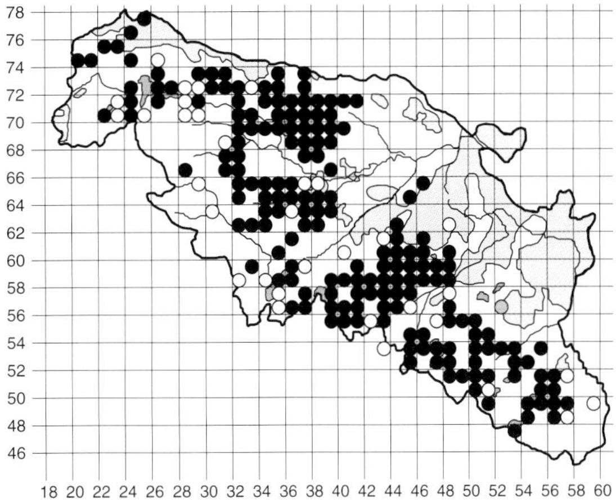

Comarum palustre

Fragaria vesca L.
Wild Strawberry

Britain: common throughout.
Native; 114/142; frequent.
1st Paisley, 1845, NSA.

Wild Strawberry is a locally frequent plant of woodlands, especially along river valleys where soils are more basic. Its range extends into the

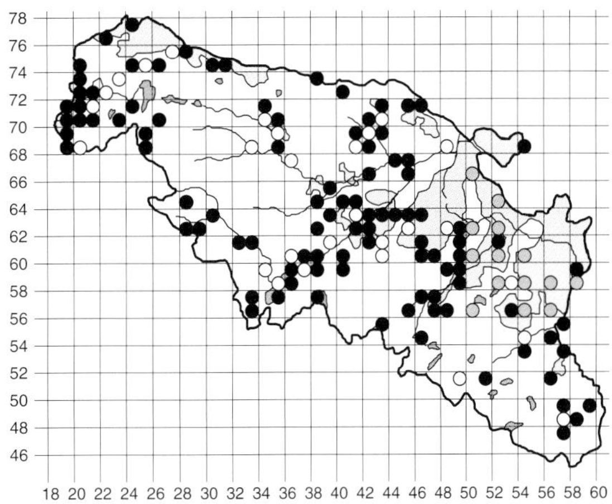

Fragaria vesca

Renfrewshire Heights, where it is found along valley sides as at Muirshiel Barytes Mine (NS2864) and along the Raith Burn (NS2962 to NS3063). It can also be found on waste ground and often grows well on mine wastes or disused railways.

Fragaria moschata (Duchesne) Weston
Hautbois Strawberry

Britain: neophyte (Europe); widely scattered, mainly in the S.
Former hortal; 0/1; extinct.
1st Cathcart, 1865, RH.

The only other mention of this species is the 1876 (FFWS) repeat of the first record.

Fragaria ananassa (Duchesne) Duchesne
Garden Strawberry

Britain: neophyte (cultivated from American parents); frequent in the S, and in lowland Scotland.
Hortal; 9/11; rare.
1st Whitecraigs Railway Station, 1926, RM.

There were four records made during CFOG surveying and a further five from the modern period, all from waste places near to housing. Four of the records are from the central north coast: Woodhall (NS3473, 1997), Whitecroft (NS3274, 1997), Langbank (NS3973, 1998) and Kelburn (NS3474, 1998); it was also recorded by the White Cart Water in central Paisley (NS4863, 2000). BWR recorded it from Glentyan House (NS3963) in 1961.

Geum rivale L.
Water Avens

Britain: common in the N.
Native; 161/191; common.
1st Gourock, 1860, GL.

A fairly common plant of the rural lowlands, but one with few urban localities and which is absent from the peaty uplands. It is most frequently seen in wet woodlands, often along watercourses, but can also be found in open fens or marshy vegetation, where not heavily grazed.

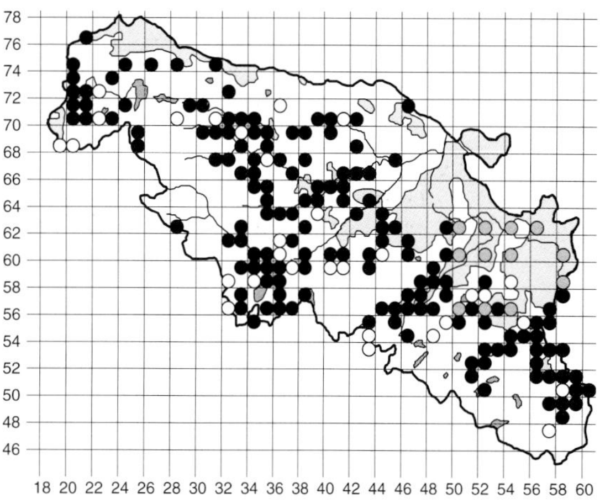

Geum rivale

Geum × *intermedium* Ehrh.
(*G. rivale* × *G. urbanum*)
Hybrid Avens

Britain: widespread, frequent in N England and C Scotland.
Native; 15/24; scarce.
1st Gourock, 1865, RH.

There are several old records scattered throughout the VC, and CFOG includes six records in the south of Glasgow. There are modern records from further south, mainly along the White Cart Water and also at Eaglesham (NS5651, 2006). Other records are from Merchiston (NS4164, 1999), River Calder (NS3459, 2003), Aird Meadow (NS3658, 2010, MG) and Devol Glen (NS3174, 2005). The two parents frequently overlap so other occurrences are very likely.

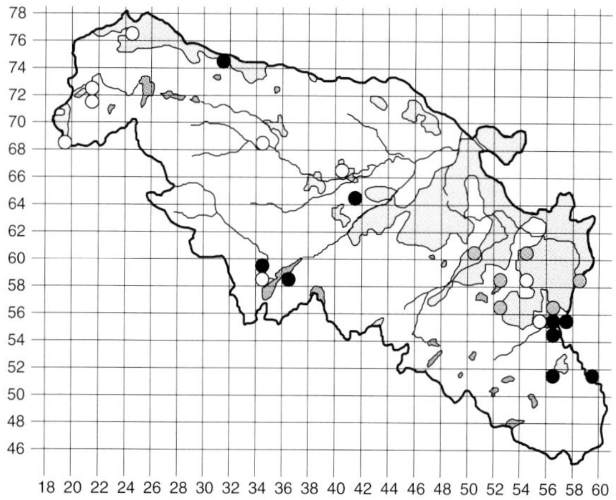

Geum × *intermedium*

Geum urbanum L.
Wood Avens

Britain: common throughout except in N Scotland.
Native; 230/249; common.
1st Paisley Canal Bank, 1865, RH.

Wood Avens is a common plant of all types of woodland, most usually seen where it is less shaded, such as at woodland edges. It is also found in scrub, along hedgerows and on open waste ground. It tends to be one of the earlier woodland colonizers of developing urban scrub woodland. Its distribution shows a strong lowland and urban-fringe pattern, being scarce on improved agricultural land and absent from the acidic uplands.

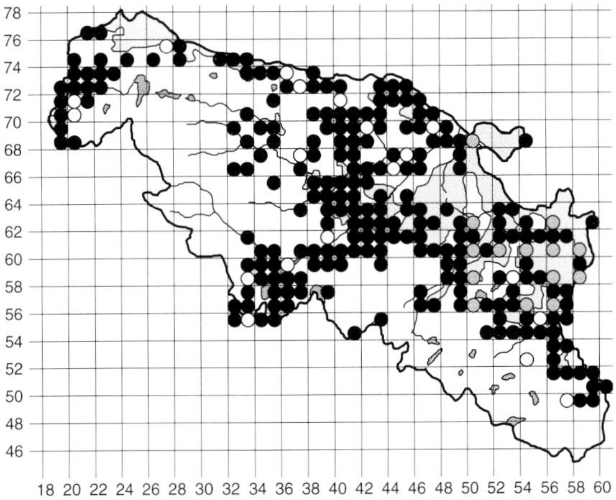

Geum urbanum

Agrimonia eupatoria L.
Agrimony

Britain: common in S to E Scotland, scattered in W Scotland.
Native and/or hortal; 1/4; very rare.
1st Renfrewshire, 1869, PPI.

Balfour recorded Agrimony along the White Cart Water (mid 19th century, CFOG), Hennedy (1891) thought it frequent, and it was listed from Gourock (1876, FFWS). It was recorded from Loch Libo in 1965 (NS4355, BWR) but not since (though note the following record). The only modern record is from the A737 roadside at Linwood (NS4363, 2005), where it likely originated from a wildflower seed mix.

Agrimonia procera Wallr.
Fragrant Agrimony

Britain: widespread and locally frequent, mainly in the W in Scotland.
Native; 0/1; extinct.
1st Loch Libo, 1956, RM.

RM was familiar with this species and recorded it in his notebook, but this appears to be the sole record for the VC.

Aremonia agrimonioides (L.) DC.
Bastard Agrimony

Britain: neophyte (S Europe); very rare, scattered in Scotland.
Former accidental; 0/1; extinct casual.
1st Castle Semple, 1965, J Lennie.

The sole record is from the edge of Castle Semple Loch.

Sanguisorba officinalis L.
Great Burnet

Britain: common in N and C England, rare alien in C Scotland.
Accidental; 0/1; extinct.
1st Paisley, 1858, PSY.

The herbarium record was highlighted in 1942 (TPNS) but there are no other reports.

Sanguisorba canadensis L.
White Burnet

Britain: neophyte (E N America); scattered in C and SW Scotland.
Hortal; 1/2; very rare.
1st Castle Semple Loch, 1965, J Lennie.

The original record was from a thicket on marshy ground to the north of Lochside Station (Lennie 1967); it is thought that this plant was seen there previously (1950), but in a non-flowering state, by Alan Crundwell. The sole modern record is from a roadside verge at Cloch (NS2075, 2010, PRG).

Poterium sanguisorba L. ssp. *sanguisorba*
Sanguisorba minor Scop.
Salad Burnet

Britain: frequent in the SE, rare in the W and Scotland.
Accidental; 0/1; doubtful record.
1st Blackbyres, 1963, JDM.

Salad Burnet was not reported by 19th-century authors and Lee considered it very rare in the Clyde area. On examination a 1977 specimen (GLAM) of Salad Burnet, from a railway cutting in Barshaw (*c.* NS5064), turned out to be Fodder Burnet. There is no specimen of the original record, which was reported by BWR (Ribbons 1964), but this may also have been Fodder Burnet, a taxa not well known at that time.

Poterium sanguisorba L. ssp. *balearicum* (Bourg. ex Nyman) Stace
Sanguisorba minor Scop. ssp. *muricata* (Gremli) Briq.
Fodder Burnet

Britain: neophyte (S Europe); scattered in the S, local in C Scotland.
Hortal; 4/5; rare casual of seed mixes.
1st Barshaw, 1977, GLAM.

The first modern record is from waste ground by sown roadside grassland at Kelburn (NS3474, 1998). Other records are from waste ground near the Calder, Lochwinnoch (NS3458, 2001), grassland – some sown with wildflowers – at Millarston (NS4563, 2002) and the sown meadow at GMRC, Nitshill (NS5160, 2008).

Acaena ovalifolia Ruiz & Pav.
Two-spined Acaena

Britain: neophyte (S America); scattered, mainly in the W.
Hortal; 2/2; very rare.
1st Ardgowan Point, 2003, IPG.

The first record is from the coast at Ardgowan (NS1972); this may be the same location as the 2007 record ('between Lunderston and Ardgowan') made by AR. More recently Two-spined Acaena was recorded from Crowhill Wood (NS2072, 2010, PRG).

Alchemilla conjuncta Bab.
Silver Lady's-mantle

Britain: neophyte (Alps); scattered, mainly in the N.
Hortal; 4/7; rare.
1st Pollok, 1952, RM.

Silver Lady's-mantle was also found at Pilmuir (NS5154, 1975) and Cunston (NS3372, 1982) by BWR. It was noted from Cathcart Castle Golf Course (1987, CFOG), and other modern localities where this garden outcast has been found are Gleniffer Braes (NS4560, 1995), Jordanhill (NS5468, AMcGS, 2002) and Dykebar Hospital (NS4961, 2004).

Alchemilla xanthochlora Rothm.
Pale Lady's-mantle

Britain: common in the N, rare in S England and N Scotland.
Native; 34/70; occasional, probably overlooked.
1st Williamwood, 1962, RM.

There are a lot of records from the 1960s–80s surveys (BWR), but none from earlier so it is difficult to assess this species' status. It is still occasionally found at grassland and waste-ground sites but may have undergone a decline in recent years, or been overlooked.

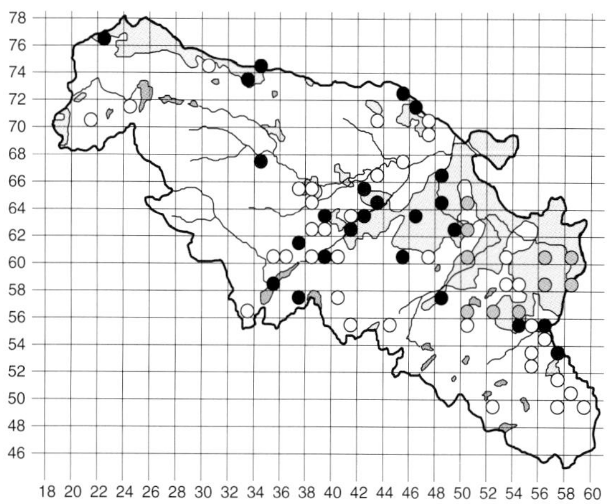

Alchemilla xanthochlora

Alchemilla filicaulis Buser ssp. *vestita* (Buser) M.E. Bradshaw
Hairy Lady's-mantle

Britain: widespread, mainly in the N.
Native; 42/55; frequent in hilly pastures.
1st Castle Semple Loch, 1887, NHSG.

This hairy-leaved lady's-mantle is an indicator of old grasslands and most of the records are from hill country pastures such as along the Gleniffer Braes (NS4460 to NS4960). It is likely to have declined although some old records may be listed as '*A. vulgaris* agg.'.

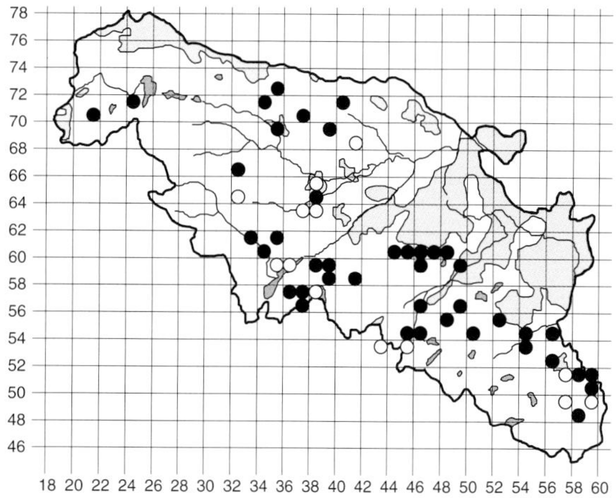

Alchemilla filicaulis ssp. *vestita*

Alchemilla glabra Neygenf.
Smooth Lady's-mantle

Britain: common in the N, very rare in the S.
Native; 225/312; common.
1st Cartside, 1854, GL.

This is the commonest lady's-mantle found in a range of grasslands; it tolerates some improvement and its presence often reflects a little extra moisture, but not

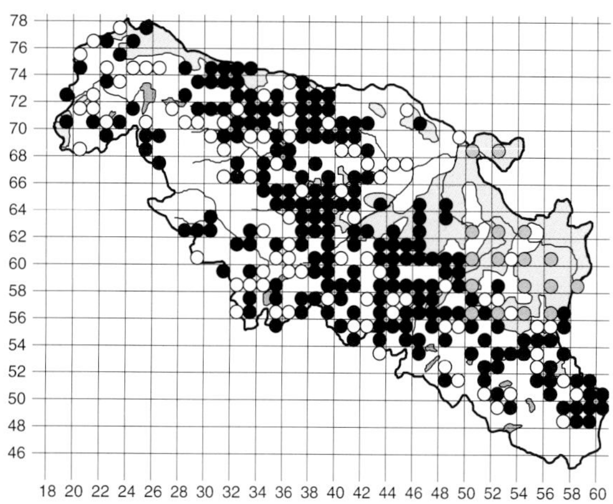

Alchemilla glabra

marshy conditions. It is also found on waste ground but is absent from intensively managed farmland or the peaty uplands.

Alchemilla mollis (Buser) Rothm.
Soft Lady's-mantle

Britain: neophyte (Carpathians); scattered throughout.
Hortal; 46/46; occasional, spreading.
1st Rosshall Farm (NS56B), 1983, CFOG.

This is a recent arrival in the area as it was not noted during the BWR surveys (1960s–80s), nor by RM. Today it is widespread and often found on waste ground or roadsides, not always near gardens. It appears well capable of local spread.

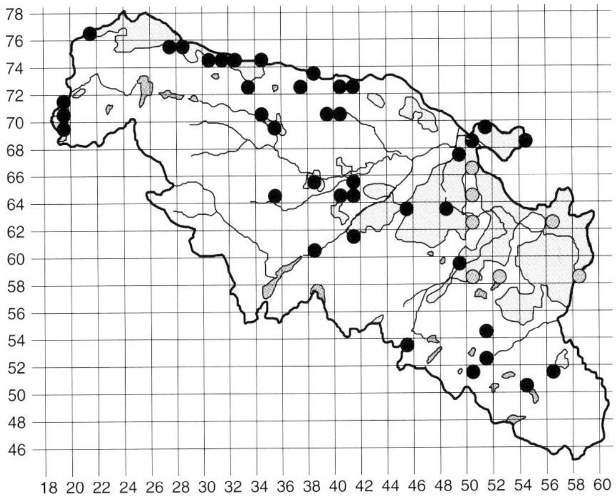

Alchemilla mollis

Aphanes arvensis agg.
Parsley-pierts

Britain: widespread except in NW Scotland.
Native; 30/41; occasional.
1st Langside, 1865, RH.

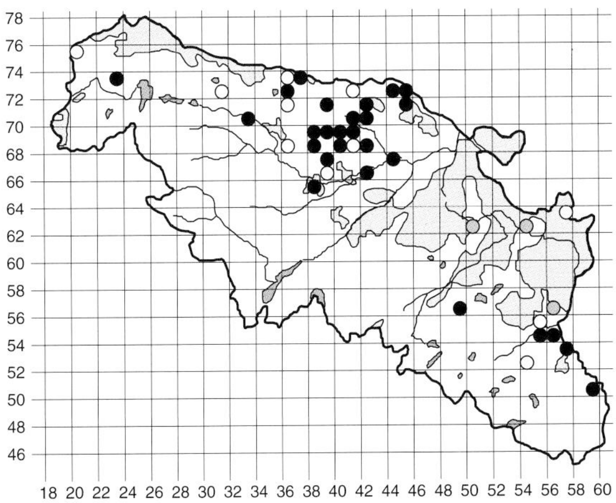

Aphanes arvensis agg.

There are several widespread records for Parsley-pierts, although for a few species-level identification is unknown. Habitats tend to be open disturbed ground along waysides and shallow soils on rocky embankments; sandy soils on golf courses were noted in CFOG.

Aphanes arvensis L.
Parsley-piert

Britain: widespread, scarce in N and W Scotland.
Native; 9/10; rare.
1st Finlaystone, 1976, AJS.

This species appears to have stronghold about the Formakin area (NS4170).

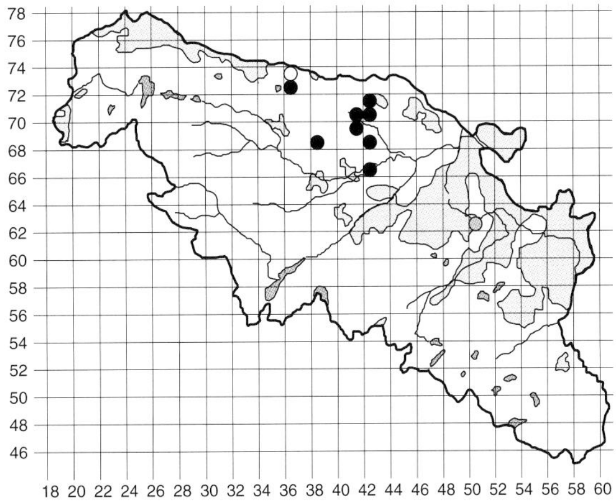

Aphanes arvensis

Aphanes australis Rydb.
A. inexspectata W. Lippert
Slender Parsley-piert

Britain: widespread but scarce locally.
Native; 17/21; scarce.
1st Pollokshields, 1872, GLAM.

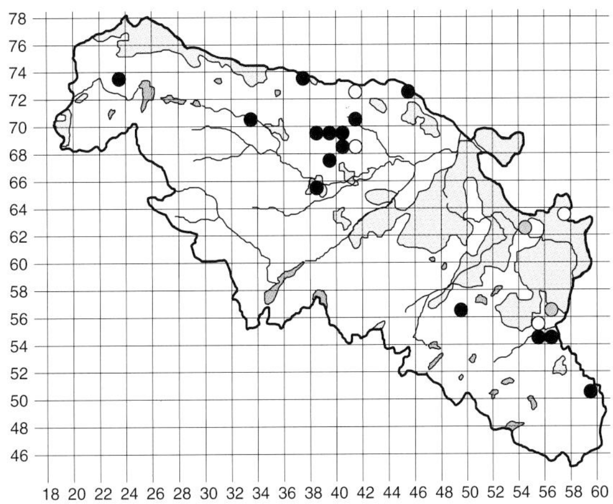

Aphanes australis

This species was also collected from Inverkip Road (1887, GRK). It appears to be more widespread than Parsley-piert and is often found on shallow soils of rocky embankments.

Rosa multiflora Thunb.
Many-flowered Rose

Britain: neophyte (E Asia); scarce and scattered.
Hortal; 5/5; rare.
1st Brock Burn, Kennishead, 1991.

The original record is from spreading plants at a planted shrubbery (NS5359) but the others are all from estates in central Renfrewshire, where Many-flowered Rose can grow rampantly, as at Formakin (NS4070, 1994), Corsliehill House (NS3969, 1998), Barochan Estate (NS4165, 1999) and Netherwood (NS3469, 1994).

Rosa arvensis Huds.
Field Rose

Britain: common in the S, local alien in S Scotland.
Hortal; 0/4; extinct or overlooked.
1st Gourock, 1865, RH.

The original record was repeated by Lee (1933) and a specimen was also collected from Gourock in 1880 (GL); there is also a record from Levan House, Ashton (1888, NHSG) and a specimen from Cloch Road (1892, GRK). More recently it was marked on a card from Jeffreystock (NS3357, 1980, BWR) but there is no other information about this find.

Rosa rugosa Thunb.
Japanese Rose

Britain: neophyte (E Asia); scattered throughout, frequent in C Scotland.
Hortal; 36/38; becoming frequent in lowlands.
1st South of Inverkip (NS1971), 1964, BWR.

Japanese Rose has been commonly used in recent

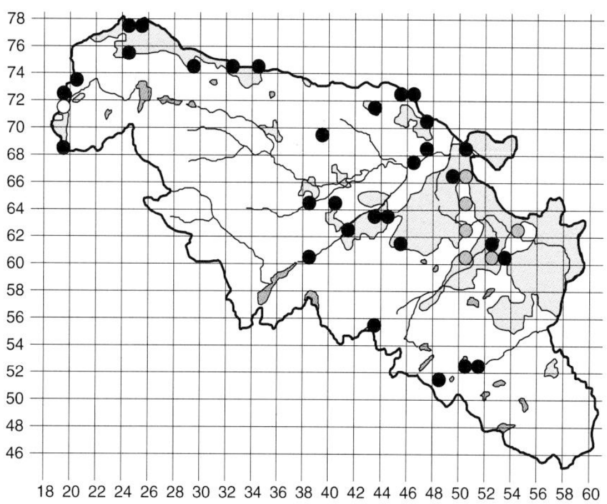

Rosa rugosa

planting schemes. It appears to be on the increase, reflecting its persistence but also spread by seed or as an outcast. Most of the records are from waste ground or disturbed roadsides, but a few may include planted specimens.

Rosa 'Hollandica'
Dutch Rose

Britain: neophyte (hybrid origin); scattered but local, frequent in C Scotland.
Hortal; 7/7; rare.
1st Kirkton Burn (NS4957), 1997.

Perhaps sometimes overlooked for the previous species, Dutch Rose is also used in similar planting schemes. The other records are from Cowdon Hall (NS4757, 1997), Woodhall (NS3473, 1997), Paisley (NS4864, 2000), Craigends Estate (NS4166, 2000), Wemyss Bay (NS1969, 2001) and Ashton (NS2276, 2001).

Rosa canina agg.
Dog-roses

There are 334 records for this aggregate, which includes Dog-rose, Hairy Dog-rose, Glaucous Dog-rose and probably some older Downy-rose records and, of course, hybrids. The map shows a very widespread distribution, but Dog-roses are absent from the uplands and some urban areas.

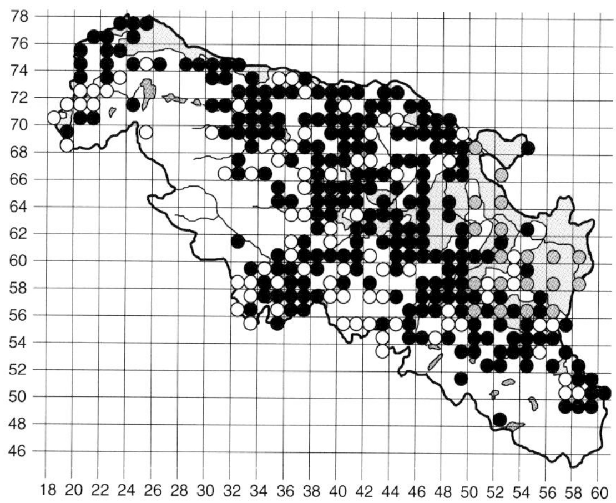

Rosa canina agg.

Rosa canina L.
Dog-rose

Britain: common in the S (except E England) and lowland Scotland.
Native; 40/41; frequent.
1st Roman Bridge, Inverkip, 1880, GL.

Records for true Dog-rose have only been gathered since the CFOG recording period, but even these were grouped in the publication. The map shows only modern records and reflects a lowland rural pattern, with Dog-rose apparently quite scarce in the west. The first record was named as 'var. *vertillicantha*'.

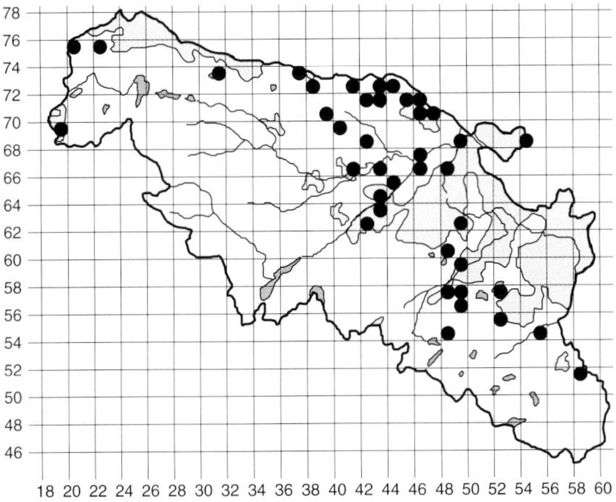

Rosa canina

Rosa caesia Sm. ssp. *caesia*
Hairy Dog-rose

Britain: local, mainly in the NE (under-recorded).
Native; 101/102; frequent.
1st Finlaystone, 1976, AJS.

Records for this dog-rose are widespread and show a lowland rural or urban-fringe pattern; Hairy Dog-rose is absent from more acidic and grazed uplands. Some sparsely hairy crosses of Glaucous Dog-rose may have been recorded as this subspecies.

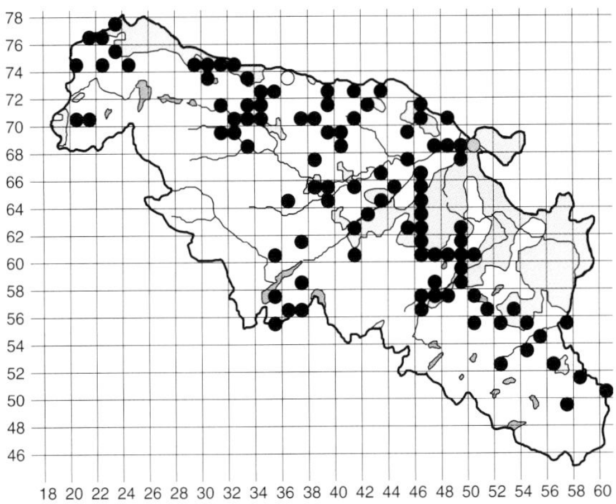

Rosa caesia ssp. *caesia*

Rosa caesia ssp. *vosagiaca* (N.H.F. Desp.) D.H. Kent
R. caesia ssp. *glauca* (Nyman) G.G. Graham & Primavesi nom. inval.
Glaucous Dog-rose

Britain: widespread, but local in the N and absent from SW and SE England.
Native; 151/153; common.
1st Finlaystone, 1976, AJS.

This is the commonest rose found in the survey area. It shows a strong lowland rural pattern and avoids the more acidic soils of the uplands.

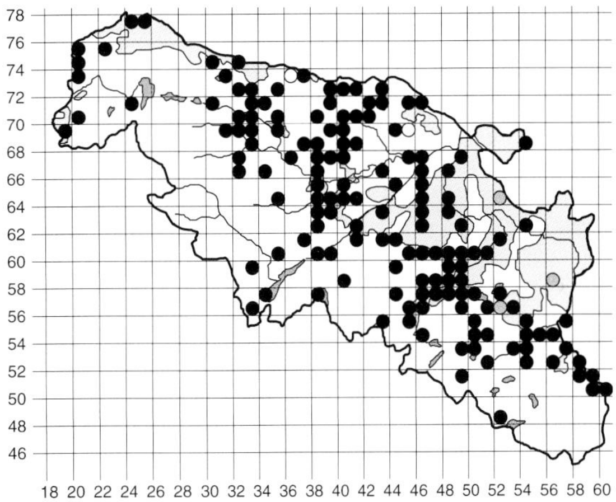

Rosa caesia ssp. *vosagiaca*

Rosa tomentosa group
Downy-roses

Native; 89/97; frequent.
1st Renfrewshire, 1834, MC.

The map shows aggregate records for the following two downy-roses of this group, and may include some hybrid records.

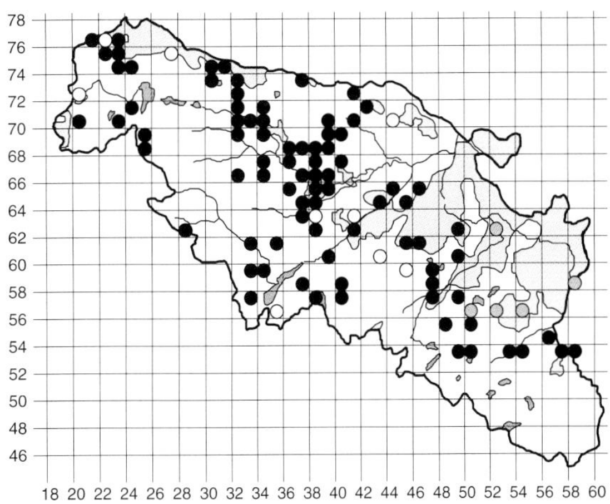

Rosa tomentosa group

Rosa sherardii Davies
Sherard's Downy-rose

Britain: frequent in the N and W.
Native; 76/79; frequent.
1st Cloch, 1892, GRK.

This is the commonest and most widespread of the downy-roses. It is found on farmland where less intensively managed, hedgerows, railway banks, roadsides and sometimes on waste ground.

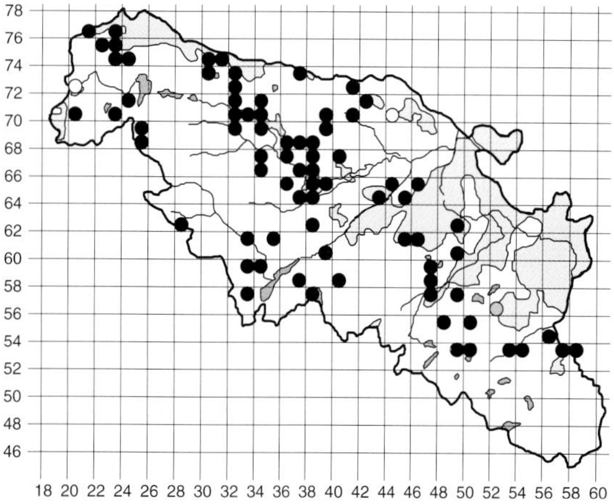

Rosa sherardii

Rosa mollis Sm.
Soft Downy-rose

Britain: local, mainly in the NE (under-recorded).
Native; 13/13; scarce.
1st Bridge of Weir, 1901, BM.

This downy-rose seems to be much rarer, with nearly all records made in the central area, notably at farmland to the north of Houston.

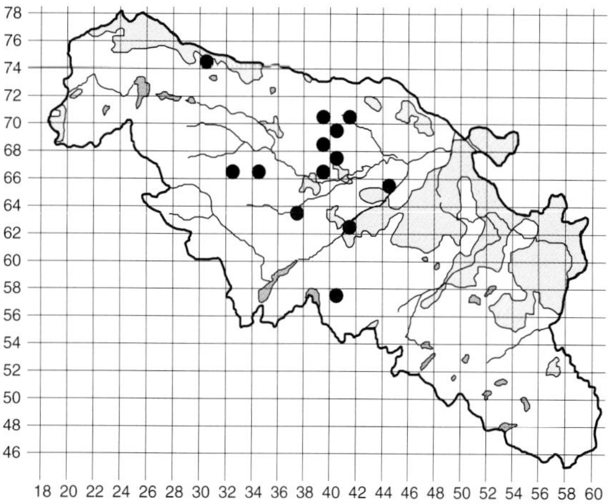

Rosa mollis

Rosa rubiginosa L.
Sweet-briar

Britain: widespread but only frequent in S England and E Scotland.
Former native or hortal; 0/1; extinct.
1st Renfrewshire, 1883, TB.

Ferguson (1915) and Lee (1933) list this rose but there is no named locality, and its native status has been questioned.

Hybrid Roses are under-recorded; tentative records include:
Rosa caesia ssp. *vosagiaca* × *R. sherardii*: Corsliehill Road (NS3969, 1996) and Humbie Burn (NS5454, 1997).
Rosa caesia ssp. *caesia* × *R. canina*: Lochwinnoch station (NS3557, 1993).

ELAEAGNACEAE

Hippophae rhamnoides L.
Sea-buckthorn

Britain: local on the E coast, widely scattered as an alien.
Hortal; 6/8; rare.
1st Renfrewshire, 1899, GC.

Sea-buckthorn was recorded at Wemyss Point in 1986 (NS1870, BWR) but has not been noted there since; it has recently been recorded at nearby Inverkip (NS2071, 1999) and from Ardgowan Point (NS1972, 1998 and NS2073, 2003, IPG). It is also found along the coast at Langbank (NS3773, 1998 and NS3873, 2008) and Battery Park (NS2577, 2012); it has presumably been introduced at all sites but has subsequently spread. It is also planted in amenity shrub beds elsewhere.

ULMACEAE

Ulmus glabra Huds.
Wych Elm

Britain: common throughout except in upland and N Scotland.
Native; 266/309; very common in lowlands.
1st Pollok Estate, 1812, AANS.

John Boyd (1908) considered the Pollok Wych Elms to be planted and noted that one of the trees had a girth of more than 10 feet. There are few other old records for this elm, which Hennedy (1891) described as very common. Wych Elm is, or perhaps more realistically was, a common tree usually preferring less acidic or richer soils as along river valleys and flood plains, but it is also common as a weedy tree in urban areas. Many of the mature trees have died over the last 30 years, but numerous seedlings and saplings still occur; however, these soon appear to succumb to Dutch Elm disease, so future seedling recruitment may become a rare event.

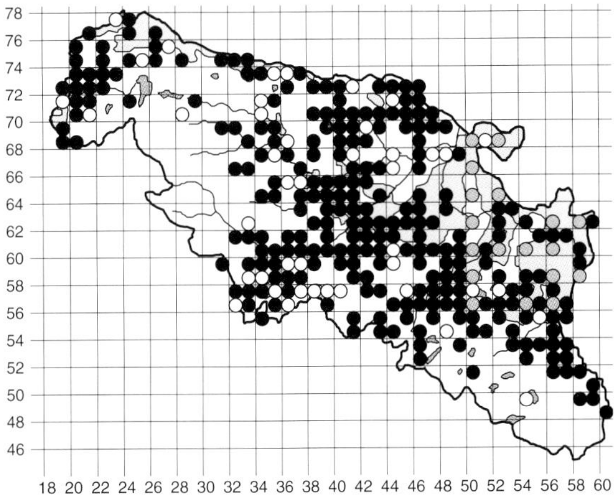

Ulmus glabra

Ulmus spp.
Elms

Several elms are planted in old estates but they have been seldom recorded and not as seeding or suckering. *Ulmus minor* Mill. (Smooth-leaved Elm) has been recorded at Ardgowan (NS2073, 1998) and Cowdon (NS4657, 2000) estates and *Ulmus procera* Salisb. (English Elm) has a few old records including Bridge of Weir, 1921 (GL) and was listed in 1883 (TB) and 1915 (TPNS).

CANNABACEAE

Cannabis sativa L.
Hemp

Britain: neophyte (SW Asia); scattered, mainly in the S.
Former accidental; 0/2; extinct casual.
1st Wayside, Mearns, 1890, AANS.

Hemp is a bird-seed casual with no modern records from the VC. The original record refers to a barley field where it was a contaminant of 'Baltic seed'. The other old record is from Paisley 1928 (TPNS) with a specimen from 'Greenock Rd, Paisley' from the same year (PSY).

Humulus lupulus L.
Hop

Britain: common in the S, frequent alien in the N.
Hortal; 6/9; rare.
1st Renfrewshire, 1872, TB.

Hop was recorded from the Levern Water, Househill House (1890, AANS), waste ground, Giffnock (1925, RM) and more recently from near Gryfeside (NS3370, 1980, BWR) and from the Formakin area (NS4170, 1987, BSBI). It can still be found at these last two sites; at Gryfeside it is well established and scrambles over the roadside hedgerow. A further record is from Paisley (NS4861, 1994, CD) and there were three Glasgow records made during CFOG surveys (NS56K, NS56L and NS56F).

URTICACEAE

Urtica dioica L.
Common Nettle

Britain: common throughout.
Native; 439/498; very common.
1st Renfrewshire, 1869, PPI.

Common Nettle is found just about everywhere and only absent from the wetter swamps or mires and acidic peaty uplands. Generally reflecting soil enrichment, it occurs in pastures about old ruins and by upland streams, and forms larger stands along alluvial fringes of lowland watercourses; it can tolerate shade and can be abundant in scrubby woodlands on rich soils, as at Georgetown (*c.* NS4468, 2005).

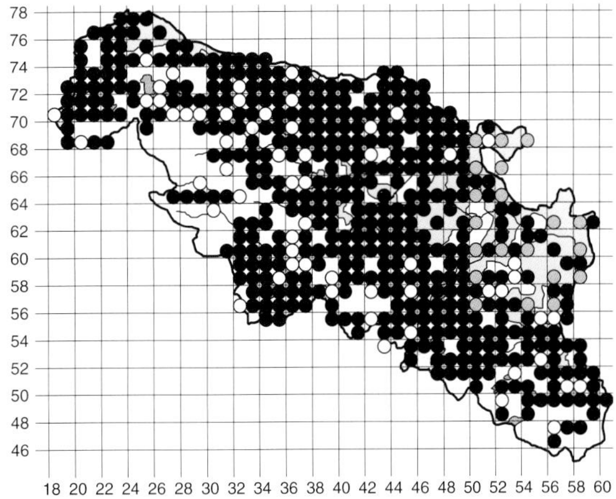

Urtica dioica

Urtica urens L.
Small Nettle

Britain: archaeophyte; common in the S, frequent in C and E Scotland.
Accidental; 1/3; very rare or overlooked.
1st Cathcart, 1865, RH.

There are only a few old records for this annual species, which Hennedy (1891) considered common. It was recorded from Barochan Estate (NS4165, 1979, BWR) and more recently from Barshaw Park (NS56C, 1985, CFOG). It is probably overlooked but even so it is likely to be a rare plant.

Parietaria judaica L.
Pellitory-of-the-wall

Britain: frequent in the S, local in E Scotland.
Former hortal; 0/2; extinct.
1st Stanely Castle, 1845, NSA.

There is only one other old record for this plant of walls and old buildings, from Roman Bridge in 1880 (NHSG). It was collected from Stanely in 1896 (PSY), a record which Lee repeated in 1933. It is now long gone from the area.

FAGACEAE

Fagus sylvatica L.
Beech

Britain: common in the S and as alien in lowland Scotland.
Hortal; 296/343; very common.
1st Renfrew, 1777, JW.

A common and widespread tree found in woodlands (plantations or semi-natural), hedges, shelterbelts, parks and gardens. It seeds freely and is naturalized in many woodlands. Although it grows well at some higher altitude shelterbelts and boundary dykes, it avoids upland grazed or peaty areas.

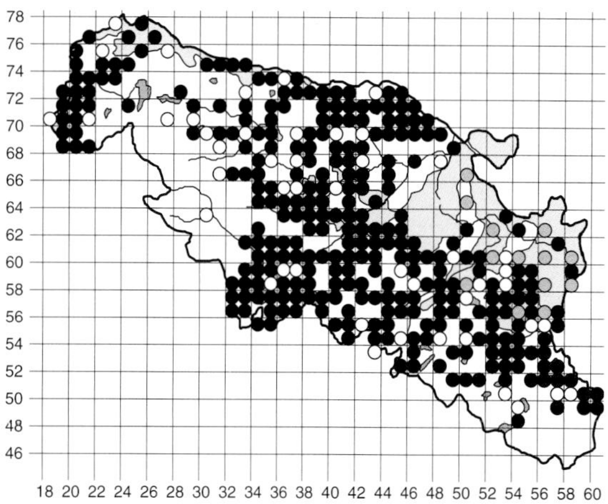

Fagus sylvatica

Castanea sativa Mill.
Sweet Chestnut

Britain: archaeophyte; common in the S, frequent in lowland Scotland.
Hortal; 14/21; occasionally planted.
1st Renfrew, 1777, JW.

Sweet Chestnut can be found at a number of old estate woodlands, often as large mature trees, mainly in the central area. There is no evidence of seedling establishment but trees can persist by suckering.

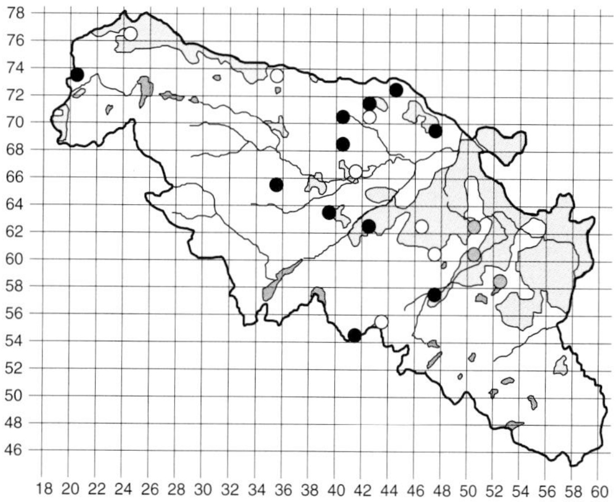

Castanea sativa

Quercus spp.
Oaks

A few exotic oaks can be found as planted trees in old estates, e.g. *Quercus ilex* L. (Evergreen Oak) grows in Ardgowan Estate (NS2073 1998) and *Quercus rubra* L. (Red Oak) was planted in Pollok Estate (1905, GL) and at Ardgowan Estate; there are no records of seedlings.

Quercus cerris L.
Turkey Oak

Britain: neophyte (S Europe); common in the S, frequent in lowland Scotland.
Hortal; 13/18; occasionally planted.
1st Paisley Cemetery, 1900, PSY.

Turkey Oak was also collected from Erskine Policies in 1915 (GL) but there are few old records for it as a planted tree; JDM noted it from Pollok Estate (NS5562, 1961) and a quarry near Finlaystone (NS3573, 1961). Modern records include Johnstone Wood (NS4262, 1993), Linwood Industrial Estate (NS4465, 1995) and Bow Hill (NS2676, 2001) and young saplings can be found in old pasture to the edge of Waulkmill Glen (NS5258, 2003), at Linwood Moss (NS4465, 1998), on a railway embankment,

Lochwinnoch (NS3558, IPG) and on coastal grassland Erskine (NS4572, 2008).

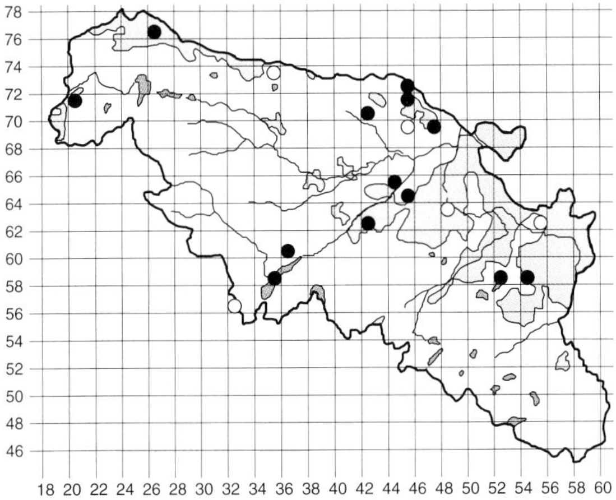

Quercus cerris

Quercus petraea (Matt.) Liebl.
Sessile Oak

Britain: throughout, but more frequent in the N and W.
Native and hortal; 30/39; occasional, scattered.
1st Renfrewshire, 1872, TB.

Considered the typical native oak in western Scotland, today this oak is by far the rarest and mainly found in estate woodlands rather than in more 'natural' old woodlands (Shielhill Glen being an exception – NS2471, 2000). Some of the records may even refer to hybrids (*Q.* × *rosacea* Bechst.), which is known from 12 1km squares and was first recorded from Dampton (1984, BWR); modern hybrid records include Tandiebrae (NS3565, 3862), Bow Hill (NS2676, 2001), Greenock Road (NS4172, 2001) and Brueacre Burn (NS2070, 2002).

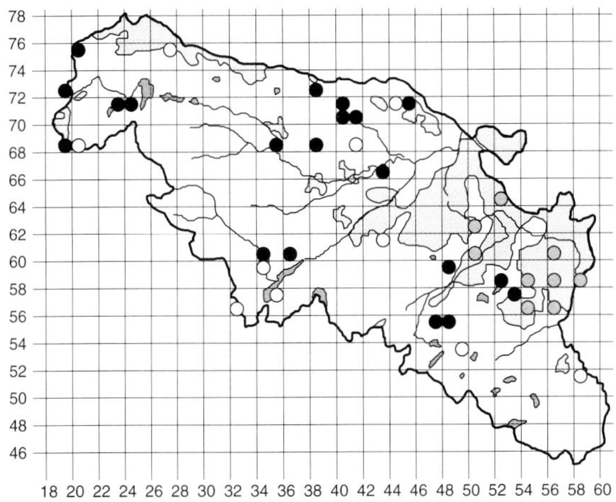

Quercus petraea

Quercus robur L.
Pedunculate Oak

Britain: common throughout except in N Scotland.
Native and hortal; 210/246; common in lowlands.
1st Renfrew, 1777, JW.

The commonest of the oaks, Pedunculate Oak is found in most local woodlands – plantation, estate and semi-natural. Many records are for mature trees but seedlings are very common. It generally avoids the wetter upland areas, as is shown markedly by the map. It is usually found on the upper slopes of valley woodlands, avoiding the richer soils below. Some records may include hybrid or the occasional Sessile Oak, but closer scrutiny of trees nearly always indicates Pedunculate Oak.

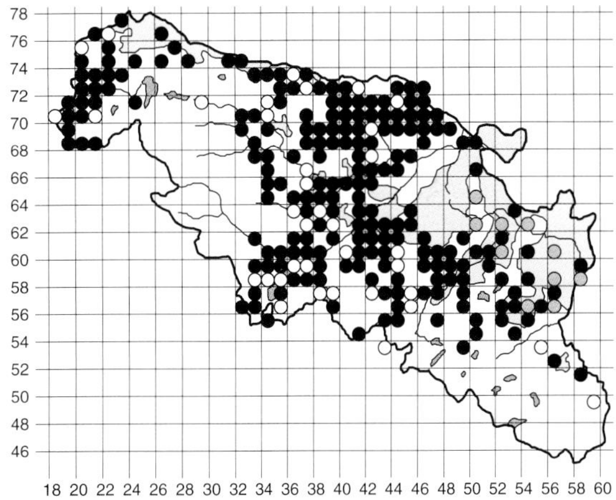

Quercus robur

MYRICACEAE

Myrica gale L.
Bog-myrtle

Britain: common in the N and W, rare and local elsewhere.
Native; 3/7; very rare.
1st Gourock, 1865, RH.

Despite extensive areas of apparently suitable acidic mires, Bog-myrtle is surprisingly absent from most of the VC, presumably due to past land management (drainage and grazing). Other old records include Loch Thom (1888, GRK), Kilmacolm (1896, NHSG), where is was noted as 'plentiful', and the banks of the Gryfe Water and marshes above Knockbuckle (1898, JMY). Today it is known from Gall Moss, Darndaff (NS2672), where it was first recorded in 1984 (BWR) occurring in a large population; unfortunately this site was recently severely, and unnecessarily, damaged by water pipeline work but some plants survived. A small population also occurs to the west side of Daff Reservoir (NS2270, 2005). At a very unexpected locality for this bog plant, there is a single plant on a roadside embankment near Inverkip (NS1971, 2000).

JUGLANDACEAE

Juglans regia L.
Walnut

Britain: neophyte (SW Asia); frequent in S and E England.
Hortal; 3/5; very rare planted tree.
1st Inchinnan and Erskine, 1845, NSA.

There are three modern records for planted trees: Whinnerston (NS3864, 1997), Boden Boo (NS4871, 1991, CB) and Pollok Park (NS5461, 2009, RW), but no seedling records. Edible fruits were reported from a tree in the district in 1926 (TPNS).

BETULACEAE

Betula pendula Roth
Silver Birch

Britain: common throughout.
Native and hortal; 138/156; common, lowland.
1st Renfrewshire, 1869, PPI.

A common tree, planted or self-sown, of most woodland types, Silver Birch shows a very strong lowland pattern; it is common on waste ground, but rare or absent on wetter and peaty soils in the uplands. Some records may refer to the hybrid with Downy Birch (*B.* × *aurata* Borkh.). The hybrid is probably more widespread but there are only two records: an old tree by the disused railway at Lilybank (NS3174, 2003) and a young tree by Daff Reservoir (NS2270, 2005).

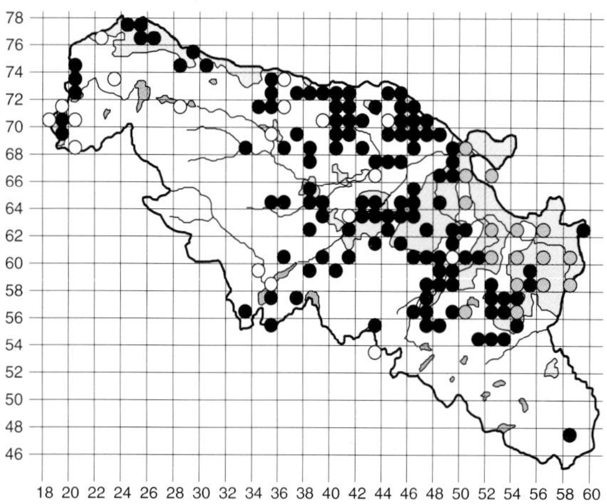

Betula pendula

Betula pubescens Ehrh.
Downy Birch

Britain: common throughout.
Native and hortal; 235/276; common.
1st Renfrewshire, 1872, TB.

The commonest of the birches, Downy Birch is found throughout the VC, usually preferring wetter soils, and is likely to be native at many sites. It occurs in various woodland types, including wet boggy areas to the margins of mires, but can also be found at disturbed waste-ground sites; a few records may refer to hybrids.

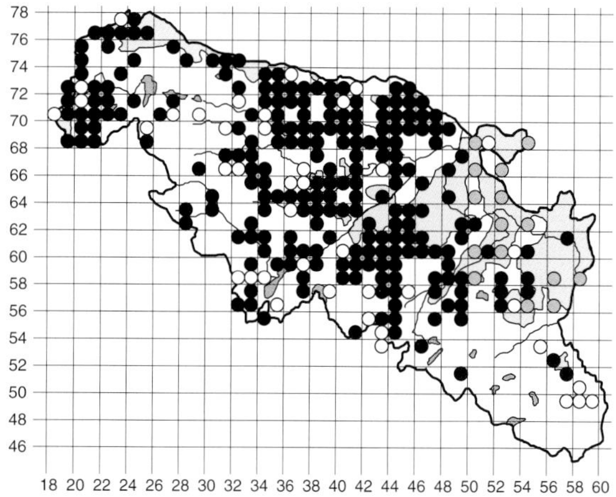

Betula pubescens

Alnus glutinosa (L.) Gaertn.
Alder

Britain: common throughout.
Native; 202/229; common.
1st Skiff Park, 1793, JW.

Alder is a common tree of wet places, especially along watercourses and the margins of wetlands, and has also been widely planted. It becomes rare in the peaty

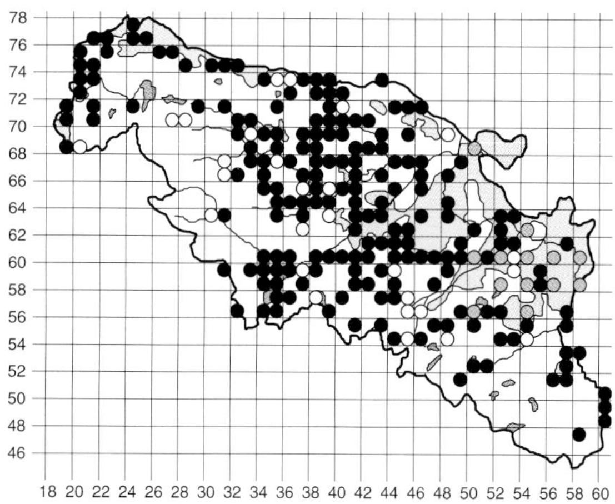

Alnus glutinosa

soils of the uplands. There are few large, wet carr woodland areas, as most stands are small or narrowly restricted to burn sides; a stand of note occurs at the small valley near Everton (NS2170 and NS2171, 2004).

Alnus incana (L.) Moench
Grey Alder

Britain: neophyte (Europe); scattered throughout, but local.
Hortal; 12/12; scarce, probably under-recorded.
1st Loch Libo, 1996.

This widely planted alder has only been recorded a few times during the survey period. It readily suckers but spread by seedlings has not been noted confidently. It is well established on the north side of Loch Libo (NS4355, 1996).

Carpinus betulus L.
Hornbeam

Britain: common in the SE, frequent in lowland Scotland (alien).
Hortal; 7/13; rare.
1st Renfrewshire, 1834, MC.

This southern species does not appear to have been widely planted and there are only a few records across the VC. Modern records include Ardgowan (NS2073, 1998), Barochan (NS4069, 1997), Glenlora (NS3358, 2006), Castle Semple (NS3660, 1990, CMRP) and a massive, sprawling tree at Dargavel House (NS4369, 2005). There were two CFOG records: from Househillwood Park, Hurlet (NS56A, 1986) and Waulkmill Glen (NS5258, 1984). Although a few saplings have been noted about Glasgow, the VC76 records are all of trees.

Corylus avellana L.
Hazel

Britain: throughout.
Native; 168/180; fairly common, widespread.
1st McInroy's Point, 1854, HMcD.

A common native of all but the more acidic or wetter woodlands, Hazel occurs throughout the lowlands. It markedly avoids higher ground – because of a combination of grazing pressure, acidic soils and wet peat – but it can be found on the steep sides of the North Rotten Burn (NS2568) and Forking of Raith (NS2862). It is also widely planted in urban areas, although these have seldom been recorded. Fruiting specimens are seldom noted, but seedlings can often be seen.

CELASTRACEAE

Euonymus europaeus L.
Spindle

Britain: common in the S, rare alien in C Scotland.
Hortal; 2/3; very rare.
1st Thornliebank, 1890, GL.

IPG recorded Spindle at the edge of a car park at Formakin House (NS4170) in 2003, but thought it probably planted. In the same year an old shrub was found in a hedgerow on farmland near Blackstone (NS4666).

PARNASSIACEAE

Parnassia palustris L.
Grass-of-Parnassus

Britain: common in the N and W.
Native; 7/10; rare, western.
1st Paisley, 1845, NSA.

Other old records are from Wemyss Bay (1845, GL), Mearns Moor (1890, AANS) and Gourock (1857, GGO). There are seven modern locations, all in the extreme western hills, from four places: Earn Hill (NS2375, 1998 and NS2275, 2003; NS2276, 2006, PB), near Kelly Reservoir (NS2268, 2001), Everton (NS2170, 2004) and Shielhill Glen (NS2471, 2007; NS2472, 1995,

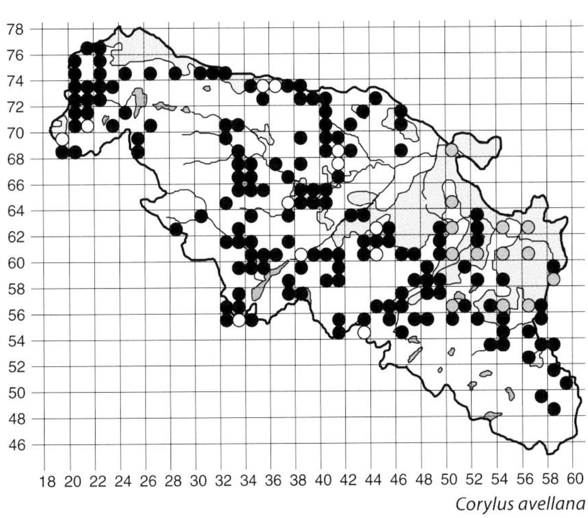

Corylus avellana

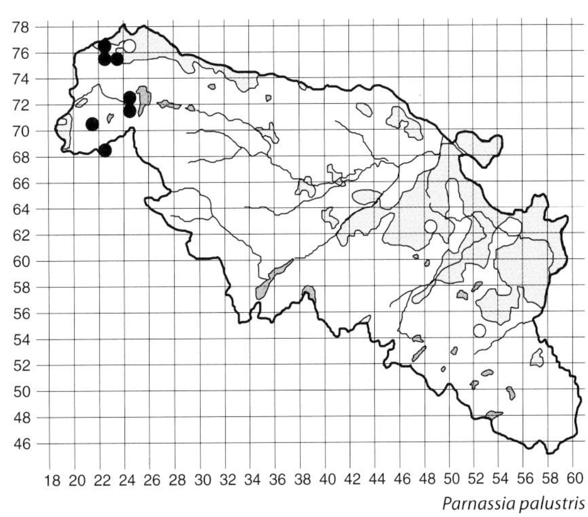

Parnassia palustris

AB). Wood (1893) knew of only 'one station' in East Renfrewshire but Hennedy (1891) described Grass-of-Parnassus as common in boggy places about Gourock; it appears that this attractive species has always been rare elsewhere, and its presence today in such small numbers at only a few sites makes its status vulnerable.

OXALIDACEAE

Oxalis corniculata L.
Procumbent Yellow-sorrel

Britain: neophyte (unknown); frequent in S England, rare in Scotland.
Hortal; 2/4; very rare.
1st Paisley, 1858, PSY.

This sorrel was also collected from an unnamed but local cemetery by T Fisher in 1888 (GRK). It was found as a weed at Greenbank (NS55T, 1984, CFOG) and more recently from Newlands South Church (NS56Q, 1998, CFOG).

Oxalis exilis A. Cunn.
Least Yellow-sorrel

Britain: neophyte (Australasia); scattered, mainly in the SE.
Hortal; 0/1; extinct.
1st Newlands, 1975, CFOG.

This garden weed is now thought extinct (PM pers. comm.).

Oxalis acetosella L.
Wood-sorrel

Britain: common throughout except in E England.
Native; 261/303; common in woodlands.
1st Wemyss Bay, 1845, GL.

There are many old records for Wood Sorrel and today it is still a fairly common plant, occurring throughout the VC and only becoming scarce in strongly urban or upland areas, or intensive agricultural land. It can extend into upland areas along the sides of burns and also occasionally about rock outcrops and boulders. It is well represented on the dry peaty soils under the mature pines on Barochan Moss (NS4268, 1994).

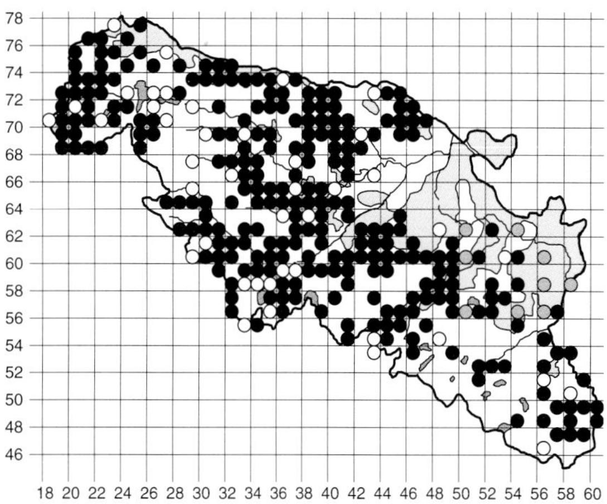

Oxalis acetosella

EUPHORBIACEAE

Mercurialis perennis L.
Dog's Mercury

Britain: common throughout except in N Scotland.
Native; 188/209; common.
1st Neilston, 1792, OSA.

Dog's Mercury is a fairly common plant of old woodlands, especially where the soils are less acidic, and occasionally of hedge banks. Its distribution shows a lowland pattern and it is absent from most relict upland burn-side woodlands. It is also absent from strongly urban areas.

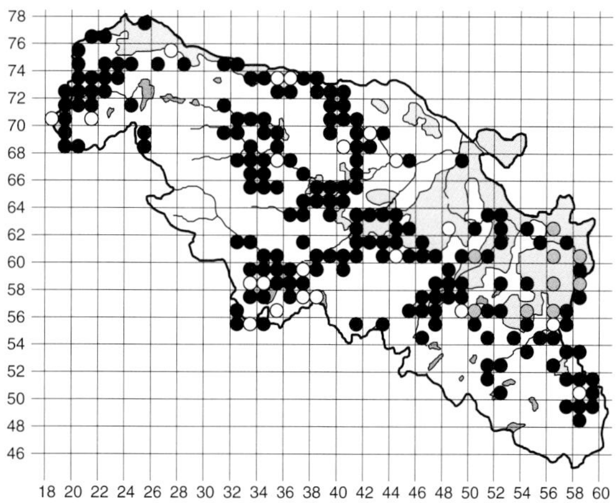

Mercurialis perennis

Mercurialis annua L.
Annual Mercury

Britain: common in SE England, scarce elsewhere.
Former accidental; 0/2; extinct.
1st Crossmyloof, 1894, GL.

This was the only record until an unusual find of it as a weed in a glasshouse near Bridge of Weir in 1966 by ERTC (Conacher 1966).

Euphorbia oblongata Griseb.
Balkan Spurge

Britain: neophyte (Balkans); rare in S England, very rare in Scotland.
Accidental; 1/1; very rare.
1st Newlands, 2004, PM.

This spontaneous weed was recently identified from PM's garden (NS5760).

Euphorbia helioscopia L.
Sun Spurge

Britain: archaeophyte; common in S and lowland Scotland.
Accidental; 12/19; scarce.
1st Renfrewshire, 1883, TB.

There are a few scattered records but only three from the 1970s–80s: Kirktonmoor (NS5551, 1977), Harelaw, Bridge of Weir (NS3863, 1984) and Formakin area (NS4070, 1987). There are a handful from CFOG and other modern records include waste ground, Clydeport (NS3174, 2001), Threeply (NS3766, 2002), Sandieston (NS3561, 2004) and Darnley (NS5259, 2006) – all these are from bare waste-ground soil – and one other record comes from a weedy field at East Fulwood (NS4567, 2007).

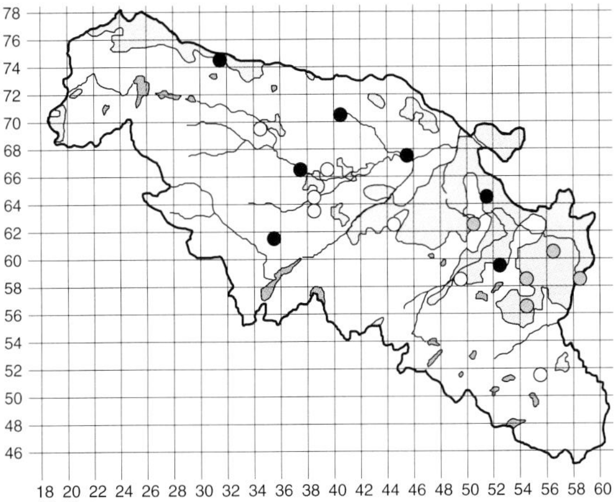

Euphorbia helioscopia

Euphorbia lathyris L.
Caper Spurge

Britain: archaeophyte; common in the SE, scarce in Scotland.
Hortal; 0/2; extinct.
1st Renfrewshire, 1869, PPI.

Caper Spurge was excluded from the Renfrewshire list (TPNS 1915) and the original source is unknown. A more recent record was made at a garden at Pollokshields in 1974 (NS56F, CFOG).

Euphorbia exigua L.
Dwarf Spurge

Britain: archaeophyte; common in the SE, rare in Scotland.
Former accidental; 0/3; extinct.
1st Cornfields, Strathbungo, 1838, GGO.

Dwarf Spurge was also collected from cornfields at Shawlands (1858, GGO) and there is a specimen from a garden in Glenburn, Paisley (1955, PSY).

Euphorbia peplus L.
Petty Spurge

Britain: archaeophyte; common in the S, extending to lowland Scotland.
Accidental; 6/10; rare but under-recorded.
1st Gourock, 1858, GGO.

There are other old records from Cathcart and Gourock (1876, FFWS), west of Paisley (1896, NHSG) and Auchentorlie, Paisley (1905, PSY) and it was collected from Bishopton in 1908 (GL). There were two records from CFOG, and more recently Petty Spurge has been recorded at four waste-ground sites: the India Works site, Inchinnan (NS4768, 1996), Linwood Moss tip (NS4465, 1998), Wemyss Bay (NS1971, 1999) and the strandline at Finlaystone Point (NS3574, 2008).

Euphorbia esula L.
Leafy Spurge

Britain: neophyte (S Europe); scarce in S England, rare in Scotland.
Former hortal; 0/1; extinct.
1st Renfrew, 1941, GL.

Euphorbia amygdaloides L.
Wood Spurge

Britain: native in S England, rare alien in Scotland.
Hortal; 1/1; very rare.
1st Barrhead, 2009, PRG.

This garden outcast (ssp. *robbiae* (Turril) Stace) was recently found under trees by the Levern riverbank (NS4959).

Euphorbia griffithii Hook. f.
Griffith's Spurge

Britain: neophyte (Himalayas); rare casual.
Hortal; 1/1; very rare garden outcast.
1st Ashton, 2001.

The sole record is of a small patch at a woodland edge near housing (NS2276).

ELATINACEAE

Elatine hexandra (Lapierre) DC.
Six-stamened Waterwort

Britain: scattered in the W.
Native; 1/2; very rare.
1st Loch Libo, 1883, MY.

The original record is accredited to 'Mr Wood', and nearly a hundred years later Six-stamened Waterwort was rediscovered at Loch Libo by AJS (NS4355, 1979), where it occurred as submerged plants in a sandy bay. Although not seen during modern visits, including the detailed SNH Loch Survey (1996), it is likely that this waterwort is still present. More recently it was found on shallow, peaty mud at the margin of Lochcraig Reservoir (NS5351, 2005).

Elatine hydropiper L.
Eight-stamened Waterwort

Britain: local in W C Scotland, rare elsewhere.
Native; 6/7; rare, but local at one site.
1st Barr Loch, 1976, JM.

The SNH Loch Survey (1996) also recorded this waterwort from Barr Loch (NS3457; NS3557; NS3558) and at Castle Semple Loch (NS3659); it is also known from the north edge of Kilbirnie Loch (NS3355, DH) just inside the VC. The first record was from the south-west corner of Barr Loch (*c.* NS3456), found when the water was low in later summer. More recently it has been found at a new site: Barcraig Reservoir (NS3956, 2003, IPG).

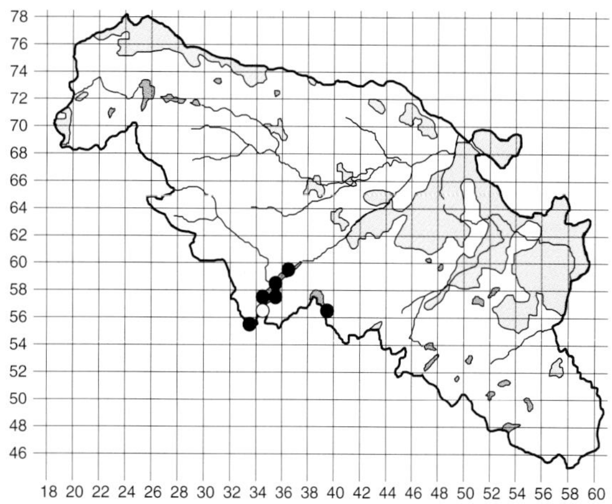

Elatine hydropiper

SALICACEAE

Populus alba L.
White Poplar

Britain: neophyte (Europe); common in S and lowland Scotland.
Hortal; 12/13; scarce but widely planted.
1st Milliken, 1905, PSY.

Also collected in 1905 from Pollok Estate, there are few modern records for this distinctive small tree, which does not appear to be establishing. Several records will be for persisting trees but White Poplar does appear to spread as at Woodhall (NS3473, 2012) and Underheugh (NS2075, 2012).

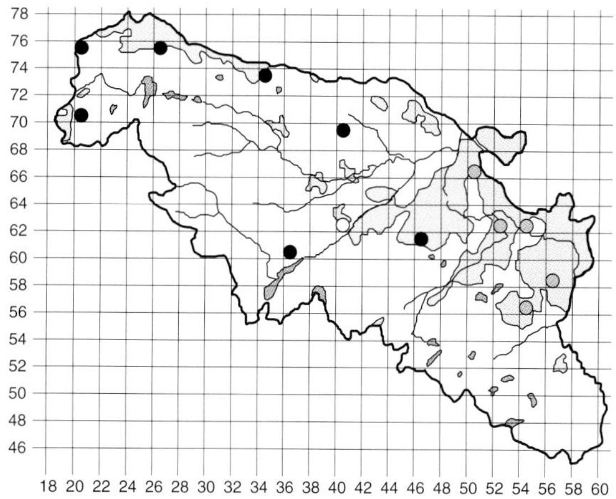

Populus alba

Populus × canescens (Aiton) Sm.
(*P. alba × P. tremula*)
Grey Poplar

Britain: neophyte (Europe); frequent in S England and C Scotland.
Hortal; 5/5; rarely recorded.
1st Kennishead (NS56K), 1986, CFOG.

Also recorded from Haggs Castle Golf Course (NS55L, 1987, CFOG), the other modern locations for Grey Poplar (all 1999) are from Houston Estate (NS4167), Stanely Reservoir (NS4661) and Dykebar Hospital (NS4961). At no place can it be considered naturalized.

Populus tremula L.
Aspen

Britain: throughout.
Native and hortal; 19/23; very rare and endangered as a native.
1st Near Bardrain, 1865, MY.

There are few named old localities for this species, although it was described as frequent in 1876 (FFWS). Some of the modern records refer to planted trees, as Aspen is often used in new broad-leaved plantation mixes. It was recorded from Loch Libo in 1984 (BWR) and today forms a large stand in the eastern fen (NS4355), but there are no old trees there. Most interest lies in the presumed native relict singletons or small stands found in four similar situations on the edges of steep rock outcrops above burns in the Renfrewshire Heights: near Muirshiel Barytes Mine (NS2864, 2003), North Rotten Burn (NS2568, 1997), Routen Burn (NS3160, 2003) and Kelly Burn (NS2268, 2006); at all sites suckering occurs but the sex is unknown. The native status of the populations about the Blacketty Water (*c.* NS3267) is unclear but all appear to be young.

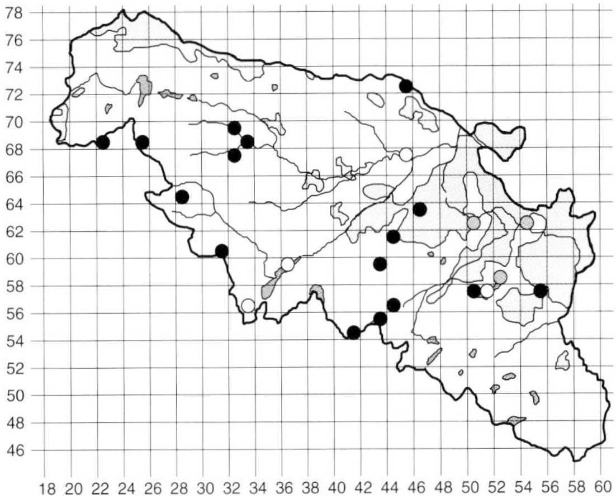

Populus tremula

Populus nigra L.
Black-poplar

Britain: frequent in S England, very rare alien in E Scotland.
Hortal; 15/20; widely planted but few records.
1st Gourock, 1850, GL.

A few records relate to young specimens as at Inchinnan (NS4768, 1996), Barochan (NS4069, 1997) and Greenock (NS2775, 2000), but it is probable that all are planted or sucker shoots (possibly outcasts). Old Black-poplars can still be found along the White Cart Water in Glasgow, although a few have recently fallen and others are threatened by flood prevention works. Hybrid Black-poplar (*P.* × *canadensis* Moench.) records are from by the Gryfe Water, Milton (NS3568, 1997) and Murdieston (NS2675, 2002), both mature trees, and a young sapling on waste ground at Woodhall (NS3473, 2012); the cultivar 'Serotina' was recorded from the Linburn area (NS4569, 1981, BWR).

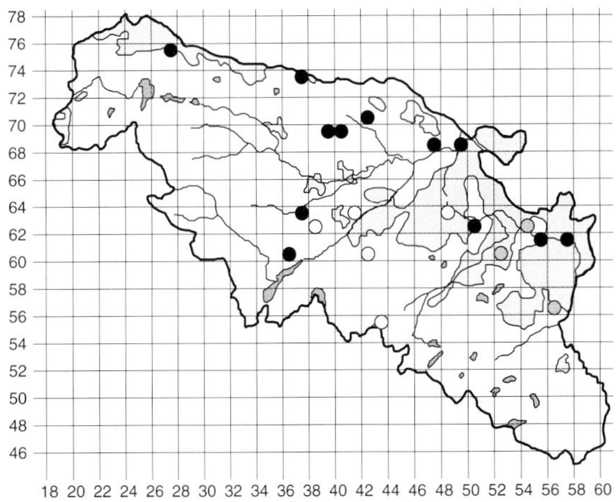

Populus nigra

Populus trichocarpa Torr. & A. Gray ex Hook.
Western Balsam-poplar

Britain: neophyte (W N America); widely scattered, occasional in C Scotland.
Hortal; 4/4; rarely planted.
1st River Calder, Lochwinnoch (NS3459), 1995.

This balsam-poplar is used in modern woodland plantings and can spread by seed, as at the original location. The other modern records are from Pilmuir Quarry (NS5154, 2000), by the White Cart Water, Porterfield (NS4967, 2001) and Castle Semple (NS3660, 1990, CMRP).

Salix pentandra L.
Bay Willow

Britain: frequent in N England and lowland Scotland, scattered alien elsewhere.
Native; 80/92; frequent, eastern.
1st Gourock, 1865, RH.

Bay Willow shows a marked preference for central and eastern parts of the VC, presumably reflecting the presence of low-lying wetlands. Usually this willow grows in mesotrophic fens, avoiding acidic or nutrient-rich sites, and also towards loch margins and along some slow-flowing watercourses.

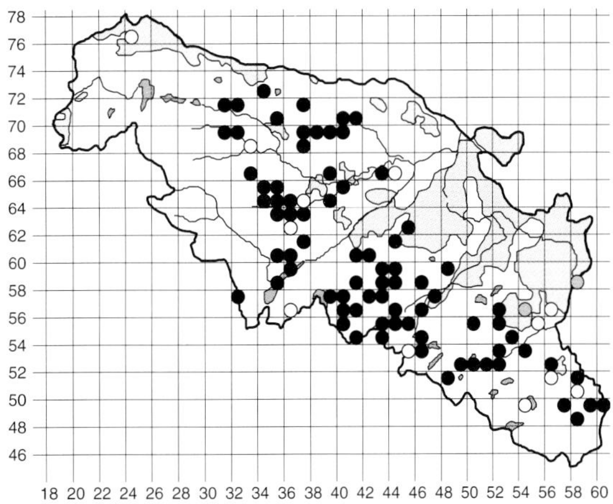

Salix pentandra

Salix fragilis L.
Crack-willow

Britain: archaeophyte; common in England and lowland Scotland.
Hortal; 80/96; frequent, eastern.
1st Renfrewshire, 1883, TB.

This tall tree willow shows a marked eastern and urban-fringe distribution, reflecting the deeper and richer riparian soils away from the upland areas. Old trees can be found along the Gryfe Water and both the White and Black Cart waters. The var. *russelliana* (Sm.) W.D.J. Koch was determined from the Yoker Burn in 1986 (CFOG) and var. *decipiens* (Hoffm.) W.D.J. Koch from Whitecraigs Golf Course (NS5557, 1987) and Foxbar (NS4561 1993); the latter is a locality from 1905 (GL).

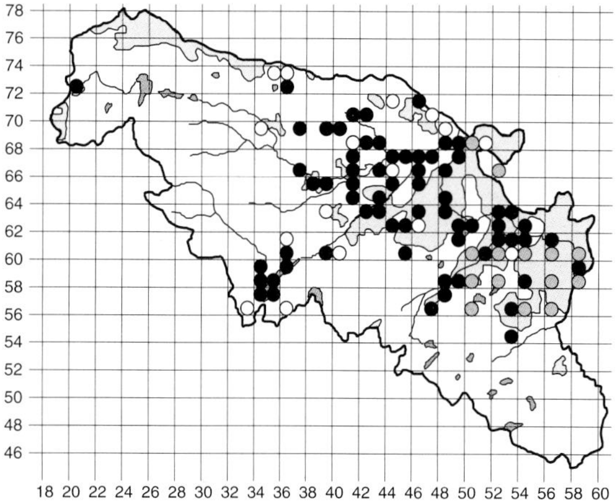

Salix fragilis

Salix alba L.
White Willow

Britain: archaeophyte; common in England and lowland Scotland.
Hortal; 40/50; frequently planted in the lowlands.
1st Gourock, 1865, RH.

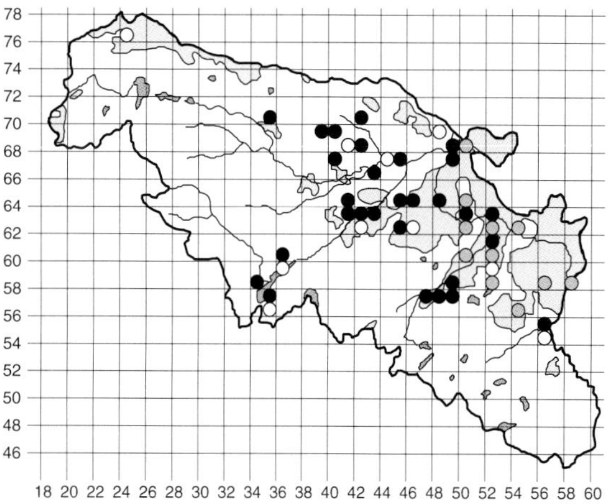

Salix alba

White Willow occurs as a planted tree, notably along riverbanks, but can spread locally and is found on drier waste ground. As with Crack Willow there is a notable absence from the west, presumably for the same reason. The var. *caerulea* (Sm.) Dumort. and var. *vitellina* (L.) Stokes were recorded from Pollok Estate in 1905 (GL), but modern recording has not noted varieties. Some records may refer to Hybrid Crack-willow (*S.* × *rubens* Schrank), which was recorded from the estate at Millikenpark (NS4161, 2001).

Salix purpurea L.
Purple Willow

Britain: widespread, rare in upland and N Scotland.
Native; 29/32; scattered, occasional.
1st Banks of the Cart, 1865, RH.

Purple Willow is typically associated with watercourses such as the Polnoon Water (NS5851, 2000), Kirkton Burn (NS4856, 1999), Roebank Burn (NS3756, 1999) and Earn Water (NS5252, 1998), and it is found at several places along the Gryfe Water and both the White and Black Cart waters. It is rare in the west, occurring at Howford Glen (NS2274, 2002), a reclaimed quarry, Underheugh (NS2075, 2012) and by the River Calder at Muirshiel (NS3163, 1996). In the south-east a young colonist was found by a trackside at Whitelee Windfarm (NS5846, 2011).

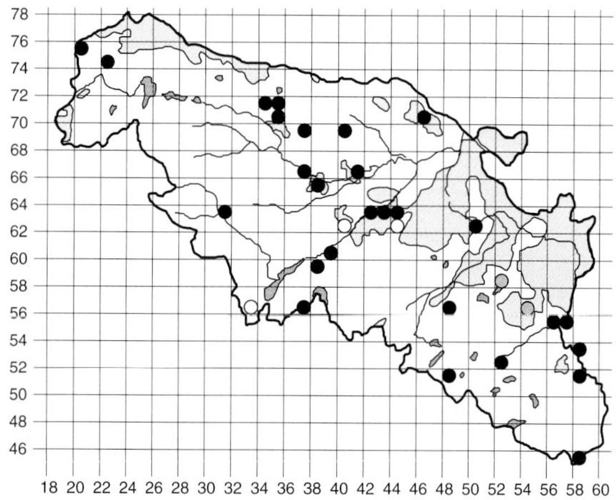

Salix purpurea

Salix × *rubra* Huds.
(*S. purpurea* × *S. viminalis*)
Green-leaved Willow

Britain: widely scattered.
Native or hortal; 18/18; scattered, occasional.
1st Auldhouse Burn, CFOG, 1988.

This hybrid willow was formerly widely planted and appears to have spread or persisted at several scattered locations across the VC, where it shows a lowland distribution pattern and, with a few exceptions, an affinity for urban areas. The margins of watercourses are favoured but some locations are in wetlands, e.g. Loch Libo (NS4355, 1996) and in a marsh at Wraes Wood (NS3868, 1996).

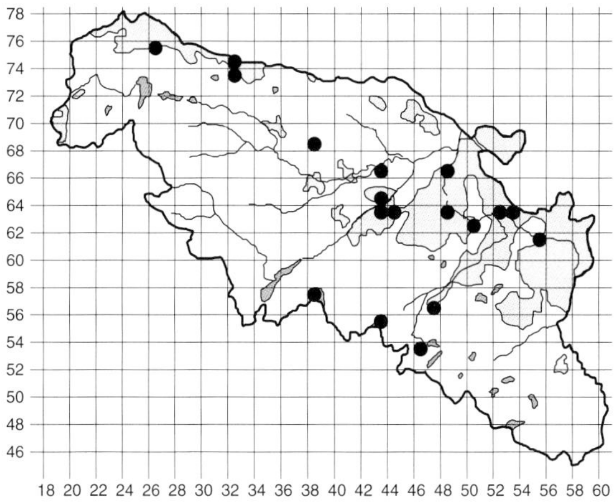

Salix × rubra

Salix viminalis L.
Osier

Britain: archaeophyte; common in England and lowland Scotland.
Hortal; 71/83; frequent in lowlands.
1st Gourock, 1865, RH.

Osier was recorded in 1906 from The Peel, Busby (GL) and in 1923 from the Ravenscraig–Gourock area (RM). It was considered very common in the 19th century by Hennedy, but he also stated that it was 'doubtfully native'. Today it is still common as it is widely planted and is also spreading, and it is well represented along watercourses and at wetter waste-ground sites.

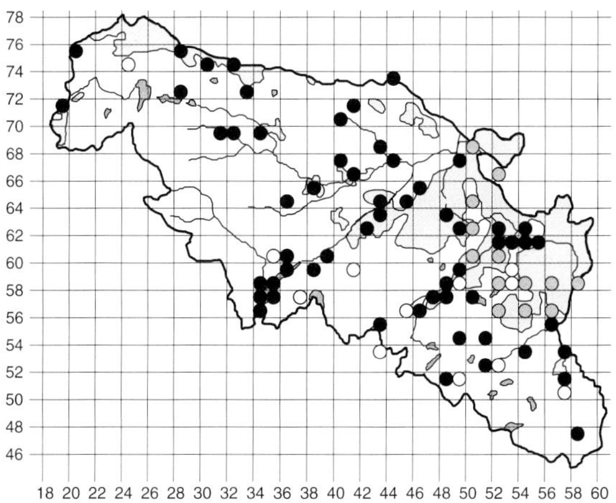

Salix viminalis

Salix × smithiana Willd.
(*S. viminalis × S. caprea*)
S. × *sericans* Tausch ex A. Kern.
Broad-leaved Osier

Britain: widely scattered and locally frequent.
Native (archaeophyte parent); 15/15; scattered and under-recorded.
1st Haggs Castle, 1987, CFOG.

Broad-leaved Osier is frequently encountered locally, occurring where both parents meet, usually in the lowlands, but also occasionally independently; a few records may be of planted varieties.

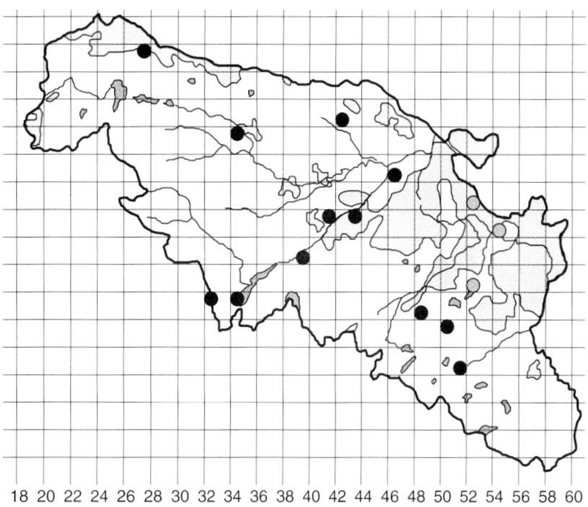

Salix × smithiana

Salix × calodendron Wimm.
(*S. viminalis × S. caprea × S. cinerea*)
Holme Willow

Britain: scattered but local, mainly in E England.
Native or hortal; 1/1; very rare.
1st Paisley, 2004.

A large specimen of Holme Willow grows by the Espedair Burn feeder to the White Cart Water in central Paisley (NS4863).

Salix × *holosericea* Willd.
(*S. viminalis* × *S. cinerea*)
S. × *smithiana* auct. non Willd.
Silky-leaved Osier

Britain: widely scattered.
Native (archaeophyte parent); 17/19; scattered and probably under-recorded.
1st Castle Semple Loch, 1934, RM.

Silky-leaved Osier is occasionally found growing with both parents in similar wet places.

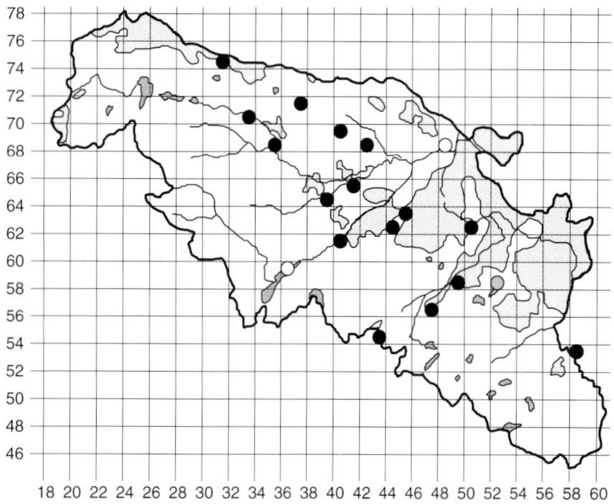

Salix × *holosericea*

Salix × *fruticosa* Döll
(*S. viminalis* × *S. aurita*)
Shrubby Osier

Britain: widely scattered.
Native (archaeophyte parent); 2/2; very rare hybrid.
1st Loch Libo (NS4355), 1996.

This hybrid willow was also recorded from Walls Loch (NS4158, 2002).

Salix caprea L.
Goat Willow

Britain: common throughout except in N Scotland.
Native; 263/306; very common in lowlands.
1st Gourock, 1865, RH.

This tall shrub or small tree can form dense scrubby woodlands in urban areas when any piece of waste ground is left undisturbed for a good few years. Goat Willow shows a strong lowland distribution pattern, with a reduction towards the more acidic and wetter uplands, but it can also be scarce at intensively farmed areas.

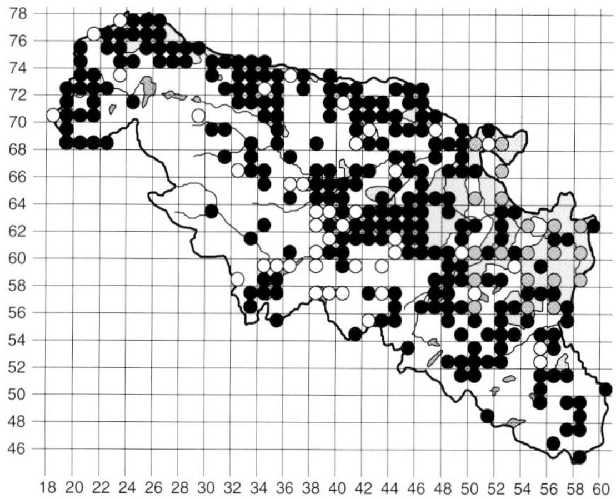

Salix caprea

Salix × *reichardtii* A. Kern.
(*S. caprea* × *S. cinerea*)

Britain: frequent but local.
Native; 6/12; rare but under-recorded.
1st The Cut, Greenock, 1965, RM.

This hybrid is very likely under-recorded as both parents often occur together, and are known to readily hybridize. There are a couple of relevant CFOG records, Hillington (NS5166, 1985) and Busby (NS55T, 1985), and four other modern records from Raith Burn (NS3063, 1997), Gryfeside (NS3370, 1998), Routen Burn (NS3160, 2003) and Threepgrass (NS4458, 2003). The plants by the upland Raith and Routen burns occur on similar remote, steep banks.

Salix × *capreola* Jos. Kern. ex Andersson
(*S. caprea* × *S. aurita*)

Britain: scarce but widely scattered.
Native; 0/1; doubtful record.
1st Flenders, 1983, BWR.

This hybrid was noted on a field card from Flenders (NS5656) but there appears to be no specimen or other verification of what is a very rare hybrid.

Salix cinerea L.
Grey Willow

Britain: common throughout.
Native; 351/383; very common.
1st Renfrewshire, 1834, MC.

This widespread willow often grows with Goat Willow in urban locations but it is typically found in wetter places, becoming the commonest willow in rural habitats, especially in fens and along burn sides and loch margins. It is scarce in the peaty uplands. Any infraspecific records refer to ssp. *oleifolia* Macreight, which is thought to be the only subspecies present in the area.

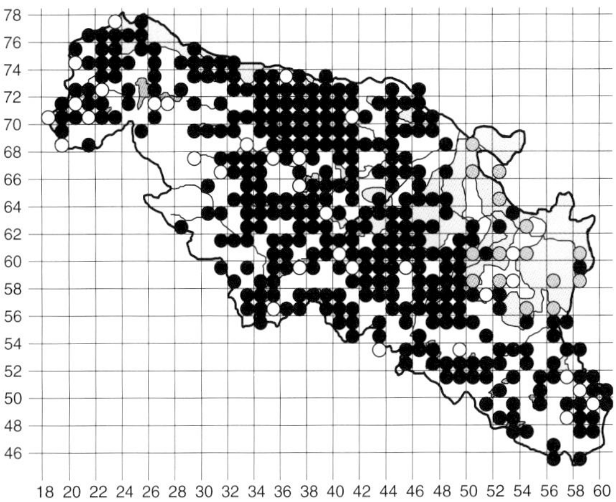

Salix cinerea

Salix × multinervis Döll
(*S. cinerea × S. aurita*)

Britain: frequent but local, mainly in the N and W.
Native; 16/21; scattered.
1st Gryfe Water, Blacksholm, 1980, BWR.

This hybrid is widely scattered and probably under-recorded, and some plants may well be backcrosses.

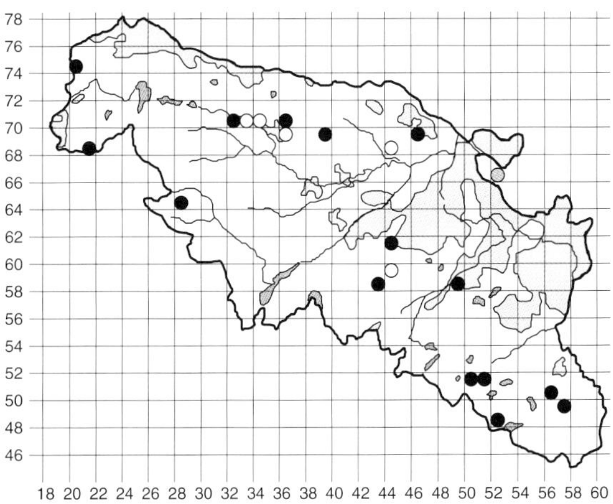

Salix × multinervis

It occurs in rural locations, usually with dampish and acidic conditions, where the two parents meet.

Salix × strepida J. Forbes non Schleich.
(*S. cinerea × S. myrsinifolia*)

Britain: rare in N Britain.
Native; 2/2; very rare.
1st Picketlaw Reservoir, 1998.

This hybrid was found to the west side of the Picketlaw Reservoir (NS5651), where it grows with both parents. A further tentative record is from the western edge of Shovelboard (NS3869, 2011), where Dark-leaved Willow was previously reported.

Salix × laurina Sm.
(*S. cinerea × S. phylicifolia*)
Laurel-leaved Willow

Britain: scattered in the N.
Native; 4/4; rare hybrid.
1st Sergeantlaw Moss, 1995.

A rare hybrid recorded from four sites, all associated with wet places: the carr to the edge of Sergeantlaw Moss (NS4459), Dunwan Burn, Netherton (NS5749, 2000), the Earn Water, Moorhouse (NS5252, 2003) and the mire below Walls Hill (NS4158, 2011).

Salix aurita L.
Eared Willow

Britain: common in the N and W.
Native; 116/160; common in upland fringes.
1st Gourock, 1865, RH.

This shrubby willow is locally common away from urban areas and heavily grazed uplands, generally reflecting wet acidic soils. It was well recorded during the 1960s–80s (BWR), but there are now some areas from which it is absent, possibly reflecting a decline caused by agricultural intensification, including an

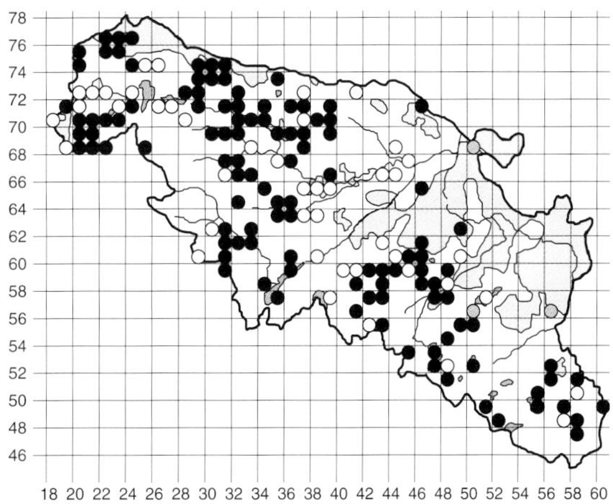

Salix aurita

increase in grazing pressure. An impressive stand occurs near Gryfe Reservoir No. 2 (NS2972).

Salix × ambigua Ehrh.
(S. aurita × S. repens)

Britain: widely scattered, mainly in the N.
Former native; 0/1; extinct.
1st Castle Semple Loch, 1887, NHSG.

This is the sole record of this hybrid and there is no further information to aid verification.

Salix myrsinifolia Salisb.
Dark-leaved Willow

Britain: local in N England and upland Scotland.
Native; 1/4; very rare.
1st North of Crawhin (NS2471), 1966, BWR.

There is no confirmation of the original record, which appears to be from near Shielhill Glen, or another record made by the same recorders from 'between Goldenlea and Bridge of Weir' (NS3965, 1970). SNH files reported it from Shovelboard Moss (NS3869, 1968, D Goode) as 'frequent', but recent searches only noted a possible hybrid. The sole modern location for this rare willow is the west side of Picketlaw Reservoir (NS5651, 1998). It was found at Braehead (VC77 border) by PM in 1985 (CFOG).

Salix phylicifolia L.
Tea-leaved Willow

Britain: frequent in N England and upland Scotland.
Native; 6/13; rare.
1st Peesweep Inn, 1905, GL.

This rare willow of fens and burn sides was also collected from the Gleniffer Braes in 1905 (GL), and later in 1917 by Lee (GL). A few years later RM recorded it from Holehill Burn, Dunwan (*c.* NS5549, 1933), Mid Dam near Eaglesham (NS5651, 1935) and the Loganswell area (NS5152, 1936); later records are from Black Loch (NS4951, 1966, JDM), Glen Moss (NS3669, 1959, BWR) and Brock Burn, Duncarnock Farm (NS5055, 1984, BWR). It can still be found at the latter (recorded in 2000), and there are two records from the Earn Water (NS5151 and NS5152, both 2005). Other south-eastern records are from Dunwan Burn, Netherton (NS5749, 2000) and the White Cart Water at Threepland (NS6049, 2011). It was collected from Lochwinnoch in 1987 (GL) and occurs in the fen near Lochlip Road (NS3558, 2011); shrubs by the nearby Calder Bridge appear to have been lost during recent gravel extraction and bridge construction.

Salix repens L.
Creeping Willow

Britain: widespread except in C England, frequent in the N and W.
Native; 7/14; rare.
1st Gourock, 1865, RH.

Today there are seven widely scattered records mainly from high-quality fen habitats, as at the Kelly Reservoir (NS2268, 1996, SNH-LS), Little Loch (NS5052, 1997), Walls Hill (NS4158, 2002), Dargavel Burn (NS3771, 2005) and Barmufflock (NS3664, 2004). The latter may be the same as the BWR 1986 'north of Barnbeth' locality, although the grid reference differs (NS3665); it may instead refer to a large, spreading stand of an apparently planted upright form at the roadside margin of Ranfurly Golf Course (NS3665, 2007). A further unexpected modern record is from flush grassland on Jock's Craig (NS3269, 2005). It has not been refound at Harelaw Dam (NS4753, 1951, RM) or from 'Shielhill' Reservoir (NS2472, 1965, JDM).

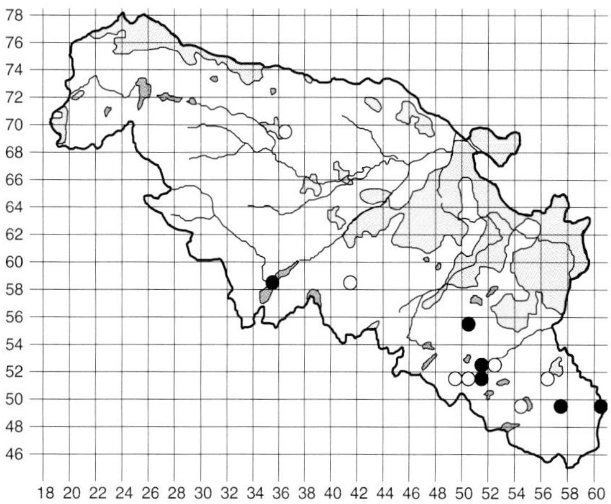

Salix phylicifolia

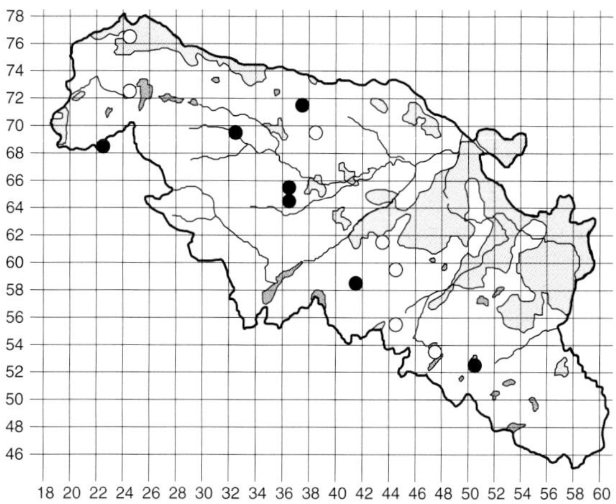

Salix repens

VIOLACEAE

Viola odorata L.
Sweet Violet

Britain: common in the S, local alien in S and E Scotland.
Former hortal; 0/3; extinct.
1st Hedge on road to Barrhead, 1869, RPB.

Sweet Violet was also known from Thornliebank in 1883 (NHSG) and more recently RM recorded it from Deaconsbank in 1926 and 1960 (CFOG), but there are no modern records.

Viola riviniana Rchb.
Common Dog-violet

Britain: common throughout.
Native; 276/336; very common.
1st Renfrewshire, 1869, PPI.

Dog-violet is a common sight in local woodlands occurring throughout the VC, only becoming rare or absent from heavily urbanized areas or improved farmland. Less acidic, well-drained woodlands are the main locations, but it can be found along scrubby burn-side slopes in upland areas and in open dry heaths or grassy embankments.

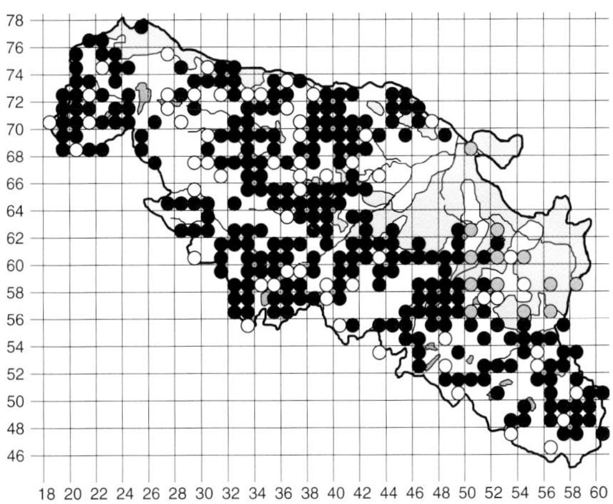

Viola riviniana

Viola canina L.
Heath Dog-violet

Britain: widespread but scattered.
Native; 0/9; extinct, but possibly overlooked.
1st Paisley, 1845, NSA.

There are no modern records for this dog-violet, which Hennedy (1891) described as frequent at heathy or sandy places (though he only named Gourock), and which was considered plentiful on the Gleniffer Braes (1869, RPB); other old records include Gourock (1869, GL), Eaglesham (1895, PSY) and the banks of the Earn Water, Muirshields Farm (1907, GL). Some very old records may represent a confusion with Common Dog-violet. However, more recently RM collected it from Harelaw Dam (1951, E) and BWR recorded it in 1982 from north of Quarriers, Bridge of Weir (NS3667) and Ryat Linn (NS5257, 1982), but queried a record on a card from a disused railway, St Brydes (NS3860, 1977), and JDM noted it at Garvock Lodge (NS2771, 1968). It may well have been overlooked during modern surveying, but if not there has been a significant decline.

Viola palustris L.
Marsh Violet

Britain: common in the N and W.
Native; 218/270; common in wet places.
1st Paisley Canal Bank, 1865, RH.

Marsh Violet shows a strong rural distribution pattern, being very rare in urban places. It is commonly found in more acidic marshes and tolerates the shade of wet woodlands; it also occurs in flushes or mires on wet heaths, especially by burn sides, and grows at over 400m, as at Queenside Muir (NS3064, 1998).

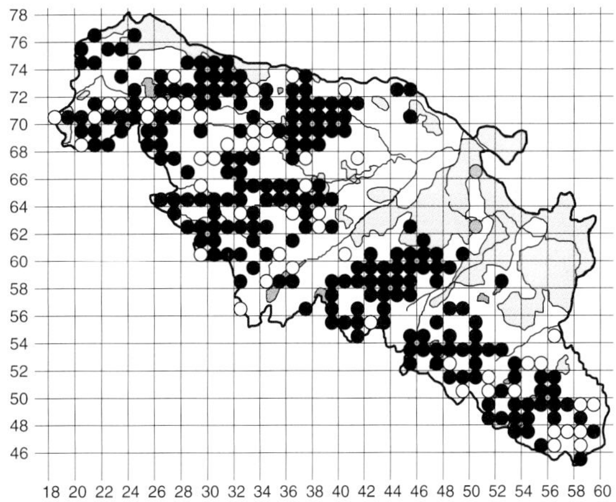

Viola palustris

Viola cornuta L.
Horned Pansy

Britain: neophyte (Pyrenees); scattered and rare, local in NE Scotland.
Hortal; 1/1; very rare.
1st Lochend, 2004.

The sole record is for a few plants growing under a fence along a lane side at Lochend (NS3764).

Viola lutea Huds.
Mountain Pansy

Britain: common in uplands, absent elsewhere.
Native; 102/143; locally common in upland fringe pastures.
1st Paisley, 1845, NSA.

Mountain Pansy can be frequent on pastures on the

hilly ground in rural areas, notably to the south-east and often in some abundance; it is not recorded from similar pasture in the extreme north-west. It avoids strongly acidic pasture and seems to tolerate some improvement.

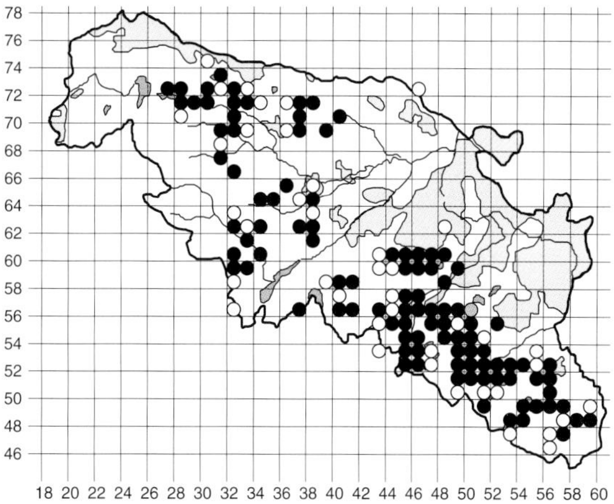

Viola lutea

Viola tricolor L.
Wild Pansy

Britain: frequent, scarcer in the W.
Native; 6/18; rare and declining.
1st Paisley, 1845, NSA.

There is some confusion about this species as it is often referred to as occurring in hill pastures but all specimens checked at such locations appear to be Mountain Pansy, although some are small flowered and perhaps may include relict hybrids. There were a few relevant CFOG records but surprisingly few modern ones, mainly from disturbed places: Gleniffer Road (NS4560, 1995), Knapps Loch (NS3668, SNH-LS, 1996) and a roadside, Roundtree (NS3562, 2001).

Viola × wittrockiana Gams ex Kappert
(*V. tricolor × V. arvensis*, and *V. altaica* Ker Gawl?)
Garden Pansy

Britain: neophyte (garden hybrid); local in the S, rare in Scotland.
Hortal; 3/4; rare outcast.
1st Finlaystone, 1976, AJS.

There are three modern records: waste ground at the India Works, Inchinnan (NS4768, 1996), Clydeport (NS3174, 2001) and Busby (NS5756, 2001).

Viola arvensis Murray
Field Pansy

Britain: archaeophyte common in the S and in NE Scotland.
Accidental; 10/17; rare.
1st Renfrewshire, 1883, TB.

There are few records for this pansy from any period, although earlier authors (e.g. Hennedy 1891) described it as common. It is likely to have been overlooked, although the reduction in arable fields and increase in herbicide use will have caused a decline. Only three modern records are from arable fields: near Knapps Loch (NS3668, 2000), South Mains (NS4266, 2003) and Cauldside (NS3270, 2005), the others being from waste ground and roadsides.

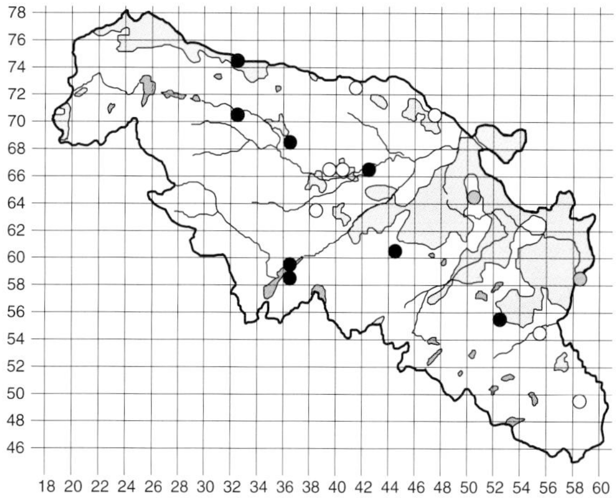

Viola arvensis

LINACEAE

Linum bienne Mill.
Pale Flax

Britain: frequent on SW coasts, rare alien inland.
Former hortal; 0/1; extinct.
1st Wasteland, Paisley, 1910, PSY.

Linum usitatissimum L.
Flax

Britain: neophyte (cultivated); frequent in the S, local in lowland Scotland.
Accidental; 2/5; very rare.
1st Renfrewshire, 1869, PPI.

Flax was in cultivation in the early 19th century (Wilson 1812) in the Lochwinnoch area, but the only named places are Langbank (1886, AANS) and Castle Semple Loch (1887, NHSG); more recently JDM noted it at Muirend railway embankment (NS5760, 1961). There are two modern records, from a motorway edge, Erskine Bridge (NS4571, 1998, PM) and the Main Street, Lochwinnoch (NS3558, 2003, IPG).

Linum catharticum L.
Fairy Flax

Britain: common throughout.
Native; 53/70; frequent.
1st Paisley Canal Bank, 1855, GGO.

Fairy Flax grows in short open grasslands, usually

reflecting some base enrichment. It can be found in more natural upland locations, usually on burn-side embankments, but it also occurs on waste ground in urban-fringe areas, e.g. at disused railways and on mine-waste shale. It is noticeably rare or absent from the south-east, but is found by the Ardoch Burn (NS5848, 2003).

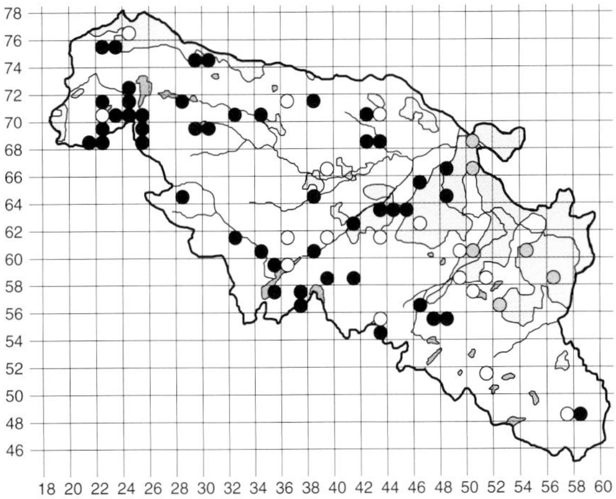

Linum catharticum

Radiola linoides Roth
Allseed

Britain: widely scattered on coasts, rare and declining inland.
Former native; 0/3; extinct.
1st Langside, 1813, TH.

Other old records are from Barr Loch (1845, NSA) with Hennedy's (1891) note of Gourock apparently the last.

HYPERICACEAE
(CLUSIACEAE *pro parte*)

Hypericum calycinum L.
Rose-of-Sharon

Britain: neophyte (SE Europe); frequent in the S, scattered in N.
Hortal; 1/4; very rare escape or outcast.
1st Gourock, 1876, FFWS.

Rose-of-Sharon was also collected from a cemetery, presumably in the Greenock area, in 1892 (GRK) and Johnstone Castle in 1904 (PSY), where it was likely planted but it does not appear to have become established. Today there is only one record from Dee Drive, Lexlands (NS4562, 2002).

Hypericum forrestii (Chitt.) N. Robson
Forrest's Tutsan

Britain: neophyte (China); scattered, very rare in the N.
Hortal; 1/1; very rare.
1st Wemyss Plantation, 2010, PRG.

The sole record is for a bush in the wood at Wemyss Bay (NS1970).

Hypericum androsaemum L.
Tutsan

Britain: native in the W, frequent alien in the E.
Native and/or hortal; 34/42; occasional.
1st Shore at Gourock, 1834, MC.

Tutsan may be native about the Clyde coast but it is difficult to state this confidently about any populations now found here due to widespread disturbance and the frequency of garden escapes. Hennedy (1891) stated that it was 'not uncommon on coast at Gourock'. Although strongly represented in the west, it is also found, undoubtedly as escapes or outcasts, further east. Some records may refer to other shrubby cultivars or hybrids.

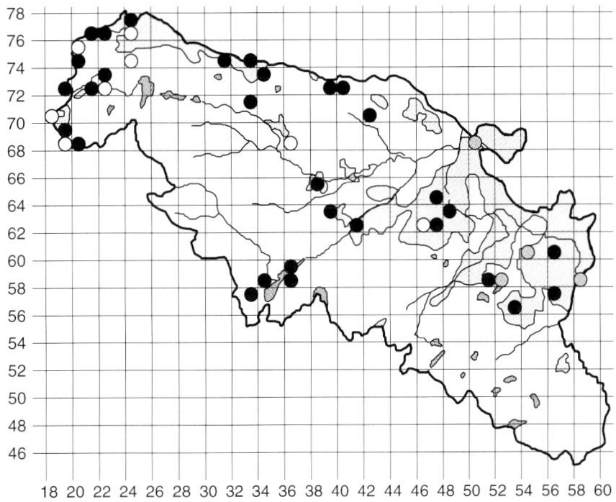

Hypericum androsaemum

Hypericum × *inodorum* Mill.
(*H. androsaemum* × *H. hircinum*)
Tall Tutsan

Britain: neophyte (SW Europe), local in S and W C Scotland.
Hortal; 3/4; very rare.
1st Hills beyond Greenock, 1860, GL.

The GL specimen (named H. *anglicum* Bert., from the D Stewart collection, but unknown collector) was exhibited by Prof. Walker Arnott in 1860 (Anon. 1869). The modern records are of a couple of plants by the cycleway near Howwood (NS3860, 2007) and from a wood in Wemyss Bay (NS1970, 2010, PRG), and a CFOG record from the King's Inch (NS5167, 1992, PM), which was just inside VC76.

Hypericum perforatum L.
Perforate St John's-wort

Britain: common in the S, extending to C and NE Scotland.
Native; 72/86; frequent in lowlands.
1st Auchinames, Kilbarchan and Paisley, 1845 NSA.

Perforate St John's-wort shows a strongly lowland urban-fringe distribution. All of the records (a few may include the hybrid) are from waste-ground places, especially disused railways, reflecting free-draining open soils. Two rural south-east records are from roadsides near Little Loch (NS5051, 1997) and near Fauldside (NS5253, 2003).

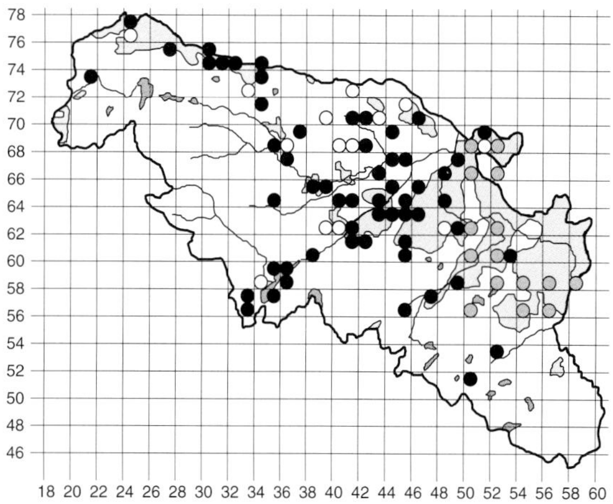

Hypericum perforatum

Hypericum × desetangsii Lamotte
(H. perforatum × H. maculatum)
Des Etangs' St John's-wort

Britain: widely scattered, extending to C Scotland.
Native; 18/18; occasional in lowlands.
1st Pollok Park, 1987, CFOG.

Although well recorded during surveys for CFOG, there was only one record from within the VC and one on the border. Today there are quite a few records for this hybrid in Renfrewshire, from waste ground and disused railway sites mostly in the centre of the VC. There is a strong overlap between the two parents' distributions, although it is unclear if any hybrids have originated in situ.

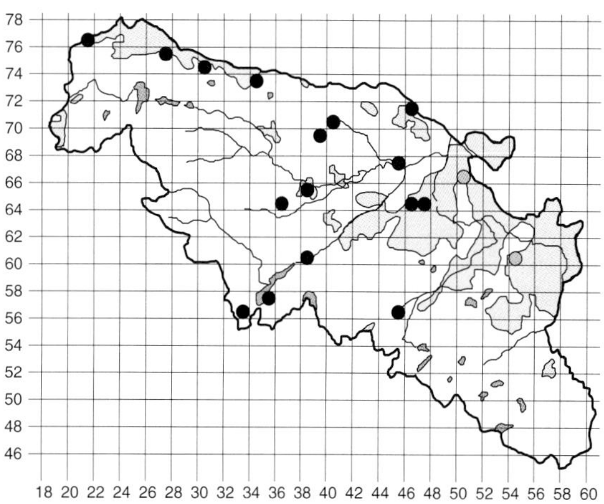

Hypericum × desetangsii

Hypericum maculatum Crantz.
Imperforate St John's-wort

Britain: common in W C England/Wales, lowland in Scotland.
Native; 11/18; scarce, scattered.
1st Paisley, 1845, NSA.

There are a few scattered old and modern records for this St John's-wort. All the records refer to ssp. *obtusiusculum* (Tourlet) Hayek, although some may represent hybrids or backcrosses, as was the case during CFOG surveys. Modern records are from various woodland, grassy and waste-ground places and include Johnstone Wood (NS4262, 1993), Arthurlie (NS4958, 1997), waste ground, Neilston (NS4857, 1998), Mill Dam (NS3469, 1999), Humbie (NS5453, 2003) and by the Gryfe Water (NS3667, 2003; NS3569, 2006).

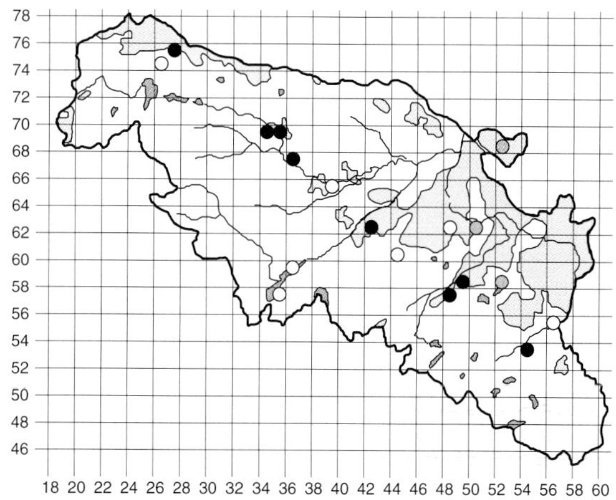

Hypericum maculatum

Hypericum tetrapterum Fr.
Square-stalked St John's-wort

Britain: throughout, rare in N Scotland.
Native; 47/66; frequent in marshy places.
1st Paisley Canal Bank and Gourock, 1865, RH.

Square-stalked St John's-wort is often found in marshy ground, often species-rich fens, at widely scattered rural lowland locations. It avoids acidic upland peaty areas although it is known from marshy flushes at Earn Hill (NS2275, 2003) and by Ardoch Burn (NS5848, 1992). It is likely to have declined, probably reflecting agricultural improvement.

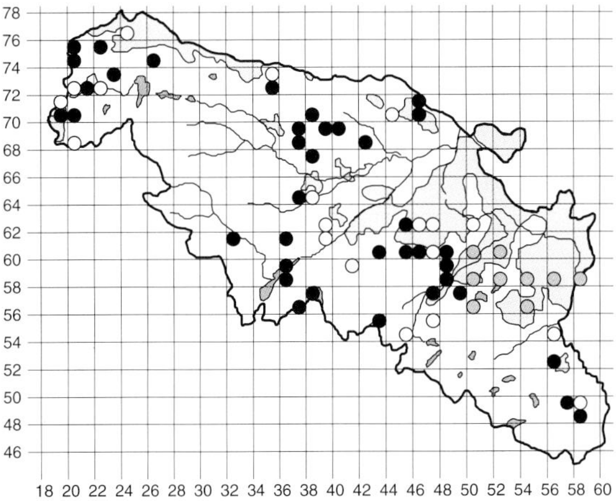

Hypericum tetrapterum

Hypericum humifusum L.
Trailing St John's-wort

Britain: common in the W, frequent in the E.
Native; 9/26; rare.
1st Paisley and Kilbarchan, 1845, NSA.

There are several old records from places such as Paisley Canal (1876, FFWS), Loch Libo and Wemyss Bay (both 1887, NHSG), Kilmacolm (1888, PSY) and Ravenscraig (1925, RM), and Hennedy (1891) stated that Trailing St John's-wort was frequent. It was also recorded a few times in the 1960s–80s, but today there are only a few records, mainly to the centre of the VC, including the Gryfe Water (NS3865, 1998), Gotter Water (NS3465, 1999), near Blood Moss (NS2168, 2001), Lochwinnoch (NS3658, 1990, CMRP), Brother Loch area (NS5153, 1991, LB), Finlaystone Estate

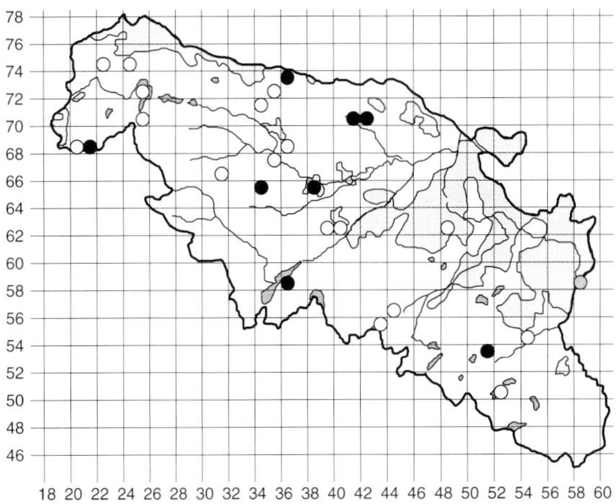

Hypericum humifusum

(NS3673, 2003, IPG), Formakin Estate (NS4171, 2003, IPG) and the Royal Ordnance Factory, Georgetown (NS4270, 2005). If not overlooked, it appears to have declined.

Hypericum pulchrum L.
Slender St John's-wort

Britain: common except in E C England.
Native; 135/185; common in rural fringes.
1st Kilbarchan and Paisley, 1845, NSA.

Slender St John's-wort was considered common in the 19th century and this is largely true today, although it is seldom seen in any great numbers. It shows a strong rural pattern and is most frequently seen on shaded rocky ledges or on the upper slopes of wooded valleys, and is also found in some open areas where the land is not heavily grazed. It can follow steep-sided burn sides to over 300m in the upland areas, e.g. Forkings of Raith (NS2862, 1997) and Dickman's Glen (NS5847, 2001).

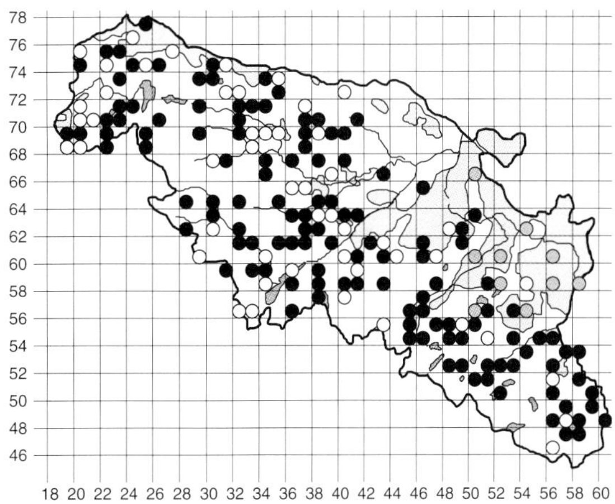

Hypericum pulchrum

Hypericum hirsutum L.
Hairy St John's-wort

Britain: common in the E but rare or absent from the W and NW.
Native; 10/13; rare.
1st Paisley, 1845, NSA.

This St John's-wort was collected from Cathcart in 1890 (GL) and recorded from Gourock and Inverkip (1876, FFWS). It was reported from riverbanks and disused railways during CFOG surveys, and can be found on what is now a cycleway at Howwood (NS3860, 2001; NS3659, 1988, CMRP), where first noted in 1977 (BWR). Other sites include waste ground, as at East Fulwood (NS4567, 2001), but also semi-natural woodlands as at Bridgewater (NS4670, 1997) and by the Green Water, Duchal Wood

(NS3368, 1994) and the Gryfe Water at Brierie Hill (NS4065, 2002).

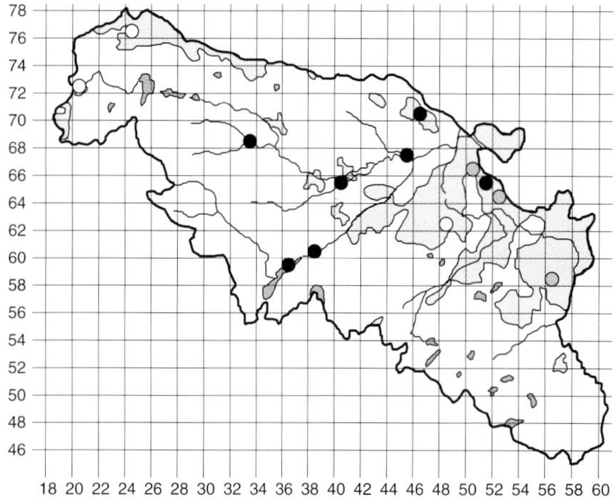

Hypericum hirsutum

GERANIACEAE

Geranium endressii J. Gay
French Crane's-bill

Britain: neophyte (Pyrenees); scattered throughout the lowlands.
Hortal; 12/12; scarce.
1st Newton Mearns (NS55N), 1985, CFOG.

There are two other CFOG localities and nine other records from the modern period, all from waste ground or wayside places. One or two records may refer to the following hybrid.

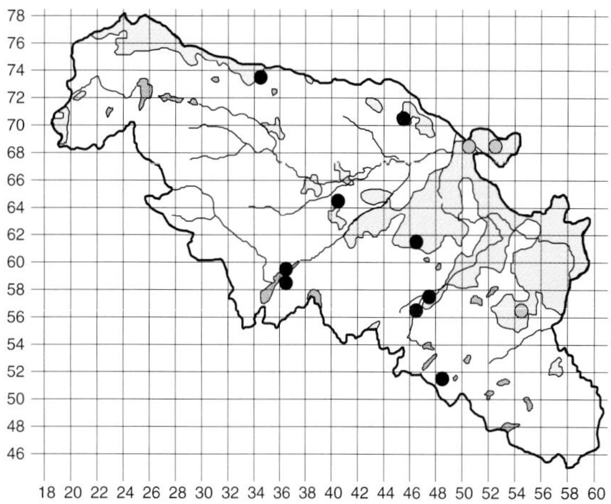

Geranium endressii

Geranium × oxonianum Yeo
(G. endressii × G. versicolor)
Druce's Crane's-bill

Britain: neophyte (garden origin); scattered in C Scotland.
Hortal; 5/5; rare.
1st Millikenpark, 2001.

This cultivated garden outcast was first recorded from old estate grounds (NS4161). Other records are from waste ground at Porterfield (NS4967, 2001), near Millarston (NS4563, 2002) and two records from the west coast were made by PRG in 2010: Trumpethill (NS2276) and Larkfield (NS2375).

Geranium versicolor L.
Pencilled Crane's-bill

Britain: neophyte (C Mediterranean); occasional in S England.
Former hortal; 0/1; extinct.
1st Near Crofts, Wallneuk St, Paisley, 1857, PSY.

Geranium sylvaticum L.
Wood Crane's-bill

Britain: common in N England and Scotland, except in the extreme N.
Native; 12/20; rare.
1st Renfrewshire, 1834, MC.

Other old records include ones from Lochwinnoch (1845, NSA), Cathcart (1876, FFWS), Crookston (1856, HMcD) and excursions to Kelly Glen and Castle Semple Loch (1887, NHSG). There are several records from the 1960s–80s period (BWR); these include south of Daff Reservoir (NS2270, 1977), Lawmarnock Wood (NS3763, 1984) and by the Gryfe Water, Hatterick (NS3567, 1971). Most modern records are from steep-sided watercourse margins affording some protection from grazing; localities include Ouse Hill (NS2670, 1997), Threeply (NS3766, 2002), Duchal (NS3269, 1999), North Rotten and Harestane burns (both

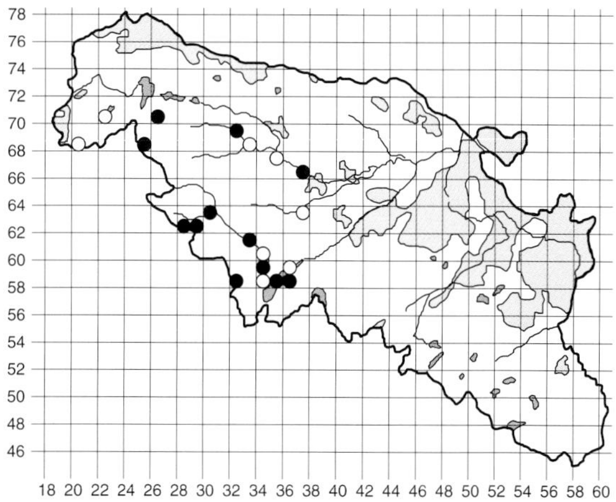

Geranium sylvaticum

NS2568, 1997), Raith Burn (NS3063, 1995), Calder Bank (NS3361, 1997) and Maich Water (NS3258, 2005). There are also a few lowland records from the Castle Semple Loch area (NS3658, 1992, CMRP).

Geranium pratense L.
Meadow Crane's-bill

Britain: common throughout, alien in N Scotland.
Native and hortal; 37/44; occasional.
1st Black Cart, Kilbarchan and Paisley, 1845, NSA.

There are several records from Castle Semple Loch, starting from 1887 (NHSG) to most recently from shingle by the River Calder (NS3558, 2003, IPG). Modern records all tend to reflect larger watercourses such as the White and Black Cart waters and Earn Water. There is an obvious absence in the west, presumably linked to the scarcity of more enriched alluvial riverbanks, the exception being alongside the Spango Burn (NS2374, 2006). A few records may refer to sown or introduced plants, as in the case of grassland to west of Millarston (NS4563, 2002), Barnbrock (NS3564, 1993, CMRP) and perhaps Bridgewater (NS4770, 1997).

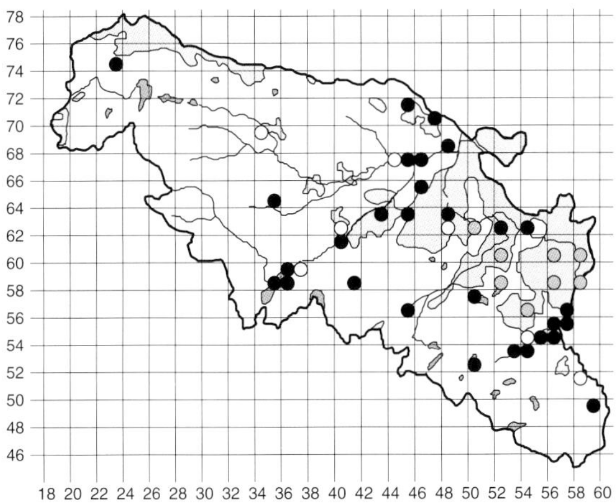

Geranium pratense

Geranium sanguineum L.
Bloody Crane's-bill

Britain: coastal in N England and Scotland, mainly alien in S England.
Hortal; 3/5; very rare.
1st Paisley, 1889, NHSG.

AMcGS recorded this crane's-bill from a grassy bank above the shore, opposite Gantocks Hospital, Gourock, in 1979 (NS2176) and it was also known from Newton Mearns (NS55N, 1985, CFOG). Today it is known as an outcast from two places: Underheugh (NS2075, 1997) and on the cycleway, Scart (NS3667, 2001).

Geranium dissectum L.
Cut-leaved Crane's-bill

Britain: archaeophyte; common throughout except in N Scotland.
Accidental; 52/65; occasional.
1st Glebe, Kilbarchan, 1845, NSA.

Cut-leaved Crane's-bill was well known to 19th-century botanists, as is evidenced by specimens (GL) from Houston (1876), Thornliebank (1890) and Giffnock (1890) and from a report from Castle Semple Loch (1887, NHSG). Today the records are mostly from urban-fringe waste ground and wayside grasslands, usually short and open, but rarely in the west.

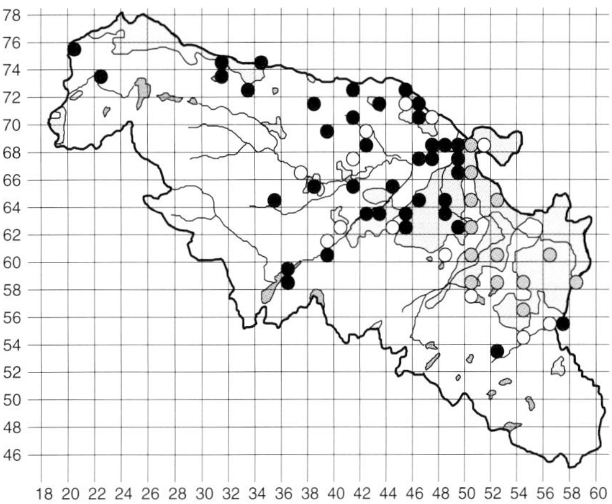

Geranium dissectum

Geranium × magnificum Hyl.
(*G. ibericum* Cav. × *G. platypetalum* Fisch. & C.A. Mey.)
Purple Crane's-bill

Britain: neophyte (garden origin); widely scattered in lowlands.
Hortal; 23/25; occasional.
1st Quarry, Lochwinnoch, 1976, AJS.

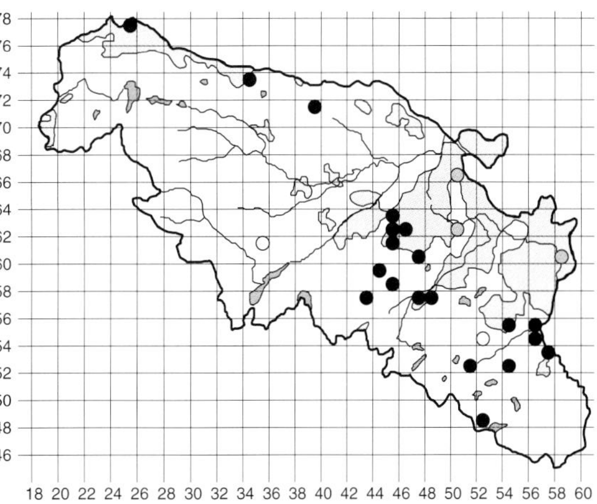

Geranium × magnificum

AJS also recorded this large-flowered crane's-bill from waste ground, Newton Mearns (NS5254), in 1978. Today it is the most widespread of the larger crane's-bills away from watercourses, and it seems capable of persisting at waste-ground sites and along roadsides and hedge banks. It is found at some remote areas such as by the roadside at Queenseat (NS5248).

Geranium pyrenaicum Burm. f.
Hedgerow Crane's-bill

Britain: neophyte (S Europe); common in the S extending to NE Scotland.
Accidental; 2/2; very rare.
1st Aitkenhead Road, 1987, CFOG.

The only other modern record is from the roadside west of Bishopton (NS4371, 2001).

Geranium pusillum L.
Small-flowered Crane's-bill

Britain: common in the S and E, extending to NE Scotland.
Native or accidental; 1/3; very rare.
1st Kilmacolm Road, near Greenock, 1866, GL.

There is also a specimen from a cornfield by Old Kilmarnock Road near Clarkston (1907, GL). Today the sole modern record is of a weed in sown grass at GMRC, Nitshill (NS5160, 2004).

Geranium molle L.
Dove's-foot Crane's-bill

Britain: common in the S, scattered in N Scotland.
Native; 15/23; scarce.
1st Cathcart, 1860, GGO.

Hennedy described Dove's-foot Crane's-bill as common but gave no localities from VC76 and apart from the first record, a 19th-century specimen from near Paisley (PSY) is the only other named place prior to Cloch in 1960 (NS2075, BWR). Today it is known from a few scattered places, usually in urban-fringes, particularly from waste ground or roadsides as along the Greenock Road (NS4172, 2002).

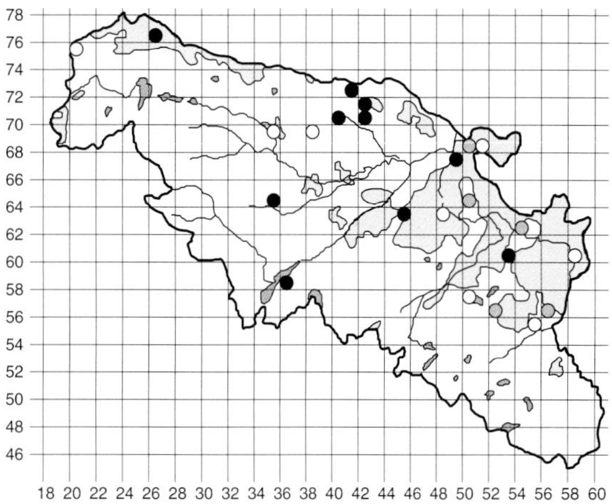

Geranium molle

Geranium macrorrhizum L.
Rock Crane's-bill

Britain: neophyte (S Europe); scattered, mainly in the S.
Hortal; 7/7; rare outcasts.
1st Merchiston, 1999.

Originally recorded from a disused railway at Merchiston (NS4164), there are six other localities for this garden outcast: waste ground, Linwood (NS4364, 2001), Forbes Place, Wemyss Bay (NS1969, 2002), Kirkland (NS5752, 2011), a roadside, Black Hill (NS4851, 2011), Planetreeyetts (NS3570, 2011) and near Spango Burn (NS2475, 2011).

Geranium lucidum L.
Shining Crane's-bill

Britain: common in the S and W, rare in N Scotland.
Hortal and former native?; 6/8; rare.
1st Renfrewshire, 1872, TB.

There are few old locations for Shining Crane's-bill, which was considered not common by Hennedy (1891), who named Devol Glen, and it is assumed that all the records since the 1950s refer to garden outcasts or escapes. It was recorded from waste ground at Newton Mearns (NS5356, 1963, BWR) and, during CFOG surveys, from Pollokshields (NS56R, 1986) and Whitecraigs (NS55N, 1987); there are four other modern records: the west side of Barcraigs Reservoir (NS3857, 1999), the White Cart Water, Millarston (NS5755, 2002), Waterfoot (NS5655, 2011) and waste ground, Wemyss Bay (NS1970, 2006).

Geranium robertianum L.
Herb-Robert

Britain: common throughout.
Native; 239/278; common.
1st Gourock, 1860, GL.

Herb-Robert is a common woodland plant but can also be found along hedge banks, on waste ground – including mine-waste shale and disused railways – and as a garden weed. It can extend into the uplands along watercourses but is absent from peaty or acid soils.

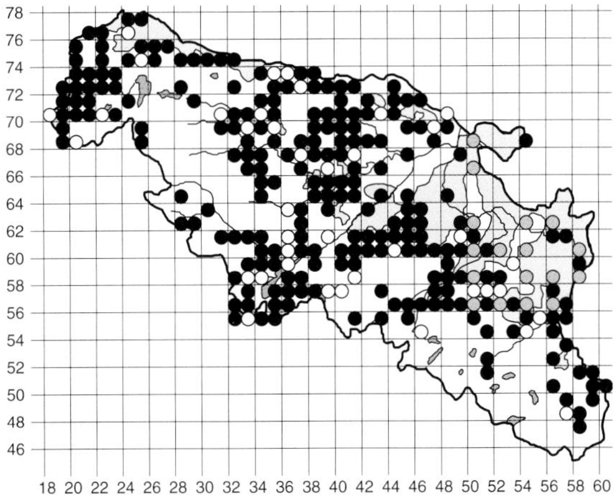

Geranium robertianum

Geranium purpureum Vill.
Little-Robin

Britain: rare on coasts in S England.
Hortal; 0/1; doubtful.
1st Neilston, 1886, GL.

The specimen collected by P Ewing is labelled as 'Neilston' but no other detail was provided. However, another Ewing specimen (collected by BM Oakeshott, 1883, GL), from Shoreham, Sussex, has an additional stem attached to the herbarium sheet labelled 'Renfrew 7/86', indicating that the Renfrew specimen was probably garden-grown from Sussex seed.

Geranium phaeum L.
Dusky Crane's-bill

Britain: frequent in England, also in C and E Scotland.
Hortal; 5/11; rare.
1st Banks of the Cart, 1821, FS.

There are three old specimens of this dark-flowered crane's-bill: from Kilbarchan (1875, GL), Inchinnan (1878, GL) and Wemyss Bay (1886, GRK). RM recorded it from Eaglesham in 1935, where it can still be seen today by the 'Common' (NS5751, 1997). Modern records are from Devol Road (NS3272, 2000), the old estate, Millikenpark (NS4161, 2001), Boylestone Quarry (NS4959, 1987, DM) and Whitecraigs (NS55I, 1983, CFOG). Other records (BWR) not seen during the modern period are from roadsides at Pomillan (NS3467, 1955) and Bishopton (NS4370, 1975).

Erodium maritimum (L.) L'Hér.
Sea Stork's-bill

Britain: native in the SW, rare inland alien.
Former accidental; 0/1; extinct.
1st Paisley Cemetery, 1858, PSY.

Erodium moschatum (L.) L'Hér.
Musk Stork's-bill

Britain: archaeophyte; scattered, mainly coastal in S England.
Accidental; 1/1; very rare casual.
1st Persistent garden weed, Newlands, 1973, CFOG.

Erodium cicutarium (L.) L'Hér.
Common Stork's-bill

Britain: common in the S, more coastal in Scotland.
Native; 2/4; very rare.
1st Clydeside, Erskine, 1883, GL.

Hennedy (1891) repeated the Erskine locality for Common Stork's-bill, noting near Erskine House, and it was collected from Cloch by T Fisher in 1892 (GRK). In 1975 AMcGS found it near Erskine Hospital on the south bank of the Clyde, the same area where the two modern records occur: coastal sandy grasslands just west of the old ferry (NS4672, 1996) and Erskine Park, west of the Erskine Bridge (NS4572, 1996).

LYTHRACEAE

Lythrum salicaria L.
Purple-loosestrife

Britain: common in the S, extending to W C Scotland.
Native and hortal; 19/25; scarce.
1st Banks of the Clyde, Renfrew, 1813, TH.

Other old records for Purple-loosestrife refer to Paisley and Lochwinnoch (1845, NSA), the Black Cart Water near Howwood (1893, AANS) and Inverkip (1933, Lee). It is still known from the White Cart Water in Paisley (NS4863, 2004) and from Castle Semple Loch (NS3659, 1996, SNH-LS), and it occurs on the banks of the Black Cart Water at Inchinnan (NS4767, 1999). Other modern records include Barr Loch (NS3557), Knapps Loch (NS3658) and Caplaw Dam (NS4358), all from the SNH Loch Survey (1996), plus Stanely Reservoir (NS4661, 2005), Whittliemuir (NS4158, 1993) and by a roadside, Eaglesham (NS5852, 2002, PM). It is often planted by ponds and a few records may be of introduced plants, but it is assumed that the riverbank and most rural wetland populations are native.

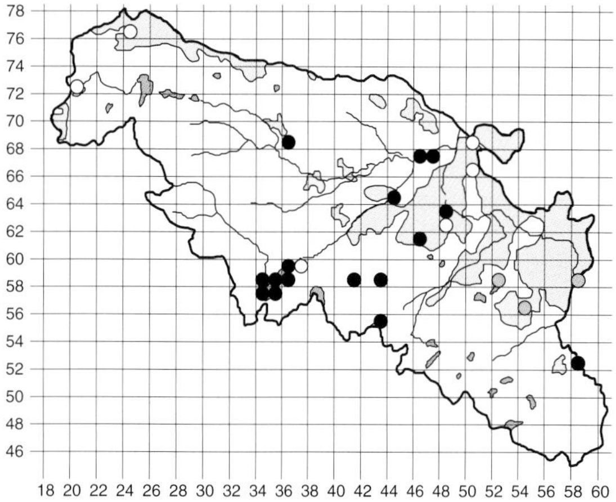

Lythrum salicaria

Lythrum portula (L.) D.A. Webb
Water-purslane

Britain: frequent in the W, rare in E England and N Scotland.
Native; 24/36; occasional.
1st Near Gleniffer, 1858, PSY.

Other old records include Loch Thom (1880, NHSG), where Water-purslane can still be found (NS2672, NS2570, 2000), and Brother Loch (1865, GGO), where it was found by the SNH Loch Survey (NS5052, 1996). Today it is occasionally found in small ponds and to the margins of lochs or reservoirs at various scattered sites. AJS, who found it at Leperstone Reservoir (NS3571, 1976) and Finlaystone (NS3673, 1976), recorded the finds as ssp. *longidentatum* (J. Gay) P.D. Sell.

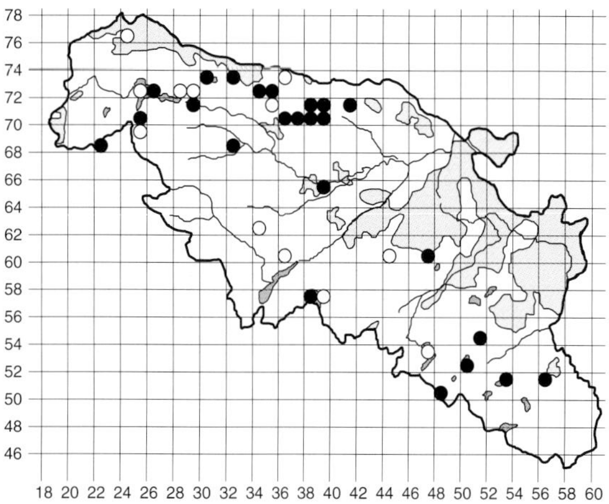

Lythrum portula

ONAGRACEAE

Epilobium hirsutum L.
Great Willowherb

Britain: common throughout, except in NW Scotland.
Native; 165/185; common.
1st Banks of the Cart, 1865, RH.

Hennedy (1865) referred to Great Willowherb as frequent but there are only a few other named locations: old canal, Hawkhead, 1896 (PSY) and Giffnock quarries, 1908 (GL). It appears to have become more common during the last century and is now often seen in urban-fringe locations, occurring in marshes, ditch sides and riverbanks, and also on waste ground; nutrient enrichment of water sources is likely to be a factor in its modern abundance, as the cluster of sites to the central lowland area indicates.

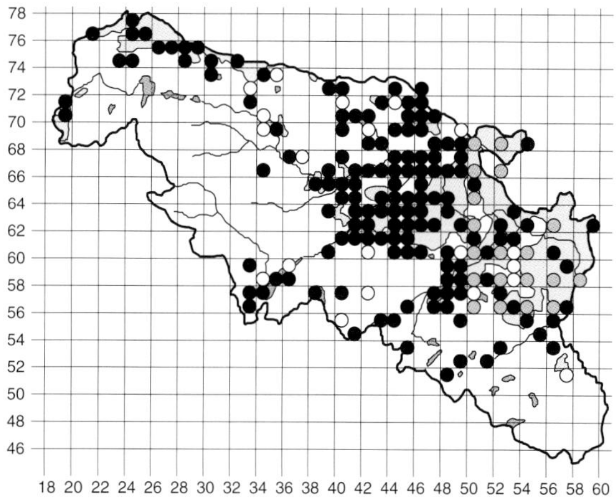

Epilobium hirsutum

Epilobium parviflorum Schreb.
Hoary Willowherb

Britain: common in the S, local and lowland in Scotland.
Native; 22/27; occasional.
1st Banks of the Cart, 1865, RH.

Other old records include Thornliebank (1891, GL), Giffnock quarries (1908, GL) and Williamwood (NS5658, 1934 and 1946, RM). Today Hoary Willowherb has a scattered, disjunct lowland distribution pattern, with many locations close to urban areas. It is always found at poorly draining ground, usually indicating some base-rich flushing, which can often occur at urban waste-ground sites.

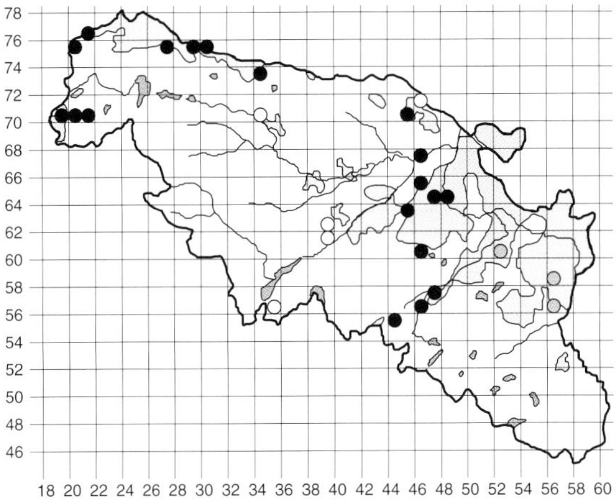

Epilobium parviflorum

Epilobium montanum L.
Broad-leaved Willowherb

Britain: common throughout.
Native; 314/375; very common.
1st Paisley, 1869, RPB.

This is a common willowherb which occurs in urban areas and extends into the uplands alongside the many burn valleys. It is often associated with some shade and is one of the first shade-tolerant species to colonize secondary scrub or woodland sites. It does also occur in open spaces, being found on waste ground at many disturbed urban-fringe marginal sites.

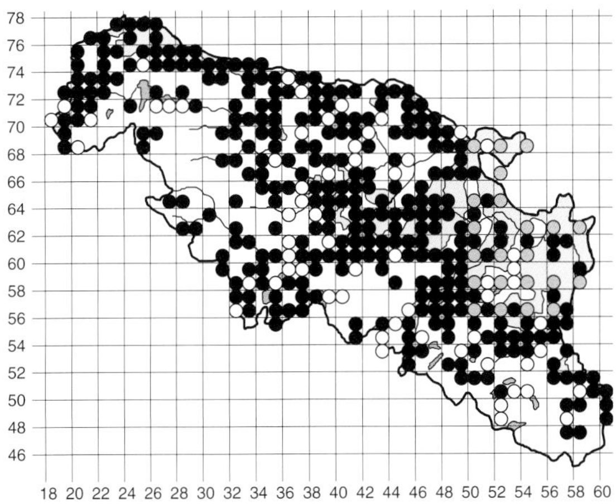

Epilobium montanum

Epilobium tetragonum L.
Square-stalked Willowherb

Britain: common in S England, absent from Scotland.
Accidental; 1/2; very rare.
1st Paisley area (NS46W), 1975, AJS.

Square-stalked Willowherb was considered frequent by earlier authors, but it is thought that most records, if not all, refer to Short-fruited Willowherb. The only specimen seen in the modern period is from the tip at Linwood Moss (NS4465, 1998).

Epilobium obscurum Schreb.
Short-fruited Willowherb

Britain: common throughout except in N Scotland and SE England.
Native; 345/364; very common.
1st Gourock, 1865, RH.

This willowherb has been confused in the past and it is thought that the old records for Square-stalked Willowherb refer to this species. It shows a similar widespread distribution pattern to that of the Broad-leaved Willowherb but is generally found in different habitats. Short-fruited Willowherb can occur on dry waste ground and as a garden weed, but is most typically associated with marshy ground or watercourse margins, where it can extend into the uplands. It often overlaps and perhaps competes with the recent American Willowherb invader and hybrids occur.

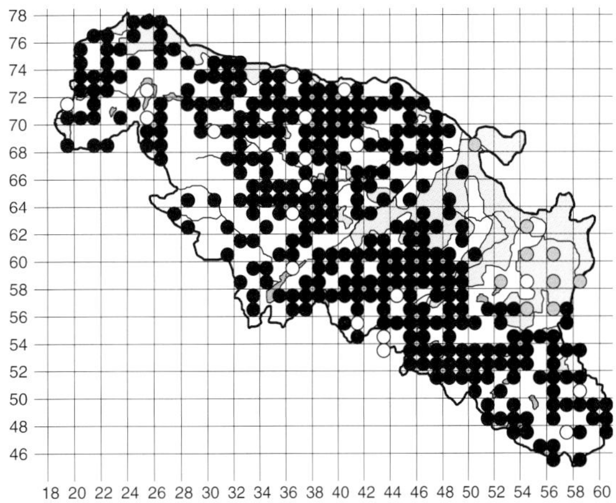

Epilobium obscurum

Epilobium roseum Schreb.
Pale Willowherb

Britain: frequent in the S but rare in the N except in C Scotland.
Accidental; 12/17; scarce.
1st Deaconsbank, 1953, RM.

This willowherb is not mentioned by earlier authors, including Lee (1933), indicating that is a recent arrival. AJS reported it from Finlaystone in 1976, where it still occurs (NS3673, 2003, IPG), and also as a weed in a car park at the University of the West of Scotland Paisley campus (NS4763, 1975). All of the records are from near habitation, and include waste ground but usually where there is some impeded drainage.

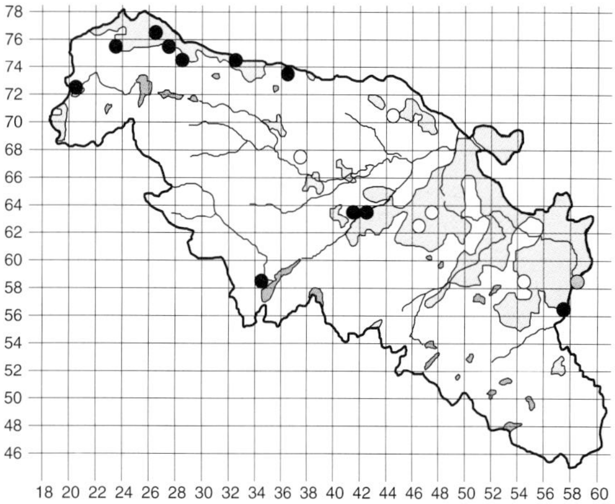

Epilobium roseum

Epilobium ciliatum Raf.
American Willowherb

Britain: neophyte (N America); common throughout except in N Scotland.
Accidental; 202/207; common, increasing.
1st Kilmacolm and Dumbreck, 1969, RM.

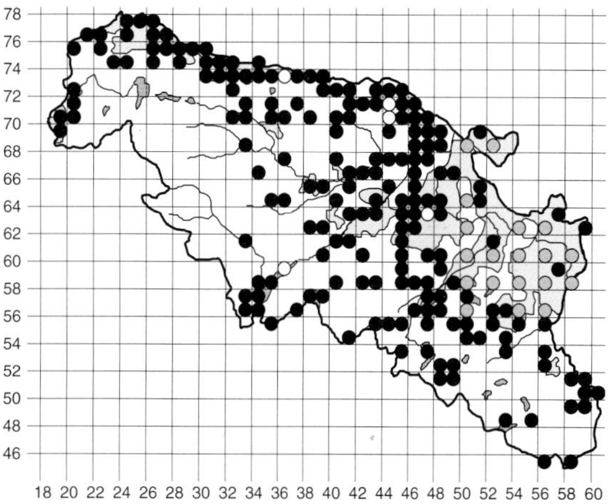

Epilobium ciliatum

Mackechnie (1974) wrote of American Willowherb that 'by the summer of 1972 it was an abundant and obvious weed in gardens and waste ground in the southern suburbs of Glasgow', indicating a dramatic early spread, which is ongoing. It is perhaps now the most common willowherb to be found at urban waste-ground sites, but, as the map shows, although it is predominantly found in the lowlands it is spreading into rural upland-fringe areas. It grows on waste ground and appears to tolerate poor drainage – it has colonized several marshy places and loch shores, as at Pilmuir (NS5154, 2000) and Harelaw (NS3073, 2000).

Epilobium palustre L.
Marsh Willowherb

Britain: common except in SE England.
Native; 266/312; common.
1st Lochwinnoch, 1845, NSA.

Marsh Willowherb shows a strongly rural pattern, becoming scarce in lowland urban areas – though less so in the west – and extending into the uplands. It is the most water tolerant of the willowherbs: it grows in marshes and extends into wetter fens and swamps, commonly with Bottle Sedge but also in less acidic bog-moss mires and to the margins of peat bogs or flushes within them.

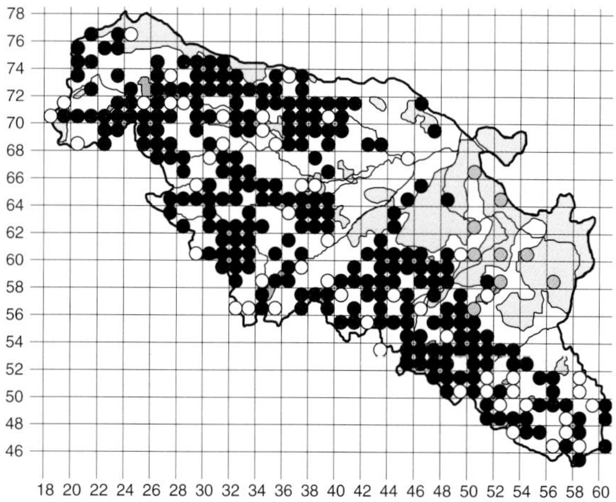

Epilobium palustre

Epilobium anagallidifolium Lam.
Alpine Willowherb

Britain: frequent on the Pennines and the high mountains of Scotland.
Accidental; 0/1; error.
1st Waste ground, near Renfrew, 1924, TPNS.

Lee (1933) cast doubt on a reported casual record for this montane species, the discovery of which, by a Mr Keir, was reported in the minutes of the TPNS (1934). However, a specimen in PSY labelled as this species (1933, from Dykebar) has been redetermined (BWR) as New Zealand Willowherb.

Epilobium brunnescens (Cockayne) P.H. Raven & Engelhorn
New Zealand Willowherb

Britain: neophyte (New Zealand); common in the N and W.
Accidental; 75/94; frequent.
1st Lochwinnoch, 1928, TPNS.

Lee (1933) considered this willowherb a rare casual about Glasgow, and it was collected from a cinder track at Jordanhill (1933, GL); in the same year it was collected from Dykebar Road, near Jenny's Well (PSY) and in 1944 reported from Loch Libo (GN). Today New Zealand Willowherb is a frequent sight, especially in the west, and is often seen on the loose soil of upland burns in the Renfrewshire Heights; Lee (1933) noted it from the old road beyond Muirshiel, perhaps the place from where it has spread over the last 70 years. It appears to be spreading in the south-east uplands along the windfarm tracks at Whitelee (NS5745, 2011).

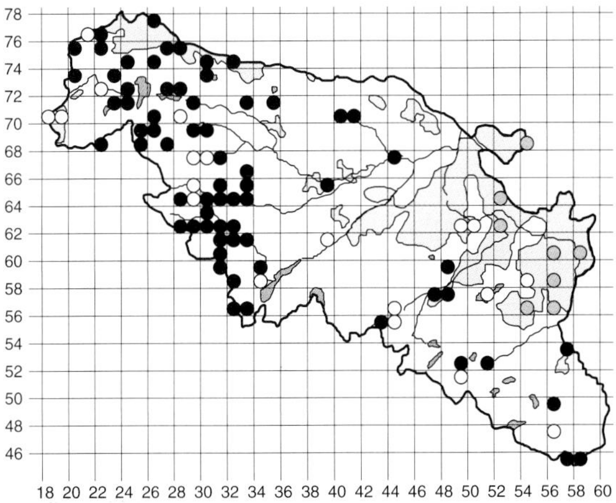

Epilobium brunnescens

Epilobium komarovianum H. Lév.
Bronzy Willowherb

Britain: neophyte (New Zealand); rare and scattered.
Accidental; 3/3; very rare.
1st Northbrae, 1994.

Bronzy Willowherb has been recorded from three locations, all about the Royal Ordnance Factory, Georgetown. The first was by an old quarry (NS4169) and the other two within the site perimeter (NS4467, 1996 and NS4268, 1999). All sites are on open, free-draining waste ground.

Hybrid Willowherbs

There are six hybrids recorded, three involving *E. ciliatum*.

Epilobium × erroneum Hausskn.
(*E. hirsutum × E. montanum*)

1/1; 1st NS47Q, 1987, BSBI.

The only record is from the BSBI database (Formakin area) but no further details are available.

Epilobium × rivulare Wahlenb.
(*E. parviflorum × E. palustre*)

0/1; 1st Williamwood (NS5658), 1934, RM.

Epilobium × interjectum Smejkal
(*E. montanum × E. ciliatum*)

5/5; 1st Rouken Glen (NS5458), 1983, CFOG.

Modern records are from Hotel Wood, Erskine (NS4671, 1997), Linwood Pool (NS4564, 2000), waste ground, Lochwinnoch (NS3458, 2001) and waste ground, Boglestone (NS3373, 2001).

Epilobium × vicinum Smejkal
(*E. obscurum × E. ciliatum*)

3/3; 1st Disused railway, Yoker (NS5068), 1985, GL.

Probably overlooked, other records for this hybrid are from a ditch, Patterton (NS5256, 1985, GL) and marsh, Walton Dam (NS4955, 1998).

Epilobium × schmidtianum Rostk.
(*E. obscurum × E. palustre*)

3/4; 1st Caplaw Dam, 1962, RM.

These two species often occur near each other in marshy places; modern records are from Loch Libo (NS4355, 1996), mire at Whinny Hill (NS3670, 1997) and Linwood Pool (NS4564, 2000).

Epilobium × fossicola Smejkal
(*E. ciliatum × E. palustre*)

2/2; 1st Phoenix site (NS4564), 1995.

The only other record is from Earn Hill, Bank (NS2275, 2002), but this hybrid is likely to be on the increase.

Chamerion angustifolium (L.) Holub
Rosebay Willowherb

Britain: common throughout.
Native and/or alien; 386/460; very common.
1st Gleniffer Glen, Barrhill and Lochwinnoch, 1845, NSA.

There are several old records for Rosebay Willowherb but none (as in CFOG) throw any light on its native status; the Gleniffer Glen record, and a later one from Nethercraigs Gorge (1896, PSY), are likely 'native' candidates, as is the report from the Rotten Burn near

Loch Thom in 1890 (NHSG). It was present at the edge of the VC on waste ground at Busby in 1885 (GL) and was collected from Castle Semple Estate in 1921 (GL). Today it can be found on some remote burn-side rock outcrops, e.g. Raith Burn (NS2862, 1997), North Rotten Burn (NS2568, 1997), Routen Burn (NS3060, 2003) and Munzie Burn (NS5848, 2001), but these may have arrived relatively recently as wind-blown seed. It is common in the lowlands, often abundantly so, on waste ground, woodland rides, waysides and neglected grasslands.

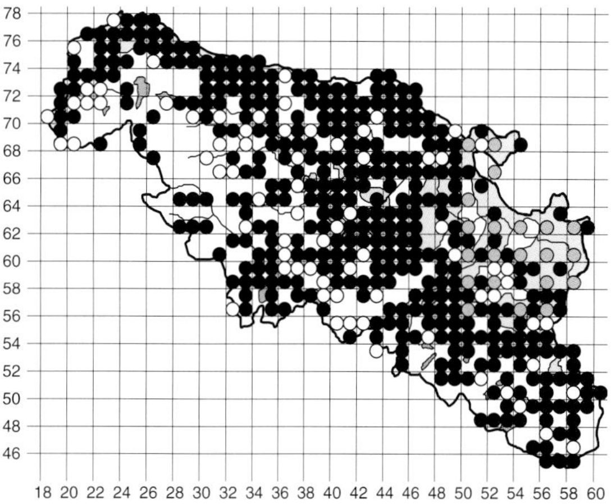

Chamerion angustifolium

Oenothera × *fallax* Renner
(*O. glazioviana* P. Micheli × *O. biennis*)
Intermediate Evening-primrose

Britain: neophyte; scattered, mainly in the S.
Hortal; 1/1; very rare.
1st Lochwinnoch, 2008.

A small population occurs about the new footpath around the wetland near Calder Bridge (NS3558).

Oenothera biennis L.
Common Evening-primrose

Britain: neophyte (N America); frequent in the S.
Hortal; 0/1; extinct.
1st Cornfield in Lochwinnoch, 1834, MC.

A somewhat surprising record, but with no specimen it is unclear which species was recorded.

Clarkia unguiculata Lindl. (Clarkia) & *Clarkia amoena* (Lehm.) A. Nelson & J.F. Macbr. (Godetia)

Britain: neophytes (W N America); scattered casuals.
Hortals; 1/1; very rare.
1st Coup, Linwood Moss, 1998.

Both these garden plants were found growing close together on waste ground to the edge of the Linwood Moss tip (NS4465).

Fuchsia magellanica Lam.
Fuchsia

Britain: neophyte (Chile and Argentina); frequent in the W.
Hortal; 23/23; occasional.
1st Port Glasgow (NS3174), 1996.

The map shows a neat pattern, reflecting a marked preference for milder, wetter coastal-fringe land near to habitation. There are a few records from further east, such as from Boylestone Quarry (NS4959, 1999), Shillford Tip (NS4455, 1999) and Neilstonside (NS4655, 1998). A few records may refer to other cultivars.

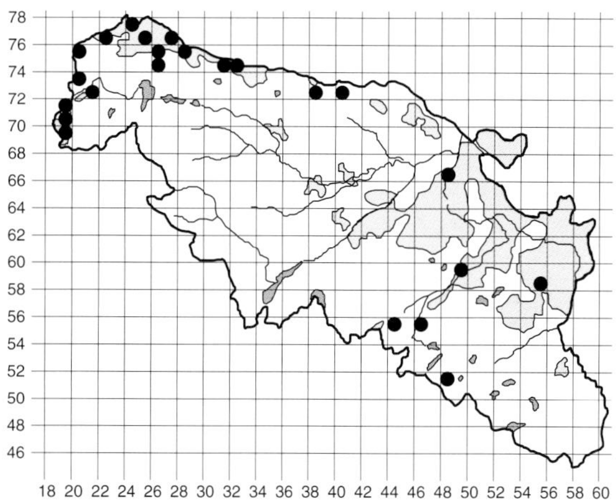

Fuchsia magellanica

Circaea lutetiana L.
Enchanter's-nightshade

Britain: common in the S, extending to C Scotland but rare in N.
Native; 86/107; frequent.
1st Paisley and Lochwinnoch, 1845, NSA.

Enchanter's-nightshade is found in mature woodlands, often where heavily shaded and also

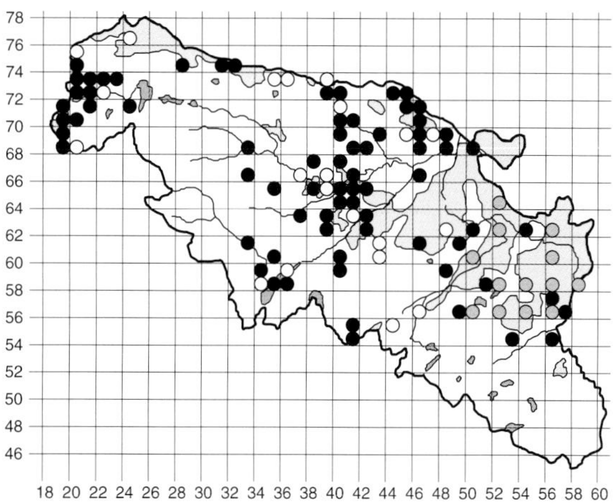

Circaea lutetiana

where the soils are not so acidic. It has a lowland distribution, remarkably similar to that of the hybrid, which is more frequent along upland burns, and it has a strong representation in the west about Inverkip.

Circaea × *intermedia* Ehrh.
(*C. alpina* L. × *C. lutetiana*)
Upland Enchanter's-nightshade

Britain: local in the W and N.
Native; 77/84; frequent.
1st Gleniffer, 1858, PSY.

The 19th-century records for '*C. alpina*' presumably refer to the hybrid although it was considered rare by authors of the day. Today it is widespread and frequently seen in semi-natural woodlands, usually on less acidic soils along watercourses, although it can occur in estate woodlands and has been recorded from disused railways, as at Bridge of Weir (NS3865, 1998) and Neilston (NS4756, 1999).

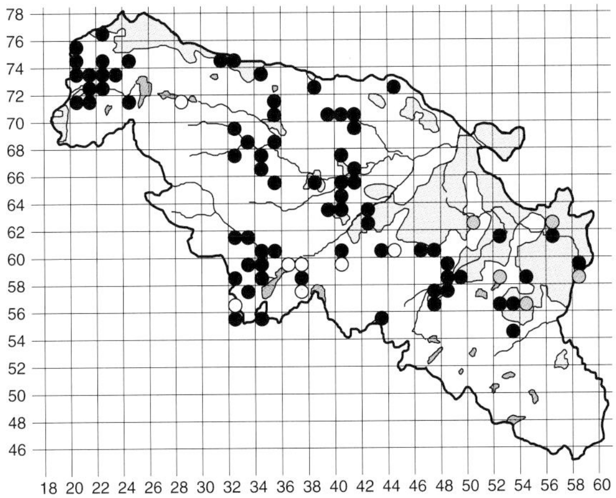

Circaea × *intermedia*

SAPINDACEAE
(HIPPOCASTANACEAE, ACERACEAE)

Aesculus hippocastanum L.
Horse-chestnut

Britain: neophyte (Balkans); common throughout except in N Scotland.
Hortal; 104/127; frequent.
1st The Park, Inchinnan, 1845, NSA.

Horse-chestnut shows a typically lowland, urban-fringe distribution, reflecting its widespread planting in parks and policy estates. It fruits readily and saplings and young trees are frequently found, but it does not appear to be naturalized as claimed by Lee (1933) and noted in CFOG. Red Horse-chestnut (*A. carnea* J. Zeyh.) has been recorded but has only been noted as a planted tree, e.g. as at Waterfoot (NS5655, 2001).

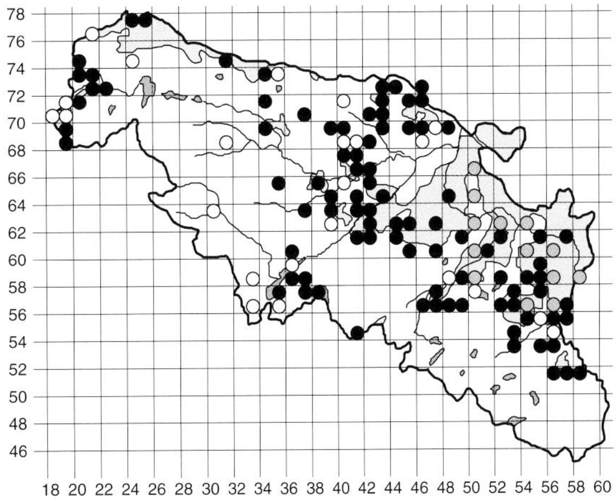

Aesculus hippocastanum

Acer platanoides L.
Norway Maple

Britain: neophyte (Europe); common or frequent, except in N Scotland.
Hortal; 77/80; frequent.
1st Wood, Paisley Road, 1842, GGO.

There are no other early references for this maple, although judging by the size of some trees it has been around for some time. BWR recorded it from Craigends House (NS4166, 1969), Georgetown (NS 4467, 1974), Chapel Farm (NS4168, 1977) and the Glenlora area (NS3358, 1995). Today Norway Maple can be found at many lowland sites, often associated with estate policies; it is commonly found as young plants and seems likely to continue to spread.

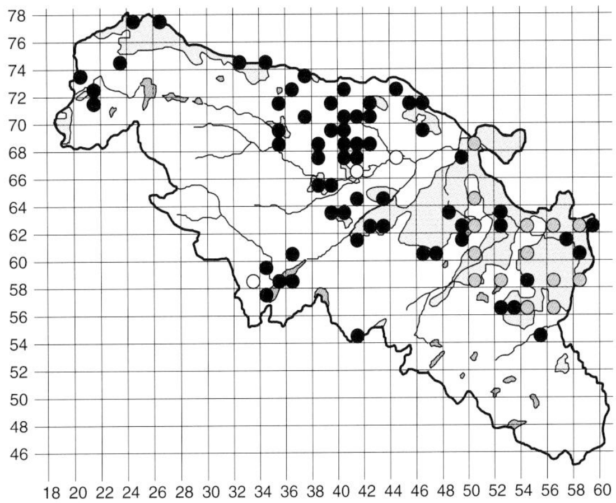

Acer platanoides

Acer cappadocicum Gled.
Cappadocian Maple

Britain: neophyte (SW Asia); scattered in E Scotland.
Hortal; 1/1; very rare.
1st Formakin Estate, 2003, IPG.

This maple is self-sown or suckering in several places around Formakin House (NS4170).

Acer campestre L.
Field Maple

Britain: common in England, local alien in the N.
Hortal; 10/14; scattered.
1st Carruth Bridge, 1889, NHSG.

There are few other old records for this non-native maple: Hawkhead (1931, PSY) and Greenfield (NS5449, 1981, BWR). There are several scattered modern records, some possibly self-sown, but probably mostly for planted trees.

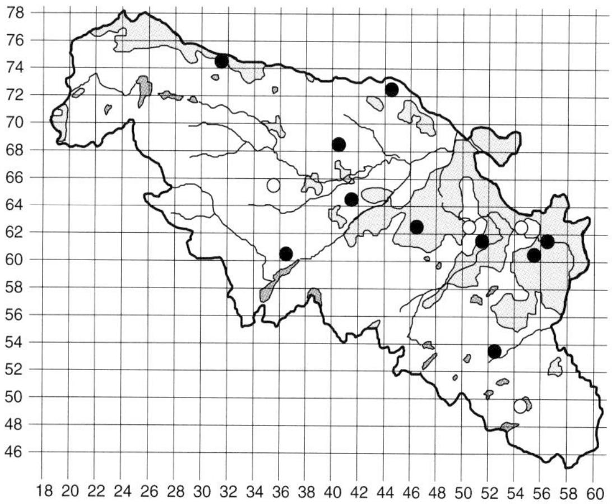

Acer campestre

Acer pseudoplatanus L.
Sycamore

Britain: neophyte (C Europe); common throughout.
Hortal; 352/401; very common.
1st Renfrew, 1777, JW.

Sycamore is a long-established and widely planted tree of parks and woodlands, occurring throughout the VC and only absent from the grazed and wet peaty uplands, but even here it has been favoured for planting near farmsteads. It is commonly found in urban places and it is likely to dominate secondary woodlands if left alone. It also appears to have benefited from the demise of Wych Elm along lowland watercourses.

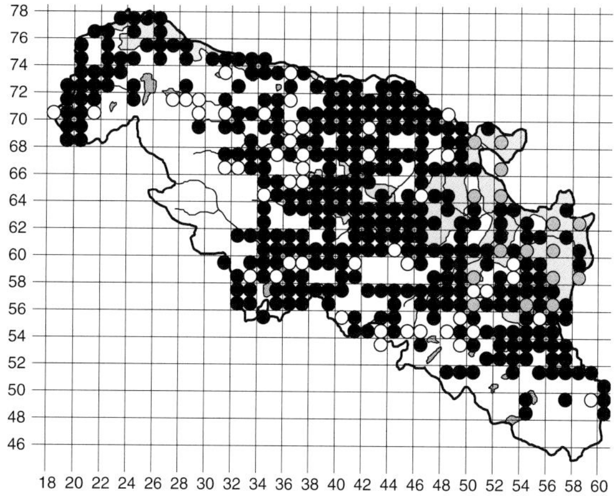

Acer pseudoplatanus

MALVACEAE
(TILIACEAE)

Malva moschata L.
Musk-mallow

Britain: common in the S, frequent alien in lowland Scotland.
Hortal; 15/17; occasional in urban-fringe areas.
1st Inverkip and Langbank, 1891, RH.

A further location was Milliken (1904, PSY), but Musk-mallow was excluded from the Renfrewshire list (TPNS 1915). Today it is known from at least five 1km squares along the A737 near Kilbarchan and Linwood (NS4062 to NS4464), presumably originating from a wildflower seed mix; it is also recorded from several urban locations including Jenny's Well (NS4962, 2001), with some populations perhaps of similar origin, and an outlier at Erskine Park (NS4572, 2011).

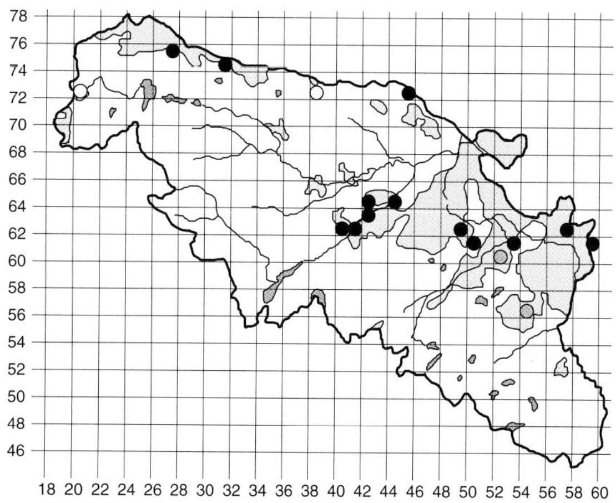

Malva moschata

Malva alcea L.
Greater Musk-mallow

Britain: neophyte (Europe); rare in the S.
Former hortal; 0/1; extinct.
1st Newlands, 1919, RG.

Malva sylvestris L.
Common Mallow

Britain: archaeophyte common in England, frequent in E Scotland.
Hortal; 8/10; rare, scattered.
1st Renfrewshire, 1883, TB.

Common Mallow was collected from Fort Matilda, Gourock, in 1909 (GL) and an undated specimen exists for Paisley Canal (presumably collected in the 19th century). There are two records in CFOG: a garden at Jordanhill (NS5468, 1987) and Auldhouse Burn (NS5560, 1988); more recent records are from South Mains farm, Houston (NS4266, 2005), Loch Libo (NS4355, 1987, DM), Inch Green Dock (NS3075, 2010, AHJ), by the White Cart Water, Howford Bridge (NS5163, 2005), a slip road off the M77 at Ryat Farm (NS5257, 2003) and Patterton area (NS5357, 1991, LB).

Malva parviflora L.
Least Mallow

Britain: neophyte (Europe); scattered casual.
Former accidental; 0/2; extinct.
1st Wasteland near Paisley, 1910, PSY.

RG (1930) also mentioned Darnley.

Malva pusilla Sm.
Small Mallow

Britain: neophyte (SW Europe); scattered casual.
Former accidental; 0/2; extinct.
1st Nitshill, 1909, GL.

This mallow was also collected near Paisley in 1913 by D Ferguson (GL).

Malva neglecta Wallr.
Dwarf Mallow

Britain: archaeophyte common in the SE, scattered in Scotland.
Former accidental; 0/1; extinct casual.
1st Slates Mill, Kilmacolm (NS3469), 1969, BWR.

There is no further information on the card for this very rare mallow.

Malva × clementii (Cheek) Stace
Lavatera thuringiaca auct. non L.
Garden Tree-mallow

Britain: neophyte (Europe); scattered in S England.
Hortal; 1/1; very rare.
1st Stanely Dam, Paisley, 2005.

The sole record is from rough ground near a road by the dam, where Garden Tree-mallow persists (NS4661).

Sidalcea malviflora (DC.) A. Gray ex Benth.
Greek Mallow

Britain: neophyte (SW N America); scattered, local in C Scotland.
Hortal; 3/3; very rare escape or outcast.
1st Disused railway, Yoker (NS56E), 1986, CFOG.

The most recent records are from waste ground at Ingliston (NS4371, 1999) and Jordanhill (NS5468, 2002, AR).

Tilia × europaea L.
(*T. cordata* Mill. × *T. platyphyllos* Scop.)
Lime

Britain: rare native hybrid, common throughout as a planted tree.
Hortal; 99/128; frequently planted.
1st Hawkhead, 1812, JW.

Lime trees have long been planted in parks, as street trees and along avenues at policy estates. At a few sites there is evidence of suckering, but no clear evidence of any lime seedlings. Small-leaved Lime (*T. cordata* Mill.) was recorded from Langside (1860, GGO) and occurs rarely as a planted tree, as does Large-leaved Lime (*T. platyphyllos* Scop.), which was listed by Ferguson (1915), but there are no modern field records.

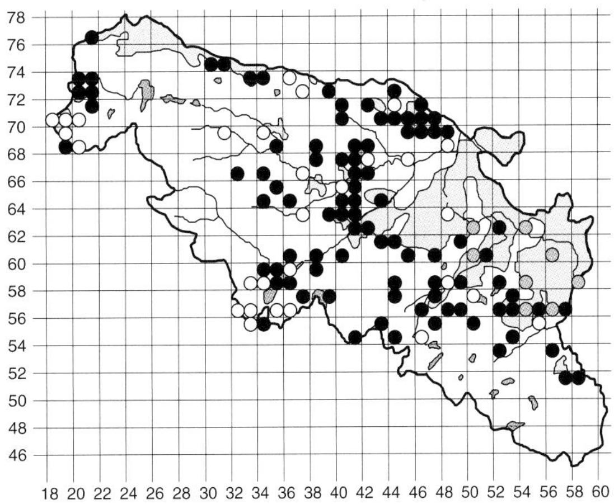

Tilia × europaea

TROPAEOLACEAE

Tropaeolum majus L.
Nasturtium

Britain: neophyte (S America); scattered, mainly in the S.
Hortal; 1/1; very rare.
1st Ladyburn (NS3075), 1998.

Possibly overlooked, this rare garden escape has very few Glasgow records (CFOG) and is unlikely to persist.

LIMNANTHACEAE

Limnanthes douglasii R. Br.
Meadow-foam

Britain: neophyte (W N America); widely scattered in NE Scotland.
Hortal; 6/7; rare.
1st Giffnock coup, 1924, RG.

There are a six modern records of this garden outcast which appears to persist locally: Braemount (NS4760, 1995), Linwood Moss tip (NS4465, 1998), waste ground, Bridge of Weir (NS3865, 1998), Castle Semple (NS3658, 1992, CMRP), Corslet Rd, Darnley (NS5256, 1998) and by the entrance gate, White House (NS4164, 2003).

RESEDACEAE

Reseda luteola L.
Weld

Britain: archaeophyte; common in England and lowland Scotland.
Hortal and accidental; 21/26; occasional.
1st Paisley, 1845, NSA.

Other old records include Cathcart and Gourock (1865, RH). Weld appears to have a local distribution, with centres about Linwood and north Paisley, Greenock and Glasgow (CFOG). There are outliers at Underheugh (NS2075, 1997), Lochwinnoch (NS3558, 2005, IPG) and a remote site at Kaim Dam (NS3462, 1998). All the records are from disturbed or waste ground sites.

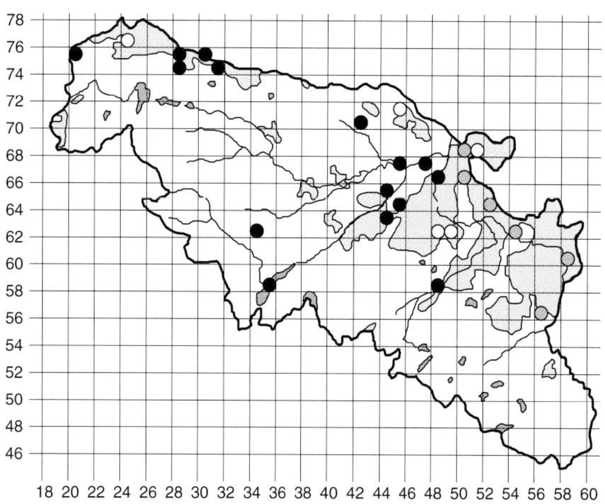

Reseda luteola

Reseda alba L.
White Mignonette

Britain: neophyte (S Europe); scattered in S England.
Former hortal; 0/1; extinct.
1st Hawkhead Estate, 1942, TPNS.

Reseda lutea L.
Wild Mignonette

Britain: common in the S and E, frequent (probably as alien) in C Scotland.
Hortal; 13/17; scarce.
1st Langbank, 1861, MY.

The native status of this species in C Scotland is questionable. Hennedy (1891) thought it alien and described it as very rare; all of its known sites are associated with disturbed waste ground. There are two old records from Crossmyloof (1908, GL) and Paisley (1910, PSY); Wild Mignonette was also recorded from the latter in 1942 (TPNS) and it is still known there today, contiguous with the sites along the Clyde (CFOG). More isolated sites occur at Port Glasgow (NS3174, 1998) and Lochwinnoch (2005, IPG).

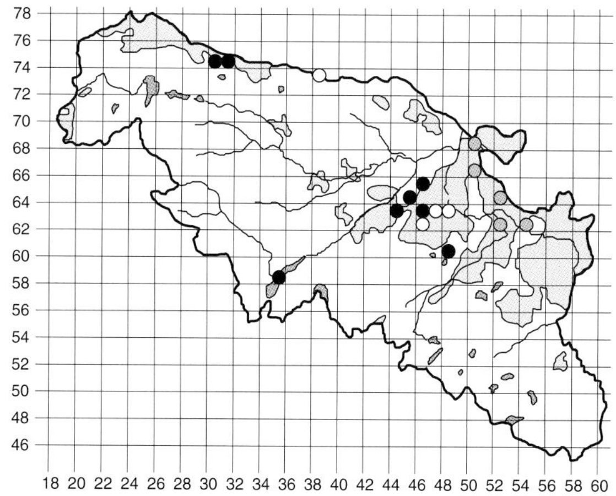

Reseda lutea

BRASSICACEAE

Erysimum cheiranthoides L.
Treacle-mustard

Britain: archaeophyte; common in the SE, scattered in the N.
Former accidental; 0/6; extinct or overlooked casual.
1st Hurlet, 1890, AANS.

Hennedy (1891) repeated the Hurlet record and there are two GL specimens from Barrhead (1891) and railway sidings, Giffnock (1901); Treacle-mustard was also collected from wasteland, Paisley (1910, PSY). In more recent times it was recorded from similar disturbed ground at Paisley (NS4765, 1977, AJS) and Lonend, Paisley (NS4863, 1980, AMcGS).

Erysimum cheiri (L.) Crantz
Wallflower

Britain: neophyte (E Mediterranean); common in S England, local in C Scotland.
Hortal; 2/3; very rare.
1st Cathcart Castle, 1889, AANS.

Today Wallflower can be found on an old wall at Kirkton (NS4857, 1999) and at a disused railway at Gallowhill (NS4864, 1999).

Erysimum repandum L.
Spreading Treacle-mustard

Britain: neophyte (S Europe); rare casual.
Former accidental; 0/1; extinct casual.
1st Paisley, 1910, DF.

This casual was noted growing throughout a waste-ground site (Ferguson 1911).

Arabidopsis thaliana (L.) Heynh.
Thale Cress

Britain: common throughout except in N Scotland.
Native; 109/140; common in lowlands.
1st Paisley, 1845, NSA.

A common, early flowering plant of waste ground, path edges, gardens and open micro-habitats, Thale Cress shows a strongly lowland, urban distribution pattern.

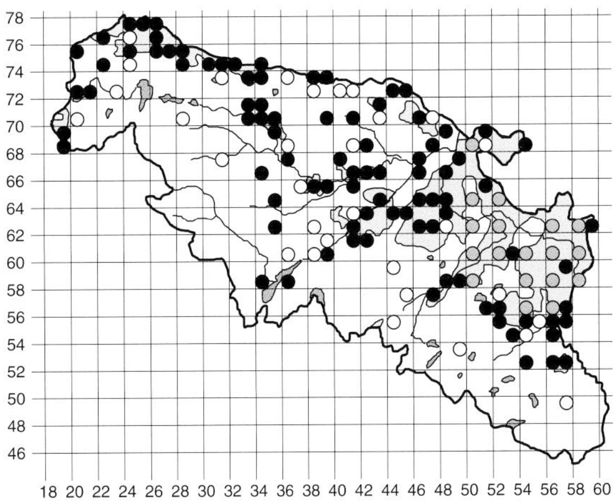

Arabidopsis thaliana

Camelina sativa (L.) Crantz
Gold-of-pleasure

Britain: archaeophyte; scattered in the SE.
Former accidental; 0/5; extinct.
1st Gourock, 1837, E.

This species was collected by W Gourlie from a flax field and was probably occasionally found as an arable weed locally; other records are from Lochwinnoch (1845, NSA), Renfrew Moor (1871, MY), Giffnock (1893, GL) and wasteland, Paisley (1910, PSY). A specimen named '*C. alyssum*' from Paisley (1910, PSY) has been redetermined as Ball Mustard.

Neslia paniculata (L.) Desv.
Ball Mustard

Britain: neophyte (Europe); rare casual.
Former accidental; 0/1; extinct casual.
1st Paisley, 1910, PSY.

The sole record is from a herbarium sheet (PSY) labelled '*Camelina alyssum*'.

Capsella bursa-pastoris (L.) Medik.
Shepherd's-purse

Britain: archaeophyte; common throughout except in N Scotland.
Accidental; 125/225; common.
1st Renfrewshire, 1869, PPI.

This is a common weedy plant of urban road and path sides, waste ground, gardens, riverbanks and arable fields, although it appears to have been under-recorded during modern surveying (notably in the extreme west). There may have been a decline in rural areas, perhaps reflecting a decrease in arable farming, although Shepherd's-purse is one of the more persistent weeds of farms and track sides.

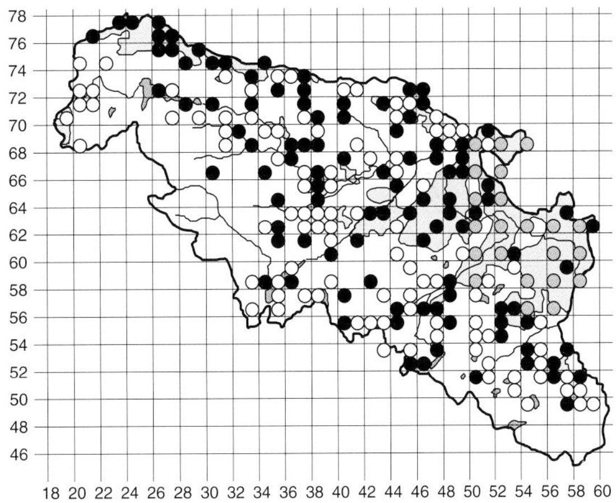

Capsella bursa-pastoris

Barbarea vulgaris W.T. Aiton
Winter-cress

Britain: common in the S and in lowland Scotland.
Native; 62/80; frequent.
1st Lochwinnoch and Paisley, 1845, NSA.

Winter-cress is a widespread species with a distribution pattern marking out several watercourses: the Gryfe, Black Cart, White Cart and Levern waters. It is frequently and readily seen in spring along riverbanks, although it becomes harder to spot later in the season; it is also found on waste ground and at roadsides.

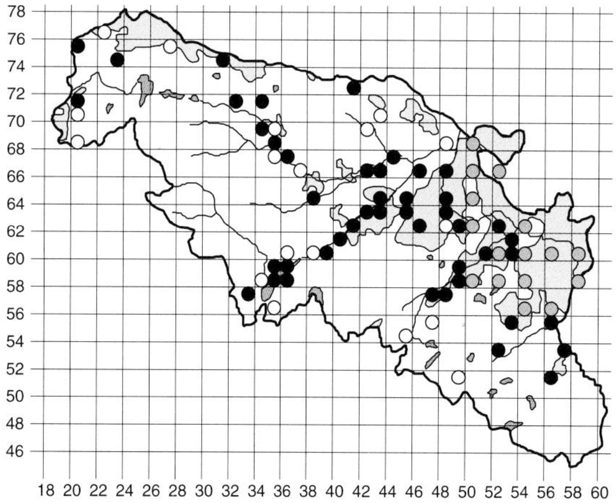

Barbarea vulgaris

Barbarea intermedia **Boreau**
Medium-flowered Winter-cress

Britain: neophyte (Europe); widely scattered but local.
Accidental; 10/11; rare but spreading.
1st Black Loch, 1966, RM.

There are two relevant CFOG records for Medium-flowered Winter-cress, which has perhaps been overlooked for Winter-cress. It appears to be spreading with several scattered records from waste-ground and roadside habitats.

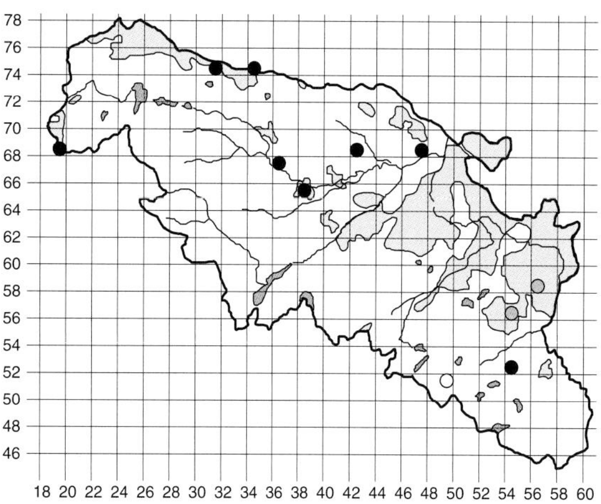

Barbarea intermedia

Barbarea verna **(Mill.) Asch.**
American Winter-cress

Britain: neophyte (Europe); widely scattered in the S, rare in Scotland.
Accidental; 1/2; very rare.
1st Bridge of Weir, 1916, GL.

The first record is of an immature specimen collected from the 'Banks of Gryfe'. Fruiting specimens were recently found growing on waste ground at Port Glasgow (NS3075, 2012).

Rorippa islandica **(Oeder ex Gunnerus) Borbás**
Northern Yellow-cress

Britain: very local, at a few sites.
Native; 2/2; very rare.
1st Lochwinnoch, 2003, IPG.

This much rarer relative of Marsh Yellow-cress, found by IPG from path sides and soil on disturbed ground near Castle Semple Loch (NS3659 and NS3558), is one of the highlights of the modern survey period; he also found it from shingle by the River Calder. Northern Yellow-cress may have arrived as a recent migrant or was perhaps accidentally brought to the area during recent earthworks. However, it could have been overlooked at other sites.

Rorippa palustris **(L.) Besser**
Marsh Yellow-cress

Britain: common in the S, extending to C Scotland, absent from N Scotland.
Native; 53/63; frequent.
1st Cart and Gourock, 1865, RH.

Marsh Yellow-cress is frequently found by the margins of lochs and dams, often in some abundance, but it is also locally common on the sandy or shingle edges of larger watercourses. There are few records from waste ground. It appears to be uncommon in the north-west.

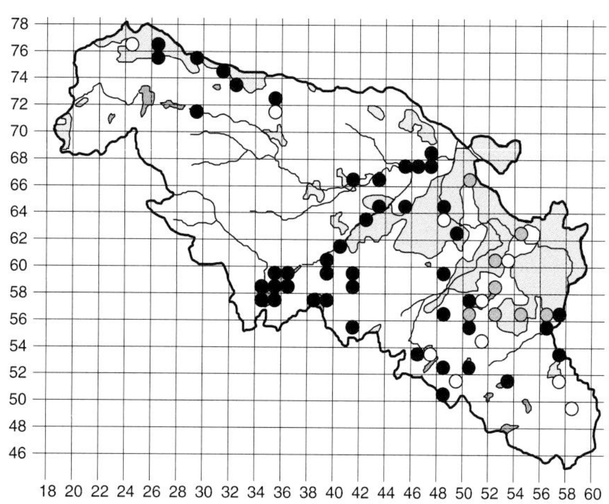

Rorippa palustris

Rorippa sylvestris (L.) Besser
Creeping Yellow-cress

Britain: widespread in the S, extending to C Scotland.
Native; 28/30; occasional.
1st Lochwinnoch, 1845, NSA.

This yellow-cress was also reported from Lochwinnoch in 1892 by J Shearer (NHSG) and collected from there in 1892 and 1916 (GL). It was additionally reported from Whiteinch in 1889 (NHSG). There are several modern records for Creeping Yellow-cress along the Black and White Cart waters, where it grows along the riverbanks, and it was commonly encountered in Glasgow (CFOG). The two westerly records are from Finlaystone (NS3673, 2003, IPG) and Cartsdyke (NS27X, 2003).

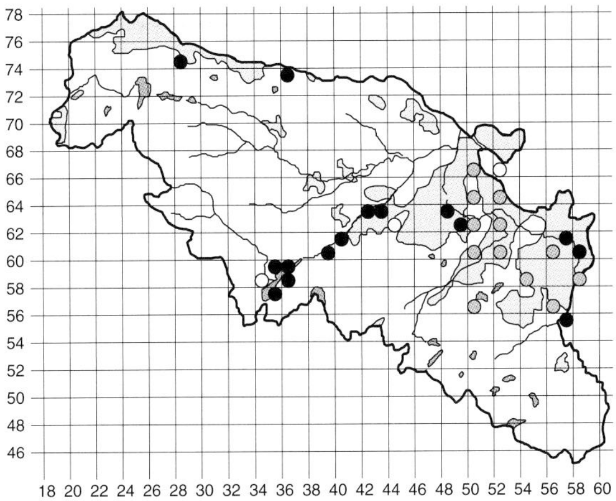

Rorippa sylvestris

Rorippa × *armoracioides* (Tausch) Fuss
(*R. sylvestris* × *R. austriaca* (Crantz) Besser)
Walthamstow Yellow-cress

Britain: rare local hybrid.
Accidental; 1/1; very rare or extinct.
1st Auldhouse Burn, Mansewood (NS5560), 1991, CFOG.

This species and the preceding have not been seen during modern recording.

Rorippa amphibia (L.) Besser
Greater Yellow-cress

Britain: common in SE England, rare alien in C Scotland.
Accidental; 0/1; probably extinct.
1st Newlands Tennis Club (NS5760), 1970, CFOG.

Nasturtium officinale agg.
Rorippa nasturtium-aquaticum agg.
Water-cresses

Britain: common throughout England and lowland Scotland.
Native or hortal; 97/139; common in wet places.
1st Paisley, 1845, NSA.

Older records do not distinguish between the following two species or their hybrid, so mapping historical distribution is not possible. Hybrid Water-cress (*N.* × *sterile* (Airy Shaw) Oefelein) [*Rorippa* × *sterilis* Airy Shaw] is by far the commonest, but it is possible that some non-flowering records may be of the two parents. The majority of the modern records shown (from 81 1km squares) refer to the hybrid, which was collected from Paisley Canal Bank in *c.* 1860 (GGO). It shows a strong rural pattern, often occurring along stream and ditch sides near dwellings.

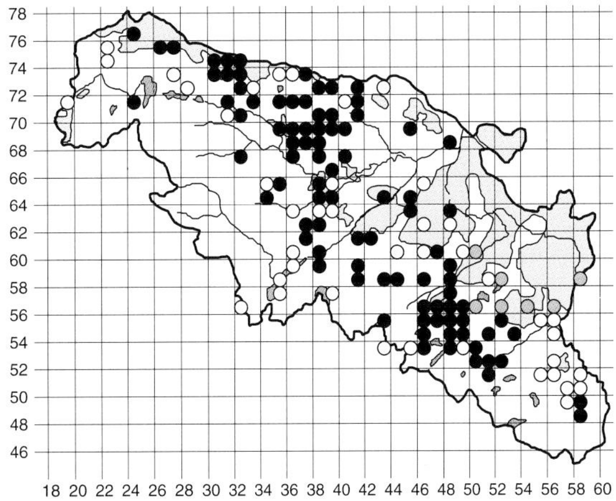

Nasturtium officinale agg.

Nasturtium officinale W.T. Aiton
Rorippa nasturtium-aquaticum (L.) Hayek
Water-cress

Britain: common in the S, scattered and local in Scotland.
Native; 4/5; rare.
1st Witch Burn, 1966, BWR.

This appears to be the rarer of the two water-cress species, but is probably overlooked, and was first recorded in the modern period from Loch Libo in 1996 (NS4355, SNH-LS). It was later found not far away at a marshy burn side near Nether Kirkton (NS4857, 1999). Two further records are from Leperstone Reservoir (NS3571, 2000) and in a conduit by the disused railway trackbed (now a cycleway) at Auchenbothie (NS3371, 2005).

Nasturtium microphyllum (Boenn.) Rchb.
Rorippa microphylla (Boenn.) Hyl. ex Á. & D. Löve
Narrow-fruited Water-cress

Britain: widespread but scattered, frequent in E Scotland.
Native; 16/17; scarce.
1st Finlaystone (NS3673), 1976, AJS.

This water-cress has been recorded from several rural locations, usually in similar places to the hybrid, but it is possibly under-recorded.

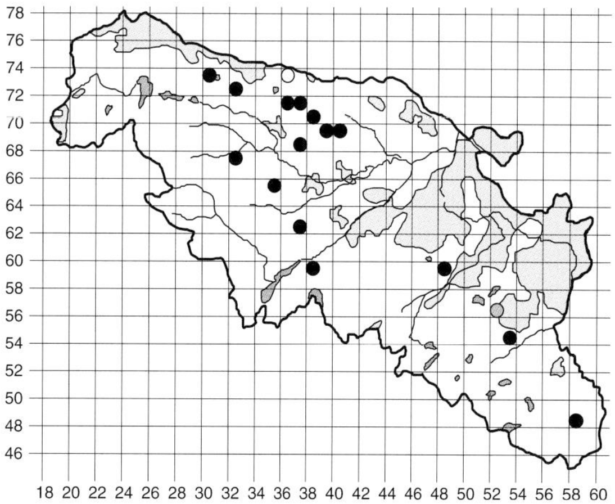

Nasturtium microphyllum

Armoracia rusticana P. Gaertn., B. Mey. & Scherb.
Horse-radish

Britain: neophyte (W Asia); common in England, occasional in C Scotland.
Hortal; 7/10; rare.
1st Castle Semple Loch, 1887, NHSG.

RM knew Horse-radish from waste ground at Giffnock (1925), JP reported it from Inverkip (1971) and there are four relevant records in CFOG from the eastern part of the VC: Yoker Ferry (NS56E, 1985), Maxwell Park (NS56R, 1986), Barshaw Park (NS56L, 1988) and east of Renfrew (NS56D, 1987). There are only three modern field records: Linwood Moss (NS4465, 1998), a roadside, East Fulwood (NS4567, 2001) and Eastwood Toll roundabout (NS5558, 1990); the last site was destroyed in the early 1990s.

Cardamine amara L.
Large Bitter-cress

Britain: common in C Britain to E Scotland, absent from NW Scotland.
Native; 143/175; common.
1st Lochwinnoch, 1845, NSA.

There are old records for this species from Shielhill Glen, Langbank, Kelly Glen, Darnley [Waulkmill] Glen, Calder Glen and Castle Semple Loch, where it can still be found today. Large Bitter-cress is a common plant of wet woodland marshes, swamps and flushes, and is frequently found along shaded burn sides and riverbanks. It shows a strong lowland pattern. It can occur in open habitats and is often noticeable as an early-flowering species beneath taller swamp dominants.

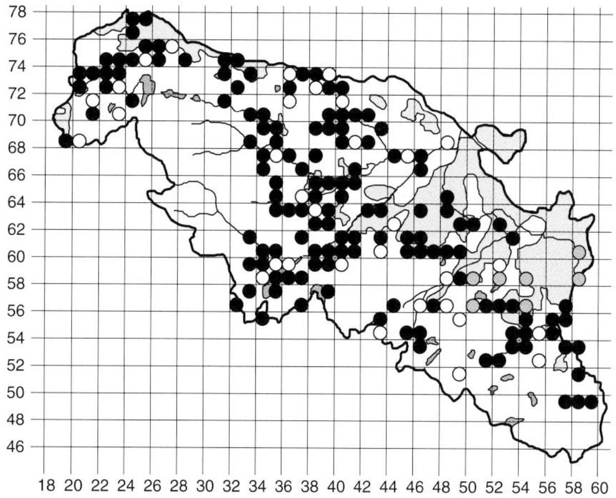

Cardamine amara

Cardamine raphanifolia Pourr.
Greater Cuckooflower

Britain: neophyte (S Europe); widely scattered but very local.
Hortal; 4/5; rare.
1st White Cart Water, Netherton, 1985, CFOG.

This showy cuckooflower was recorded in 1987 from Dargavel Burn (TPNS) and is now becoming well established on the shaded banks from Parkglen (NS3970, 1998) through to Formakin Estate (NS4070, 1994; NS4170, 2003, IPG) and on to Georgetown (NS4268, 1999).

Cardamine pratensis L.
Cuckooflower

Britain: common throughout.
Native; 363/435; very common.
1st Paisley, NSA, 1845.

Cuckooflower is a common, often abundant, and attractive flower of damp or marshy places, extending well into the uplands. It can be seen readily in springtime in rushy pastures and marshes, and also in parks and gardens on poorly draining amenity grassland or lawns – at least before the mowers arrive.

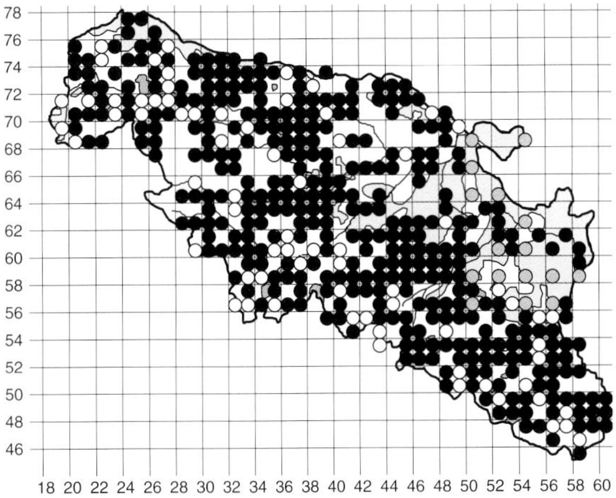

Cardamine pratensis

Cardamine flexuosa With.
Wavy Bitter-cress

Britain: common throughout.
Native; 321/389; very common.
1st Gourock, 1856, GGO.

Wavy Bitter-cress is a very common plant of burn sides, especially in woodlands, marshy places where not too wet, disturbed soil, waste ground, parks and gardens. It only becomes rare in the uplands.

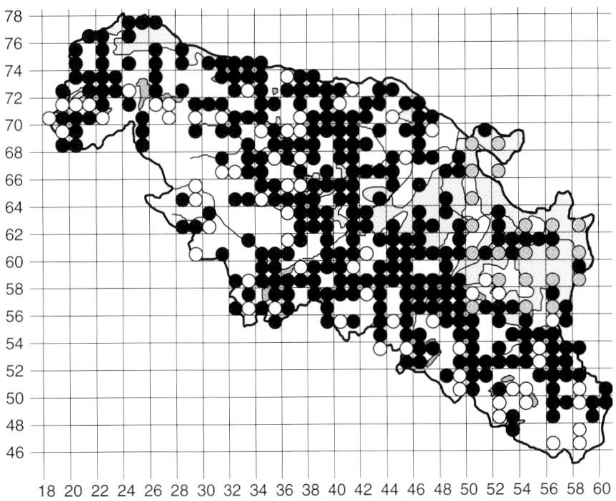

Cardamine flexuosa

Cardamine hirsuta L.
Hairy Bitter-cress

Britain: common throughout except in N Scotland.
Native; 76/133; frequent.
1st Gourock, 1860, GL.

Probably overlooked or under-recorded for the preceding, Hairy Bitter-cress is a more urban plant most frequently seen on waste ground, in gardens and on old walls. It was well recorded in rural areas during the earlier recording period (1960s–80s).

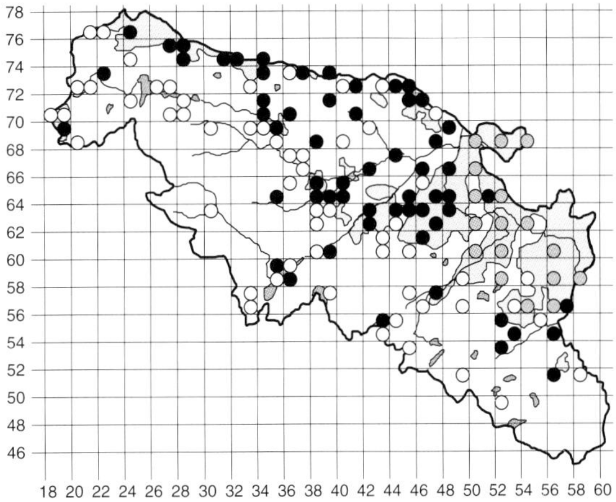

Cardamine hirsuta

Cardamine pentaphyllos (L.) Crantz
Five-leaflet Bitter-cress

Britain: neophyte (W Europe); rare casual.
Former hortal; 0/1; extinct casual.
1st Rubbish dump, Kilmacolm, 1965, ERTC.

Lepidium sativum L.
Garden Cress

Britain: neophyte (W Asia); widely scattered, mainly in the S.
Hortal; 1/5; very rare casual.
1st Paisley, 1901, FFCA.

Garden Cress was collected from Springrove Garden, Kilbarchan (1930, PSY) and Abbotsinch (NS4765, 1957, PSY); more recently it was recorded from the Finlaystone Estate (NS3973, 1976, AJS) and, most recently, from Rouken Glen Park in 1986 (NS5458, CFOG). There are no modern field records but it may well lurk in gardens and other nearby places.

Lepidium campestre (L.) W.T. Aiton
Field Pepperwort

Britain: archaeophyte; frequent in SE England, local elsewhere.
Accidental; 1/7; very rare casual.
1st Langside, 1813, TH.

This Pepperwort has only one modern record, from an old tip at Linwood Moss (NS4465, 1998), and BWR knew it from a rubbish tip at St James (NS4665, 1961). There are several other old records: Ward House, Kilbarchan, and Lochwinnoch (both 1845, NSA), Locher Water (1886, NHSG) and Bridge of Weir (1896, TPNS). There may have been confusion with the more widespread Smith's Pepperwort.

Lepidium heterophyllum Benth.
Smith's Pepperwort

Britain: widespread, locally common in the W.
Native; 25/32; occasional.
1st Dry fields, Gourock, 1879, GL.

This is the commonest of the pepperworts, found on waste ground and disused railways mainly in the north central area; it is absent from the south-east.

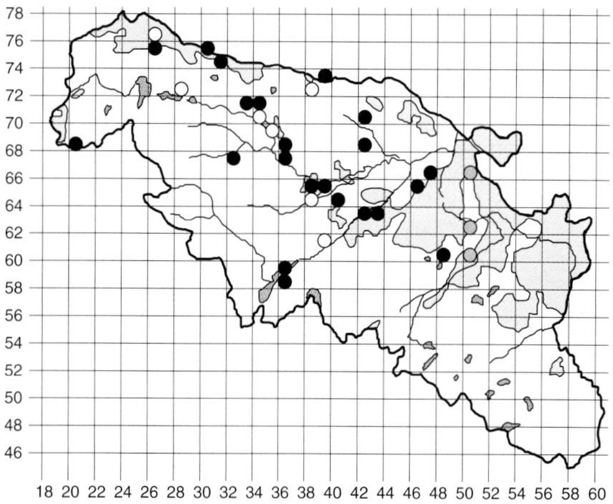

Lepidium heterophyllum

Lepidium virginicum L.
Least Pepperwort

Britain: neophyte (N America); scattered casual, mainly in the SE.
Former accidental; 0/1; extinct casual.
1st Newlands Coup, 1920, GL.

Lepidium ruderale L.
Narrow-leaved Pepperwort

Britain: archaeophyte; frequent in SE England, rare in Scotland.
Former accidental; 0/2; extinct casual.
1st Wasteland, Paisley, 1910, PSY.

There are two other specimens from 'near Paisley' in GL (1910 and 1913).

Lepidium draba L.
Hoary Cress

Britain: neophyte (S Europe); very common in the SE, scattered elsewhere.
Accidental; 3/6; very rare.
1st Near Renfrew, 1873, TGSFN.

Other old records include Hawkhead Station (1889, PSY) and Blackstone (1902, PSY); the latter is possibly by the banks of the White Cart Water, where Hoary Cress was recorded in 1981 (AMcGS) and where it can still be found locally in abundance (NS4968, 1997 and NS4967, 2001). It was also reported from Deanside (at the VC76 boundary) during CFOG surveys (NS56H, 1989).

Lepidium didymum L.
Coronopus didymus (L.) Sm.
Lesser Swine-cress

Britain: neophyte (S America); common in the S, occasional in lowland Scotland.
Accidental; 4/6; rare casual.
1st Renfrewshire, 1899, GC.

In 1979 this swine-cress was recorded from Barochan Estate (NS4168, BWR). The modern records are from waste ground at Linwood Moss (NS4465, 1998) and Inverkip (NS2072, 1999), Paisley (NS4763, 1998, AJS) and Formakin Estate (NS4170, 2003, IPG).

Subularia aquatica L.
Awlwort

Britain: frequent in NW Scotland, rare in uplands further S.
Former native; 0/1; extinct.
1st Langbank, 1869, MY.

This is the only named locality for Awlwort, which was excluded from the Renfrewshire list (TPNS 1915).

Lunaria annua L.
Honesty

Britain: neophyte (SE Europe); common in England and lowland Scotland.
Hortal; 22/25; occasional.
1st St James, Paisley, 1961, BWR.

There are only a few scattered modern records for this garden escape or outcast, usually from the vicinity of housing. It grows on waste ground and at roadsides and woodland edges and is persistent, being capable of local spread.

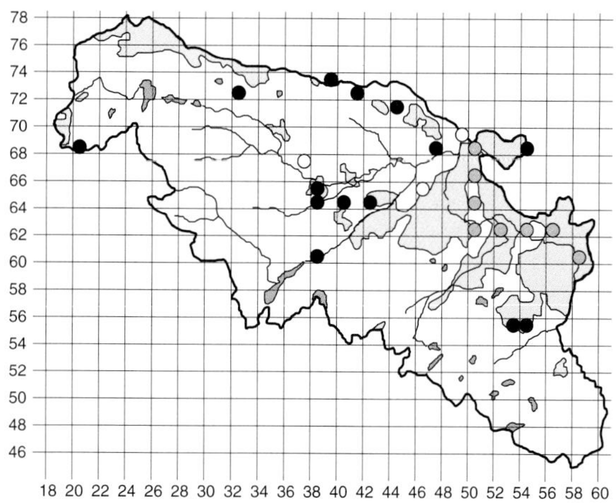

Lunaria annua

Lobularia maritima (L.) Desv.
Sweet Alison

Britain: neophyte (S Europe); locally frequent in the S, scattered in lowland Scotland.
Hortal; 3/7; very rare casual.
1st Loganswell, 1936, RM.

There are few records for this garden escape or outcast. It was noted from the University of the West of Scotland Paisley campus (NS4763, 1975, AJS) and a rubbish tip, Barr Loch (1969, AMcGS), and was recorded twice during surveying for CFOG (NS56L and NS56R). The sole modern record is from a roadside near Blackhouse Farm (NS5353, 2006).

Descurainia sophia (L.) Webb ex Prantl
Flixweed

Britain: archaeophyte; common in E England, scattered in C and E Scotland.
Former accidental; 0/2; extinct casual.
1st Cartside, 1869, PPI.

The only other record is for a single plant recorded from a waste-ground site in Paisley (Ferguson 1911).

Arabis hirsuta (L.) Scop.
Hairy Rock-cress

Britain: widespread but local, frequent in C and E Scotland.
Former native; 0/2; extinct.
1st Renfrewshire, 1872, TB.

Lee (1933) also noted Hairy Rock-cress for the VC, but there are no other named localities, past or present.

Draba muralis L.
Wall Whitlowgrass

Britain: native on limestone in England, scattered alien elsewhere.
Accidental; 2/10; very rare casual.
1st Greenock, 1880, NHSG.

Other old records for Wall Whitlowgrass include Duchal Avenue (1898, JMY), Inverkip Road (1887, GRK) and Kilmacolm (1932, PSY), and Lee (1933) noted it from Wemyss Bay as a rare plant of walls. In 1948 it was reported from Shielhill Glen (ANG excursion report) and BWR (1978) noted it from the Loanhead area (NS4267) and Bogside (NS2172). There was one garden record in CFOG (NS56R, 1998, PM) and in 1984 it was recorded from the Loch Libo area (NS4355, BWR), where the SNH Loch Survey found it growing at the side of the road in 1996.

Erophila verna (L.) DC.
Common Whitlowgrass

Britain: locally common in the S, scattered in Scotland.
Native; 51/70; frequent, scattered.
1st Paisley, 1845, NSA.

Probably under-recorded due to its early flowering, there is nevertheless a good spread of records for Common Whitlowgrass from throughout the lowlands. Many of the records are from waste-ground sites, disused railways, path edges and roadsides, although it can occur on shallow soils of rocky ground or hedge banks in more rural areas.

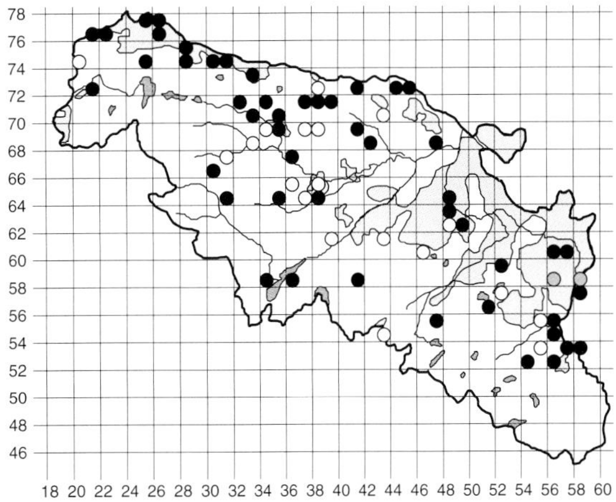

Erophila verna

Conringia orientalis (L.) Dumort.
Hare's-ear Mustard

Britain: neophyte (E Mediterranean); scarce casual, mainly in the S.
Former accidental; 0/4; extinct casual.
1st Renfrewshire, 1869, PPI.

Localized old records are from Hangingshaw (1896, NHSG) and Paisley (1910, GN), but a more recent record came from Renfrew (NS4966, 1959, JP).

Diplotaxis muralis (L.) DC.
Annual Wall-rocket

Britain: frequent in S England, local in E Scotland.
Former accidental; 0/2; extinct casual.
1st Giffnock, 1930, RG.

More recently (1977) Annual Wall-rocket was found in flower beds at the University of the West of Scotland Paisley campus (NS4863, AJS), and may be overlooked elsewhere.

Brassica oleracea L.
Cabbage

Britain: local on coasts, scattered alien inland.
Former hortal; 1/1; extinct casual.
1st Braehead Power Station (NS56D), 1984, CFOG.

There are no modern records and the original waste-ground site was developed in the 1990s.

Brassica napus L.
Oil-seed Rape

Britain: neophyte (cultivated origin); common in the S, extending to E Scotland.
Hortal; 21/27; occasional, scattered.
1st Cornfields (Renfrew), 1868, MY.

The earliest herbarium collection appears to be from the Mill Lade, Houston (1897, PSY). Probably some records are confused with Turnip, but both species can be found on waste ground, roadsides, riverbanks and towards field margins, or as relict crop weeds in arable fields.

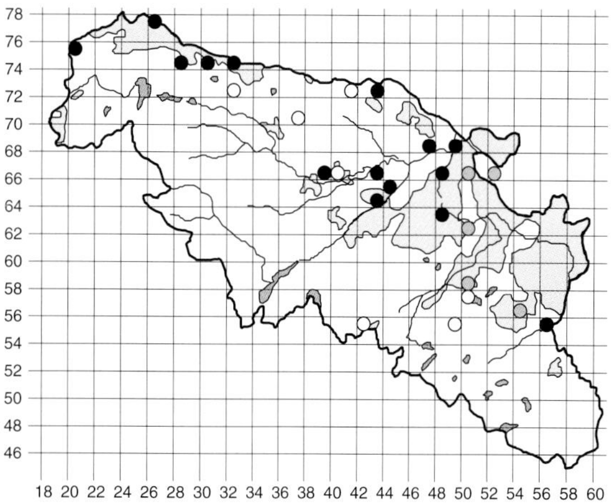

Brassica napus

Brassica rapa L.
Turnip

Britain: archaeophyte; frequent in the S, scattered in Scotland.
Hortal; 16/19; scarce, scattered.
1st Hatton, 1865, RH.

Possibly confused with the preceding, Turnip shows a similar distribution pattern and grows in similar places but it is apparently not so frequent. It was collected from the banks of the White Cart Water in 1903 (PSY).

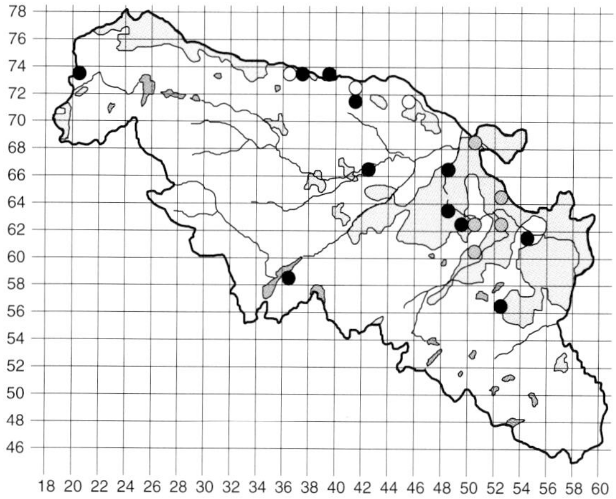

Brassica rapa

Brassica nigra (L.) W.D.J. Koch
Black Mustard

Britain: common in S England, rare in Scotland.
Accidental; 4/5; rare.
1st Rubbish tip, Barr Loch, 1969, AMcGS.

Black Mustard was also recorded from the White Cart Water, Newlands Park (NS56Q) in the 1980s (CFOG); modern field records are from waste ground at a demolished housing scheme, Ferguslie Park (NS4664, 2000), waste ground, Woodhall (NS3473, 2012) and by the White Cart Water, Paisley (NS4863, 2003).

Sinapis arvensis L.
Charlock

Britain: archaeophyte; common throughout in lowlands.
Accidental; 46/64; frequent, lowlands.
1st Renfrewshire 'weed', 1812, JW.

Charlock was collected from Cathcart (1884, GL) and there are old literature records from Loch Libo, Wemyss Bay and Castle Semple Loch (1887, NHSG) and several field records between 1950 and 1980 (BWR). Today it is frequently found on waste ground, margins of riverbanks, as a garden weed and on farmland. Its distribution shows a strong lowland urban-fringe pattern with few modern rural records.

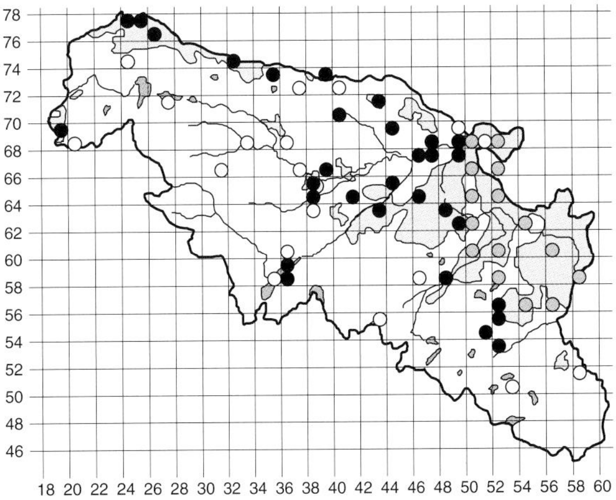

Sinapis arvensis

Sinapis alba L.
White Mustard

Britain: archaeophyte; common in the SE, local in E Scotland.
Accidental; 3/8; rare casual.
1st Cathcart, 1884, GL.

The modern records occur in the south-west of Glasgow: roadside, Corslet (NS5256, 1998), arable weed, Balgray (NS5157, 1999) and Hillington (NS56C, 1985, CFOG). BWR recorded White Mustard on four occasions, most recently (1984) at Dampton (NS3862) and Kilmacolm (NS3569).

Erucastrum gallicum (Willd.) O.E. Schulz
Hairy Rocket

Britain: neophyte (S Europe); scattered in the S.
Former accidental; 0/2; extinct casual.
1st Greenock, 1920, RG.

Grierson also collected Hairy Rocket from an ash heap at Newlands in 1930 (GL).

Hirschfeldia incana (L.) Lagr.-Foss.
Hoary Mustard

Britain: neophyte (S Europe); locally common in the S, rare in Scotland.
Accidental; 4/4; rare casual.
1st Old tip, Linwood Moss (NS4465), 1998.

Hoary Mustard was also seen in 1998 on waste ground in Paisley (NS4863) and at Weighhouse Close, Paisley (NS4763, AJS); a more recent find was from Ladyburn (NS3075, 2010, AHJ). It is considered to be spreading nationally but records from VC76 remain few.

Coincya monensis (L.) Greuter & Burdet
Isle of Man Cabbage

Britain: frequent on coast in NW England and SW Scotland.
Former native; 0/1; extinct.
1st Inverkip, *c.* 1840, BM.

Inverkip is the sole named VC76 locality (collected by W Gourlie) for this coastal species which Hennedy (1891) described as common on sandy seashores. It still occurs in nearby Ayrshire (VC75) and it was also found inland in Glasgow (VC77), where it was considered a recent accidental arrival (CFOG).

Cakile maritima Scop.
Sea Rocket

Britain: frequent on coasts throughout.
Native; 1/1; very rare.
1st Ardgowan Point, 2003, IPG.

Considered by Hennedy (1891) to be frequent on sandy shores, Sea Rocket can still be found in Ayrshire, but there were no named localities in Renfrewshire until IPG's discovery (NS1972).

Raphanus raphanistrum L.
ssp. *raphanistrum*
Wild Radish

Britain: archaeophyte (Europe); common in the SE, frequent elsewhere.
Accidental; 4/10; rare casual.
1st Canal bank, Renfrew, 1862, PSY.

There are several old herbarium (GL) records for Wild Radish: the River Cart, Inchinnan (1883), Thornliebank (1909) and Renfrew (1883), and RG knew it from Kilbarchan (1945, TPNS). AJS recorded it from the University of the West of Scotland Paisley campus (NS4763, 1976) and more recently it was collected (GL) from Dumbreck (NS5563, 1986) and Langside (NS5761, 1987) during CFOG recording. It has only been recorded twice in modern field recording, from Bankfoot (NS2173, 2010, PRG) and a car park at Castle Semple Loch (NS3558, 2011), and although possibly overlooked it must be considered rare.

Raphanus raphanistrum L.
ssp. *maritimus* (Sm.) Thell.
Sea Radish

Britain: common on W and S coasts, absent from N and E Scotland.
Native; 19/23; frequent on coast.
1st Wemyss Bay, 1865, RH.

Sea Radish is frequently encountered on the strandline sands of the west coast, from Wemyss Bay (NS1968) round to Gourock (NS2477); further east it becomes scarcer but it can be found near Erskine (NS4373, 1996).

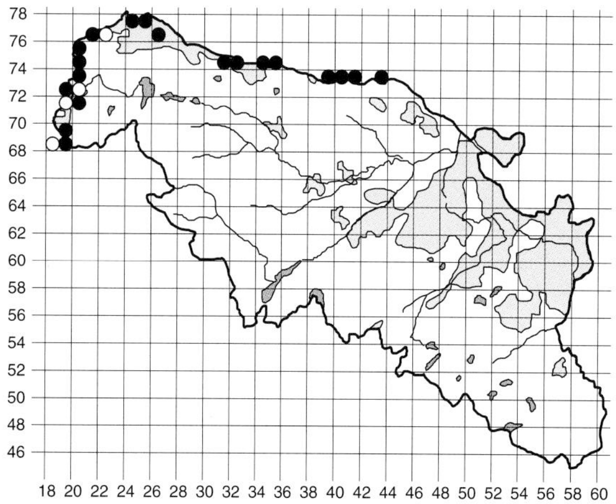

Raphanus raphanistrum ssp. *maritimus*

Raphanus sativus L.
Garden Radish

Britain: neophyte (Mediterranean?); scattered, mainly in S.
Hortal; 1/1; very rare casual.
1st Braehead Power Station (NS56D), 1985, CFOG.

Sisymbrium altissimum L.
Tall Rocket

Britain: neophyte (Europe); frequent in C England and C Scotland.
Accidental; 4/9; rare.
1st Paisley, 1913, GL.

A further old herbarium specimen of Tall Rocket is from Giffnock (1917, GL). More recently ERTC (1953) recorded it from a waste heap at Lochwinnoch and it was collected from St James' Park, Paisley, in 1975 (NS4765, PSY), and AMcGS knew it from Lonend (NS4863, 1980). There was one record from CFOG surveys, from the disused railway, Hillington (NS5166, 1985). Tall Rocket can also be found at three other locations to the west of Glasgow: waste ground, Linwood (NS4465, 1998), north Paisley (NS4866, 2000) and Ferguslie Park (NS4664, 2000).

Sisymbrium orientale L.
Eastern Rocket

Britain: neophyte (Europe); frequent in C England and C Scotland.
Accidental; 10/15; rare, scattered.
1st Hangingshaw, Glasgow, 1896, NHSG.

Eastern Rocket was also found at Kilmacolm (1932, TPNS), by a roadside, Mearns (1938, RM) and later (AJS) at Paisley (NS4764, 1975) and Gleniffer Braes (NS4560, 1976). There are a couple of CFOG records and it was recorded recently at Polmadie (NS5962, 2011). There is only one modern record from Paisley (NS4864, 2001) and one from Lochwinnoch (NS3558, 2005, IPG); all other records are from waste ground in urban Inverclyde.

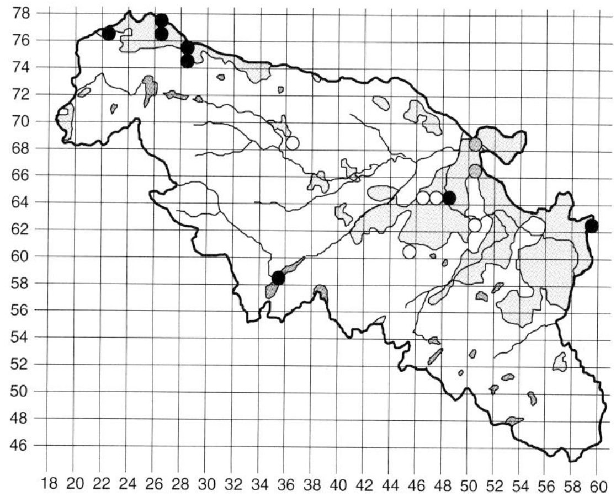

Sisymbrium orientale

Sisymbrium officinale (L.) Scop.
Hedge Mustard

Britain: archaeophyte; common throughout, lowland in Scotland.
Accidental; 45/52; frequent in urban areas.
1st Renfrewshire, 1869, PPI.

There are few old named localities for Hedge Mustard, a plant considered by Hennedy (1891) to be very common; it was collected from Haggs Castle in 1907 (GL). The distribution pattern is strongly urban, notably about Paisley and Glasgow and also along the Inverclyde conurbation. Habitats include waste ground, disused railways, roadsides and waysides. PM recorded the glabrous var. *leiocarpum* DC. from Merryvale (NS5660, 2007).

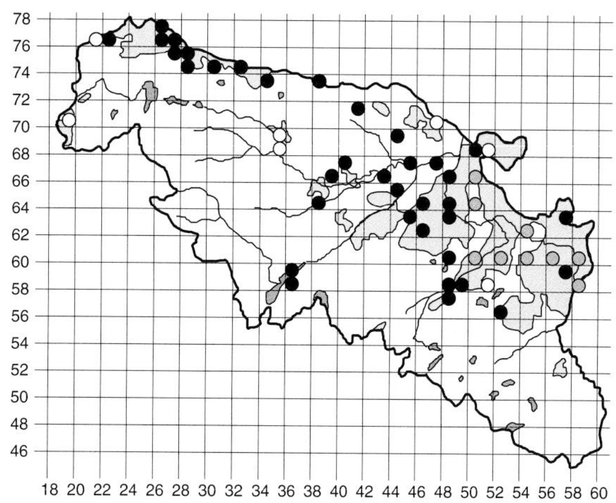

Sisymbrium officinale

Alliaria petiolata (M. Bieb.) Cavara & Grande
Garlic Mustard

Britain: common in the S, absent from much of N and W Scotland.
Native; 65/74; frequent in lowlands.
1st Gourock, 1865, RH.

The map shows a markedly peripheral lowland pattern, with Garlic Mustard absent from much of the central rural areas and upland ground; the lack of records along the Black Cart Water valley is surprising. It can be a common sight along hedgebanks and riverbanks, and in woodlands – more so towards margins; it is also found on waste ground.

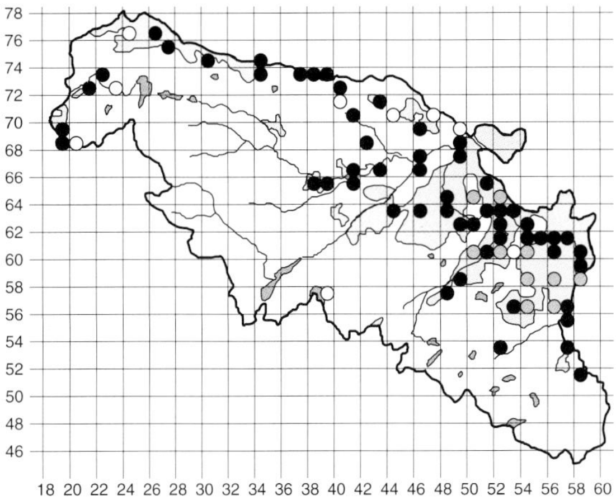

Alliaria petiolata

Teesdalia nudicaulis (L.) W.T. Aiton
Shepherd's Cress

Britain: widespread but local.
Former native; 0/1; extinct.
1st Castle Semple Loch, 1900, PSY.

This is the sole record for a plant considered very rare in the general area by Hennedy (1891).

Thlaspi arvense L.
Field Penny-cress

Britain: archaeophyte; common in the S, extending to NE Scotland.
Accidental; 7/11; rare.
1st Lochwinnoch, 1895, NHSG.

Other old records include Giffnock (1901, FFCA), Paisley (1910, PSY) and a garden, Bishopton (1908, GL). There were four CFOG records at the eastern edge of the VC (NS55Z, NS56D, NS56H, NS56I), but only three recent field records of this seemingly rare casual: waste-ground grassland, Holehouse (NS4756, 1997), waste ground, Clydeport (NS3074, 2001) and Ranfurly (NS3864, 2008, AM).

Hesperis matronalis L.
Dame's-violet

Britain: neophyte (Europe); common throughout except in upland Scotland.
Hortal; 57/66; frequent in the central lowlands.
1st Kilmacolm, 1895, GL.

Although it was present in the 19th century, both Hennedy and Lee considered Dame's-violet rare or not common, and there were few other records until the 1970s, when BWR recorded it from scattered localities such as Hatton (NS4172, 1974), Bonnyton (NS5553, 1978) and Harelaw (NS4960, 1981). Today it is frequently encountered at waste-ground sites and can be a colourful feature along many riverbanks. However, it is rare away from the central lowland urban zone, with only two sites in the north-west – by the Kip Water (NS2172, 2011) and waste ground, Port Glasgow (NS3174, 2001) – and one in the south-east at Ardoch Burn (NS5849, 1995).

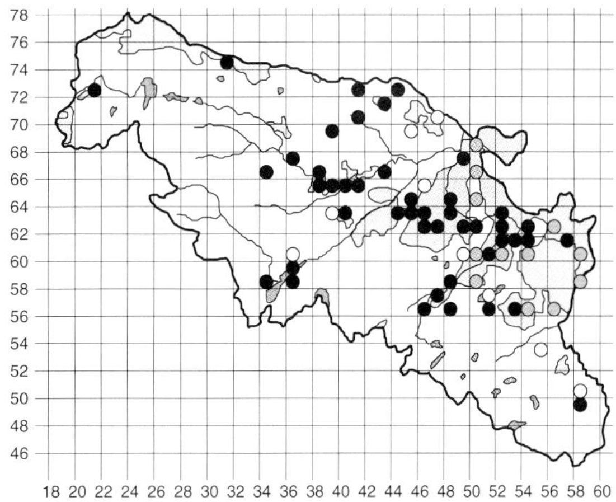

Hesperis matronalis

Cochlearia anglica L.
English Scurvygrass

Britain: frequent on coasts in the S, very rare in Scotland.
Native; 8/10; local on the Clyde coast.
1st Renfrewshire, 1915, TPNS.

A specimen was collected from the River Cart, Blythswood (1931, GL) and in more recent times AMcGS reported English Scurvygrass from the River Cart at Inchinnan (1963) and AJS recorded it from the coast at Erskine (NS4672, 1975); AJS later considered that his record was perhaps of the hybrid. Since then English Scurvygrass has been recorded as far west as Langbank (NS3773, 1998) and as far east as the Cart confluence (NS4969, 1999); it has been recorded upriver as far as Kirklandneuk (NS4866, 2000). The hybrid with *C. officinalis* (*C.* × *hollandica* Henrard) has been recorded from Erskine (NS4671, 1979,

PSY), Port Glasgow (NS3174, 2001) and Blythswood (NS5068, 2005), and some of the above records may also include hybrid plants.

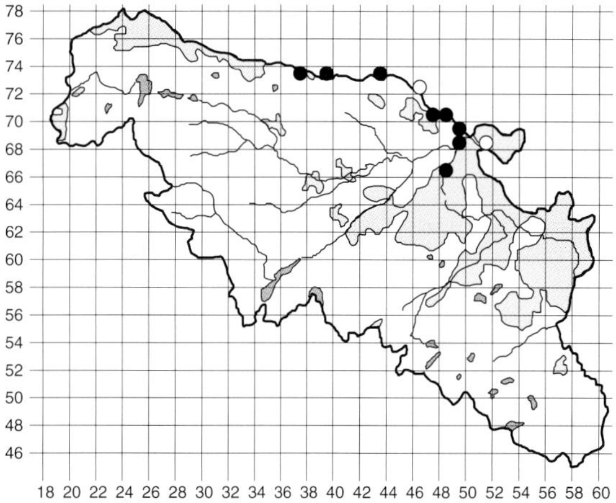

Cochlearia anglica

Cochlearia officinalis L.
Common Scurvygrass

Britain: common on coasts except in the SE, frequent in uplands.
Native; 27/33; frequent on the coast.
1st Renfrewshire, 1834, MC.

Common Scurvygrass has been recorded from many places along the muddy estuarine waters of the Cart and the Clyde, but is also found along the more exposed west coast at Lunderston Bay (NS2074, 1996) and Ardgowan Point (NS1972, 2003, IPG). A number of the records may refer to the hybrid *C.* × *hollandica* Henrard. The true identity of some specimens along the Clyde has been questioned (T Rich pers. comm.). An 1846 specimen (OXF) from Gourock has been reported as ssp. *scotica* (Druce) P.S. Wyse Jacks., but this has not been verified.

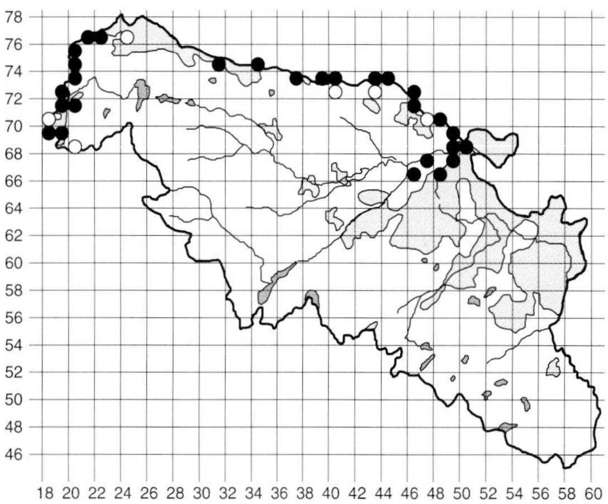

Cochlearia officinalis

Cochlearia danica L.
Danish Scurvygrass

Britain: common on coasts and frequent alien inland.
Native and/or accidental; 47/48; locally common on larger roadsides.
1st Renfrewshire, 1898, ASNH.

Over the last 20 years Danish Scurvygrass has become a common sight along local roads, often forming extensive white to purplish carpets along the central reservations of motorways. No population appears to be native, all being found by the sides of roads treated with salt, even when close to the coast. It is especially common by the M8 west of Glasgow through Paisley and Langbank, and also about the Erskine Bridge; it can also be found along other roads such as at Linwood (NS4464, 2004) and Cowglen (NS5461, 2002, PM). Outlier records include a roadside, Eaglesham (NS5652), waste ground at Cartsdyke (NS2875, 2001) and a roadside near the coast at Kelly Glen (NS1968, 2003). The earliest modern record was from a disused railway edge at Crookston (1987, CFOG).

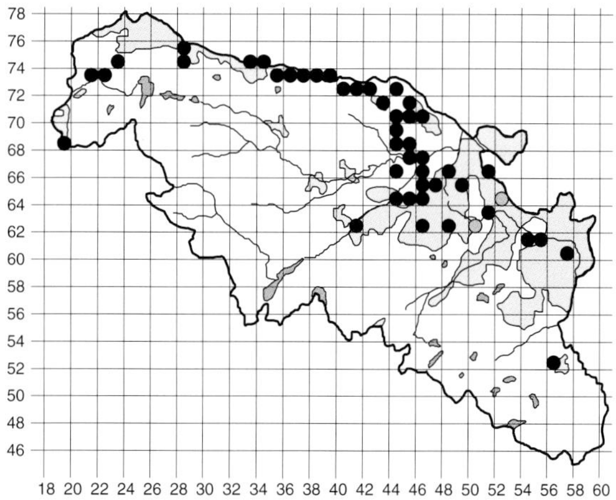

Cochlearia danica

Iberis sempervirens L.
Perennial Candytuft

Britain: neophyte (S Europe); scattered, mainly in England.
Hortal; 2/2; very rare.
1st Roadside rocks, Bow Farm, 1998.

This colourful garden escape may have been planted at Bow Farm (NS2675). It is also known as a garden escape from Jordanhill (NS5468, 2002, AMcGS).

Iberis amara L.
Wild Candytuft

Britain: local in SE England, scattered alien elsewhere.
Former hortal; 0/2; extinct casual.
1st Greenock (west), 1880, NHSG.

Wild Candytuft was also recorded from Crossmyloof (1901, FFCA).

Iberis umbellata L.
Garden Candytuft

Britain: neophyte (Europe); scattered in the S and in lowland Scotland.
Hortal; 1/1; very rare casual.
1st Williamwood, 1986, CFOG.

Erucaria hispanica (L.) Druce

Britain: neophyte (E Europe); rare casual.
Former accidental; 0/1; extinct casual.
1st Giffnock, 1926, RG.

PLUMBAGINACEAE

Armeria maritima (Mill.) Willd.
Thrift

Britain: common on coasts and some inland mountains.
Native; 11/15; locally frequent on the west coast.
1st Firth of Clyde, 1834, MC.

A common coastal plant in the 19th century, Thrift is only now easily seen on the rocky coastline in the extreme west, where it has been found from McInroys Point (NS2277, 1998) to Wemyss Bay (NS1969, 1999). The sole eastern site is from the coast at East Langbank (NS3973, 1998); it was recorded from the old parish of Erskine in 1845 (NSA).

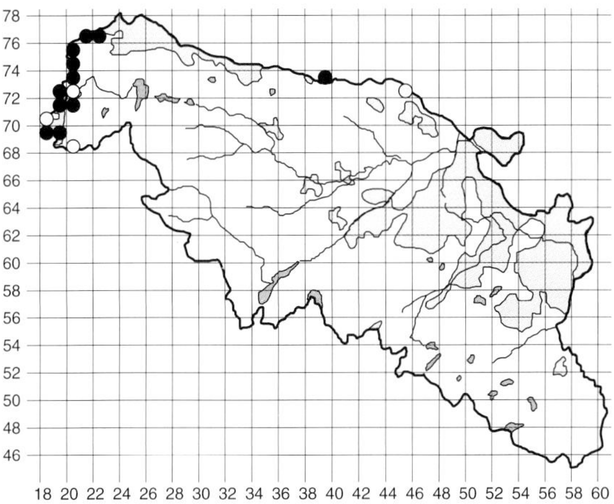

Armeria maritima

POLYGONACEAE

Persicaria alpina (All.) H. Gross
Alpine Knotweed

Britain: neophyte (Europe); rare escape, only in Scotland.
Hortal; 1/1; very rare.
1st Disused railway, Scart, 2001.

The sole record for this garden outcast is from the side of a disused railway, now a cycleway (NS3667).

Persicaria campanulata (Hook. f.) Ronse Decr.
Lesser Knotweed

Britain: neophyte (Himalayas); widely scattered, mainly in the W.
Hortal; 6/8; rare garden outcast.
1st Shore near Inverkip, 1960, BWR.

Lesser Knotweed was also recorded from Finlaystone (NS3673, 1975, AJS). Modern records include woodland edges near gardens at Skiff Wood (NS4059, 1993), Quarrelton Wood (NS4262, 1993), Uplawmoor (NS4355, 2009, PG) and Trumpethill (NS2276, 2010, PG); it has also been recorded from Ardgowan Estate (NS2073, 1998) and a roadside, Wateryett (NS3757, 2010).

Persicaria bistorta (L.) Samp.
Common Bistort

Britain: common in N and W England and in lowland Scotland.
Native and hortal; 35/57; occasional.
1st Below Greenock, 1821, FS.

Common Bistort has a widespread but scattered distribution mainly in the Glasgow conurbation or along the Black Cart Water 'valley'. It is most likely to be seen along roadsides and towards the edges of woodlands or scrub, or on waste ground, but is rarely seen on agricultural land. Wood (1893) noted, 'up all the burns which feed the Cart may be found beds of the pretty snake-weed (*Polygonum bistorta*).'

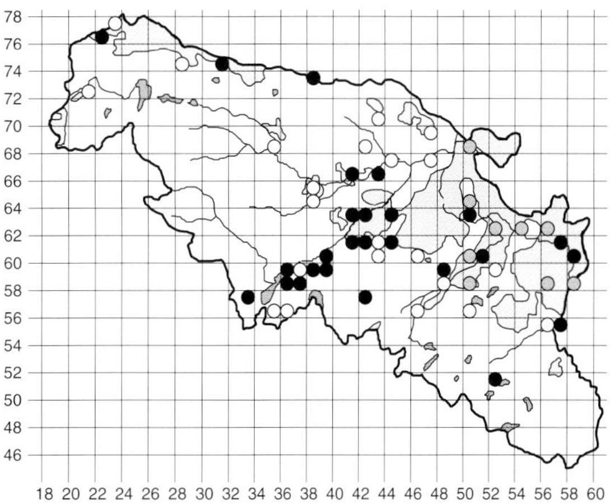

Persicaria bistorta

Persicaria amplexicaulis (D. Don) Ronse Decr.
Red Bistort

Britain: neophyte (Himalayas); widely scattered, mainly in the S and W.
Hortal; 1/2; very rare.
1st Inverkip, 1960, GL.

Collected in 1960 from the Kip estuary (A Miller), there is now only one modern location for Red Bistort: below a roadside hedge bank, near Gibblaston (NS3466, 1999), where it appears to have spread from the adjacent garden.

Persicaria wallichii Greuter & Burdet
Himalayan Knotweed

Britain: neophyte (Himalayas); widely scattered, mainly in the W.
Hortal; 3/5; very rare.
1st White House, Kilbarchan, 1976, RM.

This garden outcast, escape or relict was also recorded from Finlaystone Estate (NS3673, AJS) in the same year, and may well still occur there. Today Himalayan Knotweed is known from estates at Formakin (NS4070, 1999) and Ardgowan (NS2073, 1998), and also from the roadside, Bow Farm (NS2576, 1994, CD).

Persicaria amphibia (L.) Delarbre
Amphibious Bistort

Britain: throughout but absent from uplands.
Native; 94/116; frequent and widespread in wetter places.
1st Lochwinnoch and Paisley, 1845, NSA.

There are other 19th-century records from waterbodies such as Castle Semple Loch and Loch Libo. Amphibious Bistort has not been recorded at the latter in recent years but is known from several other open waterbodies such as Brother Loch (NS5052, 1996, SNH-LS) and reservoirs at Coves (NS2276, 1998, SNH-LS), Auchendores (NS3572, 1998), Waulkmill (NS5252, 2005) and Barcraig (NS3857, 2000). It is, however, more commonly encountered in its terrestrial form along loch shores, burn sides and on waste ground.

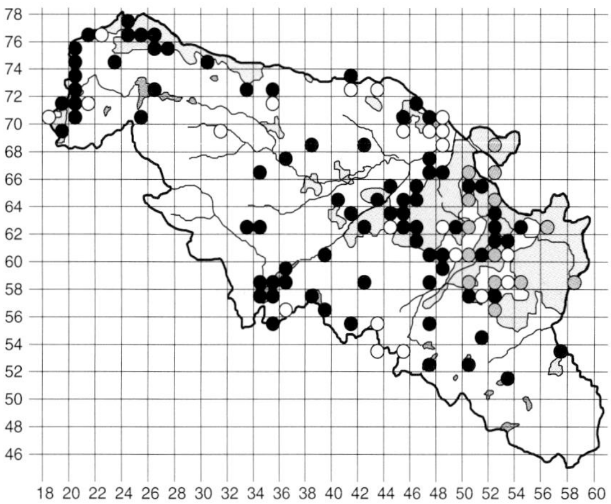

Persicaria amphibia

Persicaria maculosa Gray
Redshank

Britain: common throughout except in N Scotland.
Native; 156/241; common, widespread.
1st Paisley, 1869, RPB.

There may well have been a decline locally since the recording of the 1960s–80s period, but this species is still common, except in the uplands. The decline may reflect changes in agricultural practice, with increased permanent pastures; periodic soil disturbance may be followed by a good flush of Redshank from the seed bank.

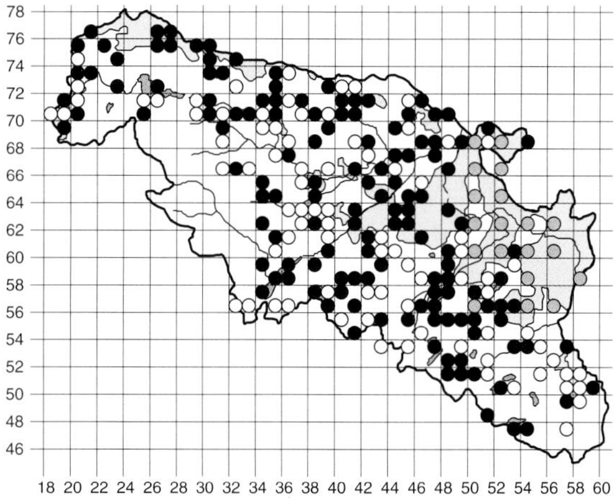

Persicaria maculosa

Persicaria lapathifolia (L.) Delarbre
Pale Persicaria

Britain: common in the S, frequent in lowland Scotland.
Native or accidental; 13/19; scarce.
1st Cathcart and Gourock, 1865, RH.

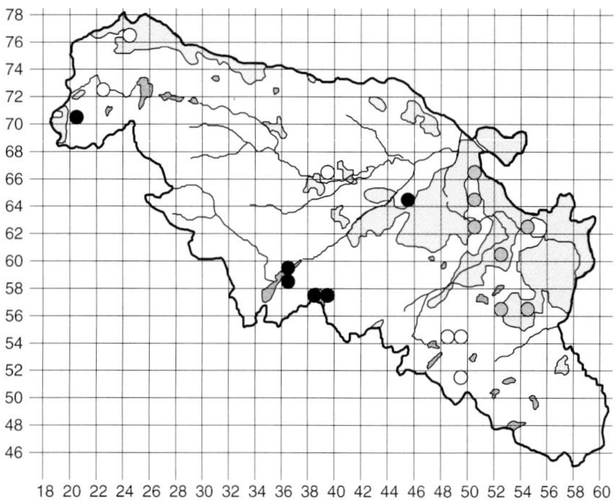

Persicaria lapathifolia

The authenticity of some records for Pale Persicaria was questioned in CFOG, and this weed of waste ground and arable fields is perhaps best considered scarce in the area, although it may be overlooked. Modern records are from Barcraig Reservoir (NS3857, 2000; NS3957, 2000, IPG), Inverkip (NS2070, 1999), Linwood Pool (NS4564, 2000) and CMRP rangers have recorded it from the Castle Semple Loch area (NS3658 and NS3659, 1990). The few BWR records include Black Loch (NS4951, 1966) and Houstonhead (NS3966, 1971).

Persicaria hydropiper (L.) Delarbre
Water-pepper

Britain: common in England, mainly in the W in Scotland.
Native; 80/92; frequent, widespread.
1st Paisley, 1845, NSA.

Water-pepper occurs throughout the VC, although it tends to favour rural and lowland areas, avoiding the uplands except near some waterbodies. It is usually found to the margins of open waterbodies, especially reservoirs, and is also commonly seen on sandy alluvium by watercourses.

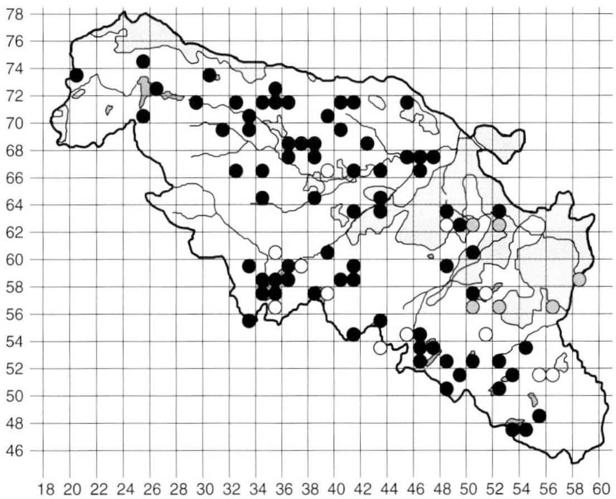

Persicaria hydropiper

Persicaria minor (Huds.) Opiz
Small Water-pepper

Britain: widely scattered and local, rare in Scotland except in the SW.
Native; 5/11; rare, declining.
1st Black Loch, 1865, GGO.

Small Water-pepper has been found over the years at several waterbodies in the south-east: Brother Loch (1876, GL), Commore (1933, Lee), Harelaw (1949, RM) and Pilmuir (1975, BWR), with the old Calder Dam the only site in the west (NS2965, 1976, RM). It is likely to have declined given the recent loss of reservoirs and the maintaining of high water levels at others, but it may well have been overlooked. The modern records are from the dams or former dams at Whittliemuir (NS4158, 1993), Gryfe Reservoir No. 2 (NS2971, 2000) and Barcraig (NS3957, IPG, 2003), with a record from river shingle near Castle Semple Loch (NS3558, 2003, IPG); it is also known from Glanderston Dam (NS5056, 1988, CFOG).

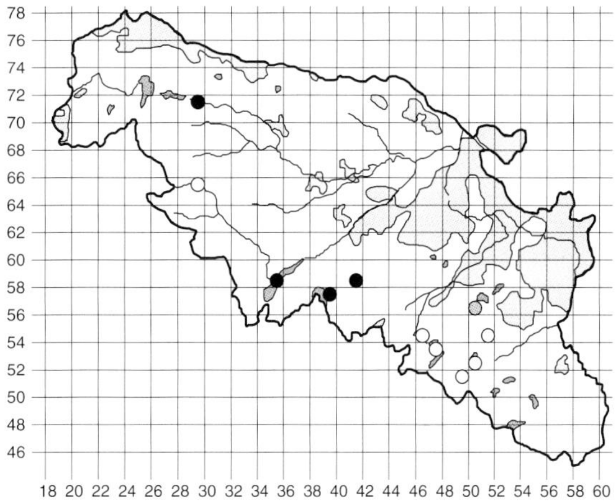

Persicaria minor

Fagopyrum esculentum Moench
Buckwheat

Britain: neophyte (Asia); frequent in the S, scarce in Scotland.
Hortal; 0/1; extinct or very rare casual.
1st Crosslee, DM, 1978.

The sole record for Buckwheat is from Home Farm (NS4166) and there have been no modern sightings.

Polygonum oxyspermum C.A. Mey. & Bunge ssp. *raii* (Bab.) D.A. Webb & Chater
Ray's Knotgrass

Britain: frequent on the W coast, rare in the E.
Native; 1/2; very rare.
1st Inverkip, *c.* 1840, BIRM.

This knotgrass was also listed by Lee (1933), and Hennedy (1891) described it as frequent on sandy shores, but there are no other named localities. The sole modern record is of a plant on the shore south of Lunderston Bay (NS2073, 2008).

Polygonum arenastrum Boreau
Equal-leaved Knotgrass

Britain: archaeophyte; common, rare in N Scotland.
Accidental; 85/85; frequent and widespread.
1st Whitecraigs area (NS55N), 1985, CFOG.

Not distinguished by earlier recorders, this small knotgrass can be found throughout the VC, although some confusion with Knotgrass is still likely. It is usually found at waste-ground and wayside locations.

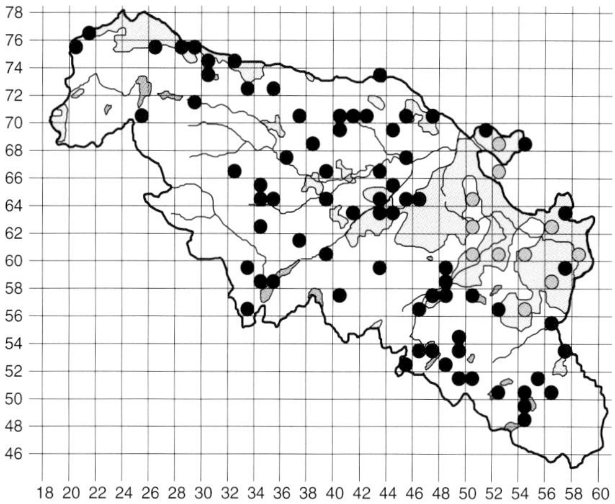

Polygonum arenastrum

Polygonum aviculare L.
Knotgrass

Britain: common throughout except in uplands or N Scotland.
Native or accidental; 177/283; common and widespread.
1st Gourock, 1837, GL.

Knotgrass is a widespread and common weedy species of waste ground, roadsides and farmyards. Some of the records may refer to Equal-leaved Knotgrass, especially those from before 1987. A specimen (E) of Cornfield Knotgrass (*P. rurivagum* Jord. ex Boreau) was collected by AMcGS at Erskine (NS4672, 1980).

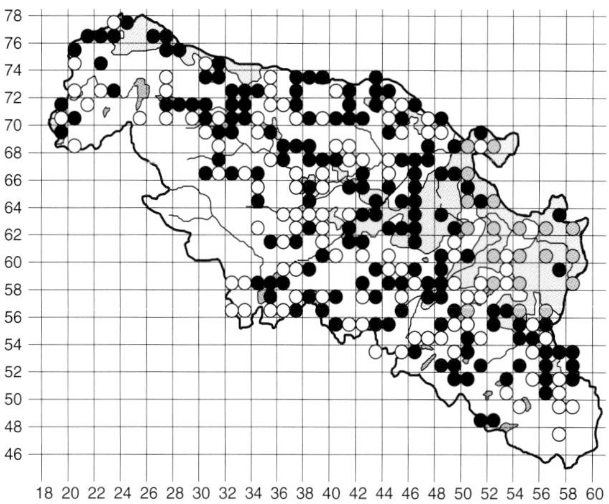

Polygonum aviculare

Fallopia japonica (Houtt.) Ronse Decr.
Japanese Knotweed

Britain: neophyte (Japan, Korea and China); common throughout except in uplands.
Hortal; 205/218; common and widespread.
1st Giffnock–Merrylee, 1926, RM.

Although present for more than 85 years, there are very few other old records locally until the 1970s, by when Japanese Knotweed appears to have been widespread. It was not mentioned by Hennedy (1891) but Lee (1933) considered it frequent; a specimen collected in 1954 (PSY) from Potterhill is annotated by the collector, W Hood, as 'now established'. Today it is certainly very well established throughout the lowland VC on waste ground and at waysides and woodland edges, often where previously dumped, and also along the more nutrient enriched riverbanks and along the coastal upper strandline, as is shown by several linear patterns on the map.

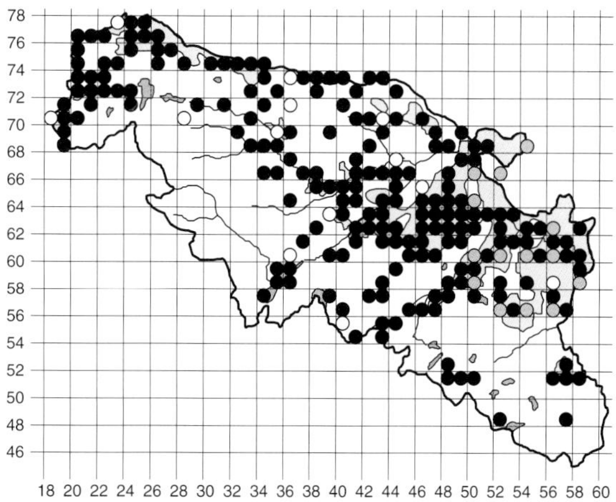

Fallopia japonica

Fallopia × bohemica (Chrtek & Chrtková) J.P. Bailey
(F. japonica × F. sachalinensis)
Conolly's Knotweed

Britain: neophyte (hybrid origin); widely scattered but local.
Hortal; 10/10; rare but increasing.
1st Killoch Water, 1999.

The hybrid, like its Giant Knotweed parent, grows in estate woodlands. At several locations it can form very large stands, as at Fulton Wood (NS4165, 1999) and by the Black Cart Water at Johnstone (NS4263, 2001). It is also found as an apparent outcast on roadsides, as at Cairn Hill (NS4851, 2002), at a disused railway at Kilbarchan (NS4264, 2003, IPG) and by burn sides at Pilmuir (NS5155, 2000), Killoch Water (NS4758, 1999) and

Auldhouse (NS5560, 2007). There is no evidence to suggest that any hybrids have arisen locally.

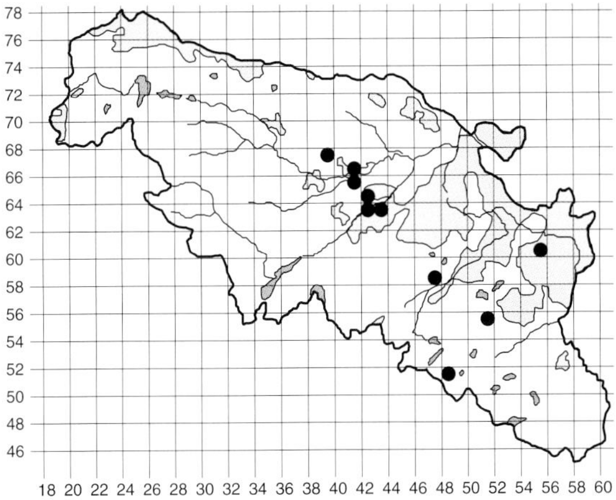

Fallopia × bohemica

Fallopia sachalinensis (F. Schmidt) Ronse Decr.
Giant Knotweed

Britain: neophyte (Sakhalin and Japan); widely scattered throughout.
Hortal; 8/10; rare, scattered.
1st Giffnock, 1926, RG.

JDM recorded Giant Knotweed at Finlaystone (NS3663, 1963), but it has not been reported subsequently from there. Like the much commoner Japanese Knotweed, this larger relative has been present in the Glasgow area for many years but by comparison has hardly spread, even though fertile. The modern records are all from estate grounds where it can form large but local stands: Ardgowan (NS2073, 1998), Barmufflock (NS3664, 1998), Caldwell (NS4154, 1999), Eaglesham (NS5651, 1991, LB) and Barochan (NS4165, 1999); it can also still be found at a couple of places in Pollok Park (NS5462; NS5561, 2005, CFOG), but it has not been recorded at Rouken Glen since 1983 (NS55N, CFOG).

Fallopia convolvulus (L.) Á. Löve
Black-bindweed

Britain: archaeophyte; common in the S and in lowland Scotland.
Accidental; 8/15; rare, declining.
1st Paisley Moss, 1858, PSY.

There are only two records for Black-bindweed from modern recording: near the Black Cart Water, Inchinnan (NS4767, 1999) and near the White Cart Water at Ford, Eaglesham (NS5753, 2006, PM), but there are quite a few relevant records from CFOG. In the 1970s it was found about the Hatton area (NS4072 and NS4172, BWR) and there are several old herbarium records: Paisley Canal Bank (1868, GLAM), above Greenock (1905, GL), a Neilston roadside (1916, GL) and Haggs (1890, GL). Both Hennedy (1891) and Lee (1933) describe this weedy plant as 'common'; recently it must have either been overlooked or undergone a dramatic decline.

Rheum × rhabarbarum L.
R. × hybridum Murray
Rhubarb

Britain: neophyte (garden origin); widely scattered, frequent in lowland Scotland.
Hortal; 20/21; occasional, scattered.
1st Rubbish tip, Barr Loch, 1969, AMcGS.

An occasional outcast, Rhubarb occurs on tracksides, roadsides and waste ground, usually near to residential areas or farm buildings. It appears to persist but has not been noticed as spreading locally. It has not been recorded growing 'wild' about the Rhubarb fields near North Mains (*c.* NS4267).

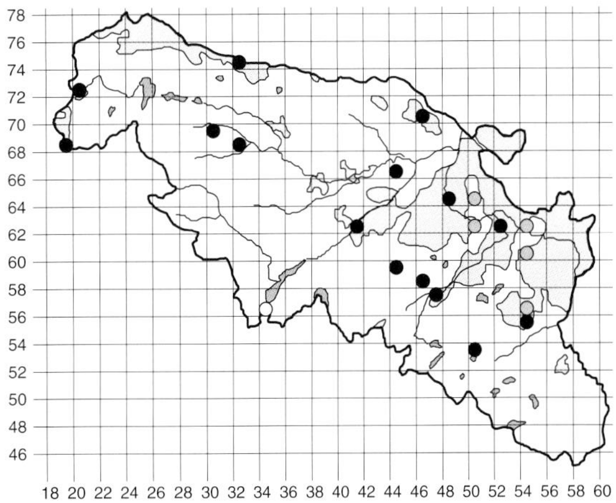

Rheum × rhabarbarum

Rumex acetosella L.
Sheep's Sorrel

Britain: common throughout.
Native; 233/333; common.
1st Paisley, 1858, PSY.

Sheep's Sorrel is a common plant of open ground such as waste ground, grasslands, gardens and waysides. It is also common on rocky outcrops and grasslands on shallow soils, and extends into the upland areas.

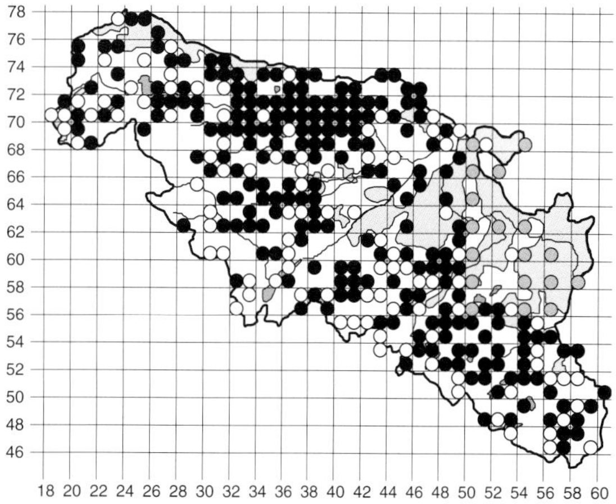

Rumex acetosella

Rumex acetosa L.
Common Sorrel

Britain: common throughout.
Native; 425/495; very common.
1st Paisley, 1869, RPB.

Common Sorrel is indeed a very common species of all but the driest or wettest grasslands, being tolerant of marshy soils and some agricultural improvement and amenity management. It also occurs at waysides, on waste ground and in scrubby areas.

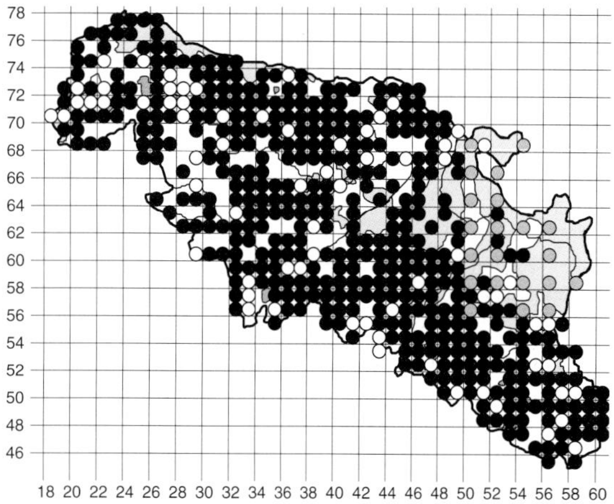

Rumex acetosa

Rumex salicifolius T. Lestib.
Willow-leaved Dock

Britain: neophyte (N America); very rare casual.
Former accidental; 0/1; extinct casual.
1st Newlands, 1920, GL.

Rumex longifolius DC.
Northern Dock

Britain: common in C and NE Scotland and upland N England.
Native or accidental; 82/114; frequent.
1st Renfrew, 1876, FFWS.

Under the name '*Rumex aquaticus*', this dock was reported from Renfrew in 1876 (FFWS) but it is scarcely mentioned by Hennedy (1891), who noted it 'below Renfrew' and previously at Scotstoun, and Lee (1933) thought it 'not common'. It appears that it was rare in the 19th century but today it is frequent on waste ground and rough grasslands and along waysides. It occurs throughout the lowlands, mainly in rural areas, and is markedly scarce in the north and west.

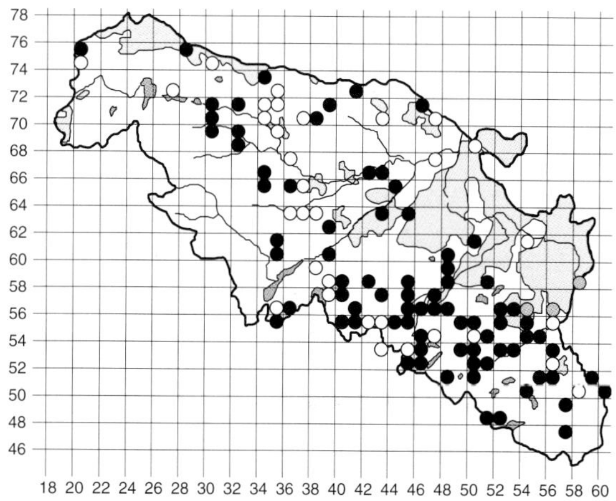

Rumex longifolius

Rumex crispus L.
Curled Dock

Britain: common throughout except in upland Scotland.
Native; 228/300; very common.
1st Renfrewshire, 1869, PPI.

A common species of waste ground, gardens, farmyards, roadsides and open grassy places, Curled Dock can be tolerant of wet places such as inundation areas to the margins of open waterbodies and some marshes. There are also 18 modern records of coastal forms occurring along the strandline from Erskine to Wemyss Bay, which appear referable to ssp. *littoreus* (J. Hardy) Akeroyd.

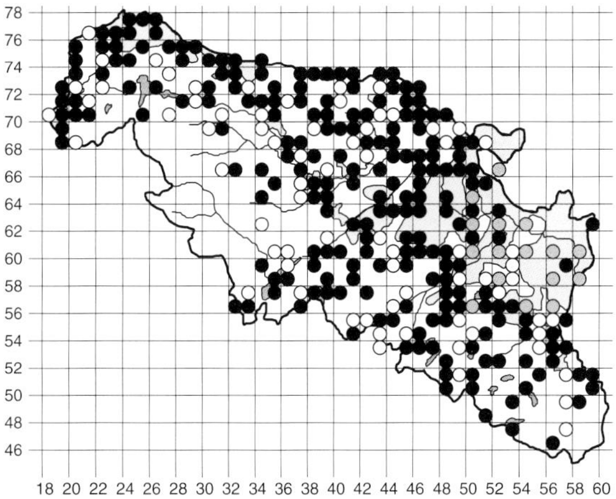

Rumex crispus

Rumex conglomeratus Murray
Clustered Dock

Britain: common in England, scattered in C Scotland.
Native; 2/3; very rare.
1st Cathcart, 1865, RH.

There are no other named old localities and only two modern records for this locally rare dock, although it may have been overlooked for Wood Dock. It is known sparingly from a ditch near Mosspark Station (NS5263, 2004), along the route of the former Paisley Canal where it may well have occurred, as it does today at the Forth & Clyde Canal in Glasgow (CFOG). It has also been found along the riverbanks of the Black Cart Water at Inchinnan (NS4667, 2002).

Rumex sanguineus L.
Wood Dock

Britain: common, rare or absent from N Scotland.
Native; 55/59; locally frequent, but scattered.
1st Renfrewshire, 1883, TB.

Wood Dock is fairly frequently found in wet woodlands or more open marshy flushes, but it is seldom found in any abundance. It is widespread, with a notable concentration in the extreme west, but avoids urban disturbance, agriculturally improved ground and the uplands. There are surprisingly few named old localities (such as Kelly Glen, Loch Libo, Castle Semple – all 1887, NHSG). The var. *sanguineus* was excluded from the Renfrewshire list (TPNS 1915). The hybrid with Curled Dock (*R.* × *sagorskii* Hausskn.) has been tentatively recorded from one place, Loch Libo (NS4355), where both parents occur, but the plant was too immature for this to be confirmed.

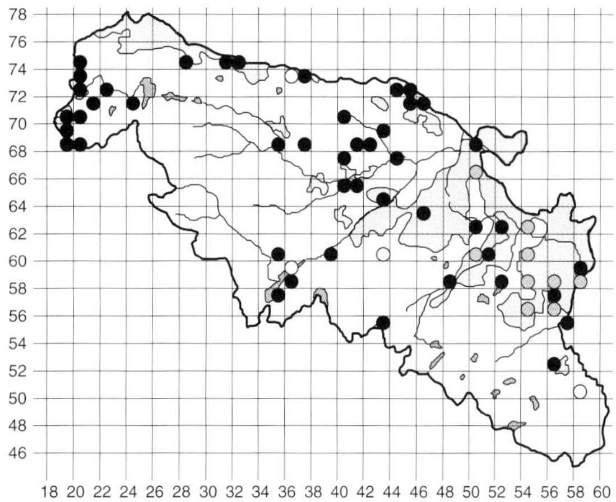

Rumex sanguineus

Rumex obtusifolius L.
Broad-leaved Dock

Britain: common throughout.
Native; 345/437; very common.
1st Renfrewshire, 1869, PPI.

This is the commonest dock, found in many places including waste ground, waysides, farmyards and fields, gardens, grasslands, marginal marshy ground, scrub and woodland edges, and frequently along riverbanks; it is only absent from the peaty uplands.

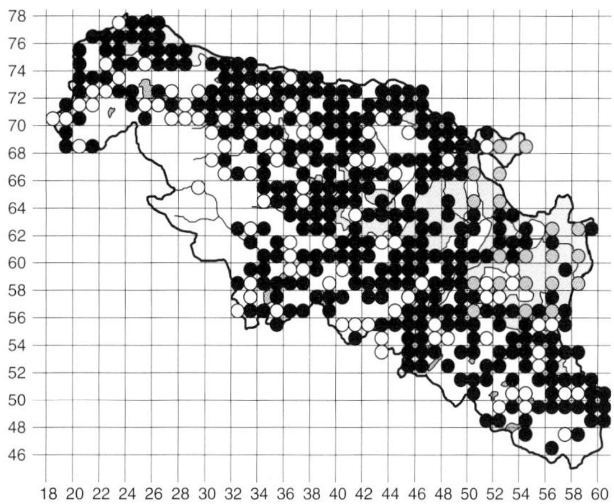

Rumex obtusifolius

Rumex obovatus Danser
Obovate-leaved Dock

Britain: neophyte (S America); very rare casual in S England.
Former accidental; 0/1; extinct casual.
1st Coups in Renfrewshire, 1920, RG.

Rumex bucephalophorus L.
Horned Dock

Britain: neophyte (S Europe); very rare casual with no modern records.
Former accidental; 0/1; extinct casual.
1st Bishopton coup, 1930, RG.

Hybrid Docks

The following hybrids have been recorded in the VC:

Rumex × hybridus Kindb.
(*R. longifolius × R. obtusifolius*)

Britain: frequent where parents overlap.
Native; 6/6; rare, overlooked.
1st Knockbartnock, 1995.

This hybrid can often be found if sites with both parents are examined closely. The modern localities are: Knockbartnock (NS3560, 1995), Neilstonside (NS4655, 1998), Walton Dam (NS4955, 1998), Carswell Hill (NS4653, 1999), Fingalton (NS5055, 2000) and Priestside (NS3271, 2008).

Rumex × pratensis Mert. & W.D.J. Koch
(*R. crispus × R. obtusifolius*)

Britain: widely scattered but local.
Native; 10/10; rare, probably overlooked.
1st White Cart, Busby, 1984, CFOG.

Likely to occur whenever the two parents coincide, which is often, this hybrid has been recorded from Deanside (NS56H, 1989, CFOG), Gleniffer Braes (NS4661, 1995), Kelburn (NS3474, 1998), Linwood (NS4564, 2000), Inverkip (NS2071, 2006) and Glentyan (NS3963, 2007). It was also found at a cluster of sites about Castle Semple in 2005 by IPG: Lochwinnoch (NS3558), Lochside Station (NS3557) and Howwood (NS3860).

Rumex × dufftii Hausskn.
(*R. sanguineus × R. obtusifolius*)

Britain: local but widely scattered, mainly in the S.
Native; 1/1; very rare, overlooked.
1st Blythswood, 2005.

Several plants occur, with both parents, along a shaded path side at Blythswood (NS5068). This hybrid probably occurs at other sites where both parents overlap.

DROSERACEAE

Drosera rotundifolia L.
Round-leaved Sundew

Britain: common in the N and W.
Native; 102/117; common in upland mires.
1st Paisley, Eaglesham and Kilbarchan, 1845, NSA.

Although it may have been lost from many of its former lowland sites, this sundew can be readily found in the bogs, acidic mires and peaty burn-side flushes of the uplands.

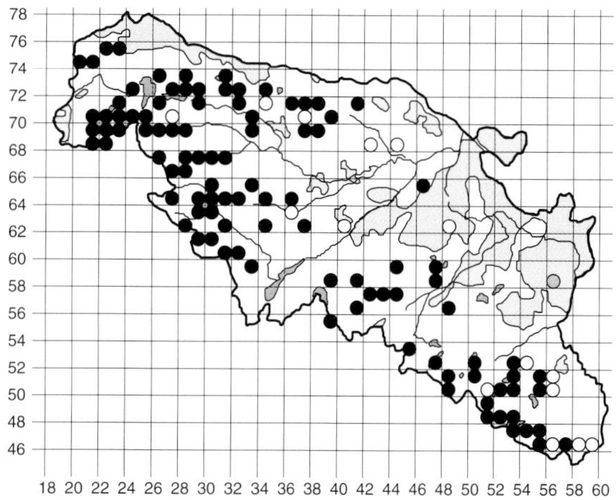

Drosera rotundifolia

Drosera anglica Huds.
Great Sundew

Britain: common in NW Scotland, local elsewhere.
Former native; 0/2; extinct.
1st Paisley, 1845, NSA.

As noted in the previous account, this species was collected from Houston Moss (1897, PSY), and Lee (1933) mentioned nearby 'Dargarvel'. More recently it was recorded from Duchal Moor (NS3067, 1992, NVC-CMRP), but there is no further information about what would be remarkable find; modern searches have failed to verify this record, which is likely to be a recording error.

Drosera intermedia Hayne
Oblong-leaved Sundew

Britain: frequent locally in the W.
Former native; 0/3; extinct.
1st Marsh between Paisley and Glasgow, 1813, TH.

Hopkirk's locality to the east of Paisley must have been destroyed a long time ago. Oblong-leaved Sundew was recorded in 1896 (TPNS), a record which Lee (1933) repeated. However, an 1897 specimen from the nearby 'Houston Moss' (PSY) appears referable to Great Sundew, casting doubt on the reliability of old records. Oblong-leaved Sundew was excluded from the Renfrewshire list (TPNS 1915).

CARYOPHYLLACEAE

Arenaria serpyllifolia L.
Thyme-leaved Sandwort

Britain: common in England and lowland Scotland.
Native or accidental; 14/22; scarce.
1st Below Battery, Greenock, 1865, RH.

Thyme-leaved Sandwort is possibly under-recorded. Modern records tend to be from open waste ground or disused railways in urban fringes, such as at Ferguslie Park (NS4664, 2000), East Fulwood (NS4567, 2001), Linwood Industrial Estate (NS4463 and NS4564, both 1995), Georgetown (NS4270, 2005) and by the roadside at Erskine Bridge (NS4572, 1998, PM). Other (BWR) records are from Bridge of Weir (NS3667, 1982) and Kilmacolm (NS3569, 1984); the most westerly record was from a track at Leapmoor (NS2170, 1981). *Arenaria leptoclados* (Rchb.) Guss. (Slender Sandwort) has records in CFOG, two of which were close to the VC border, but there are no confirmed records from within Renfrewshire.

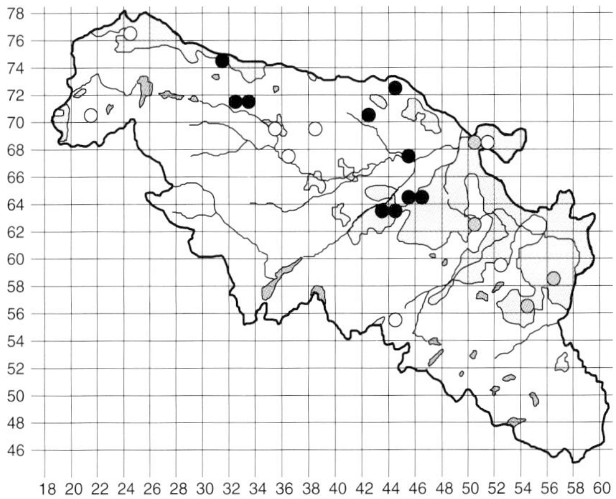

Arenaria serpyllifolia

Arenaria balearica L.
Mossy Sandwort

Britain: neophyte (W Mediterranean islands); widely scattered.
Hortal; 2/3; very rare.
1st Auchendores Reservoir (NS3572), 1969, BWR.

This Mediterranean alien has not been seen recently at its original location but it was found growing on a stone entrance post at Formakin Estate (NS4171, 1994); it was still present in 2003 (IPG). It was recorded from Barshaw Park (NS5064) in 1988 (CFOG).

Moehringia trinervia (L.) Clairv.
Three-nerved Sandwort

Britain: common in England and lowland Scotland.
Native; 46/69; frequent in woodlands.
1st Renfrewshire, 1869, PPI.

Although known 'about Glasgow' (FS 1821), there are few old locations named for this sandwort. It is perhaps an overlooked plant of generally less disturbed, semi-natural woodland, often associated with watercourses, but it is seldom found in any quantity. The distribution is scattered and tends to be in rural-fringe areas; Three-nerved Sandwort avoids the uplands, although it has been recorded from the Forkings of Raith (NS2862, 1997).

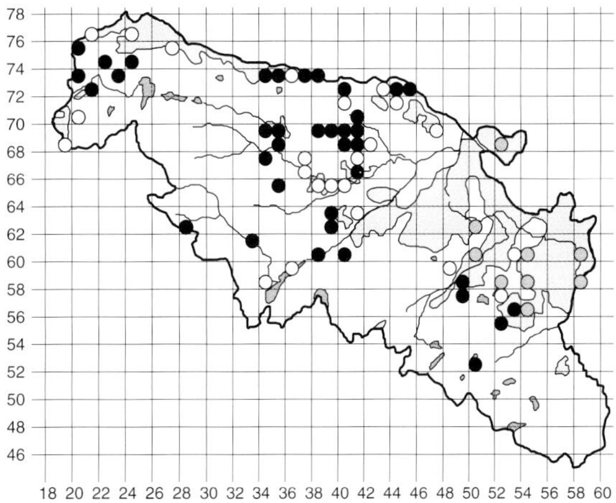

Moehringia trinervia

Honckenya peploides (L.) Ehrh.
Sea Sandwort

Britain: coastal throughout.
Native; 10/14; occasional on the coast.
1st Erskine, 1845, NSA.

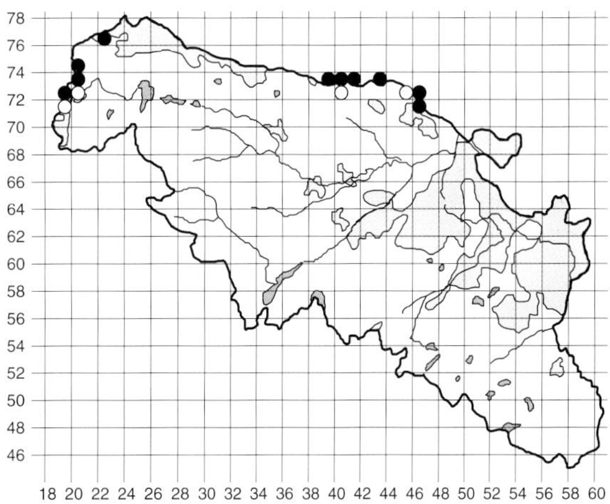

Honckenya peploides

Modern records are from coastal sands and shingles between Inverkip and Lunderston in the west and from Langbank and Erskine further up the Clyde.

Stellaria nemorum L.
Wood Stitchwort

Britain: frequent from N England to C Scotland only.
Native; 27/31; occasional along watercourses.
1st Taylor's Glen, Port Glasgow, 1834, MC.

Wood Stitchwort was reported from Lochwinnoch parish in 1845 (NSA) but the only modern record from around this area is from the River Calder (NS3459, 1987, DM). Most of the records refer to wooded or somewhat shaded riverbanks, although it can occur amongst alluvial-influenced tall herb communities. It is well represented in the west, with an 'arc' distribution pattern reflecting the burns flowing down from the Renfrewshire Heights. Curiously, there was only one record for this species made during surveying for CFOG, from the White Cart Water (Linn Park, NS55Z, 1983), and it has not been seen there since, although it does occur downriver in Paisley (NS4863, 2000). The southernmost record is from Caldwell Estate (NS4154, 1999).

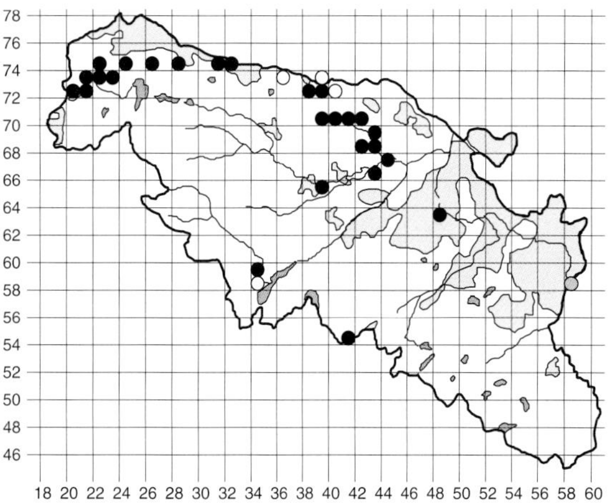

Stellaria nemorum

Stellaria media (L.) Vill.
Common Chickweed

Britain: common throughout.
Native; 255/376; very common.
1st Renfrewshire, 1869, PPI.

Found throughout the VC, Common Chickweed only becomes rare in the uplands, but even here it can be seen at nutrient-enriched sheep shelters. It occurs on waste ground and in parks and gardens, is tolerant of some scrub shade, is also common about farmyards and can be frequent in sown grass leys. Greater Chickweed (*S. neglecta* Weihe) was recorded on a field card from near Inverkip (NS2071, 1973, BWR) but there is no supporting specimen.

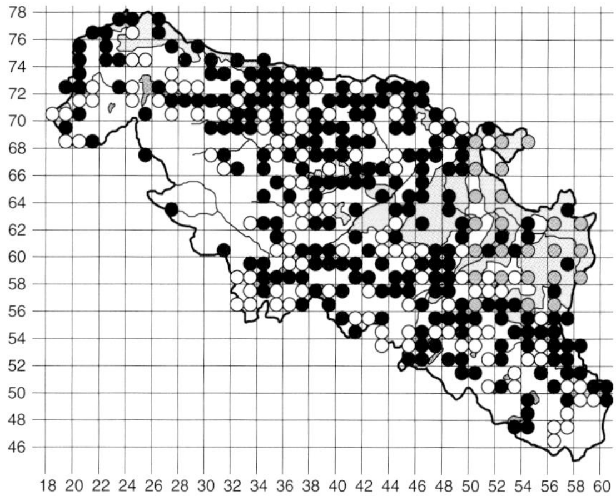

Stellaria media

Stellaria holostea L.
Greater Stitchwort

Britain: common throughout except in N Scotland.
Native; 134/166; common in wooded places and along hedge banks.
1st Wemyss Bay, 1845, GL.

Greater Stitchwort is a fairly common woodland plant, although scarce where conditions are nutrient rich or marshy, and not found in new plantations. It can also be seen along hedge banks and scrub-edge grasslands.

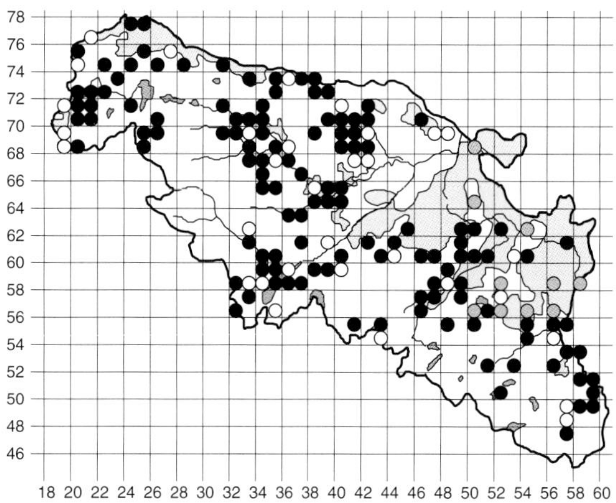

Stellaria holostea

Stellaria palustris Ehrh. ex Hoffm.
Marsh Stitchwort

Britain: widely scattered, scarce in Scotland.
Former native; 0/2; extinct.
1st Lochwinnoch, 1845, NSA.

Castle Semple Loch is a likely site but there have been no modern records from there. A JDM record from Finlaystone (NS3673, 1963) must be viewed with suspicion, as this is an unlikely site and Marsh Stitchwort can be mistaken for the following species. It was recorded from Braehead (at the VC77 boundary) in 1985 (CFOG), but the site has since been destroyed.

Stellaria graminea L.
Lesser Stitchwort

Britain: common throughout except in N Scotland.
Native; 181/274; common.
1st Paisley, 1861, PSY.

This is a common stitchwort of grassland which is also often found on waste ground. It favours neutral grasslands, although not where improved, and tolerates some inundation; its trailing habit enables it to persist in abandoned urban-fringe grasslands.

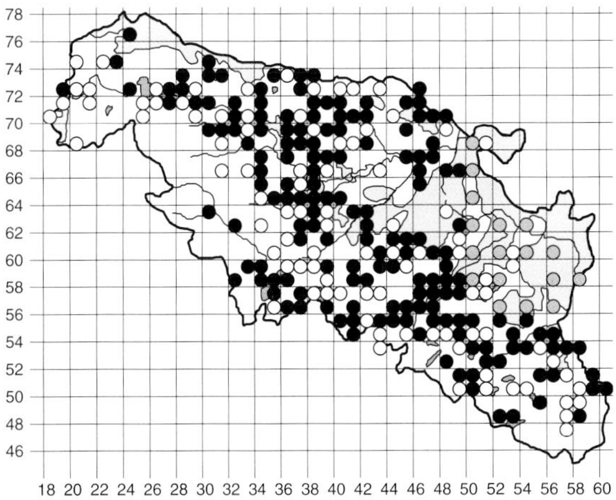

Stellaria graminea

Stellaria alsine Grimm
Stellaria uliginosa Murray
Bog Stitchwort

Britain: common throughout.
Native; 272/356; very common in wet places.
1st Paisley Canal Bank, 1865, RH.

Bog Stitchwort is a common plant found throughout the VC, but is rare or absent from urban areas. It occurs in wet to very wet places, notably in springs and flushes, and often along marshy burn sides. In spite of its common name it does not occur in true ombrotrophic bogs, although it can occur at mineral-influenced margins or channels and so is able to extend into the peaty uplands.

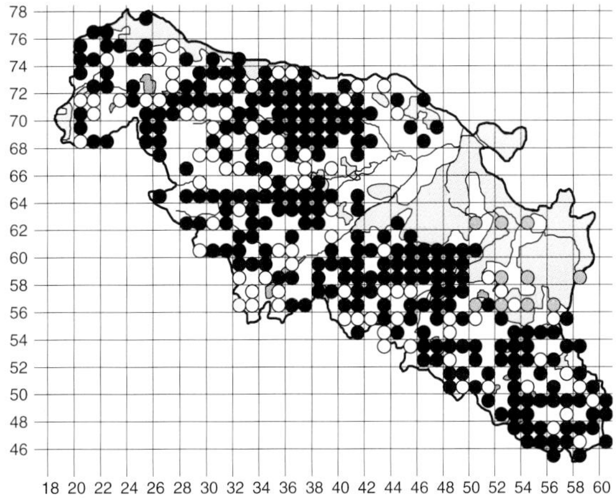

Stellaria alsine

Cerastium arvense L.
Field Mouse-ear

Britain: frequent throughout the E, rare in the W.
Native; 1/5; very rare, endangered.
1st Sea shingle, Gourock, 1879, GL.

The only modern record for Field Mouse-ear is from the sandy soils of the grassy embankment of the Clyde to the west of Erskine (NS4572, 1996), from where Lee collected it in 1938 (GL) and AMcGS reported it in 1981. It was recorded from Greenock (west) in 1880 (NHSG), and BWR recorded it from an old railway cutting at Barrhead (NS4958, 1960) and a bay south of Inverkip (NS1971, 1964), but otherwise there are no records for what must always have been a local rarity.

Cerastium tomentosum L.
Snow-in-summer

Britain: neophyte (Italy); frequent in England and lowland Scotland.
Hortal; 16/17; occasional, scattered.
1st Hatton, 1974, BWR.

This garden outcast or escape can persist on waste ground, along hedge banks and in grassy places.

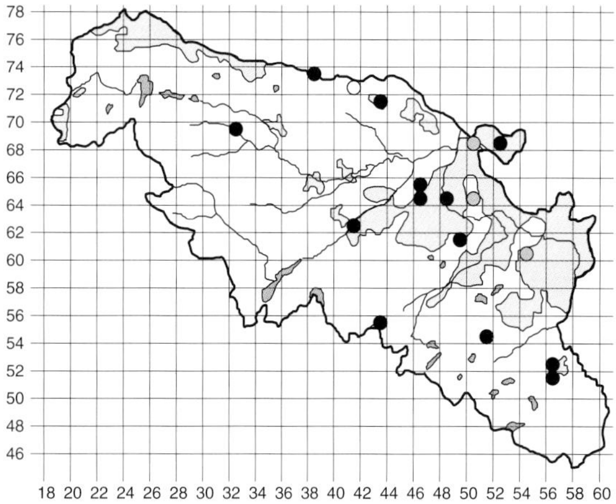

Cerastium tomentosum

Cerastium fontanum Baumg.
Common Mouse-ear

Britain: common throughout.
Native; 343/459; very common.
1st Renfrewshire, 1869, PPI.

As its name suggests, this is a common plant; it is found on waste ground and in various grasslands, tolerating some improvement and amenity management. It extends into the uplands but avoids very wet and peaty soils.

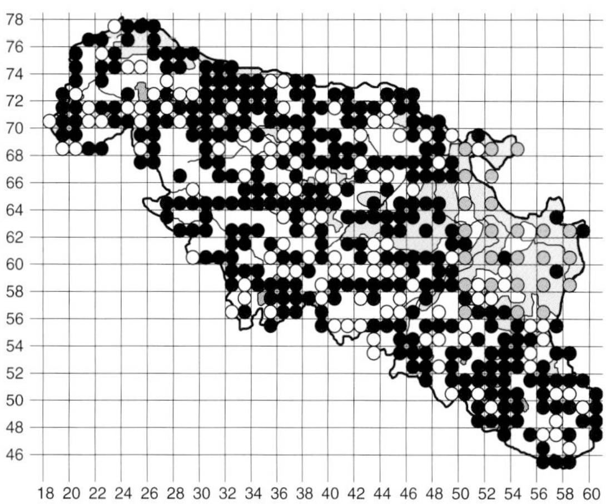

Cerastium fontanum

Cerastium glomeratum Thuill.
Sticky Mouse-ear

Britain: common throughout except in N Scotland.
Native or accidental; 175/239; common.
1st Gourock, 1857, GGO.

Sticky Mouse-ear is often encountered at urban waste-ground sites; it is often most obvious early in the year. It is also common in rural areas, especially about farmyards, gateways and disturbed pastures or reseeded leys.

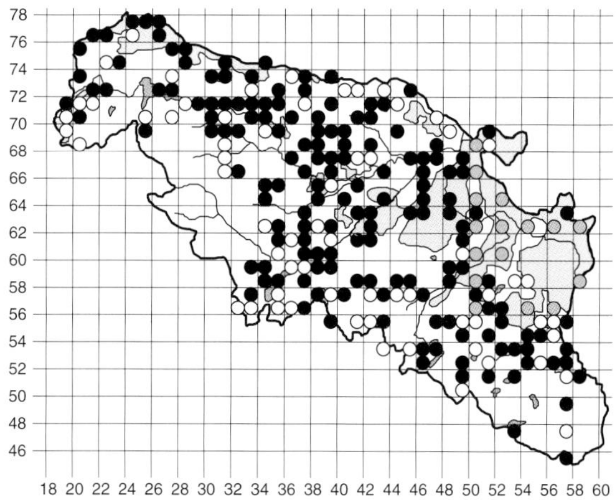

Cerastium glomeratum

Cerastium diffusum Pers.
Sea Mouse-ear

Britain: common on coasts throughout, widely scattered inland.
Native; 7/10; rare, scattered.
1st Renfrewshire, 1872, TB.

Four of the modern records are from coastal locations: Wemyss Bay (NS1969, 2001), Cartsdyke (NS27X, 2001), Port Glasgow (NS3174, 2001) and Erskine Park (NS4572, 1996); older coastal

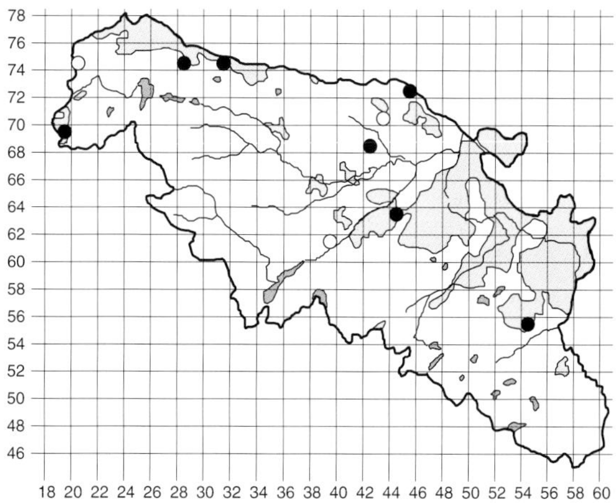

Cerastium diffusum

records include Inverkip to Ashton (1904, GL) and Lunderston Bay (1983, BWR). Inland waste-ground sites include Mearns (NS5455, 2001) and Linwood Industrial Estate (NS4463, 1995), with older records (BWR) from Drygate (NS3961, 1976) and Bishopton (NS4370, 1975); it was found near the latter at the Royal Ordnance Factory, Georgetown (NS4268, 2006).

Cerastium semidecandrum L.
Little Mouse-ear

Britain: frequent and scattered in England, mainly in the E in Scotland.
Native; 2/3; very rare.
1st Gourock, 1865, RH.

Lee (1933) repeats the original record but the only modern site is from the sandy coastal soil at Erskine Park (NS4672). Little Mouse-ear was first recorded from here in 1975 (AJS) and later by AMcGS (1995); in 1996 it was also recorded a little further to the west (NS4572).

Sagina nodosa (L.) Fenzl
Knotted Pearlwort

Britain: widely scattered, more frequent in the N.
Native; 2/10; very rare.
1st Clyde, near Renfrew, 1854, GGO.

Hennedy (1891) considered this plant frequent although there are only a few named old locations, such as Gourock (1876, FFWS), Brother Loch (1891, NHSG) and Loch Libo (1887, NHSG). It was recorded from Jenny's Well (NS4962) in 1902 (PSY) and RM found it at the Williamwood 'railway triangle' (NS5658) in 1934; more recently it was recorded (BWR) from Wemyss Point (NS1870, 1986) and an inflow at Black Loch (NS5051, 1966). The modern records are from two widely separated mire areas: Picketlaw (NS5650, 1998) and Glenshilloch (NS2269, 2011).

Sagina subulata (Sw.) C. Presl
Heath Pearlwort

Britain: frequent in Scotland, SW England and W Wales, rare elsewhere.
Native; 1/4; very rare, endangered.
1st Gourock, 1865, RH.

Considered by Hennedy (1891) as 'not common', today this plant is extremely rare. The sole modern record is for a depauperate specimen in a rock crevice on Earn Hill, above Gourock (NS2275, 2002). Lee (1933) also recorded Heath Pearlwort from Kilmacolm and in 1949 it was reported from an ANG excursion between Bishopton and Langbank.

Sagina procumbens L.
Procumbent Pearlwort

Britain: common throughout.
Native; 295/389; very common.
1st Paisley, 1869, RPB.

A very common plant, Procumbent Pearlwort is found in various places including waste grounds, as a garden weed and in parks and many grassland types. It can be common in upland areas, often along burn sides, but avoids peaty areas.

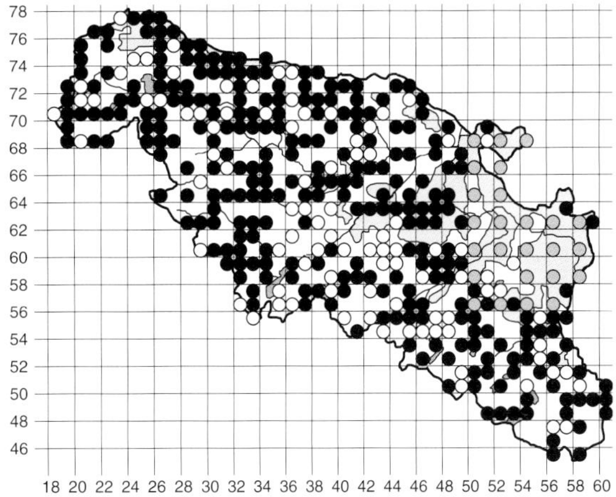

Sagina procumbens

Sagina filicaulis Jord.
Sagina apetala Ard. ssp. *erecta* F. Herm.
Slender Pearlwort

Britain: throughout England and lowland Scotland.
Native or accidental; 34/36; occasional.
1st Renfrewshire, 1883, TB.

This pearlwort did not appear in Hennedy's flora until the 5th edition (1891), and only then in the appendix, where it was described as very rare, and it was not

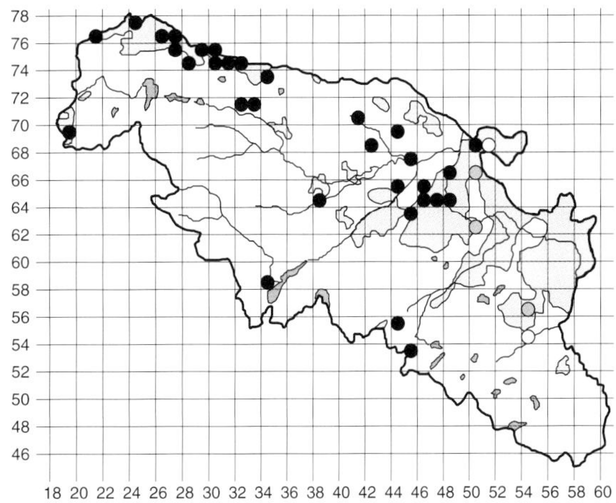

Sagina filicaulis

referenced to the VC by Lee (1933), indicating a late 19th-century arrival in the wider area. However, Hopkirk (1813) considered it 'common' in the Glasgow area and it is generally described as a native (Preston et al. 2002). There were no named locations for this diminutive plant of open waste ground until RM recorded it from a quarry at Mearns (1938, GLAM). Today Slender Pearlwort can be found at various scattered waste-ground sites. Most records, where checked, appear referable to this species although true Annual Pearlwort (*S. apetala*) is known from a few sites along the Clyde nearby in VC77 (CFOG) so may be present in Renfrewshire.

Sagina maritima Don
Sea Pearlwort

Britain: coastal throughout, and by some inland salted roads.
Native; 1/3; very rare, coastal.
1st Wemyss Bay, 1865, RH.

Sea Pearlwort was recorded from Gourock in 1876 (FFWS) but the only modern record is from a coastal saltmarsh at Erskine Park (NS4572, 1996). It may well occur at other coastal locations, but given the scarcity of its habitat it is likely to be rare. There are, as yet, no records from salted roadsides in the VC.

Scleranthus annuus L.
Annual Knawel

Britain: widely scattered in S England and lowland Scotland.
Native; 2/11; very rare, endangered.
1st Gourock, 1865, RH.

In 1908 Annual Knawel was collected at Barochan (GL); this is the same location as the two modern records, Corsliehill (NS3969, 1996) and Barochan (NS4069, 1997). At both sites it grows on open shallow soils associated with rock outcrops, presumably relying on periodic disturbance by livestock. It was collected from Bridge of Weir (1886, PSY) and RG knew it from a quarry at Howwood and also Penilee, Paisley (1942, TPNS); more recently BWR recorded it from Houstonhead (NS3966, 1971), Knocknairs Hill (NS3074, 1969), Shovelboard (NS3869, 1969) and Hatton (NS4172, 1974).

Herniaria glabra L.
Smooth Rupturewort

Britain: rare in E England, scattered alien elsewhere.
Accidental; 1/2; very rare casual.
1st Paisley, 1898, TPNS.

The original record stated 'Rupture Wort (*H. hirsuta*)', but it is assumed to be Smooth Rupturewort. Additionally, an undated specimen collected by Peter Ewing (who was very active in the 1890s), labelled '*Illecebrum verticillatum*' from 'Paisley' (GL), is Smooth Rupturewort, which proves that this species was present at the location about this time. AJS made a more recent discovery of this rarity from a flowerbed at Paisley Cross (NS4864, 1979).

Polycarpon tetraphyllum (L.) L.
Four-leaved Allseed

Britain: very rare in extreme SW, rare casual elsewhere.
Accidental; 1/1; very rare casual.
1st Newlands Garden (NS56Q), 1972, CFOG.

Spergula arvensis L.
Corn Spurrey

Britain: archaeophyte; common throughout.
Accidental; 69/133; frequent, widespread.
1st Paisley, 1869, RPB.

Corn Spurrey is scattered throughout the VC, often in some abundance, occurring on waste ground and about farms, especially towards the margins of arable fields or disturbed pastures. It can also be frequent on the dry margins of reservoirs. It was very widely recorded in the 1960s–80s period (BWR), so it may well have declined in more recent times.

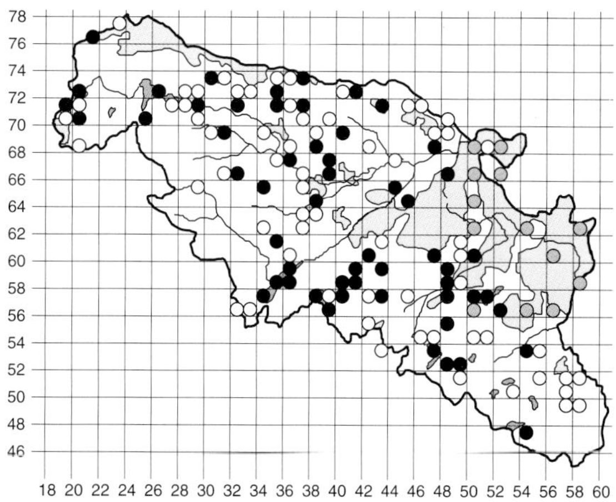

Spergula arvensis

Spergularia media (L.) C. Presl
Greater Sea-spurrey

Britain: coastal throughout.
Native; 0/3; very rare.
1st Langbank, 1885, NHSG.

A specimen of Greater Sea-spurrey was collected from the Battery Field (Greenock) in 1882 (GRK) and it was recorded from Gourock in 1901 (FFCA). Hennedy (1891) noted 'var. *marginata*' (winged seeds) as being 'not so frequent', but gave no VC76 locations.

Spergularia marina (L.) Besser
Lesser Sea-spurrey

Britain: common on coasts throughout, increasing by inland salted roads.
Native and accidental; 24/26; locally frequent on the coast, increasing inland.
1st Renfrewshire, 1834, MC.

Lesser Sea-spurrey is readily encountered along the Erskine–Renfrew stretch of the estuarine Clyde (including up the Black Cart Water at Inchinnan, NS4767), but there are surprisingly few modern records from the west. Since 1998 there have been several inland roadside records made, related to winter salting of roads, and the number is rising fast with at least ten records to date.

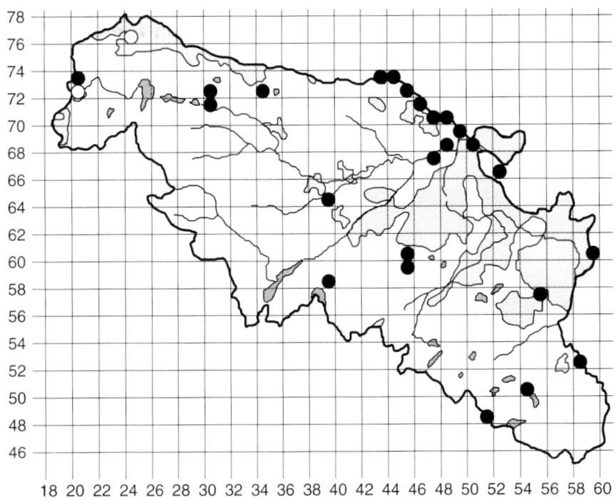

Spergularia marina

Spergularia rubra (L.) J. & C. Presl
Sand Spurrey

Britain: frequent but local throughout.
Native; 2/6; very rare.
1st Lochwinnoch, 1845, NSA.

There are several other 19th-century records: Loch Thom (1894, GRK), East Langbank coast (1885, NHSG) and Hennedy (1891) noted Greenock and Kilmacolm; the latter was a location also noted by ERTC in 1957, where Sand Spurrey was seen by 'several footpaths' in the area. The only modern records are from Corsliehill Road (NS3969, 1996) and the shore of Loch Thom (NS2672, 2000).

Agrostemma githago L.
Corncockle

Britain: archaeophyte; extinct as an arable weed, now occasional as a 'wildflower'.
Accidental; 1/6; extinct as a wild plant.
1st Paisley, 1845, NSA.

There are several old records but they are all from more than 100 years ago, so this arable weed is now considered long extinct, as discussed in CFOG; it was perhaps never as frequent in the west. In 1876 it was noted from Cathcart and Paisley Rd (FFWS) and also recorded from Greenock (1880, NHSG) and Chain Rd, Paisley (1905, PSY). The only modern record for Corncockle is as a contaminant of a meadow-seed mix at GMRC, Nitshill (NS5160, 2005).

Silene vulgaris (Moench) Garcke
Bladder Campion

Britain: common in England and lowland Scotland.
Native or accidental; 5/11; rare.
1st Paisley Canal, 1865, RH.

There are only a few other old records (Wemyss Bay, Gourock and Inverkip) for Bladder Campion, which Hennedy (1891) thought not common. In CFOG it is noted as well known along the 'Clyde corridor', with relevant records from Deanside (NS56H, 1984), Arkleston (NS56C, 1985), Renfrew (NS56D, 1984) and railway sidings at Yoker (NS56E, 1987). Surprisingly, there is only one additional modern record, from waste ground at Clydeport, Port Glasgow (NS3174, 2001).

Silene uniflora Roth.
Sea Campion

Britain: coastal, throughout.
Native; 5/8; rare, coastal.
1st Renfrewshire, 1834, MC.

The modern records for this campion are all from widespread coastal locations: Longhaugh (NS4373, 1996), Port Glasgow (NS3174, 2001), Wemyss Bay (NS1969, 2001) and Lunderston Bay (NS2074, 1996; NS2073, 1996, AB); it was recorded from the latter in 1983 (BWR) and also from nearby Cloch Lighthouse in 1903 (PSY). Lee collected it from Erskine in 1938

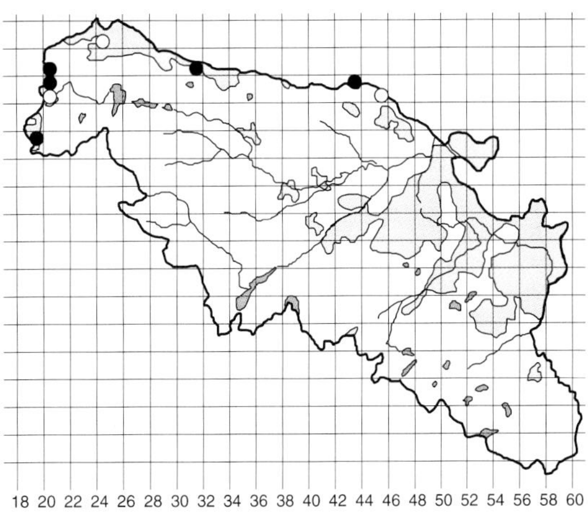

Silene uniflora

(GL). T Scott wrote an account of the floral diversity of Sea Campion populations in the Gourock area (Scott 1887).

Silene noctiflora L.
Night-flowering Catchfly

Britain: archaeophyte; frequent in SE England, very rare in SE Scotland.
Former accidental; 0/3; extinct casual.
1st Greenock, 1880, NHSG.

This catchfly was also collected from wasteland near Paisley in 1910 (PSY) and more recently from a wayside, Lochwinnoch (1959, ERTC).

Silene latifolia Poir. ssp. *alba* (Mill.) Greuter & Burdet
White Campion

Britain: archaeophyte; common in England and E and C Scotland.
Accidental; 15/18; scarce, scattered.
1st Renfrewshire, 1869, PPI.

Greenock is the only old named place (1880, NHSG) for White Campion. Today there are a few scattered records from waste grounds and roadside grasslands including Gleddoch (NS3772, 1998), Ladymuir (NS3464, 1999), by the M8 at Paisley (NS4865, 2002; NS4965, 2003), Millarston (NS4563, 2002) and Wemyss Bay (NS2068, 1998). The hybrid with Red Campion (*S.* × *hampeana* Meusel & K. Werner) was first recorded in 1913 (TPNS) from near Elderslie but the next record was from Braehead in 1985 (NS56D, CFOG), and more recently it was seen near Boglestone (NS3373, 2001).

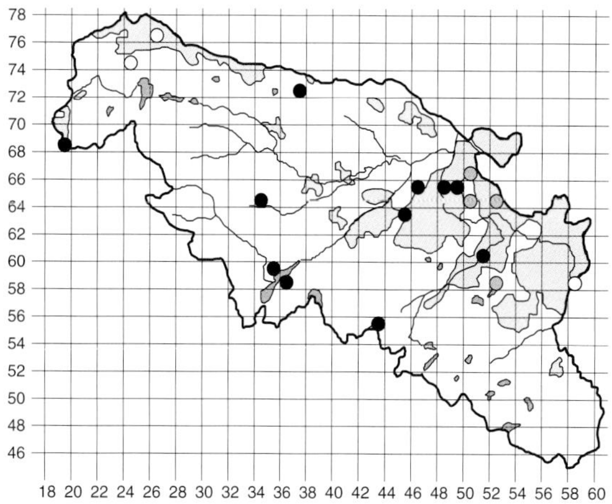

Silene latifolia ssp. *alba*

Silene dioica (L.) Clairv.
Red Campion

Britain: throughout except in N Scotland.
Native; 227/259; common.
1st Paisley, 1845, NSA.

Red Campion was also collected in 1845 from Wemyss Bay (GL) and there are plenty of old records from well-known woodland places (e.g. Loch Libo, Castle Semple and Kelly Glen, all 1887, NHSG). Today it can be commonly found in woodlands and hedge banks and can spread on to waste ground. It is markedly absent from the poorer or wetter soils of the upland areas, but shows a strong affinity for habitats by watercourses as along the White and Black Cart waters and the Gryfe and Levern waters.

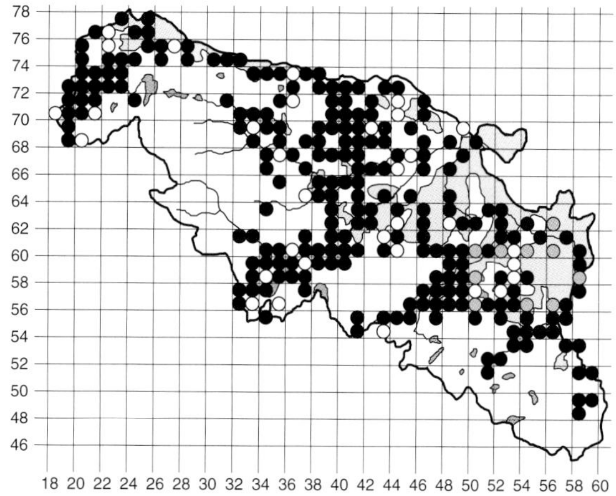

Silene dioica

Silene conica L.
Sand Catchfly

Britain: rare and scattered in the S, extinct in Scotland.
Former accidental; 0/1; extinct casual.
1st Hangingshaw, 1896, NHSG.

A specimen from the White Cart Water, New Harbour, Inchinnan (1903, PSY), is a misidentification for Cowherb; however, both these species were recorded at the original location.

Silene flos-cuculi (L.) Clairv.
Lychnis flos-cuculi L.
Ragged-Robin

Britain: throughout except in N Scotland.
Native; 174/211; common in marshy places.
1st Paisley Canal Bank, 1865, RH.

Ragged-Robin is a colourful species of marshy grasslands, fens and burn-side flushes, where not heavily grazed. It can still be found at many rural locations, but it is scarce in urban areas, on intensive farmland and in the peaty uplands.

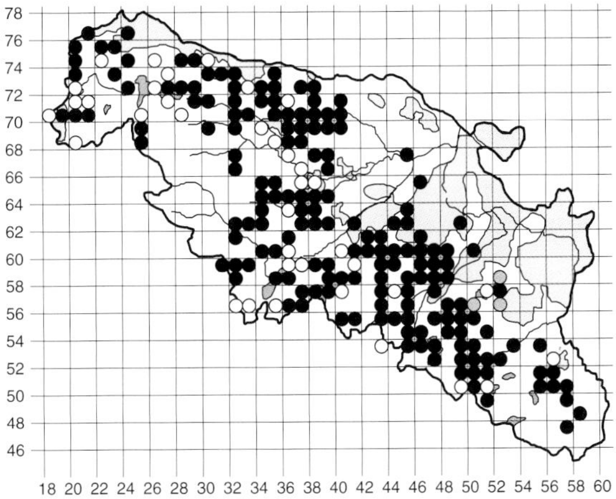

Silene flos-cuculi

Saponaria officinalis L.
Soapwort

Britain: archaeophyte (Europe); frequent in the S, scattered in lowland Scotland.
Hortal; 4/5; rare.
1st Greenlaw, Paisley, 1930, TPNS.

The modern records are from disused railways at Bridge of Weir (NS3865, 1998; NS3965, 1995, DM) and near Devol Glen (NS3174, 2005), and also from open ground at Wemyss Bay (NS1971, 1999).

Vaccaria hispanica (Mill.) Rauschert
Cowherb

Britain: neophyte (S and C Europe); scarce and scattered, mainly in the S.
Former accidental; 0/4; extinct casual.
1st Crossmyloof, 1894, GL.

Cowherb was also noted from Hangingshaw (1897, NHSG) and collected from by the White Cart Water, New Harbour, Inchinnan (1903, PSY), and wasteland, Paisley (1910, PSY); of the latter find Ferguson (1911) wrote that it 'was reported from five different localities in the neighbourhood of Paisley'. It has not been reported since.

Dianthus deltoides L.
Maiden Pink

Britain: scattered and local in England and E Scotland, rare alien elsewhere.
Hortal; 2/3; very rare.
1st Renfrewshire, 1915, TPNS.

There are two modern records, both presumably planted: one is from roadside grassland by a garden near Caplaw Dam (NS4358, 1999), and the other is from Newton Mearns (NS5254, 1988, PM).

Dianthus armeria L.
Deptford Pink

Britain: scarce in S England, rare alien in the N.
Hortal; 0/1; very rare.
1st Newlands garden, 1979, PM.

This spontaneous weed is known from a single garden (CFOG).

AMARANTHACEAE (CHENOPODIACEAE)

Chenopodium bonus-henricus L.
Good-King-Henry

Britain: archaeophyte; common in England and E Scotland.
Accidental; 1/4; very rare casual.
1st Cathcart, 1813, TH.

Good-King-Henry was collected from Hawkhead Mains (1894, PSY) and later recorded from a roadside in Eaglesham (BWR, 1956). It was seen at Darnley in 1991 (NS5258, CFOG), but has not been found there since.

Chenopodium rubrum L.
Red Goosefoot

Britain: common in S and E England, local in C Scotland.
Native or accidental; 1/3; very rare casual.
1st Renfrewshire, 1883, TB.

This goosefoot was collected from Kilbarchan in 1942 (GL). Probably overlooked, the only modern record was made by IPG at Lochwinnoch (NS3558, 2005).

Chenopodium ficifolium Sm.
Fig-leaved Goosefoot

Britain: common in S and E England, very rare in C Scotland.
Accidental; 0/1; very rare casual.
1st Paisley, 1984, AMcGS.

The sole record was from Mitchell St in Paisley.

Chenopodium album L.
Fat-hen

Britain: common in England and lowland Scotland.
Native; 46/55; frequent.
1st Renfrewshire, 1869, PPI.

Fat-hen is most frequently seen on roadsides and waste ground, but seldom found in any numbers. It is also found at the margins of arable fields and hedge banks, but records from such sites are now few.

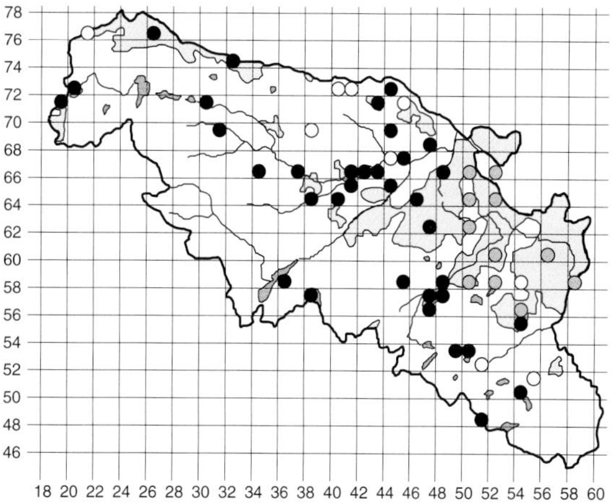

Chenopodium album

Atriplex prostrata Boucher ex DC.
Spear-leaved Orache

Britain: common in England, more coastal in Scotland.
Native; 41/43; frequent, scattered.
1st Renfrewshire, 1883, TB.

It is possible that Spear-leaved Orache has been, and perhaps still is, confused with other oraches; along shorelines this species is possibly over-recorded for Babington's Orache. Some 19th-century authors mentioned '*A. patula*' (and varieties), which presumably included this species, and although Lee (1933) separated it as '*A. hastata*', which he considered common in the area (but not coastal), he excluded it from VC76. Today Spear-leaved Orache can be found on coastal shingles and sands, extending to the strandline. It also occurs along the riverbanks of tidal rivers such as the White and Black Cart waters near Renfrew and there are a few inland records from waste-ground localities, including salted roadsides.

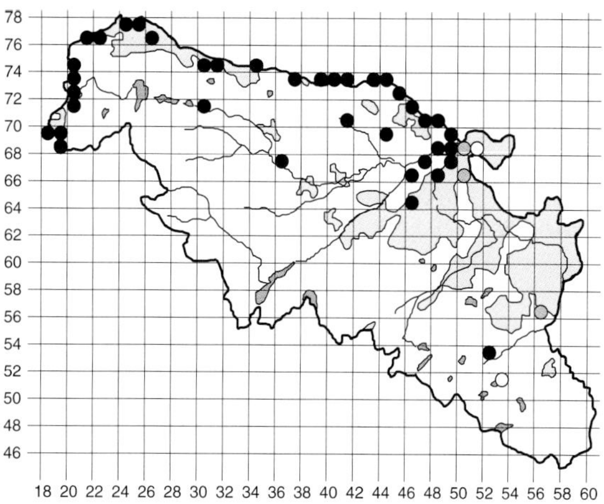

Atriplex prostrata

Atriplex glabriuscula Edmondston
Babington's Orache

Britain: coastal throughout.
Native; 8/11; occasional on the coast.
1st Renfrewshire, 1872, TB.

There are no named old localities for Babington's Orache in the VC, but Hennedy described it as frequent (although he only named the nearby island of [Great] Cumbrae, VC100). There are a couple of records made by BWR from the coast at Hatton (NS4072, 1976) and Trumpethill (NS2276, 1977). In modern times it has been under-recorded but appears widespread with records from Wemyss Bay (NS1969, 1999) to Lunderston Bay (NS2074, 2006) in the west and, in the east, from Finlaystone Point (NS3574, 2008) to the Erskine Park area (NS4572, 1996).

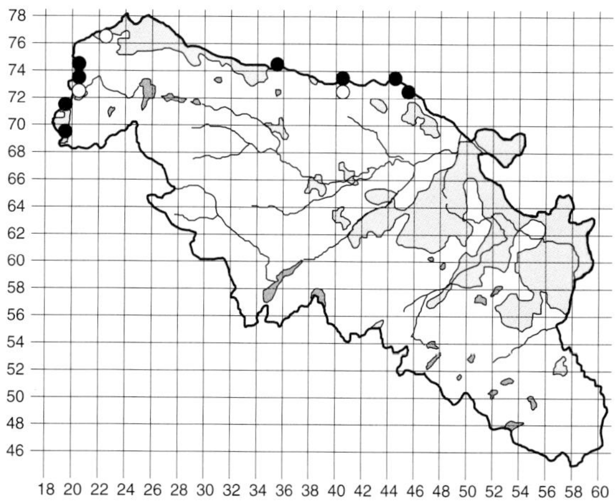

Atriplex glabriuscula

Atriplex × gustafssoniana Tascher.
(*A. longipes × A. prostrata*)
Kattegat Orache

Britain: very rare hybrid.
Native; 2/3; very rare, coastal.
1st Erskine, 1978, AJS.

This rare hybrid was first found among Common Reed at Boden Boo saltmarsh (NS4671). Kattegat Orache was also recorded further west at Erskine Park, towards Longhaugh Point (NS4473, NS4572; 1996).

One specimen collected here (NS4473) appears to be the hybrid between Long-stalked Orache and Babington's Orache *A.* × *taschereaui* Stace. Long-stalked Orache (*A. longipes* Drejer) was recorded from the Clyde in VC77 at Linthouse (NS56I, 1985, CFOG) just inside VC77, but there are no confirmed records from within Renfrewshire.

Atriplex patula L.
Common Orache

Britain: common in England and lowland Scotland.
Native; 55/69; frequent, scattered.
1st Renfrewshire, 1883, TB.

Common Orache is a frequent plant of open places, such as waste ground, waysides, reservoir margins and disturbed farmland. At no place is it abundant and it may well have declined in its traditional farmland habitats.

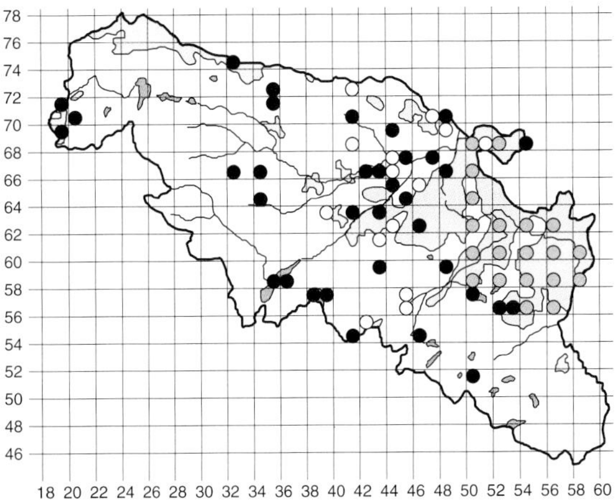

Atriplex patula

Atriplex laciniata L.
Frosted Orache

Britain: coastal throughout but local.
Native; 0/1; extinct or overlooked.
1st Renfrewshire, 1883, TB.

Hennedy (1865) described this orache as frequent 'on sandy sea beaches of the Firth', but Lee (1933) stated that it was 'not common' although he listed it as being present in Renfrewshire. Possibly overlooked, there are no named old localities and no modern sightings for this plant.

Beta vulgaris L. ssp. *maritima* (L.) Arcang.
Sea Beet

Britain: coastal in England and SW Scotland, very rare further N.
Former native; 0/1; extinct.
1st Wemyss Bay, 1901, FFCA.

Sea Beet has not been reported by any other author and is now presumed to be extinct.

Beta trigyna Waldst. & Kit.
Caucasian Beet

Britain: neophyte (SE Europe); rare casual, mainly in the S.
Accidental; 1/1; very rare casual.
1st Paisley, 2003, AJS.

The sole record appears to have been a contaminant in sown grass near Paisley Gilmour Street Station (NS4864).

Salicornia europaea agg.
Glassworts

Britain: coastal throughout but local and identification difficult.
Former native; 0/2; extinct.
1st Gourock, 1865, RH.

A specimen exists in GRK, collected from Battery Field (Greenock) in 1887, but its identity has not been confirmed. There have been no glasswort records at tidal locations since, so it is unclear which species was formerly present.

Suaeda maritima (L.) Dumort.
Annual Sea-blite

Britain: coastal throughout.
Former native; 0/1; extinct.
1st Renfrewshire, 1872, TB.

Sea-blite was collected from the Clyde at Gourock by Peter Ewing (1877, GL) but this appears to be the only named locality. However, it was considered frequent ('at Firth') by Hennedy (1865).

Amaranthus retroflexus L.
Common Amaranth

Britain: neophyte (N America); scattered, mainly in S England.
Former accidental; 0/3; extinct.
1st Near Paisley, 1910, PSY.

RG reported this waste-ground casual from Newlands (1920), and in 1959 it was found as a garden weed, also in Newlands, by F Black.

MONTIACEAE
(PORTULACAEAE *pro parte*)

Claytonia perfoliata Donn ex Willd.
Springbeauty

Britain: neophyte (W N America); frequent in C and E England, scattered in C and NE Scotland.
Hortal; 1/5; very rare.
1st Renfrewshire, 1915, TPNS.

RM noted Kilmacolm (1925) and there was a record from Paisley (NS4863) made by AJS in 1977. Springbeauty was recorded from Rosshall Park (NS5263, 1983, CFOG), but there are no other modern records.

Claytonia sibirica L.
Pink Purslane

Britain: neophyte (W N America); widely scattered, frequent in the W and N.
Hortal; 158/176; common.
1st Cloch, 1863, TGSFN.

The first record was from 'a wood between Levan House and Cloch Inn' and Pink Purslane was soon considered to be 'quite established'. There are several other late 19th-century records from the extreme north-west, as summarized by Shanks (1915), but today this long-flowering species can be found throughout the area. It is commonly found in woodlands, especially along river valleys, where it can form large stands.

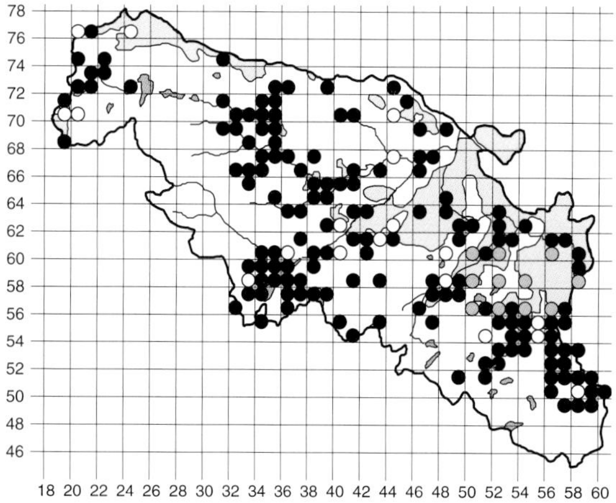

Claytonia sibirica

Montia fontana L.
Blinks

Britain: common in the N and W, local elsewhere.
Native; 181/257; common, rural.
1st Meikleriggs, 1858, PSY.

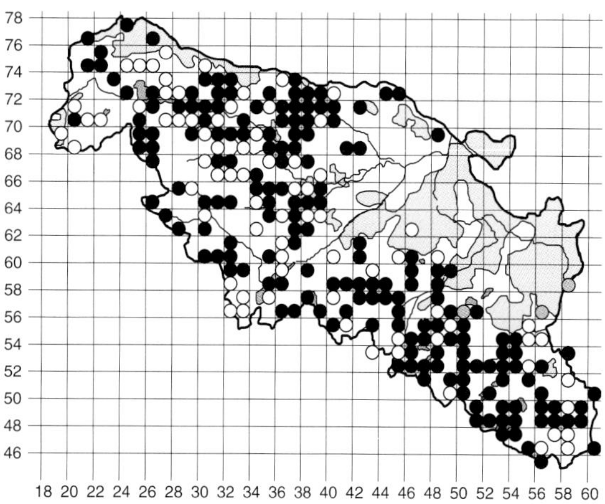

Montia fontana

Blinks is a common plant in rural areas typically found in wet flushes and along ditches or drains on agricultural land. It is also found on drier ground such as rocky outcrops, though here the plants tend to be ssp. *chondrosperma*.

Records for the subspecies include:

Montia fontana* ssp. *fontana: 25 records, probably accounting for the majority of the distribution-map sites.

***Montia fontana* ssp. *variabilis* Walters**: five modern records from the central area – Midton (NS4158, 1993), Loch Libo (NS4355, 1996), Formakin (NS4071, 1997), Marshall Moor (NS3762, 1997) and Gleddoch (NS3772, 1998). Old records are from Inverkip (1857, GL), Black Loch (1884, GL), River Calder (1909, GL) and Loganswell (1927, GL).

***Montia fontana* ssp. *amporitana* Sennen**: this subspecies was collected by RM from Ravenscraig (1925, GLAM) and there are four scattered modern records – moor, Devol Road (NS3272, 2000), Lawfield Dam (NS3769, 1997), Knocknairs Moor (NS3073, 1997) and East Moorhouse (NS5351, 1998). A somewhat intermediate specimen ('*minor-amporitana* intermediate' of Stace 2nd edn) was recorded from Dargavel Burn, Mid-Glen (NS3870, 1997).

***Montia fontana* ssp. *chondrosperma* (Fenzl) Walters**: eight records, all from the central north area – Birkmyre (NS3173, 1996), Barbeg (NS4071, 1998), Gleddoch (NS3772, 1998), Langbank (NS3773, 1998), Gryfeside (NS3370, 1998), Loch Thom (NS2672, 2000), Knockmountain (NS3671, 2007) and Barnbrock (NS3564, 2007).

CORNACEAE

Cornus spp.
Dogwoods

Britain: neophytes (E Asia/N America); scattered throughout lowlands.
Hortals; 21/25; occasional.
1st Castle Semple Loch, 1887, NHSG.

Dogwoods have been widely planted and can form large stands along the margins of ponds and sometimes along watercourses in old estate woodlands. All of the modern records have been mapped as Red-osier Dogwood (*C. sericea* L.), which was first recorded from Haggs (NS5662, *c.* 1966, JDM) and later from Finlaystone (1976, AJS), although some may refer to White Dogwood (*C. alba* L.), which was recorded from Braehead on the VC77 boundary (1983, CFOG). Lee (1933) recorded '*C. sanguinea*' from Lochwinnoch ('very abundant'), but the only dogwood now recorded here, at Calder Bridge (NS3558), is Red-osier Dogwood.

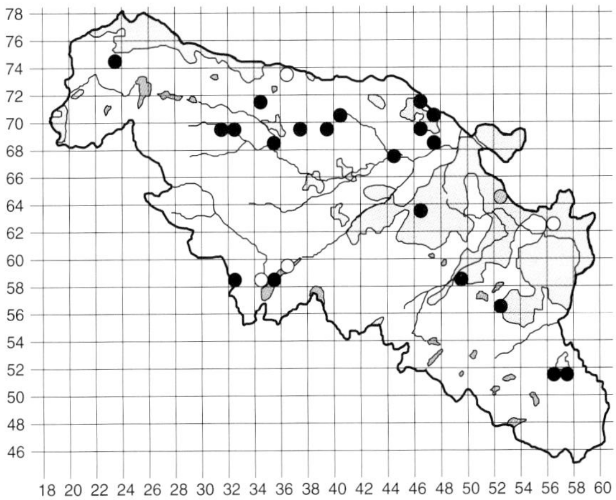

Cornus sericea

HYDRANGEACEAE

Philadelphus spp.
Mock-oranges

Britain: neophytes (Europe); scattered and local.
Hortal; 18/19; scattered.
1st Crofthead, 1976, JP.

Mock-oranges were recorded from four places during CFOG surveys. The map shows all species records, which are garden outcasts or persistent shrubs, mainly from old estates, but any spread seems to be very localized. There are seven modern records for Mock-orange (*P. coronarius* L.) but hairier specimens, referable to garden cultivars (*P.* 'Virginalis Group' or possibly *P.* 'Lemoinei Group'), have been recorded from six places (seven 1km squares): Formakin (NS4070, 1994; NS4170, 2007), a roadside at Port Glasgow (NS3075, 2012), Holehouse (NS4756, 1997), by the Gryfe Water, Kilmacolm (NS3469, 1999), and by the White Cart Water at Kirkland (NS5853, 2000) and at the Snuff Mill (NS5860, 2005).

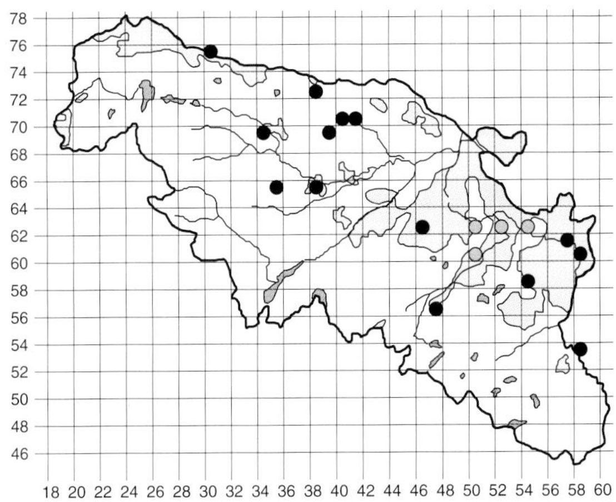

Philadelphus spp.

Hydrangea petiolaris Siebold & Zucc.
Japanese Climbing-hydrangea

Britain: neophyte (Japan); rare.
Hortal; 1/1; very rare
1st Bridge of Weir, 2005.

A large specimen grows on a disused railway and can be seen emerging over a roadside wall on the edge of Bridge of Weir (NS3965). Another outcast Hydrangea was noted from Castle Wemyss Wood (NS1970, 1999), but it was not identified to species level.

BALSAMINACEAE

Impatiens noli-tangere L.
Touch-me-not Balsam

Britain: rare and local in the W, scattered alien elsewhere.
Former hortal; 0/1; extinct.
1st Cloch, 1876, NHSG.

The first record stated '¼ of a mile west of Cloch' and there is a specimen from this location in GL (1879). Hennedy (1891) and Ewing (1901) repeated the record but there are no later sightings and Touch-me-not Balsam was excluded from the Renfrewshire list (TPNS 1915).

Impatiens glandulifera Royle
Indian Balsam

Britain: neophyte (Himalayas); common in England and lowland Scotland.
Hortal; 67/71; frequent, spreading.
1st Cart, Paisley, 1937, TPNS.

The original note (McKim 1942) considered Indian Balsam a garden outcast but added that 'it is welcomed if it beautifies an otherwise ugly spot in the river'. By 1945 the TPNS was reporting the plant well established on the 'Cartbank' from Crookston to the harbour. Although Lee (1953) noted it from the VC, the next named locality does not appear until BWR recorded it from Trumpethill (NS2276) in 1977. However, Indian Balsam appears not to have been able to establish itself in the west of the VC, perhaps deterred by the swifter flow and lack of silt and nutrients in the burns feeding off the uplands. The only other recent past record is from Glentyan (1984, BWR), where Indian Balsam still occurs today (NS3962, 1997) and appears to be spreading rapidly.

The modern distribution shows a striking pattern, being well established along the Black Cart Water and converging with the populations along the White Cart Water catchment; it is also present along the lower reaches of the Gryfe Water, but not recorded upstream of Selvieland (NS4567, 2001), though it is known further west from waste ground in Bridge of Weir (NS3866, 2006). The northern outliers are by the Gleddoch Burn (NS3972, 1998) and by the coast about Ferryhill (NS4073, 2006) and Langbank (NS3973, 2012). To the south it occurs by a small burn near Uplawmoor (NS4354, 2007, P Allan) and by the disturbed ground at Bent Bridge (NS4359, 2012). Given Indian Balsam's current notoriety as an invasive non-native, it is one to watch out for along other watercourses.

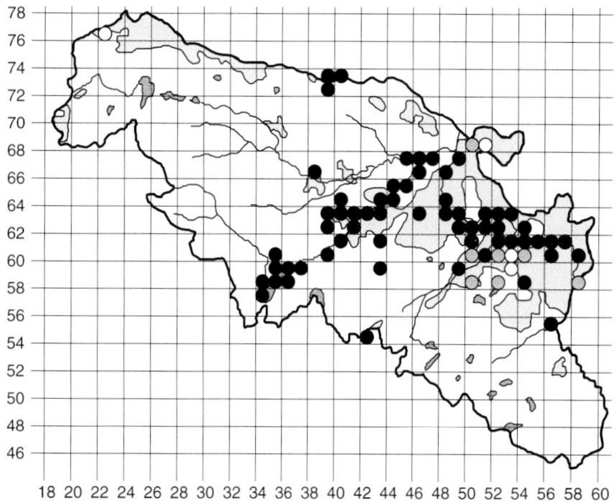

Impatiens glandulifera

POLEMONIACEAE

Polemonium caeruleum L.
Jacob's-ladder

Britain: rare native, widespread as garden outcast.
Hortal; 8/11; rare.
1st Dam at Rouken Glen, 1885, NHSG.

RG (1920) found Jacob's-ladder at a 'bank of old Paisley Canal' and Lee (1933) reported Caplerig Burn, but all other named localities of this garden outcast are from the modern period: roadside, Sergeant Law (NS4459, 1995), Park Quay, Newshot (NS4770, 1997), roadside, Greenock Road (NS4172, 2001), roadside, Carrot (NS5748, 2003), near Quarriers (NS3667, 2003), disused railway, Springhill (NS5057, 1996, SA), waste ground, Mearns (NS5152, 2006) and roadside, Black Hill (NS4851, 2011).

PRIMULACEAE

Primula vulgaris Huds.
Primrose

Britain: common throughout.
Native; 134/164; common in rural areas.
1st Neilston and Paisley, 1845, NSA.

Primrose is frequently found in local woodlands, often on steeper slopes and where the soils are less acidic; it is also found on more open scrubby banks along watercourses and on railway embankments as near Cowdon Burn (NS4556). It can extend into the uplands along steep burn sides.

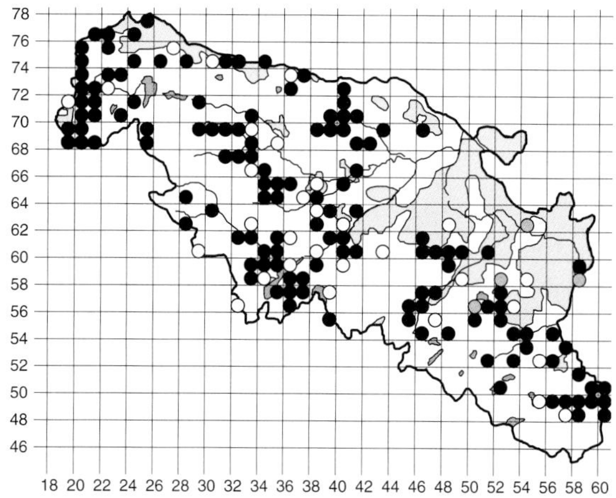

Primula vulgaris

Primula veris L.
Cowslip

Britain: common in the S, extending to E Scotland.
Hortal; 5/11; rare.
1st Paisley and Glebe, Kilbarchan, 1845, NSA.

There are a few old records for Cowslip, which earlier authors considered rare: Cardonald Place (1856, HMcD), Finlaystone and Blythswood (1869, RPB), Inverkip (1880, NHSG), Port Glasgow (1880, NHSG), Crookston (1890, GL) and Milliken (1891, AANS); in 1869 (RPB) it was stated about Cowslip, 'does not propagate and appears introduced'. More recently it was seen at Craigends (NS4166, 1969, BWR) and Abbotsinch (NS4775, 1980, AJS). Cowslip was reported (CFOG) from the railway embankment at Arkleston (NS5065, 1985) and other modern records are from short grassland, presumably sown: Kelburn (NS3471, 1998), west Millarston (NS4563, 2002) and Jenny's Well (NS4562, 2007, RW); it was also recorded from Cornalees (NS2472, 1995, CMRP).

Primula cultivars

Occasionally *Primula* cultivars are found as garden outcasts and may persist locally. Polyanthus (*P.* × *polyantha* Mill.) cultivars were recorded at Castle Semple (NS3558, 1997), Craigends (NS4166, 2002), Bow Hill (NS2577, 2004) and by the White Cart Water (NS5859, 2005). *Primula* 'Wanda' has been recorded from Barsail Wood (NS4669, 1996) and Arthurlie (NS4958, 1997).

Primula japonica A. Gray
Japanese Cowslip

Britain: neophyte (Japan); very rare in the S and W.
Hortal; 1/1; very rare escape or outcast.
1st Millbank, 2006.

A small population occurs in wet woodland by the burn below Millbank Farm (NS3357).

Lysimachia nemorum L.
Yellow Pimpernel

Britain: common throughout except in E England.
Native; 79/102; frequent, mainly in the W.
1st Paisley and Lochwinnoch, 1845, NSA.

Yellow Pimpernel is frequently found in woodlands, usually on slopes where some flushing occurs. It can be found in more open habitats such as near burn sides and, rarely, marshy flushes, but generally it is plant of at least partial shade. The map shows a marked western tendency, but it is absent from higher peaty ground.

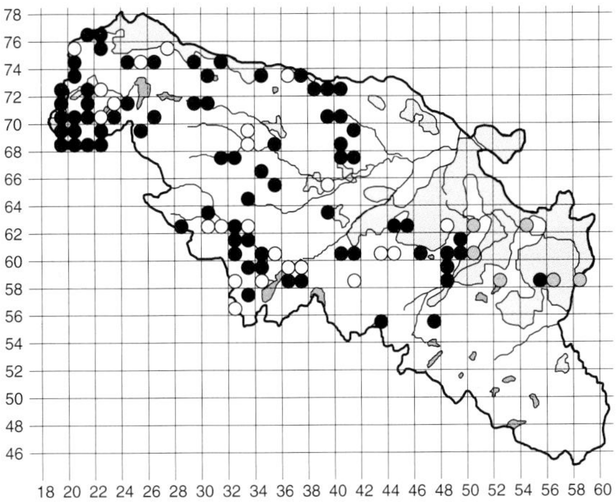

Lysimachia nemorum

Lysimachia nummularia L.
Creeping-Jenny

Britain: common in the S, scattered alien in Scotland.
Hortal; 11/16; scarce.
1st Paisley, 1845, NSA.

Creeping-Jenny was also noted at Lochwinnoch in 1887 (NHSG), a location repeated by Hennedy (1891), who added that it 'can scarcely be considered native'. It was recorded from Castle Semple in 1971 and 1973 (BWR) and there are two modern records from this area: Barr Loch (NS3557, 1995) and by the Black Cart Water, near Kenmure Hill (NS3859, 2003); other records include by the Black Cart Water, Inchinnan (NS4666, 2003), India Works site, Inchinnan (NS4768, 1996), Royal Ordnance Factory, Georgetown (NS4268, 1999) and by the White Cart Water, Kirkland (NS5853, 2000).

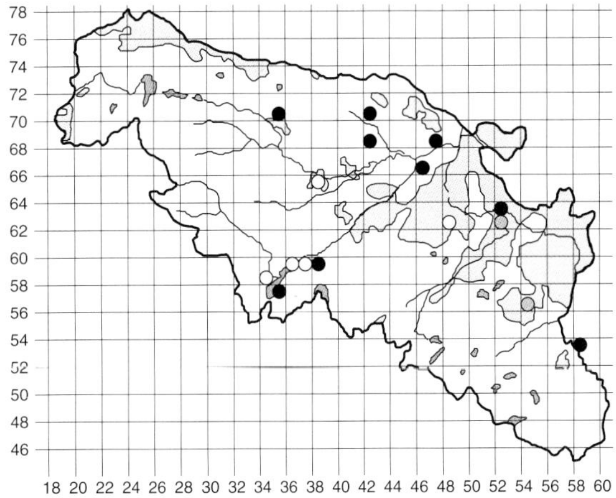

Lysimachia nummularia

Lysimachia vulgaris L.
Yellow Loosestrife

Britain: common in the S, scattered in Scotland.
Native; 1/10; very rare.
1st Lochwinnoch, 1845, NSA.

Yellow Loosestrife was recorded from Castle Semple in 1887 (NHSG) and it was also noted from the nearby Collegiate Church ponds (1893, AANS), but the sole modern record is from the adjacent Barr Loch (NS3557, 1996), made by the SNH Loch Survey, which did not relocate it at Castle Semple Loch. Other old records include Inverkip (1880, NHSG), Paisley (1910, PSY), Clydeside, Erskine (1883, GL), and RM recorded it from Braidbar (1945) and Lochside (1965).

Lysimachia punctata L.
Dotted Loosestrife

Britain: neophyte (SE Europe); frequent and widespread.
Hortal; 119/122; common, increasing.
1st Rubbish tip, Barr Loch, 1969, AMcGS.

Dotted Loosestrife was found at Gleniffer Braes (NS4459, 1974, AJS) and also recorded from St Brydes (1977), Barcraigs Reservoir (1978), Bridge of Weir (1978) and Kilmacolm (1984) by BWR, and since then appears to have increased quite rapidly; this is perhaps due more to dumping than natural colonization, although it seems capable of local spread (vegetatively only?) once at a site. It grows on disused railways, waysides, hedge banks, waste ground and scrub margins, and seems to do well in damp soils, though not in permanent wetlands. Its distribution is widespread, mostly lowland and near habitation, but a few sites are rural, e.g. along roadsides as at Queenseat (NS5248, 2001) and Melowther (NS5450, 2001).

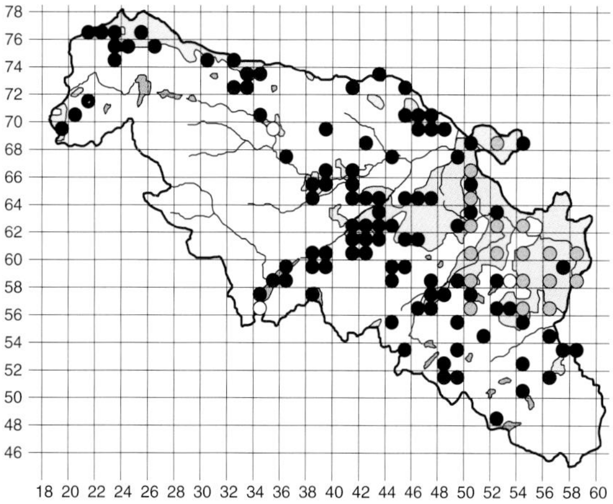

Lysimachia punctata

Lysimachia thyrsiflora L. Tufted Loosestrife

Britain: frequent in C Scotland, very rare elsewhere.
Native; 9/10; rare.
1st Castle Semple Loch, 1887, NHSG.

Castle Semple Loch appears to be the only 19th-century locality for Tufted Loosestrife and it is not recorded from Paisley Canal, a habitat favoured by this loosestrife in Glasgow (CFOG). It was collected from Glen Moss in 1959 (GL) and also found at Knapps Loch in 1975 (AJS); JDM reported it from Lawfield

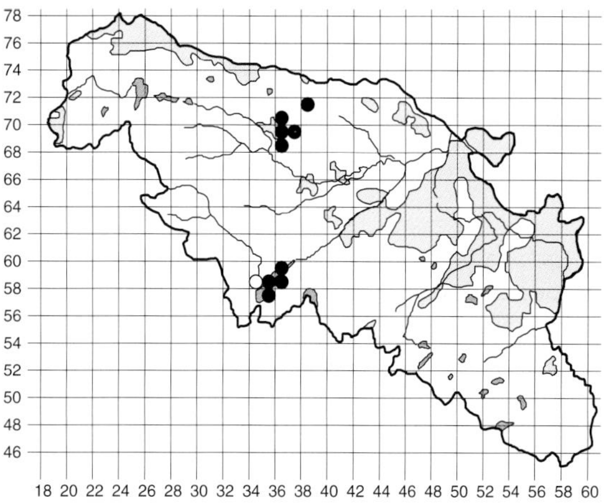

Lysimachia thyrsiflora

Dam (NS3969) in 1968. The SNH Loch Survey (1996) refound it at Castle Semple Loch (NS3559), Knapps Loch (NS3668) and Glen Moss (NS3669); other sites are a pond at Barscube Hill (NS3871, 1996), Hydro Dam, Kilmacolm (NS3670, 2002), Lawfield Dam (NS3769, 1997) and the old reservoir at Cairnkibbuck (NS3670, 2007). All the sites are swamps or loch margins, although at Glen Moss Tufted Loosestrife grows in thick carpets of bog-moss.

Glaux maritima L. Sea-milkwort

Britain: common on coasts throughout.
Native; 17/24; frequent on the coast.
1st Firth of Clyde, 1834, MC.

Sea-milkwort can still be found on the coast but it cannot be said to be very common, as Hennedy (1891) described it. Today it shows a strong disjunct pattern, occurring on the sandy shores from Gourock south to Wemyss Bay, and further east, along the muddy margins of the Clyde from Langbank to Erskine, although the neat disjunction is weakened by the recent find on the shore near Port Glasgow (NS3174, 2012).

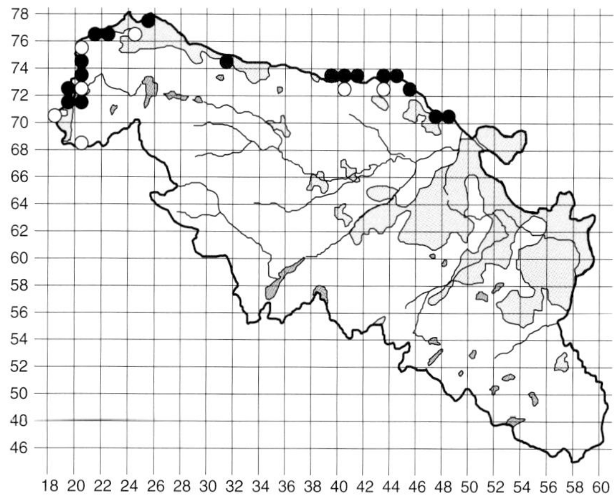

Glaux maritima

Anagallis tenella (L.) L.
Bog Pimpernel

Britain: frequent in the W, local elsewhere.
Former native; 0/2; extinct.
1st Port Glasgow, 1821, FS.

This is possibly the same record as the Greenock (parish) one from 1845 (NSA) but Hennedy (1891) mentions Gourock; Bog Pimpernel has not been recorded for well over 100 years.

Anagallis arvensis L.
Scarlet Pimpernel

Britain: common in the S, local in Scotland.
Native; 2/7; very rare.
1st Paisley, 1845, NSA.

Hennedy (1891) noted Gourock and Cathcart and in 1896 Scarlet Pimpernel was reported from Ferguslie (TPNS). The only modern records are from the quarry at Swinesglen (NS4169, 1994) and waste ground, Gourock (NS2775, 2012). Blue Pimpernel (ssp. *foemina* (Mill.) Schinz & Thell.) was collected as a garden weed in Giffnock by RM in 1972 (CFOG) and was first recorded from Paisley in 1845 (NSA) and later from Giffnock quarry in 1892 (NHSG), Hangingshaw in 1896 (NHSG) and 'near Paisley' in 1915 (TPNS 1942); no specimens exist to verify the latter records.

Trientalis europaea L.
Chickweed-wintergreen

Britain: local in NE England, common in NE Scotland.
Native; 1/3; very rare, endangered.
1st Glen, near Johnstone, 1883, NHSG.

The 'Glen' of the first record is Bardrain Glen (NS4360) and there are a number of records or specimens collected from 1883 through to 1942 (TPNS); one record (1896, PSY) refers to 'Dusky Glen, Gleniffer', presumably the same place. There is also an old record for Witch Moss, Houston (1898, JMY) and a more recent one from Torr Hall, near Bridge of Weir (NS3666, 1961) by Frances Black. Chickweed-wintergreen was still present at Bardrain in 2001 (D Parker pers. comm.).

Centunculus minimus L.
Anagallis minima (L.) E.H.L. Krause
Chaffweed

Britain: scattered in the W and S, mainly coastal, very rare elsewhere.
Former native; 0/1; extinct.
1st Marsh near Langside, 1813, TH.

Hopkirk considered Chaffweed rare. There are no other reports.

Samolus valerandi L.
Brookweed

Britain: scattered, often by coasts, rare in Scotland except in the W.
Former native; 0/1; extinct.
1st Firth of Clyde, 1834, MC.

Brookweed was considered common around the Firth of Clyde in the 19th century (Hennedy 1891), but the only named location is 'near Paisley' (1873, BIRM); it must now be considered extinct.

ERICACEAE
(EMPETRACEAE, PYROLACEAE)

Empetrum nigrum L.
Crowberry

Britain: common in the N and W.
Native; 83/100; frequent in the uplands.
1st Lochwinnoch, 1845, NSA.

Crowberry is commonly encountered in bogs and related mires in the uplands (above 250m) in the west and east. It can be found at lower altitudes, as at Burneven (NS2275, 2003), Lurg Moor (NS2973, 1998), Devol Moor (NS3272, 2000), Craig Muir (NS3770, 1997), mire at Knockmountain (NS3671, 1998) and Hartfield Moss (NS4357, 1999). It is assumed that all records belong to ssp. *nigrum*.

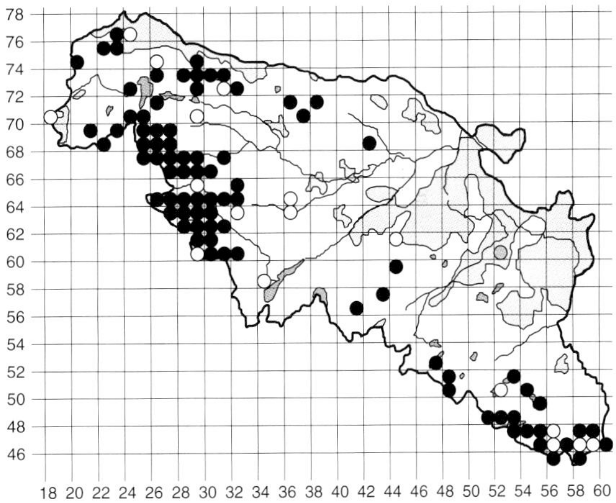

Empetrum nigrum

Rhododendron ponticum L.
Rhododendron

Britain: neophyte (SW Europe and possibly hybrid); common throughout.
Hortal; 133/143; common.
1st Barochan Moss, 1959, BWR.

Rhododendron has been present in the VC far longer than the first record indicates, as it has been widely planted in old policy estates and parks. As in other parts of Scotland, it has persisted and spread and can be abundant and problematic in woodlands, e.g. notably at Drum Estate (NS3971, 1997) and Kelly Glen (NS1968, 1995), and it is well established on Barochan Moss (NS4268, 1994). It does not appear to be spreading at upland moorland sites although it is well established at Muirshiel (NS3163, 1995) and has spread to the River Calder and Raith Burn feeder (NS3063, 1995).

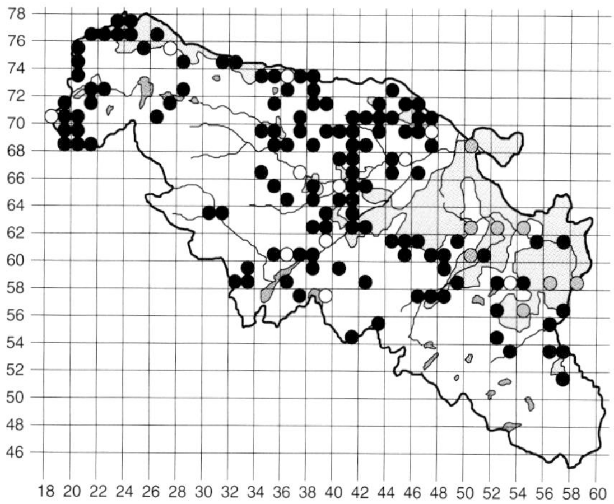

Rhododendron ponticum

Rhododendron groenlandicum (Oeder) Kron & Judd
Ledum palustre L.
Labrador-tea

Britain: neophyte (N Europe and N America); local in bogs in the N.
Hortal; 0/1; probably extinct.
1st Barochan Moss, 1976, JP.

It is unclear who first planted Labrador-tea, a species which has been introduced at other central Scottish raised bogs, but it has not been seen in the last 30 years, although it may persist in glades amid dense stands of Rhododendron. Curiously, an article about *Ledum* in Britain (Ribbons 1976) does not mention the VC76 record.

Calluna vulgaris (L.) Salisb.
Heather

Britain: common throughout except in the SE.
Native; 290/340; common.
1st Wemyss Bay, 1844, GL.

Heather is a common sight in the uplands, especially in the Renfrewshire Heights where it dominates extensive areas of moorland. In the heavier grazed or agriculturally improved foothills Heather can be frequent although often short-grazed, only being more visible on slopes or less accessible roadsides; several golf-course roughs are also good places for relict heaths. It can be found, but less abundantly, in the lowlands, including at disused railway and waste-ground sites, local rocky outcrops and also on relict peat bogs such as at Barochan Moss (NS4268, 1994).

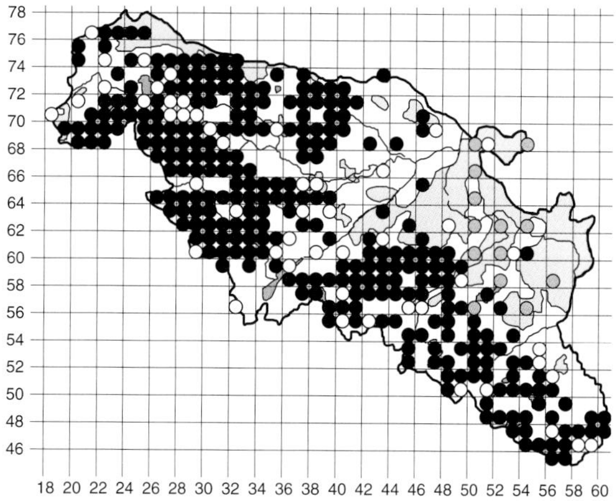

Calluna vulgaris

Erica tetralix L.
Cross-leaved Heath

Britain: common in the N and W.
Native; 163/185; common in uplands.
1st Paisley, 1845, NSA.

The map strikingly reflects the upland areas where records mark out the wetter bog areas of heathland (moorland), mire and bog sites. Cross-leaved Heath is rare or mostly absent from the lowlands, although it is found at Barochan (NS4268, 1994) and Fulwood (NS4468, 1999) mosses and a surprising waste-ground location at Port Glasgow (NS3274, 1998).

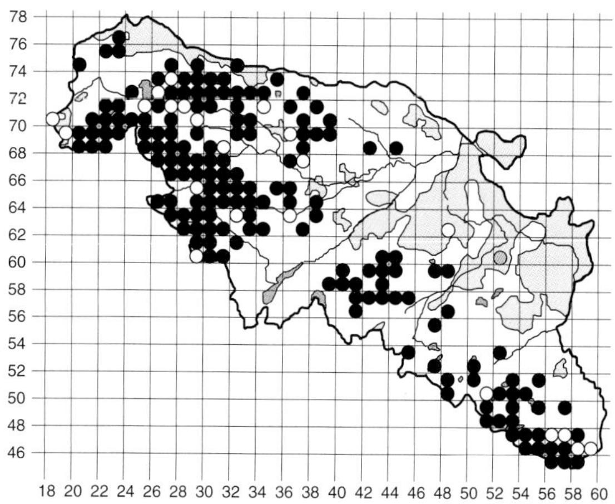

Erica tetralix

Erica cinerea L.
Bell Heather

Britain: common in the N and W.
Native; 92/111; frequent, western.
1st Paisley, 1869, RPB.

Bell Heather was considered common in the 19th century but this status can only be applied today in the Renfrewshire Heights, but even here it is scarcely in any abundance. In the south-east uplands it is notably rare, or under-recorded, with populations at Carrot Burn (NS5747, 2000), Dod Hill (NS4953, 1998) and Cairn Hill (NS4851, 2002). It is most frequently encountered on higher ground (above 200m) in the west, being found on rocky ledges or outcrops or steeper slopes and embankments, but avoiding the wetter parts of acidic peat bogs.

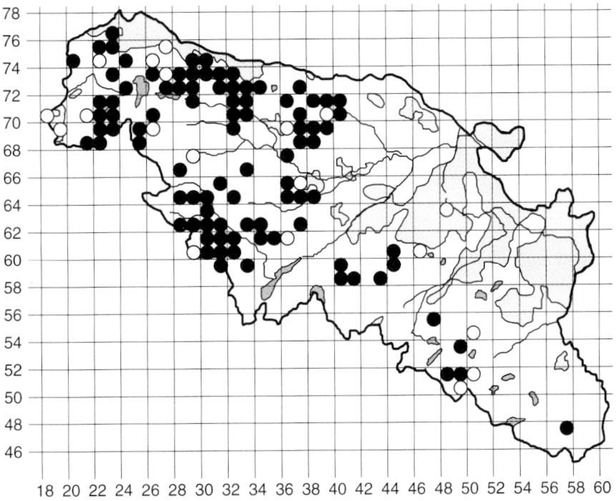

Erica cinerea

Andromeda polifolia L.
Bog-rosemary

Britain: frequent in W C Britain only.
Former native; 0/2; extinct.
1st Paisley Moss, 1821, FS.

This old record was repeated in 1845 (NSA) and 1891 (RH), although Hooker (1821) had stated that Bog-rosemary was present 'sparingly'. After years of disturbance, the modern-day Paisley Moss does not support typical raised bog habitat and this plant must be considered long gone from the site, as was thought the case in 1915 when it was excluded from the Renfrewshire list (TPNS). There was a more recent field record from Flow Moss at the south-east edge of the VC (IG pers. comm.), but the site was lost to forestry in the 1980s. Bog-rosemary still occurs nearby in a boggy glade in the Whitelee forest (VC75, NS4256, 2008), but it has not been noticed during recent surveys within VC76.

Gaultheria shallon Pursh.
Shallon

Britain: neophyte (W N America); scattered and local.
Hortal; 4/7; rare.
1st Paisley Museum, 1928, PSY.

Shallon was collected from boulders in front of Paisley Museum, where it was presumably planted. More recently it has been recorded from two estate woodlands: Ardgowan (NS2073, 1998) and Glentyan (NS3963, 1999). It is also known from a disused railway at Arthurlie (NS4958, 1997) and was recorded from Barrhead Railway Station (NS4958, 1965, RM); RM and AJS knew of Shallon from an old railway in Paisley (*c.* 1974) and it was reported from Rouken Glen (NS55N, 1983, JHD).

Gaultheria procumbens L.
Checkerberry

Britain: neophyte (E N America); rare in the S.
Hortal; 0/1; doubtful record.
1st Glentyan House, 1984, BWR.

This is the only record for Checkerberry and may be an error for Shallon, which has been recorded recently from this estate.

Gaultheria mucronata (L.f.) Hook. & Arn.
Prickly Heath

Britain: neophyte (S America); scattered and local, frequent in W C Scotland.
Hortal; 12/18; scarce, scattered.
1st Paisley, 1974, AJS.

Described as 'the pretty Patagonian shrub', this plant was noted germinating in a nursery at Pollok in 1890 (Paterson 1893) and much later JDM recorded Prickly Heath from Maxwell Park (NS5663, *c.* 1966), although its status there is unclear. It was first recorded from Glen Moss in 1976 (AJS) and other recent sites (BWR)

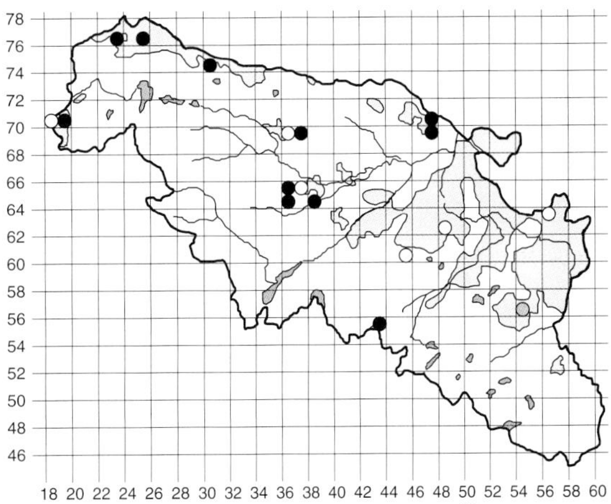

Gaultheria mucronata

include Tor Wood (1982) and Barnbeth (1986). It can still be found at Glen Moss (NS3769, 1998) and it has presumably been planted at golf courses such as Cathcart Castle (NS5557, 1987, CFOG), Ranfurly (NS3664, 1998), Ranfurly Castle (NS3864, 1999) and Kilmacolm (NS3769, 1998).

Vaccinium oxycoccos L.
Cranberry

Britain: frequent in W C Britain.
Native; 61/72; frequent in boggy areas.
1st Paisley and Lochwinnoch, 1845, NSA.

Cranberry is to be found trailing over carpets of bog-moss in the wetter mires, and is likely to be overlooked; it appears to be more frequent in the south-east, with fewer records from the extensive blanket bog in the Renfrewshire Heights. It occurs on the lowland raised bogs at Barochan (NS4268, 1994) and Fulwood (NS4468, 1999).

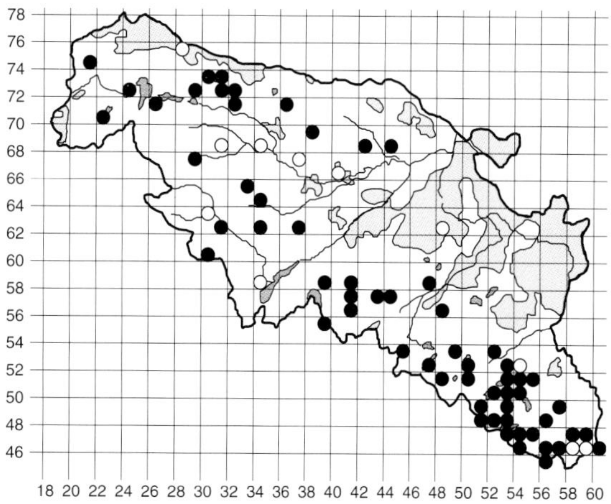

Vaccinium oxycoccos

Vaccinium vitis-idaea L.
Cowberry

Britain: common in the N and uplands.
Native; 26/32; occasional, mostly on the highest ground.
1st Gourock, 1851, GGO.

With a few exceptions, Cowberry's distribution pattern picks out the highest hills of Renfrewshire. It is particularly well represented in the Renfrewshire Heights but only above 300m altitude, where it grows in the moorland where not too wet and on rocky outcrops. In the south-east it was recorded 'south of Carrot' in 1972 (BWR) but much habitat here has been lost to forestry, although it persists at the VC77 boundary at Corse Hill (NS6046, 2002, JD). It does occur at lower altitude on a north-facing rock outcrop along Carrot Burn (NS5747, 2000). A further isolated lowland site is Broadfield Hill (NS4058, 1993), but Cowberry has not been refound at Neilston Pad where MY reported it in 1865.

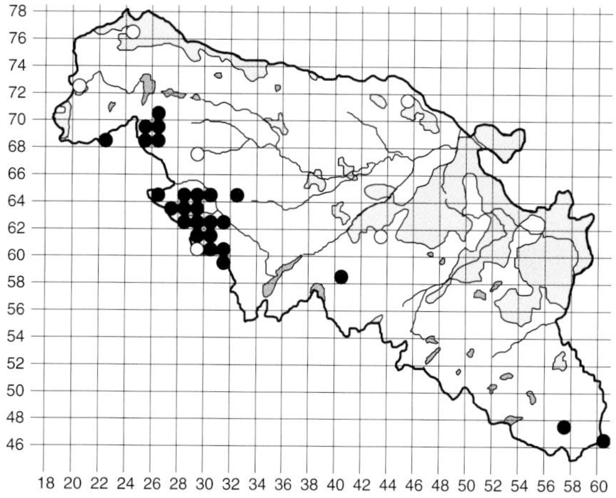

Vaccinium vitis-idaea

Vaccinium myrtillus L.
Blaeberry (Bilberry)

Britain: common in the N and W.
Native; 325/380; very common.
1st Paisley, 1845, NSA.

Blaeberry is found throughout the area, only becoming rare or absent in the strongly urban or agriculturally improved lowlands. It is commonly found at all types of heath and moorland (although not as abundantly as Heather), often most prominent on rocky ridges or outcrops, but is rare or absent from wet mires; it is also frequent in acidic, birch dominated woodlands. In the lowlands it is found on the upper slopes of watercourse valleys and on rocky outcrops, especially where north-facing.

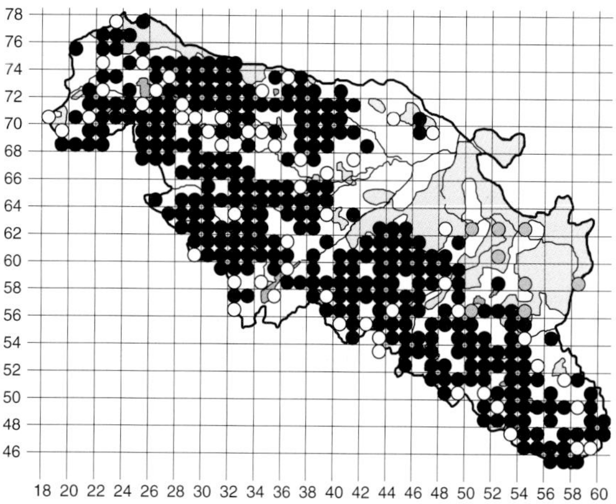

Vaccinium myrtillus

Pyrola minor L.
Common Wintergreen

Britain: frequent in the NE, rare in the S.
Native; 7/17; rare.
1st Kelly Glen, 1844, GL.

There are old records for Common Wintergreen from woods east of Gourock (1865, RH) and Lawmarnock Wood (1903, PSY); RM noted it from a marsh in the Ravenscraig–Gourock area (1923) and Lee (1933) mentions Bardrain Glen. JDM found it at three places: by the Gryfe Water (NS3766, 1961), a quarry at Craigenfeoch (NS4361, 1963) and Garple Burn (NS3458, 1967); BWR noted Glen Park (NS4760, 1974) and it occurs on the VC border at Braehead (NS55E, 1992, CFOG). Today there are seven scattered sites: Sergeant Law Moss (NS4760, 1995), Midton Wood (NS2476, 1996), mire at Knockmountain (NS3671, 1998), a disused railway, Paisley (NS4864, 2000), Inch Green dock (NS3075, 2010, AHJ), Wemyss Bay (NS1970, 2006) and a mossy wood near Fauldhead (NS4358, 2009, PG).

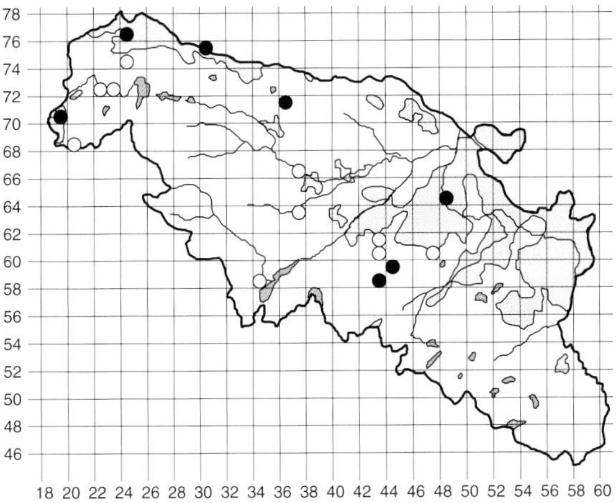

Pyrola minor

Orthilia secunda (L.) House
Serrated Wintergreen

Britain: frequent in N Scotland, rare in uplands further S.
Former native; 0/1; extinct.
1st Bardrain Glen, 1891, RH.

There is no other detail about this old record but the site has been well searched over the years, so Serrated Wintergreen must be considered long extinct.

GARRYACEAE
(CORNACEAE)

Aucuba japonica Thunb.
Spotted-laurel

Britain: neophyte (E Asia); scattered, mainly in the S.
Hortal; 2/3; very rare.
1st Rosshall (NS5263), 1966, JDM.

There are three modern records for Spotted-laurel: from Locher Bridge (NS4064, 1999), Langside Wood and by the White Cart Water, Albert Park (both NS5761, 2005).

RUBIACEAE

Sherardia arvensis L.
Field Madder

Britain: common in the S, extending to E Scotland.
Native or accidental; 3/6; very rare.
1st Cathcart Castle, 1813, TH.

Field Madder was also noted in 1845 from Paisley (NSA) and from waysides, Mearns (1890, AANS); it was considered frequent by 19th-century authors, though Wood (1893) added 'but only in small patches, and it is anything but constant'. There was one record in CFOG (NS54R, 1981) and only two modern ones: from roadside grassland at Auchenbothie (NS3471, 1998) and waste ground, Greenock (NS2776, 2000).

Asperula arvensis L.
Blue Woodruff

Britain: neophyte (Europe); casual, scattered in the S.
Former accidental; 0/3; extinct.
1st Paisley, 1865, TPNS.

Blue Woodruff was also recorded from Crossmyloof (1894, GL) and Hangingshaw, Glasgow, in 1896 (NHSG).

Galium boreale L.
Northern Bedstraw

Britain: upland areas in the N and W.
Native; 6/14; rare.
1st Lochwinnoch, 1834, MC.

Hennedy (1891) mentioned Duchal as a site for Northern Bedstraw and in 1896 it was reported from Bridge of Weir (NHSG). It was recorded about the Gryfe Water three times between 1961 and 1982 (BWR) and today can be found on rocks by the Gryfe at Craigends Bridge (NS3667, 2003). It occurs further down the river, again on rocks, at Craigends Estate (NS4166, 2000) and was reported in 1979 from the wooded riverbanks at Bridge of Weir (NS3965) by DM. A further record occurs upstream, on the Green Water feeder at Duchal (NS3269, 1999). A cluster of sites in the west is associated with rock outcrops along the North Rotten Burn (e.g. NS2569, 1997). It has not been reported recently from the River Calder (NS3361) where it was seen by AMcGS in 1969.

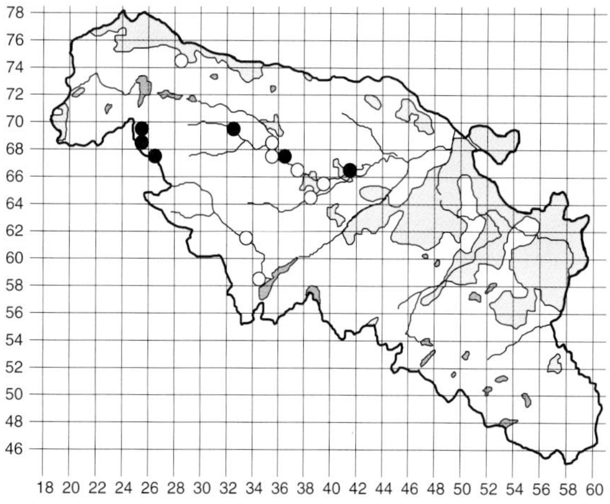

Galium boreale

Galium odoratum (L.) Scop.
Woodruff

Britain: common except in E England and N Scotland.
Native; 75/86; frequent in woodlands.
1st Paisley, 1845, NSA.

Woodruff has a widespread distribution but the localities, almost exclusively, mark out wooded watercourses, where it favours the less acidic, richer soils.

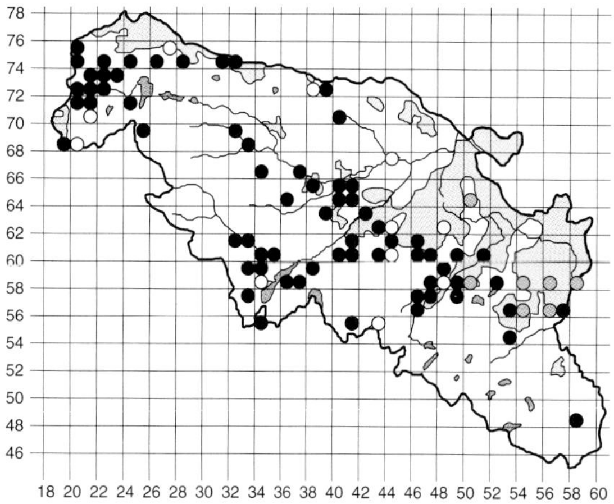

Galium odoratum

Galium uliginosum L.
Fen Bedstraw

Britain: common but absent from NW Scotland.
Native; 29/36; occasional.
1st Gourock and Paisley Canal bank, 1865, RH.

This bedstraw grows in similar-looking wet places to the much commoner Marsh Bedstraw, but is much more restricted geographically and in vegetation preference. It is local in the south-east of the VC, growing in fens and marshes which are less disturbed or enriched. There are a few old and some modern records from the west, which are viewed with some suspicion: Scart Marsh (NS3766, 1994, CD), Lady Burn, Lurg Moor (NS2974, 1992, CMRP-NVC) and two from TPNS excursions: Jock's Craig (NS3269, 1988) and River Calder (NS3959, 1990).

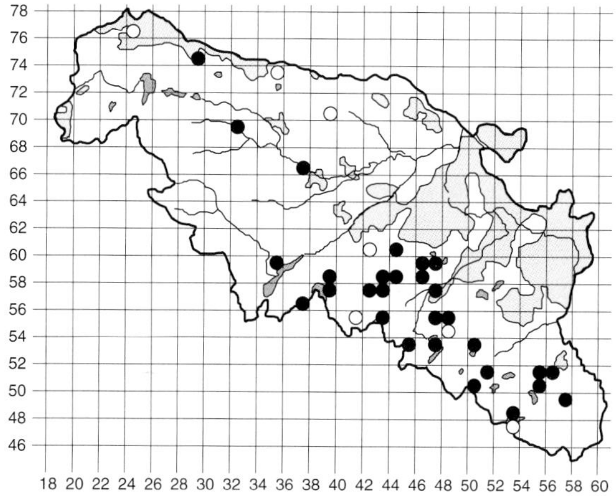

Galium uliginosum

Galium palustre L.
Common Marsh-bedstraw

Britain: common throughout.
Native; 309/347; very common in wet places.
1st Wemyss Bay, 1856, GL.

This bedstraw is one of the commonest marshland plants, found in most types of marsh and also swamps, ditches, the margins of open water and along the banks of watercourses; very rarely it can be found on waste ground colonizing poorly draining areas.

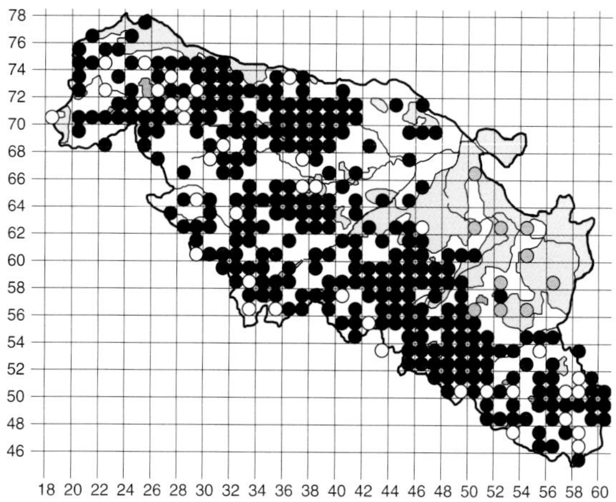

Galium palustre

Galium verum L.
Lady's Bedstraw

Britain: common throughout except locally in the W.
Native; 165/223; common.
1st Wemyss Bay, 1856, GL.

Lady's Bedstraw is typically found in free-draining, unimproved neutral to slightly acidic grasslands. It has presumably declined with agricultural improvement but can still be found on marginal grasslands, usually on embankments, road or valley sides. It is surprisingly scarce in the far west.

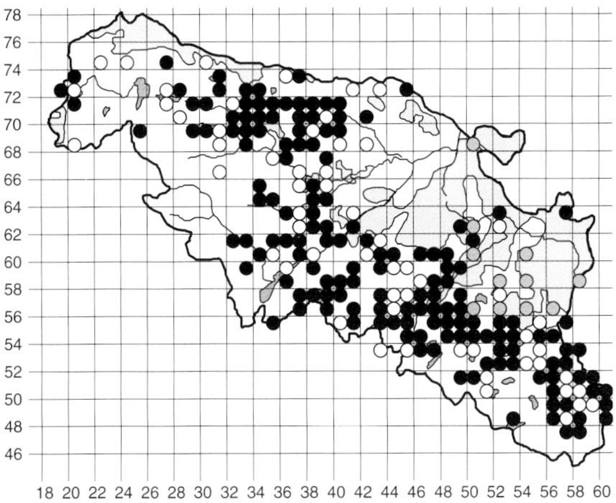

Galium verum

Galium album Miller
Galium mollugo auct. non L.
Hedge Bedstraw

Britain: common in the S, scattered in lowland Scotland.
Native or hortal; 6/11; rare.
1st Duchal Avenue, 1898, JMY.

Hedge Bedstraw was also recorded from Kilmacolm (1931, TPNS). It may have been overlooked recently, but this bedstraw appears to have always been a rare plant. BWR made no records and RM only noted two: a railway bank, Pollok (NS55K, 1959) and a roadside, Upper Enoch (NS55V,1961); the latter is close to the modern Ardoch Burn locality (NS5949, 1995). Other modern records include waste ground, Millarston (NS4563, 2002), where Hedge Bedstraw was perhaps introduced with wildflower seed, Paisley Moss (NS4665, 2003) and hedgerow at Old Mains (NS4968, 1997). Two further sites are from the roadside at Craigends of Dennistoun (NS3667, 2002) and in grassland Knapps Loch (NS3768, 2006). The hybrid with Lady's Bedstraw (*G.* × *pomeranicum* Retz.) has been reported from 'near Bull's Garage' Bridge of Weir (NS3866, 1979, BWR).

Galium saxatile L.
Heath Bedstraw

Britain: common throughout except in the SE.
Native; 359/420; very common.
1st Inverkip, 1838, GL.

Along with Tormentil, this is one of the commonest herbs to be found in acidic grassland. Although much unimproved grassland has been lost there are still extensive areas across the foothills and on local ridges elsewhere; even in improved fields Heath Bedstraw may persist at boundaries such as on old walls and along roadside ridges. It can occur in heath mosaics and lightly shading birch woodlands and can also grow among bog-moss hummocks in acidic Soft-rush mires.

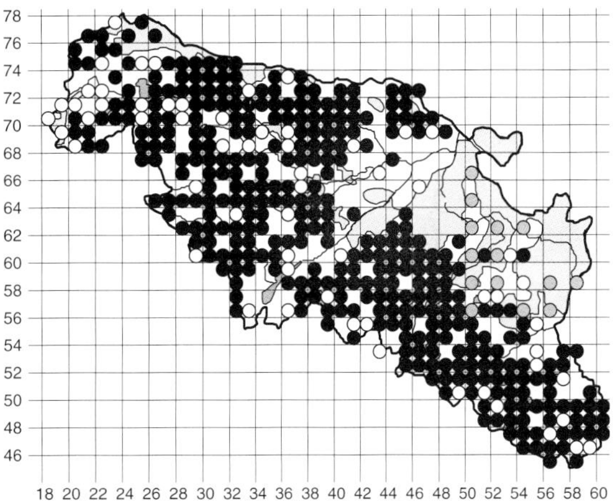

Galium saxatile

Galium aparine L.
Sticky-Willie or Cleavers

Britain: common throughout except in N Scotland.
Native; 293/359; very common.
1st Renfrewshire, 1869, PPI.

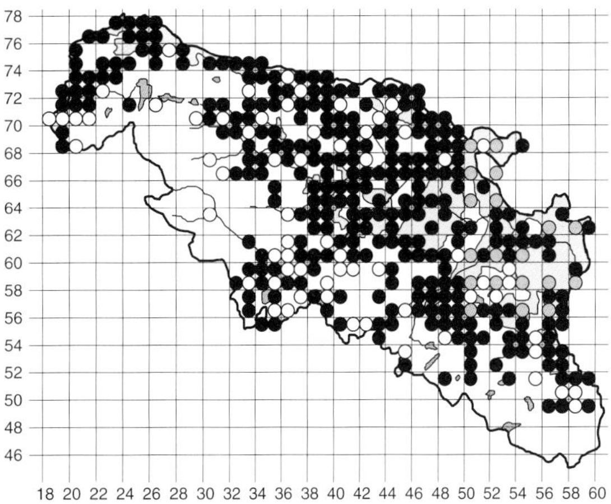

Galium aparine

Sticky-Willie is a common plant found throughout the VC and is only rare or absent from the acidic uplands. It is found in all habitats, except where permanently wet, and usually reflects some soil enrichment, e.g. along riverbanks, hedge banks, waysides, farmland, gardens, waste ground and woodlands.

Cruciata laevipes Opiz
Crosswort

Britain: common in the S but rare or absent from W and N Scotland.
Native; 29/41; occasional.
1st Paisley, 1856, GL.

Crosswort only has a few old records, such as those from Renfrew (1883, PSY) and Castle Semple Loch (1887, NHSG). Most modern records are from the east, mainly from urban-fringe areas. A western site is Kelburn (NS3474, 1998) but here it may have been introduced with wildflower sowing; another isolated record is from a roadside at Hartfield (NS4258, 1987, BWR). Habitats are marginal rough grassland by roads, hedges and riverbanks, particularly along the White Cart Water and Levern Water.

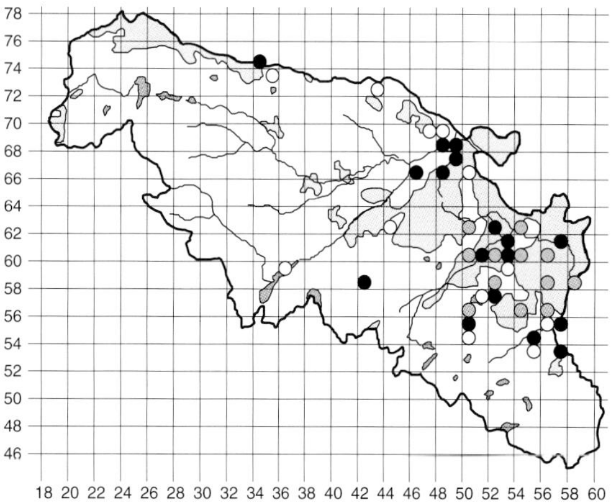

Cruciata laevipes

GENTIANACEAE

Centaurium erythraea Rafn
Common Centaury

Britain: common in the S, frequent in C and W coastal Scotland.
Native; 7/10; rare.
1st Paisley, 1845, NSA.

Both Hennedy (1891) and Lee (1933) mentioned Common Centaury at Gourock and it was recorded from Darnley by Wood in 1890, who wrote that this was the only 'station hereabout' but also noted later that it was 'common enough near the sea' (Wood 1893). Today it is known from disturbed grassland as at Wemyss Bay (NS1970, 1999) and Inverkip (NS2071, 1999), and waste-ground sites at East Fulwood (NS4567, 2002), the Royal Ordnance Factory, Georgetown (NS4270, 2005; NS4268, 2006) and the Braidbar quarry site (NS5659, 2010, J McKay). It is also found in the more natural setting of the open, species-rich grassland on top of Barscube Hill (NS3871, 1997).

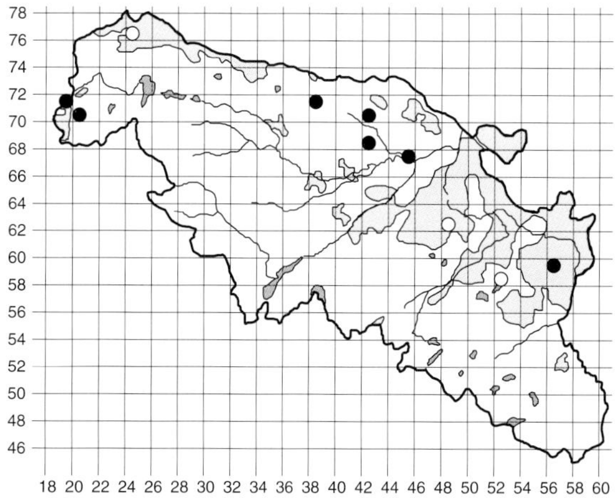

Centaurium erythraea

Gentianella campestris (L.) Börner
Field Gentian

Britain: frequent in the NW only, rare and declining elsewhere.
Native; 2/7; very rare.
1st Paisley, 1845, NSA.

Hennedy (1891) considered Field Gentian frequent in hilly pastures but only named the hills behind Lochwinnoch; other old records include Corkindale Law (1896, NHSG) and Locherside (1942, TPNS) and more recently Jock's Craig (NS3269, 1980s, DM). It may well be overlooked in short grazed pastures but with agricultural improvement it has presumably declined dramatically. Today it is only known from two sites, both ridges of species-rich grassland, which have escaped adjacent improvement at Commore Dam (NS4654, 1999) and Muirhead by Snypes Dam (NS4855, 1999).

APOCYNACEAE

Vinca minor L.
Lesser Periwinkle

Britain: archaeophyte (Europe); common in the S, frequent in C and E Scotland.
Hortal; 23/28; occasional.
1st Renfrewshire, 1869, PPI.

Other old records for Lesser Periwinkle include Cathcart (1876, FFWS), Lochwinnoch (1891, RH), Southbar (1903, PSY) and Glentyan woods (1921, GL). Today it has a scattered distribution but is occasionally found in old estate woodlands where it is capable of vegetative spread, often forming large patches.

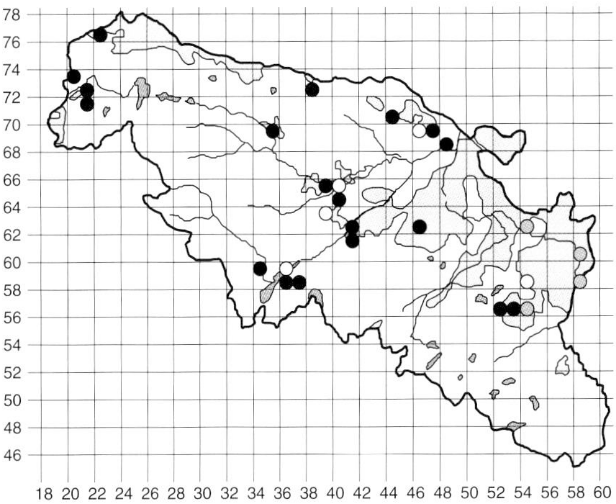

Vinca minor

Vinca major L.
Greater Periwinkle

Britain: neophyte (Mediterranean); common in the S, local in Scotland.
Hortal; 2/3; very rare.
1st Erskine, 1894, GL.

In addition to the 1983 record for Rouken Glen (NS55P, CFOG), where Greater Periwinkle may persist, the only modern record is from Jordanhill (NS5468, 2002, AR).

BORAGINACEAE (HYDROPHYLLACEAE)

Lithospermum officinale L.
Common Gromwell

Britain: frequent in the SE, very rare in Scotland.
Former accidental; 0/3; extinct.
1st Lochwinnoch and Kilbarchan, 1845, NSA.

A further record was made from Hangingshaw, Glasgow, in 1896 (NHSG).

Lithospermum arvense L.
Field Gromwell

Britain: archaeophyte; frequent in the SE.
Former accidental; 0/4; extinct.
1st Renfrewshire, 1869, PPI.

Known 'about Glasgow' (FS 1821) the only other named places for Field Gromwell are Gourock (1876, FFWS), Hangingshaw, Glasgow (1897, NHSG) and 'between Caldwell and Neilston' (1913, GL).

Echium vulgare L.
Viper's-bugloss

Britain: frequent in the S, extending to C Scotland.
Native or accidental; 12/24; scarce, scattered.
1st Lochwinnoch, 1845, NSA.

Considered rare by Hennedy (1891) the few 19th-century records for Viper's-bugloss include west of Greenock (1880, NHSG), west of Paisley (1896, NHSG) and Elderslie Station (1896, PSY). RM reported it from the Nitshill–Darnley area, where it later appeared, presumably as a part of a seed mix, at the new GMRC grounds at Nitshill in 2003 (NS5160). It can still be found at the original location (NS3558, 2005, IPG; NS3559, 2011) and the linked disused railway at Howwood (NS3961, 2001, DM) and Langslie (NS3356, 2004, IPG).

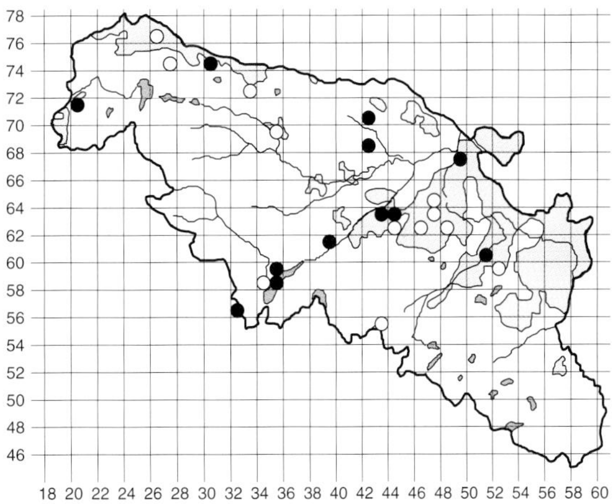

Echium vulgare

Pulmonaria officinalis L.
Lungwort

Britain: neophyte (Europe); scattered throughout lowlands.
Hortal; 12/16; scarce outcast.
1st Levernholme Wood, 1890, AANS.

Lungwort was also recorded from Kilbarchan (1905, PSY), Loch Libo (1969, JP) and more recently from a roadside at Waterfoot (1973, BWR). Modern records, which may include a few cultivars, are from urban-fringe woodlands and presumably all originate as garden outcasts, as at Uplawmoor (NS4355, 1994). There are also two records for narrow-leaved garden outcasts ('*P. angustifolia*') from Teucheen Wood (NS4868, 1996) and a roadside at Malletsheugh (NS5255, 2001).

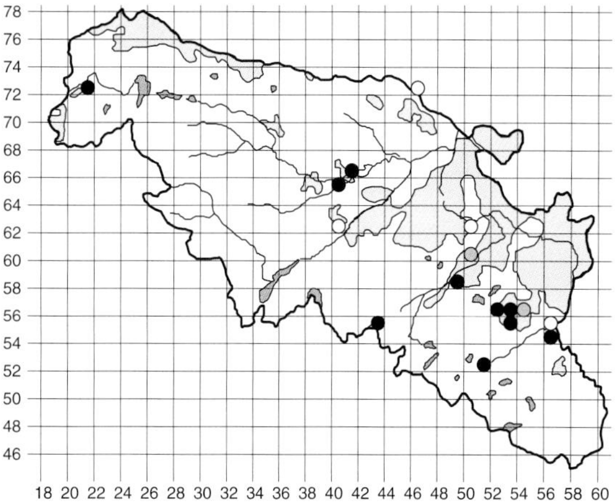

Pulmonaria officinalis

Pulmonaria rubra Schott
Red Lungwort

Britain: neophyte (SE Europe); rare local garden escape.
Hortal; 2/2; very rare.
1st Ranfurly, 1997.

A small patch of Red Lungwort occurred beneath shrubs by the side of a burn adjacent to mown grassland (NS3865) and more recently it was recorded from a roadside at Brownside (NS5058, 2009, PRG).

Pulmonaria 'Mawson's Blue'
Mawson's Lungwort

Britain: neophyte (garden cultivar); rare escape.
Hortal; 3/3; very rare.
1st Barhill Plantation (NS4671), 1996.

A small population of Mawson's Lungwort occurs in woodland by Lochside House (NS3658, 1999) and in 2003 IPG found it at the Formakin Estate (NS4170).

Symphytum officinale L.
Common Comfrey

Britain: common in the S, perhaps alien in lowland Scotland.
Native; 53/78; frequent.
1st Paisley, 1845, NSA.

This is the 'native' species which Lee (1933) said was being replaced by the introduced hybrid Russian Comfrey, but it is well represented in the local herbaria collections of that time (1891 through to 1944). Although some more recent records may include the hybrid, Common Comfrey still appears to be widespread. It has a similar distribution pattern, but is commoner to the east and apparently very rare in the west, and it is more likely to be found along riverbanks, e.g. both the Black Cart and White Cart waters.

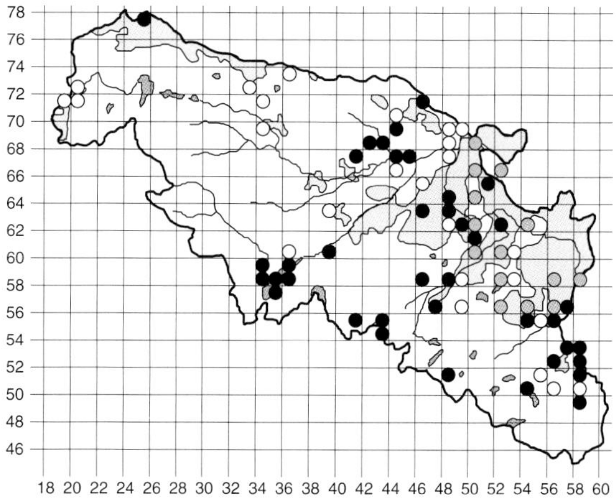

Symphytum officinale

Symphytum × *uplandicum* Nyman
(*S. officinale* × *S. asperum*)
Russian Comfrey

Britain: neophyte (cultivated origin); common in the S, extending to C and E Scotland.
Hortal; 62/69; frequent.
1st Wester Rossland, Bishopton, 1915, GL.

There are no named 19th-century localities for Russian Comfrey – and the six old Common Comfrey specimens in GL are correctly named – but by 1933 Lee stated that it had 'spread considerably and replaced the native form'. However, Hopkirk's (1813) description of 'var. *patens*' ('Plants larger and rougher … Calyx expanded') from by the White Cart Water at Cathcart and Langside, may well refer to

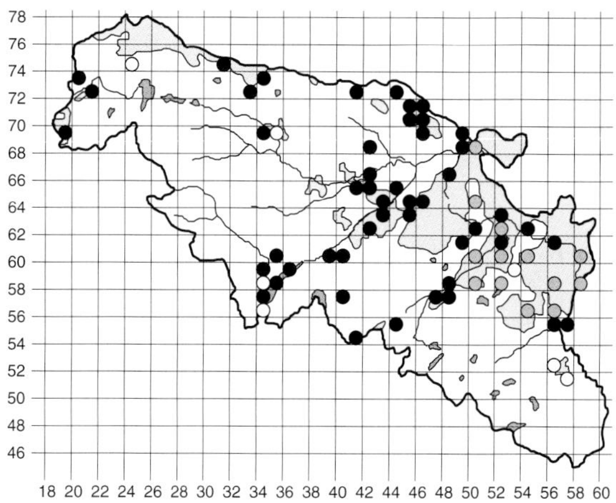

Symphytum × *uplandicum*

this hybrid. There are quite a few records from the 1960s–80s period by BWR and it was well recorded during surveying for CFOG. Today it is known from waste ground, waysides and riverbanks, mostly in the central area, but not always in urban-fringe areas.

Symphytum asperum **Lepech.**
Rough Comfrey

Britain: neophyte (SW Asia); rare and scattered.
Hortal; 1/1; very rare.
1st Langbank, 2006.

A small clump of Rough Comfrey grows to the trackside by gardens at Langbank Station (NS3873).

Symphytum tuberosum **L.**
Tuberous Comfrey

Britain: possibly native in C and E Scotland.
Native or alien; 136/172; common.
1st Castle Semple, 1834, MC.

Despite recent questioning of its native status in Scotland, Tuberous Comfrey appears to have been common in the 19th century and this is still true today. It is found in woodlands often along watercourses, generally on richer soils, and in hedge banks and along roadsides, and occasionally on waste ground. Its distribution map shows a strongly lowland pattern.

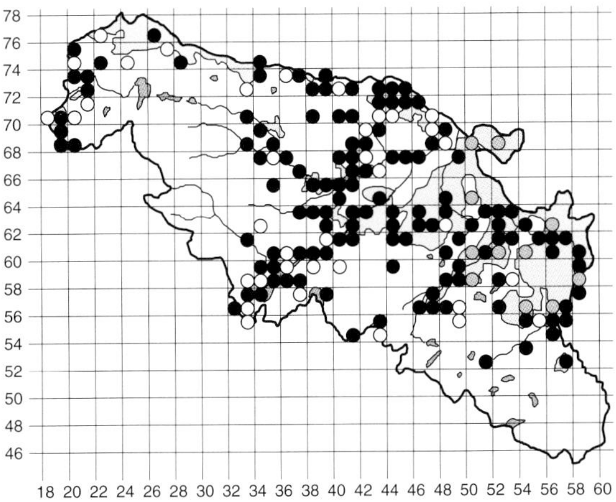

Symphytum tuberosum

Anchusa officinalis **L.**
Alkanet

Britain: neophyte (Europe); rare in the S.
Former accidental; 0/1; extinct casual.
1st Paisley, 1898, PSY.

Anchusa arvensis **(L.) M. Bieb.**
Bugloss

Britain: archaeophyte; widespread, mainly in the E.
Accidental; 1/4; very rare.
1st Gourock, 1865, RH.

Bugloss was collected from Cloch in 1893 (GRK) and in 1923 RM reported it from the Ravenscraig to Gourock area, but the only modern record is from recently disturbed ground by the sea wall at McInroy Point (NS2176, 2005).

Pentaglottis sempervirens **(L.) Tausch ex L.H. Bailey**
Green Alkanet

Britain: neophyte (SW Europe); common in the S, extending to NE Scotland.
Hortal; 14/17; scarce.
1st Renfrewshire, 1869, PPI.

RG knew Green Alkanet from near Inverkip (1930), but apart from a botanical excursion to Shielhill Glen (1948, ANG) there are no other records until AJS reported it from Finlaystone in 1976. Today it is known from several scattered places, usually on waste ground or about old estates or gardens.

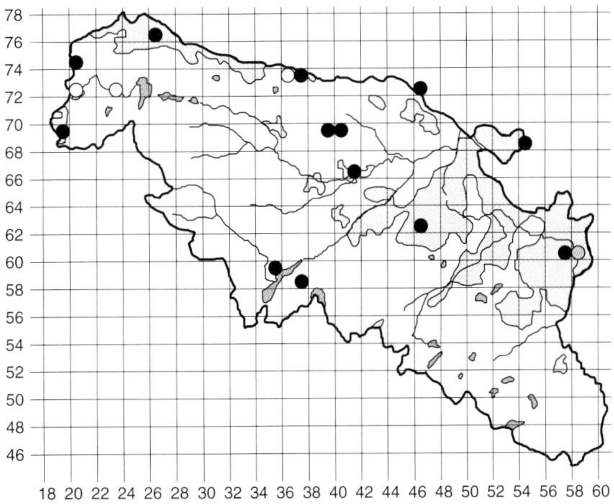

Pentaglottis sempervirens

Borago officinalis **L.**
Borage

Britain: neophyte (S Europe); frequent in the S, scarce in Scotland.
Hortal; 3/5; very rare.
1st Renfrewshire, 1895, NHSG.

Hooker (1821) had previously mentioned that Borage was to be found 'about Glasgow', although he added that this was generally near gardens, and it was collected from a field at Bishopton in 1919 (GL). Modern records are from Barhill Plantation (NS4671, 1995), as a casual at the tip at Shillford (NS4455, 1999) and from a gutter at Newlands (NS5760, 2010, PM).

Borago pygmaea (DC.) Chater & Greuter
Slender Borage

Britain: neophyte (Corsica and Sardinia); rare in S England.
Hortal; 1/1; very rare garden escape.
1st Jordanhill, 2004, AR.

The sole record for Slender Borage is from a pavement near gardens (NS5468).

Trachystemon orientalis (L.) G. Don
Abraham-Isaac-Jacob

Britain: neophyte (Caucasus and Turkey); scattered in the S, local in Scotland.
Hortal; 2/2; very rare.
1st Crookston Castle, 1982, CFOG.

More recently Abraham-Isaac-Jacob was found at Blythswood Estate (NS5068, 1997) and it can still be found on the path edge by the White Cart Water at Bonnyholm, near Crookston Castle (NS5262, 2005).

Mertensia maritima (L.) Gray
Oysterplant

Britain: local by coasts in the N.
Former native; 0/1; extinct.
1st Renfrewshire, 1834, MC.

The original record is the only one for the VC although Hennedy noted Oysterplant as frequent from the Clyde but named no location; he did note its ephemeral nature but by 1891 it was described as being 'Now rare' in the Clyde area. There are no other records of this distinctive plant.

Amsinckia lycopsoides Lehm.
Scarce Fiddleneck

Britain: neophyte (W N America); casual in the S.
Former accidental; 0/1; extinct casual.
1st Coups in Renfrewshire, 1930, RG.

Amsinckia micrantha Suksd.
Common Fiddleneck

Britain: neophyte (W N America); frequent in the E.
Accidental; 1/1; very rare casual.
1st Ranfurly Old Castle Golf Course, 2008, AM.

Common Fiddleneck was found to the edge of the fairway (NS3864); it is thought that it had arrived with imported sand.

Asperugo procumbens L.
Madwort

Britain: neophyte (Europe); rare casual.
Former accidental; 0/2; extinct casual.
1st Hangingshaw, Glasgow, 1896, NHSG.

MY noted Madwort in his list (1865) but there is some confusion over the Cartside locality. The species was excluded from the Renfrewshire list (TPNS 1915).

Myosotis scorpioides L.
Water Forget-me-not

Britain: common throughout except in N Scotland.
Native; 130/157; common.
1st Gourock, 1842, GLAM.

Old records include Paisley (1845, NSA), Castle Semple Loch and Loch Libo (1887, NHSG) and Lawmarnock (1906, PSY). Water Forget-me-not is the largest of the three 'wetland' forget-me-nots and is commonly found along watercourse margins and to the edges of waterbodies. It shares a similar pattern to Tufted-Forget-me-not and can occur with it, although it tends to prefer swampy ground rather than rushy pastures.

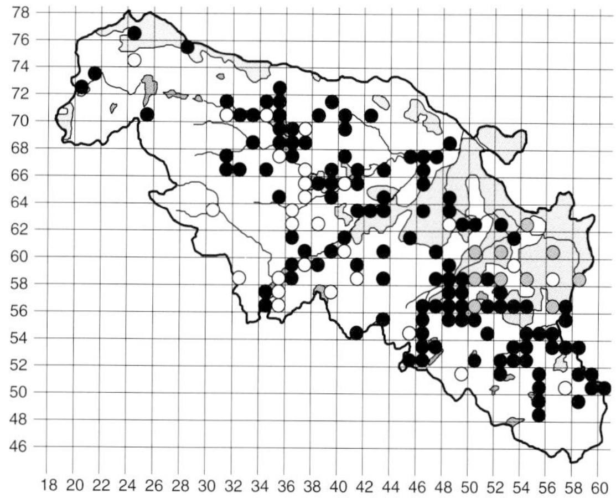

Myosotis scorpioides

Myosotis secunda Al. Murray
Creeping Forget-me-not

Britain: common in the NW.
Native; 166/204; common.
1st Mill Burn, Kilmacolm, 1874, TGSFN.

Not distinguished from the preceding species in

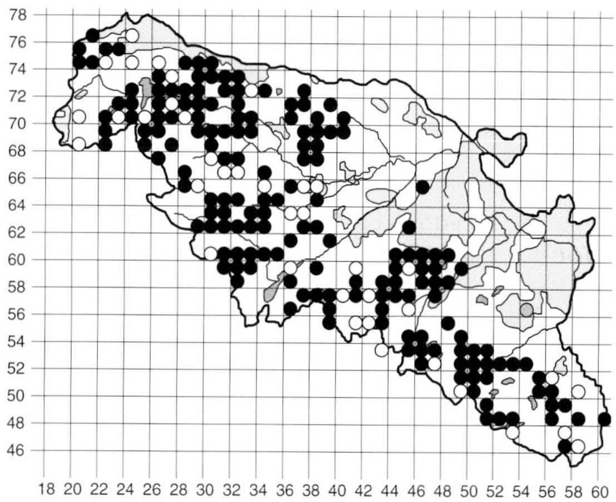

Myosotis secunda

early literature, Creeping Forget-me-not shows a very strong rural pattern that reflects its preference for low nutrient water, and is much more frequent in the west than the other two forget-me-nots of wet places. It is easily the commonest in upland locations along burn sides or in larger flushes, but is rarely found along lowland watercourse margins. It occurs in marginal fens at open waterbodies in the uplands, e.g. Gryfe Reservoirs (NS2871, 2001) and Brother and Little lochs (NS5052, 1996, SNH-LS).

Myosotis laxa Lehm.
Tufted Forget-me-not

Britain: common throughout except in N Scotland.
Native; 144/189; common.
1st Gourock, 1865, RH.

Considered frequent by 19th-century authors, there are few other old named stations of Tufted Forget-me-not except those at Castle Semple Loch (1887, GL) and Loch Libo (1887, NHSG). Today it is commonly found in marshes, often amongst Soft-rush, or to the margins of larger waterbodies and also along open river margins; it prefers less acidic conditions than Creeping Forget-me-not. An upland outlier was recorded from the Muirshiel Barytes Mine waste area in the Renfrewshire Heights (NS2865, 2003).

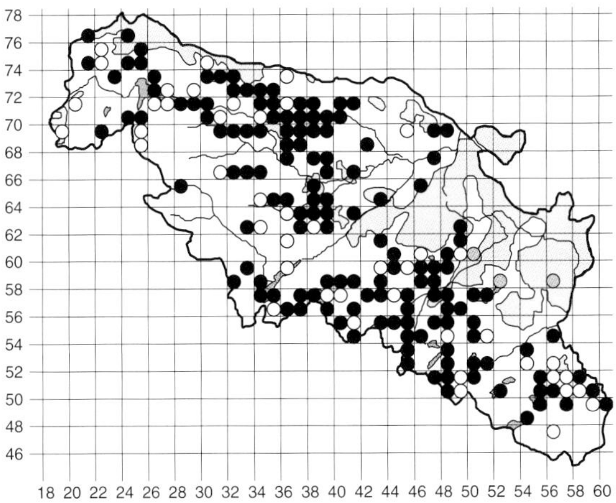

Myosotis laxa

Myosotis sylvatica Ehrh. ex Hoffm.
Wood Forget-me-not

Britain: common in the S, extending to NE Scotland.
Hortal; 38/43; frequent in lowlands.
1st Barrhead, 1901, FFCA.

There are no records for the 19th century but RM knew Wood Forget-me-not from Giffnock (1924) and an embankment near Whitecraig Station (1926, GLAM). There are very few records from the 1970s: by the Gryfe Water, Hatterick (1971, BWR), Drumcross (NS4471, 1972, BWR) and Finlaystone (NS3673, 1976, AJS). Today it is widespread, though rare in the west, and often found near gardens from where it readily escapes; it can be found on waste ground and in woodlands near habitation.

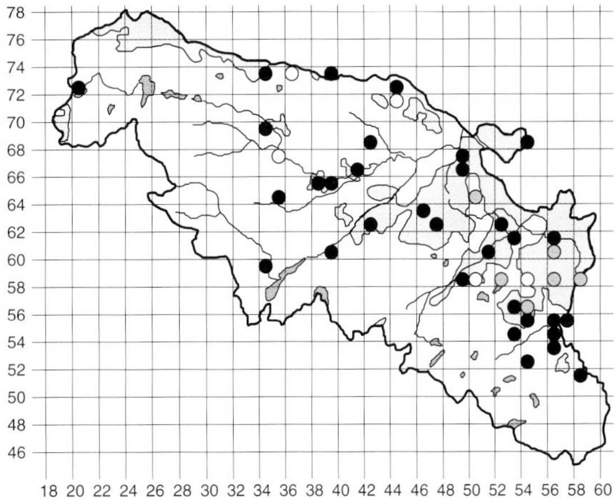

Myosotis sylvatica

Myosotis arvensis (L.) Hill
Field Forget-me-not

Britain: archaeophyte; common throughout except in N Scotland.
Accidental; 141/214; common.
1st Gourock, *c.* 1860, GL.

Field Forget-me-not is widespread and commonly found on open waste ground and at waysides, disused railways, open grasslands and the margins of agricultural fields. The var. *sylvestris* Schltdl. was recorded from Pollok Estate (NS5461) in 1979 by AJS.

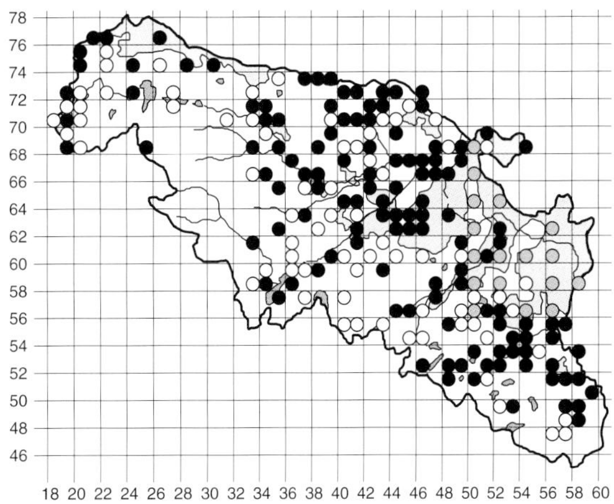

Myosotis arvensis

Myosotis discolor Pers.
Changing Forget-me-not

Britain: widespread, scarce in E England and N Scotland.
Native; 63/112; frequent.
1st Gourock, 1856, GL.

There are quite a few old records for Changing Forget-me-not from places such as Kilmacolm (1881, NHSG), Wemyss Bay (1887, NHSG), Castle Semple Loch (1887, NHSG), Langbank (1888, NHSG), Paisley (1900, PSY) and Lawmarnock (1905, PSY); there are also many records from the 1960s–80s period (BWR). Today it is still widespread but has few records from the north-west, which is possibly due to it being overlooked rather than any decline. It is often found in dry open places such as waste ground and on rocky banks, but it can also be frequent in short marshy grasslands.

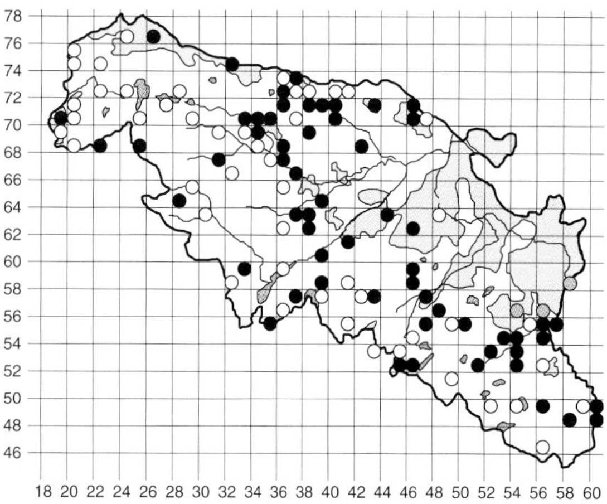

Myosotis discolor

Lappula squarrosa (Retz.) Dumort.
Bur Forget-me-not

Britain: neophyte (Europe); rare casual in the S.
Former accidental; 0/3; extinct casual.
1st Near Renfrew, 1873, TGSFN.

Further old records for Bur Forget-me-not include those from wasteland, Paisley (1910, PSY) and Paisley (1932, TPNS) but there have been no others for more than 80 years.

Omphalodes verna Moench
Blue-eyed-Mary

Britain: neophyte (SE Europe); local, mainly in S.
Hortal; 0/1; possibly extinct.
1st Thornliebank, 1883, NHSG.

The original location presumably refers to the Rouken Glen (NS5458) site from where Blue-eyed-Mary was reported in 1890 (AANS) and has been recorded for more than 100 years (CFOG); it was collected from there in 1956 (GL) and 1981 (GLAM) but has not been reported recently.

Phacelia tanacetifolia Benth.
Phacelia

Britain: neophyte (California); scattered, mainly in the S.
Hortal; 1/1; very rare.
1st Lyoncross, 1999.

Phacelia was found, presumably cultivated, in some abundance in an arable field near Balgray (NS5157). RG (1930) saw it several times but there are no named localities (except for 'near Glasgow').

CONVOLVULACEAE (CUSCUTACEAE)

Convolvulus arvensis L.
Field Bindweed

Britain: common in the S, frequent in C Scotland.
Native; 14/25; scarce.
1st Cathcart, 1813, TH.

Away from urban Glasgow (CFOG), Field Bindweed appears to be a rare plant; Hennedy (1891) considered it to be very rare. Other 19th-century records include the White Cart Water near Paisley (1869, GLAM), Castle Semple Loch (1887, NHSG) and wasteland near Paisley (1910, PSY). In addition to those in CFOG, modern records include: a roadside, West Fulton (NS4265, 1987, BWR), a railway line, Netherhouse (NS3556, 1995), Uplawmoor (NS4355, 2000, JC) a disused railway, Auchenbothie (NS3471, 2001; NS3271, 2005) and Cartsdyke (NS2975, 2003).

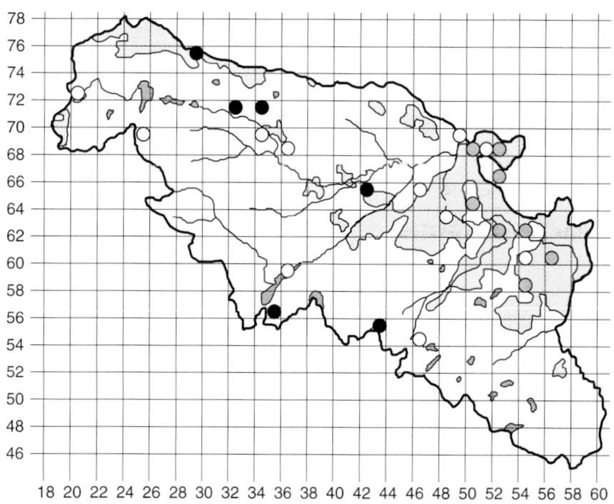

Convolvulus arvensis

Calystegia sepium (L.) R. Br.
Hedge Bindweed

Britain: common in the S, scarce in N Scotland.
Native; 117/135; common in lowlands.
1st Paisley, Kilbarchan and Lochwinnoch, 1845, NSA.

Some of the records may refer to the other bindweeds as many were for non-flowering plants (including some CFOG records), although the majority will refer to Hedge Bindweed. It is a common plant in urban areas, found on waste ground, neglected margins of open ground, disused railways and in hedge banks and gardens, but it can also occur in rural areas along roadsides or on rich soils along riverbanks. The pink form (f. *colorata* (Lange) Dorfl.) was noted on a fence at Glasgow Airport (NS4765, 2012, IPG).

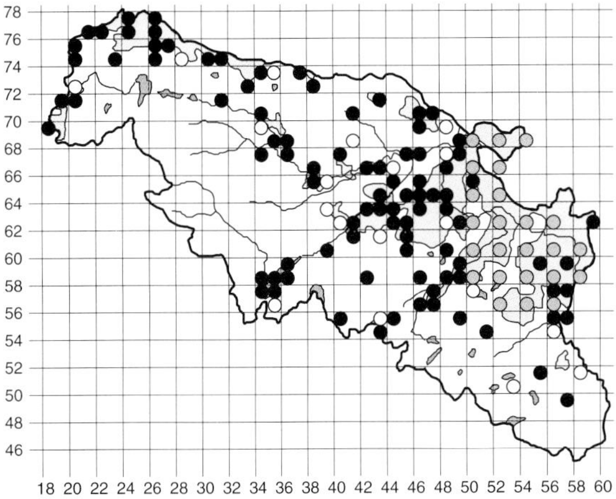

Calystegia sepium

Calystegia × *lucana* (Ten.) G. Don
(C. sepium × C. silvatica)

Britain: spontaneous hybrid; local in the S.
Hybrid; 2/2; very rare.
1st Disused railway, Johnstone, 2003, IPG.

There is also a tentative record from Jordanhill (NS5468, 2002, AR) but the identity of the plant has not been confirmed.

Calystegia pulchra Brummitt & Heywood
Hairy Bindweed

Britain: neophyte (uncertain origin); widespread in lowlands.
Hortal; 16/18; occasional.
1st Disused railway, Kerse, 1981, BWR.

This garden outcast or escape was only recorded twice before the modern period, but today Hairy Bindweed is known from several scattered sites. It is usually found on neglected corners of waste ground or railway embankments, all near habitation.

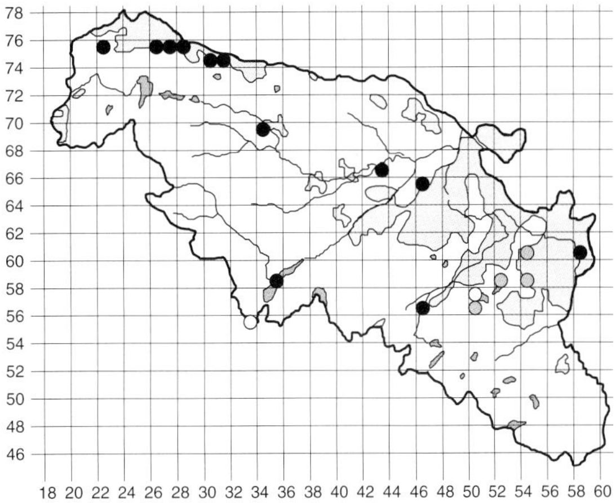

Calystegia pulchra

Calystegia silvatica (Kit.) Griseb.
Large Bindweed

Britain: neophyte (S Europe); common in England, extending to lowland Scotland.
Hortal; 22/24; occasional, under-recorded.
1st Bridge of Weir, 1980, BWR

There are no old records and surprisingly few modern ones for this large-flowered bindweed, in contrast with how often it was found during CFOG surveys;

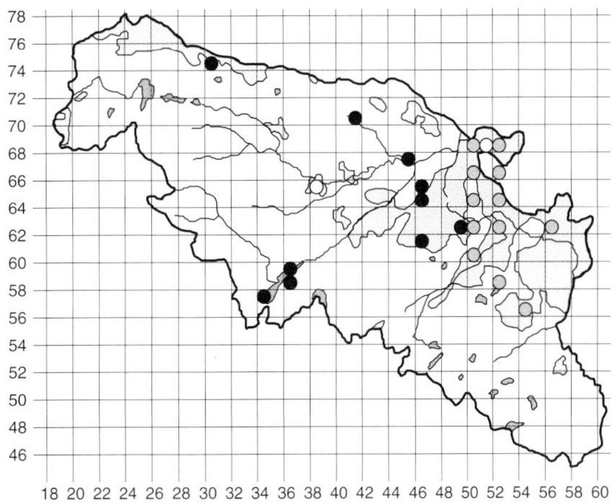

Calystegia silvatica

several non-flowering records may be included under Hedge Bindweed. It grows in similar disturbed grassy places, but generally in urban-fringe areas.

Cuscuta europaea L.
Greater Dodder

Britain: rare in the extreme SE.
Former accidental; 0/1; extinct.
1st Nether Auldhouse, 1944, RM.

The first noted occurrence, reported as being 'on leeks', is a surprising and unusual record. There are sporadic records from western Scotland for Flax Dodder (*C. epilinum* Weihe), which Hennedy considered the most likely species to occur locally, but he named no localities for VC76. It is safe to assume both species were always rare locally but are now extinct.

SOLANACEAE

Atropa belladonna L.
Deadly Nightshade

Britain: frequent in the SE, rare alien in C Scotland.
Former hortal; 0/1; extinct.
1st Cathcart, 1813, TH.

Both Hopkirk and Hennedy (1891) quoted Dr Brown: 'Banks of Cart near Cathcart Mill', but this is the only record.

Hyoscyamus niger L.
Henbane

Britain: archaeophyte (Europe); frequent in the SE, extending to E Scotland.
Former hortal; 0/2; extinct.
1st Cathcart, 1821, FS.

In 1845 (NSA) the original record was refined to Cathcart Castle. Henbane was collected from Jenny's Well near the White Cart Water in 1900 (PSY).

Datura stramonium L.
Thorn-apple

Britain: neophyte (America); frequent in the S, rare in the N.
Former accidental; 0/2; extinct casual.
1st Wasteland, near Paisley, 1910, PSY.

Of the original site Ferguson (1910) wrote that there were only six plants 'none of which flowered'. This species was also noted as a spontaneous weed from a garden in Craigielea, west Paisley (1933, TPNS).

Solanum nigrum L.
Black Nightshade

Britain: common in the S, scattered alien in Scotland.
Hortal; 0/4; extinct casual.
1st Greenock, 1880, NHSG.

Black Nightshade was collected from Lochwinnoch (NS5935) in 1955 (PSY) and also recorded as a garden weed from Bridge of Weir (1957, ERTC) and Pollokshields (1978, CFOG).

Solanum dulcamara L.
Bittersweet

Britain: common in the S and in C Scotland, rare in N Scotland.
Native; 93/104; frequent.
1st Lochwinnoch and Paisley, 1845, NSA.

Considered 'not common' in the past (Hennedy 1891), today Bittersweet can be frequently found in lowland-urban-fringe areas. There are a cluster of sites around Paisley between the Gryfe and Levern Water, and it occurs sparsely along the north coast but is rare in the west; it was well recorded during surveying for CFOG. Habitats vary: it is often found in wet places, often where partially shaded, but can also be found on dry waste ground, especially in the north-west, e.g. at Ladyburn (NS3074, 1998), Port Glasgow (NS3274, 1998) and Greenock (NS2775, 2000).

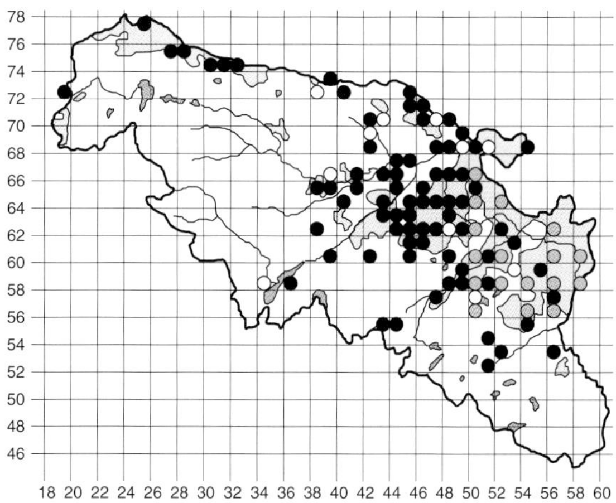

Solanum dulcamara

Solanum tuberosum L.
Potato

Britain: neophyte (S America); scattered in lowland areas.
Hortal; 3/4; very rare garden outcast.
1st Rubbish tip, Barr Loch, 1969, AMcGS.

Outcast plants derived from potato tubers were recorded a couple of times during CFOG surveys but the sole modern record for Potato is from the edge of Teucheen Wood (NS4869, 1996).

Solanum lycopersicum L.
Lycopersicon esculentum Mill.
Tomato

Britain: neophyte (C and S America); scattered and local throughout.
Hortal; 8/9; rare casual.
1st Rubbish tip, Barr Loch, 1969, AMcGS.

Grierson (1930) stated that 'Tomatoes are among the commonest plants on our coups', but did not name a VC76 site. Modern records are from scattered urban-fringe locations: Patterton (NS55N, 1985, CFOG), Old Mains (NS4969, 1997), Jordanhill (NS5468, 2002, AR), Crossmyloof (NS5762, 2009, PM) and Inch Green dock (NS3075, 2010, AHJ); there are also three from the west coast: Underheugh (NS2075, 1997), Wemyss Bay (NS1971, 1999) and Lunderston Bay (NS2073, 2008).

Nicotiana tabacum L.
Tobacco

Britain: neophyte (S and C America); rare casual in the S.
Former hortal or accidental; 0/1; extinct casual.
1st Giffnock, 1926, RG.

OLEACEAE

Forsythia* × *intermedia Zabel
(*F. suspensa* (Thunb.) Vahl × *F. viridissima* Lindl.)
Forsythia

Britain: neophyte (garden origin); scattered, mainly in the S.
Hortal; 3/3; very rare.
1st Ranfurly (NS3855), 1997.

Two other sites are Ardgowan Estate (NS2073, 1998) and a record from stonework by the White Cart Water at Cathcart (NS5860, 2005).

Fraxinus excelsior L.
Ash

Britain: common throughout except in N Scotland.
Native; 346/394; very common.
1st Renfrew, 1777, JW.

Ash is a very common tree throughout the lowlands. It is found in all but the most acidic woodlands, although it is most numerous in valley woodlands occurring along watercourses; it tolerates flushing but is not found in wet carr woodlands. It is also common in the urban environment and readily colonizes waste ground and other neglected areas.

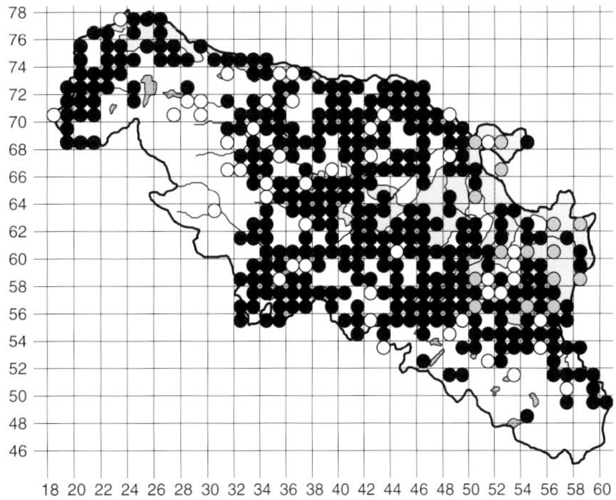

Fraxinus excelsior

Syringa vulgaris L.
Lilac

Britain: neophyte (SE Europe); common in the S, frequent in C and E Scotland.
Hortal; 21/23; occasional.
1st Giffnock Quarry (NS5659), 1966, JDM.

Lilac is usually seen in urban-fringe areas, persisting as relict plantings, outcasts or self-sown individuals. It is found on waste ground or waysides near habitation, but can occur in remote areas, such as a roadside at Dodside (NS5053, 1998).

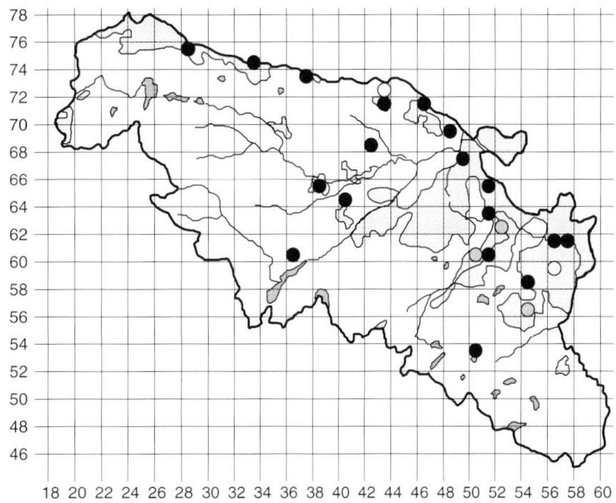

Syringa vulgaris

Ligustrum vulgare **L.**
Wild Privet

Britain: common in the S, frequent alien in C and S Scotland.
Hortal; 60/76; frequent.
1st Lochwinnoch, 1845, NSA.

Hennedy described Wild Privet as much planted and 'questionably native' but common. There are few other records from before the 1960s, but it was collected from Kilbarchan (1921, GL), near Greenock (1904, GL) and behind Giffnock Quarry (1909, GL). This is the commonest privet recorded and is more frequently found in old estate woodlands where, although presumably first planted, it appears capable of local spread.

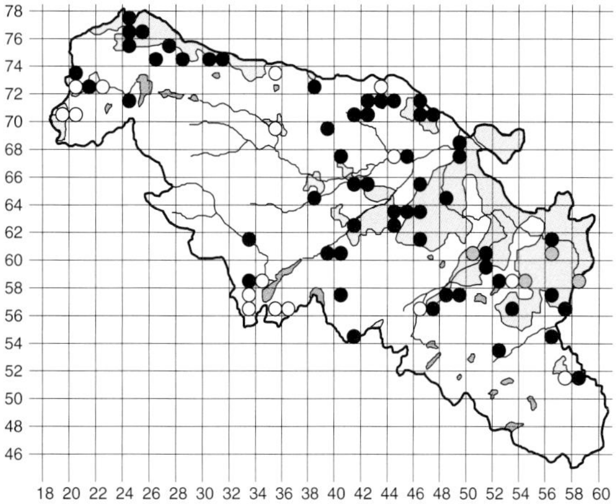

Ligustrum vulgare

Ligustrum ovalifolium **Hassk.**
Garden Privet

Britain: neophyte (Japan); common in the S and in C Scotland.
Hortal; 43/43; frequent.
1st Linn Park (NS55Z), 1981, CFOG.

As its name suggests, Garden Privet is a common hedging shrub in urban gardens and many of the records refer to relict shrubs or outcasts, although some are likely to be self-sown. However, although persistent it cannot be described as naturalized.

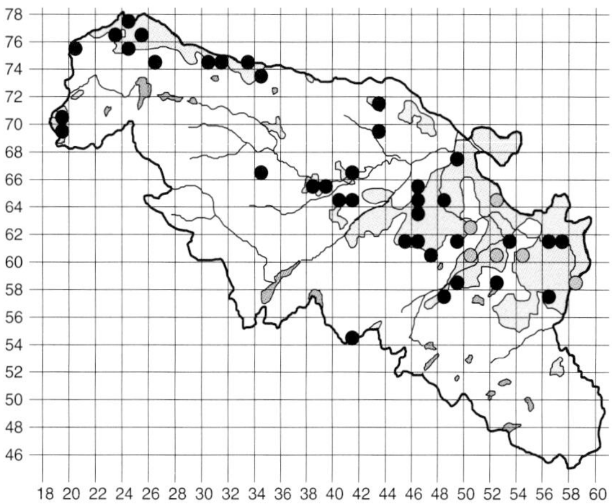

Ligustrum ovalifolium

VERONICACEAE
(SCROPHULARIACEAE *pro parte*)

Digitalis purpurea **L.**
Foxglove

Britain: common throughout.
Native and hortal; 349/397; very common.
1st Neilston, 1845, NSA.

Foxglove is found throughout the area and may be commoner than in the 19th century (as noted in CFOG). It is found in woodlands and on hedge banks by waysides, along old walls, on rocky outcrops and along watercourse margins, only becoming rare or absent in the wetter peaty uplands. It is often equally common in urban areas where it is found in similar places and on waste ground, disused railway lines and also in gardens; some of the records refer to introduced garden material (CFOG).

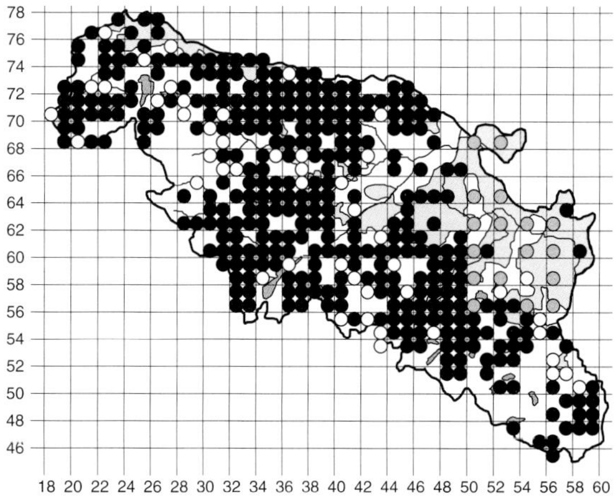

Digitalis purpurea

Erinus alpinus **L.**
Fairy Foxglove

Britain: neophyte (SW Europe); scattered, frequent in W C Scotland.
Hortal; 11/12; scarce.
1st Glentyan House, 1984, BWR.

Fairy Foxglove was recorded during surveys for the BSBI Monitoring Scheme (1987) from the Formakin Estate, where it was refound (NS4170) in 2003 by IPG. Other records include Johnstone Wood (NS4262, 1993), Duchal Wood (NS3368, 1994), Kilbarchan (NS4063, 1999) and Polnoon Water (NS5851, 2000). It has also been found, as an outcast, on waste ground at Greenock (NS2776, 2000), Gourock (NS2477, 2001) and Cartsdyke (NS2875, 2001).

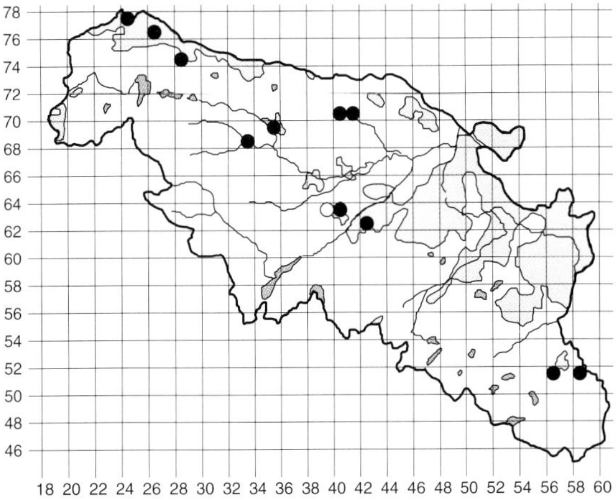

Erinus alpinus

Veronica officinalis L.
Heath Speedwell

Britain: common throughout except in E C England.
Native; 104/143; frequent in rural areas.
1st Renfrew, 1845, NSA.

Heath Speedwell was considered very common by Hennedy (1891) and there are quite a few named old localities, and it was also well recorded in the 1960s–80s (BWR). It is likely to have undergone a decline due to intensive farming and heavy grazing in the foothills. However, it is still widespread, although populations may now be restricted in scale. It is associated with unimproved grasslands, usually on the shallow soils of rocky outcrops or embankments and valley sides; it is seldom found on acidic heaths but may occur as part of habitat mosaics, and it can ascend high into the uplands.

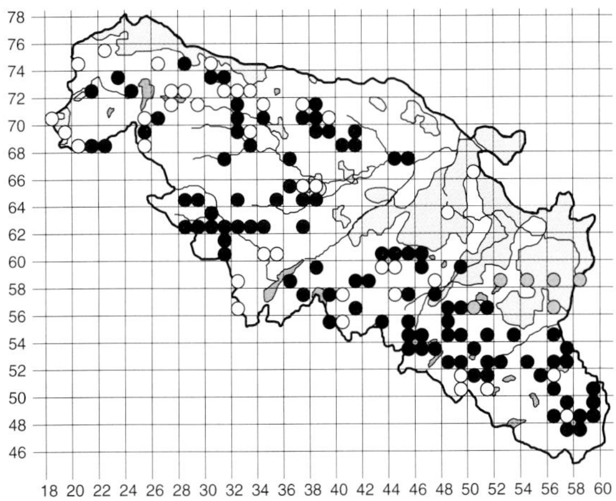

Veronica officinalis

Veronica montana L.
Wood Speedwell

Britain: common except in N Scotland and E England.
Native; 41/49; frequent in valley woodlands.
1st Langside Wood, 1813, TH.

Wood Speedwell is exclusively found on the richer soils of woodlands, often associated with watercourses. The map shows a widespread pattern, but with a cluster of sites in the west picking out the wooded valleys descending from the Renfrewshire Heights; it is also well represented along the White Cart Water in Glasgow.

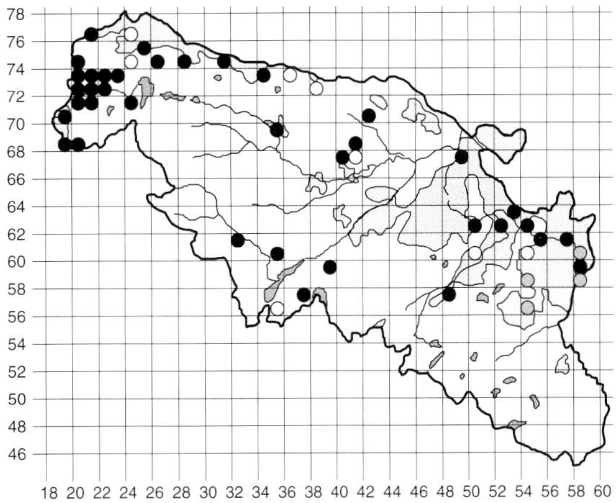

Veronica montana

Veronica scutellata L.
Marsh Speedwell

Britain: common in the N and W.
Native; 73/100; frequent in marshes.
1st Lochwinnoch, 1845, NSA.

Marsh Speedwell was considered frequent in the 19th century, but there are few named old localities:

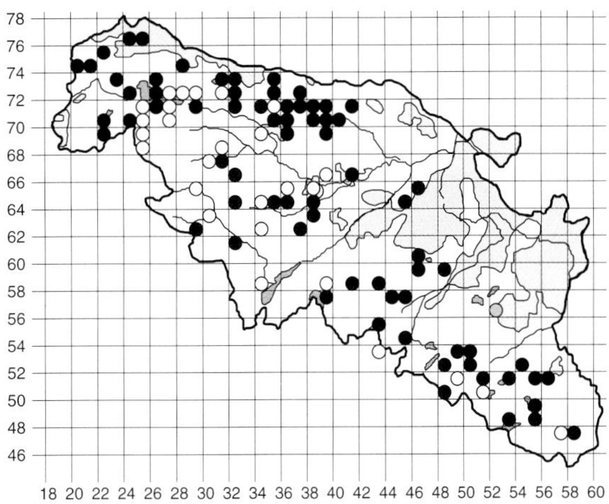

Veronica scutellata

Hennedy (1891) mentioned the Gourock area and it was collected from Kilmacolm Reservoir in 1890 (GL). It was well recorded in the 1960s–80s period (BWR) and has also been well recorded in the modern period, including during the SNH Loch Survey. It is a rural species, ascending into the uplands, and is usually found, often well hidden, in wet places amongst rushes or sedges and is tolerant of fairly deep swamp water.

Veronica beccabunga L.
Brooklime

Britain: common throughout except in the NW.
Native; 180/227; common in wet places.
1st Gourock, 1851, GL.

Considered common in the past, Brooklime has retained this status. Today it is found throughout the area, mostly in the rural lowlands; it can occur in the uplands but is discouraged by acidic peaty conditions. It is found at the margins of ditches or slow-flowing burns, along inundated riverbanks, and in marshes and at the margins of waterbodies.

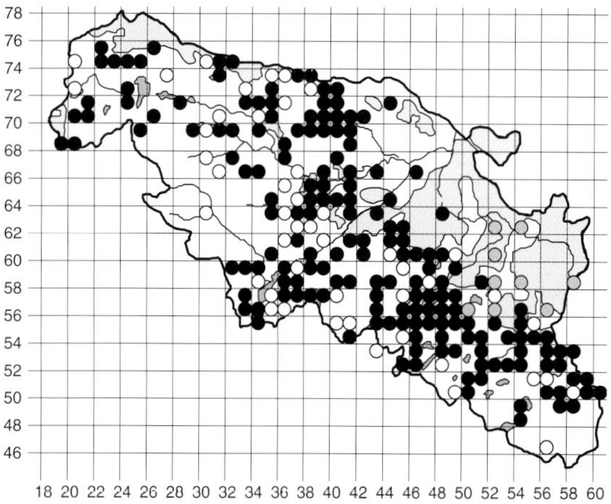

Veronica beccabunga

Veronica peregrina L.
American Speedwell

Britain: neophyte (N and S America); scattered but local.
Accidental; 7/8; rare, probably overlooked.
1st Camphill Gardens, Glasgow, 1908, CFOG.

More recently PM recorded American Speedwell from a Newlands garden (NS56Q, 1982), JHD collected it from the walled garden at Rouken Glen (NS55P, 1978, GL) and AJS found it at Finlaystone (1975). Today it is still present at the latter (NS3673, 2003, IPG) and also known from waste ground, Phoenix site (NS4564, 1995), a pavement in Gourock (NS2477, 2001), Wemyss Bay (NS1971, 2010, PRG) and Formakin Estate (NS4170, 2003, IPG).

Veronica serpyllifolia L.
Thyme-leaved Speedwell

Britain: common throughout.
Native; 237/317; common.
1st Paisley, 1858, PSY.

This speedwell is readily found on waste ground, in gardens and in farm fields and other grasslands, and tolerates some improvement or enrichment. It can often appear in other habitats, including woodlands, where local disturbance has opened up soils and at inundation zones along rivers. A few records are from higher altitudes (up to *c.* 400m) along burn sides in the Renfrewshire Heights.

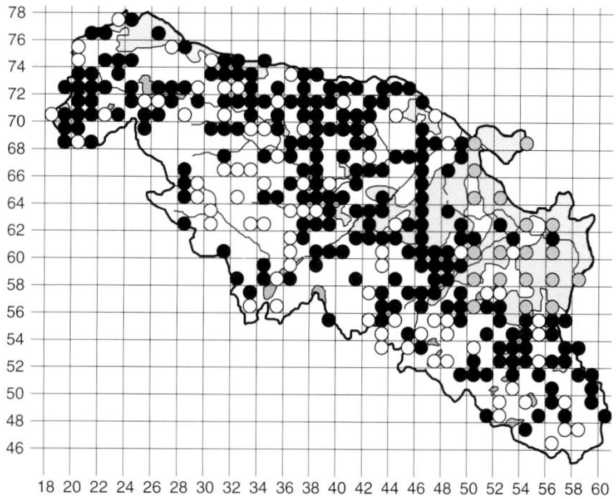

Veronica serpyllifolia

Veronica hederifolia L.
Ivy-leaved Speedwell

Britain: archaeophyte; common in the S and E, mostly eastern in Scotland.
Accidental; 19/23; scarce.
1st Jenny's Well, 1858, PSY.

Ivy-leaved Speedwell was considered a common weed of cultivated ground and disturbed places in the 19th century but there are few named localities; surprisingly it was only reported twice by BWR, from Houston (NS4066, 1969) and Kilbarchan (NS3762, 1971). Perhaps it never has been common in the VC or else it has undergone a strong decline. Today most records are associated with watercourses, notably the Gryfe Water from Duchal (NS3568, 2000), and especially about Craigends (NS4166, 2002), and the White Cart Water from Busby (NS5756, 2001) to Hawkhead (NS4166, 2005). A couple of sites are from disused railways: Arthurlie (NS4958, 1997) and Brodie Park (NS4762, 2003). The original record is for ssp. *hederifolia*, as is the specimen from Fort Matilda, Greenock (1888, GRK), but several modern specimens, if not all, from both river systems named above, belong to ssp. *lucorum* (Klett & Richt.) Hartl, first recorded from Finlaystone in 1976 (NS3673, AJS).

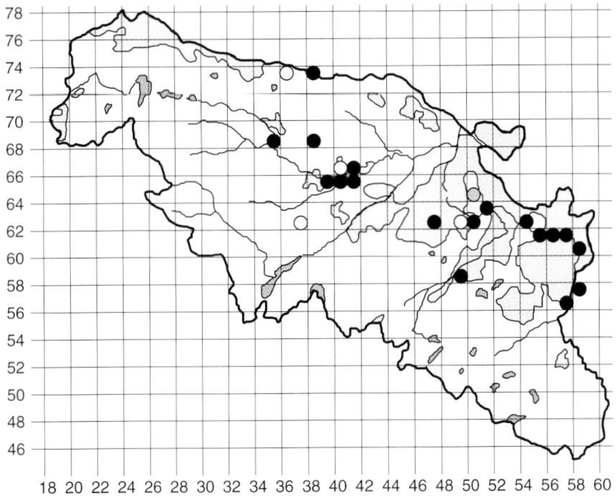

Veronica hederifolia

Veronica filiformis Sm.
Slender Speedwell

Britain: neophyte (Turkey and Caucasus); common throughout except in the NW.
Accidental; 37/47; frequent.
1st Bridge of Weir, 1940, GL.

Other old records for Slender Speedwell include Cathcart (1942, GL) and BWR recorded it from a number of locations between 1960 and 1984, including Craigenfeoch (NS4361, 1960) and Willowbank, Newton Mearns (NS5356, 1963). Today this colourful weed of lawns and other amenity cut grasslands is widespread, and is very likely under-recorded from urban areas.

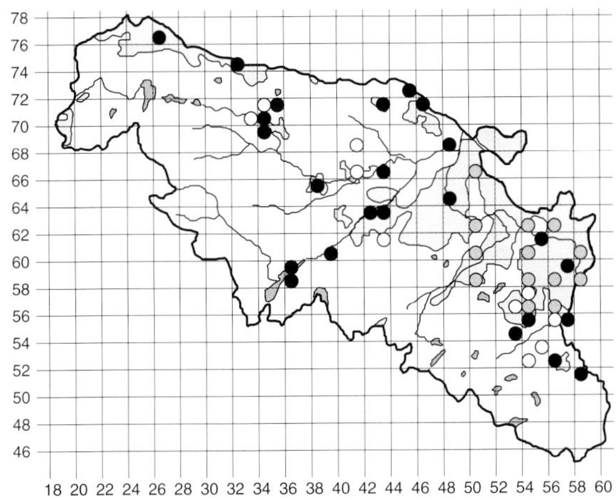

Veronica filiformis

Veronica agrestis L.
Green Field-speedwell

Britain: archaeophyte; frequent throughout except in the NW.
Accidental; 9/11; rare, overlooked.
1st Renfrewshire, 1883, TB.

This speedwell has probably been overlooked during modern recording as there are few modern records, although it is likely that it has declined along with its weedy cornfield habitat. Hennedy (1891) described it as very common, although this is more likely to have applied to the eastern part of the VC. There were two relevant CFOG localities, Newton Mearns (NS55N, 1987) and 'The Cunyon' (NS56L, 1986), and other modern records are from waste ground, Inverkip (NS2072, 1999), waste ground, Linwood (NS4364, 2001), arable weed, South Mains (NS4266, 2003), Finlaystone (NS3673, 2003, IPG), waste ground, Cartsdyke (NS2975, 2003), Lochwinnoch (NS3558, 2005, IPG) and Yoker (NS5169, 2011).

Veronica polita Fr.
Grey Field-speedwell

Britain: neophyte (Europe); common in the SE, local in Scotland.
Accidental; 1/6; very rare, possibly overlooked.
1st Kilbarchan, 1845, NSA.

Not distinguished by most 19th-century authors, it is thought that Grey Field-speedwell is the rarest of the field-speedwells locally, although Lee (1933) described it as common. There is only one modern record, from Finlaystone (NS3673, 2003, IPG), where AJS first found it in 1976, so it has either declined or has been overlooked. Old records include Paisley in 1859 (TPNS, 1942) and the Clarkston area (1933, RM); AJS also knew it from Paisley in 1975.

Veronica persica Poir.
Common Field-speedwell

Britain: neophyte (SW Asia); common except in NW Scotland.
Accidental; 8/11; rare, overlooked.
1st Renfrewshire, 1869, PPI.

As discussed in CFOG, this speedwell started to increase in the late 19th century; old records include Hurlet (1890, AANS), Thornliebank (1890, GL), near Newton Mearns (1907, GL), a roadside at Bridge of Weir (1916, GL), railway track, Clarkston (1929, GLAM) and allotments, Netherauldhouse (1948, RM). It was the most common of the field-speedwells in Glasgow in the 1980s, with a handful of CFOG VC76 records. However, it may well be overlooked as a farm weed or in gardens and allotments, as it has only been noted a few times in the most recent recording: South Mains (NS4266, 1987, BWR), Carriagehill Ave, Paisley (NS46R, 1997, GL) and Formakin Estate (NS4170, 2003, IPG).

Veronica chamaedrys L.
Germander Speedwell

Britain: common throughout.
Native; 336/406; very common.
1st Cart, Jenny's Well, 1858, PSY.

This is the commonest of the speedwells; it occurs in most dry habitats, except those with strongly acidic soils. It is most frequently associated with woodlands, although it prefers edges or glades, and it is often found in scrub, along hedgerows and in grassland that is not too heavily grazed or improved. It is only rare or absent from the peaty uplands.

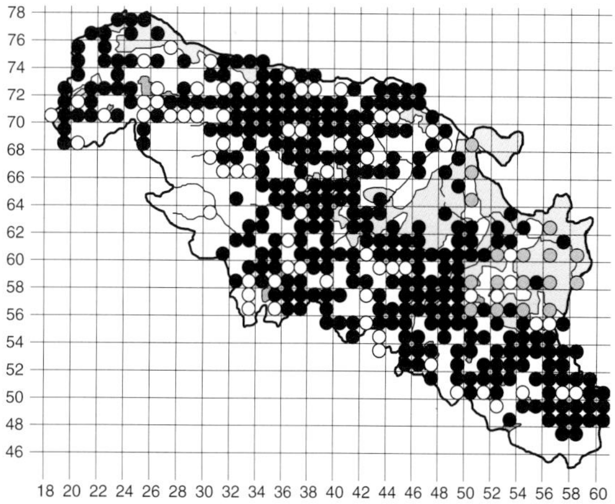

Veronica chamaedrys

Veronica arvensis L.
Wall Speedwell

Britain: common throughout.
Native; 108/158; common in open places.
1st Gourock, 1860, GL.

Wall Speedwell can be commonly found in urban areas as a weed of gardens and on waste ground, walls, tracks and pavements, and has likely been overlooked during modern recording. It also can occur in more natural, rural locations, usually on disturbed or shallow soils often associated with rock outcrops and ridges.

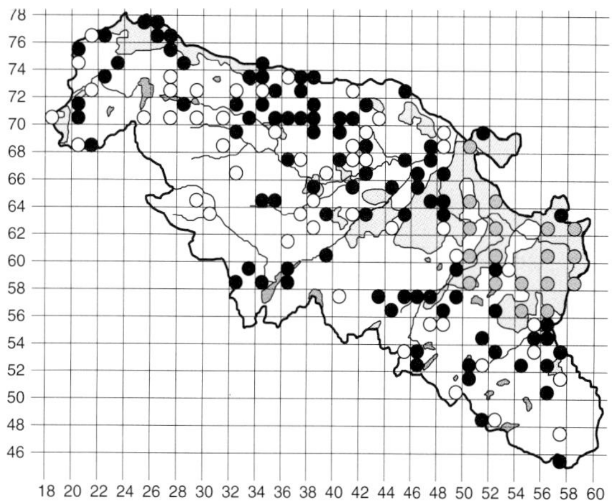

Veronica arvensis

Sibthorpia europaea L.
Cornish Moneywort

Britain: local in the SW, very rare alien elsewhere.
Former hortal; 0/1; extinct.
1st The Cottage, Paisley, 1890, PSY.

The only specimen is presumably of a garden-grown plant.

Antirrhinum majus L.
Snapdragon

Britain: neophyte (SW Europe); frequent in the S and in C Scotland.
Hortal; 2/2; very rare.
1st Crookfur, 1986 (NS55N), CFOG.

The only other record is of a garden outcast on waste ground, Greenock (NS2073, 1998).

Chaenorhinum minus (L.) Lange
Small Toadflax

Britain: archaeophyte (Europe); common in the S, rare in N Scotland and uplands.
Accidental; 29/38; occasional, scattered.
1st Old wall, Renfrew, 1873, GLAM.

Other old records for Small Toadflax refer to railways at Nitshill (1890, AANS) and Howwood (1895, NHSG) and many of the recent and modern records are from now disused railways sites, or those converted to cycleways (where the liberal use of tarmac usually destroys most of the open, gravelly trackbed habitat). A few other records refer to waste-ground sites, as at Port Glasgow (NS3074, 1998) and Inverkip (NS2072, 1999).

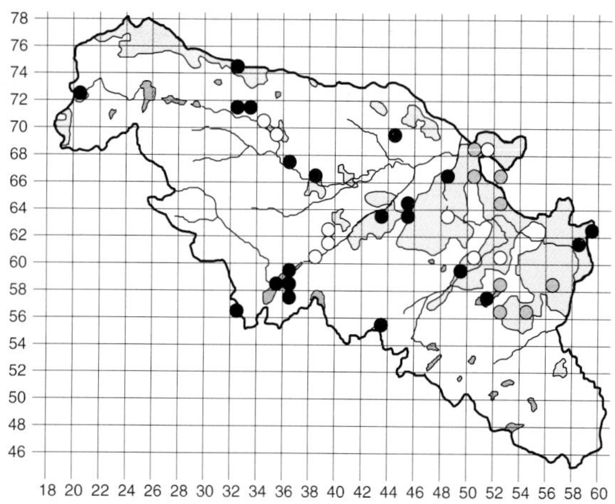

Chaenorhinum minus

Misopates orontium (L.) Raf.
Weasel's-snout

Britain: archaeophyte (Europe); frequent in the S.
Accidental; 1/1; very rare casual.
1st Nitshill, 2005.

A surprising singleton was found in the disturbed but seeded meadow of GMRC, Nitshill (NS5160); it has not reappeared.

Asarina procumbens Mill.
Trailing Snapdragon

Britain: neophyte (SW Europe); scattered in England.
Hortal; 1/1; very rare.
1st Glentyan, 2007.

Trailing Snapdragon is well established on the walls of the old walled garden at Glentyan Estate (NS3963).

Cymbalaria muralis P. Gaertn., B. Mey. & Scherb.
Ivy-leaved Toadflax

Britain: neophyte (S Europe); common throughout except in upland and N Scotland.
Hortal; 56/66; frequent, urban.
1st Renfrewshire, 1869, PPI.

Old named localities for Ivy-leaved Toadflax include West of Greenock (1880, NHSG), Wemyss Bay (1887, GRK), the Collegiate Church at Parkhill (1891, AANS), Castle Semple (1890, GL), Erskine Church (1904, PSY) and old wall, Kilbarchan (1921, GL). Today its pattern reflects long-established human habitation; it can be found on old walls but it is also occasionally found on waste ground.

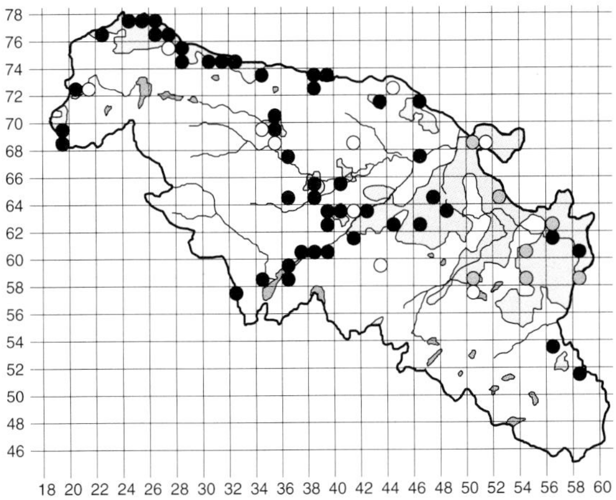

Cymbalaria muralis

Cymbalaria pallida (Ten.) Wettst.
Italian Toadflax

Britain: neophyte (Italy); scattered in C Britain.
Hortal; 2/2; very rare.
1st Glebe Road, Newton Mearns, 2000, PM.

The original locality (NS5355) is from the side of a lane; Italian Toadflax is also recorded from a similar location in Jordanhill (NS5468, 2002, AR).

Linaria vulgaris Mill.
Common Toadflax

Britain: common in the S, rare in N Scotland.
Native; 72/93; frequent.
1st Paisley, 1845, NSA.

Common Toadflax was considered frequent in the 19th century and it appears to have a similar distribution today. It is widespread in the lowlands, usually near habitation and generally associated with disturbed places such as waste ground, disused railways, waysides and along some riverbanks.

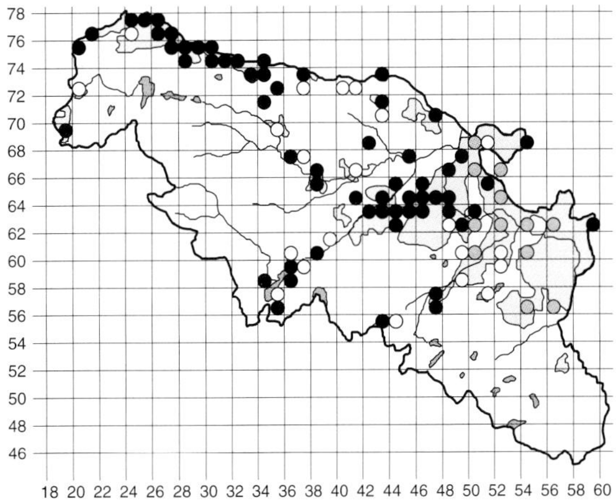

Linaria vulgaris

Linaria × sepium G.J. Allman
(*L. repens × L. vulgaris*)

Britain: scattered in the S.
Spontaneous hybrid; 3/3; very rare.
1st Scotstoun Station, 1988, CFOG.

More recently this hybrid was found growing with both parents on waste ground below an old shed at Cartsdyke, Greenock (NS2975, 2003), and from similar waste ground by the Clyde at Battery Park (NS2575, 2012).

Linaria purpurea (L.) Mill.
Purple Toadflax

Britain: neophyte (Italy); common in the SE, frequent in C and E Scotland.
Hortal; 8/11; rare.
1st Kilmacolm, 1931, TPNS.

There are no other old localities for this garden outcast or escape except a disused railway at Harelaw (NS4960, 1981, BWR). In CFOG it is noted from a couple of sites and there are six modern records: from waste ground at Cathcart (NS5860, 1996, AMcGS), Ladyburn (NS3074, 1998), Port Glasgow (NS3174, 2001) and Cowglen (NS5360, 2005, LR), and from walls at Kilmacolm (NS3569, 2006) and Greenock (NS2776, 2007).

Linaria × dominii Druce
(*L. purpurea × L. repens*)

Britain: scattered in the S.
Spontaneous hybrid; 1/1; very rare.
1st Muirend, 2011, PM.

A single plant was found in a gutter at Earlspark Avenue (NS5761).

Linaria repens (L.) Mill.
Pale Toadflax

Britain: scattered, frequent, only in S and W.
Hortal or accidental; 22/27; occasional.
1st Renfrewshire, 1915, TPNS.

Pale Toadflax was collected from by the White Cart Water, Inchinnan (1929, PSY) and RG knew it from Kilmacolm (1931). It was found on disused railways at Kilmacolm and Auchenbothie (1984, BWR), where it can still be found today (NS3470, 2010, AM). Other records include Loch Libo (NS4355, 1965, JDM) and waste ground at Springhill (NS3264, 1986, BWR). The modern records are all from scattered waste-ground sites or railway sides.

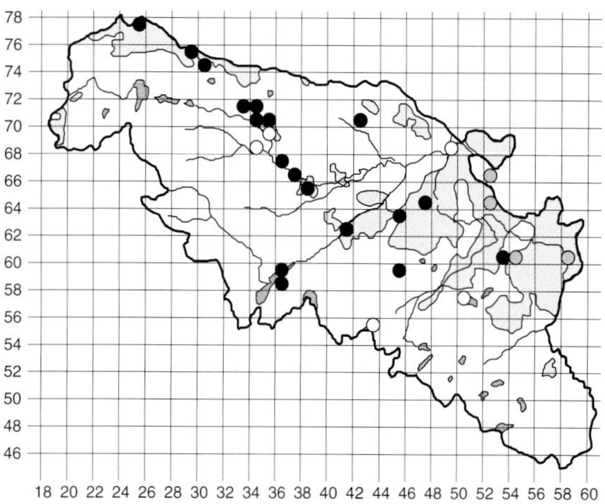

Linaria repens

Veronicastrum virginicum (L.) Farw.
Culver's-root

Britain: neophyte (E N America); rare garden escape.
Hortal; 1/1; very rare.
1st White Loch, 2003.

This garden outcast occurs to the edge of the car park at White Loch (NS4952).

PLANTAGINACEAE

Plantago coronopus L.
Buck's-horn Plantain

Britain: common by coasts and inland in the SE.
Native; 1/5; very rare.
1st Renfrewshire, 1869, PPI.

Hennedy considered Buck's-horn Plantain common along the sea shore but named no localities; it was collected from Wemyss Bay (1893, GL). It is likely to have declined greatly with development along the coast. It was recorded (BWR) from the Clyde Estuary, Trumpethill (NS2276, 1977) and at Wemyss Point (NS1870, 1986). The only modern record is from stony waste ground by the edge of the Clyde at Cartsdyke (NS2975, 2003).

Plantago maritima L.
Sea Plantain

Britain: common all round coast, and in some uplands.
Native; 24/29; frequent on coast.
1st Renfrewshire, 1834, MC.

Apart from Balfour's curious 1859 inland White Cart Water record (CFOG) all other records for Sea Plantain are coastal. It is one of the commoner plants of a few relict coastal grassland sites and is occasionally found on more exposed tidal mud or rocks. As with other coastal plants, the populations are split into two zones by the industrial developments in and around Greenock.

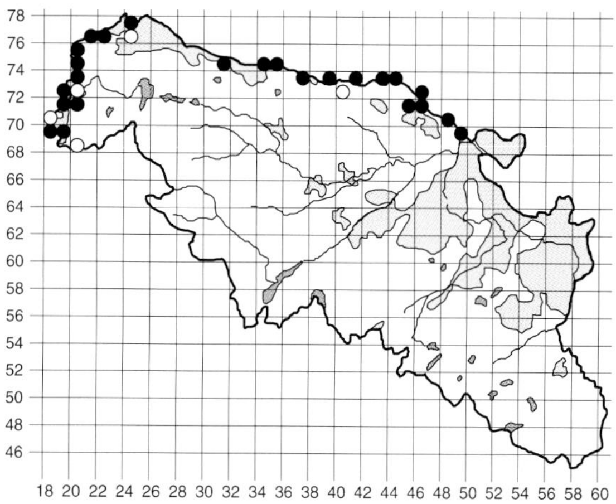

Plantago maritima

Plantago major L.
Greater Plantain

Britain: common throughout.
Native; 286/425; very common.
1st Renfrewshire, 1869, PPI.

Greater Plantain occurs in both urban and rural locations. At the latter it occurs on farmland, including disturbed improved pasture, but generally along tracks and paths and about gateways, reflecting its trample tolerance; it is also common on waste ground and along riverbanks. The ssp. *intermedia* (Gilib.) Lange has been recorded from the stony margins of Balgray Reservoir (NS5057, 1999) and Formakin Estate (NS4170, 2003, IPG).

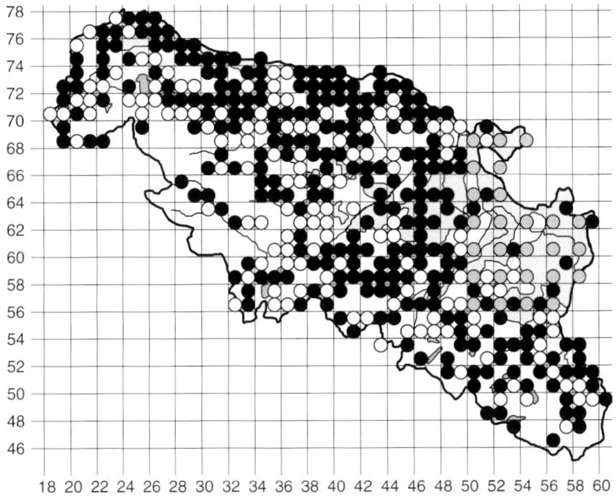

Plantago major

Plantago media L.
Hoary Plantain

Britain: common in the S and E, rare (mainly as alien) in Scotland.
Former accidental; 0/4; extinct.
1st Dennistoun, Kilmacolm, 1889, RH.

Hennedy (1891) added that this plantain was 'only found as an introduced plant with grass seeds, not retaining its place so as to become permanent in pasture.' There are also specimens of Hoary Plantain from Pollokshaws (1905, GLAM) and Ferguslie (1913, PSY) and RG (1930) noted it from near Johnstone.

Plantago lanceolata L.
Ribwort Plantain

Britain: common throughout.
Native; 376/467; very common.
1st Near Hurlet, 1842, GLAM.

Ribwort Plantain is a very common species of grassland except where poorly drained, strongly acidic or heavily improved. It is also commonly found on waste ground, to waysides and as a garden weed.

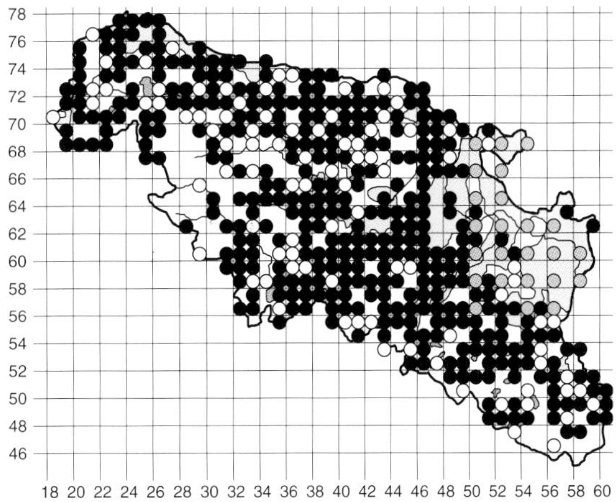

Plantago lanceolata

Littorella uniflora (L.) Asch.
Shoreweed

Britain: common in the N and W.
Native; 55/74; common in open waterbodies.
1st Stanely Reservoir, 1869, RPB.

Shoreweed was considered common in the past and it can still be found at many old localities. Today it is widespread and commonly seen in the shallow margins of most upland lochans and reservoirs; it is

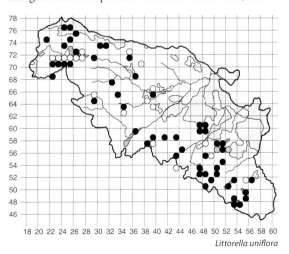

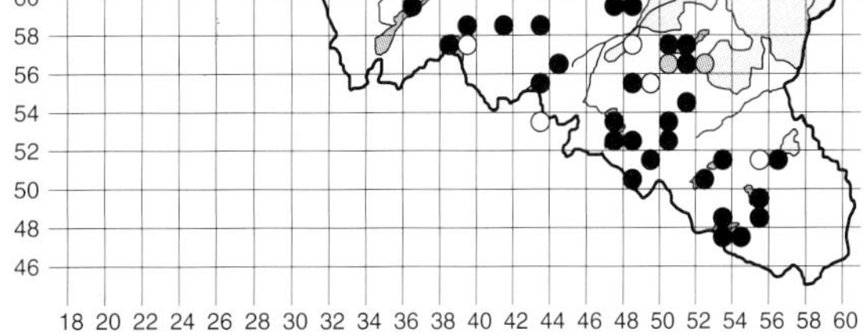

Littorella uniflora

also recorded from lowland sites such as Loch Libo and Castle Semple Loch (both 1996, SNH-LS).

HIPPURIDACEAE

Hippuris vulgaris L.
Mare's-tail

Britain: frequent in the E, including C and NE Scotland.
Native; 11/20; scarce.
1st Castle Semple, 1813, TH.

Hennedy (1891) considered Mare's-tail to be not common, repeated Castle Semple and also noted it from the 'hills beyond Gourock'; the latter may be the same as the 'Loch above Ravenscraig' reported in 1880 (NHSG). There are several other records for Castle Semple Loch and also for Loch Libo (1887, NHSG), but none recently from the latter. In 1897 (TPNS) it was found at Craighall Dam, Neilston Pad, where it still occurs (NS4755, 1994). It is also known from a scattering of other swamps and margins of waterbodies, and it was found in pools near the Muirshiel Barytes Mine (NS2864, 1998), which may be the same location as RM's 1934 Mistylaw Muir record.

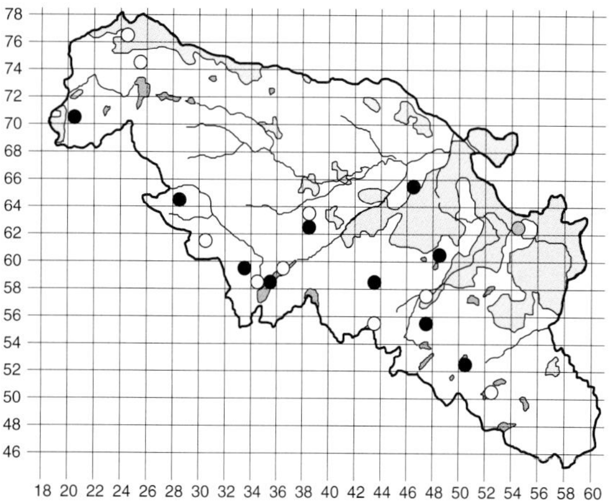

Hippuris vulgaris

CALLITRICHACEAE

Callitriche hermaphroditica L.
Autumnal Water-starwort

Britain: scattered in the N.
Native; 15/24; rare.
1st Renfrewshire, 1915, TPNS.

This water-starwort does not appear to have been specifically recorded in the 19th century locally but later RM made several records, including Eaglesham Dam (1927, GLAM), Pilmuir Dam (1944), Walton Dam (1952), Knapps Loch (1958), dams on Kirkton Burn, Neilston (1952) and Barr Loch (1974); whether this represents a recent spread of Autumnal Water-starwort (by birds or anglers?) is unclear. It is not mentioned by Hennedy (1891) and Lee (1933) considered it very rare. Today it is still present at Barr Loch (several 1km squares) and Walton Dam (NS4955, 1998); other records include Knapps Loch (NS4951, 1996, SNH-LS), Black Loch (NS4951, 1997), Harelaw Dam (NS4753, 1997), Harelaw Reservoir (NS4859, 1999) and the pond in Murdieston Park (NS2675, 2002).

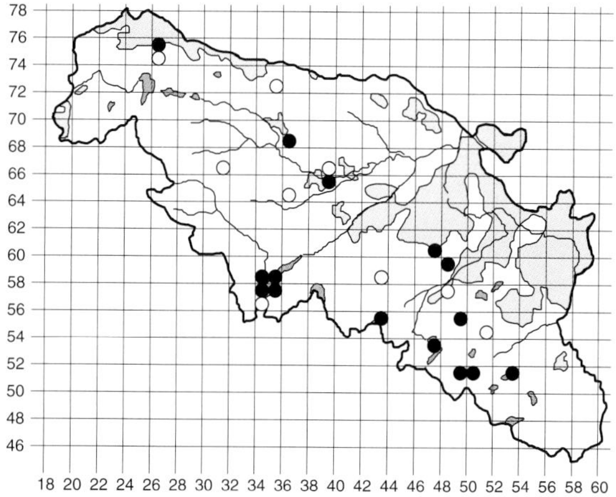

Callitriche hermaphroditica

Callitriche stagnalis s.l.

Britain: common throughout.
Native; 113/221; common.
1st Paisley Canal bank, 1865, RH.

The map shows all records, including those from CFOG, and includes Various-leaved Water-starwort, which is difficult to distinguish in the absence of fruit – which is often the case. The records are from lochs, ponds, slow-flowing parts of watercourses and marginal inundation areas, and ditches and pools in marshes.

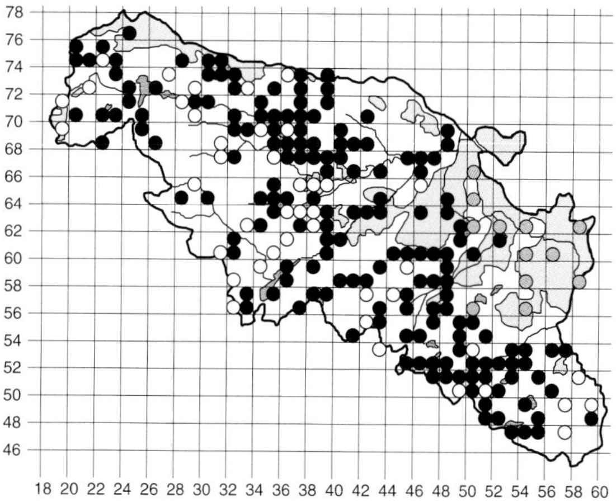

Callitriche stagnalis s.l.

Callitriche stagnalis Scop.
Common Water-starwort

Native; 33/34; frequent.
1st Gourock, 1879, GL.

The original record for Common Water-starwort was expertly determined but most other records are based on the presence of distinctly winged fruit, or occasionally vegetative characters. It seems likely that Various-leaved Water-starwort is the commoner, but more detailed study is required.

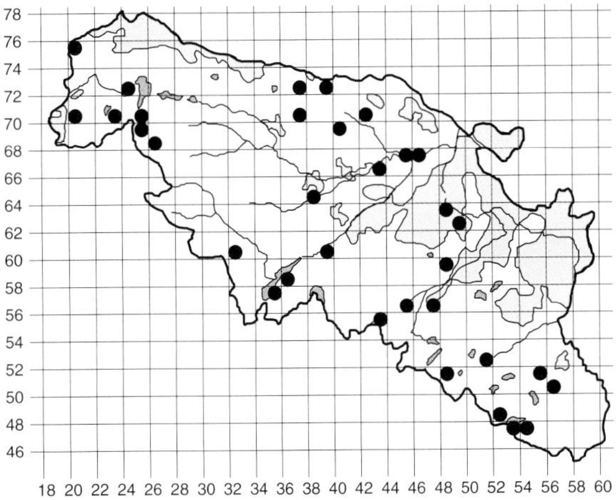

Callitriche stagnalis

Callitriche platycarpa Kütz.
Various-leaved Water-starwort

Britain: locally common in the S and lowlands including C Scotland.
Native; 55/58; frequent.
1st Renfrewshire, 1883, TB.

There is no way of knowing the true identity of the first records, but this species appears to be the commonest water-starwort in the lowlands. Various-leaved Water-starwort is usually found in ditches and in pools and marshes, although a few records are from loch margins.

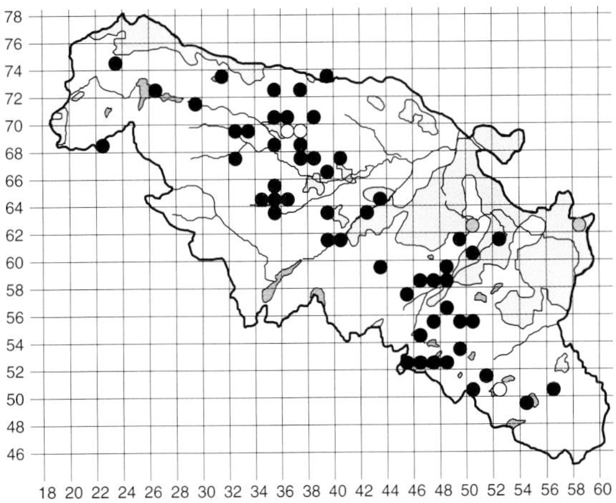

Callitriche platycarpa

Callitriche brutia Petagna
ssp. *hamulata* (Kütz. ex W.D.J. Koch) O. Bolòs & Vigo.
Callitriche hamulata (Kütz. ex W.D.J. Koch) A.R. Clapham
Intermediate Water-starwort

Britain: common in the N and W.
Native; 36/47; frequent.
1st Burn in wood, Kilmacolm, *c.* 1860, GGO.

Lee thought this water-starwort was common and overlooked but there are no other precise VC records. More recently RM found it at Walton Dam (NS4955, 1968) and the old Calder Dam (NS2965, 1976) and AJS noted it from Finlaystone (NS3673, 1976). All the other records come from the modern period and mostly from open waterbodies. The SNH Loch Survey (1996) found it at Caplaw Dam (NS4755), Greenock Cut Reservoir No. 2 (NS2373), Loch Libo (NS4355), Houston Estate (NS4466), North Loch Drum Estate (NS3971), Coves Reservoir (NS2476) and Castle Semple Loch (NS3669). Many of the records are from upland (i.e. lower-nutrient) locations, but not all.

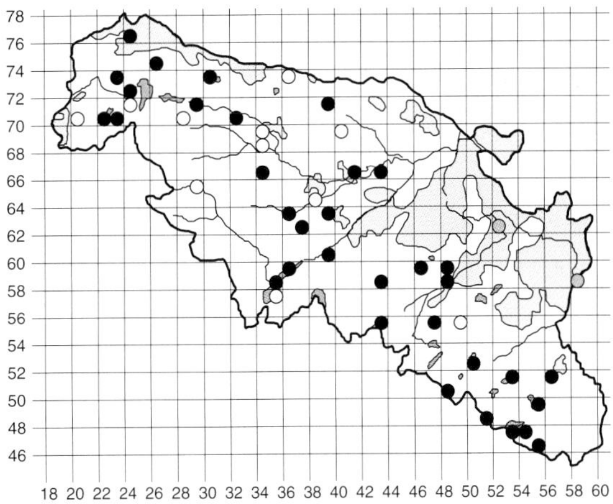

Callitriche brutia ssp. *hamulata*

SCROPHULARIACEAE (BUDDLEJACEAE)

Verbascum blattaria L.
Moth Mullein

Britain: neophyte (Europe); scattered in the SE.
Former hortal; 0/1; extinct.
1st Greenock, 1880, PSY.

Verbascum thapsus L.
Great Mullein

Britain: common throughout lowlands
Native or hortal; 21/32; occasional.
1st Banks of Clyde below Renfrew, 1813, TH.

Great Mullein was well recorded in the 19th century, although considered not common by Hennedy (1891); localities included Crossflats, Kilbarchan (1845, NSA), Plantation, Wemyss Bay (1880, NHSG), Calder, Lochwinnoch (1890, AANS) and Kilmacolm Station (1896, NHSG). Today there are scattered records, all from freely draining waste-ground locations.

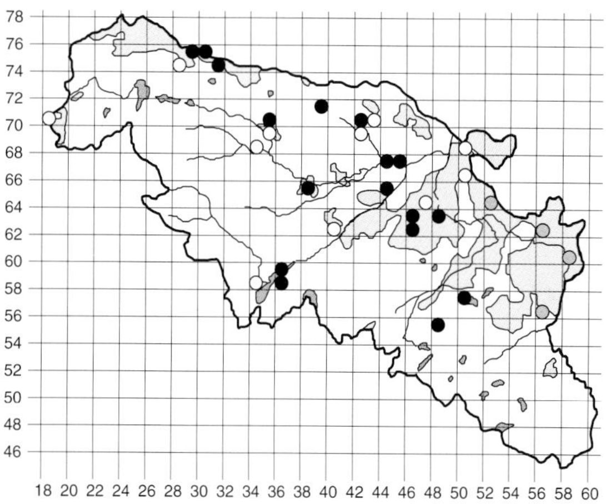

Verbascum thapsus

Verbascum nigrum L.
Dark Mullein

Britain: frequent in the SE, scattered elsewhere as alien.
Hortal; 1/4; very rare.
1st Thornliebank, 1930, RG.

A specimen of Dark Mullein was collected from Paisley (1930, PSY) and in 1975 AJS recorded it from the University of the West of Scotland Paisley campus (NS4763). Today it can be found along the edges of gravelly tracks at the Royal Ordnance Factory, Georgetown, and on the adjacent railway line just outside (NS4467, 2005).

Verbascum lychnites L.
White Mullein

Britain: rare in the S, occasional alien elsewhere.
Former hortal; 0/1; extinct.
1st Greenock, 1942, TPNS.

The spontaneous arrival of White Mullein in a neglected garden was reported by the Rev. JG McCulloch.

Scrophularia nodosa L.
Common Figwort

Britain: common throughout except in N Scotland.
Native; 158/224; common.
1st Paisley, 1845, NSA.

Common Figwort is commonly found throughout the lowlands and occurs at both rural and urban locations, although it is seldom found in abundance – many records are of individuals, some perhaps occurring sporadically from persistent seed banks. It is found in woodlands and other shaded places, but also along hedge banks, riverbanks, disused railways and roadsides and can be found on waste ground.

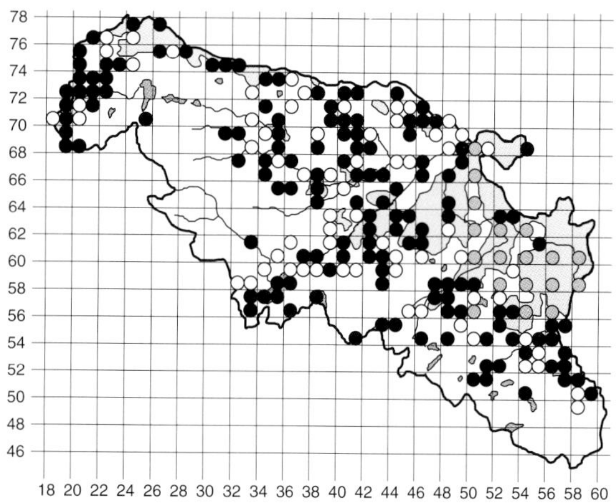

Scrophularia nodosa

Scrophularia auriculata L.
Water Figwort

Britain: common in the S and E, rare in Scotland.
Native or alien; 4/5; very rare.
1st Cartside, 1865, MY.

The original record may well be an error for Green Figwort, so there is little evidence for this species' native status in the VC. The first modern record for Water Figwort is from by a stream in the Duchal Estate (NS3568, 1994, CD); Alan Wood found it in a garden near Gourock (NS2776) in 2000 and it occurs by a track at Larkhall (NS2375, 2012). A few individuals were recently spotted by members of the ranger service (REM) near the visitor centre at Whitelee Windfarm (NS5349, 2010); it was seen recently by tracks further south (in VC75, *c.* NS5543, 2008), so it may be spreading in this area.

Scrophularia umbrosa Dumort.
Green Figwort

Britain: doubtful native with local distribution, frequent in C Scotland.
Native or alien; 9/13; rare.
1st Paisley, 1845, NSA.

Green Figwort was collected from Inchinnan in 1854 (GGO) and 1878 (GL) – Hennedy (1891) refined these records as 'Banks of the Cart at Inchinnan Bridge', but added that it was 'apparently extinct'; it has not been recorded from there in recent times. However, it can be found at the 1890 AANS locality 'Aurs–Brock burn confluence', where it still occurs at Darnley (NS5258, 2004). Wood (1893) wrote that it was plentiful along the Levern and White Cart waters downstream; at the former it is found about Crookston (NS5361, NS5262, 2005) and at the latter along the banks in the centre of Paisley (NS4863, 2000 and NS4864, 2001). It is also found along the Dargavel Burn at Formakin (NS4070, 1994). An isolated, and somewhat different-looking plant, occurred as a colonist on waste ground at Linwood Moss tip (NS4465, 1998).

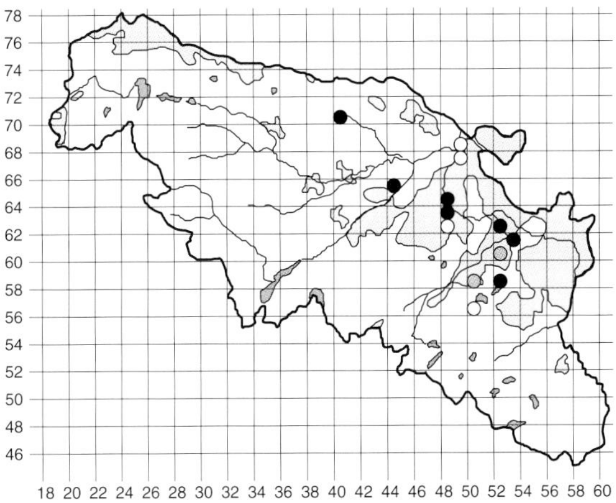

Scrophularia umbrosa

Scrophularia vernalis L.
Yellow Figwort

Britain: neophyte (Europe); scattered in the SE, locally frequent in E C Scotland.
Hortal; 1/5; very rare.
1st Paradise, Finlaystone Estate, 1889, RH.

Hennedy (1891) also listed Barrhead and Kilmacolm; Lee (1933) noted Carruth Bridge. An excursion to Finlaystone in 1963 (JDM) failed to find Yellow Figwort but AJS relocated it in 1976 and it may still occur there. It can still be found at the old churchyard, St Fillan, Kilallan (NS3866, 2004), where it has been known for more than 100 years (JMY, 1898).

Phygelius capensis E. Mey. ex Benth.
Cape Figwort

Britain: neophyte (S Africa); very scattered in the W.
Hortal; 1/1; very rare.
1st Wemyss, 2010, PRG.

The sole record is of a patch in a wood at Wemyss Bay (NS1970).

Limosella aquatica L.
Mudwort

Britain: scattered, local in C Scotland.
Native; 2/2; very rare, but possibly overlooked.
1st Balgray Reservoir, 1975, JM.

A population of this, then considered very rare, Scottish plant was found when water at the Balgray Reservoir was low (Mitchell 1976) and, in similar circumstances, later in 1984 (CFOG). It was still present in some abundance in 2000 (NS5057) but recently the water level has been retained high throughout the year. In 1984 Mudwort was recorded from 'Rowbank' [Barcraigs] Reservoir (NS3957, BWR) where, in 2003, IPG refound it.

Buddleja davidii Franch.
Butterfly-bush

Britain: neophyte (China); common in S, frequent in C Scotland.
Hortal; 45/45; frequent, spreading.
1st Braehead (NS56D), 1985, CFOG.

Butterfly-bush appears to be enjoying a modern-day spread, with no records before a few for VC76 in CFOG. It shows a disjunct distribution pattern, and is quite frequent on waste ground, in old quarries and on buildings in the north-west, and extends to the rural fringe about Spango (NS2374 and NS2474, 1997). It grows in similar places further east. It was first reported from Lochwinnoch in 1992 (NS3658, CMRP) and later in 2005 (NS3557, IPG).

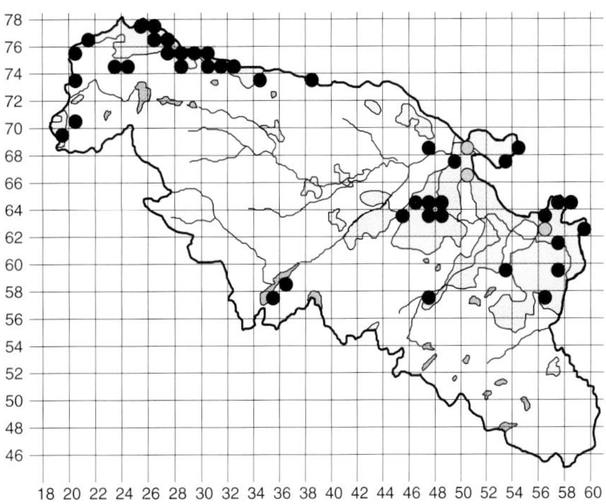

Buddleja davidii

Buddleja globosa Hope
Orange-ball-tree

Britain: neophyte (S America); scattered and local.
Hortal; 1/1; planted and persisting.
1st Ardgowan Estate (NS2073), 1998.

LAMIACEAE

Stachys sylvatica L.
Hedge Woundwort

Britain: common throughout except in N Scotland.
Native; 217/273; common.
1st Paisley, 1845, NSA.

Hedge Woundwort has been a common lowland plant since recording first began. It is absent from most of the upland Renfrewshire Heights, although there are a few records in the south-east uplands such as at High Myres (NS5646, 2000) and Ardoch Burn (NS5849, 1982, BWR). Its typical habitats are woodlands, hedge banks, riverbanks – open or shaded – waysides and also waste ground.

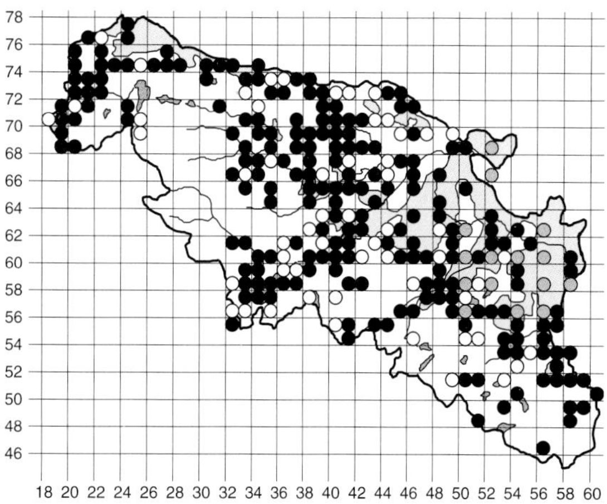

Stachys sylvatica

Stachys palustris L.
Marsh Woundwort

Britain: common throughout except in N Scotland.
Native; 124/152; common.
1st Near Hurlet, Glasgow, 1842, GLAM.

Marsh Woundwort is a common plant of wet or inundated places such as marshes (but not swamps), loch margins and along watercourses, generally reflecting more nutrient-enriched conditions; however, it is also frequently found at better-drained sites such as roadsides, waste ground and hedge banks.

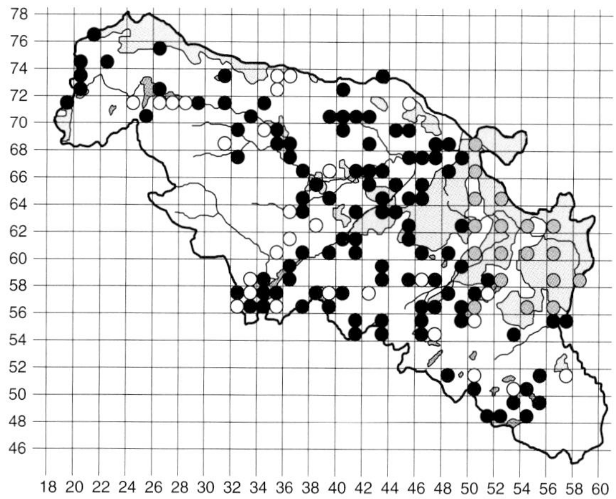

Stachys palustris

Stachys × ambigua Sm.
(*S. sylvatica × S. palustris*)
Hybrid Woundwort

Britain: frequent, mainly in the W.
Native; 64/71; frequent.
1st Renfrewshire, 1834, MC.

Considered frequent in the 19th century, there are few named local sites for Hybrid Woundwort until

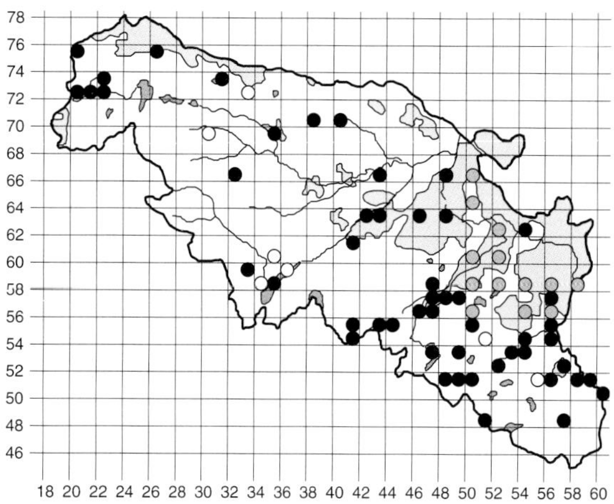

Stachys × ambigua

the middle of last century when RM knew it from Loganswell (1936), the Mearns area (1945) and Loch Libo (1945). Today it is widespread, though commoner in the central and urban south-east, and is found in the rough grasslands of hedge banks, waysides, riverbanks and waste ground.

Stachys arvensis (L.) L.
Field Woundwort

Britain: archaeophyte; common in the S and W, local in Scotland.
Accidental; 0/6; possibly extinct or overlooked
1st Gourock, 1865, RH.

Field Woundwort was also collected from Gourock (1872, GLAM) and from Blackland (1900, PSY). It may well survive at arable field margins or even as a garden weed, but it must be very rare as there are no modern records. It was recorded some 50 years ago by BWR: Neilston (NS4656, 1960), Knapps Loch (NS3668, 1963) and West Dougliehill (NS3173, 1963), and more recently AMcGS knew it from waste ground at Erskine (NS4672, 1980).

Betonica officinalis L.
Stachys officinalis (L.) Trevis
Betony

Britain: common in the S, rare in Scotland.
Former native and hortal; 1/3; very rare.
1st Paisley, 1845, NSA.

Ten years after the first record Betony was collected from 'west of Paisley' (PSY) and in 1865 it was noted at Glen Patrick (MY), perhaps the same place. It is possibly a very rare native in Lanarkshire but apart from the Paisley records there is little evidence to support this status in Renfrewshire. A small population occurs on the motorway embankment in Paisley (NS4965, 2010), presumably planted.

Ballota nigra L.
Black Horehound

Britain: archaeophyte; common in S and E extending to E Scotland.
Former accidental; 0/3; extinct.
1st Cathcart Castle, 1813, TH.

Black Horehound was also found at Stanely Reservoir in 1893 (PSY) and a 1938 Lee specimen (GL) from borderland Anniesland is annotated as VC76.

Lamiastrum galeobdolon (L.) Ehrend. & Polatschek
Yellow Archangel

Britain: common in the S, scattered alien in Scotland.
Hortal; 2/6; very rare.
1st Pollokshaws, 1887, GL.

Not recorded in the earlier editions of *The Clydesdale Flora*, Hennedy (1891) described Yellow Archangel as rare and reported it from hedges and old houses, but referred to luxuriant growth – over many years – at the side of a small stream near 'Netherauldhouse, Pollockshaws' (there are two other GL specimens from there); Hennedy also noted Barscube Old Mill, Erskine (1890). Lee collected a specimen from Neilston in 1923 (GL) and RM collected another from near Patterton (1952, E). AMcGS recorded it (as ssp. *montanum* (Pers.) Ehrend. & Polatschek) from Capelrig House (NS5457) in 1986, from where it was first noted in 1963 (Mackechnie 1964). More recently a small population was found growing along the margin of woodland by the house at Corsliehill (NS4069, 1999).

Lamiastrum galeobdolon ssp. *argentatum* (Smejkal) Stace

Britain: neophyte (cultivated origins); common in the S, frequent in C Scotland.
Hortal; 61/61; frequent, spreading.
1st Greenbank (NS55T), 1984, CFOG.

Today this garden outcast or escape is the only Yellow Archangel that can be readily seen. It has a widespread distribution, occurring in many rural as well as urban locations, often in abundance. It tolerates shade and can be seen in various woodlands but it can also be found on waste ground and railway banks.

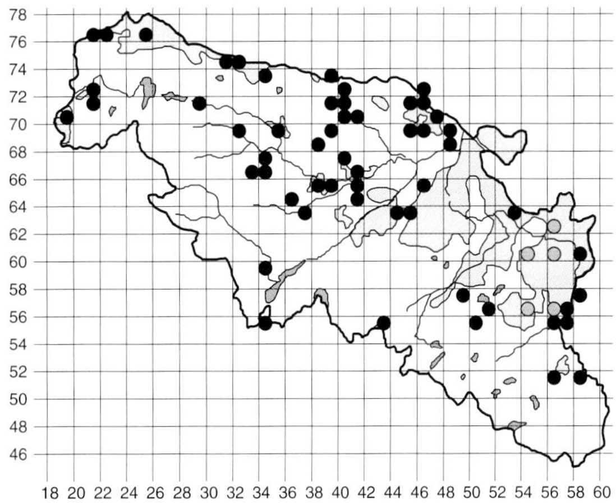

Lamiastrum galeobdolon ssp. *argentatum*

Lamium album L.
White Dead-nettle

Britain: archaeophyte; common in England and in E Scotland.
Accidental; 13/24; scarce.
1st Paisley, 1845, NSA.

The only other old localities for White Dead-nettle are 'Eaglesham–Ballageich' (1874, TGSFN), side of road Houston (1921, GL) and Eaglesham (1941, GL).

It was well recorded in the 1960s–80s (BWR) with ten records and a further five during surveying for CFOG: all to the eastern half of the VC. Today White Dead-nettle is known from Barnbrock (NS3564, 1994, AB), Barochan Estate (NS4168, 1997), Wester Fulwood (NS4366, 2001) and Houston (NS4067, 2002), as well as from a few sites along the Levern Water and White Cart Water to the edge of Glasgow.

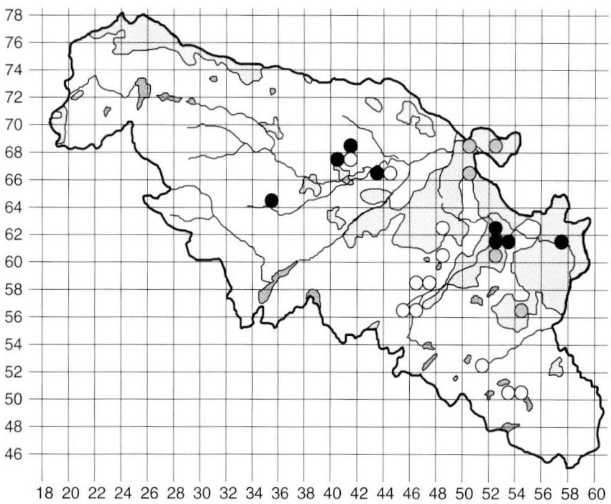

Lamium album

Lamium maculatum (L.) L.
Spotted Dead-nettle

Britain: neophyte (Europe); frequent in the S, local in C Scotland.
Hortal; 3/7; very rare.
1st Near Paisley, 1930, PSY.

Lee (1953) added Glen Killoch, JDM reported Spotted Dead-nettle from Craighall Dam (1964) and AJS knew it from a roadside on the Gleniffer Braes (NS4459, 1975). There are two records for this garden outcast in the Johnstone area: Floors Burn (NS4262, 1993) and near Auchingreoch (NS4260, 1993), and it was also recorded from Barsail Wood, Erskine (NS4669, 1996).

Lamium purpureum L.
Red Dead-nettle

Britain: archaeophyte; common throughout except in N Scotland.
Accidental; 25/40; occasional.
1st Paisley, 1845, NSA.

Other old records for Red Dead-nettle include Paisley Canal Bank (1885, PSY), Gourock (1880, GL) and Bishopton (1919, GL). There are quite a few records from the 1960s–80s (BWR) and a few more were made during CFOG surveys. It is the commonest dead-nettle, absent from the west and upland fringes, but like the others probably under-recorded; habitats include waste ground, waysides, disturbed soil and arable weed, as at Woodneuk (NS4858, 1998) and South Mains (NS4266, 2003). It has recently appeared at a couple of disturbed places near the visitor centre at Whitelee Windpark (NS5248, 2010, REM).

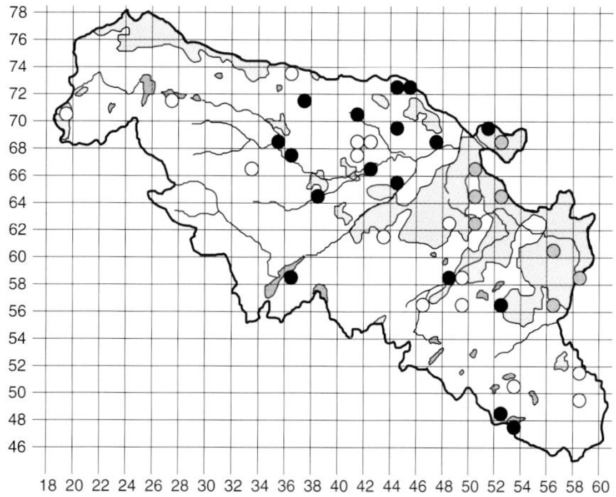

Lamium purpureum

Lamium hybridum Vill.
Cut-leaved Dead-nettle

Britain: archaeophyte; common in C and E England and E Scotland.
Accidental; 3/6; very rare or overlooked.
1st Renfrewshire, 1834, MC.

This dead-nettle was recorded from Bridge of Weir in 1975 (GMTC) and by AJS in 1976 from the University of the West of Scotland Paisley campus (NS4863), and there are two CFOG records of it as garden weed from Yoker (NS56E, 1985) and Jordanhill (NS56N, 1984). Another recent record is as an arable weed, near Inchinnan Bridge (NS4968, 1988, AMcGS).

Lamium confertum Fr.
Northern Dead-nettle

Britain: archaeophyte; very rare in the S, frequent in lowland coastal Scotland.
Accidental; 0/2; extinct or possibly overlooked.
1st Cathcart and Gourock, 1865, RH.

Considered a frequent weed of arable fields in the past (but 'not common' by Lee in 1933), it would be surprising if Northern Dead-nettle was actually extinct, but there have been no records for it made during the last 50 years.

Lamium amplexicaule L.
Henbit Dead-nettle

Britain: archaeophyte; common in the SE, extending to E Scotland.
Accidental; 1/4; very rare or overlooked.
1st Lochwinnoch, 1845, NSA.

Henbit Dead-nettle was collected from a rubbish

heap, Crookston in 1919 (GL) and Lee (1933) reported Gourock, but there are no modern records; it was recorded for CFOG from Mosspark (NS56G, 1985). It is likely that this plant has been overlooked during modern recording.

Galeopsis speciosa Mill.
Large-flowered Hemp-nettle

Britain: archaeophyte; frequent in C England and C and E Scotland.
Accidental; 24/33; occasional.
1st Renfrewshire, 1834, MC.

Considered very common about Glasgow by Hooker (1821) and similarly so by Hennedy (1891), there are not many recent or modern records for Large-flowered Hemp-nettle. It was well recorded in CFOG but only a handful of times in the 1970s (BWR), so although it may have been overlooked recently (when not in flower) it appears to have undergone a marked decline. The map shows one record from the north-west, an arable field margin at Cauldside (NS3270, 2005), perhaps indicating that this species has always preferred the drier soils, and arable farming, to the east.

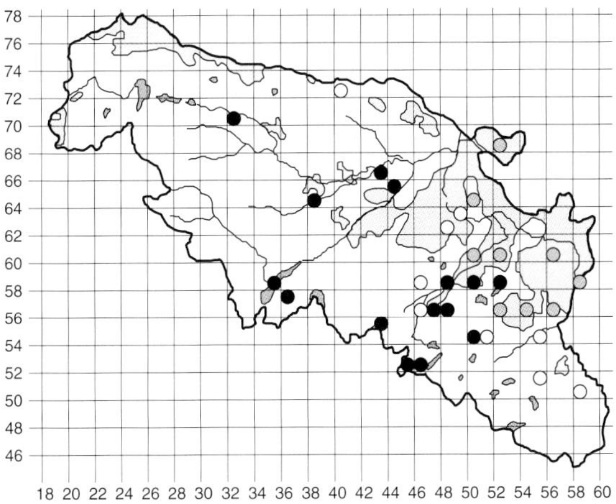

Galeopsis speciosa

Galeopsis tetrahit s.l.
Hemp-nettles

Britain: common throughout except in N Scotland.
Native; 139/232; common.
1st Renfrewshire, 1869, PPI.

Previous recorders did not distinguish between the following two species so the map includes all records; it also includes some modern records for non-flowering plants. Hemp-nettles may have undergone a decline, but not one as dramatic as the map suggests.

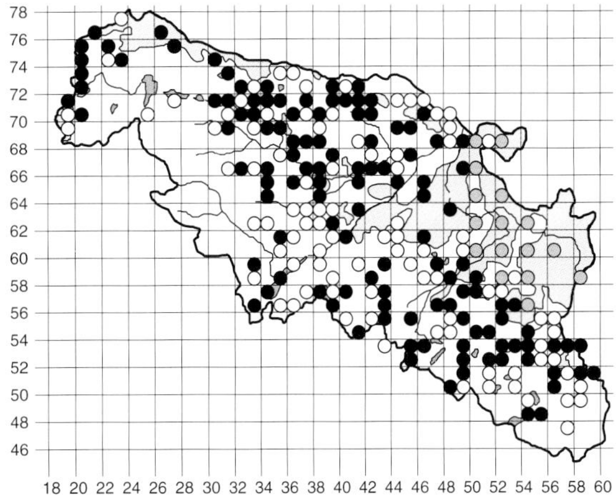

Galeopsis tetrahit s.l.

Galeopsis tetrahit L.
Common Hemp-nettle

Britain: common in the S and in C and E Scotland.
Native; 42/44; frequent.
1st Renfrewshire, 1899, GC.

Common Hemp-nettle has probably been under-recorded with many records (old and new) included above. The first herbarium sheet is from Port Glasgow (1956, GL) and the first reliable field record is from

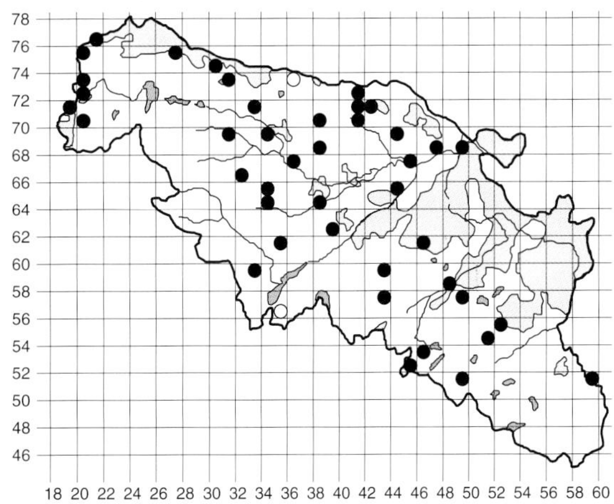

Galeopsis tetrahit

Finlaystone (1976, AJS). Today it is known from widespread localities which are generally open waste-ground sites including waysides and farmland.

Galeopsis bifida Boenn.
Bifid Hemp-nettle

Britain: frequent but local in lowlands.
Native; 49/50; frequent.
1st Kilmacolm, 1894, GL.

Bifid Hemp-nettle is often encountered in open and poached damp grasslands, sometimes with bracken, and tolerates rushy pastures; it can also be found at drier open waste-ground sites. It is strongly rural and scarce in the north-west. The map shows only modern (and CFOG) records.

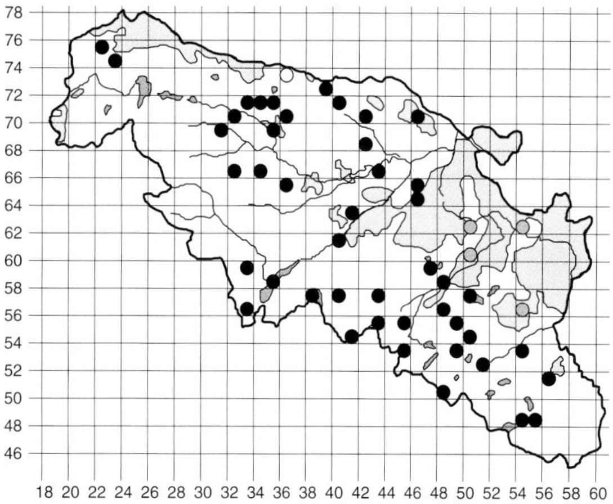

Galeopsis bifida

Scutellaria galericulata L.
Skullcap

Britain: common in the S, common in W Scotland.
Native; 8/17; rare.
1st Gourock, 1842, GLAM.

There are quite a few past records for Skullcap including Castle Semple Loch (1971, BWR), where it was earlier recorded in 1845 (NSA); however, there are no modern records from former localities on the west coast, e.g. Wemyss Bay (1887, NHSG), Gourock (1857, GGO) and Lunderston Bay (1978, BWR). It was recorded in 1887 (NHSG) from Loch Libo, and it can still be found there today (NS4355, 1996). The SNH Loch Survey (1996) found it in marginal swamps at Knapps Loch (NS3668) and Brother Loch (NS5052); Hennedy reported the latter in 1891. The other modern records are from a marshy drain, near the Black Cart Water, Howwood (NS3960, 1994, CD), Barcraigs Reservoir margins (NS3857, 2001), Reed Canary-grass swamp, Knowes (NS3756, 2003) and a burn side, near Barrance Hill (NS5153, 2008). It appears that Skullcap is quite rare but may well be hidden in thick swamp vegetation at other locations.

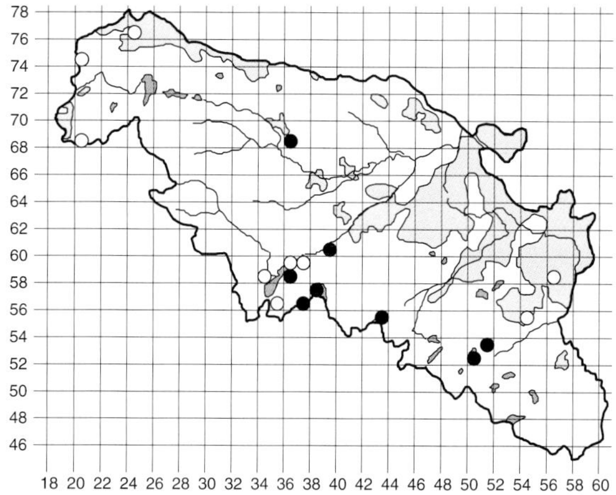

Scutellaria galericulata

Teucrium scorodonia L.
Wood Sage

Britain: common throughout except locally in the E.
Native; 174/209; common in rural areas.
1st Paisley, 1845, NSA.

Wood Sage is a fairly common plant in the rural lowlands, although seldom found in any abundance, and can extend into the uplands along valley sides. It is most frequently encountered on rocky slopes or outcrops in woodlands, along river valleys, on open grasslands and dry heaths, and often on steep disused railway embankments.

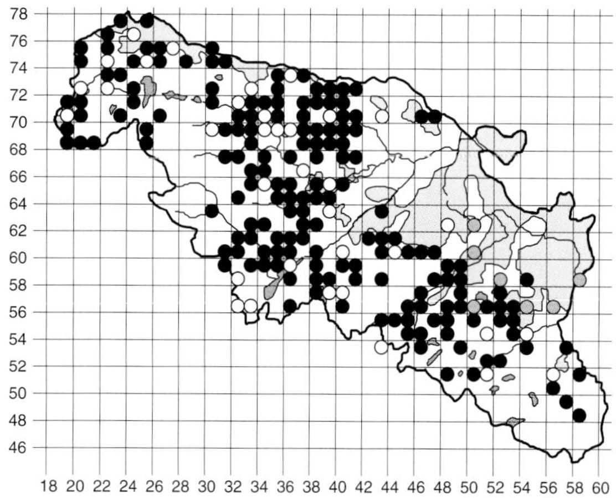

Teucrium scorodonia

Ajuga reptans L.
Bugle

Britain: common throughout except in the extreme N of Scotland.
Native; 152/208; common in damp places.
1st Inverkip, 1837, GL.

Bugle is commonly found in damp places but usually where there is at least partial shade as in wet woodlands or along riverbanks. It can occur in more open habitats, such as wet woodland glades or rides and often along upland burn sides and at some loch margins, but it is not a feature of rush pastures or marshes. It shows a rural pattern, extending into the upland zone along burn sides. A copper-coloured cultivar was recorded from a shelter roof at Auchenbothie (NS3571, 1998).

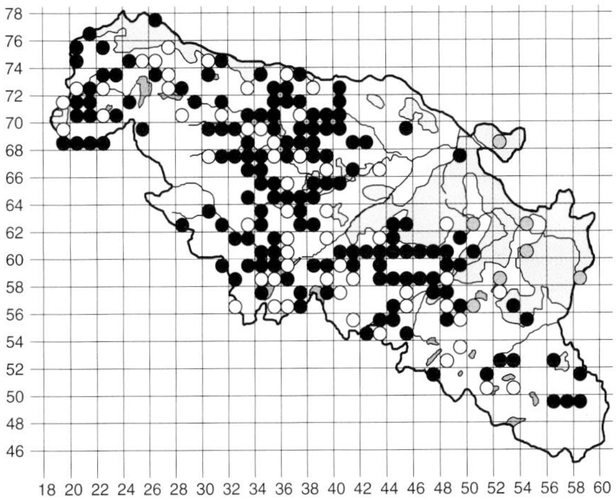

Ajuga reptans

Nepeta cataria L.
Cat-mint

Britain: archaeophyte; frequent in the S, rare in Scotland.
Former hortal; 0/2; extinct.
1st Paisley, 1845, NSA.

Cat-mint was recorded in 1966 from Greenhags Coup (NS5152, BWR). Garden Cat-mint (*N.* × *faassenii* Bergmans ex Stearn) is well established on stonework at Pollok House (NS5461, 1996, CFOG).

Glechoma hederacea L.
Ground-ivy

Britain: common in the S, extending to C and E Scotland.
Native; 88/123; frequent, central.
1st Paisley, 1848, GL.

Ground-ivy exhibits a very strong central lowland and rural pattern, but is surprisingly absent from much of the south-east and the west, except for a cluster about Inverkip. It is frequently encountered along riverbanks, is usually tolerant of some shade, and can also be found in woodlands and along waysides and hedge banks.

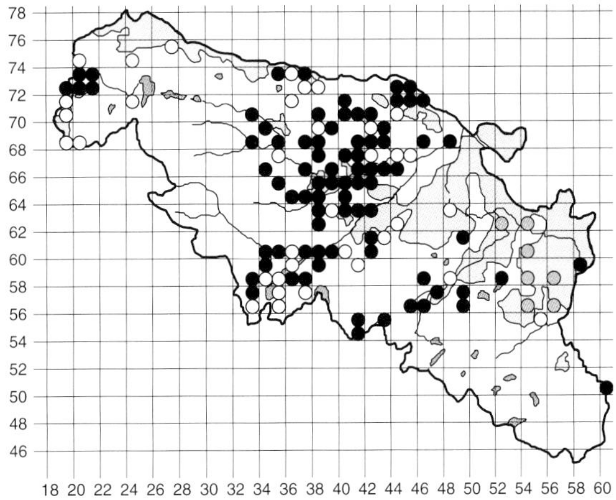

Glechoma hederacea

Prunella vulgaris L.
Selfheal

Britain: common throughout.
Native; 321/382; very common.
1st Paisley, 1845, NSA.

Selfheal is typically found in short, neutral grasslands, including amenity-managed types; it tolerates some dampness, although not true marshy conditions, and can be common on compacted, clayey soils of waste grounds. It has a very widespread distribution, extending into the peaty uplands along burn sides or marking out less acidic and often flushed soils on mineral ground, e.g. at Queenside Loch (NS2964, 2003), North Rotten Burn (NS2667, 1997), Calder Water (NS2764, 1993), Maich Water (NS3060, 2001) and Dickman's Glen (NS5847, 2001). A curious small, pale-flowered form occurs by the Gryfe Water at Cauldside (NS3270, 2005).

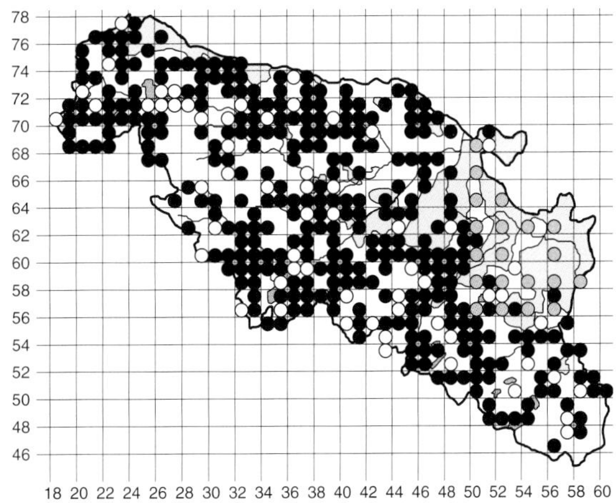

Prunella vulgaris

Clinopodium vulgare L.
Wild Basil

Britain: common in the S, frequent in E Scotland.
Former native; 0/9; extinct.
1st Barrhill, Kilbarchan, 1845, NSA.

There are quite a few old records for Wild Basil: Paisley (1858 and 1860, PSY), Paisley Canal bank (1865, RH), Jenny's Well (1865, MY), Castle Semple Loch (1887, NHSG), Darnley (1890, GL), Barrhead (1901, FFCA) and Lee (1933) noted it at Polnoon and Busby. Despite all these old locations it has not been seen for more than 60 years, since RM collected it from the riverbanks of the Earn Water, Mearns, in 1945 (GLAM).

Clinopodium acinos (L.) Kuntze
Basil Thyme

Britain: frequent in the SE, very rare in Scotland.
Former accidental; 0/1; extinct casual.
1st Wasteland, near Paisley, 1910, PSY.

Origanum vulgare L.
Wild Marjoram

Britain: common in the S, frequent in E Scotland.
Hortal; 4/6; rare garden escape or outcast.
1st Bishopton Station, 1893, PSY.

Lee (1933) noted Wild Marjoram at Polnoon and also repeated Bishopton. The modern locations are: waste ground, Ladyburn (NS2975, 1998), roadside, Jordanhill (NS5468, 2002, AR), waste ground, Millarston (NS4563, 2002) and Daff Burn, Inverkip (NS2072, 2003).

Thymus polytrichus A. Kern. ex Borbás
Wild Thyme

Britain: common throughout except in C and E England.
Native; 51/73; frequent, rural.
1st Kelly Glen, 1854, HMcD.

Wild Thyme shows a strongly rural distribution, with virtually no records close to urban areas, except in the extreme west. It extends into upland areas where it grows on exposed rock outcrops or similar along watercourses. Even in less improved, lowland pastures it is typically associated with the less acidic, shallow soils of rocky features.

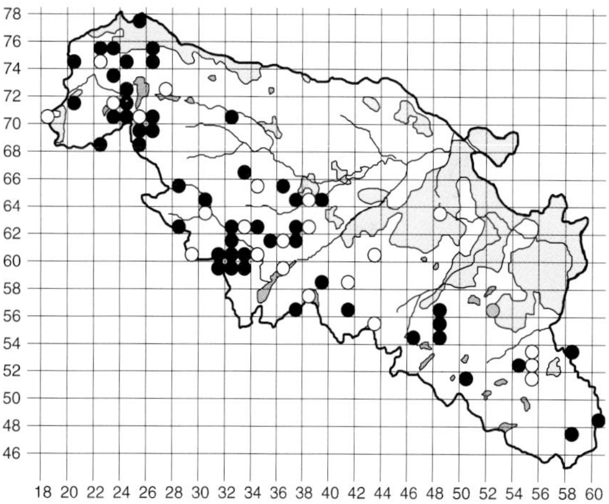

Thymus polytrichus

Lycopus europaeus L.
Gipsywort

Britain: common in the S, frequent in W Scotland
Native; 1/4; very rare.
1st Kelly, 1834, MC.

Hennedy (1865) and Lee (1933) noted Gipsywort at Gourock and it was recorded from Cloch Lighthouse in 1880 (NHSG), but there were no further records until 1981 when it was seen at Paisley Moss (BWR), where it can still be found (NS4665, 2000).

Mentha arvensis L.
Corn Mint

Britain: common in the S, frequent in Scotland except in the N.
Native; 4/14; rare, overlooked.
1st Renfrewshire, 1869, PPI.

Hennedy (1891) considered this mint to be 'very common in corn fields' but named old localities are few; it was collected from Bridge of Weir (1886, PSY) and GL records include Darnley (1890), Kilmacolm (1895), Eastwood Cemetery (1906) and Clarkston Road (1907). There are ten records by BWR from the 1960s–80s but today Corn Mint is only known from four sites: three at reservoirs – the margins of Barcraigs Reservoir (NS3857, 2000) and the Compensation Reservoir, Loch Thom (NS2472, 2001), and the drained margins of Leperstone Reservoir (NS3571, 2001) – and a flushed area at Lorabar (NS3958, 2010). It appears to have undergone a decline but it is also probably overlooked.

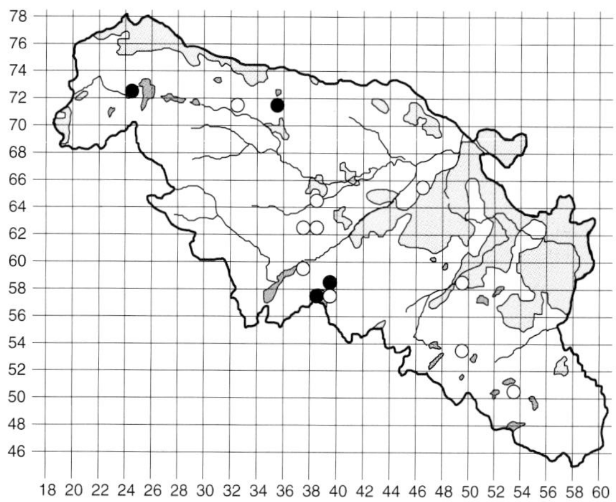

Mentha arvensis

Mentha × *verticillata* L.
(*M. arvensis* × *M. aquatica*)
Whorled Mint

Britain: frequent throughout except in N Scotland.
Native; 38/46; frequent, scattered.
1st Inchinnan Bridge and Gourock, 1865, RH.

This hybrid mint has a widespread distribution and may be overlooked for Water Mint when non-flowering (and not sniffed!). It grows in similar lowland rural places but extends into the uplands at Dunwan Dam (NS5548, 2001). It was also considered frequent by Hennedy (1891).

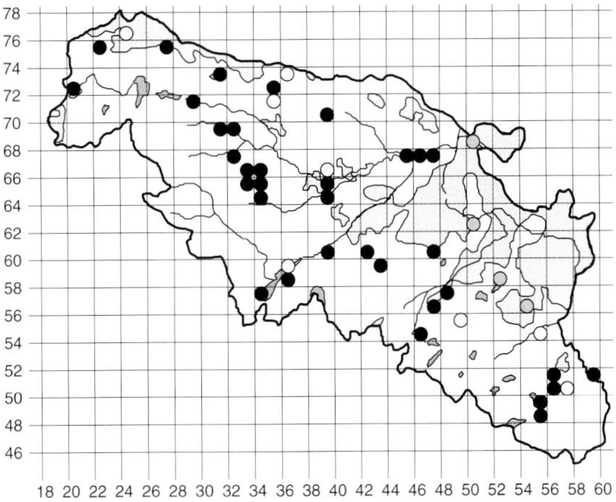

Mentha × verticillata

Mentha × *smithiana* R.A. Graham
(*M. arvensis* × *M. aquatica* × *M. spicata*)
Tall Mint

Britain: neophyte (C Europe); scattered in the S, rare in Scotland.
Hortal; 1/1; very rare.
1st Dumbreck, 1986, CFOG.

This sole record is from soil on waste ground, at Springkell Avenue, Dumbreck (NS55L).

Mentha × *gracilis* Sole
(*M. arvensis* × *M. spicata*)
Bushy Mint

Britain: spontaneous hybrid (native × alien); scattered in C Scotland.
Hortal; 16/18; scattered.
1st Clarkston–Giffnock area, 1933, RM.

Bushy Mint was collected from Giffnock (1937, GLAM), although this is predated by a less convincing specimen from Newlands in 1921 (GL). There are no other records until PM recorded it from a burn side at Lochwinnoch (NS3558, 1971). Today it is quite widespread, mainly in rural fringes, and may well have been overlooked; several records are from watercourses, notably along the White Cart Water south of Glasgow. A CFOG specimen from Busby (1984, NS55T – but from the VC77 side) was determined as this species by R Harley who added that it was 'the large form which approximates to *M.* × *smithiana*. The distinction is unclear'. Others records refer to waste ground, including the two western outliers at Overton (NS2674, 2002) and Larkfield (NS2476, 2010, PRG).

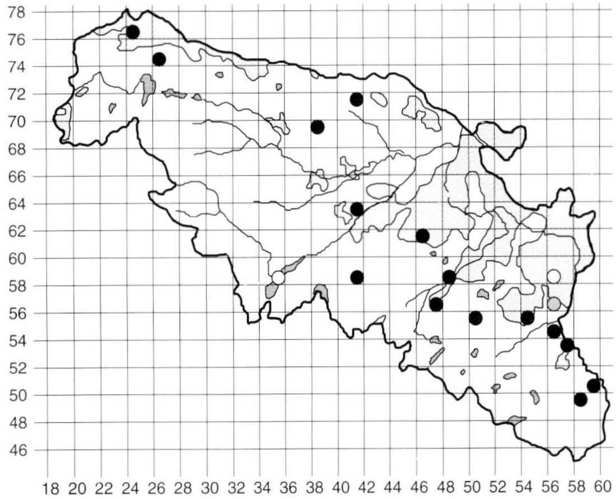

Mentha × gracilis

Mentha aquatica L.
Water Mint

Britain: common throughout except in N Scotland.
Native; 168/218; common in wet places.
1st Gourock, 1860, GL.

There are plenty of old literature and herbarium records for Water Mint, which was described by Hennedy (1891) as frequent. Today it is the commonest mint encountered although some records may refer to non-flowering Whorled Mint (or other hybrids). Its pattern shows a strong

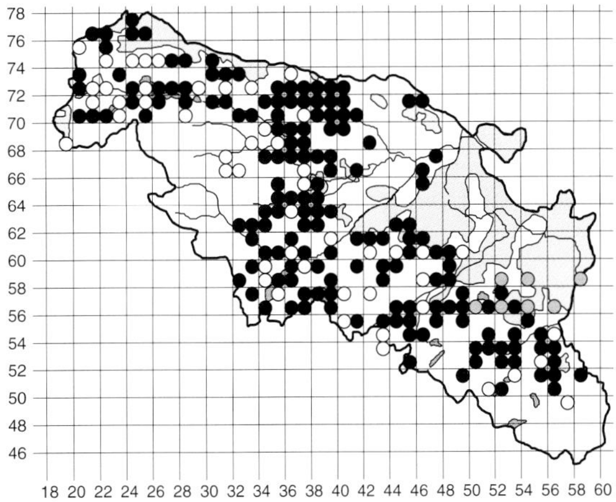

Mentha aquatica

lowland rural distribution; it becomes rare in urban or heavily improved farmland and it is also absent from the peaty uplands. Habitats include open waterbodies, watercourse and ditch margins, fens and marshes.

Mentha × *piperita* L.
(*M. aquatica* × *M. spicata*)
Peppermint

Britain: spontaneous hybrid and garden escape; scattered throughout.
Hortal; 15/20; occasional.
1st Elderslie and Paisley Canal, 1883, GL.

Other old records for Peppermint include the road to Waas (Walls) Hill (1895, PSY), the Earn Water, Newton Mearns (1909, GL) and a wet meadow, Clarkston (1929, RM); it was only noted a couple of times by BWR. The modern records are grouped about the central area, with western outliers at Gryfe No. 1 Reservoir (NS2771, 2001) and Larkfield (NS2376, 2011). It is mainly found along burn sides near habitation, plus at the margins of open waterbodies and tips such as Shillford (NS4455, 1999).

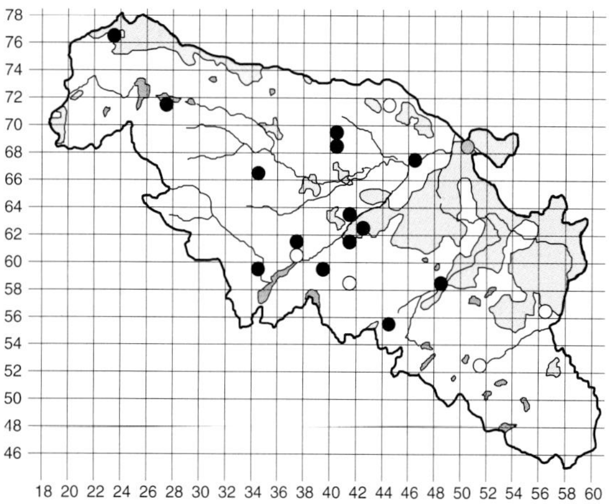

Mentha × *piperita*

Mentha spicata L.
Spear Mint

Britain: neophyte (garden origin?); common in the S, frequent in lowland Scotland.
Hortal; 19/22; occasional.
1st Polnoon, 1957, RM.

Hennedy (1891) referred to Spear Mint being cultivated near dwellings but there are no named old localities in the VC. BWR recorded it from Trumpethill (NS2276, 1977) and Ardoch Burn (NS5849, 1982), but all other records are modern. Localities include waste ground, roadsides and loch margins as at Black Loch (NS4951, 1997). The pattern shows a south-eastern, rural distribution.

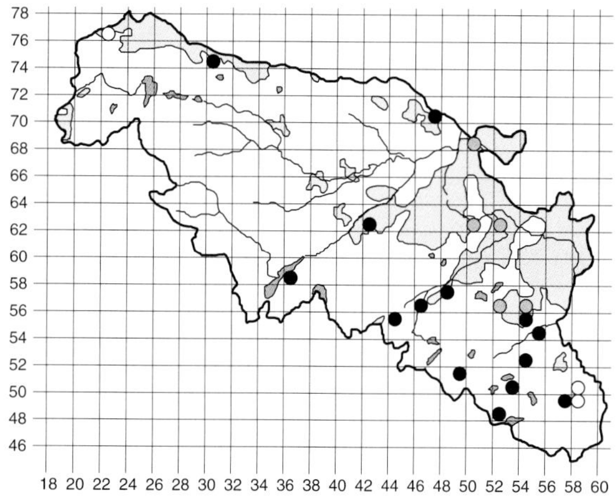

Mentha spicata

Mentha × *villosonervata* Opiz
(*M. spicata* × *M. longifolia* (L.) Huds.)
Sharp-toothed Mint

Britain: neophyte (garden origin); scattered in lowlands.
Hortal; 2/2; very rare.
1st Yoker, 1985, GL.

The original record was determined by RH Harley; there is a tentative record from a roadside at Neilston (NS4757, 2004).

Mentha × *villosa* Huds.
(*M. spicata* × *M. suaveolens* Ehrh.)
Apple-mint

Britain: neophyte (garden origin); frequent and scattered in C Scotland.
Hortal; 36/40; frequent.
1st Loganswell, 1936, RM.

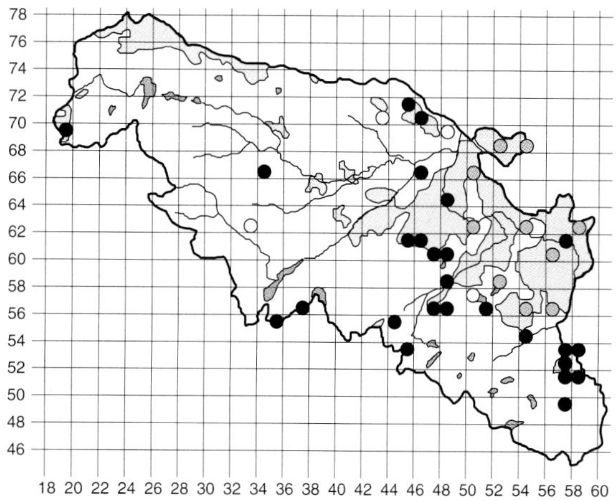

Mentha × *villosa*

Lee (1933) may well have referred to this species from Langbank (as *M. alopecuroides*, 'Horse-mint'). More recently BWR recorded it from Kaim Dam (NS3362, 1971), Bishopton Station (NS4370, 1975), Northbar (NS4869, 1977) and Springhill (NS3264, 1987), and AMcGS knew it from Parklea (NS3574, 1984). Today this is the commonest of the hairy mints, although some records may include other cultivars. Like Spear Mint, it is frequent in the east in urban-fringe areas, generally found on waste ground, but can persist in long, neglected grasslands, and can also occur at watercourse margins. Several, if not many, records may refer to Garden Apple-mint (var. *alopecuroides* (Hull) Briq.), which has been noted from Eaglesham Common (NS5751, 1997), Kirkland (NS5752, 2011) and roadside, near Beith (NS3555, 2011).

Mentha × rotundifolia (L.) Huds.
(*M. longifolia × M. suaveolens*)
False Apple-mint

Britain: neophyte (Europe); widely scattered (uncertain).
Hortal; 9/10; rare.
1st Clarkston–Giffnock area, 1933, RM.

False Apple-mint is one of several critical hairy mint taxa but not all the modern field records have been expertly verified. It was collected from Darnley in 1984 (NS55I, GL) during surveying for CFOG and modern reports are from Nether Johnstone (NS4163, 1998), Paisley (NS4864, 2000), Pilmuir Quarry (NS5154, 2000), a roadside, Roundtree (NS3562, 2001), Millarston (NS4663, 2001), Smiddyhill (NS4556, 2003), a roadside, Black Hill (NS4851, 2011) and waste ground, Port Glasgow (NS3174, 2012).

Mentha requienii Benth.
Corsican Mint

Britain: neophyte (Corsica and Sardinia); rare and scattered in the S.
Hortal; 1/1; probably extinct.
1st Barrhead Station (NS45Z), 1985, CFOG.

This rare hortal has not been reported for more than 25 years.

Salvia verbenaca L.
Wild Clary

Britain: frequent in the SE, very rare (mainly alien) in Scotland.
Former hortal; 0/1; extinct.
1st Clyde floating dock, Greenock, 1881, NHSG.

Salvia verticillata L.
Whorled Clary

Britain: neophyte (S Europe) scattered, mainly in the S.
Former hortal; 0/1; extinct.
1st Walkinshaw oil works, 1901, PSY.

Sideritis romana L.

Britain: neophyte (S Europe); rare casual.
Former accidental; 0/1; extinct.
1st Paisley, 1913, GL.

There are two specimens in GL (named as '*S. purpurea*'), both collected by DF in 1913. One was from a 'rubbish heap', and the other has the location noted as Stonefield, Paisley.

Dracocephalum parviflorum Nutt.
American Dragon-head

Britain: neophyte (N America); rare casual.
Former accidental; 0/1; extinct casual.
1st Giffnock coup, 1930, RG.

PHYRMACEAE
(SCROPHULARIACEAE *pro parte*)

Mimulus moschatus Douglas ex Lindl.
Musk

Britain: neophyte (W N America); scattered.
Hortal; 2/4; very rare.
1st Pokeston, Mearns, 1890, AANS.

Wood (1893) thought that Musk might 'overrun' the 'small syke' at the original location, but it does not appear to have become established there or elsewhere. There are only two modern records: a damp trackside at Muirshiel (NS3163, 2000) and disused railway, Kilbarchan (NS4265, 2003, IPG). In 1964 it was reported from Dykefoot, near Gryfe Reservoir (NS2971, JDM).

Mimulus **spp.**
Monkeyflowers

Britain: neophytes (N and S America); common except in N Scotland and E England.
Hortals; 111/142; common.
1st Bank of Gryfe, near Houston Station, 1879, PSY.

There appears to have been a rapid spread of Monkeyflowers in the late 19th century: records include a ditch, Inverkip Road (1888, GRK), River Calder, Lochwinnoch (1890, AANS), the Black Cart Water, Howwood (1891, AANS), Ardgowan Estate (1891, AANS) and Risk Burn, Castle Semple (1891, RH). Wood (1893) reported it from a 'marshy piece of ground' on the Aurs Burn below Glanderston, which was in season 'a golden mass of Wild Musk'. Monkeyflower was also well recorded during the 1960s–80s period (BWR). Today it is common, showing a rural pattern extending into urban areas, although perhaps surprisingly it does not appear to have penetrated the uplands along the numerous feeder burns. Localities are nearly always associated with watercourses or margins of open waterbodies, usually reservoirs.

It is thought that most records refer to *M. guttatus* DC. but a few may be referable to 'unblotched' plants of Hybrid Monkeyflower (*M.* × *robertsii* Silverside). The latter (as '*M. luteus*') was first recorded in 1960 from Brockburn, near Commore Dam (1960, GL) and earlier field notes (1963, JDM) include Dykefoot, near Gryfe Reservoir (NS2971) and Finlaystone (NS3673). Today the hybrid has been recorded from five sites: the Gryfe Water at Auchenfoyle (NS3171, 2001) and Cauldside (NS3270, 2005, confirmed AJS), Glenbrae (NS2872, 1999, confirmed AJS), Spateston (NS4161, 1993) and Dodside (NS5053, 1998).

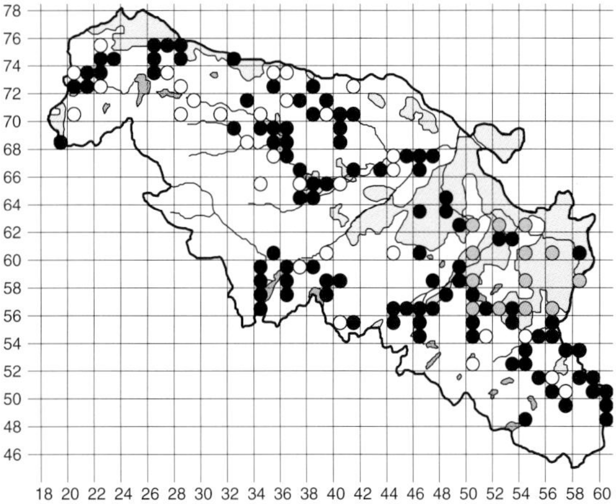

Mimulus spp.

OROBANCHACEAE
(SCROPHULARIACEAE *pro parte*)

Melampyrum pratense L.
Common Cow-wheat

Britain: common, mainly in the N and W.
Native; 4/13; rare.
1st Gourock, 1842, GL.

Other old records for Common Cow-wheat include Lochwinnoch (1845, NSA), near Port Glasgow (1866, GL), Kelly Glen (1887, NHSG) and Neilston Pad (1890, NHSG); the latter was considered the only station in East Renfrewshire by Wood (1893). It was known from Bardrain Glen in 1930 (Stewart 1930), and was recorded on a handful of occasions by BWR: Blacketty Water (NS3368, 1969), west of Bennan Hill (NS5150, 1982), River Calder (NS3459, 1964), Mill Burn (NS3366, 1965) and above Inverkip (NS2171, 1985). Common Cow-wheat may be overlooked but appears to have declined, presumably due to agricultural improvement and heavy grazing, with only four modern records: Devol Glen (NS3174, 1996), rocky edges of Dod Hill (NS4953, 1998), grassy ridges at Kirkton (NS4856, 1999) and Neilston Pad (NS4755, 1996, DM).

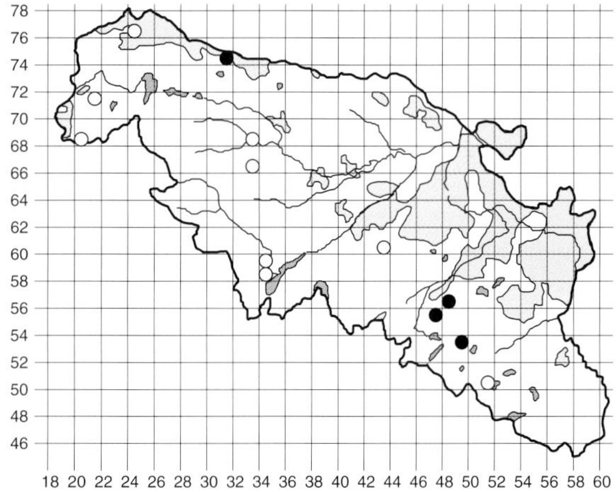

Melampyrum pratense

Euphrasia officinalis agg.
Eyebrights

Britain: common except in C and E England.
Native; 191/238; common.
1st Paisley, 1845, NSA.

Eyebrights are commonly found in unimproved pasture, both lowland and upland, but also some urban grasslands, notably at disused railways and spoil heaps. They form a critical group of microspecies and hybrids and accurate recording relies on expert determinations. The modern recording has been chiefly at the aggregate level, with some tentative field records for the more distinctive members. Dr Alan Silverside (AJS) has examined more than 30 specimens collected from the VC, which, along with earlier determinations for CFOG, has provided the foundation for establishing the taxa found in the

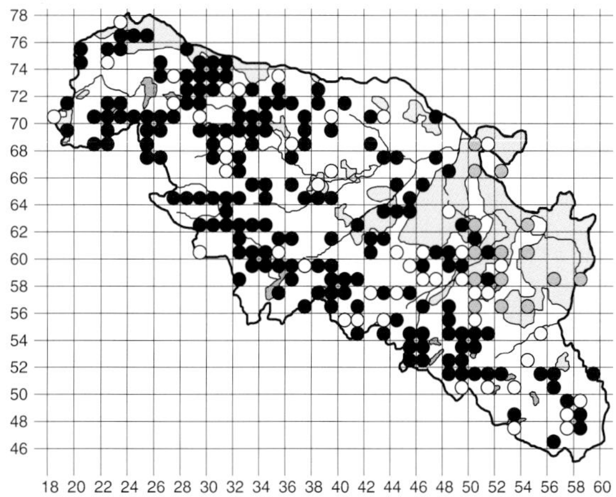

Euphrasia officinalis agg.

area. The map shows the distribution of records at the aggregate level.

Euphrasia arctica Lange ex Rostrup ssp. *borealis* (F. Towns.) Yeo
Arctic Eyebright

Native; 5/6; probably frequent and widespread.
1st Nitshill–Darnley area, 1929, RM.

This taxon (as '*E. brevipila*') was recorded by RM but there is no specimen to confirm the identity. It is thought to be one of the commoner taxa, although several samples have proved to be hybrids. There have been five determinations from the modern survey: Whinny Hill (NS3770, 1997), Hartfield Moss (NS4358, 1999), Knowes (NS3756, 2003), Waulkmill Reservoir (NS5257, 1990) and Harelaw Mire (NS3863, 1996); the specimen from the latter is considered somewhat untypical and worth further investigation.

Euphrasia nemorosa (Pers.) Wallr.
Common Eyebright

Native; 0/1; very rare, overlooked.
1st Loganswell, 1928, GLAM.

This collection from pastures (*c.* NS5152) was originally determined by Pugsley and later confirmed by AJS, but is the only record at the species level. However, Common Eyebright is a frequent parent in hybrid collections.

Euphrasia confusa Pugsley
Confused Eyebright

Native; 1/2; very rare, overlooked
1st Nitshill–Darnley area, 1929, RM.

There is no voucher specimen for the original record. This eyebright was collected from species-rich grassland at Cauldside, Strathgryfe (NS3270, 2005), but although it commonly occurs as hybrid material this is the only record at the species level.

Euphrasia micrantha Rchb.
Slender Eyebright

Native; 3/4; very rare, possibly occasional in heathy places.
1st Moorland, Whitelees (NS2073), 1976, BWR.

There is no voucher specimen for the original record and although there are five modern field records, the only two samples that have been determined were considered hybrids with Scottish Eyebright (*E.* × *electa*), resulting in no confirmed records for the species. Unverified field records, of strongly coloured plants on heathy embankments, are from a trackside at Blood Moss (NS2168, 2001), Jock's Craig (NS3269, 2002) and Daff Reservoir (NS2271, 2005). Two hybrids with *E. arctica* have also been collected (see Eyebright hybrids below).

Euphrasia scottica Wettst.
Scottish Eyebright

Native; 29/29; frequent in the western hills.
1st Raith Burn, 1993, AMcGS.

This slender eyebright has a strong western distribution pattern and it is usually found along open flushes by watercourses, usually at higher altitudes and marking out some mineral influence. Four specimens, including the original, have been confirmed by AJS; the other three are from small flushes at Glenwood Hill (NS3163, 1995), Earn Hill (NS2375, 1998) and Crawhin Reservoir (NS2470, 2004). The map shows all field records but given the species' ability to hybridize it should be viewed with caution.

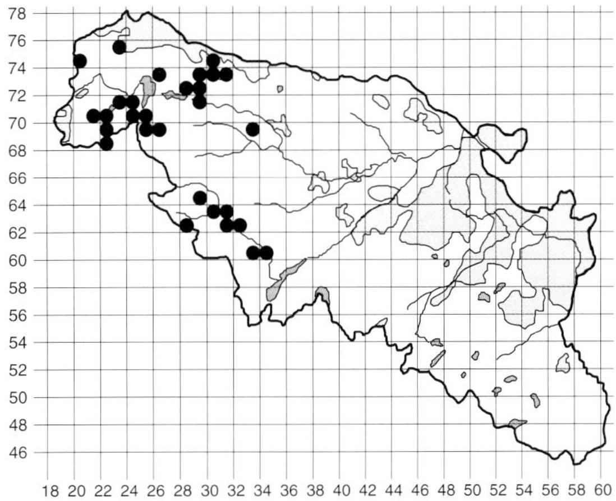

Euphrasia scottica

Eyebright hybrids

Euphrasia arctica ssp. *borealis* × *E. nemorosa*

Native; 1/2; probably frequent in lowland urban areas.
1st Giffnock quarries, 1931, GLAM.

This RM specimen was determined by Pugsley and confirmed by AJS. Its original location fits with its modern-day distribution along disused railways and mine wastes about Glasgow (CFOG). There is only one determined modern record, from Picketlaw (NS5651, 1998), but this may reflect a bias of sampling towards less vigorous plants of species-rich grasslands.

Euphrasia arctica ssp. *borealis* × *E. confusa*

Native; 7/7; overlooked, perhaps frequent in old grasslands.
1st Knocknairshill (NS3074), 1998.

This hybrid appears to be frequent at species-rich grassland sites. There are six other determined records: Earn Hill (NS2376, 1998), Gotter Burn (NS3264, 1998), Dyke Hill (NS4856, 2002), Burneven Hill (NS2275, 2003), Turnave Hill (NS3260, 2003) and Whinnerston (NS3864, 2003).

Euphrasia confusa × *E. nemorosa*

Native; 5/5; possibly frequent in old grasslands.
1st Levern Water, Harelaw (NS4654), 1999.

This hybrid also appears to be frequent at species-rich grassland sites. There are four other determined specimens: Kelly Reservoir (NS2268, 2001), Muirhead (NS4855, 1999), Snypes Dam (NS4855, 2002) and Turnave Hill (NS3260, 2003).

Euphrasia × *electa* F. Towns.
(*E. micrantha* × *E. scottica*)

Native; 2/2; very rare, overlooked
1st Lyles Hill (NS3064), 1998.

This hybrid was identified from material collected (as Slender Eyebright) along the track near Clyde Muirshiel (NS3064); the same redetermination occurred with a specimen collected from a heathy embankment at Snypes Dam (NS4855, 1999).

Euphrasia arctica ssp. *borealis* × *E. micrantha*

Native; 2/2; very rare, overlooked.
1st Duchal Bridge (NS3368), 1999.

This hybrid was also identified from a collection made at Jock's Craig (NS3269, 2002).

Euphrasia arctica ssp. *borealis* × *E. nemorosa* × *E. confusa*

Native; 2/2; very rare, overlooked.
1st Neglected lawn, Cowdon Hall (NS4757), 1999.

This triple combination was also determined for specimens collected at Lochend (NS3764, 2003).

Euphrasia nemorosa × *E. confusa* × *E. scottica*

Native; 1/1; very rare, overlooked.
1st Knowes (NS3756), 2003.

This triple hybrid was collected from species-rich grassland.

Odontites vernus (Bellardi) Dumort.
Red Bartsia

Britain: common except in N Scotland.
Native; 111/133; common in lowlands.
1st Paisley, 1845, NSA.

Red Bartsia appears to have remained common since the 19th century (Hennedy 1891). Today it is quite widespread, showing a lowland pattern occurring in rural and urban locations, but absent from the uplands. It is nearly always associated with waste ground and is tolerant of trampling or soil compaction, occurring also on waysides, disused railways, reservoir shores and about farms.

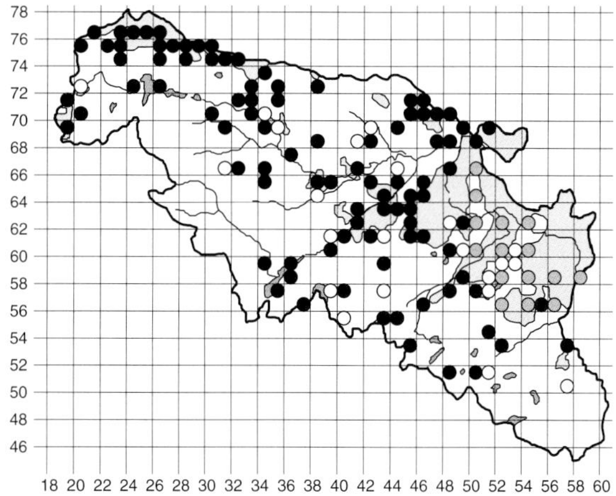

Odontites vernus

Parentucellia viscosa (L.) Caruel
Yellow Bartsia

Britain: local in the SW, scattered alien elsewhere.
Native or accidental; 3/5; very rare.
1st Greenock Battery, 1821, FS.

Yellow Bartsia was collected from Gourock in 1858 (GGO), from where there are also three other specimens (all 1879, GL) and in 1880 it was recorded from Ashton, Gourock (NHSG), which likely refers to the same place. A single plant was found at South Pollok (NS5360, 1994, CFOG), a site soon after destroyed. In 1999 Yellow Bartsia was found in some numbers on the exposed, open gravels at Balgray Reservoir when the water level was low (NS5057 and NS5157); the level has been kept high since 2000.

Rhinanthus minor L.
Yellow-rattle

Britain: common throughout.
Native; 120/161; common.
1st Gourock and near Hurlet, 1842, GLAM.

Yellow-rattle is a hemiparasite found in unimproved grasslands, generally avoiding more acidic types, but capable of tolerating some impeded drainage. It is widespread, found throughout the lowlands but becomes rare in urban or intensively farmed areas; habitats are grasslands including roadsides, hedge banks and riverbanks. Larger plants found in more marshy conditions may be referable to ssp. *stenophyllus* (Schur) O. Schwarz, which was first recorded at Caplaw Dam (NS4358, 1962, BWR), and has been recorded at nine modern sites including Harelaw Mire (NS3863, 1996), Lunderston Bay (NS2074, 1996), Corsliehill (NS3969, 1996) and Mid-Glen (NS3870, 1997).

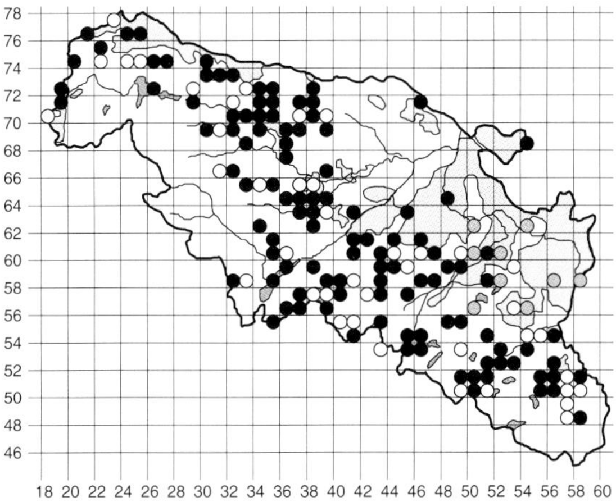

Rhinanthus minor

Pedicularis palustris L.
Marsh Lousewort

Britain: common in the N and W.
Native; 90/112; frequent.
1st Paisley, 1845, NSA.

Marsh Lousewort was considered a frequent plant of marshes and wet meadows in 1891 (RH) and there are quite a few old named stations including Wemyss Bay (1842, GL), near Hurlet (1842, GLAM), Harelaw Dam (1858, PSY), Gourock (1859, GLAM) and Castle Semple Loch (1887, NHSG). Today it is frequently found in mires or fens and by some burn sides in rural areas, usually where nutrients are low, and usually marking out flush lines. It can be found above 300m but it is absent from acidic bogs or moorland, apart from where mineral seepage occurs.

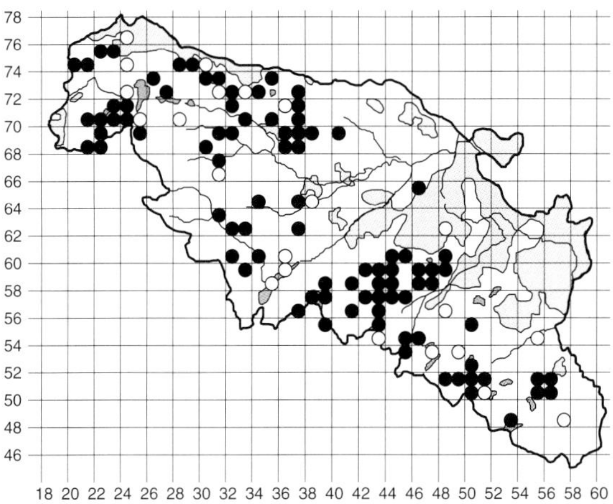

Pedicularis palustris

Pedicularis sylvatica L.
Lousewort

Britain: common in the N and W.
Native; 114/153; common in upland fringes.
1st Meikleriggs Dam, 1858, PSY.

Lousewort is quite often found on less-improved acidic grasslands, heaths or mosaics, on hilly ground fringing the Renfrewshire Heights and Eaglesham Moors, and also isolated hills such as on the Gleniffer Braes; it is now virtually absent from urban areas. It is likely that this species will have declined with the increasing agricultural improvement of acidic grasslands. A white-flowered form is occasionally seen, as at Witch Burn (NS4658, 1999).

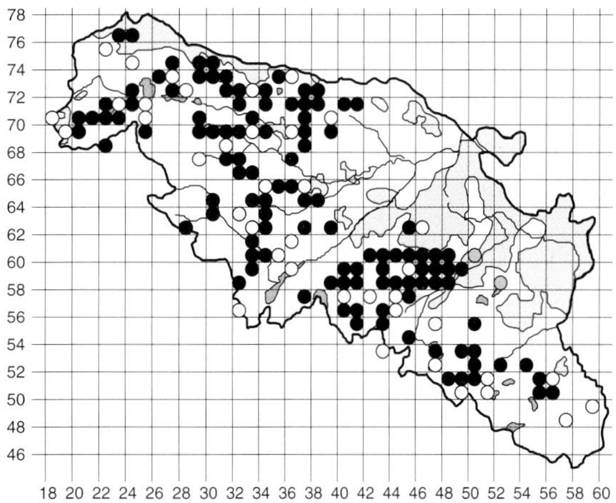

Pedicularis sylvatica

Lathraea squamaria L.
Toothwort

Britain: frequent in the S, absent from N Scotland.
Native; 4/4; rare.
1st Cathcart, 1821, FS.

Toothwort can still be found by the White Cart Water in Linn Park (NS5859), where PM noted it in 1980, perhaps in the same place as the first record. There are two other White Cart riverbank localities: Langside (NS5761, 2000, CB), where it was first recorded in 1889 (Hennedy 1891), and Pollok Park (NS5561), where it was first found in 1974 (Mackechnie 1975). All these populations are still extant and have been counted recently during Glasgow LBAP monitoring, with a maximum count of more than 3000 flower spikes in one year. An isolated record is from the ranger service at Finlaystone Estate (NS3673, 1999); Toothwort was first recorded from there (by T Anderson) in 1907, from the east gate (GN).

Lathraea clandestina L.
Purple Toothwort

Britain: neophyte (S Europe); scattered in the S, scarce in C Scotland.
Hortal; 2/2; very rare.
1st Gryfe Water, Kilmacolm, 1997.

This colourful parasitic plant grows under willows at a couple of places along the banks of the Gryfe Water to the south of Kilmacolm (NS3569, 1997; NS3568, 2003).

Orobanche alba Stephan ex Willd.
Thyme Broomrape

Britain: local in the NW, very rare elsewhere.
Former accidental; 0/1; extinct.
1st Newlands garden (NS5760), PM, 1976.

This surprising find grew in a potted plant of *Thymus* × *citriodorus* (Pers.) Schreber, bought in 1963, but although monitored it never reappeared (Macpherson and Macpherson 1978).

Orobanche minor Sm.
Common Broomrape

Britain: frequent in the S, rare alien in Scotland.
Accidental; 1/2; very rare.
1st Sidings, Crossmyloof Station, 1909, GL.

The early record was overlooked in CFOG but this appears to be the only Glasgow-area record for Common Broomrape apart from its occurrence on the north bank of the Clyde at Bowling from 1964 to the 1980s (VC99, JP). However, in 2001 during a survey of waste ground at Clydeport, Port Glasgow (NS3174), Alan Wood found a large well-established colony; the site has been subsequently developed.

LENTIBULARIACEAE

Pinguicula vulgaris L.
Common Butterwort

Britain: common in the N and W.
Native; 40/58; frequent in boggy places.
1st Lochwinnoch and Paisley, 1845, NSA.

Common Butterwort has a strongly rural distribution with many of the records from above 200m. It is found in boggy places, usually where the peat is exposed and flushed as along the embankments of upland burns. Most of the records are in the Renfrewshire Heights, with a few in the south-east uplands, but it can occur closer to urban-fringe areas as at Killoch Hill (NS4759, 1999), Knocknairshill (NS3074, 2001) and Earns Hill (NS2275, 1998).

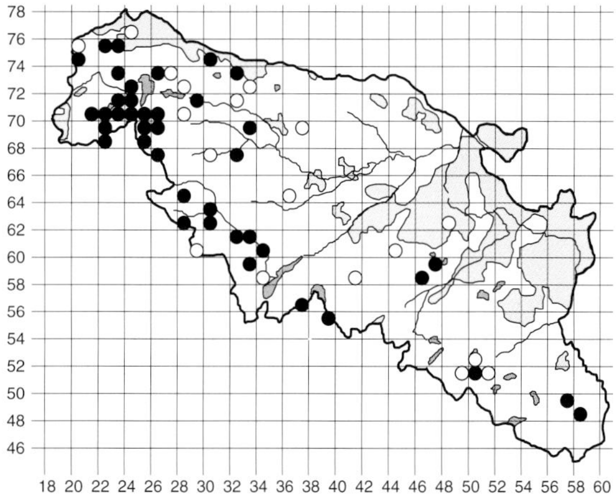

Pinguicula vulgaris

Utricularia vulgaris group
Greater Bladderwort/Bladderwort

Britain: frequent in NW Scotland, scattered elsewhere (declining).
Native; 1/3; very rare.
1st Curling Pond, Kilmacolm, 1888, RH.

The original record stated 'Greater Bladderwort' (*U. vulgaris* L.) but there is no specimen to confirm the species. 'Greater Bladderwort' was also recorded from a 'pool in a marsh at Knockmountain' in 1898 (JMY). The SNH Loch Survey (1996) recorded '*U. vulgaris* group' from the original location, Glen Moss (NS3669); however, a file note from ERTC (1958) states '*U. neglecta*' at Glen Moss but again there is no supporting specimen. A specimen collected from Glen Moss by RM (1969, GLAM) has no flowers to enable reliable identification, but the quadrifid hairs point to *U. australis* R. Br. A field record from Ballageich Hill (NS5350, 1959, BWR) for '*U. vulgaris* sens. str.' must be viewed with doubt, as there is no other comment or information. It is therefore unclear whether just one or both species have been seen in the VC.

Utricularia minor L.
Lesser Bladderwort

Britain: frequent in the N and W uplands, rare elsewhere.
Native; 3/4; very rare.
1st Barmufflock Dam, 1953, ERTC.

The original file note on this find stated 'in some quantity in a peat pool and flowered in 1955 and 1957'; Lesser Bladderwort was also seen there in 1961 (NS3664, BWR) but has not been reported since. It was seen at Glen Moss in 1969 (NS3669, BWR) and 1996 (SNH-LS). Nearby, a small population was also found, in flower, in a small pool to the edge of Elphinstone Wood (NS3769, 1997). A further population occurs in a mire to the south-east of Corsliehill Wood (NS3869, 1998).

VERBENACEAE

Verbena bonariensis L.
Argentinian Vervain

Britain: neophyte (S America); rare in the S.
Hortal; 1/1; very rare escape.
1st Path-edge wall, Jordanhill (NS5468), 2004, AR.

This is the sole record for this garden escape.

AQUIFOLIACEAE

Ilex aquifolium L.
Holly

Britain: common except in N Scotland.
Native; 160/176; common.
1st Renfrewshire, 1869, PPI.

Holly is a common sight in lowland areas, occurring in parks, at some waste-ground sites or neglected places and in plantation or policy woodlands, and also in more semi-natural settings and along hedgerows. Some records may refer to cultivars or hybrids, such as 'Ferox' from the hotel grounds near Erskine Harbour (NS4671, 1997).

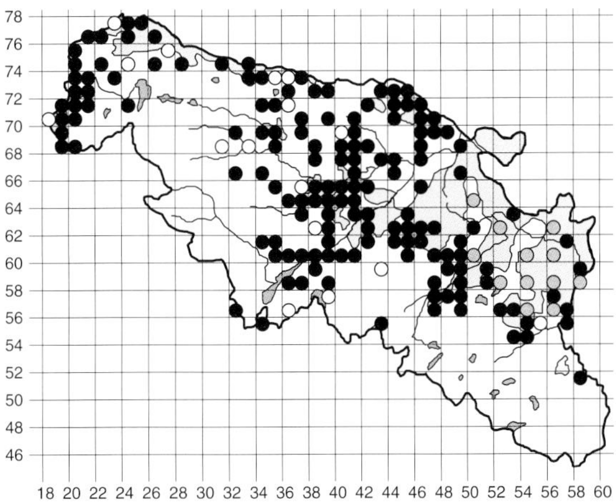

Ilex aquifolium

Ilex × *altaclerensis* (Loudon) Dallim.
(*I. aquifolium* × *I. perado* Aiton)
Highclere Holly

Britain: neophyte (garden origin); local, mainly in the S and W.
Hortal; 18/19; scarce.
1st Capelrig House (NS55N), Crookfur, 1986, AMcGS.

All of the other records are from the modern period and the distribution pattern reflects that of estate woodlands. Highclere Holly has been widely planted but it appears to be spreading locally, e.g. Glen Park (NS4760, 1995), Gleddoch Estate (NS3872, 1998), Ferryhill Plantation (NS3972, 1998) and Ardgowan Estate (NS2073, 1998). The rural outlier at Kaim Burn (NS3461, 1998) was from near a house.

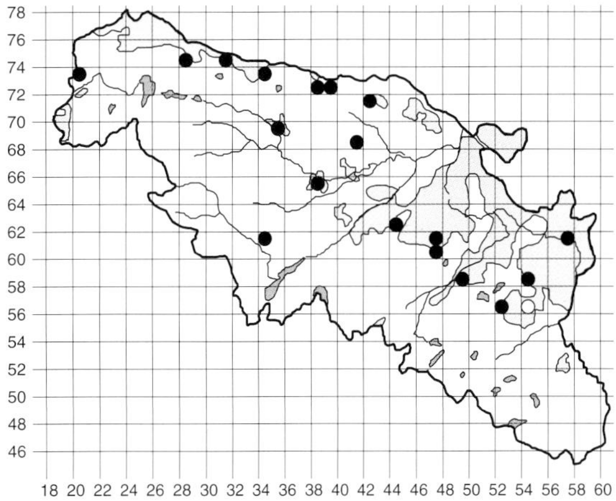

Ilex × *altaclerensis*

CAMPANULACEAE

Campanula lactiflora M. Bieb.
Milky Bellflower

Britain: neophyte (SW Asia); scattered in E Scotland.
Hortal; 1/1; very rare.
1st Quarrelton, 1993.

A well-established clump of Milky Bellflower occurs by a path in mature woodland (NS4262).

Campanula persicifolia L.
Peach-leaved Bellflower

Britain: neophyte (Europe); scattered in lowlands.
Hortal; 0/4; probably extinct.
1st Cloch and Ardgowan, 1887, GL.

RM knew Peach-leaved Bellflower from the bottom of a hedge at Deaconsbank Golf Course (1926), and more recently it was recorded from a roadside hedge near Skiff Wood (NS4060, 1983, PM).

Campanula poscharskyana Degen
Trailing Bellflower

Britain: neophyte (W Balkans); frequent in the S and W, rare in Scotland.
Hortal; 3/3; very rare.
1st Old stonework, Pollok House, 1996.

Other records for Trailing Bellflower as an outcast or escape are from waste ground, Greenock (NS2776, 2000) and a roadside and walls, Jordanhill (NS5468, AR, 2002).

Campanula latifolia L.
Giant Bellflower

Britain: frequent except in the extreme S and N.
Native; 17/26; scarce.
1st Lochwinnoch, 1845, NSA.

This tall, attractive bellflower was considered a frequent plant of moist woods by Hennedy (1891) and old localities include Wemyss Bay (1848, GL), Loch Libo (1887, NHSG), Castle Semple Loch (1887, NHSG), by the Black Cart Water near Howwood (1893, AANS) and Linn Park (1893, AANS). It was last seen at Loch Libo in 1984 (BWR), but can still be found at Castle Semple Loch (NS3658, 1992, CMRP) and Linn Park (NS5859, 2004). In addition to occurring on the banks of the White Cart Water, Giant Bellflower is also known from by the Gryfe Water about Bridge of Weir (NS3766 and NS4064, 2002), Dargavel Burn (NS4270, 2005) and by the Black Cart Water at Johnstone (NS4263, 2001). Other records are from Rouken Glen (NS5458, 2005) and Gleddoch Estate (NS3872, 1998), but the latter may well be a recent introduction.

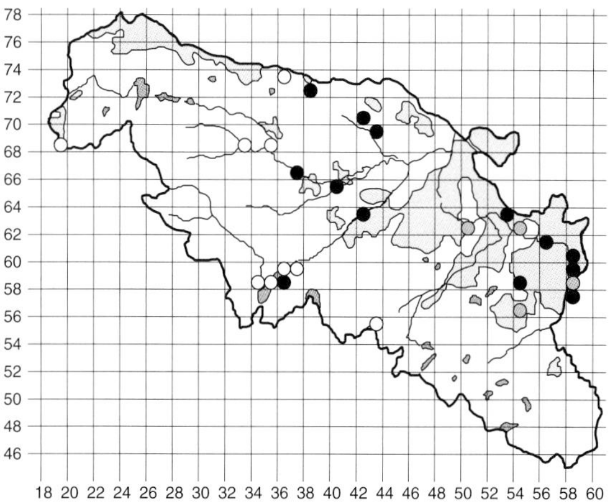

Campanula latifolia

Campanula trachelium L.
Nettle-leaved Bellflower

Britain: frequent in the S, rare alien in Scotland.
Hortal; 2/4; very rare.
1st Colinslee (Paisley), 1904, PSY.

Nettle-leaved Bellflower was also reported from Georgetown in 1932 (TPNS 1942). In 1996 PM found it at Cathcart (NS5860) and in 2000 it was found on waste ground in Greenock (NS2776).

Campanula rapunculoides L.
Creeping Bellflower

Britain: neophyte (Europe); scattered throughout, mainly in the S and E.
Hortal; 1/4; very rare.
1st Renfrew Moor, 1865, MY.

This bellflower was also reported from near Langside (1872, TGSFN) but was excluded from the Renfrewshire list (TPNS 1915). A more recent record is from the side of Nithsdale Road, Pollokshaws (NS5663, 1982, CFOG). The sole modern record of Creeping Bellflower is of a large patch, presumably from dumped material, which had established on waste ground at Clydeport (NS3174, 2001), a site now developed.

Campanula rotundifolia L.
Harebell

Britain: common except in the extreme NW and SW.
Native; 176/243; common.
1st Renfrew and Paisley, 1845, NSA.

Harebell was a very common plant in the 19th century and is still common today in suitable habitats, but there are quite a few gaps in the distribution map, reflecting a decline due to agricultural improvement. It can still be found at dry unimproved grassland sites, generally where not very acidic, and usually on embankments such as hillsides and along watercourse valleys. Often it persists about ridges and old boundary walls by improved field margins or roadside banks.

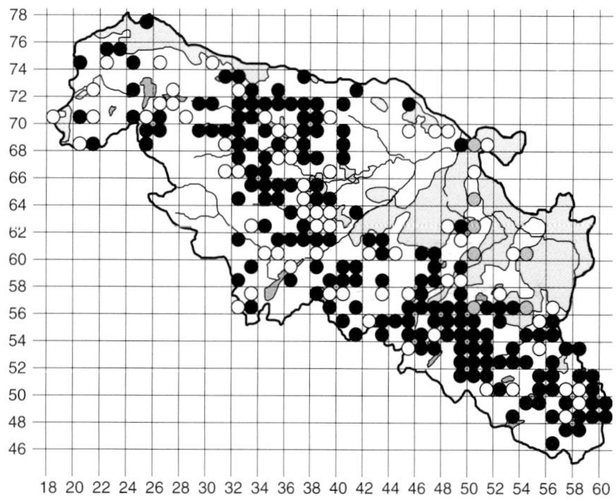

Campanula rotundifolia

Wahlenbergia hederacea (L.) Rchb.
Ivy-leaved Bellflower

Britain: frequent in the SW, absent from most of the E and N.
Former native or accidental; 1/3; extinct as a native.
1st Neighbourhood of Greenock and Ardgowan, 1821, FS.

Hooker (1821) wrote that Dr Brown, the discoverer,

found Ivy-leaved Bellflower to the west of Cloch lighthouse and that it was present on the lawns at Ardgowan. A number of old records from the west coast refer to it and there are no less than 11 specimens in GL and GLAM combined. Hennedy (1891) mentioned Inverkip and Cloch, although most specimens are from either the latter (mainly about the lighthouse) or Ardgowan; a specimen label (1842, GL) indicates that it was considered abundant. The last records are from Cloch (1904, GL) and Gourock (1911, GL), but during modern searches this small, south-western species has not been rediscovered. It was recorded from a lawn in Dumbreck at the edge of the VC in 1998 (NS5663, CFOG).

Jasione montana L.
Sheep's-bit

Britain: frequent in the W, rare and declining elsewhere.
Native; 1/6; very rare, endangered.
1st Gourock, 1837, GL.

Considered frequent in the 19th century on dry sandy soils, other old records for Sheep's-bit include Inverkip Road (1886, GRK), Paisley and Lochwinnoch (1845 NSA) and a bank of Newton Wood (1865, MY). There are few other named localities and BWR reported it just once from south of Leitchland (NS2274) in 1974. Today the sole record is from the disused railway (cycleway) at Craigends of Dennistoun (NS3667, 2001), where it grows on the rocky embankment and a large population occurs on nearby trackside gravels; a 1957 specimen in PSY (collected by Magnus Park) from a 'railway bank near the trout farm' (with grid reference NS362678) indicates this population has been there for more than 50 years.

Lobelia erinus L.
Garden Lobelia

Britain: neophyte (S Africa); frequent in the S.
Hortal; 1/2; very rare.
1st Disused railway, Kilmacolm, 1984, BWR.

Garden Lobelia probably occurs as a sporadic weed in urban areas but the only modern record is from a disused railway near Linwood (NS4264, 2003, IPG).

MENYANTHACEAE

Menyanthes trifoliata L.
Bogbean

Britain: common in the N and W.
Native; 92/113; frequent.
1st Erskine and Paisley, 1845, NSA.

Bogbean is commonly found in marginal fen or swamp vegetation of open waterbodies where it grows with sedges, notably Bottle Sedge. It also occurs in upland mires and in the wetter parts of rushy pastures.

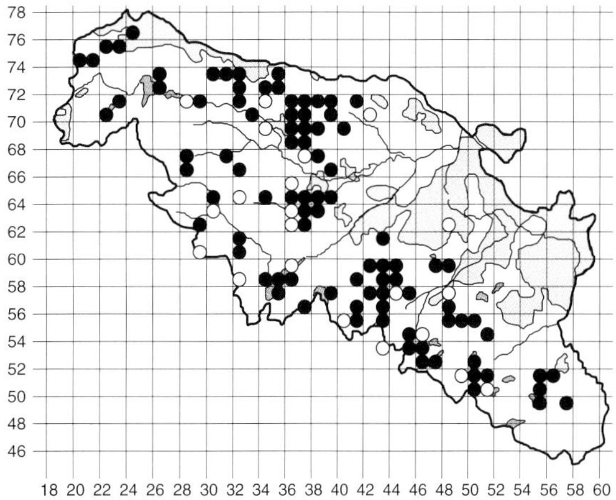

Menyanthes trifoliata

ASTERACEAE

Echinops sphaerocephalus L.
Glandular Globe-thistle

Britain: neophyte (S Europe); scattered in the S.
Hortal; 0/1; extinct.
1st Trumpethill, 1977, BWR.

The sole record is for a 'garden origin' plant recorded by the coast west of Gourock (NS2276).

Echinops exaltatus Schrad.
Globe-thistle

Britain: neophyte (SE Europe); scattered in the S.
Hortal; 1/2; very rare escape or outcast.
1st White Cart (NS5661), 1978, CFOG.

There is one other more recent CFOG record for this globe-thistle, from Cowglen Golf Course (NS5461, 1989).

Arctium nemorosum Lej.
Arctium minus (Hill) Bernh. *pro parte*
Wood Burdock

Britain: distribution uncertain, but more common in N Britain.
Native; 74/106; frequent.
1st Paisley, 1845, NSA.

It is thought that most if not all records refer to Wood Burdock (the presumed common taxa in the north), although most recording has been for the formerly more widely circumscribed 'Lesser Burdock'. Wood Burdock was recorded from Formakin Estate (NS4170, 1987, BSBI; 2003, IPG) and Ardgowan Estate (NS1972, 2003, IPG). 'Burdock' was considered frequent by Hennedy though only Gourock is mentioned, but there are other old records from

Castle Semple Loch (1887, NHSG) and Jenny's Well (1900, PSY). Additionally, earlier authors used the scientific name '*A. lappa*', adding to the confusion, but there is no authenticated record of Great Burdock. Today Wood Burdock is still frequent and widespread, but mostly lowland, and often found in urban-fringe areas. It grows along waysides, riverbanks and woodland edges and about farm margins and occasionally on waste ground.

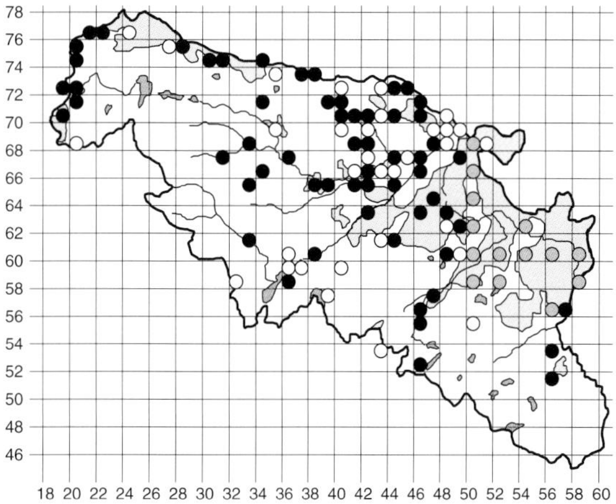

Arctium nemorosum

Carduus crispus L.
Welted Thistle

Britain: common in the S and E, rare in the W.
Native; 7/10; rare.
1st Cathcart, 1865, RH.

There are also old specimens of Welted Thistle from Pollokshields (1872, GLAM) and Cartside east (1882, PSY), and it was recorded from Scotstoun (1883, NHSG), but there are no other named old localities and it was excluded from the Renfrewshire list in 1915 (TPNS). It was quite frequently found on waste-ground sites in Glasgow during CFOG surveys, including at a handful in VC76. There are only two more recent field records: from Inverkip (NS2071, 2002) and by a trackside at the Royal Ordnance Factory, Georgetown (NS4268, 2005). It may have been overlooked in modern times but has a strong eastern bias and has always been a rare Renfrewshire plant.

Carduus nutans L.
Musk Thistle

Britain: common in the S and E, very rare in Scotland.
Former accidental; 0/2; extinct.
1st Newlands coup, 1920, RG.

A more recent record was from a field near Kilmacolm (1960, ERTC).

Cirsium eriophorum (L.) Scop.
Woolly Thistle

Britain: frequent in C S England only.
Former accidental; 0/1; extinct.
1st Field, Kilbarchan, 1921, GL.

The specimen collected by CW Wardlaw is annotated 'not indigenous'.

Cirsium vulgare (Savi) Ten.
Spear Thistle

Britain: common throughout.
Native; 254/368; very common.
1st Renfrewshire 'weed', 1812, JW.

The map indicates a decline since recording in the 1960s–80s, which would be surprising so may represent modern under-recording in rural areas. Spear Thistle is widespread, but sometimes only present in low numbers, occurring on waste ground, agricultural land and rough grasslands; it extends into the uplands where associated with upland farms, fields or tracks.

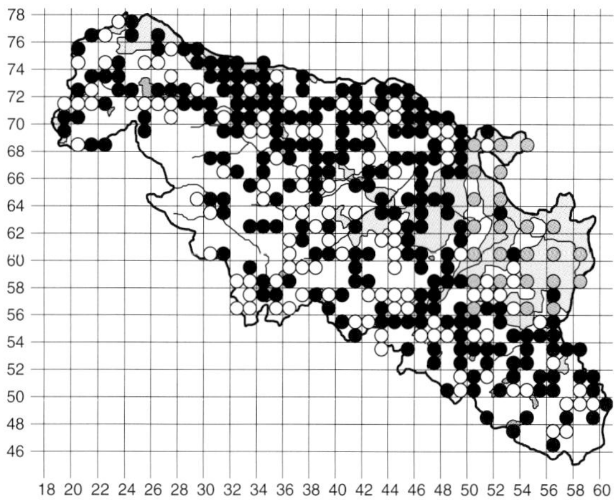

Cirsium vulgare

Cirsium heterophyllum (L.) Hill
Melancholy Thistle

Britain: frequent in Scotland and N England.
Native or hortal; 6/12; rare.
1st Cloak, 1834, MC.

Melancholy Thistle was recorded from Lochwinnoch (1845, NSA) and the River Calder (1891, NHSG) and more recently JP reported it from Pollok (1959) and the River Calder at Muirshiel (1974); AJS also found it at Finlaystone Estate in 1976. It was recorded during surveying for CFOG from Blythswood Estate (NS56E, 1985) and a railway bank, Williamwood (NS55U, 1978). There are four more recent field records, all probably hortals: the edge of estate woodland, Barmufflock (NS3664, 1998), north of Neilston (NS4757, 1999), a roadside ditch, Old Ranfurly Golf Course (NS3864, AM, 2012) and Uplawmoor (NS4455, 2009, RW); the last site may be the same as RM's 1963 Uplawmoor record (Mackechnie 1964).

Cirsium palustre (L.) Scop.
Marsh Thistle

Britain: common throughout.
Native; 362/430; very common.
1st Gourock, *c.* 1840, GLAM.

Marsh Thistle is a very common plant of marshy grasslands, mires, flushes and loch and burn margins. It is found throughout the area but becomes rare in the heavily urbanized lowlands – although it can be found on waste ground – and on peaty upland moorland, but even here it can extend to high elevations along burn sides or in local flushes.

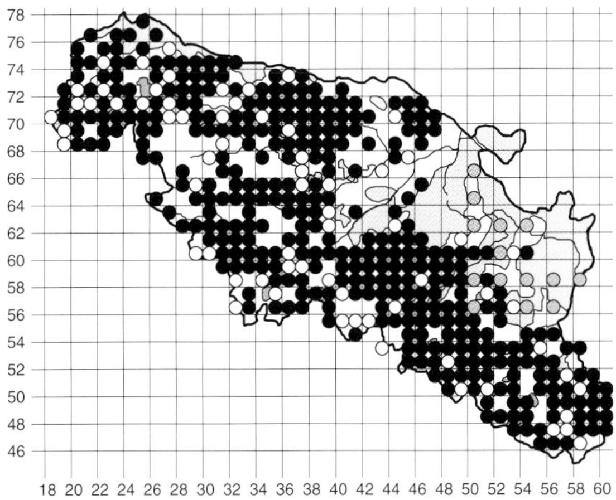

Cirsium palustre

Cirsium arvense (L.) Scop.
Creeping Thistle

Britain: common throughout.
Native; 369/452; very common.
1st Renfrewshire 'weed', 1812, JW.

Creeping Thistle is a very common plant of neglected pastures and meadows, and waste-ground grasslands where it can eventually become abundant along with tall grasses. It can be well represented in tall herb stands along riverbanks and also occurs as a weed of arable fields. Its plumed seeds are capable of spreading to suitable disturbed or enriched sites in the uplands. The 'var. *setosa*' was collected from wasteland, Paisley (1910, PSY), recorded from Giffnock (1918, RG) and collected from Braidbar by RM (1946, E).

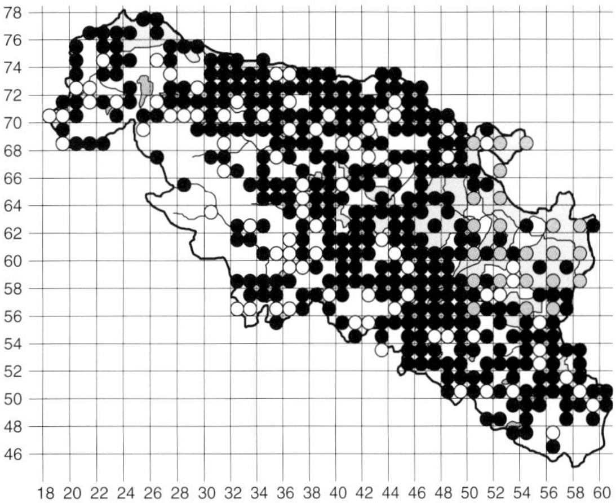

Cirsium arvense

Onopordum acanthium L.
Cotton Thistle

Britain: archaeophyte (Europe); frequent in the SE, scattered in E Scotland.
Former hortal; 0/1; extinct.
1st Renfrewshire, 1915, TPNS.

Cotton Thistle was only mentioned in the county's 'Excluded' list (TPNS 1915); there is no other record.

Centaurea scabiosa L.
Greater Knapweed

Britain: common in the SE, very local in E and N Scotland.
Former accidental; 0/3; extinct.
1st Renfrewshire, 1872, TB.

Greater Knapweed was reported from the railway at Arkleston (1896, NHSG) and from Johnstone (1936, TPNS). Some erroneous reports have been made because of confusion with rayed forms of Common Knapweed.

Centaurea montana L.
Perennial Cornflower

Britain: neophyte (C and S Europe); scattered, frequent in lowland Scotland.
Hortal; 50/52; frequent.
1st Lochside, 1965, RM.

There are few other records for Perennial Cornflower from before the 1980s: roadside, Inchinnan (NS4867, 1977, BWR) and roadside, Trumpethill (NS2276, 1977, RM). However, it was well established by the CFOG recording period. Today this widely grown garden plant can be found on rough grassland on waste ground and often along roadsides and disused railways, where it seems capable of persisting and spreading.

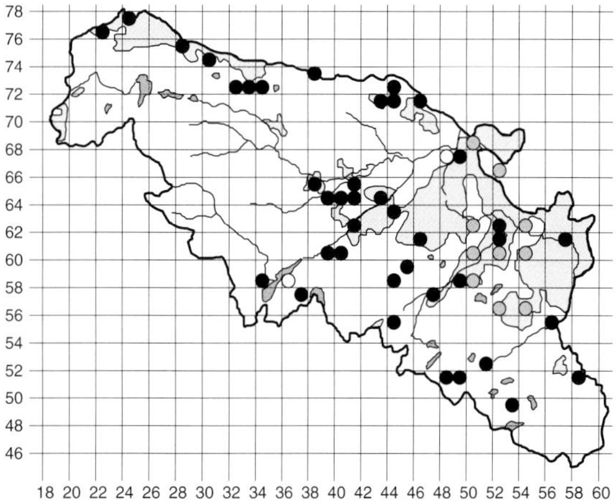

Centaurea montana

Centaurea cyanus L.
Cornflower

Britain: archaeophyte (Europe); frequent in the S, local in Scotland.
Accidental; 1/3; very rare.
1st Paisley, 1845, NSA.

There is an old record for Kilmacolm (1876, FFWS) and in 1869 (RPB) Cornflower was described as 'not very common'. In 1891 Hennedy considered it rare (around Glasgow); it may well have occurred in the lowland farmland to the west of Glasgow at this time, but there are no records. Today the only record is for a solitary plant, presumably a contaminant of seed mix, from the grounds of GMRC at Nitshill (NS5160, 2004); however, it is a common constituent of wild-flower seed mixes so it is likely to be more frequent.

Centaurea calcitrapa L.
Red Star-thistle

Britain: archaeophyte (Europe); rare casual in the S.
Former accidental; 0/1; extinct casual.
1st Crossmyloof, 1894, GL.

Centaurea melitensis L.
Maltese Star-thistle

Britain: neophyte (S Europe); very rare casual in the S.
Former accidental; 0/1; extinct.
1st Newlands, 1919, RG.

Centaurea jacea L.
Brown Knapweed

Britain: neophyte (Europe); former casual.
Former accidental; 0/1; extinct.
1st Ferguslie, 1903, PSY.

This knapweed was reported in the TPNS (Smith 1915) with comments that there were only 'three or four records' in the British Isles. It has not been seen since.

Centaurea nigra L.
Common Knapweed

Britain: common throughout except in N Scotland.
Native; 316/396; very common.
1st Near Hurlet, 1842, GLAM.

Common Knapweed has remained a very common plant since recording began, although it is often now marginal on farmland. It is usually found in less acidic, freely draining grasslands, most notably at neglected sites and along waysides, and is often found on waste ground in urban situations. Rayed forms have been found in rough grassland at Paisley Moss (NS4765, 2002) and Nitshill (NS5260, 2003).

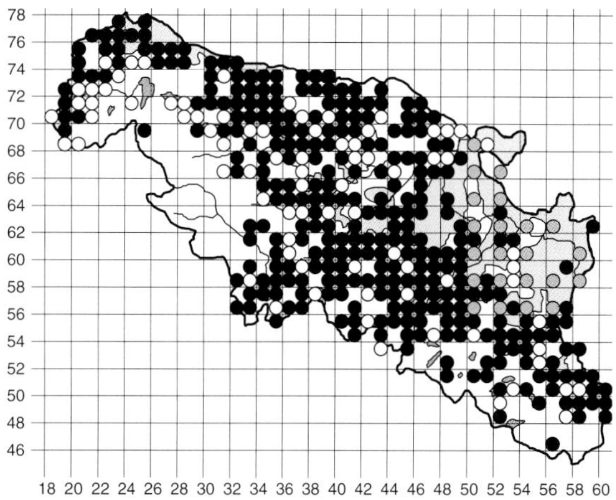

Centaurea nigra

Carthamus lanatus L.
Downy Safflower

Britain: neophyte (S Europe); rare casual in S England.
Former hortal; 0/1; extinct casual.
1st Newlands coup, 1922, RG.

Cichorium intybus L.
Chicory

Britain: archaeophyte (Europe); frequent in the S and E, local in Scotland.
Hortal; 1/5; very rare.
1st Inverkip, 1889, RH.

There are also old records for Chicory from wasteland, near Paisley (1910, PSY), Hayfield, Bishopton (1911, GL) and waste ground, Giffnock (1925, GLAM). The sole modern record is from an embankment by the Black Cart Water at Barnsford Bridge (NS4667, 2002).

Lapsana communis L.
Nipplewort

Britain: common throughout except in N Scotland.
Native; 182/231; common.
1st Renfrewshire, 1869, PPI.

Nipplewort was a very common plant in 1891 (RH) and today it can still be described as such in the lowlands. The map shows a very strong lowland pattern with only sporadic upland records; the latter includes a roadside near Brownside (NS5051, 2001). Habitats tend to be marginal disturbed ground, often shaded (woods, scrub or hedges), but also open waste ground.

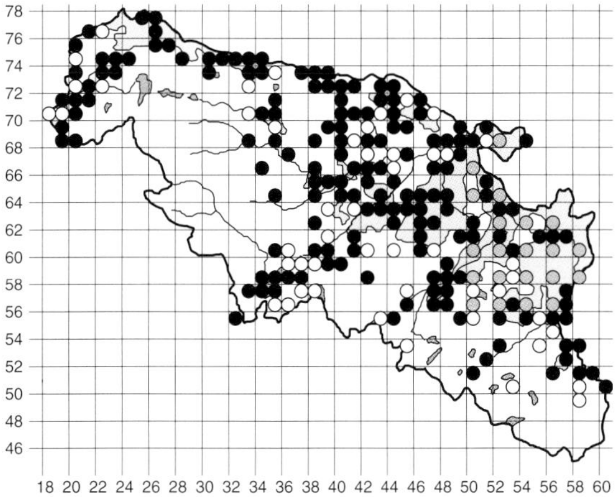

Lapsana communis

Hypochoeris radicata L.
Cat's-ear

Britain: common throughout.
Native; 291/364; very common.
1st Paisley Canal bank, 1865, RH.

Cat's-ear can be readily found in open grassy places throughout the area. It is tolerant of grazing pressure, although it does not cope well with agricultural improvement, and can persist in amenity cut grassland, most often on embankments. It is found on neutral to acidic soils, but becomes rare on more acidic peaty upland soils. It can also be frequent at waste-ground grassland in urban areas.

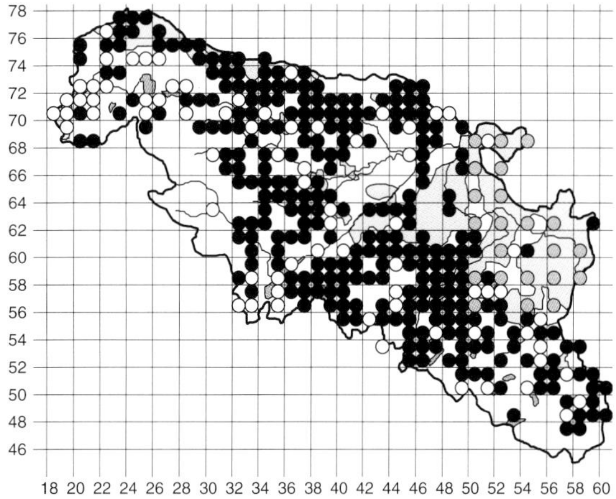

Hypochoeris radicata

Scorzoneroides autumnalis (L.) Moench
Leontodon autumnalis L.
Autumn Hawkbit

Britain: common throughout.
Native; 218/291; common.
1st Paisley, 1845, NSA.

Autumn Hawkbit is commonly found throughout the survey area although the map shows a few gaps, some related to peaty moorlands or improved farmland. It is found in grasslands, usually neutral rather than acidic, and tolerates some trampling and dampness, being found readily in short grass by burns, waterbodies and some footpaths, in amenity parkland and on lawns; it can be found in grassy places at high altitudes but also colonizes open waste ground in the lowlands.

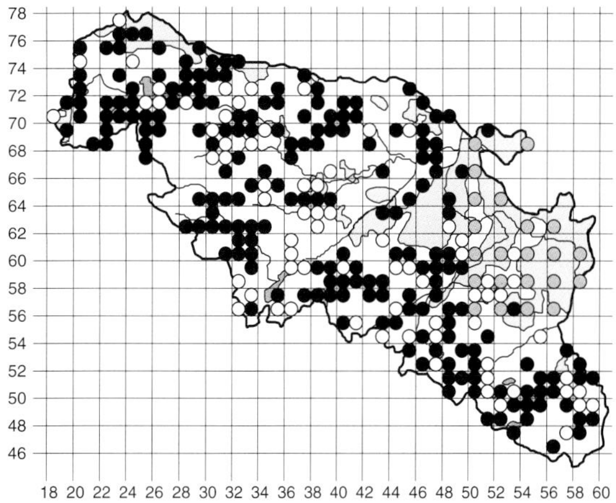

Scorzoneroides autumnalis

Leontodon hispidus L.
Rough Hawkbit

Britain: common up to SE Scotland, absent from the W and N.
Native or accidental; 2/3; very rare.
1st Paisley, 1845, NSA.

This hawkbit was supposedly 'common' in Clydesdale in the 19th century (Hennedy 1891), but there are no other named VC76 localities, and Lee (1933) was perhaps more accurate by stating 'not common'. In CFOG there is a record for 'east of Renfrew' on the VC border (NS56C, 1987). The only other modern record is from rough grassland at Paisley Moss (NS4665, 2003).

Picris hieracioides L.
Hawkweed Oxtongue

Britain: frequent in the SE, very rare in Scotland.
Former accidental; 0/1; extinct or error.
1st Paisley, 1845, NSA.

This is an unlikely record for a species not mentioned by any other author.

Helminthotheca echioides (L.) Holub
Picris echioides L.
Bristly Oxtongue

Britain: archaeophyte (S Europe); common in the SE, rare in Scotland.
Former accidental; 0/1; extinct casual.
1st Crossmyloof, 1894, GL.

Tragopogon pratensis L.
Goat's-beard

Britain: common in the S, rare in Scotland except in the C lowlands.
Native or accidental; 27/32; occasional.
1st Paisley Cemetery, 1869, RPB.

There are few old records for this species, considered by Hennedy (1891) to be 'not common'; it was recorded from the canal bank in Paisley in 1902 (PSY). Goat's-beard was well recorded during CFOG surveys and it has been recorded from widespread, usually urban, locations more recently. All old specimens and modern records (including CFOG) refer to ssp. *minor* (Mill.) Wahlenb.

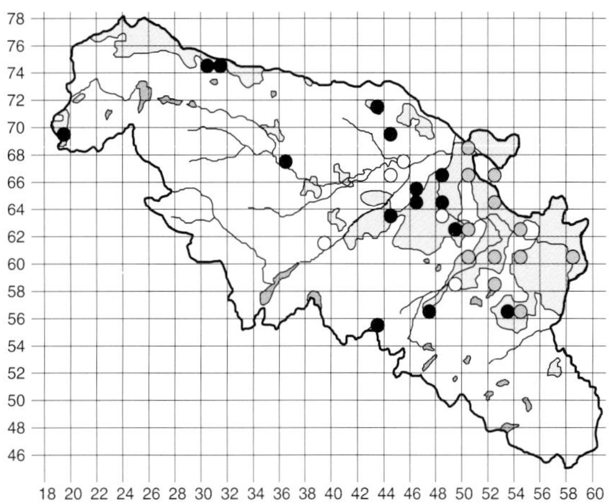

Tragopogon pratensis

Tragopogon porrifolius L.
Salsify

Britain: neophyte (Mediterranean); occasional in the S, very rare in Scotland.
Former hortal; 0/1; extinct.
1st Renfrew, 1865, RH.

Hennedy considered this plant as 'very rarely found except where grown in gardens and escaping, such as near Renfrew'.

Sonchus arvensis L.
Perennial Sowthistle

Britain: common in the S and lowland Scotland.
Native; 41/50; frequent on the coast.
1st Cathcart, 1865, RH.

Hennedy (1891) considered Perennial Sowthistle common and noted it from 'corn fields'; it was collected from Cardonald in 1888 (PSY). Today it is widespread but not common, although it is frequent along the coastal strandline, a habitat not mentioned by Hennedy but observed by Lee (1933). It also occurs inland at wayside and waste-ground locations; the three south-eastern records are all from roadsides: High Dam (NS5551, 2002), Eaglesham (NS5753, 2002) and Queenseat (NS5248, 2001).

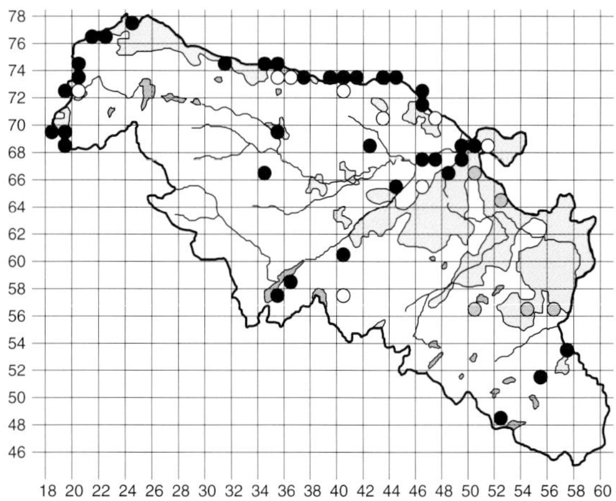

Sonchus arvensis

Sonchus oleraceus L.
Smooth Sowthistle

Britain: common in the S and in lowland Scotland.
Native; 65/87; frequent, urban.
1st Paisley Canal bank and Cathcart, 1865, RH.

Hennedy (1891) considered this sowthistle a 'common' plant of 'waste ground and cultivated fields', but there are no other records until the 1970s. Smooth Sowthistle shows a strong urban preference. It is usually found on waste ground and to waysides.

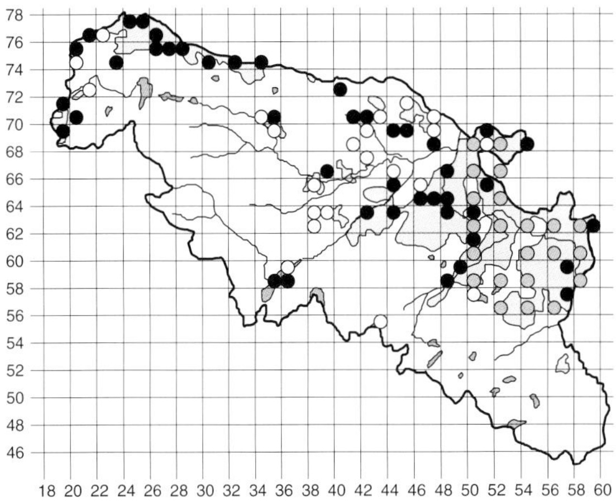

Sonchus oleraceus

Sonchus asper (L.) Hill
Prickly Sowthistle

Britain: common throughout except in upland Scotland.
Native or accidental; 139/161; common in lowlands.
1st Renfrewshire, 1869, PPI.

As for the previous sowthistle, Hennedy (1891) also noted 'common' and 'waste ground and cultivated fields', but there are no other named localities other than Giffnock (1890, GL) prior to the 1960s.

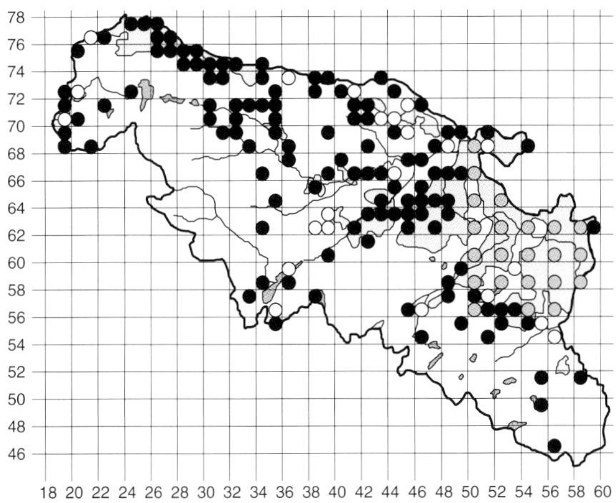

Sonchus asper

Today Prickly Sowthistle is commonly encountered at roadside and waste-ground locations, and less frequently along burn sides and riverbanks.

Lactuca serriola L.
Prickly Lettuce

Britain: archaeophyte; common in S England, very rare in Scotland.
Accidental; 1/1; very rare casual.
1st King's Inch, 2012.

There are very few old or modern records for this lettuce, which seems to be appearing more frequently in the Glasgow area. The sole record is for a single plant at path-edge waste ground by the Clyde at Braehead (NS5168).

Cicerbita macrophylla (Willd.) Wallr.
Common Blue-sowthistle

Britain: neophyte (Urals); throughout but local.
Hortal; 21/22; occasional.
1st Potterhill, Paisley, 1936, TPNS.

AM Stewart reported the first record for this species, which was later collected from the original location (Southfield Avenue) by the finder, WH Thomson, in 1941 (GL – as '*C. plumieri*'). Reported from a few tetrads during surveying for CFOG, Common Blue-sowthistle occurs at a few localities in the lowlands, but it has not been recorded in the west since 1970 ('Kelly Glen', NS1968, BWR). It is usually found on waste ground, where it is capable of spread and persisting, often doing well on scrubby embankments.

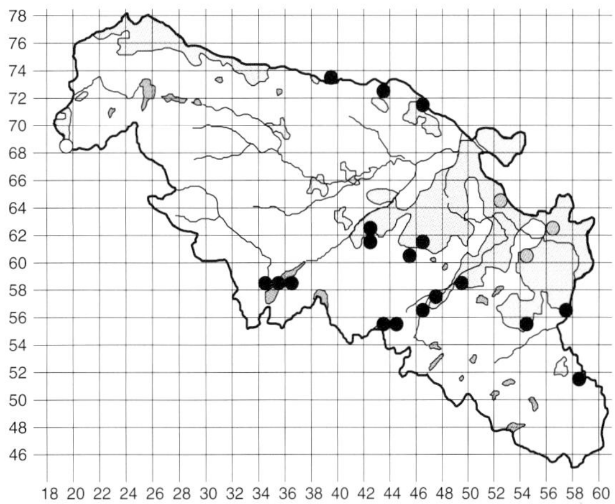

Cicerbita macrophylla

Cicerbita plumieri (L.) Kirschl.
Hairless Blue-sowthistle

Britain: neophyte (W C Europe); scattered and rare.
Former hortal; 0/2; extinct.
1st Renfrew and near Paisley, 1953, GN.

There are no specimens to verify the original records,

which were quoted by Peter Sell (1986), although Hairless Blue-sowthistle is known from the east of Glasgow (CFOG).

Mycelis muralis (L.) Dumort.
Wall Lettuce

Britain: common in the S except in E England, lowland in Scotland.
Hortal or accidental; 8/12; very rare.
1st Kilmacolm Railway Station, 1947, GN.

Wall Lettuce was recorded on several occasions about Kilmacolm (BWR), and there are three records for Wall Lettuce in CFOG: Rouken Glen (1982) – which Lee had noted in 1953 – Aurs Road, Barrhead (1985) and Whitecraigs (1987). Four of the modern records are from the north-west: waste ground at Ladyburn (NS3075, 1998), on walls by the main road at Wemyss Bay (NS1969, 2002), Inverkip (NS2072, 2006) and walls in central Greenock (NS2776, 2007). The sole central record is from a shaded, scrubby embankment in Newton Wood (NS4462, 2005).

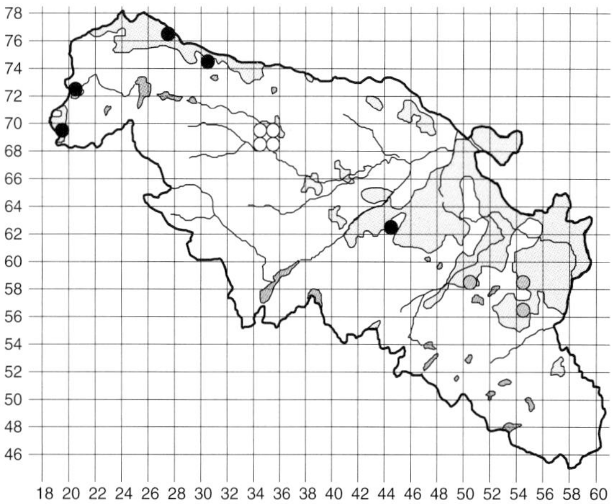

Mycelis muralis

Taraxacum spp.
Dandelions

Britain: natives and aliens; common throughout.
Native and accidental; 351/447; very common.
1st Paisley, 1845, NSA.

This difficult genus has many microspecies and needs expert determination for accurate recording. There are several named localities from the CFOG period and also recorders such as AJS and AMcGS, but mostly from the 1970s and 1980s. Most of these records have been expertly determined or confirmed by Professor AJ Richards or Chris Haworth. Bert Reid kindly supplied data from the Taraxacum Database which has helped confirm a number of records. The microspecies names follow those used in *Dandelions of Great Britain and Ireland* (Dudman and Richards 1997).

There are old references to dandelions but species (or microspecies) data are lacking, except for the occasional mention of 'var. *palustre*' (more likely *T. faroense*). Hennedy (1865) wrote 'Common everywhere; were it not so, it would be highly prized on account of its beauty'. The following map shows all records and reflects how widespread dandelions are; many records refer to 'weeds' on waste ground or park grassland, but there are a few from more natural places, some of which refer to native taxa.

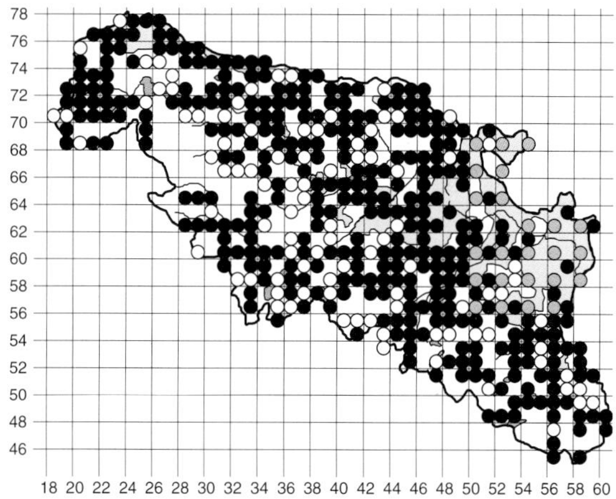

Taraxacum spp.

Section *Erythrosperma* (H. Lindb.) Dahlst.

T. brachyglossum (Dahlst.) Raunk.

This dandelion of sandy soils was first recorded from Erskine Harbour (NS4672) by AJS in 1976. AMcGS also recorded it from west of Erskine Bridge (NS4672) in 1982.

T. lacistophyllum (Dahlst.) Raunk.

Near Erskine Bridge (NS4572, 1981, AMcGS) and St James, Paisley (NS4865, 1982, AMcGS).

T. scoticum A.J. Richards

Sandy riverbank, below Erskine Bridge (NS4572, 1981, AMcGS).

Section *Spectabilia* (Dahlst.) Dahlst.

T. faeroense (Dahlst.) Dahlst.
This dandelion has been recorded, with some confidence, from 29 scattered, mostly rural, sites in upland-fringe areas. It is found in marshy or flushed grasslands. It was collected from Gourock in 1860 (GL); AMcGS recorded it from Newshot Island (NS4870, 1970), Shielhill Glen (NS27, 1971) and Loch Libo (NS4355, 1979).

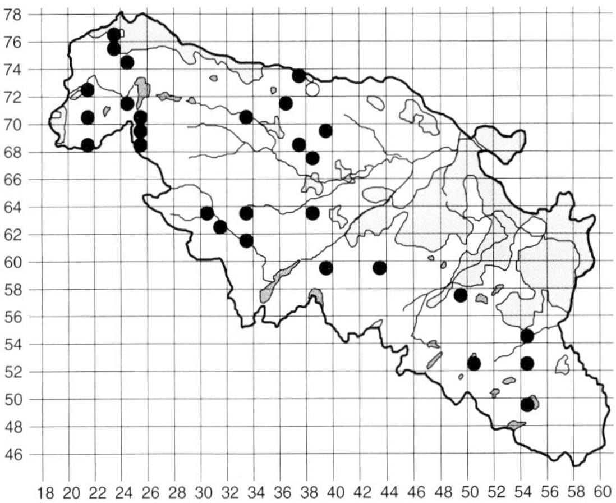

Taraxacum faeroense

Section *Naevosa* M.P. Christ

The following map shows 39 field observations (35 modern) of members of this section (i.e. those with distinctly purple blotched leaves), which are likely to be native taxa but most have not been expertly determined to species level; several will refer to *T. maculosum* but there are four other determined species.

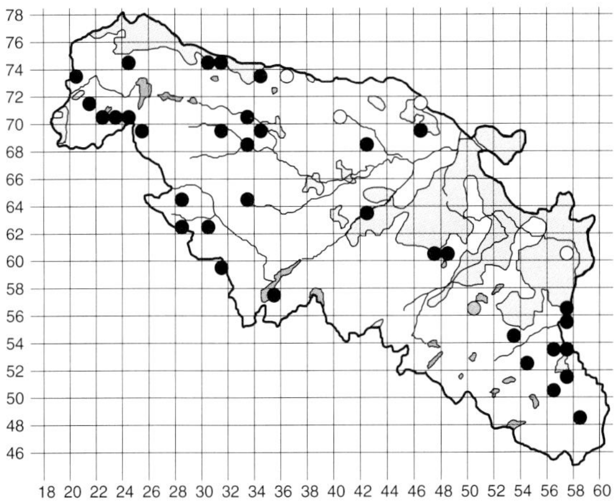

Taraxacum Section *Naevosa*

T. euryphyllum (Dahlst.) Hjelt.
Newshot Island (NS4870, 1970, AMcGS) and Shielhill Glen (NS27, 1971, AMcGS).

T. maculosum Dahlst.
Shielhill Glen (NS27, 1971, AMcGS), Broadfield Hill (NS409599, 1979, AJS), Finlaystone (NS368735, 1979, AJS) and Waterside, Balgray (NS5056, 1992).

T. naevosiforme Dahlst.
Near Loch Libo (NS4355, 1982, AMcGS).

T. pseudolarssonii A.J. Richards
Erskine (NS4672, 1981, AMcGS) and as a garden weed, Newlands (NS5760, 1983, PM).

T. stictophyllum Dahlst.
Near Formakin (NS47, 1981, AMcGS) and Waterside, Balgray (NS5056, 1992).

Section *Celtica* A.J. Richards

T. bracteatum Dahlst.
Scotstoun (NS5267, 1981, AJS), Inchinnan Bridge (NS4967, 1982, AMcGS), Waterside, Balgray (NS5056, 1992), Park Holdings, Elderslie (NS4669, 1996) and Ardgowan Estate (NS2073, 1998).

T. duplidentifrons Dahlst.
This dandelion has 37 unconfirmed modern field records generally from rural areas, often where there is some shade. Older confirmed records include near Loch Libo (NS4355, 1979, AMcGS), near Erskine Bridge (NS4572, 1982, AMcGS), Blythswood (NS4969, 1977, AJS), Finlaystone (NS367735, 1979, AJS) and Paisley (NS5063, 1988, AJS).

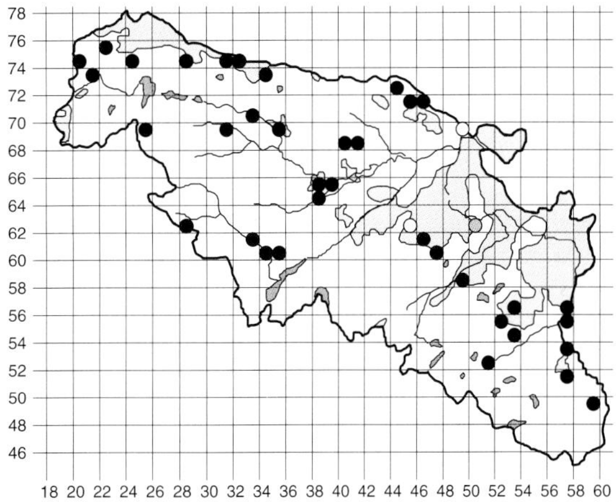

Taraxacum duplidentifrons

T. fulvicarpum Dahlst.
Near Erskine Bridge (NS4572, 1981, AMcGS).

T. landmarkii Dahlst.
Shielhill Glen (NS27, 1971, AMcGS).

T. nordstedtii Dahlst.
Wet cart track, Finlaystone Estate (NS3673, 1976, AJS), Finlaystone (NS371736, 1977, AJS), a roadside, Barnaigh (NS3661, 1979, OS), Black Loch, Mearns (NS4951, 1979, AMcGS), Paisley (NS4763, 1981, AJS) and Erskine Ferry (NS4572 and NS4672, 1995, AMcGS).

T. unguilobum Dahlst.
Turf on coastal rocks, Gourock (NS2276, 1976, AJS), the shore, Lunderston (NS2173, 1981, AMcGS), Finlaystone (NS3773,1977, AJS), Erskine Bridge (NS4672, 1976, AJS), Shielhill Glen (NS27, 1971, AMcGS) and Erskine Ferry (NS4572 and NS4672, 1995, AMcGS).

Section *Hamata* H. Øllg.

T. boekmanii Borgv.
This appears to be one of the commoner members of this section, with 12 unconfirmed field records from urban-fringe areas. Confirmed records include Inchinnan Bridge (NS4967, 1982, AMcGS), Lunderston (NS2173, 1981, AMcGS), Polmadie (NS5962, 1988, AJS), Paisley (NS4861, 1983, AJS) and Paisley (NS5063, 1988, AJS).

T. fusciflorum H. Øllg.
The sole VC record is unlocalized (NS45, 1981, AJS).

T. hamatiforme Dahlst. in Lindm.
Roadside, Bridge of Weir (NS3865, 1979, OS) and near Erskine Bridge (NS4572, 1982, AMcGS).

T. hamatulum Hagend., Soest & Zevenb.
Midbarsay, Bridge of Weir (NS3865, 1979, OS) and Neilston (NS4755, 1979, AJS).

T. hamatum Raunk.
There are several records from AJS in 1976 from the Paisley–Erskine area (NS4864, NS4672, NS4672, NS4671, NS4668) so this species is likely to be locally frequent. Other records include Paisley (NS4863, 1979, AMcGS) and near Erskine Bridge (NS4572, 1982, AMcGS).

T. hamiferum Dahlst.
Neilston (NS4755, 1979, AJS) and Hillington (NS5165, 1988, AJS).

T. lamprophyllum M.P. Christ
Roadside, Johnstone (NS4463, 1976, AJS), St James, Paisley (NS4865, 1982, AMcGS) and at the VC border at Braehead (NS5167, 1985, CFOG).

T. pseudohamatum Dahlst.
Roadside, south of Bridge of Weir (NS3865, 1979, OS), waste ground, Paisley (NS46, 1980, AMcGS) and Ralston (NS5063, 1988, AJS).

T. quadrans H. Øllg.
Paisley (NS4563, 1976, AJS), Paisley (NS4864, 1976, AJS), Linwood Moss (NS4465, 1976, AJS), Renfrew (NS4968, 1977, AJS) and near Erskine Bridge (NS4572, 1982, AMcGS).

Section *Ruderalia* Kirschner, H. Øllg. & Stepanek

T. adiantifrons Ekman ex Dahlst.
The sole record for the VC is unlocalized (NS56, 1980, AJS).

T. alatum H. Lindb.
Drive-side grass, Finlaystone Estate (NS3773, 1977, AJS) and Paisley (NS4863, 1980, AJS).

T. angustisquameum Dahlst. ex H. Lindb.
Paisley (NS5063, 1988, AJS).

T. aurosulum H. Lindb.
Rough grass by the banks of the White Cart Water, Inchinnan (NS4968, 1977, AJS).

T. cordatum Palmgr.
Underwood Rd, Paisley (NS46, 1983, AMcGS) and Ralston (NS5063, 1988, AJS).

T. croceiflorum Dahlst.
Paisley (NS4863, 1980, AJS).

T. dahlstedtii H. Lindb.
Near Erskine Bridge (NS4572, 1980, AMcGS), Inchinnan Bridge (NS46, 1982, AMcGS) and Jordanhill, Glasgow (NS5468, 1982, AMcGS).

T. ekmanii Dahlst.
Ralston (NS5063, 1988, AJS).

T. exacutum Markl.
There are 14 unconfirmed modern field records, mainly from urban waste ground. Determined records include Polmadie (NS5962, 1988, AJS) and Paisley (NS5063, 1988, AJS).

T. exsertum Hagend., Soest & Zevenb.
Garden, Potterhill (NS4861, 1983, AMcGS).

T. fasciatum Dahlst.
Paisley (NS4863, 1980, AJS) and waste ground, St James, Paisley (NS4865, 1982, AMcGS).

T. interveniens G.E. Haglund
Polmadie (NS5962, AJS, 1988).

T. laticordatum Markl.
Paisley (NS4863, 1976, AJS) and Langside (NS5660, 1989, CFOG).

T. latissimum Dahlst.
The sole record for the VC is unlocalized (NS56, 1980, AJS).

T. lepidum M.P. Christ
Polmadie (NS5962, AJS, 1988).

T. oblongatum Dahlst.
Shielhill Glen (NS27, 1971, AMcGS).

T. pannucium Dahlst.
Waste ground, Paisley (NS4563, 1976, AJS).

T. piceatum Dahlst.
Inchinnan Bridge (NS4967, 1980, AMcGS).

T. polyodon Dahlst.
Coastal rocks, McInroys Point (NS2276, 1976, AJS), Wemyss Bay (NS17, 1980, AJS), Blythswood (NS4969, 1977, AJS), Polmadie (NS5962, AJS, 1988) and Paisley (NS4763, 1998, AJS).

T. sellandi Dahlst.
Roadside, Abbotsinch (NS4967, 1979, AJS), Glasgow Airport (NS46, 1983, AMcGS) and Paisley (NS46, 1983, AMcGS).

T. xanthostigma H. Lindb.
Paisley (NS4863, 1976, AJS) and near Erskine Bridge (NS4572, 1981, AMcGS).

Crepis paludosa (L.) Moench
Marsh Hawk's-beard

Britain: common in Scotland and N England.
Native; 179/220; common.
1st Renfrewshire, 1834, MC.

Marsh Hawk's-beard shows a strong rural pattern extending up into the high ground along upland burns, and similarly to urban-fringe areas along wooded burn sides. It is plant most likely to be found along shaded burn sides, sometimes among marginal rocks but usually in wet flushes or inundation areas.

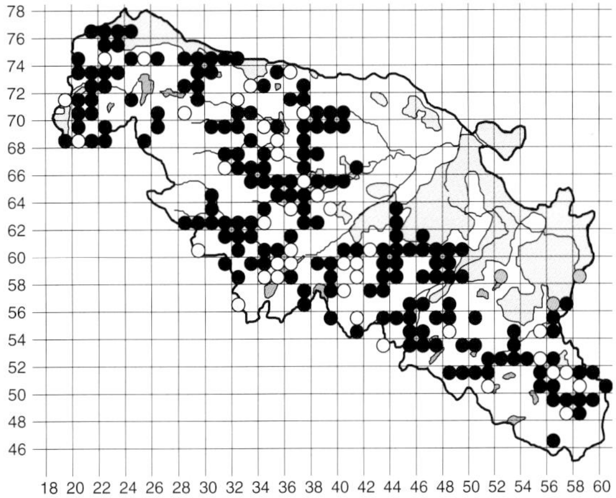

Crepis paludosa

Crepis capillaris (L.) Wallr.
Smooth Hawk's-beard

Britain: common throughout except in uplands and N Scotland.
Native; 27/42; occasional, lowland.
1st Cathcart, 1865, RH.

Smooth Hawk's-beard cannot be described as common as stated in 1891 (RH). It is now much rarer and very much an urban plant from freely draining urban-fringe waste ground. It occurs on the disused railway (now cycleway) near Scart (NS3667, 2001) and Pennytersal (NS3271, 2005).

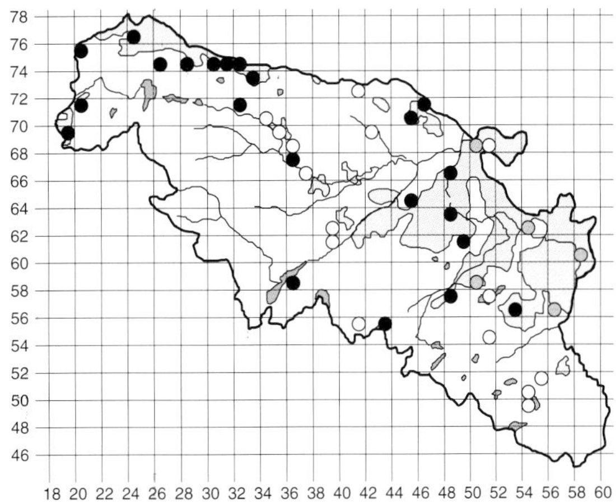

Crepis capillaris

Pilosella officinarum F.W. Schultz & Sch. Bip.
Mouse-ear-hawkweed

Britain: common throughout.
Native; 183/230; common.
1st Gourock, 1860, GL.

Mouse-ear-hawkweed was a common plant in the 19th century but there are few named localities; it was collected from Lochwinnoch in 1876 (GLAM). Today it is quite often found on lowland waste ground, such as gravels along disused railways, but it is most often encountered at unimproved grasslands, usually associated with embankments and rocky outcrops; on the latter it can extend into the peaty uplands.

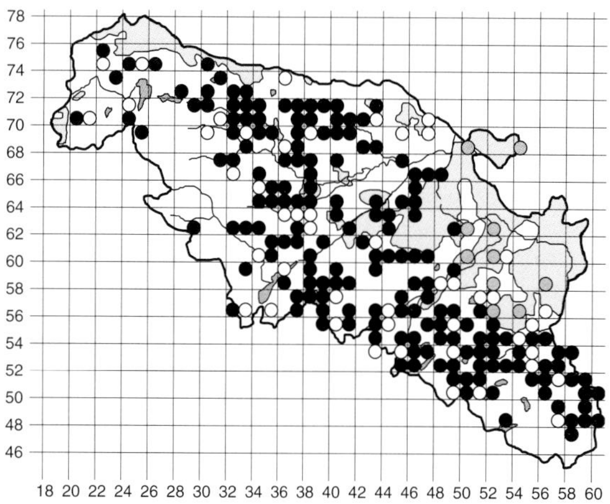

Pilosella officinarum

Pilosella flagellaris (Willd.) P.D. Sell & C. West ***ssp. flagellaris***
Spreading Mouse-ear-hawkweed

Britain: neophyte (C and E Europe); local, but frequent in E C Scotland.
Accidental; 2/2; very rare.
1st Port Glasgow, 2001.

A small population occurs on waste ground at Clydeport, Port Glasgow (NS3174) and Spreading Mouse-ear-hawkweed was also found by the Levern Water at Househill (NS5261, 2005).

Pilosella aurantiaca (L.) F.W. Schultz & Sch. Bip.
Fox-and-cubs

Britain: neophyte (N and C Europe); frequent throughout except in N Scotland.
Hortal; 35/44; occasional.
1st Kilbarchan, 1905, PSY.

In 1921 Fox-and-cubs was collected from Kilbarchan as a 'garden outcast' (GL) and RM found it on a railway bank at Clarkston (1942, GLAM). Today it is widespread but scattered and generally associated with waste ground and disused railways, although it is capable of persisting at many sites with rough grassland; it can also occur at remote areas, associated with habitation or garden dumping such as at a roadside, Plymuir (NS4357, 1999) and the margins of Black Loch (NS4951, 1997). All records are thought to refer to ssp. *carpathicola* (Nägeli & Peter) Soják.

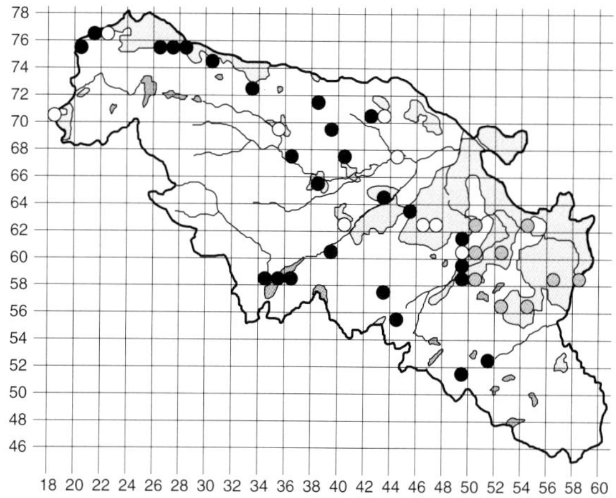

Pilosella aurantiaca

Hieracium **spp.**
Hawkweeds

Several members of this critical group have been identified from the VC but the frequency and distribution of most is unclear; in total there are 204 records shown in the following aggregate map. There are few herbarium sheets that confirm old records and it is difficult to interpret old literature records; there also appears to have been very few records made in the 1960s–80s period (BWR). Only a small number of specimens have been expertly determined (mostly by David McCosh, and some by Peter Sell), but this does verify that 16 taxa are definitely present (and one

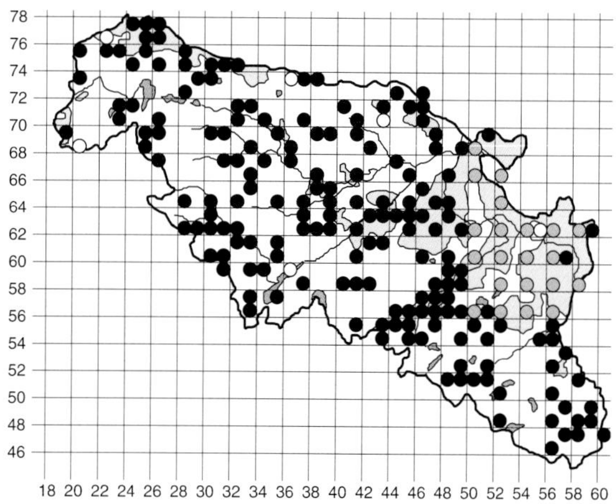

Hieracium spp.

queried – *H. 'dipteroides'*). The following accounts list the confirmed records but distribution maps are based mostly on unconfirmed field records. Common names follow those used in the *Atlas of British and Irish Hawkweeds* (McCosh & Rich 2011).

Section *Hieracoides* Dumort.

Hieracium umbellatum L.
Umbellate Hawkweed

Native or accidental; 0/2; probably extinct.
1st Renfrewshire, 1915, TPNS.

This species seems to be rare, with the most recent record made by AMcGS from waste ground at Williamwood (NS5657, 1969, E) and with no modern sightings.

Section *Sabauda* (Fr.) Arv.-Touv.

Hieracium sabaudum L.
Autumn Hawkweed

Native; 30/32; occasional.
1st Renfrewshire, 1887, NHSG.

This leafy hawkweed, with prominent stem hairs, is presumed to be native and was formerly considered widespread (as *H. boreale* Fries). Its distribution shows a strong lowland urban-fringe pattern. It can occur on open waste ground or rough grassland but is often associated with the shade of scrub or woodland edges. The forma *bladonii* (Pugsley) P.D. Sell was collected from Maxwelltown, Paisley (NS4663, 1961, AMcGS in E).

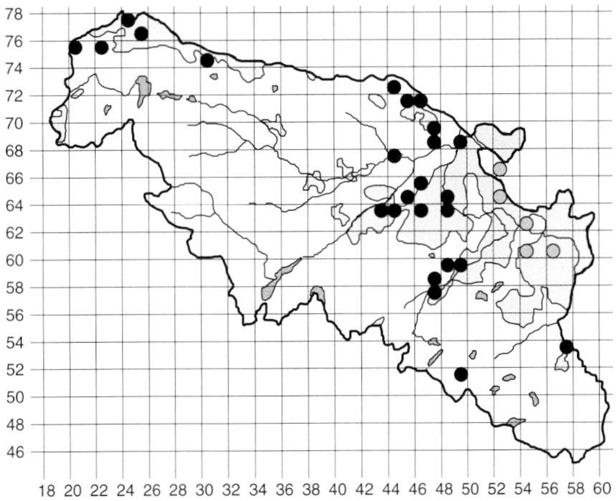

Hieracium sabaudum

Hieracium vagum Jord.
Glabrous-headed Hawkweed

Native or accidental; 4/6; rare.
1st Scotstounhill, 1890, GL.

This broad-leaved hawkweed was collected (E) by AMcGS from waste ground at Elderslie (NS4563, 1981) and was also found in Glasgow at Scotstoun in 1991 (CFOG). There are three unconfirmed modern field records: Lex Wood (NS4562, 1993), Thornley Dam (NS4860, 2005) and waste ground near Gourock Station (NS2477, 1997).

Hieracium salticola (Sudre) P.D. Sell & C. West
Bluish-leaved Hawkweed

Accidental; 21/22; occasional.
1st Elderslie, Paisley, 1981, AMcGS (E).

This hawkweed, with a glabrous involucre, is noted in CFOG to be possibly the commonest of the leafy species. There are a few widely scattered records from the modern period, so it is perhaps under-recorded; most records are from urban waste ground such as by the disused railway at Hurlet (NS5061, 1986, GLAM) but it was also collected from rocky scree at the more rural Walls Hill (NS4158, 1993, GLAM).

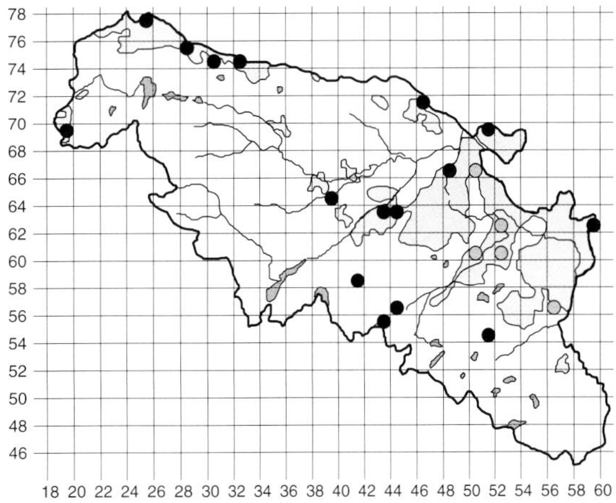

Hieracium salticola

Hieracium virgultorum Jord.
Long-leaved Hawkweed

Accidental; 5/7; rare.
1st Waste ground, White Cart Water, Pollok, 1967, AMcGS (E).

AMcGS also collected this tall hawkweed from Williamwood Railway Station (NS5658, 1969, CGE). Long-leaved Hawkweed was found at several places near the VC76 boundary during CFOG surveys, but there are only five widely scattered modern records: from the railway station, Lochwinnoch (NS3557, 1993), waste ground at Linwood Industrial Estate (NS4463,

1995), Ladyburn (NS3075, 1998), Clydeport (NS3174, 2001) and Neilston (NS4757, 2004).

Section *Foliosa* (Fr.) Arv.-Touv.

Hieracium latobrigorum (Zahn) Roffey
Yellow-styled Hawkweed

Native or accidental; 13/16; scarce.
1st Near Scotstounhill, 1921, GL.

This hawkweed, with sessile leaves and yellow stigmas, was collected from a roadside, Lochwinnoch (1969, AMcGS) and from near Neilston (NS4855, 1972, AGK), both CGE, and was frequently found at urban sites during CFOG surveys. There are a few records from the outskirts of Glasgow, e.g. by the Levern Water (NS4654, 1998) and White Cart Water, Waterside (NS5655, 2001), but it is presumably more widespread, and under-recorded, as the two north-western records indicate: Underheugh (NS2075, 1997) and waste ground, Greenock (NS2776, 2000).

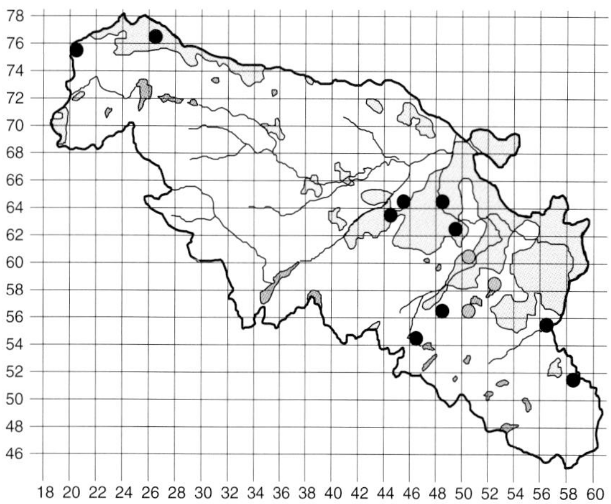

Hieracium latobrigorum

Hieracium subcrocatum (E.F. Linton) Roffey
Dark-styled Hawkweed

Native or accidental; 0/2; very rare.
1st Bing, Nitshill, 1929, RM (RNG).

The only other record is from near Erskine Bridge (NS4672, 1981, AMcGS in E).

Section *Lanatella* (Arv.-Touv.) Zahn

Hieracium rionii Gremli
Marbled Hawkweed

Accidental; 1/1; very rare.
1st Garden weed, Newlands, 1990, CFOG.

This rare hawkweed (recorded as *H. pollichiae* in CFOG) persists in PM's garden (NS5760) where it was presumably introduced along with other plant material.

Section *Hieracium*

Hieracium exotericum group

Accidentals; 16/18; scarce.
1st Wemyss Bay, 1982, AMcGS (GLAM).

There are probably four or five species in this aggregate group characterized by large basal rosette leaves and glandular involucres. *H. grandidens* was recorded several times during surveying for CFOG (though not all of these were expertly determined), so may have included other members of this group. The map, based on field records for the group, shows a scattered picture but with two main centres: about Greenock and south-western Glasgow. Habitats are waste ground and disused railways. The original record has been tentatively identified as *H. koehleri* Dahlst. by D McCosh. The confirmed records include those following the map on p. 251.

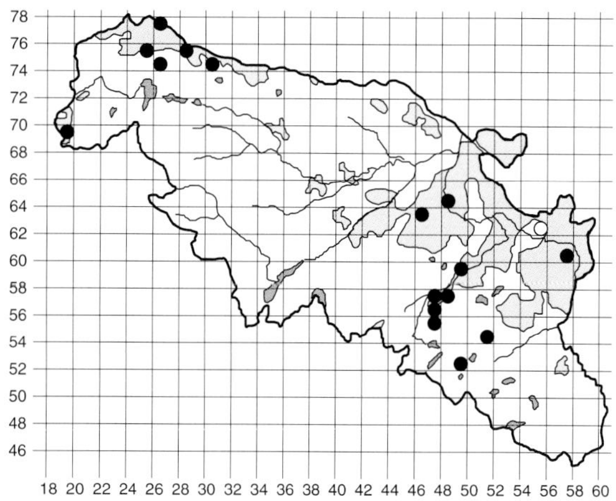

Hieracium exotericum group

H. grandidens **Dahlst.**
Grand-toothed Hawkweed

First recorded from Cowdon Hall (NS4757, 1997) and subsequently from eight other field sites, but it has only been confirmed from Bannerbank Quarry (NS4952, 2011).

H. subaequiatulum **Hyl.**
Black-bracted Hawkweed

This hawkweed occurred in small numbers on a railway bank by the Kirkton Burn, Neilston (NS4857, 1999, GLAM).

H. subcrassum **(S. Almq. ex Dahlst.) Johanss.**
Trackway Hawkweed

This hawkweed was collected from Langside Drive in 1988 (NS5760).

H. sublepistoides **(Zahn) Druce**
Grey-bracted Hawkweed

AMcGS collected this hawkweed from a shrubbery in Pollok Estate (NS5562, 1970, E) and more recently it was collected from a rocky bank in Craig Wood, Neilston (NS4755, 1999).

Section *Vulgata* (Griseb.) Willk. & Lange

Hieracium vulgatum **Fries**
Common Hawkweed

Native; 164/169; common.
1st Cathcart, 1840, GL.

This is by far the commonest rosette hawkweed, although the map may include a few other rosette hawkweeds. Old literature records presumably include those from Renfrewshire (1834, MC) and Loch Libo, Wemyss Bay and Lochwinnoch (1887, NHSG), and other confirmed herbarium records (GL) include Bardrain Glen (1907) and Lochwinnoch (1921). It is recorded throughout the VC, from the uplands – where it can be found on rocky outcrops, notably along burn sides – and urban areas, such as waste ground, disused railway banks and walls. The var. *sejunctum* (W.R. Linton) P.D. Sell was collected from Lochwinnoch (1921, GL), Cathcart (1840, GL) and Bardrain Glen (1907, GL).

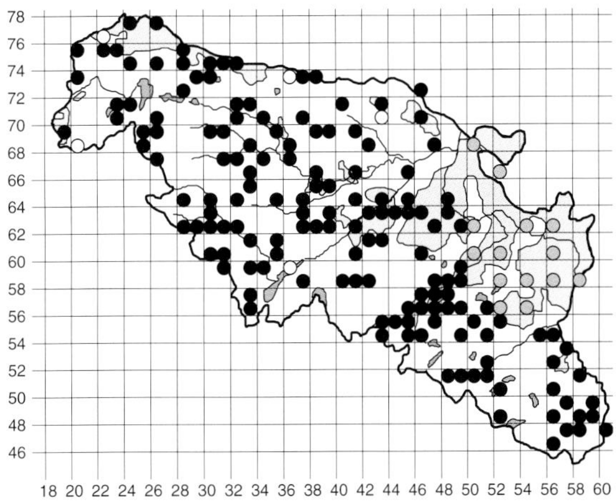

Hieracium vulgatum

Hieracium rubiginosum **F.J. Hanbury**
Rusty-red Hawkweed

Native; 0/1; extinct.
1st Lochwinnoch, 1921, GL.

This collection, made by Lee, is the sole record for the VC.

Hieracium **'dipteroides'**

Native; 0/1.
1st Shielhill Glen (NS2372), 1960, BWR

AMcGs also collected a specimen (CGE) from the same place in 1967. However, the plant from Shielhill is no longer regarded as matching true *H. dipteroides* Dahlst. and is currently without a name (D. McCosh pers. comm.).

Section *Oreadea* (Fr.) Dahlst.

Hieracium chloranthum **Pugsl.**
Green-flowered Hawkweed

Native; 0/1; extinct or overlooked.
1st Lochwinnoch, 1964, D McClintock.

This record from by a roadside wall was confirmed by Dr C West as reported by RM (Mackechnie 1966).

Filago vulgaris **Lam.**
Common Cudweed

Britain: frequent but local, in the E in Scotland.
Former native; 0/2; extinct.
1st Paisley, 1858, PSY.

Hennedy also stated Gourock (1865) but there are no other named places for this weed of open places.

Filago minima **(Sm.) Pers.**
Small Cudweed

Britain: scattered, mainly in the E in Scotland.
Former native; 0/1; extinct.
1st Paisley, 1845, NSA.

Although considered frequent in Clydesdale by Hennedy (1891), there are no other named places in the VC, where it may well have been a rarity.

Antennaria dioica (L.) Gaertn.
Mountain Everlasting

Britain: frequent in the N and W and uplands.
Native; 5/14; rare.
1st Renfrewshire, 1834, MC.

There are at least four 19th-century specimens of Mountain Everlasting from Gourock, but no modern records, although it may persist in the hills above. In 1845 it was recorded from Paisley and Erskine and was also recorded from Fereneze Braes (1896, PSY); more recently it was collected from heathy pasture at Kilmacolm marsh (Glen Moss?) by RM in 1925 (GLAM). The five modern records are from rocky banks by the River Calder (NS3361, 1997), a small rocky bank by the Witch Burn (NS4658, 1999), a similar bank just to the south of Commore Dam (NS4654, 1999), near Dunconnel Hill (NS3359, 2001) and in some local profusion by Snypes Dam (NS4855, 2002).

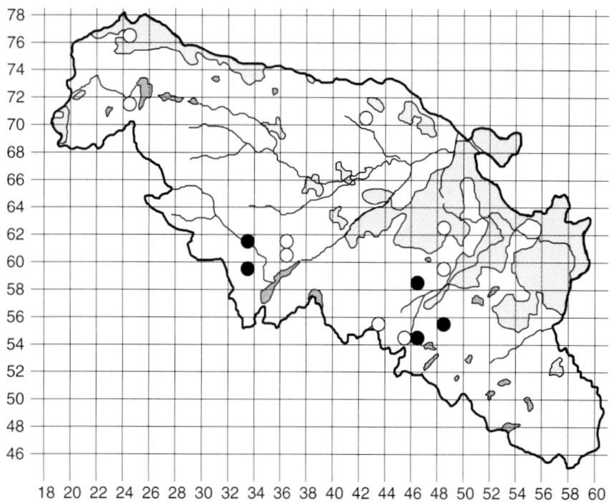

Antennaria dioica

Anaphalis margaritacea (L.) Benth.
Pearly Everlasting

Britain: neophyte (E Asia/N America); scattered casual, more in the W.
Hortal; 1/3; very rare casual.
1st Thornliebank, 1922, RG.

JP knew Pearly Everlasting from beside a wall at Blythswood (1968), but the only modern find is from waste-ground grassland in Cowglen (NS5360, 2005, LR).

Gnaphalium sylvaticum L.
Heath Cudweed

Britain: frequent in N Scotland, rare and declining in the S.
Former native; 0/8; extinct.
1st Lochwinnoch, 1845, NSA.

Other old records for Heath Cudweed are from Cathcart and Gourock (1865, RH) and Loch Libo and Caldwell (1887, NHSG). It was collected from Shielhill Glen and Inverkip Road (*c.* 1890, GRK) and from Erskine (1882, GL). There are no records since the 19th century, indicating a fairly dramatic decline.

Gnaphalium uliginosum L.
Marsh Cudweed

Britain: common throughout except in NW Scotland.
Native; 104/126; widespread.
1st Gourock, 1865, RH.

Marsh Cudweed was considered a frequent plant in the 19th century, which seems an appropriate description today, as it shows a widespread, scattered distribution. It is often associated with seasonally inundated ground, including reservoir and loch marginal mud and gravel, and can also be found at poached farm gateways and sometimes on waste ground.

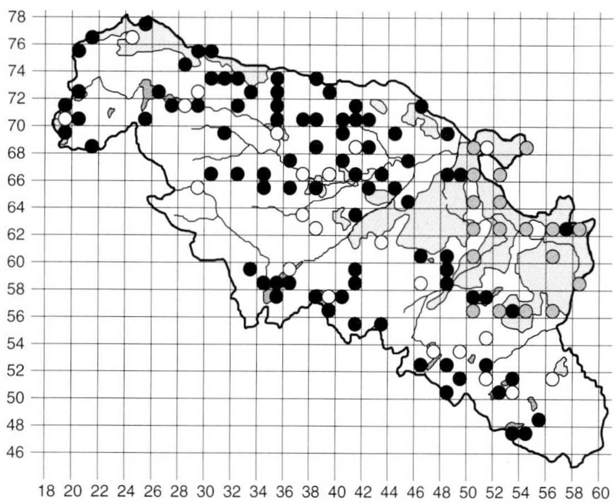

Gnaphalium uliginosum

Helichrysum bracteatum (Vent.) Andrews
Strawflower

Britain: neophyte (Australia); rare casual or escape.
Hortal; 0/2; probably extinct casual.
1st Giffnock, 1926, RG.

In 1981 AMcGS found this rare casual growing on waste ground with garden rubbish, by St James' Playing Fields, Paisley (NS4665).

Inula helenium L.
Elecampane

Britain: archaeophyte (W Asia); scattered in the S, local in Scotland.
Hortal; 3/4; very rare.
1st Inverkip, 1982, AMcGS.

The original record for Elecampane was from a roadside verge near Inverkip Power Station (NS1970). Recently a large plant was found nearby on rough grassland by a ramp down to the shore (NS2071, 2002). A further record is from between Lunderston and Ardgowan (*c.* NS2073, 2007, AR). It was recorded from Cowglen (1989, PM), where one plant persists but is struggling beneath dense scrub (NS5460, 2011).

Pulicaria dysenterica (L.) Bernh.
Common Fleabane

Britain: common in the S, very rare in Scotland.
Accidental; 2/2; very rare.
1st Paisley Moss, 2003.

A small population of Common Fleabane, in rough grassland, was found at the east edge of Paisley Moss (NS4656); a second population was found in disturbed grassland to the side of Spango Burn (NS2374, 2006).

Telekia speciosa (Schreb.) Baumg.
Yellow Oxeye

Britain: neophyte (C and SE Europe); scattered in England and Scotland.
Hortal; 3/3; very rare.
1st West Tandlemuir, 1996.

The first site (NS3361) was a roadside population next to a garden (not seen in 2011), a very similar situation to the stand at Corsliehill House (NS3969, 2004), which does still occur (2011). Yellow Oxeye is also recorded from Ardgowan Estate (NS2073, 1999), although the occurrence of Elecampane nearby casts some doubt on this identification.

Solidago virgaurea L.
Goldenrod

Britain: common in the W and N, rare in the E.
Native; 30/40; occasional.
1st Gourock, 1837, GL.

There are quite a few 19th-century records for Goldenrod, which was then considered frequent, e.g. Lochwinnnoch (1845, NSA) and Wemyss Bay (1887, NHSG), and it was collected from Kilmacolm in 1894 (GL). Today it is widespread and rural, being well represented in the west; it is usually found on rocky ground mainly along wooded valleys, although generally where not heavily shaded.

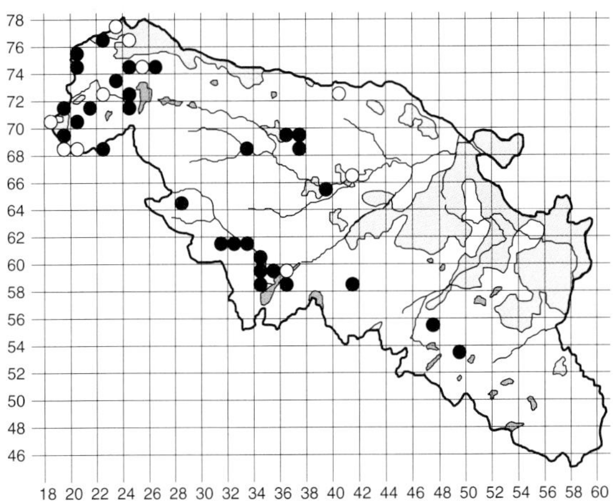

Solidago virgaurea

Solidago rugosa Mill.
Rough-stemmed Goldenrod

Britain: neophyte (N America); very rare.
Hortal; 1/1; very rare.
1st Lochwinnoch Railway Station, 1993.

A specimen collected from disturbed ground by station car park (NS3557) was determined by Eric Clement as this goldenrod.

Solidago canadensis L.
Canadian Goldenrod

Britain: neophyte (N America); common in the S, rare in Scotland except in the C lowlands.
Hortal; 41/48; frequent.
1st Polnoon, 1957, RM.

Canadian Goldenrod was considered well established 'all over the shire' in 1942 (TPNS) and RG (1930) considered it 'established and frequent', but Polnoon seems to be the first named locality. It was recorded by BWR from six sites in the 1960s–80s. It was well recorded in urban Glasgow during the 1980s (CFOG) but not so commonly in urban Renfrewshire during more recent surveying. It is widespread though, but probably under-recorded, being found on waste ground, roadsides and disused railways, and is capable of persisting in rank grassland.

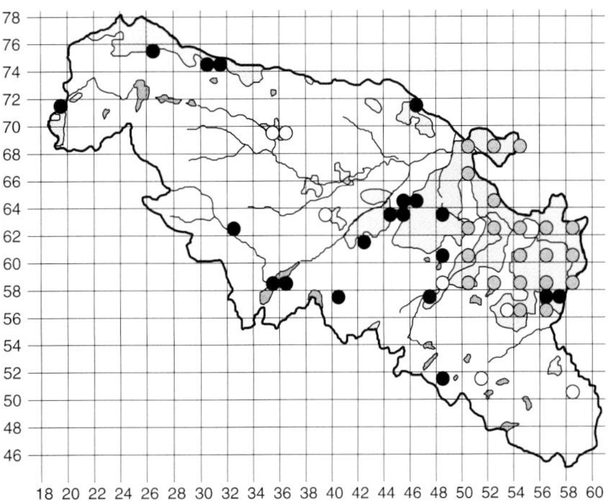

Solidago canadensis

Solidago gigantea Aiton
Early Goldenrod

Britain: neophyte (N America); scattered in N to C Scotland.
Hortal; 15/16; scarce.
1st Finlaystone Estate, 1976, AJS.

Like the previous, and possibly confused with it, this goldenrod grows in similar waste ground, roadside and disused railway localities, but appears to be less frequent, certainly about the Glasgow and Paisley conurbations.

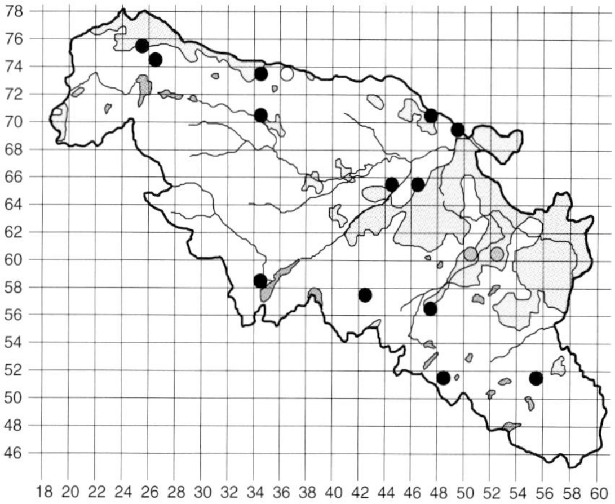

Solidago gigantea

Aster nova-angliae L.
Hairy Michaelmas-daisy

Britain: neophyte (N America); local in S England, very rare in Scotland.
Hortal; 1/1; very rare.
1st Lochwinnoch (NS3558), 2003, IPG.

Aster × *versicolor* Willd.
(*A. laevis* L. × *A. novi-belgii*)
Late Michaelmas-daisy

Britain: neophyte (garden origin); scattered and local.
Hortal; 1/2; very rare.
1st Roadside, Gleniffer Braes (NS4459), 1975, AJS.

A small population occurs to the edge of woodland at Bargarran, Erskine (NS4671, 1997).

Aster novi-belgii L.
Confused Michaelmas-daisy

Britain: neophyte (N America); scattered, frequent in C Scotland.
Hortal; 78/84; frequent.
1st Fields, Gourock, 1916, GL.

This is thought to be the commonest Michaelmas-daisy in the area, but the records may include some for Common Michaelmas-daisy (*A.* × *salignus* Willd.) or other cultivars (CFOG). Two old records from Lochwinnoch and Langbank by Lee (1933) refer to '*A. longifolius*' and as he also mentions '*A. novi-belgii*' – these may represent a different taxon. More recently there were seven records from the 1960s–80s (BWR). Today the map shows a strong lowland and urban-fringe distribution, avoiding cultivated land and the acidic uplands. It is often found on waste ground and rough grassland, where it seems well capable of persisting, especially if the soils are damp; occasionally it can spread into marshes.

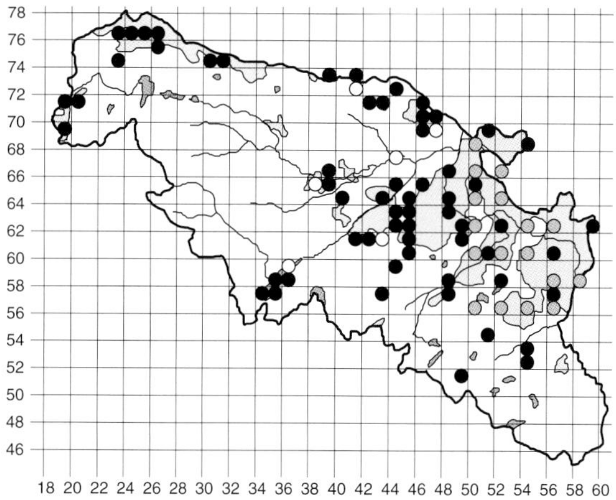

Aster novi-belgii

Aster schreberi Nees
Nettle-leaved Michaelmas-daisy

Britain: neophyte (E N America); only one British record.
Former hortal; 0/1; extinct.
1st 'Lochside Station' [Lochwinnoch], 1922, RG.

Nettle-leaved Michaelmas-daisy is one of the famous Renfrewshire discoveries which Lee (1933) referred to as 'perfectly naturalised'. It was last reported in 1965 by RM and has not been recorded since despite frequent searches.

Aster tripolium L.
Sea Aster

Britain: common on coasts throughout except in N Scotland.
Native; 27/31; frequent on the coast.
1st Firth of Clyde, 1834, MC.

Sea Aster remains a frequent sight along the coastal mudflats and rocks although it has decreased at several points due to development as indicated by Hennedy (1891) when he noted that the 'banks of the river below Renfrew have been built up'. It can extend inland along the estuarine Black Cart and White Cart waters towards Inchinnan (NS4767, 1999) and Paisley (NS4866, 2000) respectively, and up the Clyde to beyond Blythswood (NS5068, 2005).

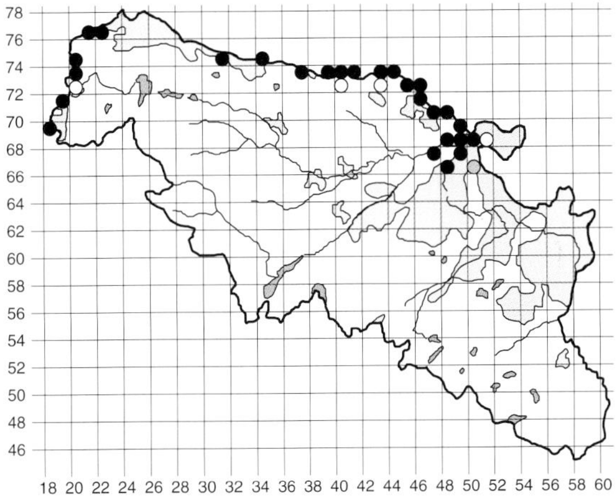

Aster tripolium

Conyza canadensis (L.) Cronquist
Canadian Fleabane

Britain: neophyte (N America); common in the S and E, rare in Scotland.
Accidental; 1/1; very rare.
1st Nitshill, 2004.

A single plant was found at the grounds of GMRC at Nitshill (NS5160).

Conyza bonariensis (L.) Cronquist
Argentine Fleabane

Britain: neophyte (tropical America); rare casual.
Accidental; 1/2; very rare.
1st Giffnock, 1926, RG.

A curious modern find is from the bodywork of a locomotive when stored at GMRC in Nitshill (NS5160, 2008). The engine had recently come from South Africa and photographs taken there indicate that Argentine Fleabane grew as a weed about the locomotive's engine sheds, suggesting that the plants represent very-long-distance stowaways.

Bellis perennis L.
Daisy

Britain: common throughout.
Native; 332/431; very common.
1st Inverkip, 1841, GL.

Daisy is one of the commonest plants in the area and it will be under-recorded from many urban parks and gardens. It can extend into the uplands, being found at 300m where the grass is less acidic or along tracksides, and can also be found on waste ground.

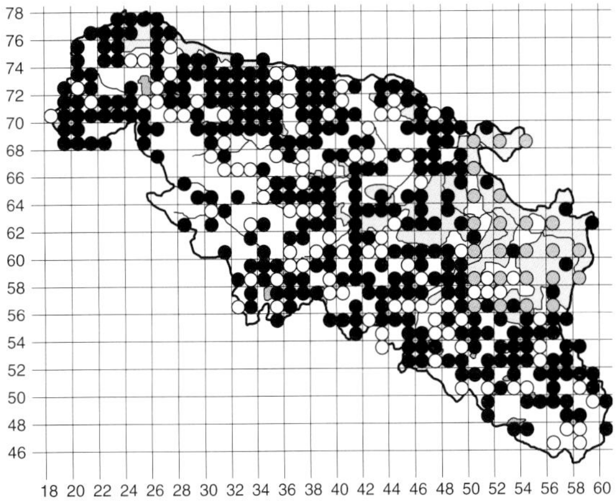

Bellis perennis

Tanacetum parthenium (L.) Sch. Bip.
Feverfew

Britain: archaeophyte (Balkans); common in the S and in lowland Scotland.
Hortal; 45/60; frequent.
1st Paisley and Kilbarchan, 1845, NSA.

Hennedy (1891) considered Feverfew frequent, and often found near farmhouses, and mentioned Gourock and Cathcart. Today it is widespread and

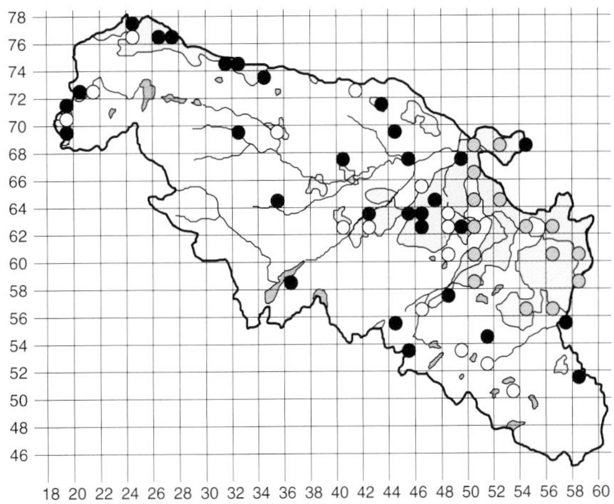

Tanacetum parthenium

generally associated with habitation, although it can be seen at remote sites, occurring at waysides, gardens, waste ground and woodland edges.

Tanacetum vulgare L.
Tansy

Britain: archaeophyte; common in the S and in C and E Scotland.
Hortal; 27/36; frequent.
1st Paisley, 1845, NSA.

Tansy was considered frequent in the 19th century but the only named locations are Cathcart and Gourock (1876, FFWS); it was collected from the banks of the White Cart Water in 1905 (PSY) and RM collected it from Giffnock in 1925 (GLAM), and Lee (1933) added Eaglesham. Today there are few records away from the Glasgow conurbation, with a scattering of isolated sites such as at Kelburn (NS3474, 1998), Greenock (NS2775, 2000) and Lochall Bridge (NS3558, 2001).

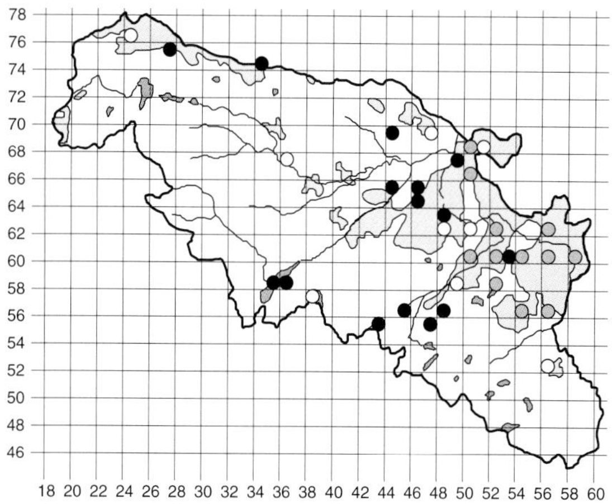

Tanacetum vulgare

Artemisia vulgaris L.
Mugwort

Britain: archaeophyte (Europe); common in the S and in lowland Scotland.
Accidental; 78/95; frequent.
1st Paisley, 1845, NSA.

Mugwort was considered frequent in the past and can still be described as such. It shows a very strong lowland, urban-fringe pattern, notably in the north-west. There are two records from the south-eastern uplands: a roadside at Melowther (NS5450, 2001) and a trackside near Myres Hill (NS5646, 2000). It is often found on waste ground or at waysides, including disused railways, and can often be seen on riverbanks such as by the Black Cart Water.

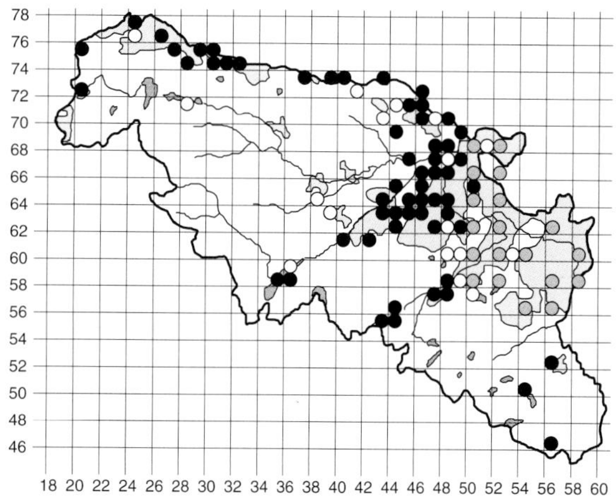

Artemisia vulgaris

Artemisia absinthium L.
Wormwood

Britain: archaeophyte (Europe); frequent in C England, local in E Scotland.
Accidental; 1/1; probably extinct.
1st Meadowside Street, Renfrew (NS5068), 1985, CFOG.

There has been no further sighting of Wormwood.

Achillea ptarmica L.
Sneezewort

Britain: common throughout except in SE England.
Native; 221/307; common in wet places.
1st Paisley, 1869, RPB.

Sneezewort has few old named stations but appears to have always been a common plant. Today it is widespread, occurring throughout the lowlands and extending to the uplands, but rare in the north-central agricultural area. It is generally found in marshy grasslands with Soft-rush or Tufted

Hair-grass, or towards wetland margins, but can occasionally be found on waste ground.

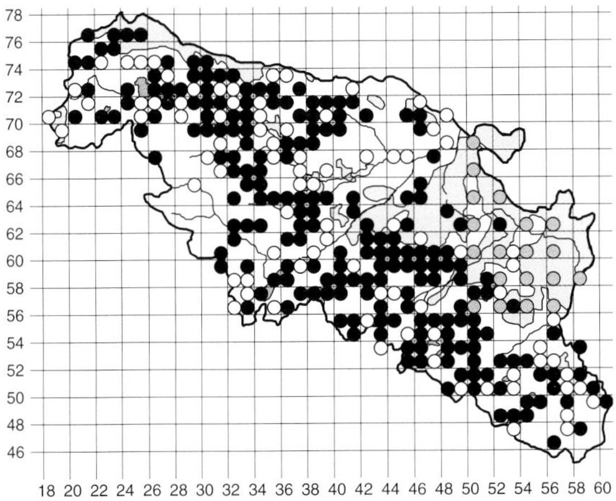

Achillea ptarmica

Achillea millefolium L.
Yarrow

Britain: common throughout.
Native; 344/430; very common.
1st Near Hurlet, 1842, GLAM.

Yarrow is a very common plant of grasslands found throughout the lowlands and extending up into the uplands in semi-improved pastures and along farm tracks. It prefers neutral soils, avoiding acidic or wet types, but can tolerate some improvement or amenity-type cutting. It can also be found in urban areas in gardens and parks, at waysides and on waste ground.

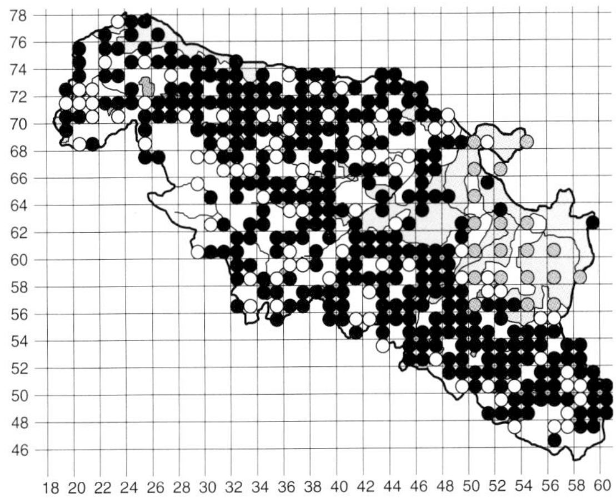

Achillea millefolium

Achillea tomentosa L.
Yellow Milfoil

Britain: neophyte (SW Europe); rare casual.
Former hortal; 0/1; extinct.
1st Neighbourhood of Paisley, 1813, TH.

This garden escape was claimed as a new British record by Hopkirk. Hooker (1821) stated that Yellow Milfoil was 'on hills near Paisley' and added 'discovered by Mr Hugh Ross'. Hennedy (1891) did not mention this location but stated 'Cultivated and not native'. There are no specimens or further details of this unusual record.

Anthemis arvensis L.
Corn Chamomile

Britain: archaeophyte (Europe); frequent in SE, local in lowland Scotland.
Former accidental; 0/4; extinct.
1st Near Paisley, 1860, PSY.

In 1888 Corn Chamomile was reported from a field to the east of Gourock (NHSG) and from Langbank in 1891 (NHSG); Hennedy (1891) mentioned Kilmacolm and considered it 'very rare'. There are no further records.

Anthemis cotula L.
Stinking Chamomile

Britain: archaeophyte (Europe); common in the SE, very rare in Scotland.
Accidental; 0/3; extinct.
1st Near Cathcart, 1865, RH.

Hennedy (1891) considered Stinking Chamomile very rare. It was reported in 1892 from Crosssmyloof (NHSG) and collected from wasteland near Paisley (1910, PSY).

Glebionis segetum (L.) Fourr.
Chrysanthemum segetum L.
Corn Marigold

Britain: archaeophyte (Europe); frequent throughout lowlands.
Accidental; 10/15; rare.
1st Renfrewshire 'weed', 1812, JW.

Corn Marigold, or 'Gule', was listed as one of the four most common 'prevailing weeds' of cornfields (Wilson 1812) and later it was considered 'so plentiful' that Causewayside was called 'Gooleyland' (1869, RPB). It was frequent in the late 19th century, but by then considered to be in decline (CFOG); other named locations include Cartside (1856, PSY), Cathcart and Pollokshields (1876, FFWS), Kilmacolm (1896, GL) and Banks of Earn, Newton Mearns (1908, GL), and in 1942 RM knew it from Houston and Kilbarchan. More recently, it was found (BWR) west of Bridge of Weir (NS3667) in 1982 and at a roadside West Fulton (NS4265, 1987), and three relevant records were made during surveying for CFOG. The other modern records are from near Elderslie (NS4462, 1993), a gateway near Barr Loch (NS3558, 1995), a roadside near Brownsfield (NS4667, 2000) and a field margin, Wester Fulwood (NS4366, 2001), and from GMRC, Nitshill (NS5160, 2003) as a seed contaminant. It is increasingly used in wildflower seed mixes so can appear sporadically, as along the new section of the M74 at Polmadie (NS5962, 2011, PM).

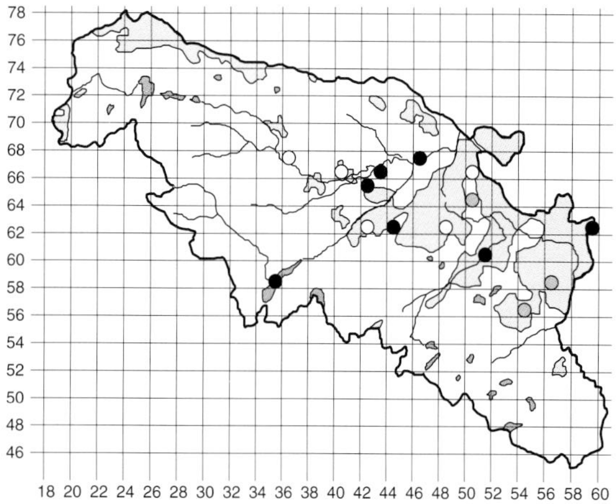

Glebionis segetum

Leucanthemum vulgare Lam.
Oxeye Daisy

Britain: common throughout except in N Scotland.
Native and hortal; 123/164; common, lowland.
1st Renfrewshire 'meadows', 1812, JW.

Oxeye is a common plant in the lowlands and the map shows an absence from the uplands and also upland fringe pastures; one exception is the track approaching High Myres (NS5646, 2000). It is most likely to be found in free-draining neutral grasslands; today such sites are represented by waste ground, waysides and disused railways rather than 'old meadows', and some records may reflect origins from wildflower mixes.

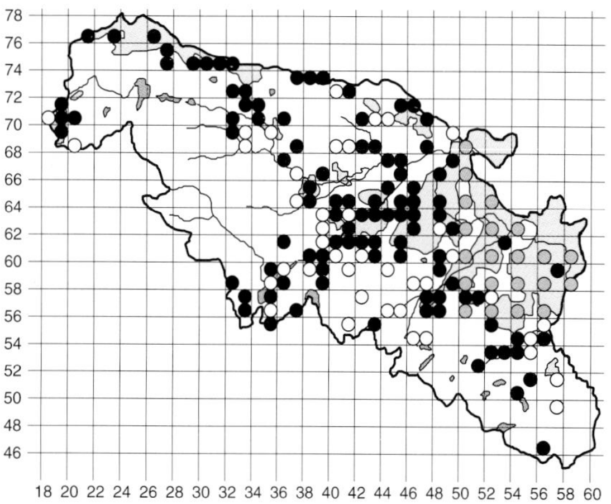

Leucanthemum vulgare

Leucanthemum × superbum (Bergmans ex J.W. Ingram) D.H. Kent
Shasta Daisy

Britain: neophyte (hybrid origin); scattered and local in C Scotland.
Hortal; 24/26; occasional.
1st Uplawmoor, 1972, RM.

Shasta Daisy is an occasional outcast, found widely scattered throughout the area along the urban fringe. Most, if not all, records refer to garden outcasts. A couple of rural records include roadside, Cairn Hill (NS4851, 2002) and nearby White Loch (NS4862, 2000). A specimen presumed to be Autumn Oxeye (*Leucanthemella serotina* (L.) Tzvelev) from a roadside at Muirhouse Farm (NS4256, 1999) has subsequently been redetermined as Shasta Daisy.

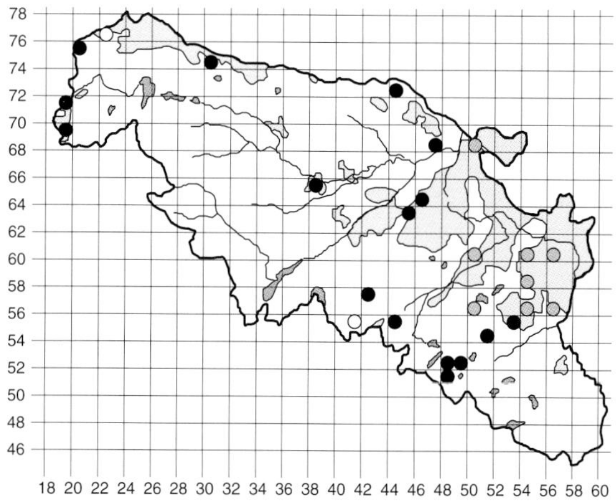

Leucanthemum × superbum

Matricaria chamomilla L.
Matricaria recutita L.
Scented Mayweed

Britain: archaeophyte (Europe); common in the S, scattered in lowland Scotland.
Accidental; 3/6; very rare.
1st Giffnock, 1892, GL.

Scented Mayweed is annotated in the MY list with '1865' stated but no location is given, and there are no other literature records for a plant considered rare by Hennedy (1891) and excluded from the

Renfrewshire list (TPNS 1915). It was collected from Kilmacolm in 1933 (PSY) and later found along the Greenock Railway (NS4665, 1956, PSY). During surveying for CFOG it was known from Renfrew (NS56D, 1983, AMcGS) and Scotstounhill (NS56J, 1988, JHD). More recently it was recorded from weedy fields at Balgray (NS5157, 1999).

Matricaria discoidea DC.
Pineappleweed

Britain: neophyte (Asia/N America); common throughout.
Accidental; 189/306; common.
1st Lochwinnoch and Inverkip, 1910, GL.

Other specimens of Pineappleweed were collected (GLAM) from Giffnock Quarries (1925) and Clarkston (1931) and it was recorded from Lochwinnoch in 1956 (BWR). By the 1960s–80s recording (BWR) it was obviously well established, with 174 records made. The map shows a widespread distribution, although it may have been under-recorded in the modern period rather than be in any decline. It is typically a plant of weedy places: waste ground, paths, tracks, roadsides, farm gateways and field margins.

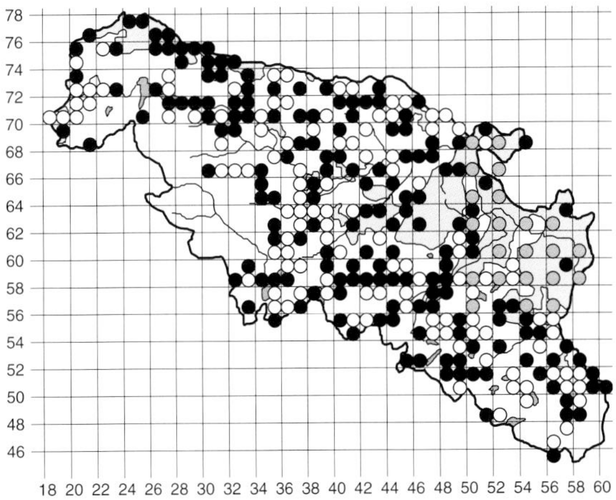

Matricaria discoidea

Tripleurospermum maritimum (L.) W.D.J. Koch
Sea Mayweed

Britain: common all round the coast.
Native; 16/17; frequent on the coast.
1st Firth of Clyde, 1834, MC.

Apart from Kelly Glen (1887, NHSG) there are no other old localities but Hennedy (1891) referred to the var. '*maritima*' ('merely a fleshy state') growing on the sea coast; it was also not distinguished from the following species during the 1960s–80s period, although there are some coastal records for the aggregate. During the modern period Sea Mayweed has been recorded at several coastal locations, usually on shingles, sandy shores, walls and rocks.

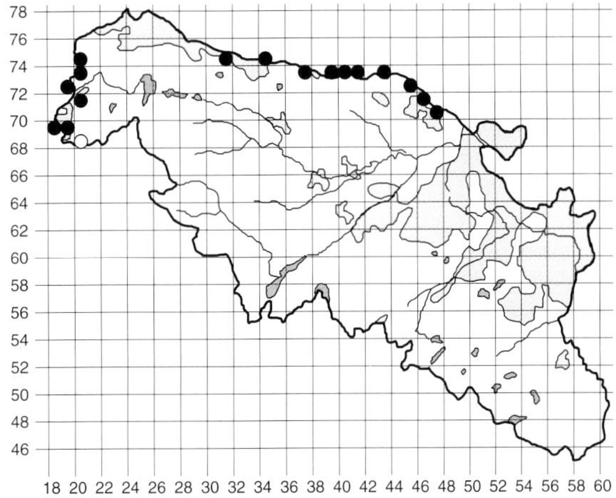

Tripleurospermum maritimum

Tripleurospermum inodorum (L.) Sch. Bip.
Scentless Mayweed

Britain: archaeophyte (Europe); common throughout except in uplands.
Accidental; 102/166; common at open places.
1st Paisley, 1845, NSA.

There are several old records for this common weed of 'fields and waste places' (Hennedy, 1891), although some of the coastal records may refer to Sea Mayweed, notably during the 1960s–80s period. Today Scentless Mayweed is still commonly found in the lowlands on waste ground, but also as a farm weed and, often extending into the uplands, along tracks and roadsides.

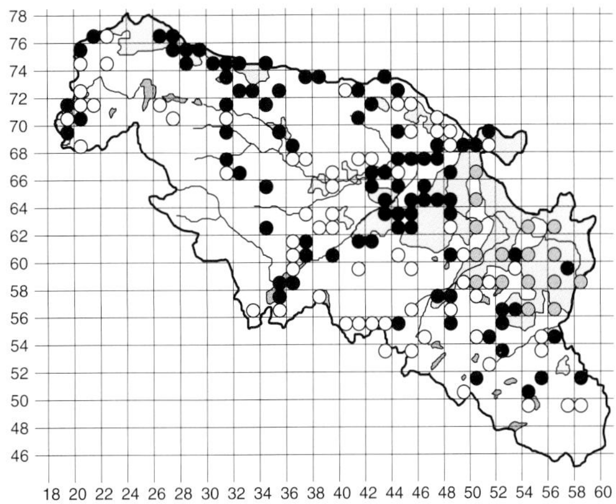

Tripleurospermum inodorum

Senecio cineraria **DC.**
Silver Ragwort

Britain: neophyte (S Europe); scattered in the S.
Hortal; 2/2; very rare.
1st Cathcart, 1994, CFOG.

The original site is from a grassy embankment above the road at Manse Brae (NS5860) and more recently it was found on waste ground as an outcast in Gourock (NS2477, 2001). PM collected the hybrid with Common Ragwort (*S.* × *albescens* Burb. & Colgan) from on top of a roadside wall at the original location (1994) and also from a roadside at Clarkston (NS5757, 1999).

Senecio jacobaea **L.**
Common Ragwort

Britain: common throughout.
Native; 304/391; very common.
1st Near Hurlet, 1842, GLAM.

Common Ragwort was noted as very common by Hennedy (1865) – he quipped, 'a good crop with bad farmers'. Today it remains very common, occurring in neutral grasslands, at roadsides and other waysides, and on waste ground, often in some abundance. It can extend to grassy places in the uplands, as at Muirshiel Barytes Mine (NS2864, 2003), but generally it avoids wet or acidic soils.

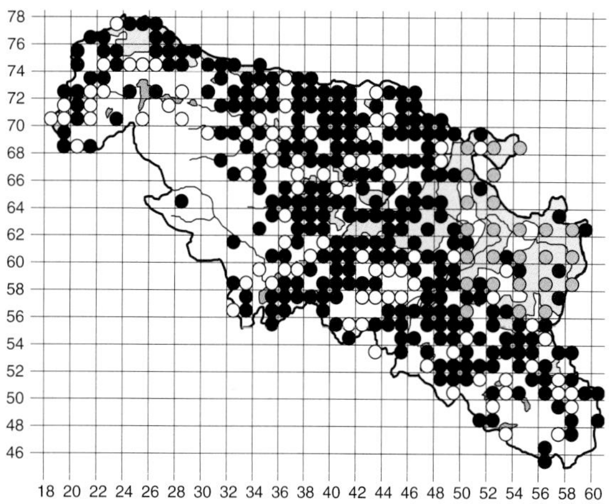

Senecio jacobaea

Senecio aquaticus **Hill**
Marsh Ragwort

Britain: common but rare in C N Scotland and E England.
Native; 224/266; common.
1st Renfrew, 1865, RH.

Marsh Ragwort shows a strong rural pattern, avoiding many urban zones, but is also absent from the acidic peaty uplands. It is commonly encountered along burn sides and some rivers and loch shores, and is often found in marshy rush pastures. The hybrid with Common Ragwort (*S.* × *ostenfeldii* Druce) was recorded from Barochan (NS4069, 1998) and from by Green Water, Duchal Bridge (NS3368, 2007), and is likely to be more widespread; an immature 19th-century specimen in GRK, from 'Inverkip Rd', appears to be this hybrid.

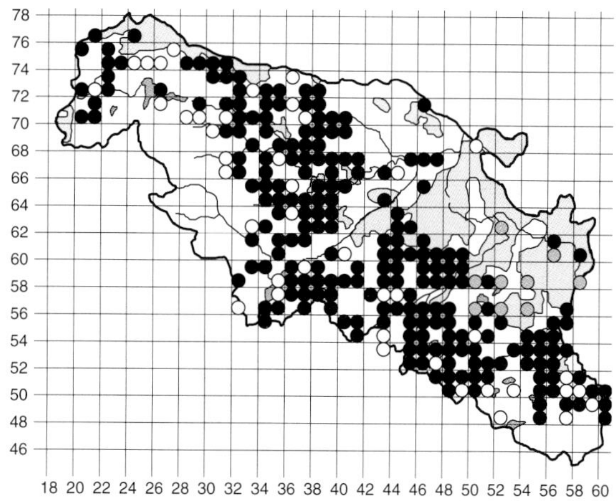

Senecio aquaticus

Senecio sarracenicus **L.**
Senecio fluviatilis Wallr.
Broad-leaved Ragwort

Britain: neophyte (Europe); scattered, local in NW England and C Scotland.
Hortal; 4/20; rare.
1st Cloak, 1834, MC.

Broad-leaved Ragwort was recorded from Lochwinnoch in 1845 (NSA), collected from Kilmacolm in 1880 (GL) and reported from the Locher Water in 1886 (NHSG), and Hennedy (1891) knew it from Lawmarnock and Gryfeside, near Kilmacolm; a more recent record was from Farm Road near Kilbarchan (1942, TPNS). It was recorded nine times between 1969 and 1986 (BWR) and it was collected from by the White Cart Water east of Pollok House in 1974 (GLAM), where it still occurs today (NS5561, 2005). There are only three other modern records, all from water-side stations:

by a burn at Tandlemuir (NS5561, 1997), Green Water, Stepend Bridge (NS3467, 1999) and by the Gryfe Water at Houstonhead (NS3965, 2002). Broad-leaved Ragwort has been around for nearly 200 years, and is vigorous at its modern stations, but appears to have disappeared from several former locations.

Senecio squalidus L.
Oxford Ragwort

Britain: neophyte (S Europe); common, rare in Scotland except in the C and SE.
Accidental; 33/34; locally frequent.
1st Rubbish dump, St James, Paisley (NS4665), BWR, 1961.

Oxford Ragwort appears to have been a recent arrival in the area but it has spread locally. It was collected from James Watt Dock, Greenock (1965, GLAM) and recorded by AJS from behind Paisley Gilmour Street Station in 1974 and from waste ground at Port Glasgow in 1975. It is well noted in CFOG and, outside Glasgow, there are two main urban centres: Greenock and Paisley; isolated colonies are developing at Bridge of Weir (NS3865, 2005) and Kilmacolm (NS3670, 2007).

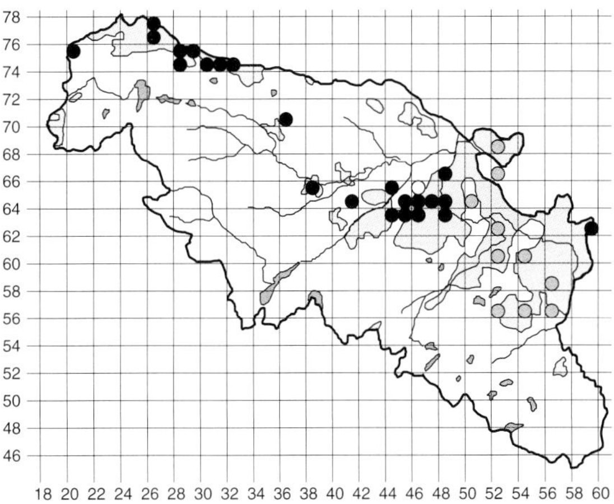

Senecio squalidus

Senecio vulgaris L.
Groundsel

Britain: common throughout except in N Scotland.
Native; 156/233; common.
1st Paisley, 1869, RPB.

There are few named old localities for this weedy species, considered very common in the 19th century. Today Groundsel is widespread, occurring in lowland rural and urban areas, on waste ground, in gardens, at disturbed waysides and on farmland.

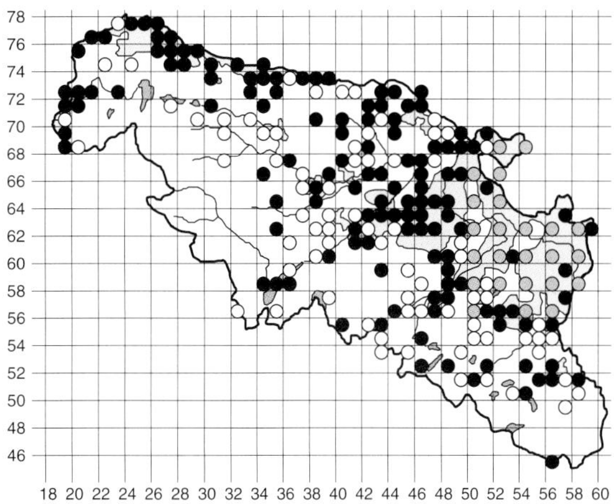

Senecio vulgaris

Senecio vulgaris L. var. *hibernicus* Syme
'Rayed Groundsel'

Britain: hybrid origin; unclear distribution pattern.
Native or accidental; 12/13; scarce.
1st Inchinnan, 1970, JP.

This species was recorded a few times during CFOG surveys, the earliest of these records being from Williamwood (NS56U, 1978). The more recent records are from waste ground and urban sites,

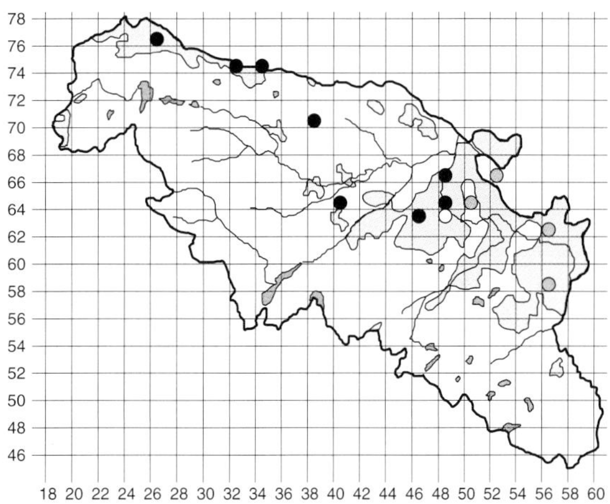

Senecio vulgaris var. *hibernicus*

except for a remote roadside record from Mid Glen (NS3870, 1997).

Senecio sylvaticus L.
Heath Groundsel

Britain: frequent except in N Scotland.
Native; 5/10; rare.
1st Lochwinnoch, 1845, NSA.

Hennedy (1891) thought Heath Groundsel 'not common' and mentioned Gourock and there is also a record from Greenock (west) (1880, NHSG). More recently it was recorded from Black Loch (NS4951, 1964, BWR) and the disused railway, Kilmacolm (NS3569, 1984, BWR), and was recorded at the VC boundary at Braehead (NS5167, 1985, CFOG). The four modern records are from a felled plantation, North Barlogan (NS3768, 1998), a bracken-covered embankment, Muirhead (NS4855, 2002), a track side, Kilmacolm Golf Course (NS3769, 2003) and the edge of a rock outcrop, Nether Cairn (NS4852, 2008).

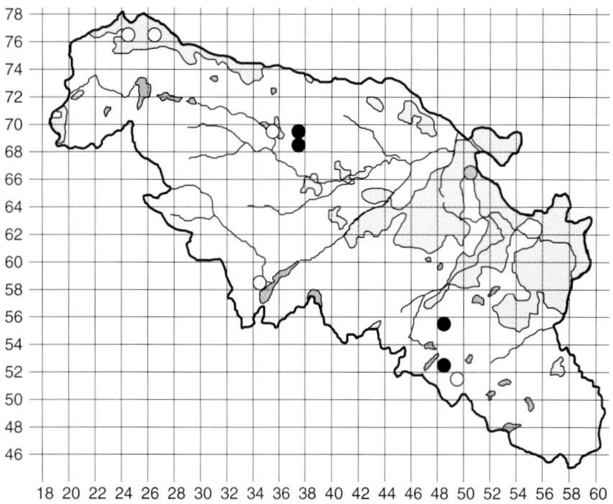

Senecio sylvaticus

Senecio viscosus L.
Sticky Groundsel

Britain: possibly native; common throughout except in N Scotland.
Accidental; 50/61; frequent, urban.
1st Renfrewshire, 1834, MC.

There are only a few old records for Sticky Groundsel, which Hennedy thought uncommon locally in 1865: West of Greenock (1880, NHSG), Kilmacolm (1888, PSY) and Nitshill (1916, GL), and RM knew it from Giffnock (1929) and Nitshill (1929). Today it is fairly well distributed, mainly occurring on urban waste ground; remote sites include Kaim Dam (NS3462, 2001) and a roadside, Brownside (NS5051, 2001).

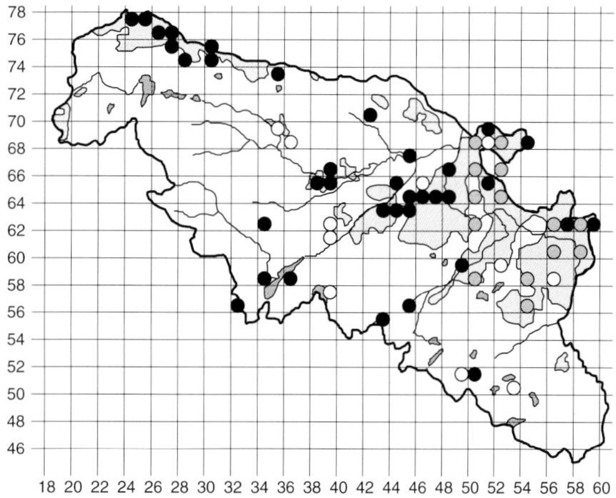

Senecio viscosus

Brachyglottis × jubar P.D. Sell
Brachyglottis 'Sunshine'
Shrubby Ragwort

Britain: neophyte (garden origin); scattered in the S.
Hortal; 1/1; very rare.
1st Langbank, 1998.

This widely planted, colourful shrub is well established by the motorway at Langbank (NS3873).

Sinacalia tangutica (Maxim.) B. Nord
Chinese Ragwort

Britain: neophyte (China); rare and scattered.
Hortal; 3/5; very rare.
1st Banks of Cart, Newlands, 1926, RG.

RM collected Chinese Ragwort from a disused quarry at Mearnskirk (1966, E) and during surveying for CFOG this garden outcast was recorded from three places in VC76 where it has persisted into the modern survey period: Rouken Glen (NS5458, 1996), a roadside, Springhill (NS5057, 1999) and a roadside, Cowglen (NS5460, 2005).

Ligularia sibirica (L.) Cass.

Britain: neophyte (E Asia); very rare as an escape.
Hortal; 1/1; very rare garden escape.
1st Banks of White Cart, Rosshall, 1991, CFOG.

This plant appears to have been short lived at its riverbank site near to the gardens at Rosshall (NS5263).

Doronicum pardalianches L.
Leopard's-bane

Britain: neophyte (W Europe); scattered throughout, rare in W Scotland.
Hortal; 5/12; rare.
1st Paisley, 1845, NSA.

Considered a rare plant in the 19th century, there are records for Leopard's-bane from west of Greenock (1888, NHSG), Crookston (1873, MY), Duchall, Kilmacolm (1890, AANS), the Inverkip area (1891, AANS) and Milliken, Kilbarchan (1891, AANS). It was recorded from the Giffnock area (NS5669) in 1966 (JDM) and there are a couple of roadside records from BWR: Waterfoot (1973) and Knowes (1975); the latter presumably the same as the roadside at the Knowes House locality (NS3656, 1999). It still appears to be a rare plant in the area, recorded from north-east of Arkleston Cemetery (NS56C, 1988, CFOG), Locher Water (NS4064, 1999), Barhill Wood (NS4163, 1999) and Houston Estate (NS4067, 2002).

Doronicum × *willdenowii* (Rouy) A.W. Hill
Willdenow's Leopard's-bane

Britain: neophyte (W Europe or garden origin); scattered.
Hortal; 2/4; very rare.
1st Shielhill, 1888, GRK.

A further GRK specimen from Inverkip Road (1889) also appears referable to this hybrid. It was found along the Gryfe Water at Duchal House (NS3568) in 2000; it is possible that old records from here (as '*D. plantagineum*') may refer to this hybrid, in which case it has persisted for more than 100 years. A further record is from the riverside at Waterfoot (NS5655, 2001).

Doronicum plantagineum L.
Plantain-leaved Leopard's-bane

Britain: neophyte (W Europe); local, scattered in E Scotland.
Former hortal; 0/2; extinct.
1st 'Duchall', 1889, AANS.

This first record 'between Craigbet and Kilmacolm' may be an error for the hybrid (see above), but there is a specimen collected from Waulkmill [Darnley] Glen in 1914 (GL) which appears correct; it does not occur there now.

Tussilago farfara L.
Colt's-foot

Britain: common throughout.
Native or accidental; 306/393; very common.
1st Paisley, 1869, RPB.

Hennedy (1891) wrote that Colt's-foot was 'very common and troublesome' and today it can still be said to match the first part of his description, though perhaps not the latter. It can be found throughout the VC, including some upland areas. It is most often found on waste ground and can persist in secondary grasslands, but it is also found along waysides, on farmland, along watercourse margins and sometimes in woodlands.

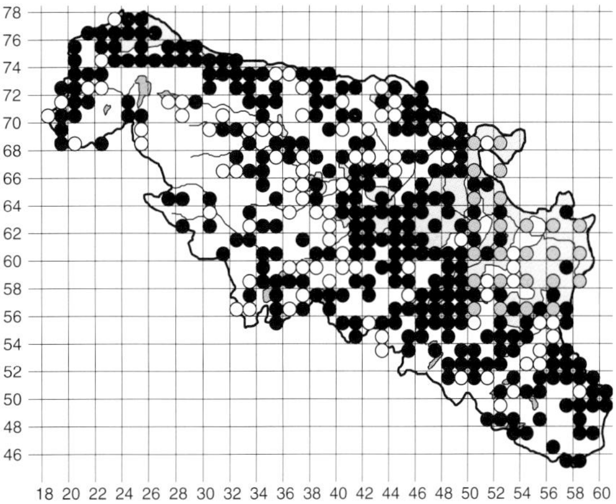

Tussilago farfara

Petasites hybridus (L.) P. Gaertn., B. Mey. & Scherb.
Butterbur

Britain: common except in N Scotland.
Native; 50/61; frequent.
1st Lochwinnoch and Paisley, 1845, NSA.

In 1887 (NHSG) Butterbur was recorded from Loch

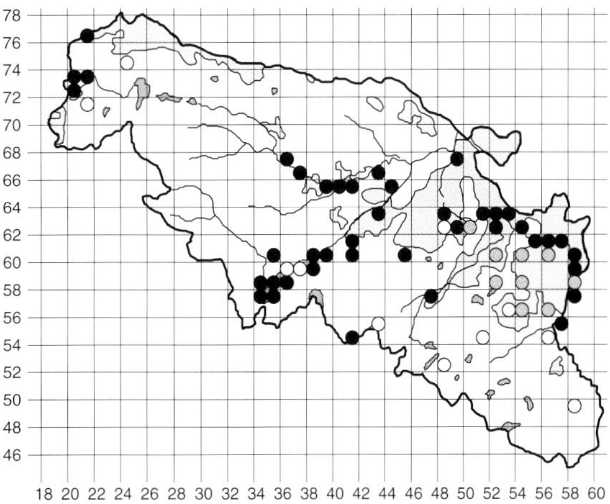

Petasites hybridus

Libo and Castle Semple Loch, where it is still present, and it was considered common in 1891 (Hennedy) and recorded from by the White Cart Water; it was collected from Hawkhead Mill in 1904 (PSY). In 1869 (RPB) it was noted from 'sand at water edge', a habitat where it can be readily found today along the larger lowland riverbanks such as the Gryfe, Black Cart, Levern and White Cart waters; it is also to be found in the west associated with the Kip Water (NS2173, 2003). It was reported from Ardoch Burn in the south-east in 1982 (NS5449, BWR).

Petasites albus (L.) Gaertn.
White Butterbur

Britain: neophyte (Europe); frequent in C and NE Scotland, scarce in the S.
Hortal; 9/12; rare.
1st Neighbourhood of Barrhead, 1880, NHSG.

A later record is from Paradise, Finlaystone (1890, AANS), where White Butterbur can still be found (NS3673, 2011, P&TNT), and a further old record is from Locher Mill in 1902 (PSY), from where the plant can also be found at Locher Water (NS4064, 2011); JDM reported it from a quarry near Port Glasgow (NS3573) in 1961. RG first reported it from Darnley in 1920 and RM noted it there in 1956, and it can be still seen today at several places along the hedgerow on Corslet Road (NS5259, 2005). There are three other CFOG records from the south-east and three other modern ones: West Lady Wood (NS5256, 2000), Pilmuir Reservoir (NS5154, 2000) and Auldhouse Burn, Glasgow (NS5560, 2005). Even though it grows quite vigorously locally and has been in the VC for a long time, it appears to have not spread widely.

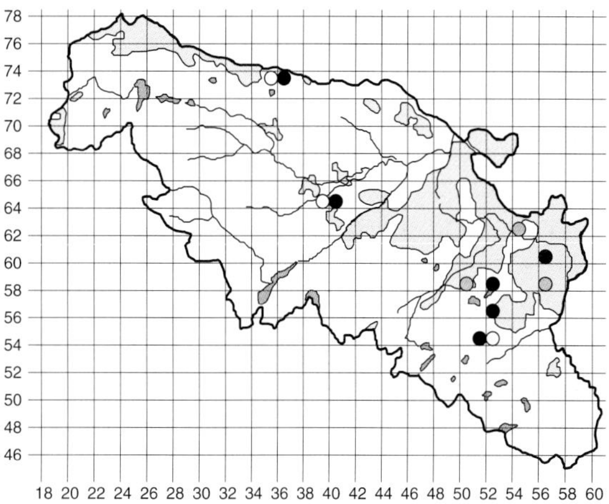

Petasites albus

Petasites fragrans (Vill.) C. Presl
Winter Heliotrope

Britain: neophyte (N Africa); common in the S, lowland in Scotland.
Hortal; 7/8; rare.
1st Newlands, 1920, GL.

The next record was from a roadside at Knowes (1975, BWR), where Winter Heliotrope can still be seen today (NS3656, 1999); it was also found at Darnley Mains in 1990 (NS5358, CFOG), although the site was destroyed by developers in spring 2007. Other records include roadside grassland, West Tandlemuir (NS3361, 1997), Shillford (NS4456, 1999), Gourock (NS2477, 2001), Mill Burn, Pomillan (NS3466, 2006) and Craigends Estate (NS4166, 2002).

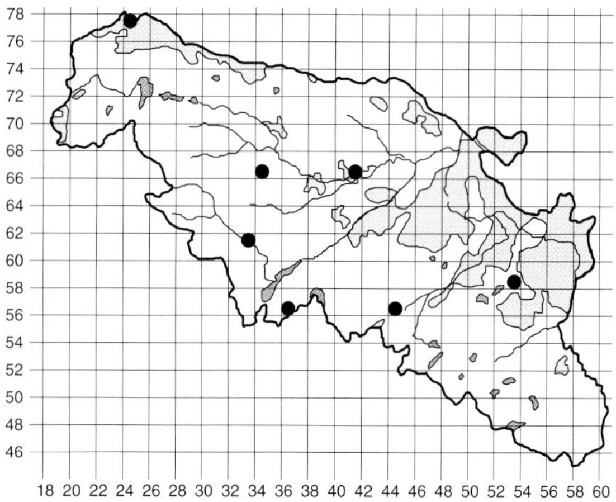

Petasites fragrans

Ambrosia artemisiifolia L.
Ragweed

Britain: neophyte (N America); scattered casual, rare in Scotland.
Accidental; 1/3; very rare casual.
1st Crossmyloof, 1894, GL.

Ragweed was also collected from wasteland near Paisley (1910, PSY) and more recently appeared as a garden weed in Greenock (NS27, 2003, Paul Matthews).

Ambrosia trifida L.
Giant Ragweed

Britain: neophyte (N America); rare casual.
Former accidental; 0/2; extinct casual.
1st Crossmyloof, 1894, GL.

Like Ragweed, this species was also collected from wasteland near Paisley in 1910 (PSY).

Ambrosia maritima L.

Britain: neophyte (S Europe); very rare casual.
Former accidental; 0/1; extinct casual.
1st Giffnock, 1892, GL.

Rudbeckia laciniata L.
Coneflower

Britain: neophyte (N America); scattered and local.
Former hortal; 0/1; extinct casual.
1st Newlands, 1921, RG.

Helianthus annus L.
Sunflower

Britain: neophyte (N America); casual, frequent in the S.
Hortal; 1/2; very rare.
1st Rubbish tip, Barr Loch, 1969, AMcGS.

The only modern record is from Lochwinnoch (NS3558, 2005, IPG).

Helianthus × laetiflorus Pers.
(*H. rigidus* (Cass.) Desf. × *H. tuberosus* L.)
Perennial Sunflower

Britain: neophyte (hybrid origin); scattered, occasional in the S.
Former hortal; 0/1; extinct.
1st Newlands, 1919, RG.

Galinsoga quadriradiata Ruiz & Pav.
Shaggy-soldier

Britain: neophyte (S America); frequent in the S.
Accidental; 2/2; very rare.
1st Crossmyloof, 2009, PM.

Recorded elsewhere in Glasgow on a few occasions (CFOG), the only VC records are from Boleyn Road (NS5762) and Quarrelton Paisley (NS4262, 2012, LP).

Bidens cernua L.
Nodding Bur-marigold

Britain: frequent in S England, rare in lowland Scotland.
Native; 4/4; rare.
1st Renfrewshire, 1883, TB.

The first named location is Glen Moss (1957, ERTC) and refers to var. *radiata* DC. Nodding Bur-marigold can still be found at Glen Moss (NS3871, 1996, SNH-LS), and has also been recorded from a pond on Barscube Hill (NS3669, 1994, CD) and at two sections of strandline at Barr Loch (NS3457 and NS3558, 1996, DH).

Bidens tripartita L.
Trifid Bur-marigold

Britain: frequent in the S, local in lowland Scotland.
Native; 13/15; scarce.
1st St Bride's Mill, Kilbarchan, 1845, NSA.

There is a specimen from 'Renfrew' in GL (1857) and another one is the first record from Balgray Reservoir (1901, GL), where Trifid Bur-marigold was seen in 1975 (NS5157, BWR) and can still be found (NS5057, 2000; NS5156, 2006); other old records are from Stanely Reservoir (1869, RPB), Donald's Wood in Paisley (1942, TPNS) and Barr Loch (NS3557, 1969, BWR). Trifid Bur-marigold was known from three sites in south-east Glasgow (CFOG) and other modern records include Barr Loch (NS3557, SNH-LS), Midton Loch (NS4158, 1993), Craighall Dam (NS4755, 1994), Auchendores Reservoir (NS3572, 1998), Linwood Pool (NS4564, 2000) and Craigmarloch (NS3472, 2006).

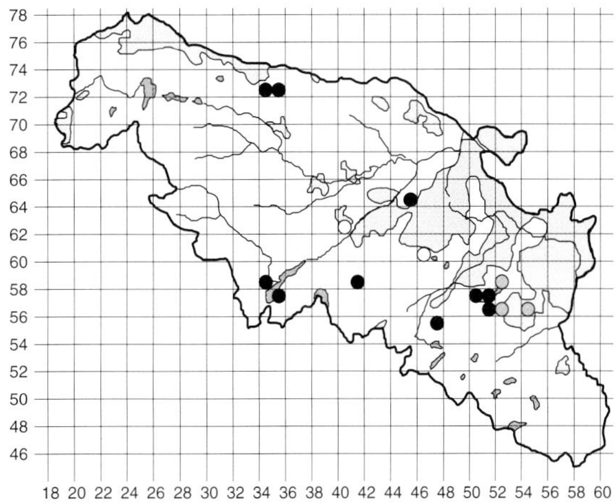

Bidens tripartita

Tagetes sp.
Marigold

Britain: neophyte (C America); rare casual.
Former hortal; 0/1; extinct casual.
1st Rubbish tip, Barr Loch, 1969, AMcGS.

It is not clear what species was found at the original location, but there have been no further records of this garden outcast.

Eupatorium cannabinum L.
Hemp-agrimony

Britain: common in the S, mainly coastal in Scotland.
Native; 9/12; rare, western.
1st Gourock and Wemyss Bay, 1865, RH.

Lee (1933) repeated both first localities for Hemp-agrimony and it was collected from Gourock in 1880 (GL). MY (1865) noted it from Holland Bush (near Dykebar). Today there are a cluster of records mainly associated with the coastal (cliff) woodlands between Wemyss Bay (NS1969, 2001) and Lunderston Bay (NS2074, 1996). There are two records from further east, both perhaps for recent invasives: disturbed grassland, Boden Boo, Erskine (NS4572, 1997) and rough grassland, Paisley Moss (NS4665, 2000).

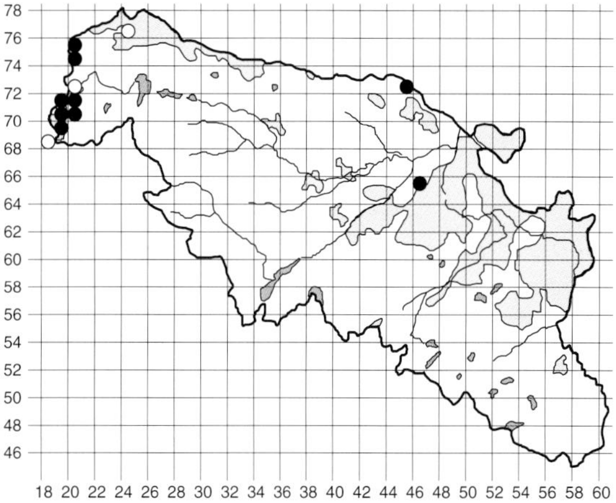

Eupatorium cannabinum

Heliopsis helianthoides (L.) Sweet
Rough Ox-eye

Britain: neophyte (N America); very rare casual.
Former hortal; 0/1; extinct.
1st Newlands, 1921, RG.

Dendranthema × *grandiflorum* group
Florist's Chrysanthemum

Britain: neophyte (cultivated); rare casual.
Former hortal; 0/1; extinct casual.
1st Giffnock, 1922, RG.

ESCALLONIACEAE

Escallonia macrantha Hook. & Arn.
Escallonia

Britain: neophyte (Chile); scattered, mainly in the SW.
Hortal; 5/5; rare.
1st Crookston, 1986, CFOG.

The original record for Escallonia was from the relict gardens at the site of Crookston House (NS56G) and there are four modern records of persistent bushes: near gardens to the edge of Earn Hill, Banks (NS2375, 2002), by the coast path at Lunderston Bay (NS2374, 2006), waste ground, Trumpethill (NS2176, 2012) and a solitary relict by the old walled garden at Waterside (NS5160, 2005).

ADOXACAEAE

Adoxa moschatellina L.
Moschatel

Britain: common throughout except NW Scotland.
Native; 42/55; occasional.
1st Lochwinnoch, Kilbarchan and Paisley, 1845, NSA.

There are quite a few old records and specimens collected, indicating that Moschatel was widespread in the past, which is probably still the case today, although its old woodland habitat has probably diminished. It is present in woodlands along many watercourses, notably the White Cart Water, but probably overlooked being small and early-flowering. There is no modern record from the River Calder, although there is one for Castle Semple Loch (NS3658, 1992, CMRP) and Belltrees (NS3758, 1999). It can also occur at roadside hedge banks as at Corsliehill Road, Barochan (NS4168, 1997).

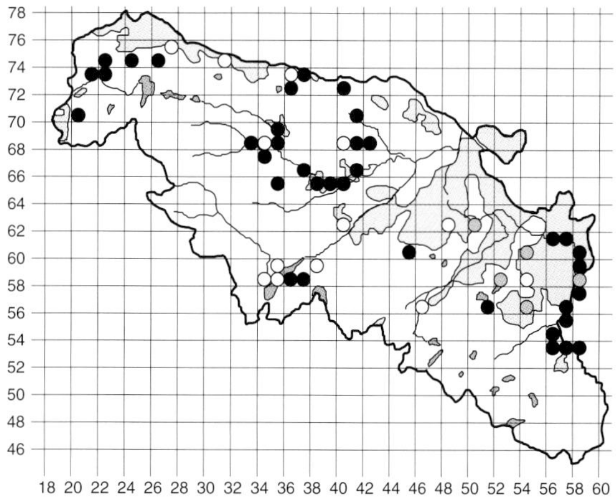

Adoxa moschatellina

CAPRIFOLIACEAE

Sambucus racemosa L.
Red-berried Elder

Britain: neophyte (N temperate zone); frequent in S and E Scotland, rare in England.
Hortal; 7/14; rare.
1st The Cottage, Paisley, 1902, PSY.

There are no other old records before the 1960s–80s, when it was recorded on eight occasions, including from estate woodlands at Gleddoch (NS3872, 1971), White House, Kilbarchan (NS4163, 1976), Barochan (NS4168, 1984) and Glentyan (NS3963, 1984). Red-berried Elder may still be present at these but apart from two CFOG records, the only other modern records are from Formakin Estate (NS4170, 1994), woodland by the Gryfe Water, Kilmacolm (NS3569,

1997), Wemyss Bay (NS1968, 2003), Castle Semple (NS3660, 1992, CMRP) and Glen Moss (NS3669, 1993, TNT).

Sambucus nigra L.
Elder

Britain: common throughout except in N Scotland.
Native; 259/302; very common in lowlands.
1st Cathcart, 1865, RH.

A very common shrub or small tree showing a strong lowland pattern, found in rural and urban situations. Elder is rare in the uplands where the soils are more acidic and its lowland presence tends to mark out richer soils, e.g. along watercourses and about farms, plus also shady plantations, scrub, hedgerows, disused railway banks and waste ground.

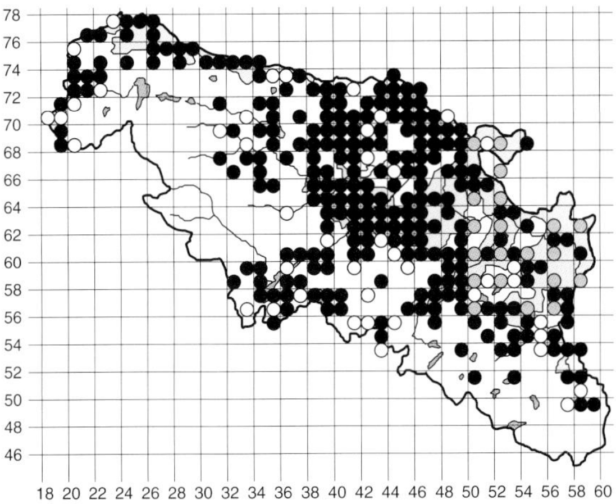

Sambucus nigra

Sambucus ebulus L.
Dwarf Elder

Britain: archaeophyte (Europe); frequent in the S, rare in Scotland.
Hortal; 0/3; very rare.
1st Eaglesham, 1899, NHSG.

The first report was made by J Orr. Later Lee collected specimens of Dwarf Elder from Polnoon in 1925 – presumably the same place as RM's 1928 'roadside, south of Eaglesham' record – and there is an undated specimen, probably from the late 19th century, from Millhall (at 'Polnoon'), collected by George Horn (GLAM). The most recent record from this area is that of PM in 1973 (NS5851). Dwarf Elder is also marked on cards (BWR) from Wemyss Bay (NS1970, 1974) and Lunderston Bay (NS2074, 1983), but there is no supporting literature and it has not been reported before or since from these places.

Viburnum opulus L.
Guelder-rose

Britain: common in the S, frequent in C Scotland.
Native and hortal; 19/27; scarce.
1st Lochwinnoch, 1845, NSA.

Other old localities include Castle Semple Loch (1887, NHSG), Langside (1858, GGO) and Bridge of Weir (1906, PSY); Guelder-rose is still found at the first site but has not been recorded along the Gryfe Water catchment in recent times. It can be found in woodlands along watercourses such as the White Cart Water (CFOG), Brock Burn (NS5258, 2004), Swinetrees Burn (NS4160, 1993), the River Calder (NS3459, 1995) and the Black Cart Water (NS4263, 2001); the northern outlier is at Devol Glen (NS3174, 1995). Some of the CFOG records may refer to introduced plants; two other modern records are for planted or outcast material: a disused railway, Brimstone (NS4656, 1998) and Jenny's Well (NS4962, 2001).

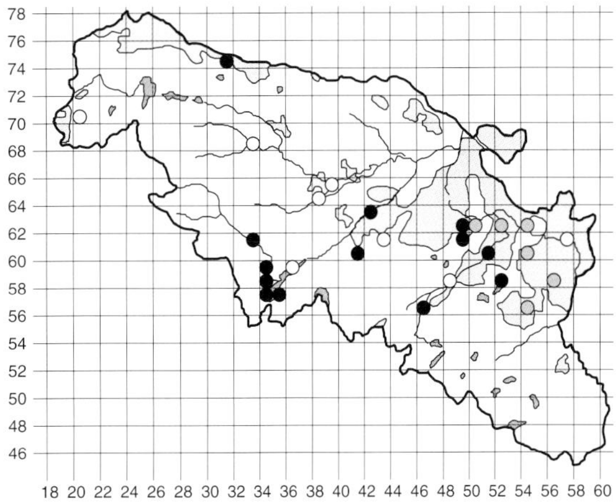

Viburnum opulus

Viburnum lantana L.
Wayfaring-tree

Britain: common in the S, local alien in the N.
Hortal; 2/6; very rare.
1st Devol Glen, 1884, GL.

Several authors repeat the first record but there have been no recent or modern sightings from Devol Glen. Wayfaring-tree was also collected from Pollok Estate in 1905 (GL) and a single shrub was reported from a hedge at Darnley [Waulkmill] Glen in 1890 (AANS); there are no modern records from either (CFOG). More recently it was reported from the old bridge (Bridgend) on the River Calder in 1948 (ANG), but not since. Today it is known from a few spots along the Gryfe Water, east of Bridge of Weir (NS3965, 1997) and the coast at Langbank (NS3773, 1998), presumably recently planted at the latter. An as yet unidentified non-flowering shrubby *Viburnum* grows under mature willow scrub at the edge of Paisley Moss (NS4665, 2004).

Symphoricarpos albus (L.) S.F. Blake
Snowberry

Britain: neophyte (W N America); common except in N and W Scotland.
Hortal; 124/138; common.
1st Glentyan Estate, 1961, BWR.

There are a quite a few other records for Snowberry from the 1960s–80s (BWR) but none before; Lee (1933) first noted it as an unlocalized Renfrewshire plant but earlier authors do not mention it as a wild plant. Today it can be readily seen in lowland woodlands, particularly estate types, where it was probably planted in the past but where it is capable of local spread; Snowberry can also be found on waste ground or along disused railways, where the plants are presumably more recent garden outcasts or colonists.

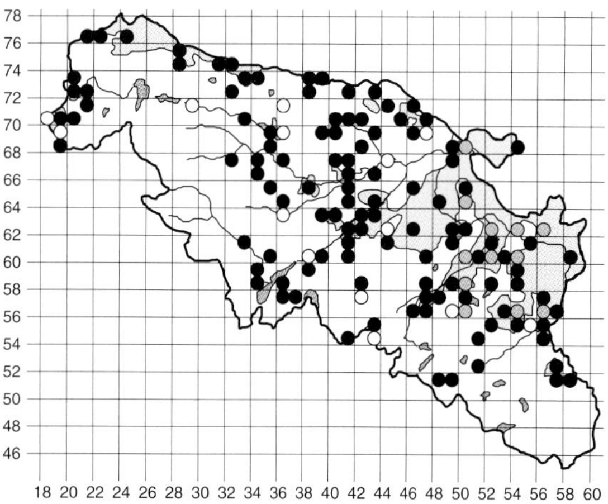

Symphoricarpos albus

Symphoricarpos × *chenaultii* Rehder
Chenault's Coralberry (Pink Snowberry)

Britain: neophyte (garden origin); scattered and local in the S.
Hortal; 7/7; very rare.
1st Whitecroft, Port Glasgow (NS3274), 1997.

This hybrid, including perhaps other related cultivars, has been recorded as outcast material or spreading relicts of past planting from a few sites: waste ground near Spango (NS2375, 1997), Shillford Tip (NS4455, 1999), Wemyss Castle woods (NS1970, 1999), the cycleway, Johnstone (NS4263, 2001), disused railway near Devol Glen (NS3174, 2005) and by the White Cart Water, Albert Park (NS5761, 2005).

Leycesteria formosa Wall.
Himalayan Honeysuckle

Britain: neophyte (Himalayas); frequent in the S, scattered and local elsewhere.
Hortal; 4/5; very rare.
1st Levanne, Gourock, 1984, BWR.

There are four modern records for Himalayan Honeysuckle, from waste ground at Port Glasgow (NS3274, 1998), Jordanhill (NS5468, 2002, AR), Ranfurly Castle Golf Course (NS3864, 2008, AM) and Lochwinnoch Station (NS3557, 2005, IPG).

Lonicera pileata Oliv.
Box-leaved Honeysuckle

Britain: neophyte (China); scattered and local in the S.
Hortal; 8/8; rare.
1st Hotel grounds, Erskine (NS4671), 1997.

There are seven other widespread records for Box-leaved Honeysuckle from waste ground, woodland edges near gardens and other marginal scrubby areas; localities range from Daff Burn, Inverkip (NS2172, 2002), Spango Burn (NS2475, 2011) and Levan Burn (NS2176, 1999) in the west, to Lochwinnoch (NS3558, 2005, IPG) in the centre and, further east, Barskiven (NS4563, 2002), Paisley (NS4864, 2000) and by the White Cart Water, Busby (NS5756, 2005).

Lonicera nitida E.H. Wilson
Wilson's Honeysuckle

Britain: neophyte (China); frequent in the S and W, occasional in C Scotland.
Hortal; 21/21; occasional.
1st Glen Park, 1995.

This shrubby honeysuckle is quite often encountered along shaded burns or scrubby woodlands near housing.

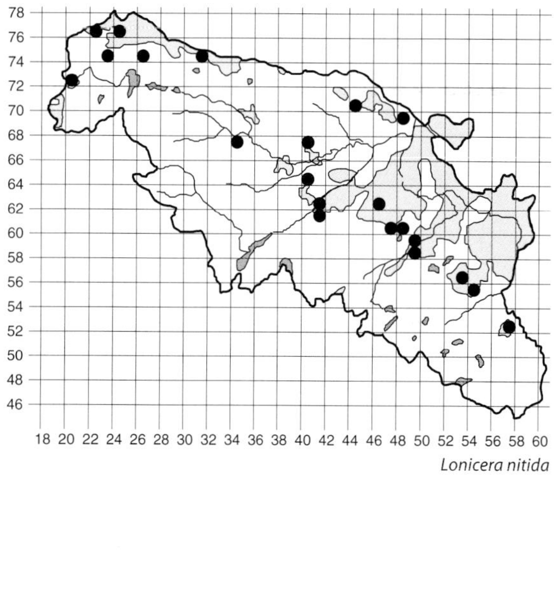

Lonicera nitida

Lonicera involucrata (Richardson) Banks ex Spreng.
Californian Honeysuckle

Britain: neophyte (W N America); scattered in the S.
Hortal; 5/5; rare.
1st Disused railway, Yoker, 1986, CFOG.

The four modern records for Californian Honeysuckle are from urban-fringe areas: Limecraig (NS4661, 1996), burn side, Capelrig Burn (NS5356, 1997), a hedgerow, Craigmuir (NS4469, 2001) and Blackstone (NS4666, 2003).

Lonicera xylosteum L.
Fly Honeysuckle

Britain: neophyte (Europe); scattered, mainly in the S.
Hortal; 4/5; rare.
1st Monkwood, 1864, PSY.

Fly Honeysuckle has been recorded from four modern sites: Whitecroft, Port Glasgow (NS3274, 1997), Wemyss Castle woods (NS1970, 1999), by the White Cart Water, Kirkland (NS5853, 2000) and, possibly planted but spreading, at Pilmuir Quarry (NS5154, 2000).

Lonicera periclymenum L.
Honeysuckle

Britain: common throughout.
Native; 183/211; common.
1st Kelly Glen, 1854, HMcD.

Honeysuckle is a common plant of woodlands, including those of old estates and semi-natural valleys, and it can follow watercourses into upland areas. It is also found in hedgerows and can occasionally occur in more open conditions such as on rocky outcrops. Some lowland urban records may refer to garden forms.

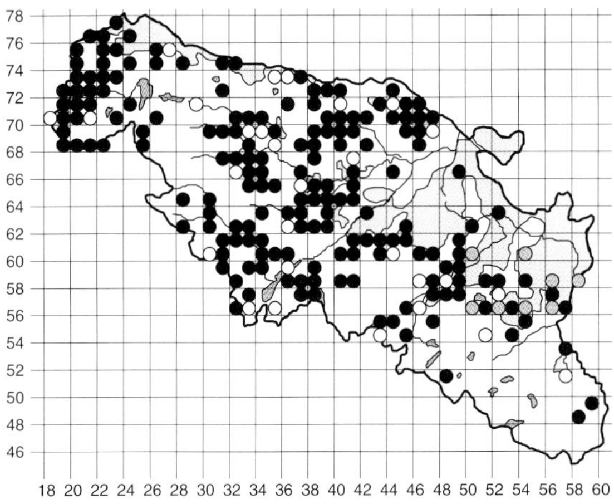

Lonicera periclymenum

Lonicera caprifolium L.
Perfoliate Honeysuckle

Britain: neophyte (S Europe); scattered in the S.
Hortal; 1/2; very rare.
1st Castle Semple, 1834, MC.

Garden cultivars can occasionally be seen creeping over walls from gardens, as at Eaglesham (NS5651, 2006).

VALERIANACEAE

Valerianella locusta (L.) Laterr.
Common Cornsalad

Britain: frequent in the S, more coastal in the N.
Native; 4/8; rare, overlooked.
1st Gourock, 1856, GGO.

There are few named old locations for this cornfield weed, which was previously considered frequent: Wemyss Bay (1887, NHSG), Gourock (1879, GL) and Langbank (1888, PSY). A further Langbank record was made by RM in 1925 and in 1960 a specimen was collected from there (GLAM); recently it was refound close by on gravels by the railway (NS3773, 1998) and from near the station (NS3873, 2006). Other records are from a roadside embankment at Brodie Park (NS4762, 2003) and roadside at Wemyss Bay (NS1968, 2011).

Valeriana officinalis L.
Common Valerian

Britain: common throughout.
Native; 206/245; common.
1st Lochwinnoch and Paisley, 1845, NSA.

Common Valerian, as its name suggests, was common in the 19th century and can still be readily found today in wet places. It is widespread but rare or absent from urban areas or the more acidic peaty uplands. It is generally found in fens and marshes where

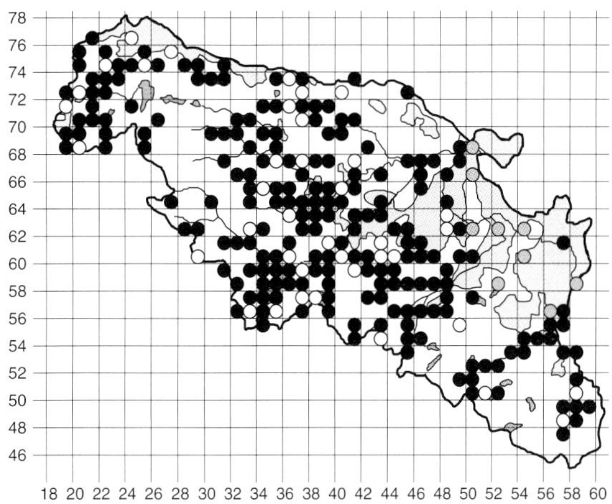

Valeriana officinalis

grazing pressure is limited, and at the margins of open waterbodies and along riverbanks; it graphically marks out the Earn Water and upper White Cart Water in the south-east.

Valeriana pyrenaica L.
Pyrenean Valerian

Britain: neophyte (Pyrenees); frequent in C Scotland, rare elsewhere.
Hortal; 30/39; occasional.
1st Langside, 1813, TH.

Other old records for this species include Lochwinnoch (1876, FFWS), Inverkip Road (1886, GRK), Carruth Bridge (1890, AANS), Ardgowan (1894, NHSG), Bridge of Weir (1903, PSY), the Ravenscraig–Gourock area (1923, RM) and Bridgend, River Calder (1948, ANG); more recently BWR reported it from seven sites including Castle Farm, Mearns (1976) and the disused railway, Kilmacolm (1984). Today Pyrenean Valerian is widespread: there are several populations about Lochwinnoch and the White Cart and Levern waters in Glasgow, but it tends to be scattered elsewhere, e.g. at Wattiston Burn (NS3657, 1999), Caldwell-law (NS4355, 1999), Roebank Burn (NS3455, 1999) and by the Gryfe Water at Craigends (NS4166, 2002). In the north-west it occurs at Howford Burn (NS2274, 2002), Bankfoot (2173, 2010, PRG) and by the Gryfe Water, Auchenfoyle (NS3171, 2001), but also on waste ground at Clydeport (NS3174, 2001).

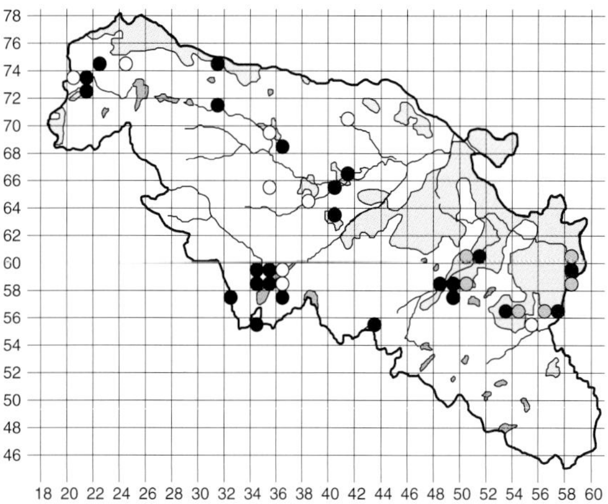

Valeriana pyrenaica

Centranthus ruber (L.) DC.
Red Valerian

Britain: neophyte (SW Europe); common in the S, rare in Scotland except in the SE.
Hortal; 3/3; very rare.
1st Whiteinch, 1986, CFOG.

Modern survey records for Red Valerian are from a roadside to the edge of Houston (NS4067, 2002) and a white form has colonized a roadside wall at Threeply (NS3765, 2008).

DIPSACACEAE

Dipsacus fullonum L.
Wild Teasel

Britain: common in the S, frequent in C and E Scotland.
Accidental; 14/22; scarce.
1st Langside, 1813, TH.

Considered a rare plant of waste places by Hennedy (1891), there are a few named old records for Wild Teasel: fields near Cathcart (1821, FS), Arden Quarry, near Thornliebank (1865, RH), Battery Field, Gourock (1888, NHSG) and Inchinnan Bridge (1891, RH); surprisingly, it was excluded from the Renfrewshire list (TPNS 1915). Wood (1893) described it as occasional from Arden to Paisley, but 'gradually dying out', linking its occurrence to old cloth mills, citing the 'wack-mill' at Darnley [Waulkmill] Glen. There are only two records from the 1960s–80s period (BWR). Today it is found at quite a few scattered places near Paisley, in grassy waste places and along roadsides. A remote population is at a path edge, Castle Semple (NS3659, 2011).

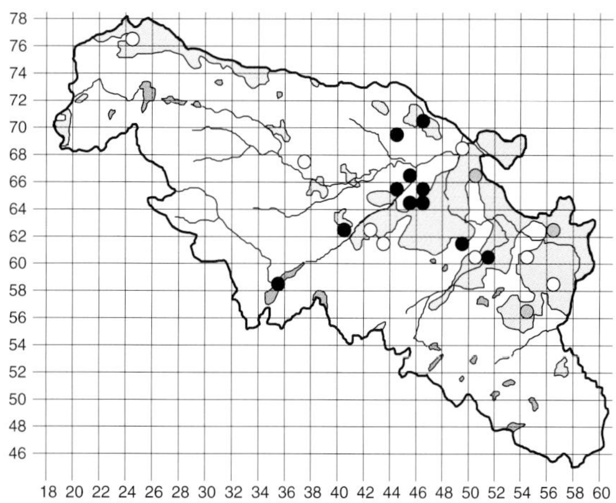

Dipsacus fullonum

Knautia arvensis (L.) Coult.
Field Scabious

Britain: common in the S and in SE Scotland.
Former native and hortal; 5/13; very rare.
1st Kilbarchan, 1845, NSA.

Field Scabious appears to have been an uncommon plant in the 19th century but there are a few named locations, including Lochwinnoch and Clochodrick (1845, NSA), a railway bank, Pennilee (1859, PSY), Giffnock (1897, GLAM), Dennistoun (1898, JMY) and

Kilmacolm (1933, Lee); Hooker (1821) described it as 'Rare in meadows and pastures about Glasgow'. It was recorded near the Earn Water, Burnhouse, in 1983 (NS5554, BWR) but today the only records are from the roadsides of the A737 between Kilbarchan and Linwood (NS4162, NS4263, NS4363 and NS4464) and the grounds of GMRC, Nitshill (NS5160, 2006) where it arrived as part of a wildflower seed mix.

Succisa pratensis Moench
Devil's-bit Scabious

Britain: common throughout except in SE England.
Native; 274/328; very common.
1st Lochwinnoch, 1845, NSA.

Devil's-bit Scabious appears to have always been a common plant and today its bluish-purple heads can be readily seen in late summer in rural areas. It is likely to have declined due to agricultural improvement, as it avoids nutrient-enriched places and often reflects some flushing or impeded drainage, though not really wet marshes or bogs; it is most commonly associated with species-rich grasslands and heaths, often on embankments, including roadsides.

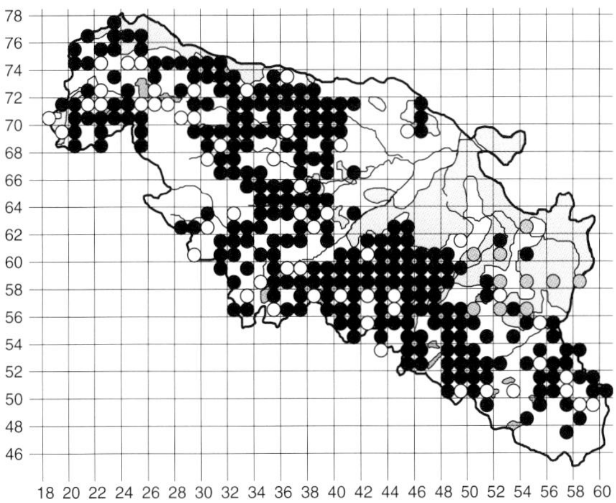

Succisa pratensis

ARALIACEAE

Hedera colchica (K. Koch) K. Koch
Persian Ivy

Britain: neophyte (Caucasus); scattered in C Scotland.
Hortal; 5/5; very rare.
1st Crosshill Station, 1985, CFOG.

This large-leaved ivy was also recorded during CFOG surveys from Cathcart Station (NS5860, 1985). The other modern records are from Glen Park (NS4760, 1995), a hedgerow near the lodge, Drums (NS4072, 1998) and Barcraigs Reservoir (NS3957, 1999).

Hedera helix L.
Common Ivy

Britain: common throughout except in N Scotland.
Native; 237/261; common.
1st Kelly Glen, 1854, HMcD.

Old records do not distinguish the now widespread 'Hibernica' cultivar so the map includes old records for both taxa. Ivy was considered common in the 19th century and this remains the case today. Common Ivy is found in woodlands but also in the open on north-facing rock outcrops or along steep-sided upland burns; along the latter it is able to penetrate the upland zones.

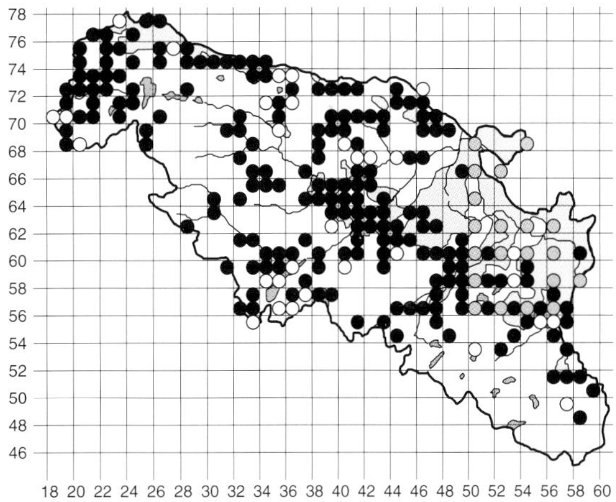

Hedera helix

Hedera 'Hibernica'
Irish Ivy

Britain: common in the SW, frequent alien elsewhere except in N Scotland.
Hortal; 95/95; frequent.
1st Crosshill Station (NS56W), 1986, CFOG.

Not distinguished by earlier authors, this often

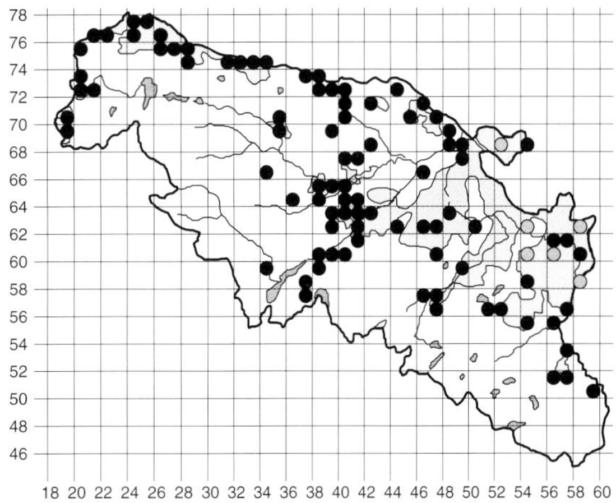

Hedera 'Hibernica'

rampant cultivar has presumably been present in the VC for a long time. It can be difficult to distinguish from Common Ivy, and it grows in similar woodland places but is usually associated with old estate woodlands, parks and gardens; it shows a strong preference for coastal locations in the north and west.

Aralia racemosa L.
American-spikenard

Britain: neophyte (N America); very rare casual.
Former accidental; 0/1; extinct.
1st Banks of Cart, Newlands, 1926, RG.

HYDROCOTYLACEAE
(APIACEAE *pro parte*)

Hydrocotyle vulgaris L.
Marsh Pennywort

Britain: widespread, mainly in the W.
Native; 81/97; locally frequent.
1st Stanely Reservoir, 1869, RPB.

Marsh Pennywort is likely to have declined in the lowlands throughout the last century, but today it can still be found in rural mires and close to loch shores, often in abundance, but sometimes hidden beneath taller vegetation. It is well represented in the fringes of the Renfrewshire Heights but appears to be limited to open waterbodies in the south-east. It is typically a fen species, usually growing in species-rich vegetation.

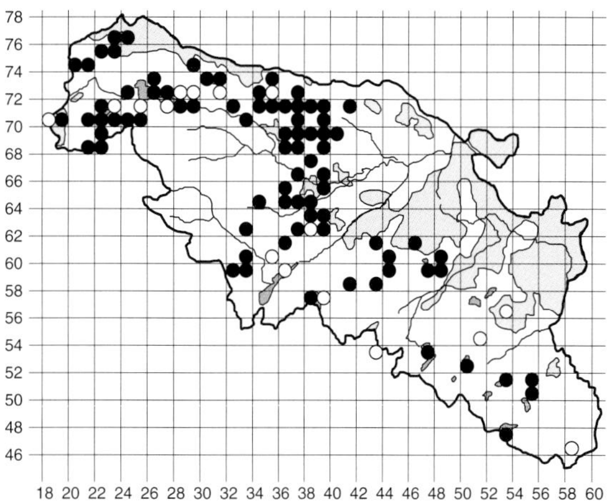

Hydrocotyle vulgaris

APIACEAE

Sanicula europaea L.
Sanicle

Britain: common except in N Scotland.
Native; 34/44; occasional.
1st Lochwinnoch and Paisley, 1845, NSA.

Sanicle is strongly represented in the south of Glasgow along the White Cart Water and tributaries (CFOG) but there are no records from further upstream, past or present. All of the records are from mature woodlands. It is often found in the wooded valleys (cleughs) cutting down from the Renfrewshire Heights: Kelly Glen (NS1968), Daff Burn (NS2172), Shielhill Glen (NS2471), Auchmountain Glen (NS2874) and Devol Glen (NS3174), but less frequently recorded to the east side: Garpel Burn (NS3459), River Calder (NS3560) and Gotter Water (NS3466). There are also a few sites about the Gleniffer Braes (*c.* NS4660) and an isolated site at Swinetree Glen (NS4060).

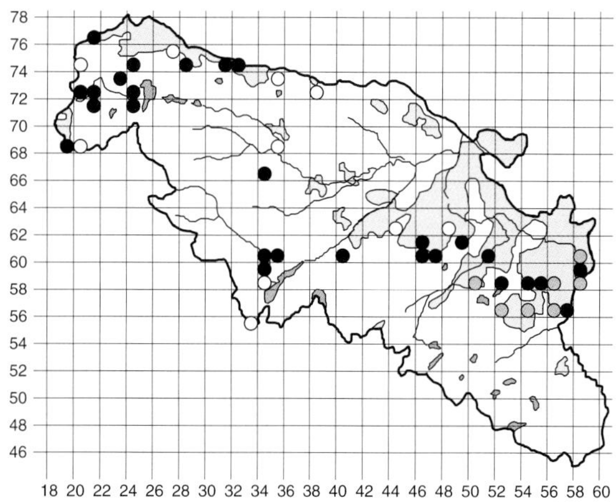

Sanicula europaea

Astrantia major L.
Astrantia

Britain: neophyte (Europe); occasional, scattered.
Hortal; 7/10; rare.
1st Castle Semple Loch, 1887, NHSG.

Astrantia was excluded from the Renfrewshire list (TPNS 1915) but RM (1934) and BWR (1971) both reported it from Castle Semple Loch, where it is still to be found (NS3658, 1992, CMRP). There are two records in CFOG and it was found at 'Levern Glen' in 1965 (NS4655, BWR) and is known nearby today from a disused railway track at Neilstonside (NS4655, 1998) and by the Levern at Holehouse (NS4756, 1997). The other modern records are from Shillford Tip (NS4455, 1999) and Millbank Burn (NS3357, 2006).

Eryngium maritimum L.
Sea-holly

Britain: coastal, mainly in the S but extending to NW Scotland.
Former native; 0/2; extinct.
1st Inverkip, 1901, FFCA.

Later Lee (1933) reported this plant of sandy shores present 'Sparingly along the coast from Inverkip southwards'. Given all the development in recent years and lack of suitable habitat, Sea-holly must now be considered extinct.

Chaerophyllum temulum L.
Rough Chervil

Britain: common in England, extending to C and E Scotland.
Native; 1/2; very rare, overlooked.
1st Renfrewshire, 1891, TB.

The sole modern record is from the Formakin Estate (2003, IPG). Considered a common or very common plant of hedges and thickets by both Hennedy (1891) and Lee (1933), Rough Chervil must, given the paucity of modern records, have either undergone a dramatic decline or be frequently overlooked. However, it was not recorded during the 1960s–80s and was found to be strongly eastern during surveying for CFOG, so perhaps it may never have been so frequent in the western part of the Clyde area.

Anthriscus sylvestris (L.) Hoffm.
Cow Parsley

Britain: common throughout except in N Scotland.
Native; 320/382; very common.
1st Renfrewshire, 1869, PPI.

Cow Parsley is a very common sight along wayside hedge banks in early summer and also in tall, usually neglected, meadows, along riversides and also on waste ground. It is common throughout and only absent from the acidic upland soils.

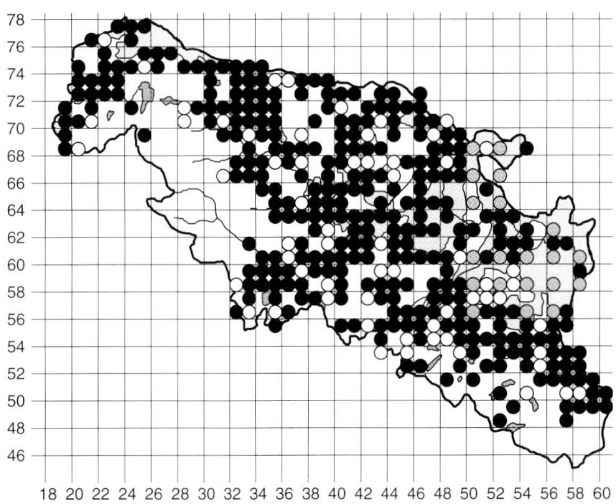

Anthriscus sylvestris

Anthriscus caucalis M. Bieb.
Bur Parsley

Britain: local, mainly in the S and E, occasional in E Scotland.
Former accidental; 0/2; extinct.
1st Loch Libo and Caldwell area, 1887, NHSG.

In 1865 Hennedy wrote that Bur Parsley was sometimes found on rubbish around Glasgow, but there are no named places. This species is also marked on a 1971 BWR card from by the Gryfe Water (NS3567), but there is no other information about what would have been an interesting local find.

Scandix pecten-veneris L.
Shepherd's-needle

Britain: archaeophyte; local in SE England, now absent from the N.
Former accidental; 0/5; extinct.
1st Cathcart Castle, 1813, TH.

CFOG describes the decline of this arable weed in the Glasgow area. It was recorded from Erskine (1886, AANS), Langbank (1891, RH), Hangingshaw in 1896 (NHSG) and collected from the cottage, Paisley (1902, PSY), but appears long extinct.

Myrrhis odorata (L.) Scop.
Sweet Cicely

Britain: neophyte (Europe); common in N England and C and E Scotland.
Hortal; 75/99; frequent.
1st Renfrewshire, 1834, MC.

Sweet Cicely has been present for a long time in the VC, with records from Erskine and Paisley dating from 1845 (NSA). Today it is widespread across urban-fringe areas, although seldom in any great abundance, and most notably in the east. It occurs along riverbanks (open or shaded), hedge banks and waysides and is capable of colonizing waste ground.

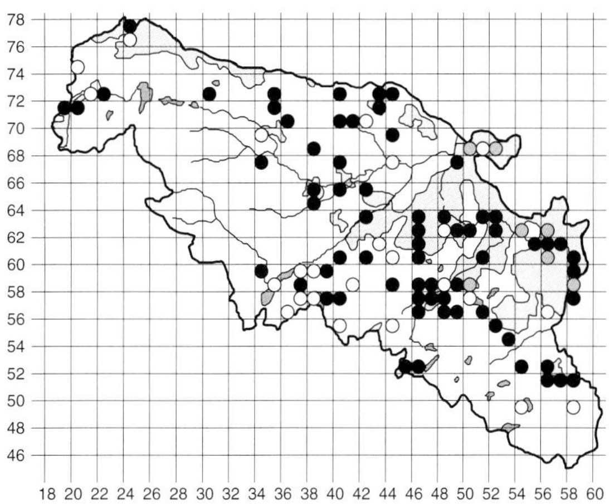

Myrrhis odorata

Coriandrum sativum L.
Coriander

Britain: neophyte (E Mediterranean); scattered in the S.
Former hortal; 0/2; extinct casual.
1st Renfrewshire, 1895, NHSG.

There are no modern records for this casual and no further details about the original. It was noted as a casual from Battlefield (NS56V) in 1978 (PM and ERTC).

Smyrnium olusatrum L.
Alexanders

Britain: archaeophyte (Europe); common near coasts in S England, very rare in the N.
Former hortal; 0/4; extinct.
1st Renfrewshire, 1899, GC.

There are no old named locations for Alexanders until Lee (1933), who noted near Renfrew, but curiously there were a number of sightings reported in 1942 (TPNS) from Hawkhead Estate, Kilbarchan and Howwood. There are no modern records.

Conopodium majus (Gouan) Loret
Pignut

Britain: common throughout.
Native; 279/339; common.
1st Renfrewshire, 1869, PPI.

Pignut is a common plant occurring in pastures and scrub or woodland edges throughout the VC, only becoming scarce where wet or peaty and in urban areas where it shuns disturbed waste ground. It can be quite abundant and noticeable in early summer at some of the less improved pastures in the foothills.

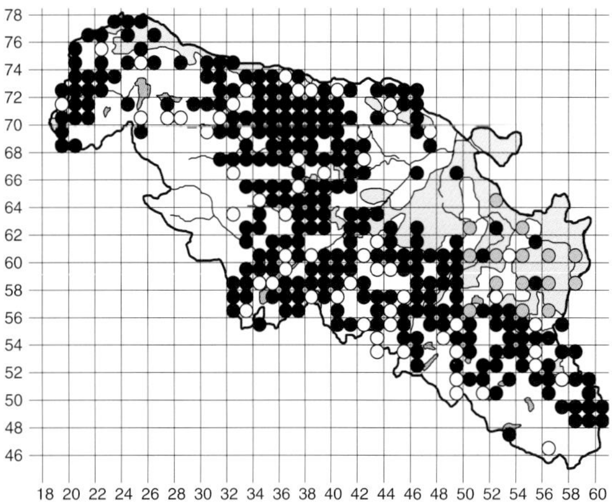

Conopodium majus

Pimpinella major (L.) Huds.
Greater Burnet-saxifrage

Britain: common in C England, very rare alien in Scotland.
Accidental; 0/1; extinct.
1st Garden weed, Pollokshields, 1980, CFOG.

Pimpinella saxifraga L.
Burnet-saxifrage

Britain: common, rare or absent from NW Scotland.
Native; 18/27; scarce.
1st Gourock and Inverkip, 1865, RH.

Burnet-saxifrage was considered frequent by earlier authors and was recorded on several occasions in the 1960s–80s period (BWR). It has declined due to agricultural improvement and heavy grazing, although the latter may result in it being overlooked. Populations occur at two disjunct areas, all relatively unimproved pastures associated with rocky ridges. It is found along the Earn and White Cart waters in the south-east and there is a cluster of records along the Gryfe Water, e.g. at Strathgryfe (NS3370, 1998), and nearby grassy ridges at Matherneuk (NS3371, 2005), Kilmacolm (NS3569, 2003) and Craigmarloch (NS3471, 2006); an isolated record is from the species-rich grassland at Muirhead (NS4856, 1999).

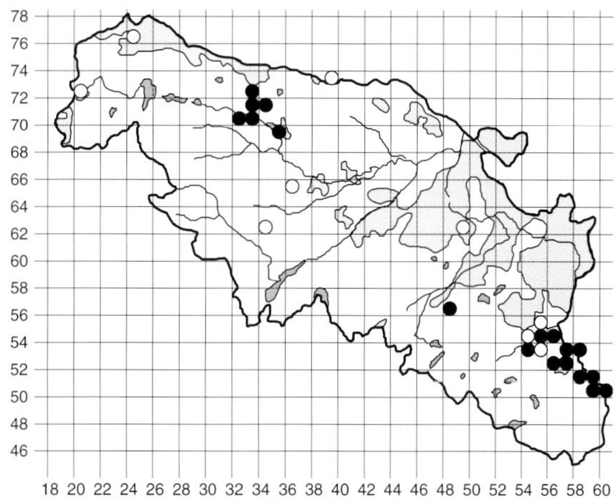

Pimpinella saxifraga

Aegopodium podagraria L.
Ground-elder

Britain: archaeophyte; common throughout except in N Scotland.
Hortal; 284/342; very common.
1st Renfrewshire, 1869, PPI.

Ground-elder was considered very common in the 19th century (Hennedy 1891) and today this is still an apt description; it occurs throughout urban and rural areas, except the uplands. It is found in woodlands, often along riverbanks where it appears perfectly naturalized; it frequently marks areas of past woodland disturbance such as rubble dumping. It can also be found in open rank grassland areas along watercourses and hedge banks, and colonizes waste ground and is well known as a garden weed.

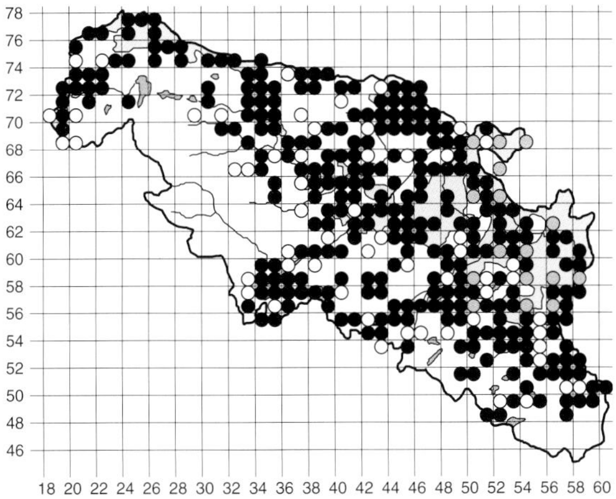

Aegopodium podagraria

Berula erecta (Huds.) Coville
Lesser Water-parsnip

Britain: common in the S, uncommon but scattered in Scotland.
Native; 11/11; rare.
1st Commore Dam, 1890, AANS.

Lesser Water-parsnip was recorded by Lee (1933) and then RM (1945) and BWR (1976) at Loch Libo, and is still there today (NS4355, 2007). It is also still present in ditches close to Commore Dam (NS4654, 1999). The species occurs in three areas: locations include a burn at Dampton (NS3863, 1997), the Gryfe Water, Kilmacolm (NS3568, 2000 and NS3589, 2003), a ditch at Milliken Park (NS4062, 1995, AR) and at Merchiston (NS4164, 1999). Old records for Greater

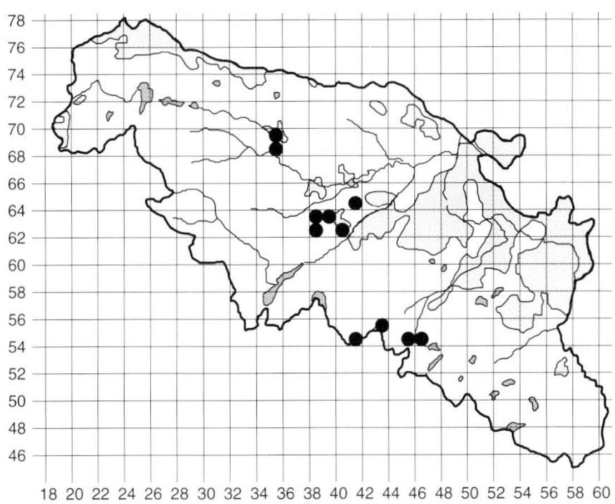

Berula erecta

Water-parsnip (*Sium latifolium* L.) may be errors for this species (Lee, 1933), although both species are included in the Renfrewshire list (TPNS 1915).

Oenanthe fistulosa L.
Tubular Water-dropwort

Britain: frequent in S England, very rare in E Scotland.
Native or accidental; 0/5; extinct.
1st Greenock, 1821, FS.

McKay (FFWS 1876) questions the Greenock record for Tubular Water-dropwort, which Hooker (1821) described as 'common along coast below Greenock', but it was repeated by Hennedy (1891) and Ewing (1901). Lee (1933) considered it very rare, adding 'wet ground near the sea' and listing Inverkip as a site; it was reported as being seen, not in flower, during a visit to Shielhill (1893, AANS). MY noted it from Loch Libo (1865), but this is the only record from that well botanized locality, so probably an error. There are no specimens of this distinctive southern species, which has not been reported for a very long time.

Oenanthe pimpinelloides L.
Corky-fruited Water-dropwort

Britain: restricted to S England.
Accidental; 0/1; doubtful record.
1st Greenock, 1821, FS.

This species is not mentioned by any other author, and is a highly questionable record for a plant well outside its native range. The name was used in the past for the following species (Baker & Newbold 1883), which is the most likely explanation of the record.

Oenanthe lachenalii C.C. Gmel.
Parsley Water-dropwort

Britain: frequent by coasts except in N and E Scotland.
Native; 1/3; very rare.
1st 'All around shores of Firth', 1865, RH.

Gourock was named in 1901 (FFCA) but, if as Hennedy implied, Parsley Water-dropwort was frequent along the Inverclyde coast, it has subsequently shown a dramatic decline. Today it is only known from one site: marshy grassland by the Clyde to the west end of Erskine Park (NS4473, 1996). It is likely that the old record (1821, FS) for Corky-fruited Water-dropwort refers to this species.

Oenanthe crocata L.
Hemlock Water-dropwort

Britain: common in the W, scattered and local elsewhere.
Native; 119/137; common in wet places.
1st Lochwinnoch, 1845, NSA.

Hemlock Water-dropwort shows a dramatic disjunct pattern with a cluster of sites in the extreme west isolated from the core central area; it is virtually absent from the south-east apart from the Little Loch record (NS5052, 1996, SNH-LS). It is typically found in open marshy ground but can occur in shaded wet woodland and is frequently seen along riverbanks, where it can also tolerate some shading.

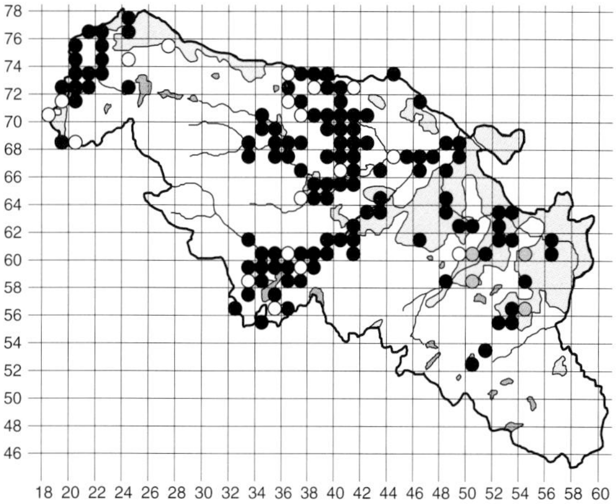

Oenanthe crocata

Aethusa cynapium L.
Fool's Parsley

Britain: common in England, local in S Scotland.
Native or accidental; 11/13; scarce.
1st Gourock and Cathcart, 1865, RH.

Fool's Parsley was recorded from a disused railway in Kilmacolm in 1984 (NS3965, BWR) and there are several locations for it in Glasgow (CFOG), but it was never found in any abundance. More recently it has been recorded from five places to the west of Glasgow: rough grassland to the edge of Paisley Moss (NS4665, 2003), just east of Glasgow Airport (NS4866, 2006), Linwood (NS4465, 1997, SA), Formakin Estate (NS4170, 2003, IPG) and Carriagehill Avenue, Paisley (NS46R, 1994, GL).

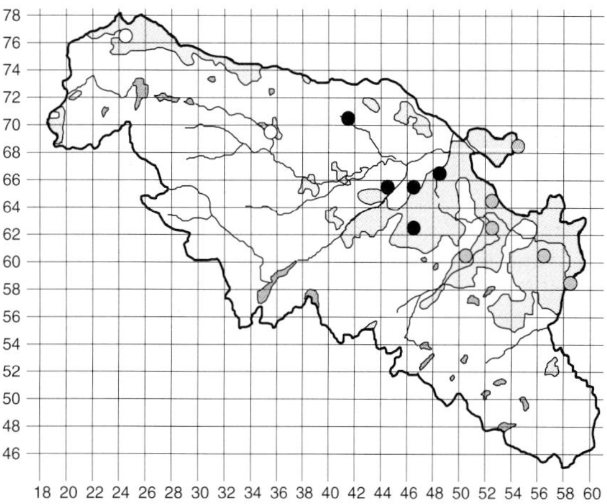

Aethusa cynapium

Meum athamanticum Jacq.
Spignel

Britain: local in S and C Scotland, very rare or absent elsewhere.
Native; 57/76; locally frequent.
1st Lochwinnoch, 1834, MC.

There are plenty of old records for this Renfrewshire speciality: FFWS (1876) lists Kilmacolm, Lochwinnoch, Greenock, Inverkip Road and Cloch. Today Spignel is still frequent in the foothills west of Bridge of Weir and to a lesser extent in the Neilston area. It was recorded from Law Hill (NS5751, BWR) in 1982, but there are no modern records from the extreme south-east. It seems to have survived agricultural improvement by persisting in species-rich grass on rocky embankments and other steep slopes, but its population sizes will have decreased. There are only two records now for the extreme west, from grasslands above Spango (NS2474, 1997) and Coves (NS2476, 2011), although it was recorded from Banks (NS2275, BWR) in 1977. The 1981 record from Erskine (NS4671, BWR) must be viewed with suspicion.

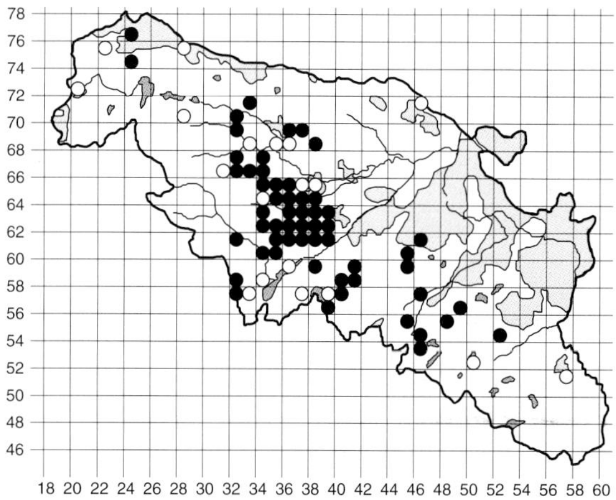

Meum athamanticum

Conium maculatum L.
Hemlock

Britain: archaeophyte; common in England and E Scotland, scattered elsewhere.
Accidental; 5/9; rare.
1st Paisley, Neilston and Kilbarchan, 1845, NSA.

It appears that Hemlock was better known in the mid 19th century than today. It was recorded from Inchinnan Bridge (NS4967, BWR) in 1984 and there are a couple of relevant Glasgow localities (CFOG). There are only three nearby records from more recent recording: waste ground about Ferguslie Park (NS4764, 1997; NS4665, 1999) and waste ground by the White Cart Water north of Porterfield (NS4967, 2001).

Bupleurum subovatum Link ex Spreng.
False Thorow-wax

Britain: neophyte (S Europe); scattered in the S.
Former accidental; 0/2; extinct.
1st Crossmyloof, 1892, TPNS.

A specimen of False Thorow-wax was also collected from Crossmyloof in 1894 (GL) and a further old record is from wasteland, near Paisley (1910, PSY). It is thought that most, if not all, old literature records for *B. rotundifolium* refer to this species, which herbarium sheets, when available, confirm (CFOG).

Apium graveolens L.
Wild Celery

Britain: scattered, mainly coastal in England, rare alien in Scotland.
Former hortal; 0/2; extinct.
1st Paisley, 1898, PSY.

Another herbarium sheet is from the 'marshy bank of the Clyde, Langbank' (1908, GL). There are no modern records.

Apium nodiflorum (L.) Lag.
Fool's-water-cress

Britain: common in the S, rare in Scotland.
Native; 1/3; very rare.
1st Gourock, 1865, RH.

There are very few old records and no specimens of Fool's-water-cress from the area. It was recorded in 1984 (BWR) from Dampton (NS3862), but Lesser Water-parsnip occurs there now so this may well have been an error. However, the SNH Loch Survey found it at the east end of Loch Libo (NS4355, 1996) – apparently growing with Lesser Water-parsnip; both these species were also recorded from here in 1976 (JP).

Apium inundatum (L.) Reichb. f.
Lesser Marshwort

Britain: scattered, mainly in the W.
Native; 33/42; occasional.
1st Lochwinnoch, 1845, NSA.

There are a few named old locations for Lesser Marshwort, such as Gleniffer Braes (1905, GL), Lochwinnoch (1936, GL) and Hawkhead policies (1936, PSY). RM recorded it from Little Loch (1936), Snypes Dam (1952), Harelaw Dam (1951) and Bennan Loch (1956), and it has been seen recently at all of these places. It is perhaps overlooked as it tends to grow in deep water or hidden amongst the taller vegetation of waterbodies or, rarely, slow burns and ditches as at Lochend (NS3664, 2004). The pattern is distinctly rural, tending towards the upland fringes.

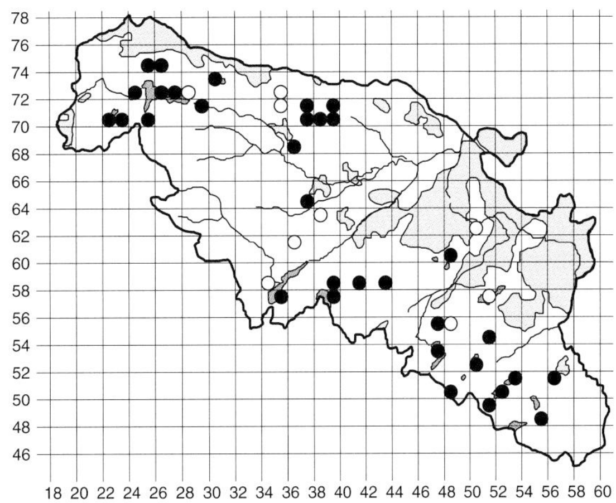

Apium inundatum

Cicuta virosa L.
Cowbane

Britain: scattered and local, occasional in lowland Scotland.
Native; 3/3; very rare.
1st Loch Libo, 1887, NHSG.

Cowbane has long been celebrated at Loch Libo – although interestingly it is not named from there by Hennedy (1891) – and it can still be found in the swampy fen vegetation to the north and south margins of the loch (NS4358, 2004). In 1996 the SNH Loch Survey recorded it from Caplaw Dam (NS4358) but did not note it from Aird Meadow (NS3558), where it was found in 2010 by MG while doing RSPB survey work. It is surprising that this distinctive species has not been noted here previously.

Carum carvi L.
Caraway

Britain: archaeophyte; scattered in NE Scotland.
Accidental; 1/3; very rare.
1st Holland Bush, 1865, MY.

A further old record is from near Elderslie (1904, PSY). The sole modern record for Caraway is for a single plant beside a path edge with recently sown grass at Porterfield (NS4967, 2001).

Carum verticillatum (L.) W.D.J. Koch
Whorled Caraway

Britain: locally frequent in W Scotland, rare elsewhere.
Native; 133/164; common in west.
1st Greenock, 1821, FS.

There are plenty of old records for Whorled Caraway, which Hennedy (1891) described as 'very abundant in low pastures all round the shores of the Firth'. The modern distribution shows a striking western pattern, with only one outlier in the south-east uplands at Little Loch (NS5052, 1996, SNH-LS). Its preferred habitat is flushed slightly acidic grassland, often with rushes, and also some more permanently wet marshes; it is absent from the more acidic upland peat.

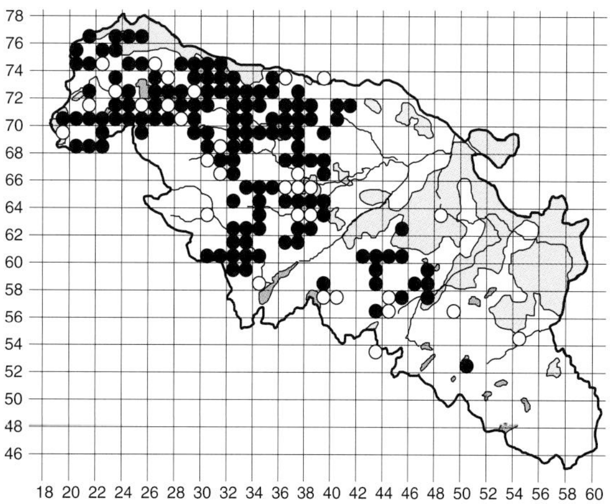

Carum verticillatum

Ligusticum scoticum L.
Scots Lovage

Britain: frequent by coasts in Scotland, absent from England.
Former native; 0/1; extinct.
1st Trap (basalt) dyke, near Wemyss Bay, 1845, GL.

Angelica sylvestris L.
Wild Angelica

Britain: common throughout.
Native; 298/350; very common.
1st Banks of Cart, 1865, RH.

Wild Angelica remains a common plant of marshy places, including some wet scrub woodlands and watercourse margins; it occurs throughout the VC, only becoming scarce in the heavily urbanized lowlands or upland peaty moorlands.

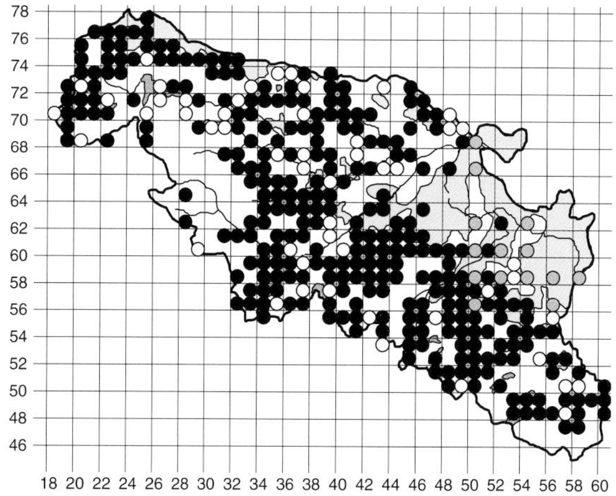

Angelica sylvestris

Angelica archangelica L.
Garden Angelica

Britain: neophyte (N and E Europe); local in the S, very rare in Scotland.
Hortal; 2/2; very rare.
1st Braehead, 1984, GL.

Garden Angelica's distribution along the Clyde in Glasgow was noted in CFOG; it just crept into Renfrewshire at Braehead, where it grew on stonework. The only modern record is from waste ground, close to the Clyde, at Clydeport, Port Glasgow (NS3174, 2001), a site subsequently developed.

Imperatoria palustre (L.) Raf.
Peucedanum ostruthium (L.) W.D.J. Koch
Masterwort

Britain: archaeophyte (C and S Europe); scattered in the N.
Hortal; 5/19; rare.
1st Lochwinnoch, 1845, NSA.

There are quite a few old records for Masterwort, which include Eaglesham (1874, TGSFN), Kilmacolm (1888, NHSG), Ardgowan Estate (1889, AANS), the road up to Misty Law, Muirshields (1900, PSY), Polnoon (1933, Lee) and Mearns–Eaglesham (1938, RM); Wood (1893) noted it near Loch Libo and Peesweep, and also at Commore and a nearby roadside with a 'magnificent bed'. Today it is only known from five sites, all to the south-east. It can still be found on the roadside at Bonnyton (NS5553, 2003) where it was first recorded in 1978 (BWR), and also by a roadside at Carrot (NS5748, 2003); the other localities are burn-side scrub, Mackiesmill (NS4461, 1993), Eaglesham Common (NS5751, 1997) and Taphead, near Cowdenmoor (NS4455, 1999). There are a few unlikely old references to 'Hog's Fennel' (*Peucedanum palustre/officinale*), presumably due to misidentification or nomenclature confusion, but either way it was 'excluded' from the Renfrewshire list (TPNS 1915).

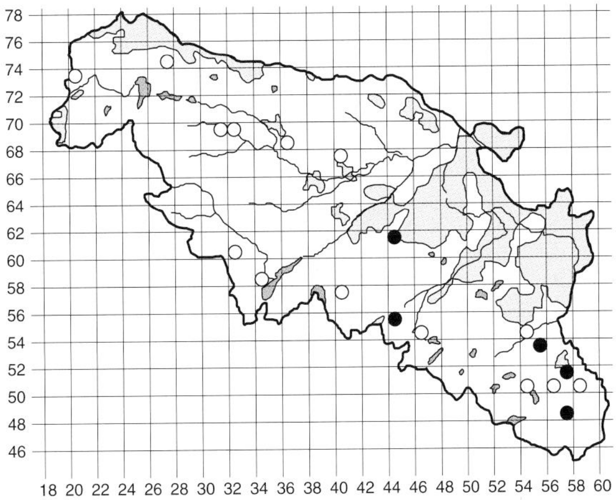

Imperatoria palustre

Pastinaca sativa L.
Wild Parsnip

Britain: common in the SE, local alien in C Scotland.
Accidental; 1/3; very rare.
1st Waste ground near Whitecraigs Railway Station, 1926, RM.

Wild Parsnip was not mentioned from the area by earlier authors and BWR only recorded it once, from Barcraigs Reservoir (1978); there is only one modern record for the VC, from waste ground at Cartsdyke (NS2975, 2003). It was reported along the M8 in Glasgow (CFOG) but so far appears not to have spread further west.

Heracleum sphondylium L.
Hogweed

Britain: common throughout.
Native; 323/414; very common.
1st Banks of Cart, 1865, RH.

Hogweed was a common plant in the 19th century and remains so today, being very tolerant of disturbance, readily colonizing waste ground and also persisting in rank grass and secondary scrub at neglected sites and along waysides. It is rare in intensively cultivated ground and absent from the uplands except by roadsides.

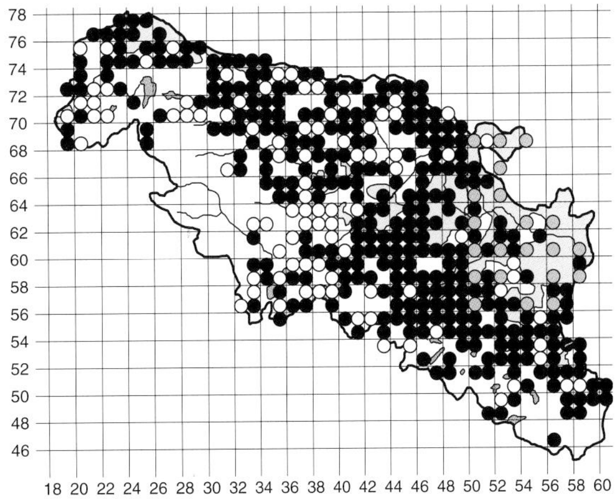

Heracleum sphondylium

Heracleum mantegazzianum Sommier & Levier
Giant Hogweed

Britain: neophyte (SW Asia); frequent throughout lowlands.
Hortal; 35/38; locally frequent on riverbanks.
1st Streamside, Eastwood Golf Course, 1925, GLAM.

There are only a few records from the 1960s–80s (BWR), such as from waste ground, Georgetown (NS4467, 1974) and Barochan Estate (NS4168, 1979), indicating a perhaps slow initial spread. Today this notorious but impressive alien is now well established along the White Cart Water, where most of the modern records are from, and from where AJS described it as 'abundant on the riverbanks' back in 1974; the first record from the White Cart Water is that from Pollokshaws in 1961 (JDM). Giant Hogweed can occur at waste-ground sites away from the river in western Glasgow (CFOG) and other modern records

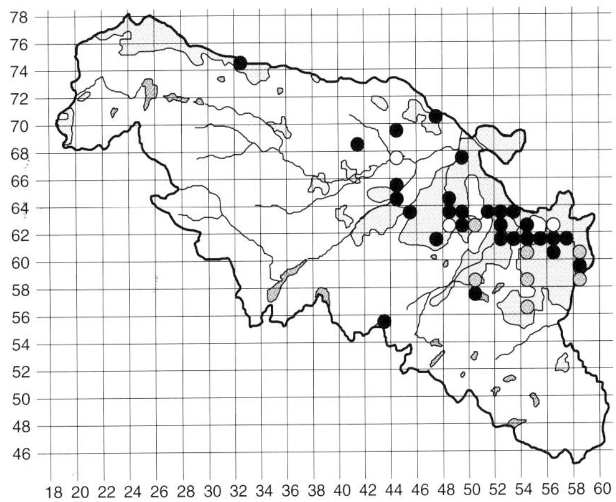

Heracleum mantegazzianum

include Barrangary Tip (NS4469, 2000) and outliers at Port Glasgow (NS3175, 1998) and Uplawmoor (NS4355, 2000, JC).

Torilis japonica (Houtt.) DC.
Upright Hedge-parsley

Britain: common in the S extending to C and NE Scotland.
Native; 31/41; occasional.
1st Gourock, 1850, GL.

Considered common in the 19th century, there are few named stations for Upright Hedge-parsley in the literature but it was collected from Elderslie (1880, GL). Today there are a number of records from waysides and hedgerows in the central lowlands, notably about Bishopton, but it is either rare or overlooked from large areas, particularly to the south-east.

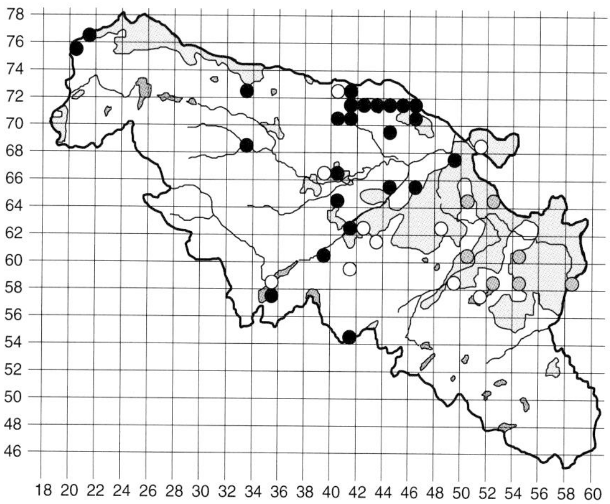

Torilis japonica

Torilis nodosa (L.) Gaertn.
Knotted Hedge-parsley

Britain: frequent in the SE, very rare in SE Scotland.
Former accidental; 0/2; extinct.
1st Queen's Park, 1874, GL.

There is also another specimen of Knotted Hedge-parsley from Crossmyloof (1894, GL).

Daucus carota L.
Wild Carrot

Britain: common in the SE, mainly coastal in Scotland.
Native, hortal or accidental; 8/10; rare.
1st Gourock, 1865, RH.

Wild Carrot was noted as being present 'plentifully in the neighbourhood [Paisley]' in 1869 (RPB) and recorded from Wemyss Bay (1887, NHSG). The old records may refer to native populations but the only records today are from waste ground, disused railways or roadside grassland, mainly about the Black Cart Water corridor. It was recorded from both sides of the A737 (NS4363) in 1994 (SNH Grassland Survey), where it was presumably sown as part of a roadside wildflower mix, but its population size has decreased over the years. It was also found along the Black Cart Water, near Barnsford (NS4667, 2002), as a contaminant of a seed mix in the grounds of GMRC, Nitshill (NS5160, 2004) and on waste ground, Port Glasgow (NS3174, 2012).

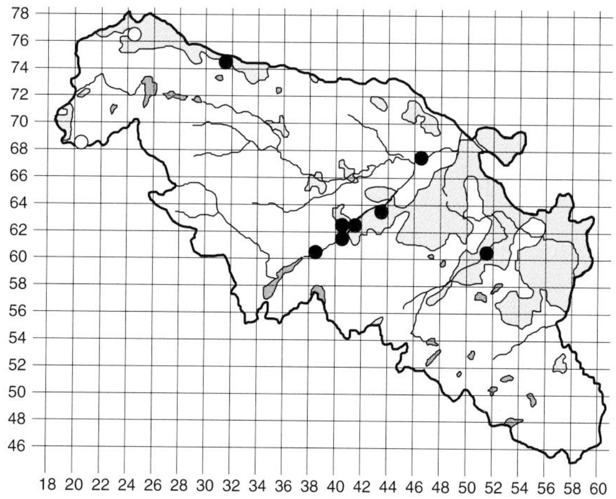

Daucus carota

Caucalis platycarpos L.
Small Bur-parsley

Britain: neophyte (S Europe); rare casual.
Former accidental; 0/1; extinct.
1st Wasteland near Paisley, 1910, PSY.

There was only one specimen according to Ferguson (1911).

Turgenia latifolia (L.) Hoffm.
Greater Bur-parsley

Britain: neophyte (Europe); rare casual.
Former accidental; 0/1; extinct.
1st Paisley, 1910, GL.

There were only half a dozen plants according to Ferguson (1911); two of them ended up in GL.

MONOCOTS (MONOCOTYLEDONS)

ACORACEAE
(ARACEAE *pro parte*)

Acorus calamus L.
Sweet-flag

Britain: neophyte (Asia/N America); frequent in the S and E, rare in the N.
Hortal; 1/2; very rare.
1st Castle Semple, 1834, MC.

This record was repeated in 1865 (RH) but there are

no other reports of Sweet-flag from this well-known site by more recent recorders. However, it is well established in the small loch at Glentyan (NS3963, 2007), where it was presumably planted.

ARACEAE

Lysichiton americanus Hultén & H. St. John
American Skunk-cabbage

Britain: neophyte (N America); scattered, mainly in the W and S.
Hortal; 6/7; rare.
1st Whitecraigs, 1987, CFOG.

American Skunk-cabbage is found in wet woodlands at several places along the Dargavel Burn and tributaries, from where it has been known since 1987 (TPNS). Today it occurs from Parkglen (NS3970 and NS4070, 1998), through Formakin (NS4070, 1994; NS4170, 2003, IPG) to the Royal Ordnance Factory, Georgetown (NS4270 and NS4467, 2005); it also occurs in wet woodland to the edge of Barmufflock (NS3664, 2004).

Arum maculatum L.
Lords-and-Ladies

Britain: common in England, frequent in lowland Scotland (where considered alien).
Native or hortal; 31/41; occasional.
1st Crookston Castle, 1847, GGO.

There are several old stations for Lords-and-ladies, considered not common by Hennedy (1891), from places such as Thornliebank (1883, NHSG) and, reported in the AANS (1893), from Levernholme Wood (1890), Finlaystone (1890) and North Barr (1892); all of these are old estates rather than 'wild woods', adding weight to doubts over this species' native status in Scotland. Today it is scattered around the central lowlands and also along wooded river

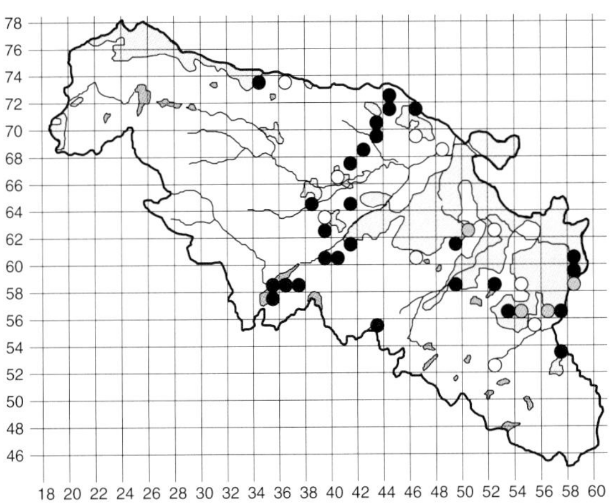

Arum maculatum

valleys to the south of Glasgow, e.g. by the White Cart Water at Linn Park (NS5859, 2005), but very rare or absent from the west.

Arum italicum Mill.
Italian Lords-and-Ladies

Britain: neophyte (S Europe); local in the S, rare elsewhere.
Hortal; 7/7; rare.
1st Barshaw Park, 1988, CFOG.

Italian Lords-and-ladies appears to be a recent arrival in the area, but to be spreading. It occurs in the central area at a couple of places in Barochan Estate (NS4168, 1997), at Fulton Wood (NS4165, 1997) and by a roadside at Cleaves (NS4067, 1998). It was recorded (as ssp. *italicum*) from two places in Lochwinnoch (NS3459 and NS3558, 1996, IPG), and as a clump in woodland at Lochfield (NS4162, 2009, PRG).

LEMNACEAE

Lemna minor L.
Common Duckweed

Britain: common, but rare in NW Scotland.
Native; 79/87; frequent.
1st Renfrewshire, 1869, PPI.

There are no other old named localities for this duckweed, though it was considered common in the 19th century. Today Common Duckweed is widespread, occurring in lowland waterbodies such as lochs, ponds, ditches and some swamps; it can colonize secondary open waterbodies, such as pools along disused railways, presumably spread by birds.

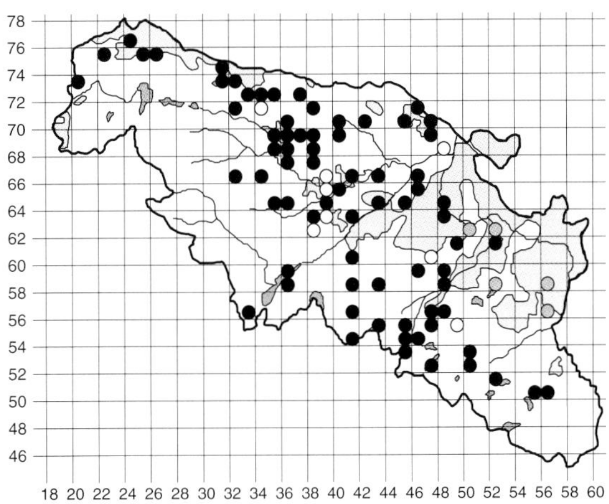

Lemna minor

Lemna trisulca L.
Ivy-leaved Duckweed

Britain: common in the S and E, frequent in C Scotland.
Native; 14/15; occasional.
1st Renfrewshire, 1915, TPNS.

Unlike the previous species, Ivy-leaved Duckweed was considered rare in the 19th century; the first named location was Bishopton (1931, PSY) and RM collected it from Braidbar Quarry (1946, E). Today it is only known from a few sites mainly in the central lowlands. It was found at six sites by the SNH Loch Survey but may have been overlooked during more general surveys.

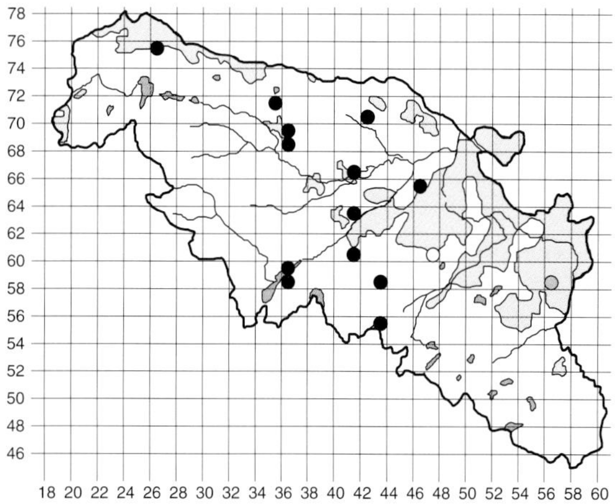

Lemna trisulca

ALISMATACEAE

Sagittaria sagittifolia L.
Arrowhead

Britain: frequent in S England, rare alien in C Scotland.
Former native or hortal; 1/2; very rare.
1st Paisley, 1834, MC.

The earliest known specimen is from the White Cart Water at Paisley, collected in 1836 by G McNab (E). Arrowhead's native status was questioned by Hennedy (1865) when he recorded it from the banks of the White Cart Water, Inchinnan Bridge, and it had been considered long extinct in the area until its discovery near the River Calder inflow at Castle Semple Loch (NS3558, 2003, AB). This species has not been recorded in the past from this well-visited, though recently disturbed, location, so it likely represents a recent arrival.

Baldellia ranunculoides (L.) Parl.
Lesser Water-plantain

Britain: scattered, declining.
Former native; 0/1; extinct.
1st Paisley, 1845, NSA.

The FFCA (1901) and Lee (1933) both mention 'Renfrewshire' but there are no other site details.

Alisma plantago-aquatica L.
Water-plantain

Britain: common except in NW Scotland.
Native; 35/43; frequent in central lowlands.
1st Lochwinnoch and Paisley, 1845, NSA.

There are quite a few old records for Water-plantain. It was collected from the Paisley Canal (1845, PSY; 1890, GL) and later from 'the works' dam' at Thornliebank (1909, GL). Other records (NHSG) are from Loch Libo (1887) and Castle Semple Loch (1887), and it has been recorded from both these lochs in modern times. Today it is found in the central lowland area, occurring in several lochs or ponds, and also along the riverbanks of the Black Cart, White Cart and Gryfe waters; it is notably absent from the west.

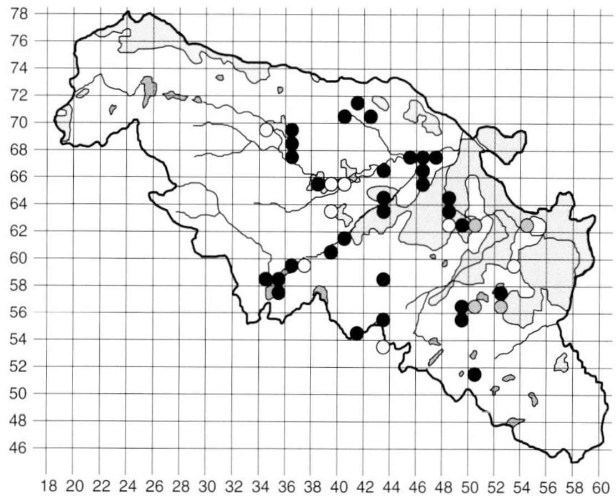

Alisma plantago-aquatica

Alisma lanceolatum With.
Narrow-leaved Water-plantain

Britain: common in SE England, rare in Scotland (mainly along the Forth & Clyde Canal).
Former native and hortal; 3/4; very rare.
1st Lochwinnoch, 1845, NSA.

The original record implies Narrow-leaved Water-plantain has been present in the VC for a long time but there are no other records, nor any mention for the area in Hennedy (1891), so there is some doubt about its local status. It has been recorded from ponds at Pollok Country Park (NS5562, CFOG),

Nether Johnstone (NS4163, 1998) and Glentyan Estate (NS3973, 2007), but it may well have been introduced at all these sites.

BUTOMACEAE

Butomus umbellatus L.
Flowering-rush

Britain: common in S England, scattered hortal in Scotland.
Former hortal; 0/1; extinct.
1st Jenny's Well, White Cart, 1900, PSY.

The minutes for 1939 in the TPNS (1942) reported that Jenny's Well was to be filled in, and further, that Flowering-rush was introduced from Duddingston Loch by 'the late Dr Cullen's father'. There are no other records for this distinctive plant.

HYDROCHARITACEAE

Elodea canadensis Michx.
Canadian Waterweed

Britain: neophyte (N America); common except in NW Scotland.
Accidental; 59/71; frequent.
1st Canal at Crookston, 1890, GL.

Canadian Waterweed may have been present in Paisley Canal prior to the first record, as Hennedy (1865) stated that it was 'now becoming a pest in canals'. It was also recorded from Bishopton (1930, PSY) and RM knew it from Knapps Loch (1938) and Pilmuir (1944). It is now quite widespread, and can be abundant, but does not seem to be such a pest as formerly feared. It is recorded from many lochs, ponds, dams and reservoirs, all in the lowlands although two are from above 200m: Picketlaw Reservoir (NS5651, 1998) and Black Loch (NS4951, 1997). A few records may refer to the following species.

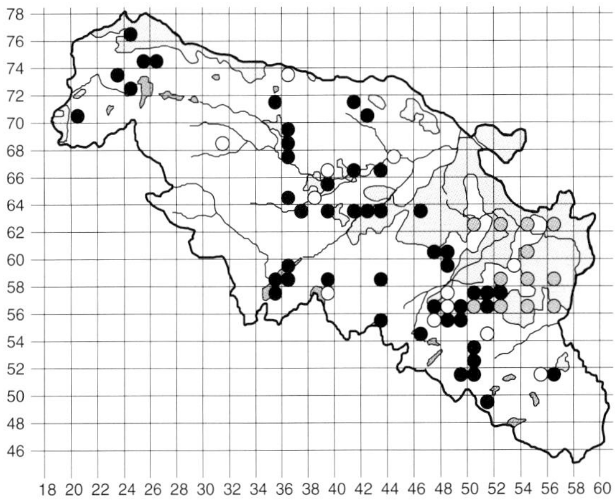

Elodea canadensis

Elodea nuttallii (Planch.) H. St. John
Nuttall's Waterweed

Britain: neophyte (N America); common in the S, spreading.
Accidental; 12/12; scarce but spreading.
1st Castle Semple Loch, 1996, SNH-LS

This waterweed is a newcomer to the area but seems to be spreading. It is found in Barr Loch (NS3558, 2001) and the Black Cart Water at Johnstone (NS4363, 2000), Linwood (NS4364, 2000) and Inchinnan (NS4767, 1999), and can also be found in the Gryfe Water at Wester Fulwood (NS4366, 2002) and Selvieland (NS4567, 2001). It occurs at Stanely Reservoir (NS4661, 2005) and two other open-water records are from the south-east: Harelaw Dam (NS4753, 2006) and Pilmuir Reservoir (NS5154, 2000) – the same place as RM's 1944 Canadian Waterweed field record.

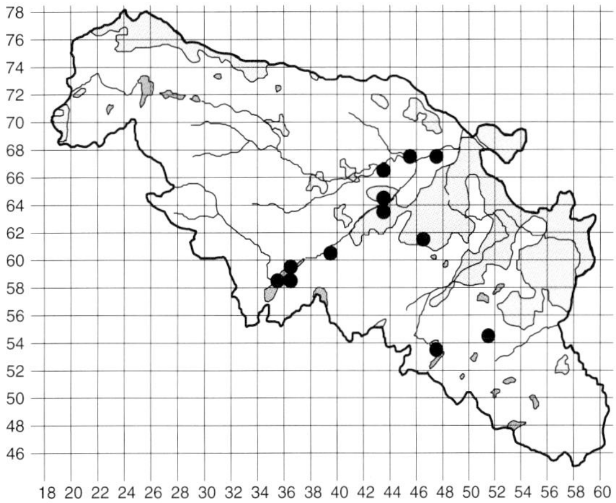

Elodea nuttallii

Lagarosiphon major (Ridl.) Moss ex V.A. Wager
Curly Waterweed

Britain: neophyte (S Africa); frequent in the S, spreading.
Accidental; 1/1; very rare.
1st Near ford of the White Cart, near Eaglesham (NS5753), 1995, PM.

This is the only record although Curly Waterweed may be overlooked elsewhere.

JUNCAGINACEAE

Triglochin palustris L.
Marsh Arrowgrass

Britain: common, declining in the S.
Native; 75/112; frequent.
1st Lochwinnoch, 1845, NSA.

Other old records include Gourock (1876, FFWS) and Neilston Pad (1891, GL). Today Marsh Arrowgrass is widely distributed, shows a strong rural pattern and can extend into the uplands, e.g. Queenside (NS3064, 1998) and Kirktonmuir (NS5551, 2002). It is most often encountered, but still likely overlooked, in marshes, flushes or mires with some mineral seepage. One unusual record was from a crack in the tarmac of a farmyard entrance at Mossneuk (NS4558, 1999).

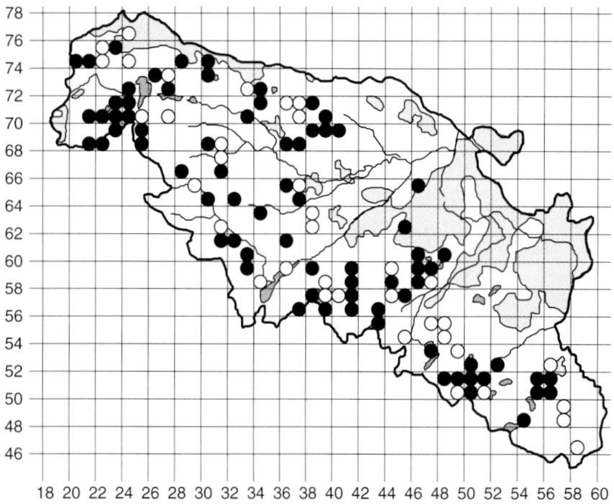

Triglochin palustris

Triglochin maritima L.
Sea Arrowgrass

Britain: common on all coasts.
Native; 15/19; locally frequent on the coast.
1st Gourock and Greenock, 1834, MC.

Sea Arrowgrass was collected from the saltmarshes at Gourock (1859, GL), which are now long gone, but it can still be found along the coast, at other small areas of relict coastal marsh. Like many other coastal species in the VC it shows a disjunct pattern.

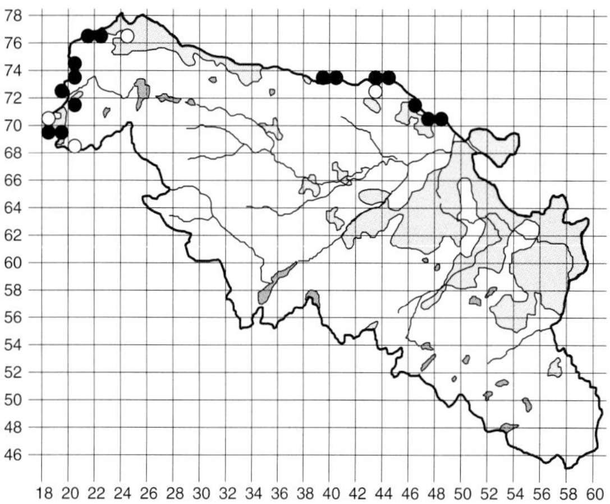

Triglochin maritima

ZOSTERACEAE

Zostera marina L.
Eelgrass

Britain: coastal, scattered, western.
Former native; 0/1; extinct.
1st Firth of Clyde, 1834, MC.

Eelgrass was also reported from Langbank in 1885 (NHSG). Although considered very common by Hennedy (1891), there are no modern sightings.

Zostera noltei Hornem.
Dwarf Eelgrass

Britain: coastal, scattered.
Native; 1/2; very rare.
1st West Ferry, Langbank, 1960, AMcGS.

Hennedy (1891) was aware of *Z. nana* Roth. locally but did not mention any VC locality. AMcGS more recently noted Dwarf Eelgrass from Langbank (NS3773) in 1980. Today it is only recorded from shallow tidal mud at Longhaugh Point (NS4374, 1996), but may well be present at similar muddy sites elsewhere.

POTAMOGETONACEAE (ZANNICHELLIACEAE)

Potamogeton natans L.
Broad-leaved Pondweed

Britain: common throughout.
Native; 74/88; common in open waterbodies.
1st Renfrewshire, 1872, TB.

Broad-leaved Pondweed was also present in Paisley Canal 'before draining' (1892, NHSG) and 40 or so years later RM reported it from several places: Kilmacolm Marsh (1924, GLAM), dams at Eaglesham (1927, GLAM), Knapps Loch (1938), Williamwood

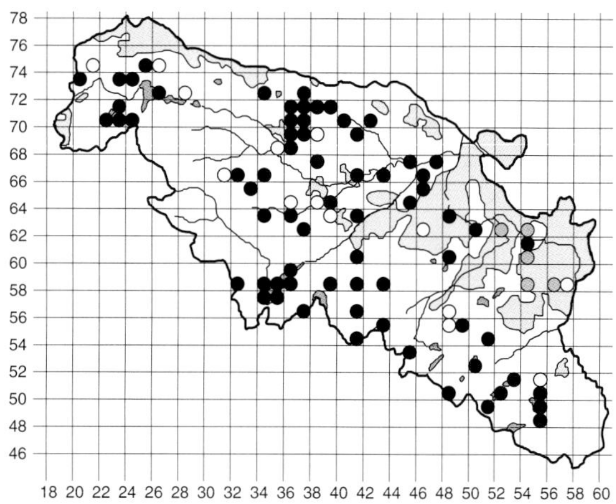

Potamogeton natans

Golf Course (1945), Braidbar Quarry (1945), Pilmuir Dam (1945), Snypes Dam (1952) and Barmufflock (1955). Today it is widespread, usually in more enriched waters, but can occur in peaty upland sites, e.g. Dunwan Dam (NS5548, 2001) and Crawhin Reservoir (NS2470, 2004). It was found at all of the 10 VC76 sites sampled by the SNH Loch Survey team (1996).

Potamogeton polygonifolius Pourr.
Bog Pondweed

Britain: common in the N and W, rare in S and E England.
Native; 113/132; common in peaty wetlands.
1st Mearns, 1865, GGO.

Bog Pondweed was considered common in the 19th century although Hennedy (1865) only named Gourock; it was collected from Brother Loch, Mearns in 1890 (GL) and from Meikleriggs in 1882 (PSY). It is now the commonest pondweed recorded, in part because it is often found in the accessible shallower water of ditches and open waterbodies; it also grows well in flushed peaty mires, often picking out mineral-enriched channels, and is readily encountered in the uplands.

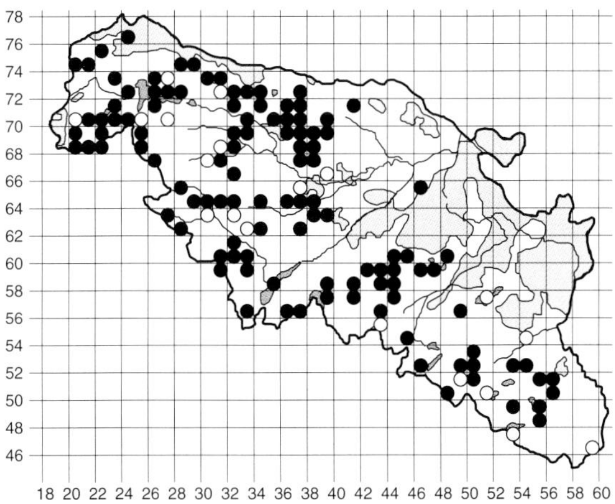

Potamogeton polygonifolius

Potamogeton lucens L.
Shining Pondweed

Britain: common in the SE, scattered in the N.
Former native; 0/1; extinct.
1st Renfrewshire, 1872, TB.

Although considered common by Hennedy (1891) there are no local sites named for Shining Pondweed apart from Barrhead (1901, FFCA), and it has not been recorded in the VC by anyone else.

Potamogeton gramineus L.
Various-leaved Pondweed

Britain: northern, scattered.
Native; 7/11; rare.
1st Lochwinnoch, 1845, NSA.

Considered common in 1876 (FFWS) or, perhaps more realistically, frequent by Hennedy in 1891, who mentioned the 'Cart'; there are no other named old locations for Various-leaved Pondweed. More recently RM found it at Daff Reservoir in 1973 and collected it from 'Howwood-Head Dam' in 1971 (GLAM); and in 1976 Mrs J Howitt reported it from Loch Thom. BWR (1973) knew it from Houstonhead Dam, where it can still be found (NS3965, 2006). The SNH Loch Survey (1996) found this pondweed at Caplaw Dam (NS4368) and Coves Reservoir (NS2476), AJS recorded it from Castle Semple Loch (NS3659, 1997) and it was recently seen at the western edge of Brother Loch (NS5052, 2011). The find at Balgray Reservoir (NS5057, 2000 and NS5156, 2006) restored Various-leaved Pondweed to the Glasgow area (CFOG).

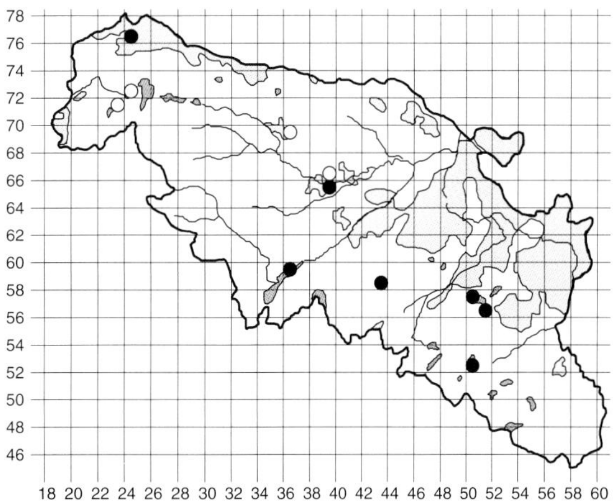

Potamogeton gramineus

Potamogeton × *nitens* Weber
(*P. gramineus* × *P. perfoliatus*)
Bright-leaved Pondweed

Britain: northern, occasional.
Native; 8/8; rare.
1st Daff Reservoir, 1977, BWR.

This hybrid pondweed was refound at Daff Reservoir (NS2270) in 2001, and was recorded from Black Loch (NS4951, 1997), the east end Corsehouse Reservoir (NS4850, 1997), the north-east edge of Brother Loch (NS5053, 2006) and the Black Cart Water near Linwood (NS4363, 2005). The SNH Loch Survey (1996) found it at Coves Reservoir (NS2476) and Knapps Loch (NS3668), and AJS found it at Castle Semple Loch (NS3659, 1997).

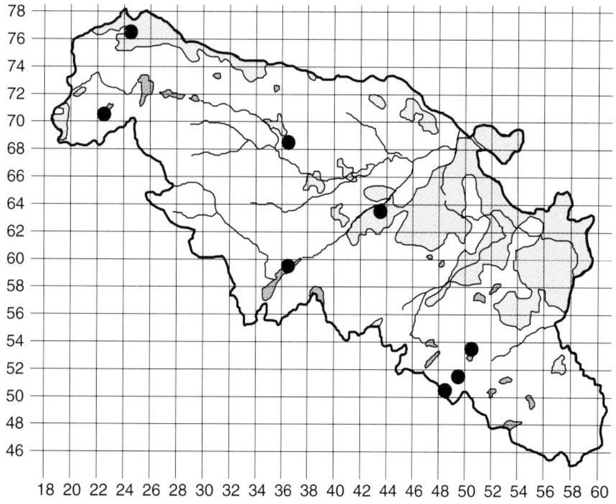

Potamogeton × nitens

Potamogeton alpinus Balb.
Red Pondweed

Britain: frequent but scattered in the N, rare in the S.
Native; 8/17; rare.
1st Renfrewshire, 1872, TB.

A 1907 (GL) specimen of this pondweed from Foxbar was incorrectly named as Long-stalked Pondweed (*P. praelongus* Wulfen). There are no named 19th-century localities for Red Pondweed, which Hennedy (1891) thought 'not common'. However, it was collected from a dam at Thornliebank (1909, GL) and RM knew it quite well with records from 'dams near Eaglesham' (1927), Bennan Loch (1929, GLAM), Braidbar Quarry, Giffnock (1945, BM), Knapps Loch (1938), a pool on Williamwood Golf Course (1945) and the Loganswell area (1957). In the recent past (BWR) it was found at a lochan above Lunderston (NS2174, 1967) and

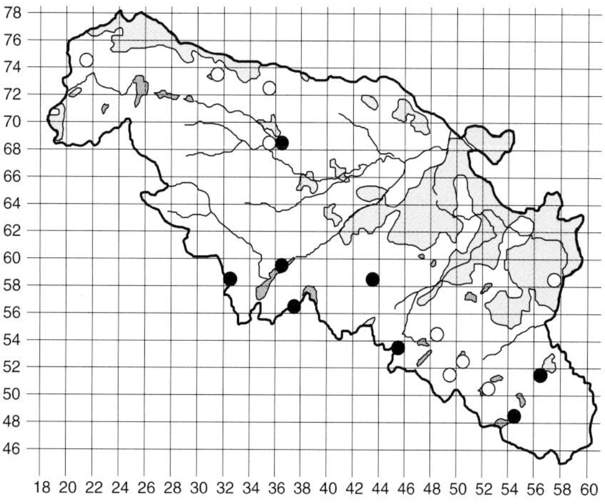

Potamogeton alpinus

Auchendores Reservoir (NS3572, 1969), and it was collected from Loch Thom by R Howitt (1976, CGE). The SNH Loch Survey (1996) found Red Pondweed on three occasions: at Castle Semple Loch (NS3659), Knapps Loch (NS3668) and Caplaw Dam (NS4358); other records are from mire at Knockmade (NS4553, 1999), a drain at Blackwoodhill (NS5448, 2001), Glenlora, Ladyland (NS3258, 2005) and a slow-moving burn at Knowes (NS3756, 2003).

Potamogeton perfoliatus L.
Perfoliate Pondweed

Britain: widespread, frequent.
Native; 31/37; frequent in open waterbodies.
1st Paisley Canal, 1843, GLAM.

Old literature accounts refer to Perfoliate Pondweed as common but there were no named locations, other than Lochwinnoch (1891, GL), until RM noted it at the dam outlet, Eaglesham (1927), Bennan Loch (1929), Pilmuir Dam (1940) and Walton Dam (1952). Today it is one of the commoner pondweeds found at a number of rural open waterbodies, mainly in the central south area; the western outliers are a quarry pool at Underheugh (NS2075, 2004) and Reservoir No. 4, Greenock Cut (NS2574, 2007). It has only been recorded from two rivers: the Gryfe Water at Wester Fulwood (NS4366, 2002), and the Black Cart Water at Barnsford (NS4667, 2002) and Linwood (NS5250, 2005).

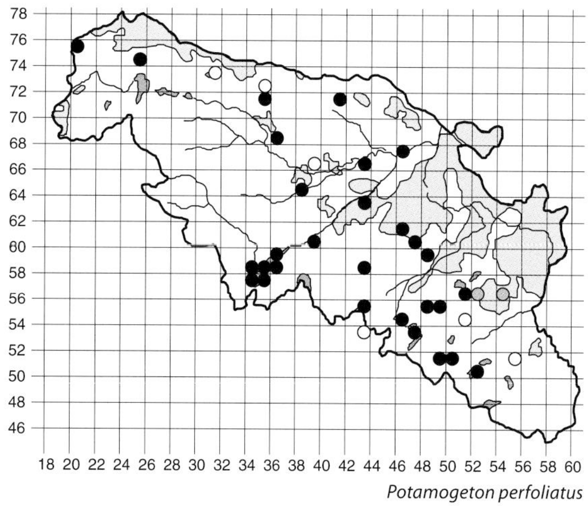

Potamogeton perfoliatus

Potamogeton obtusifolius Mert. & W.D.J. Koch
Blunt-leaved Pondweed

Britain: scattered, frequent in the N.
Native; 23/26; occasional.
1st Knapps Loch, 1938, RM.

There are no Renfrewshire localities prior to RM's for this pondweed, which was earlier considered rare by both Hennedy (1891) and Lee (1933). RM also noted it at Pilmuir Dam (1944) and Walton Dam (1955) and BWR recorded it from Auchendores Reservoir (1969); AJS found it at both Finlaystone Estate and Leperstone Reservoir in 1976. It was found at the latter (NS3571, 2000), but is likely to struggle as the dam wall has now been breached. The other sites are all standing open waterbodies in rural areas; Blunt-leaved Pondweed does not appear to relish the more peaty upland sites, although it is known from Reservoir No. 4 near Greenock Cut (NS2574, 2007).

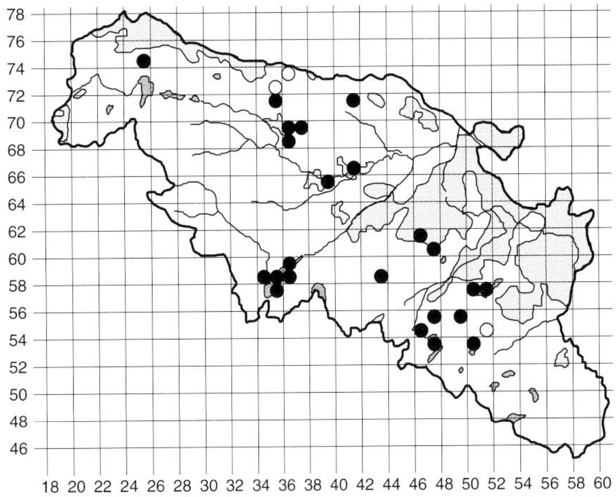

Potamogeton obtusifolius

Potamogeton pusillus L.
Lesser Pondweed

Britain: widespread, southern.
Native; 6/8; rare, but probably overlooked.
1st Paisley Canal bank, 1865, RH.

Apart from an undated 19th-century herbarium sheet from Greenock (GL), there are no other records for Lesser Pondweed until Queen's Park (1985, CFOG). Without specimens it is unclear whether old literature statements such as 'frequent' (Hennedy 1891) refer to this species or Small Pondweed. The SNH Loch Survey (1996) found Lesser Pondweed three times: at Barr Loch (NS3557), Castle Semple Loch (NS3659) and Loch Libo (NS4355); other records are from Walton Dam (NS4955, 1998) and pond at Murdieston Park (NS2675, 2002).

Potamogeton berchtoldii Fieber
Small Pondweed

Britain: common throughout.
Native; 20/24; occasional.
1st Pool on Williamwood Golf Course, 1945, GL.

Not distinguished from Lesser Pondweed by earlier authors, the above collection (by RM) is the first verifiable record; Small Pondweed is, however, the commoner of the two pondweeds today, so Hennedy's (1865) 'frequent' may well refer to this pondweed. JDM (*c*. 1966) recorded it from Maxwell Park (NS5663), RM collected it at Castle Semple Loch (1969, E) and AJS recorded it from Knapps Loch and Finlaystone Estate in 1976. All other records are from the modern period: the SNH Loch Survey reported it from five waterbodies – but not Castle Semple Loch, although AJS has reported it here (NS3659, 1997); other locations include the Gryfe Water at Fulwood (NS4366, 2002), a small marsh with open water on a disused railway (cycleway) at Elderslie (NS4366, 2000) and Kelly Cut (NS2471, 2007).

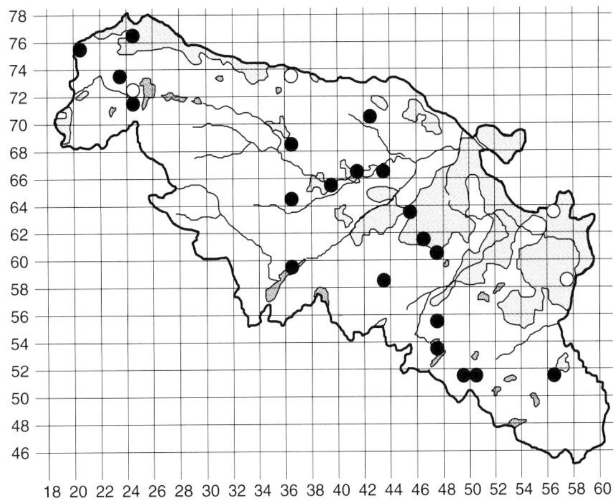

Potamogeton berchtoldii

Potamogeton crispus L.
Curled Pondweed

Britain: common except in NW Scotland.
Native; 14/20; occasional.
1st Lochwinnoch, 1845, NSA.

Considered common in 1876 (FFWS), the only other

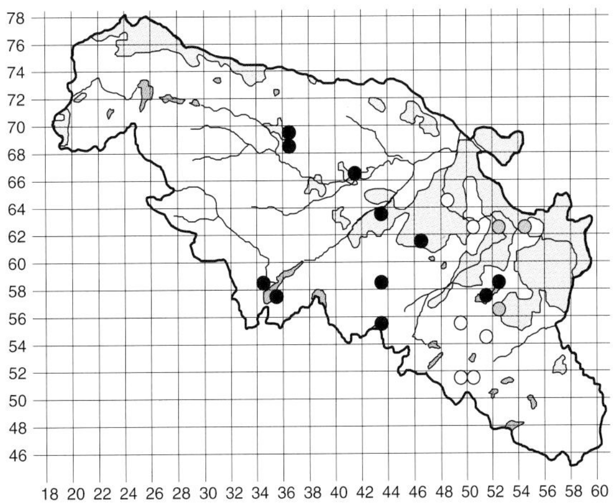

Potamogeton crispus

named location for Curled Pondweed is the Paisley Canal where it was first collected in 1859 (GL); much later (1941) RM recorded it from Pilmuir Reservoir. Modern records include Caplaw Dam (NS4358, 1993), Balgray Reservoir (NS5157, 1999), Stanely Reservoir (NS4661, 2005) and Loch Libo (NS4355, 2004), and the SNH Loch Survey found it at another four sites. It has also been seen in the Brock Burn at Waulkmill Glen (NS5258, 1991, CFOG) and in the Black Cart Water above the weir at Linwood (NS4346, 2005).

Potamogeton pectinatus L.
Fennel Pondweed

Britain: common in the S, frequent in C Scotland and NW islands.
Native; 3/4; very rare.
1st Dam, Eaglesham, 1927, RM (E).

Fennel Pondweed may be a recent arrival in the Clyde area (CFOG). Today it appears well established along the Black Cart Water at Inchinnan (NS4767, 1999) and further downstream at Colin's Isle (NS4968, 1997). A more isolated occurrence is at Caplaw Dam (NS4358, 1993), but it was not reported from there, or other local sites, during the detailed SNH Loch Survey.

Zannichellia palustris L.
Horned Pondweed

Britain: southern, common, occasional in C Scotland.
Former native; 0/2; extinct.
1st Renfrewshire, 1883, TB.

Ewing named Neilston (1901, FFCA) as a site for Horned Pondweed, but there are no modern records for this species in the VC.

RUPPIACEAE

Ruppia maritima L.
Beaked Tasselweed

Britain: coastal, scattered.
Native; 1/3; very rare.
1st Langbank, 1885, NHSG.

Beaked Tasselweed was first recorded at the end of the 19th century but there are no other named old localities. It was reported from brackish pools at Langbank by JP in 1968 and more recently it was found in shallow tidal mud at nearby Longhaugh Point (NS4373, 1996).

NARTHECIACEAE
(LILIACEAE *pro parte*)

Narthecium ossifragum (L.) Huds.
Bog Asphodel

Britain: common in the N and W.
Native; 182/207; common on peaty ground.
1st Gourock, 1842, GLAM.

A common plant of the uplands, there are plenty of old records for Bog Asphodel from the 19th century and today the map still shows a strong rural and upland pattern. It is absent from many lowland places but can still be found at the low-lying Paisley Moss (NS4865, 2000) and at Darnley (NS5258, 2007). It is typically found in acidic peaty soils, usually where there is some flushing or water movement.

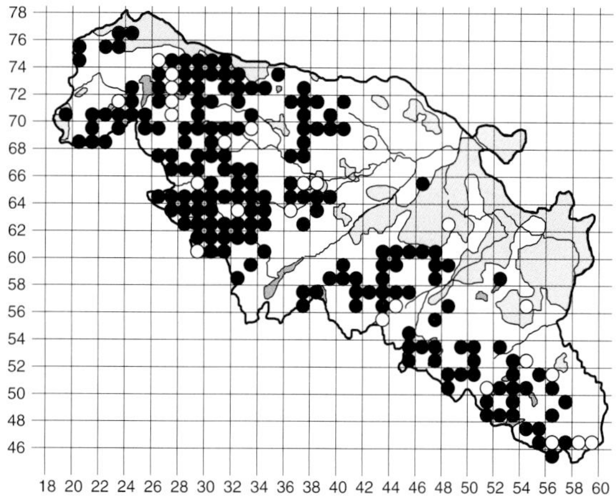

Narthecium ossifragum

MELANTHIACEAE
(LILIACEAE *pro parte*)

Paris quadrifolia L.
Herb-Paris

Britain: southern, declining, rare in Scotland.
Former native; 0/5; extinct.
1st Banks of Cart opposite mill, Cathcart, 1813, TH.

There are a few old records for Herb-Paris from wooded glens but despite much searching it has not been recorded from any site for more than 100 years. The last date for it is 1912 (GL) for Bardrain Glen, where first recorded in 1859 (PSY); a TPNS botanical report on the glen (Thomson 1944) states, 'not seen for a long time'. Records for Darnley go back to 1883 (GL) but stop with Lee (1933); it was still considered vigorous there in 1890 (Wood 1893), but he also noted that although it set seed readily they never germinated. Herb-Paris was known in the parish of Lochwinnoch in 1845 (NSA) and

Wood (1893) described the colony from the Calder (above Lochwinnoch) as the 'finest I have ever seen'. There is also an undated, but presumably 19th-century, herbarium sheet (GL) from 'Williamwood, Shawlands'.

ALSTROEMERIACEAE (LILIACEAE *pro parte*)

Alstroemeria aurea Graham
Peruvian Lily

Britain: neophyte (Chile); occasional, scattered.
Hortal; 1/1; very rare outcast.
1st Bridge of Weir, 2005.

A small clump was found on waste ground near the cycleway (NS3865).

COLCHICACEAE (LILIACEAE *pro parte*)

Colchicum autumnale L.
Meadow Saffron

Britain: declining in the S, rare alien in Scotland.
Hortal; 1/1; very rare.
1st Birkmyre Park, Port Glasgow (NS3174), 1996.

There is only one record for this garden outcast.

LILIACEAE

Tulipa sylvestris L.
Wild Tulip

Britain: neophyte (S Europe); mainly southern.
Former hortal; 0/2; extinct.
1st Arden House, 1886, GL.

Wild Tulip was well known from an old lime quarry station (near Darnley) in the 19th century (CFOG). Wood wrote (1893), 'there used to be plenty' but 'its beautiful, conspicuous yellow flowers have proved its ruin'. It was also collected from Bridge of Weir in 1905 (PSY), a record which was repeated by Lee (1933). There are no modern records.

Tulipa gesneriana L.
Garden Tulip

Britain: neophyte (garden origin); scattered.
Hortal; 4/4; rare.
1st Teucheen Wood, 1996.

The only records for this tulip are for garden outcasts, none of which are known to persist: Levern Water, Arthurlie (NS4958, 1997), Woodhall (NS3473, 1997) and Busby Glen (NS5756, 2005).

Lilium martagon L.
Martagon Lily

Britain: neophyte (Europe); scattered.
Hortal; 1/1; very rare.
1st Durrockstock (NS4662), 1991.

Lilium pyrenaicum Gouan
Pyrenean Lily

Britain: neophyte (Pyrenees); scattered.
Hortal; 2/3; very rare.
1st Finlaystone, 1975, AJS.

Pyrenean Lily, found as a garden outcast in the VC, has also been recorded from a roadside at Plymuir (NS4357, 1999) and near the Mill Dam at Kilmacolm (NS3469, 1999).

ORCHIDACEAE

Epipactis helleborine (L.) Crantz
Broad-leaved Helleborine

Britain: frequent in the S, rare in Scotland except in the C lowlands.
Native; 64/73; frequent.
1st Pollok Wood, 1842, GLAM.

Hennedy (1891) considered Broad-leaved Helleborine to be frequent and there are plenty of old literature and herbarium records from places such as Renfrew, Paisley and Lochwinnoch (1845, NSA), woods east of Gourock (1865, RH), Newton Woods (1900, PSY), Darnley (1890, GL), near Clarkston (1908, GL) and Rouken Glen (1926, GLAM). Its distribution in the Glasgow area was discussed in CFOG and the modern map shows it extends to the west of Paisley and down the Black Cart Water valley to Lochwinnoch. It appears to be much rarer in the west but was recently recorded from woodland at Wemyss Bay (NS1970, 2010, PRG) and by the railway at Spango (NS2374, 2010, PRG).

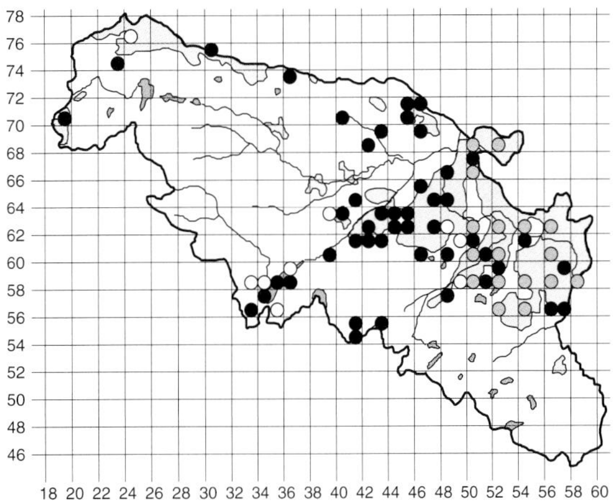

Epipactis helleborine

Neottia ovata (L.) Bluff & Fingerh.
Listera ovata (L.) R. Br.
Common Twayblade

Britain: common in the S, local in N Scotland.
Native; 26/36; occasional.
1st Gourock, 1842, GLAM.

Other old records include Lochwinnoch, Paisley and Renfrew (1845, NSA), Shielhill Glen (1881, NHSG), Castle Semple Loch (1887, NHSG), Kelly Glen (1887, NHSG) and Bardrain (1902, PSY). There are only a handful of records from the 1960s–80s period (BWR) but Common Twayblade appears to be a widespread species, and it is well capable of colonizing waste ground as at Ladyburn (NS3075, 1996), Boglestone (NS3373, 2001) and the disused railway (cycleway) near St Brydes (NS3860, 2001).

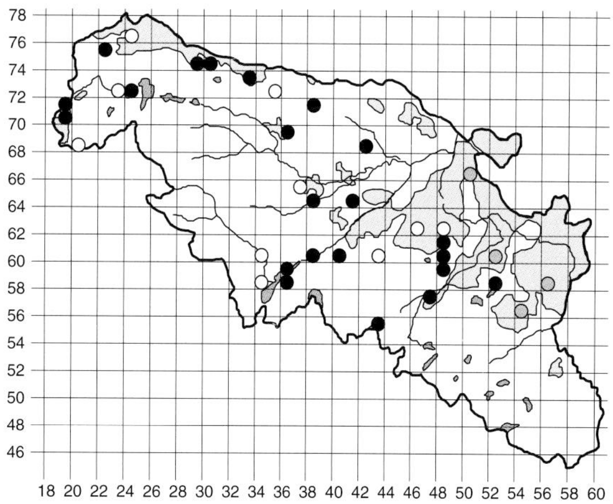

Neottia ovata

Neottia cordata (L.) Rich.
Listera cordata (L.) R. Br.
Lesser Twayblade

Britain: frequent in the N and W.
Native; 26/32; occasional in upland moors.
1st Misty Law, 1834, MC.

Lochwinnoch and Renfrew parishes are noted in 1845 (NSA) as sites for Lesser Twayblade, and Hennedy (1891) considered it to be frequent and mentioned its presence on the hills above Gourock; it was collected (GL) from Gourock in 1856 and Port Glasgow in 1884. It was found several times in the 1960s–80s (BWR) and today it shows a disjunct pattern, well represented in the Renfrewshire Heights but also known from a few sites on Eaglesham Moor. It grows in moorland, usually on low bog-moss hummocks under leggy heather.

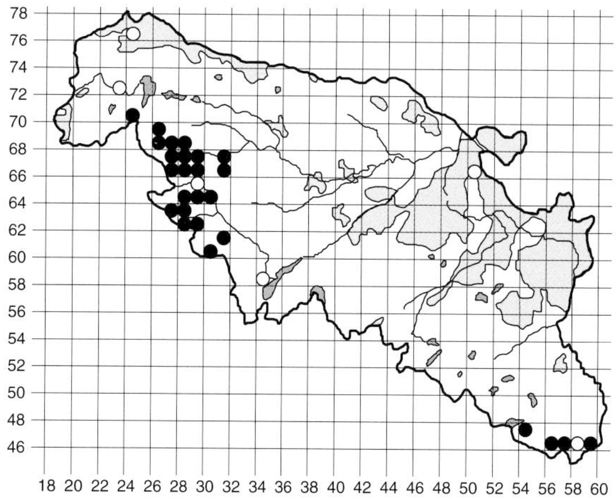

Neottia cordata

Neottia nidus-avis (L.) Rich.
Bird's-nest Orchid

Britain: scattered, local in Scotland.
Native; 1/3; very rare.
1st Wood near Kennels, Inverkip, 1858, GGO.

Hennedy first reported Bird's-nest Orchid from Cloch lighthouse (1865) and considered it to be rare. JDM reported it from the lower Kelly Glen (NS1968) in 1959, but it has not been seen there since. The only modern record is from the RSPB reserve at Lochwinnoch (NS3658, 1999, ERTC), perhaps the same locality as AMcGS's earlier 'Lochwinnoch' record (1955).

Hammarbya paludosa (L.) Kuntze
Bog Orchid

Britain: local, mainly in the N and W.
Former native; 0/1; extinct.
1st Lochwinnoch, 1876, FFWS.

This orchid has not been recorded from the 'Linthills' above Lochwinnoch since Hennedy's time (1891).

Corallorhiza trifida Châtel.
Coralroot Orchid

Britain: local in NE Scotland, rare in S Scotland.
Native; 5/6; rare.
1st Bent (NS4359), 1982, BWR.

Coralroot Orchid was unknown to 19th-century recorders in Renfrewshire although Lee (1933) knew it from Kilmarnock (VC75). It can still be found at Glen Moss (NS3669, 1998), where it was first seen in 1985 (BWR), and it also occurs at three places nearby: Lawfield Dam (NS3769, 1990, TNT), small ('crater') mires at Corsliehill (NS3969, 1996), and Shovelboard (NS3869, 2004). A further site to the south is at the edge of Sergeantlaw Moss (NS4459, 1995), not far from the original site (Bent). At all sites it grows in bog-moss under willow carr or wet birch woodland.

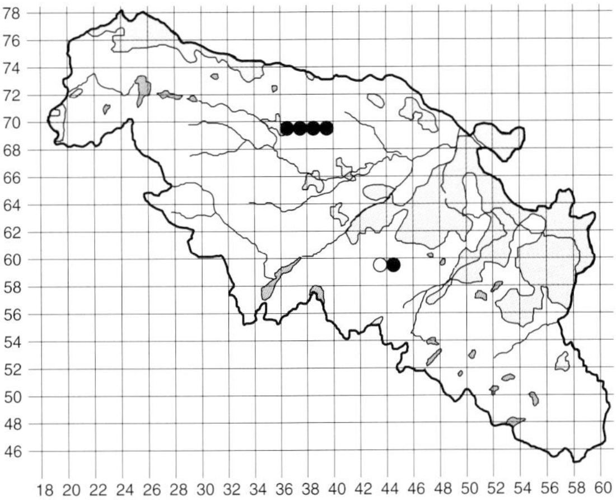

Corallorhiza trifida

Platanthera chlorantha (Custer) Rchb.
Greater Butterfly-orchid

Britain: widespread but local, mainly in the W and C in Scotland.
Native; 88/99; frequent.
1st Gourock, 1865, RH.

There are only a few old localities for this orchid, which Hennedy (1891) thought frequent: hills behind Gourock (1880, NHSG), damp wood near Cloch Lighthouse (*c.* 1880, GLAM) and heath near Kilmacolm (1925, RM). Today Greater Butterfly-orchid is widespread in the lowlands but mainly central. It is typically found in species-rich neutral or slightly acidic pastures although it can occur in neglected grasslands, where it appears capable of colonizing or persisting, and can be found on roadside grassland at Parklea (NS3573, 2007).

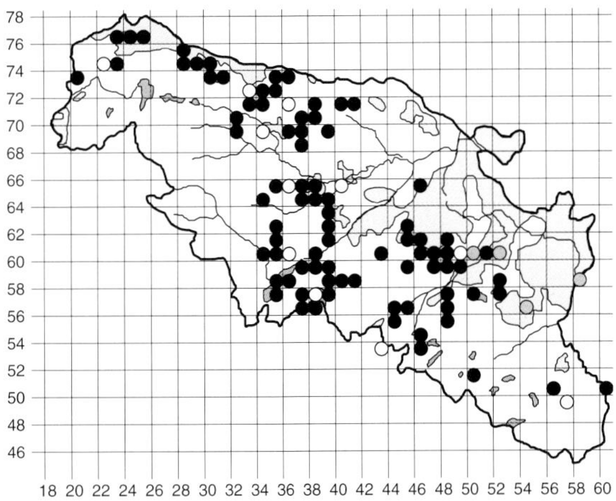

Platanthera chlorantha

Platanthera bifolia (L.) Rich.
Lesser Butterfly-orchid

Britain: scattered and local, mainly in the W and N.
Native; 2/9; very rare.
1st Cowglen, 1858, GL.

Hennedy (1865) described Lesser Butterfly-orchid as frequent and named Gourock and Paisley, but there are few other old localities; it was collected from Newton Wood (1901, PSY) and reported from Bardrain Glen (Stewart 1930). In the recent past it was recorded (BWR) from Knocknairshill (NS3074, 1969), Dargavel Burn (NS3771, 1971) and Airds Meadow, Castle Semple Loch (1971, SNH files). Today, despite many searches, it is only known from the species-rich grassland at Knocknairshill SSSI (NS3074, 2001) and a wet heath at Newhouse (NS3757, 2011); at both sites it grows with or very near to Greater Butterfly-orchid.

Pseudorchis albida (L.) Á. Löve & D. Löve
Small-white Orchid

Britain: scattered in NW Scotland, rare elsewhere.
Former native; 0/10; probably extinct.
1st Lochwinnoch, 1834, MC.

Small-white Orchid was considered rare in the 19th century but there are quite a few records: Greenock (1837, MANCH), Gourock (1842, GGO), Paisley (*c.* 1840, GL), Inverkip (1851, GL), Mill Burn, Kilmacolm (1874, TGSFN), Tower Hill, Gourock (1880, NHSG), Ranfurly (1902, PSY) and Bardrain Glen (1930, TPNS). More recently it was found at Ranfurly Golf Course (NS3665, 1969, BWR) and from near Craigenfeoch (NS3472, 1971) by J Blackwood; the latter reported more than 20 spikes annually between 1971 and 1974 (Blackwood 1976). It has not been refound during modern searches of these places.

Gymnadenia borealis (Druce) R.M. Bateman, Pridgeon & M.W. Chase
Gymnadenia conopsea (L.) R. Br.
Heath Fragrant-orchid

Britain: widespread but only frequent in uplands and the NW.
Native; 7/17; rare.
1st Gourock, 1837, GL.

It is assumed that Heath Fragrant-orchid, previously recorded as Fragrant Orchid, is the taxa occurring in the VC, but this has not been studied in detail. 'Fragrant Orchid' was described as frequent by Hennedy (1891), who added that it was 'plentiful all about the heathy shores of the Firth'. Old localities include Paisley and Lochwinnoch (1845, NSA), Bridge of Weir (1902, PSY) and Kilmacolm (1925, RM). There are a few recent past records (BWR) from places such as Ranfurly Golf Course (NS3665, 1969), Daff Reservoir (NS2371, 1973), Bardrainny (NS3472, 1975), Gibblaston (NS3465, 1977) and Bridge of Weir (NS3865, 1980). It may well have been overlooked during the modern survey, but it is also likely that it has declined with agricultural improvement. There are seven modern records for Heath Fragrant-orchid, all from species-rich grassland: Whinnerston (NS3864, 1996), Cornalees (NS2371, 1996, CMRP), Cauldside (NS3270, 2005), north of Craigmarloch (NS3472, 2008), Gryfe Reservoir No. 2 (NS2972, 2008), Greenside (NS3761, 2008) and Pulpit Rock (NS2376, 2011). At Whinnerston and Cauldside it occurs with Heath Spotted-orchid as well as scented hybrids, presumably × *Dactylodenia evansii* (Druce) Stace. At the species-rich embankment at Greenside, it grows with Northern Marsh-orchid, where one individual of × *Dactylodenia varia* (T. & T.A. Stephenson) Aver. was seen.

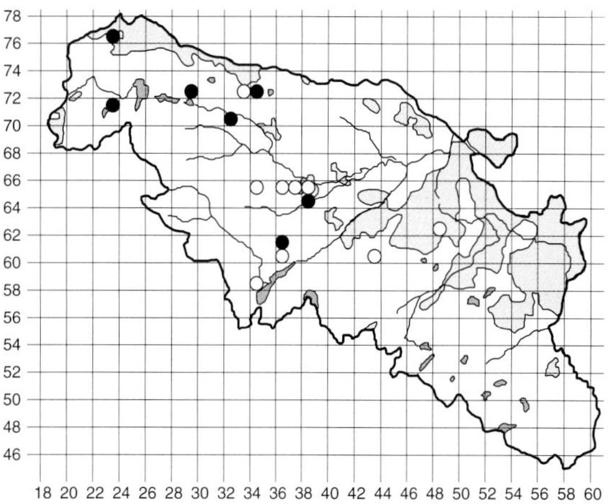

Gymnadenia borealis

Coeloglossum viride (L.) Hartm.
Frog Orchid

Britain: widespread but scattered and declining.
Native; 4/14; rare.
1st Paisley, 1845, NSA.

Frog Orchid was considered not common in the 19th century but Wood (1893) said it was plentiful at Neilston Pad and by the River Calder above Lochwinnoch; other named stations include Eaglesham (1865, GGO), Brother Loch (1891, RH), Bridge of Weir (1893, GL) and behind Peesweep (1896, PSY), and, more recently, heath near Kilmacolm (1925, RM) and Locherside (1942, TPNS). Today it is only known from five nearby sites (in four 1km grid squares), all species-rich pastures: Knockenae (NS4554, 1994), Levern Water, Harelea (NS4654, 1998), Carswell Hill (NS4653, 1999), Commore Dam (NS4654, 1999) and Muirhead (NS4855, 1999).

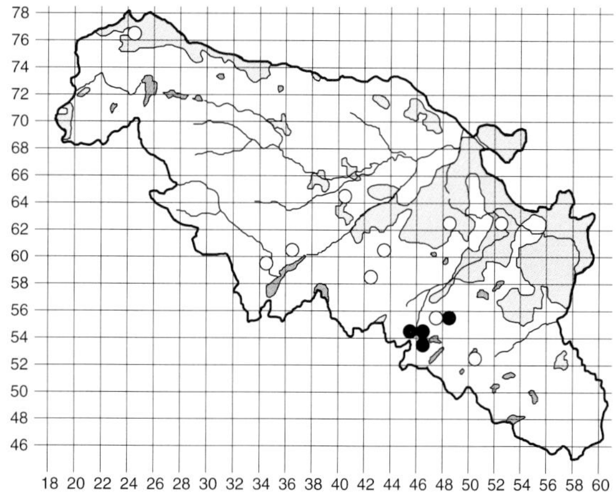

Coeloglossum viride

Dactylorhiza fuchsii (Druce) Soó
Common Spotted-orchid

Britain: common throughout except in N Scotland.
Native; 148/167; common.
1st Paisley, 1845, NSA.

'*Orchis maculata*' was considered common in the 19th century, although this name was used for both Common and Heath Spotted-orchid (all local herbarium sheets checked are of the latter). Today Common Spotted-orchid has a widespread distribution, occurring throughout the lowlands and upland fringes, but is rare or absent from the peaty uplands. It is found in less-acidic marshes and marshy grasslands; it is also a common feature of urban waste ground, notably when on poorly draining compacted clayey soils.

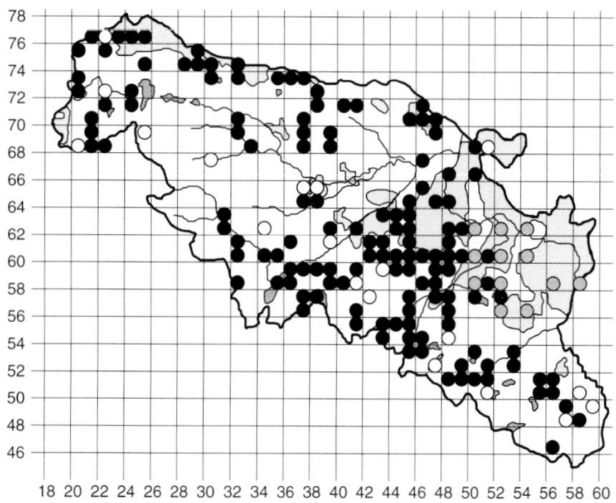

Dactylorhiza fuchsii

Dactylorhiza* × *transiens (Druce) Soó
(*D. fuchsii* × *D. maculata*)

Britain: scattered.
Native; 2/2; very rare, probably overlooked.
1st Harelaw Brae (NS4959), 1999.

This hybrid, which is likely to be more frequent than the number of records suggests, is also recorded from Jock's Craig (NS3269, 2002).

Dactylorhiza* × *venusta (T. & T.A. Stephenson) Soó
(*D. fuchsii* × *D. purpurella*)

Britain: occasional in N to C Scotland.
Native; 11/11; scarce.
1st Williamwood (NS55U), 1978, CFOG.

This hybrid has been recorded on several occasions during modern surveys, and is likely to be widespread. It has been found at waste ground, Ladyburn (NS3075, 1998), Jenny's Well (NS4962, 2001), Barscube Hill (NS3871, 2002), waste ground, Whinhill (NS2875, 2003), Humbie (NS5353, 2003), Lochend (NS3664, 2003), Harelaw Reservoir (NS3073, 2007) and at a roadside, Parklea (NS3573, 2007).

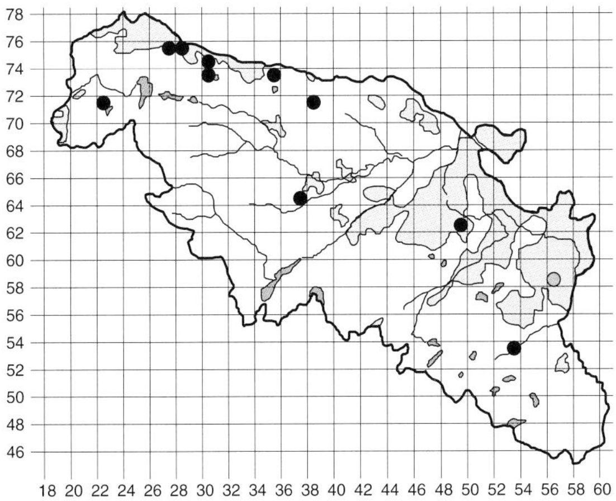

Dactylorhiza × venusta

Dactylorhiza maculata (L.) Soó
Heath Spotted-orchid

Britain: common in the N and W, local elsewhere.
Native; 106/137; common in upland fringe.
1st Gourock, 1837, GL.

There are other old herbarium sheets for Heath Spotted-orchid from Inverkip Road (1886, GRK), Black Loch (1891, GL) and Newton Woods (1901, PSY), and literature records for Paisley (1845, NSA) and Castle Semple Loch (1887, NHSG), but it was not distinguished by Hennedy from Common Spotted-orchid. Today it is widespread but shows a strong rural upland-fringe pattern. It is found in marshy acidic grasslands or wet heaths and the drier parts of mires, often in some abundance.

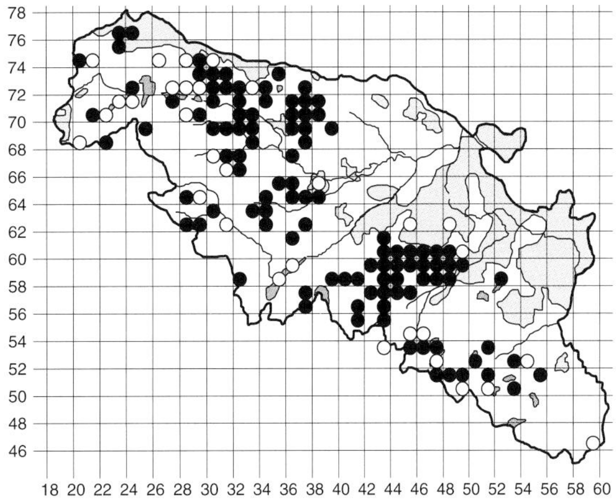

Dactylorhiza maculata

Dactylorhiza* × *formosa (T. & T.A. Stephenson) Soó
(*D. maculata* × *D. purpurella*)

Britain: scattered in the N.
Native; 11/12; scarce.
1st South of Leitchland, 1974, BWR.

This hybrid has been recorded from several sites including Corsliehill (NS3969, 1996), mire at Dargavel Burn (NS3771, 1997), Knocknairs Moor (NS3073, 1997), Witch Burn (NS4557, 1999; NS4658, 2007), Kirktonmoor (NS5551, 2003), Midhill (NS3573, 2007), Lochend (NS3664, 2003), Cauldside (NS3270, 2005) and Harelaw Reservoir (NS4859, 2007).

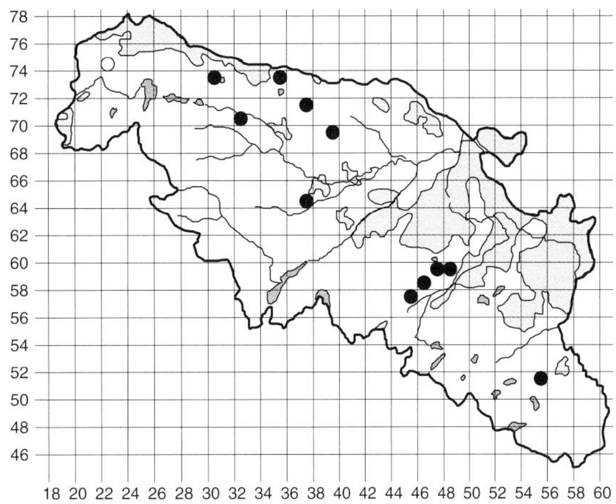

Dactylorhiza × formosa

Dactylorhiza incarnata (L.) Soó
Early Marsh-orchid

Britain: widespread but local.
Native; 1/1; very rare.
1st Dargavel Burn, 1997.

The sole record for Early Marsh-orchid, of the pale form of ssp. *pulchella* (Druce) Soó, is from the flushed margins of the species-rich mire by the Dargavel Burn (NS3771). A specimen from this location has been tentatively identified as the hybrid with Northern Marsh-orchid (*D.* × *latirella* (P.M. Hall) Soó).

Dactylorhiza purpurella (T. & T.A. Stephenson) Soó
Northern Marsh-orchid

Britain: frequent in the N and W, absent in the S.
Native; 74/93; frequent.
1st Gourock, 1837, GL.

Northern Marsh-orchid (as '*Orchis latifolia*') was a common plant in the 19th century and may have undergone a decline in rural areas, although it is capable of colonizing urban waste ground. It has a scattered distribution, mainly lowland and rural; it avoids the uplands. It is typically found in less acidic mires and flushed marshy grasslands, and also on wet waste ground and at disused railways.

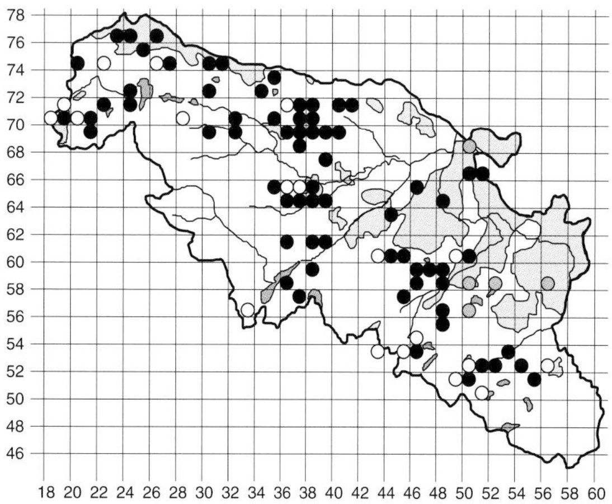

Dactylorhiza purpurella

Orchis mascula (L.) L.
Early-purple Orchid

Britain: widespread but local.
Native; 7/15; rare.
1st Gourock, 1865, RH.

Although considered frequent by Hennedy (1891), he named no other local stations apart from that of the first record. There are herbarium sheets (GL) from Devol Glen (1884), Brandy Burn, Paisley (1905), Old Patrick Water (1917) and Locher Burn, Bridge of Weir (1921). Early-purple Orchid was recorded on several occasions during the 1960s–80s period (BWR), e.g. Locher Water (1969), Calder Glen (1969) and Bridge of Weir (1980). Today it is known from a few scattered sites, but may well be overlooked later in the season. It is found in woodlands as by the River Calder (NS3361, 1997) and Shielhill Glen (NS2471, 1999, AB), but also on burn-side ledges as at Munzie Burn (NS5848, 2003) and North Rotten Burn (NS2568, 1997), roadside grass at West Tandlemuir (NS3361, 1997) and pastures at Knowes (NS3756, 2003), Whinnerston (NS3864, 1996) and Holehouse (NS5753, 2000).

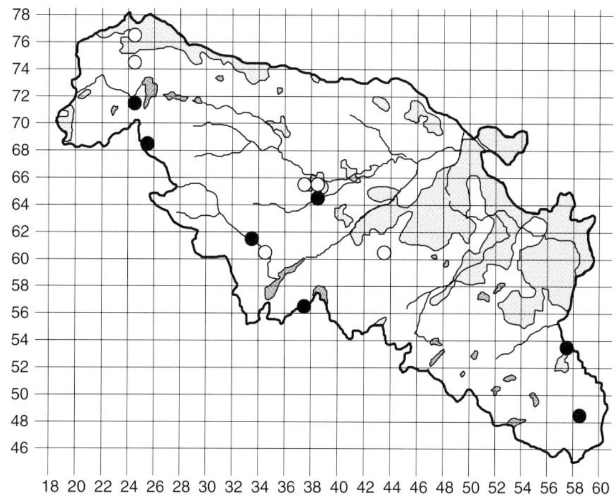

Orchis mascula

IRIDACEAE

Sisyrinchium montanum Greene
American Blue-eyed-grass

Britain: neophyte (N America); occasional, western.
Hortal; 5/7; rare.
1st Giffnock, 1919, RG.

American Blue-eyed-grass was also reported from Bishopton in 1931 by WH Thomson (TPNS 1942). Grierson's quarry locality and RM's 1940 record from the railway 'triangle' at Williamwood were noted in CFOG; other CFOG locations included Hillington West Station (NS5165), where it has not been seen recently. It is still present at Williamwood (NS5658, 1991) and was also found at Malletsheugh (NS5255, 2001), a roadside at Black Hill (NS4851, 2011) and Gleniffer Braes (NS46T, 1997, SA).

Iris pseudacorus L.
Yellow Iris

Britain: common throughout except in upland N Scotland.
Native; 223/257; common.
1st Paisley, 1845, NSA.

Yellow Iris was considered very common in the 19th century and today it can certainly be described as common, being widespread and only absent from some urban areas and the peaty uplands. It can be found in a variety of wet places, including pond or loch margins, by ditches or watercourses and in marshes or wet woodlands. A few other Iris species or cultivars (neophytes) have been recorded as garden outcasts, and may persist locally, but none have been expertly determined or appear to be established. *I. versicolor* L. (Purple Iris – or cultivar) grows in the old estate grounds at Milliken (NS4161, 2001).

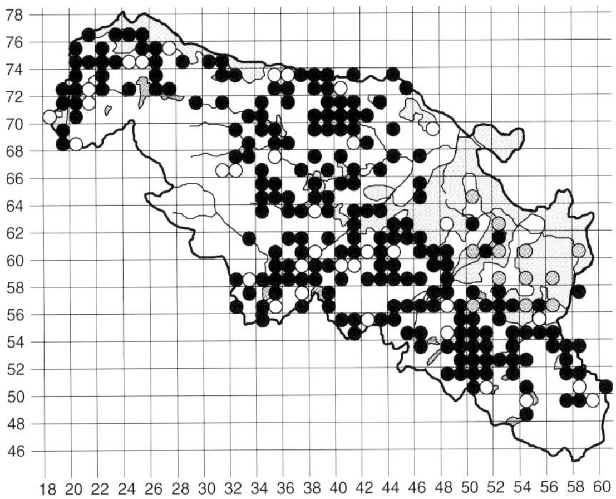

Iris pseudacorus

Crocus vernus (L.) Hill
Spring Crocus

Britain: neophyte (S Europe); scattered.
Hortal; 10/12; rare.
1st Gleniffer Braes (NS4459), 1974, AJS.

The first record was of a singleton growing in bog-moss. AJS also recorded Spring Crocus from Finlaystone Estate in 1976. There are only a few modern records for crocuses and these likely include more than one species or cultivars – Early Crocus (*C. tommasinianus* Herb.) was recently recorded from a wood in Uplawmoor (NS4355, 2009, PRG). Localities for Spring Crocus are generally riversides or wayside grasslands or woodland edges. It is not clear how well these plants persist, but they are a frequent feature of lawns and park amenity grasslands. However, they have not been well recorded as 'wild' plants.

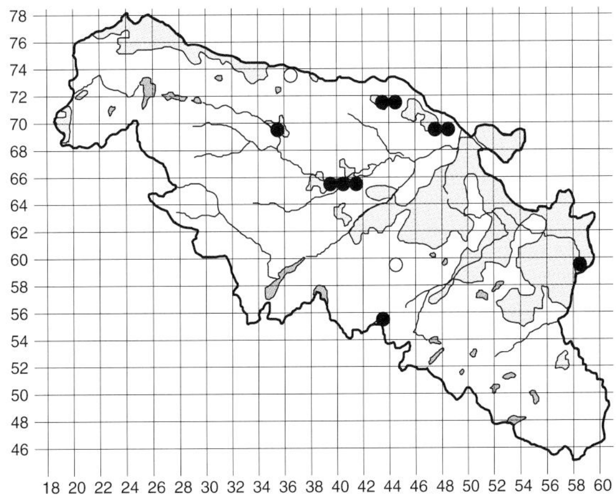

Crocus vernus

Crocosmia paniculata (Klatt) Goldblatt
Aunt-Eliza

Britain: neophyte (S Africa); occasional.
Hortal; 17/17; scarce.
1st Spateston Burn, 1993.

There are no records for this larger-leaved relative of Montbretia before the modern period, and some records may include hybrids or similar cultivars. Aunt-Eliza appears to be quite widespread, with records near the west coast but also scattered around the central area; populations though are generally small. Habitats are usually woodland edges, roadsides or hedge banks.

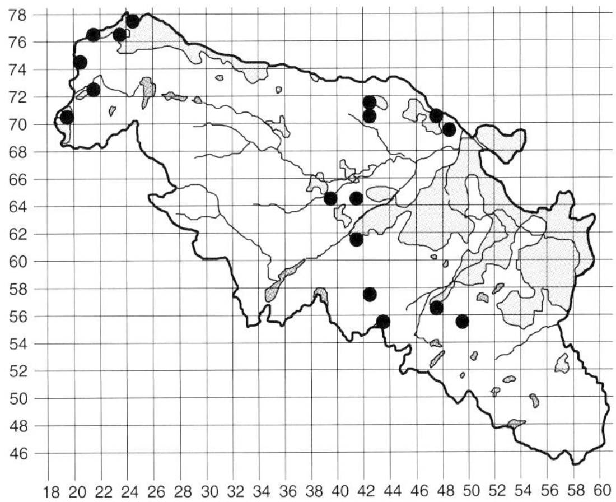

Crocosmia paniculata

Crocosmia pottsii (Macnab ex Baker) N.E. Br.
Potts' Monbretia

Britain: neophyte (S Africa); rare in W Scotland.
Hortal; 3/3; very rare.
1st Inverkip, 1999.

Possibly overlooked for Montbretia, Potts' Monbretia has been recorded from three sites: Inverkip (NS2070), a road verge, Aston (NS2276, 2010, PRG) and waste ground, near sewage works, Renfrew (NS4866, 2000).

Crocosmia × *crocosmiiflora* (Lemoine) N.E. Br.
Montbretia

Britain: neophyte (hybrid from S African parents); common especially in W.
Hortal; 87/89; frequent, established.
1st Riverside, Gryfe, Hattrick, 1971, BWR.

There are only a handful of records (BWR) from before the CFOG period, from widespread places mainly in the west, including Wemyss Bay (NS1870, 1986), Levanne (NS2176, 1984) and Trumpethill (NS2276, 1977). Today Montbretia can be frequently seen in lowland urban-fringe areas, where it grows on waysides, in rough grasslands and on waste ground. The distribution shows a strong coastal pattern, but a few records occur inland or at remote places, such as by a roadside, Queenseat (NS5248, 2001) and Black Loch (NS4951, 1997), representing dumping.

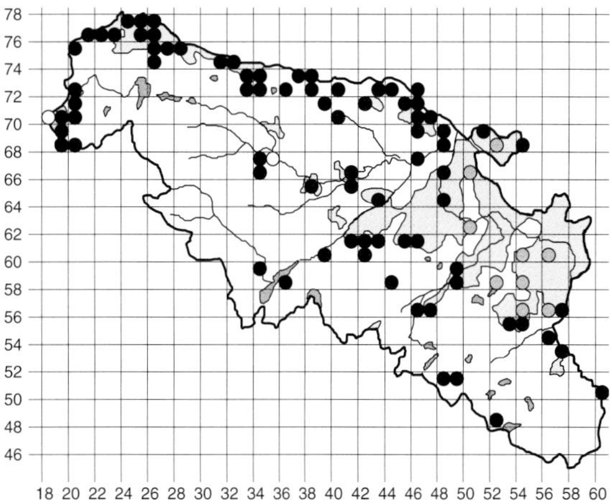

Crocosmia × crocosmiiflora

XANTHORRHOEACEAE
(LILIACEAE *pro parte*)

Hemerocallis fulva (L.) L.
Orange Day-lily

Britain: neophyte (garden origin); scattered, mainly in the S.
Hortal; 1/1; very rare.
1st Caplaw Bridge (NS4358), 2009, PRG.

This garden outcast was recorded from the edge of some willows.

ALLIACEAE
(AMARYLLIDACEAE)

Allium roseum L.
Rosy Garlic

Britain: neophyte (Mediterranean); spreading in S England.
Hortal; 1/2; very rare.
1st Paisley, 1956, PSY.

The original record for Rosy Garlic is from a hedge on the Greenock railway near St James' Park (NS467655) where it was collected by a Mr Spearing. A small clump of this outcast occurred to the edge of Teucheen Wood (NS4869), with other dumped garden waste, in 1996.

Allium subhirsutum L.
Hairy Garlic

Britain: neophyte (Mediterranean); occasional in S England.
Hortal; 1/1; very rare.
1st Teucheen Wood, 2007.

A small patch of a garden outcast occurs to the edge of the wood near gardens (NS4869).

Allium paradoxum (M. Bieb.) G. Don
Few-flowered Garlic

Britain: neophyte (Caucasus); scattered, frequent in C Scotland.
Hortal; 11/11; local and spreading.
1st White Cart, Linn Park, 1985, CFOG.

This recent invasive alien has been found along the White Cart Water at Linn Park (NS5859, 2005), Lochar Park (NS5857 2005) and Cathcart (NS5661, 2005), and downstream on the River Cart at Blythswood (NS4968, 2007). Few-flowered Garlic is also known from Tod Burn, Brownside (NS4846, 1995), Teucheen Wood (NS4869, 1996) and Castle Semple (NS3558, 2010, AB). Given its recent spread elsewhere in Glasgow, it is likely to increase over the coming years.

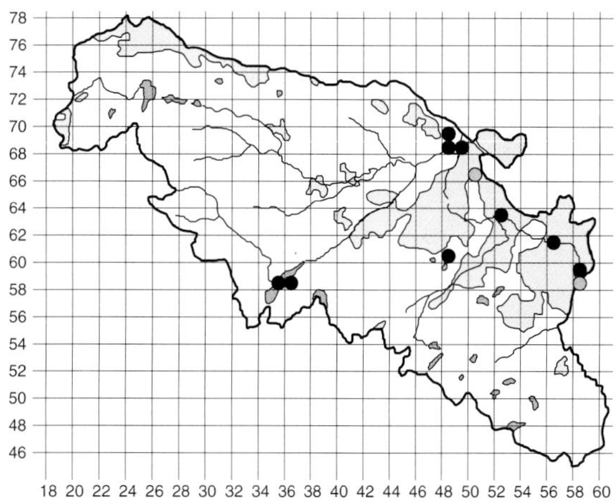

Allium paradoxum

Allium ursinum L.
Ramsons

Britain: common throughout except in N Scotland.
Native; 85/97; frequent in woodlands.
1st Gourock, 1856, GL.

Hennedy (1891) considered Ramsons to be a frequent plant of moist woodlands and hedge banks; old locations include Kelly Glen (1887, NHSG), Calder Glen (1903, NHSG) and Castle Semple Loch (1904, PSY). Today it is found in woodlands, often in profusion, particularly on the damper lower slopes or level ground of wooded watercourse margins.

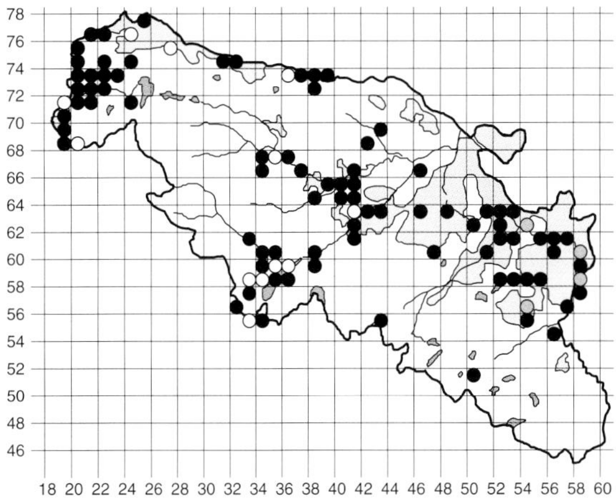

Allium ursinum

Allium vineale L.
Wild Onion

Britain: widespread, more common in the S.
Native; 11/12; scarce.
1st Gryfe, Bridge of Weir, 1957, ERTC.

Although known to Hennedy (1891) and Lee (1933) there are no VC records for Wild Onion until the find more than 50 years ago along the Gryfe Water, with a later one from Inchinnan (1969, JP), presumably indicating a recent arrival. Today it is well established by the Gryfe Water, with records from Duchal House (NS3568, 2000) downstream to Craigends Estate (NS4166, 2002), and by the Black Cart Water from Johnstone (NS4263, 2005) down to Inchinnan (NS4868, 2005). It is also found by the River Cart at Blythswood (NS4968, 2007) and by the White Cart Water at Porterfield (NS4967, 2001) and upriver at Crookston (NS5262, 2005).

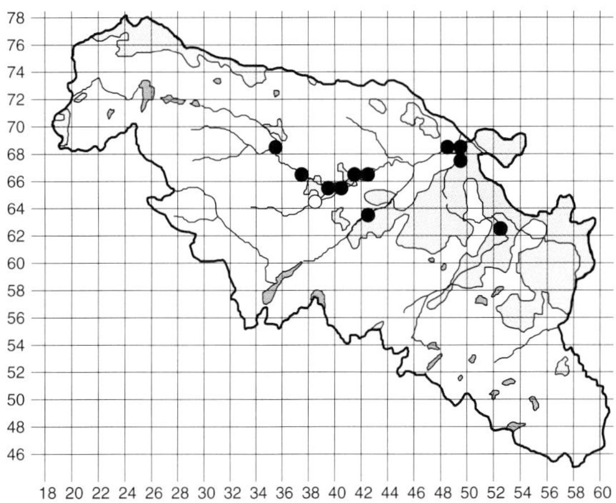

Allium vineale

Galanthus nivalis L.
Snowdrop

Britain: neophyte (S Europe); common except in the NW and uplands.
Hortal; 62/64; frequent, established.
1st Castle Semple, 1834, MC.

Snowdrop was known to be present 'plentifully' in 1869 (RPB) and Hennedy (1891) described it as 'perfectly naturalised', an opinion which appears to have been upheld, as it is now well established in woodlands and along riverbanks, e.g. along the White Cart Water; however, surprisingly, there are only three records from the 1960s–80s period. Given its early flowering period it is still probably under-recorded.

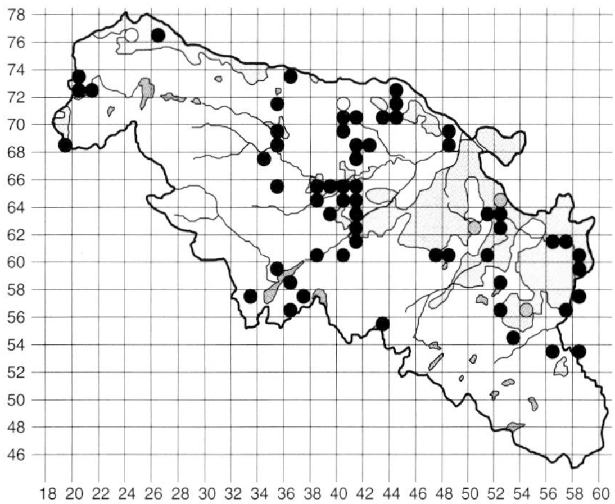

Galanthus nivalis

Narcissus spp.
Daffodils

Britain: native and neophytes (Europe); common in the S, scattered alien in Scotland.
Hortals; 105/110; common in lowlands.
1st Near Neilston, 1871, GLAM.

Hennedy (1891) makes very little reference to daffodils and even Lee (1933) considered them 'probably never established'. Today daffodils of varying types can be readily seen in urban areas, in parks and gardens and along riverbanks and in woodlands, where they appear to be established. Many of the records have not been at the species level and very few specimens have been expertly determined. However, Daffodil (*N. pseudonarcissus* L.) comprises the majority of the records (82), and members of this group are the most likely daffodils to be encountered. Spanish Daffodil, *N. hispanicus* Gouan (*N. pseudonarcissus* ssp. *major* (Curtis) Baker), was recorded from Barrhead (1901, FFCA) but there is no supporting specimen.

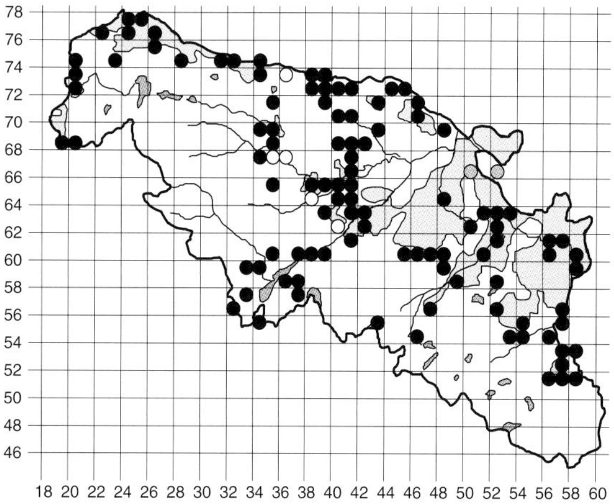

Narcissus spp.

Narcissus poeticus L.
Pheasant's-eye Daffodil

Britain: neophyte (S Europe); scattered.
Hortal; 10/11; scarce.
1st Gourock, 1972, AJS.

Modern records for Pheasant's-eye Daffodil are from widespread places such as Daff Burn, Inverkip (NS2171, 2006), Strathgryfe (NS3370, 1998), Georgetown (NS4268, 1999), Duchal House (NS3568, 2000), Houstonhead (NS3965, 2002) and three locations along the White Cart Water. Primrose Peerless (*N.* × *medioluteus* Mill. – *N. poeticus* × *N. tazetta* L.) was recorded from Finlaystone in 1976 by AJS.

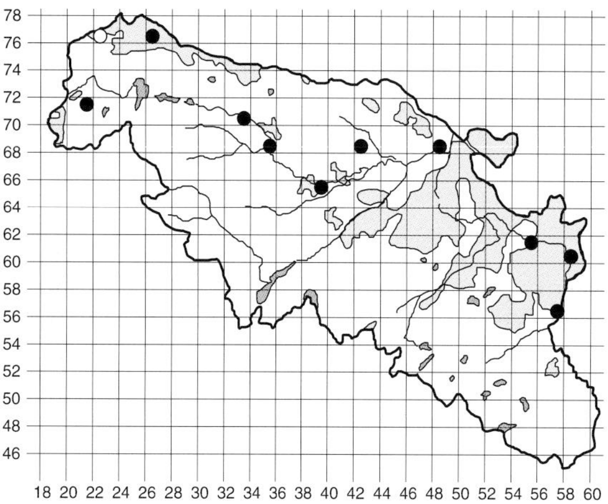

Narcissus poeticus

Narcissus × *incomparabilis* Mill.
(*N. poeticus* × *N. pseudonarcissus*)
Nonesuch Daffodil

Britain: neophyte hybrid; scattered and local.
Hortal; 18/19; scarce.
1st Finlaystone Estate, 1976, AJS.

Probably overlooked, Nonesuch Daffodil is quite widespread, particularly along the White Cart Water; however, the records will include several cultivars.

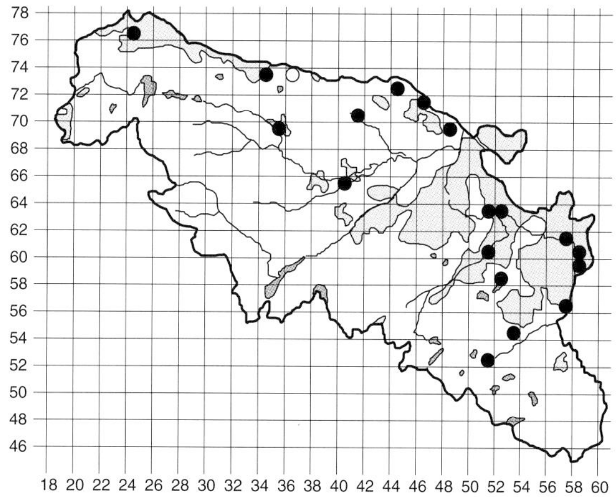

Narcissus × *incomparabilis*

ASPARAGACEAE
(LILIACEAE *pro parte*)

Convallaria majalis L.
Lily-of-the-valley

Britain: frequent in the S, local in Scotland (where mostly alien).
Hortal; 6/10; rare.
1st Castle Semple, 1834, MC.

Other old records are from Birkmyre, Thornliebank (1886, PSY), Giffnock (1901, FFCA), Shawlands (1908, GL) and a garden, Bridge of Weir (1921, GL). Lily-of-the-valley was found locally on a few occasions during CFOG recording, as at Deaconsbank Golf Course (NS55P, 1983), Renfrew Golf Course (NS56E, 1987), Mosspark (NS56L, 1986) and Renfrew (NS56D, 1983). There are only two more recent field records, both for garden outcasts: rough grass, Foxbar (NS4561, 1995) and waste ground, Ladyburn (NS3075, 1998).

Polygonatum multiflorum (L.) All.
Solomon's-seal

Britain: local in the S, scattered alien further N, rare in Scotland.
Former hortal; 0/3; extinct.
1st Carruth Glen, 1890, AANS.

A large-flowered, old specimen from Cathcart (1856, GGO) appears to be the hybrid Garden Solomon's-seal, so the first record is from the literature. Much later JDM reported it from two places (1964, GN): Southbar and Rouken Glen. It is not possible to verify the identity of these records, which are presumably of garden origin.

Polygonatum × hybridum Brügger
(*P. multiflorum* × *P. odoratum* (Mill.) Druce)
Garden Solomon's-seal

Britain: neophyte (hybrid); scattered.
Hortal; 30/38; occasional.
1st Cathcart, 1856, GGO.

A presumed hybrid (labelled '*P. odoratum*') was collected from woods, Cardonald, in 1912 (GL). Today Garden Solomon's-seal is readily seen, especially in the central area and about Glasgow; an outlier in the west is from Ardgowan Estate (NS2073, 1998). It is usually found in woodlands near to houses, or in old estates, and occasionally on more open ground along riverbanks or hedge banks, where it can form quite large clumps.

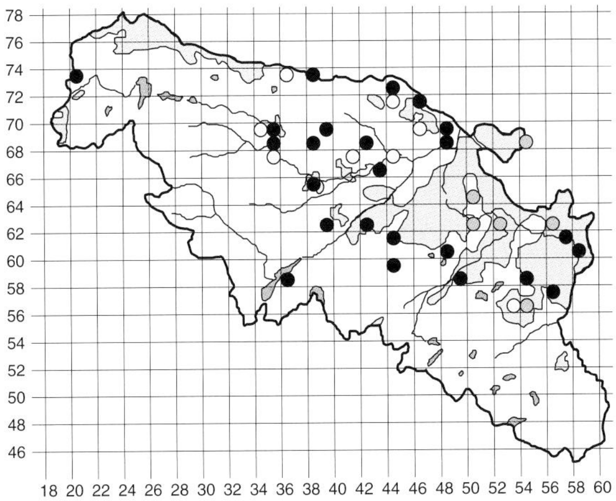

Polygonatum × hybridum

Ornithogalum umbellatum **L. ssp. *campestre*** Rouy
Ornithogalum angustifolium Boreau
Star-of-Bethlehem

Britain: neophyte (S Europe); widespread.
Hortal; 14/23; scarce.
1st Banks of Cart, 1865, RH.

Star-of-Bethlehem was considered a rare escape by Hennedy (1891) and he only mentioned the unlocalized 'banks of Cart'; Inverkip is noted in 1891 (AANS) and Kilmacolm in 1899 (PSY), both repeated by Lee (1933). JDM reported it from near St Fillan's Church (NS3838, 1963) and Langbank (NS3772, 1966). Today it appears well established in the north central area, where it is most often encountered along roadsides and hedge banks, but not in any great number (although it is possibly overlooked due to early flowering). A southern outlier is recorded from a roadside at Plymuir (NS4357, 2003).

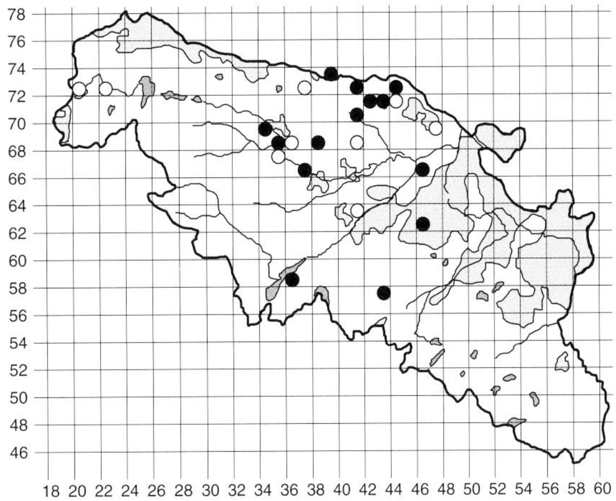

Ornithogalum umbellatum ssp. *campestre*

Scilla forbesii (Baker) Speta
Chionodoxa forbesii Baker
Glory-of-the-snow

Britain: neophyte (Turkey); scattered.
Hortal; 2/2; very rare.
1st Kelly Glen, 1995.

This presumed garden outcast grew by a wall near the road at Kelly Glen (NS1968). A further record is from a grassy area by the White Cart Water, Cathcart (NS5860, 2006).

Hyacinthoides non-scripta (L.) Chouard ex Rothm.
Bluebell

Britain: common throughout except in NE Scotland.
Native; 275/308; very common.
1st Neilston and Paisley, 1845, NSA.

Bluebell was reported from many woods in the 19th century and Hennedy (1891) called it 'very plentiful'. Today it is widespread, occurring throughout the area but absent from the uplands, apart from along some burn sides, and also from the farmland north of Houston (*c.* NS4466). It can be found in urban areas, although it is often present, and interbreeding, with hybrids or Spanish outcasts. It is most often found in broad-leaved woodlands or occasionally in conifer plantations, if not too heavily shading, and its presence generally indicates an old woodland site. Bluebell can also occur in hedge banks, on rock outcrops, under bracken, often in profusion as on the Gleniffer Braes, and occasionally in open grasslands if not too heavily grazed or improved.

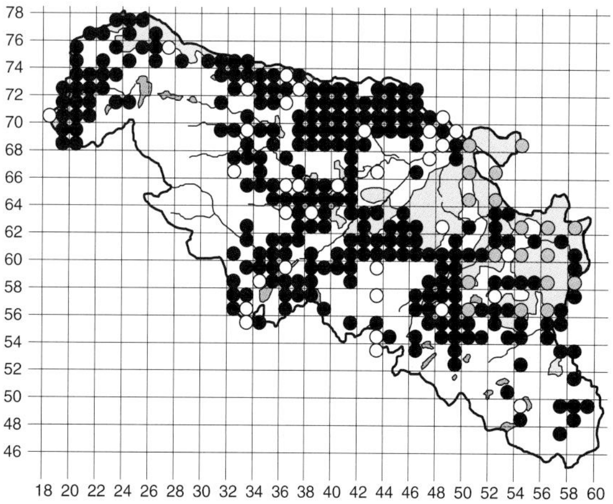

Hyacinthoides non-scripta

Hyacinthoides hispanica (Mill.) Rothm.
Spanish Bluebell

Britain: neophyte (Spain and Portugal); widespread.
Hortal; 61/65; occasional, established.
1st Paisley (NS4963), 1974, AJS.

There are quite a few modern records for Spanish Bluebell, but distinguishing it from hybrids is not always straightforward, as is noted in CFOG; modern records here refer to the distinctly broad-leaved cultivar that is thrown out, or occasionally escapes, from gardens. The map shows that it is now widely found, usually in woodlands or scrubby areas, often on riverbanks, and occasionally on grassy waste ground or roadsides; it occurs most frequently in urban areas.

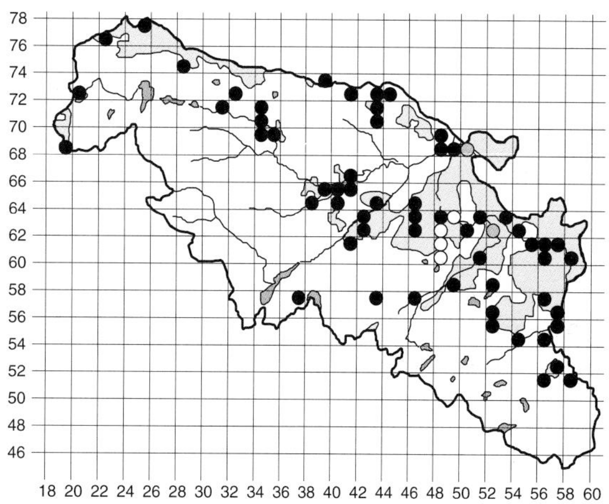

Hyacinthoides hispanica

Hyacinthoides* × *massartiana Geerinck
(*H. non-scripta* × *H. hispanica*)

Britain: garden grown and spontaneous hybrid; widespread.
Spontaneous hybrid or hortal; 84/87; frequent.
1st Paisley (NS4861), 1974, AJS.

AJS also found this hybrid at Finlaystone Estate in 1976 but all other records are from the CFOG period or more recent surveying; records for Spanish Bluebell and possibly some for the native species may include hybrid plants. The hybrid shows a similar distribution pattern, and occurs in similar places, to Spanish Bluebell, often growing with it or with both parents, and several of these populations appear to be spontaneous hybrids and include backcrosses.

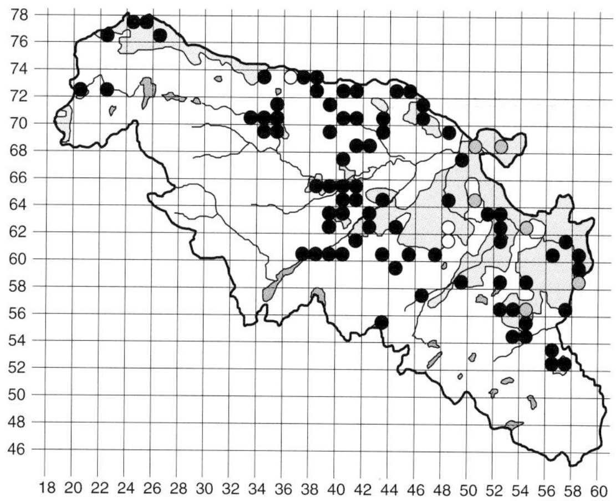

Hyacinthoides × *massartiana*

Muscari armeniacum Leichtlin ex Baker
Garden Grape-hyacinth

Britain: neophyte (Balkans to Caucasus); scattered, common in the S.
Hortal; 3/3; very rare outcast.
1st Braehead, 1987, CFOG.

Garden Grape-hyacinth is also reported from a woodland at Kilmacolm (NS3569, 1997) and from Jordanhill (NS5468, 2003, AR).

Asparagus officinalis L.
Garden Asparagus

Britain: archaeophyte (Europe); common in the SE.
Former hortal; 0/1; extinct.
1st Erskine House, 1894, GL.

This record is presumably for a garden-grown plant.

Ruscus aculeatus L.
Butcher's-broom

Britain: native and frequent in the extreme S, scattered alien in the N.
Hortal; 3/4; very rare.
1st Cathcart, 1845, GL.

Today Butcher's-broom has been recorded from two estate woodlands, at Ardgowan Estate (NS2073, 1998) and Lochside House (NS3658, 1999). It also occurs near the mansion house in Linn Park (NS5858, 2003), where it grows with Spineless Butcher's-broom (*R. hypoglossum* L.); presumably both were planted there.

Hosta spp.
Hostas

Britain: neophytes (E Asia); scattered.
Hortal; 5/5; rare outcasts.
1st Brock Burn (NS5360), 1986, CFOG.

Clements and Foster (1994) reported *H. fortunei* (Baker) L. Bailey from Gourock but there is no further detail. There have been three other modern field records of garden outcasts, though these have not been identified to species level: Teucheen Wood (NS4869, 1996), Whinhill (NS2875, 2003) and Darnley (NS5258, 2006).

TYPHACEAE
(SPARGANIACEAE)

Sparganium erectum L.
Branched Bur-reed

Britain: common throughout except in N Scotland.
Native; 126/135; common in lowland wetlands.
1st Paisley and Lochwinnoch, 1845, NSA.

This is by far the commonest of the bur-reeds, as was the case in the 19th century. Branched Bur-reed shows a very strong lowland pattern, being common in the

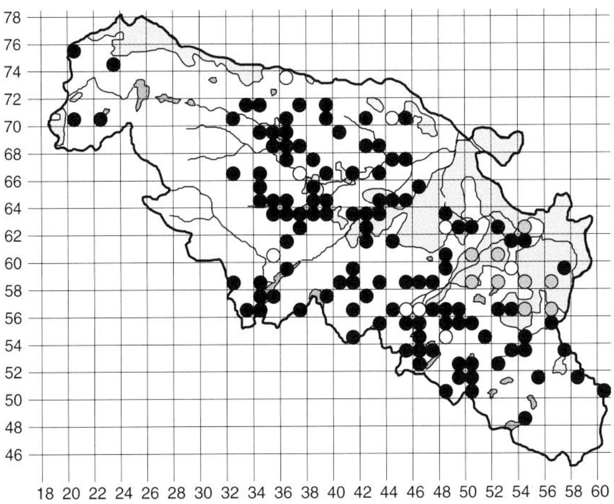

Sparganium erectum

central area and avoiding the peaty uplands; it is also scarce in the extreme west, presumably due to a lack of slow-moving water. It is found in ditches, at loch or pond sides and along slower stretches of burns or larger rivers. It is thought that many records refer to ssp. *neglectum* (Beeby) K. Richt. The ssp. *erectum* was recorded from Caplaw Dam (NS4358) by AJS in 1983.

Sparganium emersum Rehman
Unbranched Bur-reed

Britain: widespread, frequent in C Scotland.
Native; 41/45; frequent in deeper water.
1st Paisley Canal, *c.* 1850, GL.

Hennedy (1891) described Unbranched Bur-reed as frequent but did not name any locations in the VC. It was collected from near Eaglesham in 1891 (GL) and it was still to be found in the 'old canal' at Hawkhead in 1896 (PSY). Today it has been found at a wide range of sites, mainly central and southern; a few records are tentative as access is often limited by deep water. It has been recorded from several lochs, including Loch Libo, Little Loch and Barr Loch (1996, SNH-LS) and also slow-flowing rivers including the Black Cart Water, Howwood (NS3960, 2006), the Gryfe Water at Selvieland (NS4567, 2001), the Earn Water, north of Moorhouse (NS5152, 2005) and the White Cart Water in Pollok Country Park (NS5561, 2005) and Paisley (NS4863, 2004).

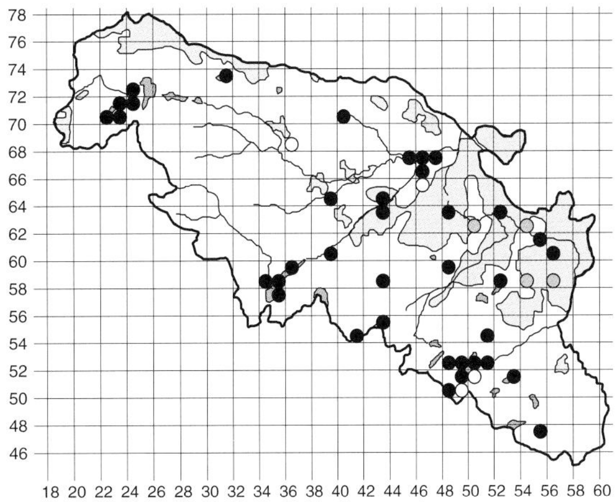

Sparganium emersum

Sparganium angustifolium Michx
Floating Bur-reed

Britain: common in the N and W, very rare in the S.
Native; 5/5; rare.
1st Reservoir 4, Greenock Cut, 1992, CMRP-NVC.

This bur-reed was not distinguished by 19th-century authors and considered rare by Lee (1933), with no VC76 records until the modern period. Floating

Bur-reed has been recorded from two sites in the south: Caplaw Dam (NS4358, 1996, SNH-LS) and Brother Loch (NS5052, 2011). It is also known from three places in the north-west: Kelly Reservoir (NS2268, 1996, SNH-LS), Crawhin Reservoir (NS2470, 2000) and can still be found at Gryfe Reservoir No. 4 (NS2574, 2007). Access is not always easy, so it may be present but not yet observed at other upland waterbodies.

Sparganium natans L.
Least Bur-reed

Britain: scattered, mainly in the N and W.
Native; 3/10; very rare.
1st Lochwinnoch, 1845, NSA.

Least Bur-reed may have been confused in the past with Floating and Unbranched Bur-reed. It was considered frequent in the 19th century with stations at Gourock (1865, RH), Meikleriggs Dam (1865, MY and 1882, PSY), Castle Semple Loch and Loch Libo (both 1887, NHSG). More recently it was recorded from Jenny's Well in Paisley (1942, TPNS) and from Daff Reservoir (NS2371, 1973, RM). In the modern period it was found by the SNH Loch Survey (1996) at North Loch, Drum Estate (NS3971) and Glen Moss (NS3669). It has also been noted from a peaty ditch near the Craig of Todholes (NS3164, 1998).

Typha latifolia L.
Bulrush

Britain: common, rare in Scotland except in the C and S.
Native; 67/70; frequent in lowlands.
1st Lochwinnoch and Paisley, 1845, NSA.

Bulrush was also recorded from Loch Libo and Inchinnan (1876, FFWS) and in 1891 (AANS) it was reported from the collegiate church near Castle Semple Loch; however, Hennedy (1891) described Bulrush as 'very rare', which is not the case today, as discussed in CFOG. Interestingly, there are only four records from the 1960s–80s period. Today Bulrush shows a very strong lowland rural pattern, notably in the central area, presumably representing fairly recent spread; it is absent from the peaty uplands, except for a record from the Kelly Cut (NS2370, 2005). It is generally found in the deeper water of loch or pond margins and has colonized ditches and quarry pools as at Pilmuir (NS5154, 2000) and Underheugh (NS2075, 2004).

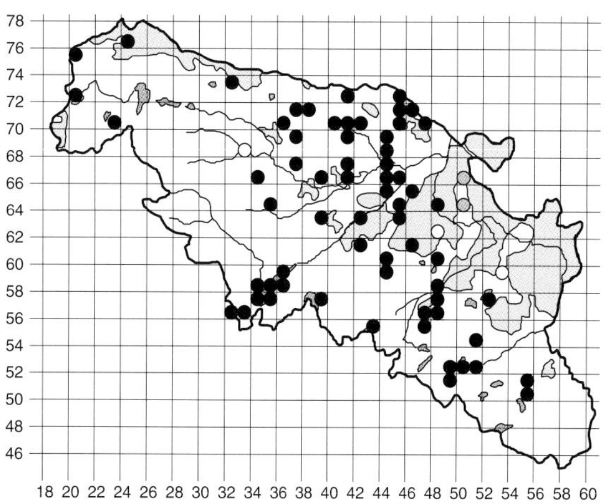

Typha latifolia

Typha angustifolia L.
Lesser Bulrush

Britain: frequent in the SE, local elsewhere.
Hortal; 2/3; very rare.
1st Loch Libo, 1865, MY.

There may be doubts about the sole old record and Lesser Bulrush was excluded from the Renfrewshire list (TPNS 1915). There were no other named localities until Keith Futter found it at Town Reservoir, Greenock (NS2675, 1992), during a SWT survey. It is also found in ponds at the Gleddoch Estate (NS3872, 1998), where it was presumably planted.

JUNCACEAE

Juncus subnodulosus Schrank
Blunt-flowered Rush

Britain: frequent in the S and E, rare and local in S Scotland.
Native or accidental; 1/2; very rare.
1st Greenock–Inverkip, 1865, RH.

The original record ('In bog on road from Greenock to Inverkip') was questioned in Hennedy's 1891 edition. However, this rush was found by RM at marshy ground near Clarkston in 1928 and collected from there in 1940 and 1961 (GLAM); it was last recorded there, at the marsh isolated by three railway lines (the 'triangle') at Williamwood, in 1992 (NS5658, CFOG).

Juncus articulatus L.
Jointed Rush

Britain: common throughout.
Native; 245/276; common in wet places.
1st Gourock, 1865, RH.

Seldom in such abundance as the following rush, Jointed Rush is, however, widespread and is more likely to be found in lowland or urban areas, where it appears more tolerant of higher nutrients or minerals. It is found at pond and stream margins and also marshes, but seldom forms dominant stands like some rushes; it can also colonize wet waste ground, particularly where previously compacted by vehicles.

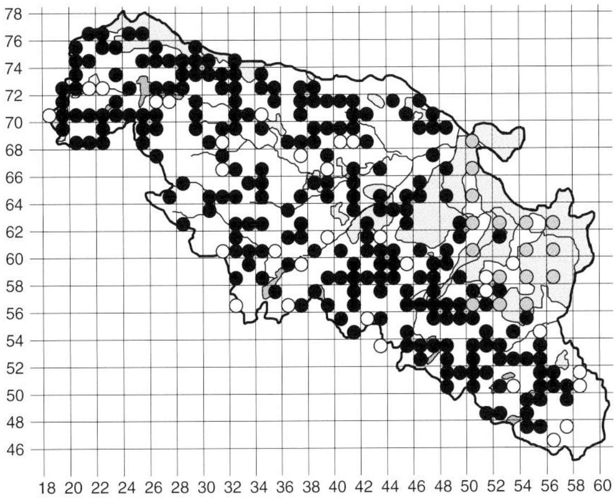

Juncus articulatus

Juncus acutiflorus Ehrh. ex Hoffm.
Sharp-flowered Rush

Britain: common throughout.
Native; 357/395; locally abundant.
1st Gourock, 1872, GLAM.

The only other old named locality is Loch Libo (1887, NHSG), but Sharp-flowered Rush was described as very common in 19th-century literature. Today it can still be described as such, away from urban encroachment. It can be a dominant feature of poorly drained marshy grasslands or mires, especially where there is some flushing. It extends into the uplands, although it is absent from the blanket mires except at marginal flushes.

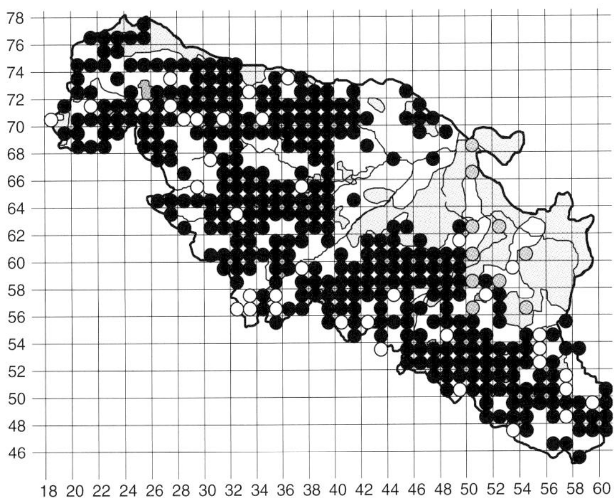

Juncus acutiflorus

Juncus bulbosus L.
Bulbous Rush

Britain: common, northern.
Native; 167/206; common in uplands.
1st Gourock, 1837, GL.

Other old localities include Meikleriggs Dam (1882, PSY), Loch Libo (1887, NHSG) and near Johnstone (1905, GL). Today Bulbous Rush is widespread but strongly rural and is well represented in the uplands. It is found in wet peaty places, although usually where there is some flushing, and it is also a common feature, though often non-flowering, of the margins of upland waterbodies.

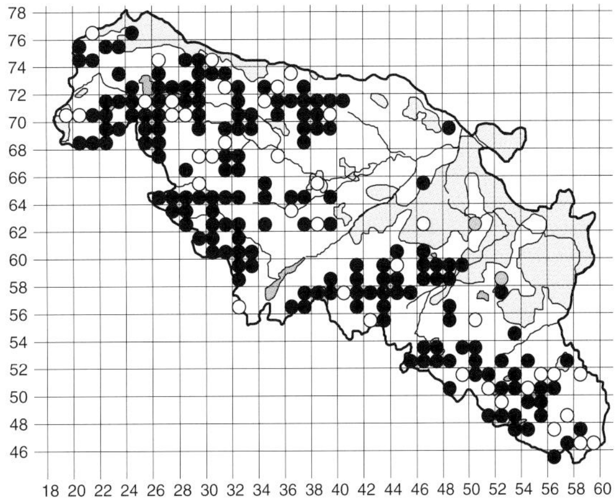

Juncus bulbosus

Juncus maritimus Lam.
Sea Rush

Britain: frequent in the W and on SE coasts.
Former native; 0/2; extinct.
1st Renfrewshire, 1872, TB.

Lee (1933) thought this rush not common, but retains Renfrewshire as a location; there has never been a named local site and Sea Rush is thought to be long extinct.

Juncus squarrosus L.
Heath Rush

Britain: common in the N and W.
Native; 221/263; common.
1st Neilston Pad, 1858, PSY.

Heath Rush was very common according to Hennedy (1891), occurring on 'all moors and heaths'. Today it is still widespread and common in the uplands, being found in acidic grasslands and as part of heath mosaics. It is strongly rural and will have been lost from many grasslands following agricultural improvement.

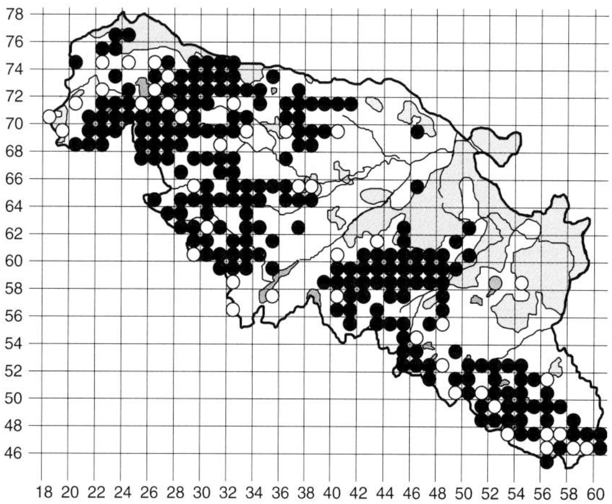

Juncus squarrosus

Juncus tenuis Willd.
Slender Rush

Britain: neophyte (N and S America); widespread but mainly in the W.
Accidental; 53/69; frequent.
1st Dennistoun, Kilmacolm, 1863, NHSG.

There are several 19th-century reports from the original location but no other localities are named for Slender Rush before RM found it at Clarkston in 1933. There were more records made in the 1960s–80s period (BWR) and it was well recorded by the time of surveying for CFOG. Today it is quite widespread, occurring on trampled ground and waste-ground grassland and appears to be spreading in rural areas. It can be found in the uplands, as along the track at Muirshiel (NS3064, 2003).

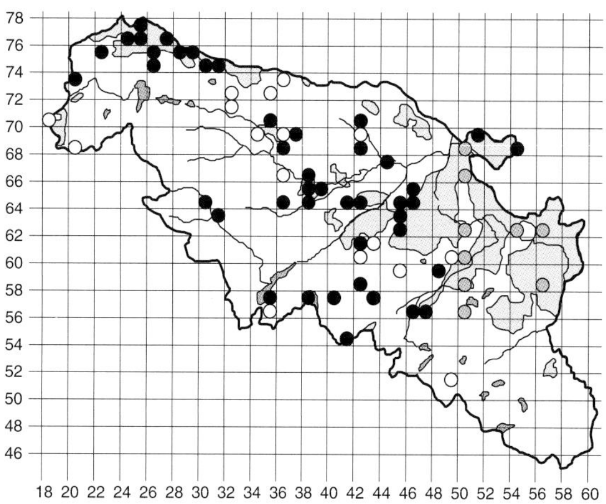

Juncus tenuis

Juncus compressus Jacq.
Round-fruited Rush

Britain: frequent in the S.
Native or accidental; 2/3; very rare.
1st Hills above Gourock, 1865, RH.

The native status for Round-fruited Rush is questionable as there are no other old records and the modern sites are from disturbed places: school grounds, Quarrelton (NS4262, 1993) and waste ground, Greenock (NS2775, 2000).

Juncus gerardii Loisel.
Saltmarsh Rush

Britain: common all around the coast.
Native; 17/22; frequent on the coast.
1st Gourock, 1865, RH.

Saltmarsh Rush has probably declined, certainly around the urban Greenock–Gourock area where its habitat has gone; small patches were found by launch ramps at Clydeport, Port Glasgow (NS3174, 2001). It can still be found on the west coast and further east at the Cart–Clyde confluence at Blythswood (NS4969, 1986, CFOG).

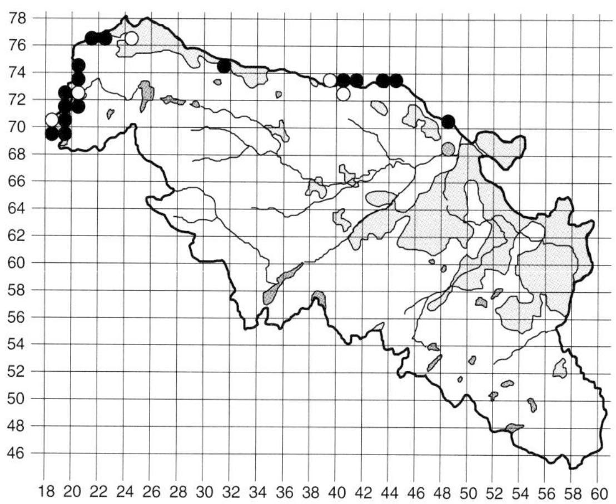

Juncus gerardii

Juncus bufonius L.
Toad Rush

Britain: common throughout.
Native; 167/239; common.
1st Renfrewshire, 1869, PPI.

This was a very common rush in the 19th century; the few named localities include Meikleriggs Dam (1882, PSY), Loch Libo (1887, NHSG) and Commore Dam (1881, GL). Today Toad Rush is found throughout the area, generally in wet places – though often where seasonally dry, such as open margins of waterbodies and streams – and also on waste ground and at roadside margins.

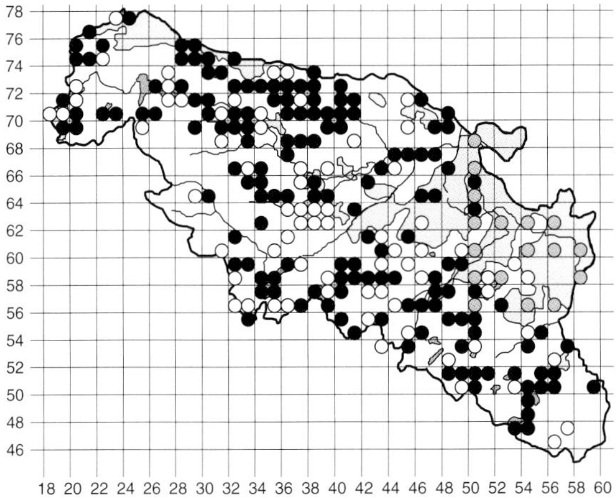

Juncus bufonius

Juncus foliosus Desf.
Leafy Rush

Britain: western, occasional.
Native; 6/6; rare, overlooked.
1st Undercraig (NS3772), 1998.

Leafy Rush has been recorded from a few sites and may occur elsewhere. However, a number of specimens when closely checked do not match Leafy Rush for all the diagnostic characters, e.g. leaf width and distinct striate seeds. Recorded localities are from Barbeg Hill (NS4071, 1998), Corselet (NS5256, 1998), marshy slope, Dod Hill (NS4953, 1998), Harelaw Brae (NS4959, 1999) and on bare ground near the River Calder (NS3558, 2003, IPG).

Juncus ranarius Songeon & E.P. Perrier
Juncus ambiguus auct. non Guss.
Frog Rush

Britain: coastal, scattered throughout, under-recorded.
Native; 1/1; very rare, overlooked.
1st Newshot Island, 2006.

Frog Rush was found in estuarine-influenced grassland at Newshot Island LNR (NS4870). It may well occur at other brackish sites elsewhere along the coast.

Juncus filiformis L.
Thread Rush

Britain: rare in Scotland and at a few sites in NW England.
Native; 6/9; very rare.
1st West shore of Auchendores Reservoir, 1969, BWR.

Thread Rush appears to be a recent arrival in the area, and seems to have spread in recent years. It was reported (BWR) from Gryfe Reservoir No. 2 (1984) and Barcraigs Reservoir (1978), and can still be found at both. The modern localities are all from mud at reservoir margins: Auchendores Reservoir (NS3572, 2000, JC), Rotten Burn inflow Loch Thom (NS2570, 1997), near Killochend (NS2672, 2000), Gryfe Reservoir No. 2 (NS2971, 2000) and Barcraigs Reservoir (NS3957, 2003, IPG; NS3857, 2000).

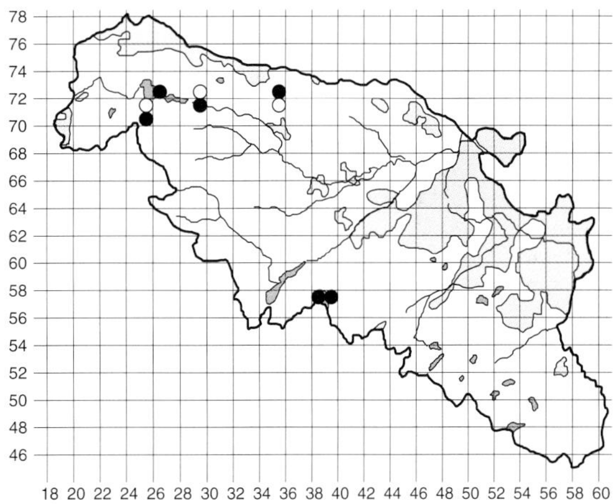

Juncus filiformis

Juncus inflexus L.
Hard Rush

Britain: common in the S, rare or absent in Scotland except in the SE.
Native or accidental; 6/8; rare.
1st Gourock, 1865, RH.

There appear to be few old references to Hard Rush in the area, the second record being for Williamwood in 1962 (NS5658, RM). The modern records are from disturbed waste-ground sites, usually with compacted clayey soils, as at 'The Cunyon', Pollokshields (NS5563, 1991, CFOG); the other localities are Park Quay, Newshot Island (NS4770, 1997), Linburn Junction (NS4570, 1997), roadside, near Star and Garter (NS5253, 2003), fields at Darnley (NS5258, 2007) and Harelaw (NS4859, 2000, SWT).

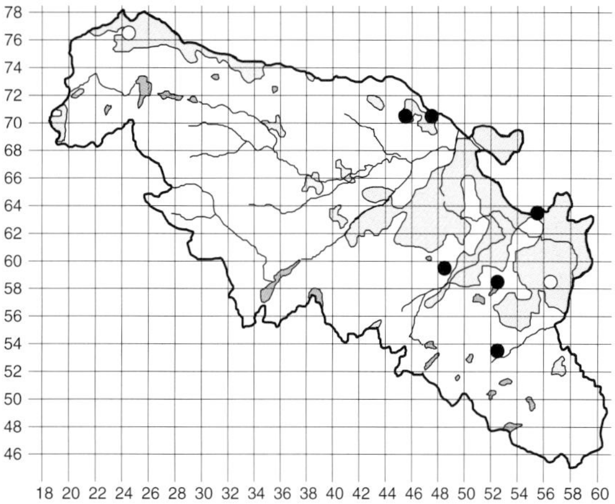

Juncus inflexus

Juncus effusus L.
Soft-rush

Britain: common throughout.
Native; 479/525; abundant.
1st Gourock, 1837, GL.

Soft-rush is a very common sight throughout the VC, found on wet waste ground and in wet pastures, mires, wet woods, open water margins and along watercourses. It can dominate large areas of 'rush pasture' on poorly draining farmland, and can often be seen marking out drainage lines. It can also be found high in the uplands to the edges of blanket mires along drains and flushes.

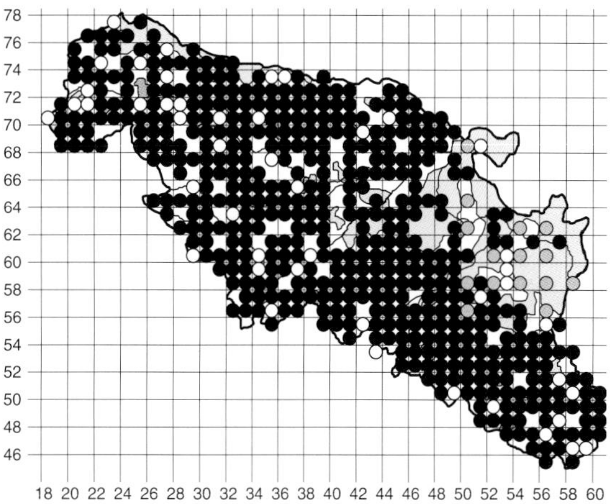

Juncus effusus

Juncus conglomeratus L.
Compact Rush

Britain: common throughout.
Native; 173/228; common.
1st Renfrewshire, 1869, PPI.

This rush (as 'var. *conglomeratus*') was considered common by earlier authors but there are few named localities. Today Compact Rush is widespread, occurring on waste ground, often where compacted and moist, but also extending into the uplands along burn valleys; it can be found in wet heaths where the peat is shallow. The hybrid with Soft-rush (*J.* × *kern-reichgeltii* Jansen & Wacht. ex Reichg.) was seen at Darnley (NS5259, 2003), with both parents, and is likely to occur at other similar sites.

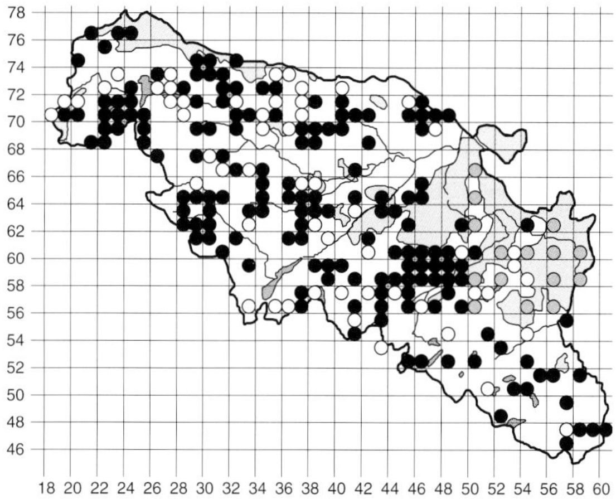

Juncus conglomeratus

Luzula pilosa (L.) Willd.
Hairy Wood-rush

Britain: common throughout except in E England.
Native; 102/129; common in upland fringe.
1st Kilmarnock Road, near the Cart, 1843, GLAM.

This wood-rush was considered frequent by Hennedy (1891) but there are few old local records: Kelly Glen (1887, NHSG), Kilmacolm (1891, GL) and Killoch Glen (1902, PSY). The modern map shows a widespread but rural distribution, more frequent in the west. Hairy Wood-rush occurs on less fertile soils, often along the upper slopes or upland stretches of watercourse valleys, and can occur on open heathy acidic grassland embankments.

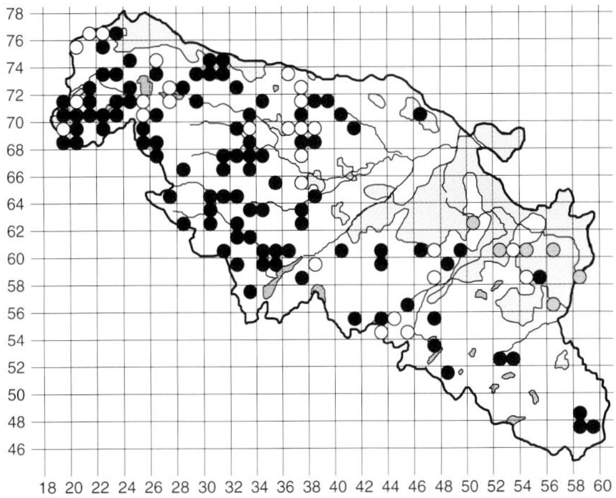

Luzula pilosa

Luzula sylvatica (Huds.) Gaudin
Great Wood-rush

Britain: common in the N and W.
Native; 180/205; common.
1st Renfrewshire, 1869, PPI.

Great Wood-rush has a widespread distribution; it is commoner in the west and tends to be rural. It is encountered, often in abundance, on usually heavily shaded, steep wooded valley slopes. It follows steep-sided burns into the peaty uplands and can occur in some open places amongst grass or heath.

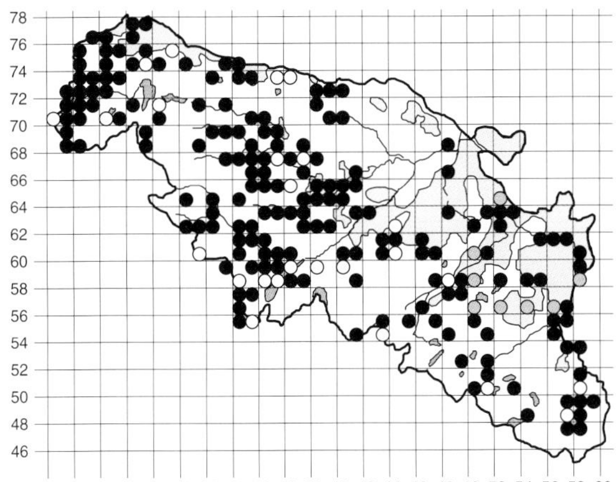

Luzula sylvatica

Luzula luzuloides (Lam.) Dandy & Wilmott
White Wood-rush

Britain: neophyte (C Europe); scattered, mostly in the N.
Hortal; 1/2; very rare.
1st Duchal Estate, 1956, ERTC.

There are no modern records from the Kilmacolm area but White Wood-rush was found at Rouken Glen in 1982 (NS55P, CFOG) and was last collected from there in 1990 (GL).

Luzula campestris (L.) DC.
Field Wood-rush

Britain: common throughout.
Native; 269/344; common.
1st Gourock, 1858, GLAM.

Field Wood-rush is a diminutive plant, locally very common in short grasslands, and very likely overlooked later in the year. It is a common feature of less improved bent-fescue pastures; it is also found in urban park grasslands such as on less-enriched embankments, margins of golf courses and some old lawns.

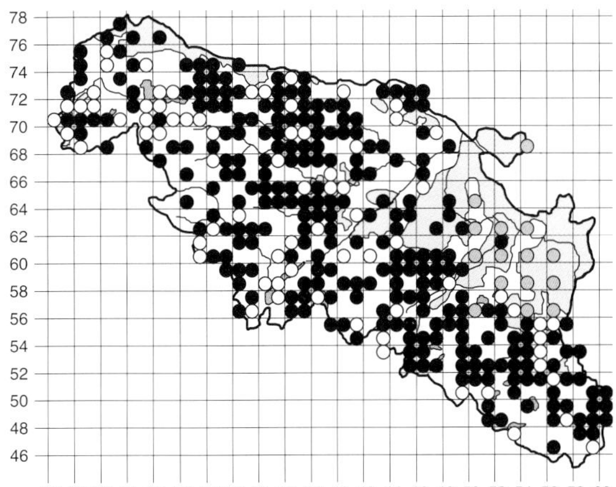

Luzula campestris

Luzula multiflora (Ehrh.) Lej.
Heath Wood-rush

Britain: common throughout especially in the N and W.
Native; 223/301; common.
1st Moor above Eaglesham, 1871, GLAM.

Early authors did not distinguish this species from Field Wood-rush, although they described var. *congesta* as common. Today Heath Wood-rush is common – and was presumably so in the 19th century – and readily found in rural areas extending into the uplands. It occurs in more acidic and poorly draining soils than its smaller relative Field Wood-rush, often in wet heaths, but there are a few records from lowland waste ground and also on disused railways.

The ssp. *congesta* (Thuill.) Arcang. has been recorded from 24 widespread localities.

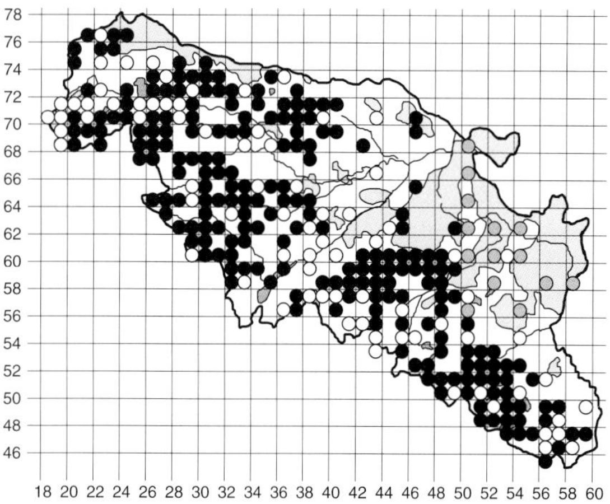

Luzula multiflora

CYPERACEAE

Eriophorum angustifolium Honck.
Common Cottongrass

Britain: common in the N and W.
Native; 189/220; common in uplands.
1st Paisley, 1845, NSA.

Common Cottongrass is a common plant of upland wet peaty places. Its distribution shows a strong rural upland pattern, and it is absent from more agriculturally improved lowland zones and also urban areas. It is well represented in the blanket-bog moorland and peaty pools in the uplands and can also occur in wetter parts of acidic mires elsewhere.

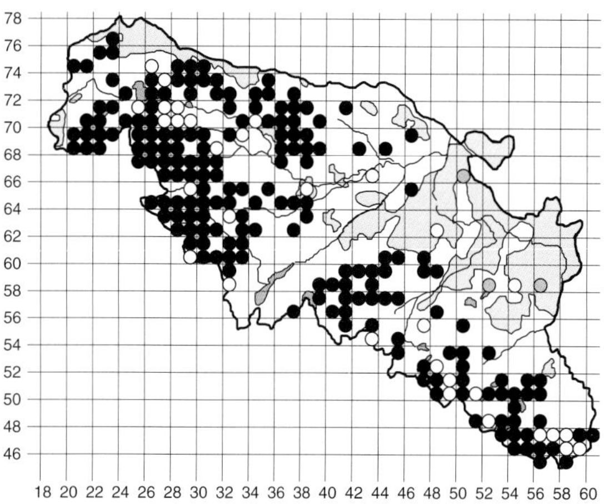

Eriophorum angustifolium

Eriophorum latifolium Hoppe
Broad-leaved Cottongrass

Britain: frequent in the NW and in S Scotland, local in the S.
Native; 2/2; very rare.
1st South of Daff Reservoir, 1977, BWR.

Broad-leaved Cottongrass is a rare plant in the region, with no old records for the VC. It is found in a series of base-rich flushes along the Glenshilloch Burn, to the south of Daff Reservoir (NS2269, 2001 and NS2268, 2007), where it grows with Yellow Saxifrage.

Eriophorum vaginatum L.
Hare's-tail Cottongrass

Britain: common in the N and W.
Native; 153/177; common in peaty uplands.
1st Renfrewshire, 1834, MC.

Hare's-tail Cottongrass has a very strong upland distribution. It is only found on the deep peat of bogs, often abundantly so as in the blanket bogs of the uplands and at a few lowland raised mires such as about Barochan Moss (*c.* NS4268).

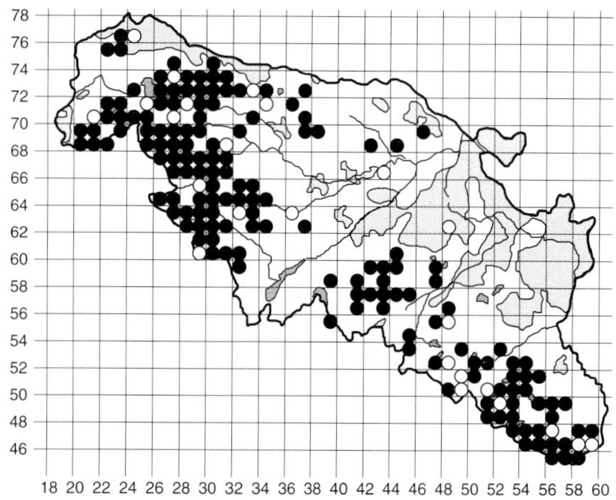

Eriophorum vaginatum

Trichophorum germanicum Palla
Trichophorum cespitosum ssp. *germanicum* (Palla) Hegi
Deergrass

Britain: common in the N and W.
Native; 156/181; common in uplands.
1st Gourock, 1850, GGO.

Hennedy (1891) described Deergrass as being very common and 'on all our heaths'. Today it is common, but only in the upland areas, being absent from lowland rural or urban areas. It is found on bogs and most commonly on grazed wet heaths, notably in the upland fringes of the Renfrewshire Heights. It occurs in the lowlands on the relict bog at Fulwood (NS4468, 1999). Proliferous forms have been recorded from

Brownhill (NS2369, 2004), Kelly Reservoir (NS2268, 2007) and Lochgoin Reservoir (NS5447, 2000).

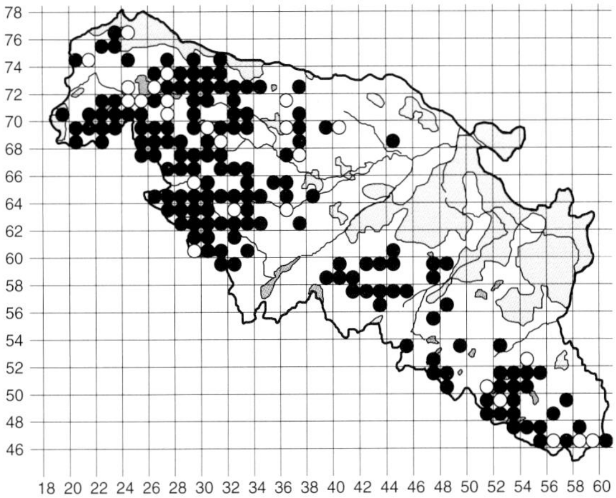

Trichophorum germanicum

Bolboschoenus maritimus (L.) Palla
Sea Club-rush

Britain: coastal, throughout except in extreme N Scotland.
Native; 13/16; local on the coast.
1st Wemyss Bay, 1865, RH.

Hennedy described Sea Club-rush as frequent in saltmarshes and mentioned western locations; from these there is only one modern record, from Lunderston Bay (NS2074, 2008). It is still fairly well represented in the brackish tidal mud of the Clyde and Cart margins, where it can form dense stands, adjacent (seaward) to the Common Reed swamp. It is found from Langbank (NS3973, 1998) eastward to near Blythswood (NS4968, 1997), and extends up the Black Cart Water as far as Inchinnan (NS4868, 2005).

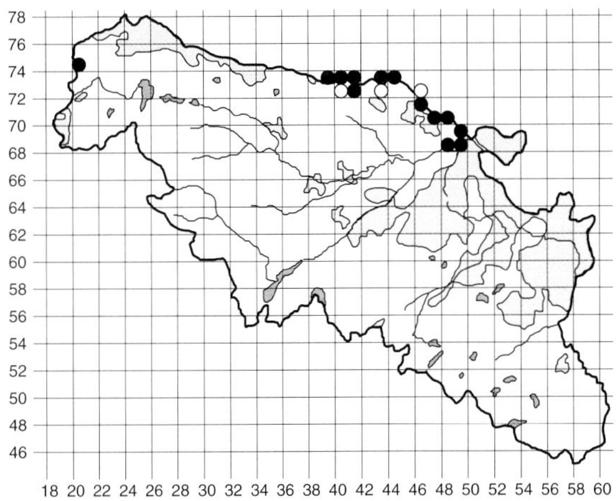

Bolboschoenus maritimus

Scirpus sylvaticus L.
Wood Club-rush

Britain: frequent, rare in N Scotland and E England.
Native; 21/23; occasional.
1st Renfrewshire, 1869, PPI.

Wood Club-rush was considered frequent in the 19th century and named places include Inchinnnan and Busby (1876, FFWS), and later Blackland, Paisley (1932, PSY) and Lochside–Howwood (1934, RM). Today it is found mainly to the south-east, notably along the riverbanks of the White Cart Water and the Levern Water, and tributaries such as the Brock Burn (NS5055, 2000). It was recorded from Castle Semple Loch in 1996 (SNH-LS) and had previously been recorded from there in 1971 (BWR). The western outlier is from wet woodland at Ravenscraig Wood (NS2575, 1996).

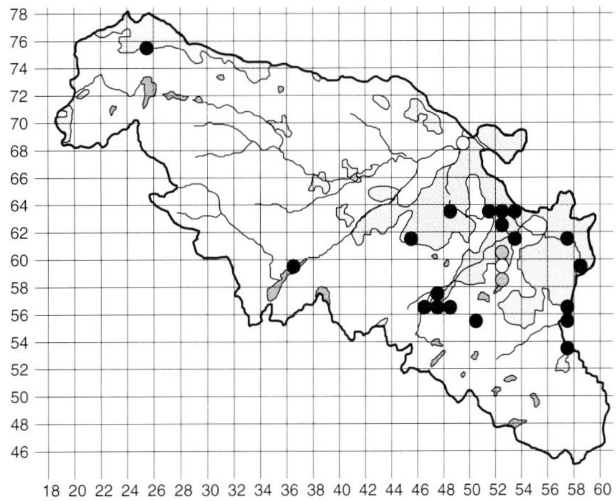

Scirpus sylvaticus

Schoenoplectus lacustris (L.) Palla
Common Club-rush

Britain: widespread, common in the SE, locally frequent in W Scotland.
Native; 8/9; rare.
1st Lochwinnoch, 1845, NSA.

There are few other old records for Common Club-rush: Brother Loch (1879, GL) and RM noted it from 'nearby' Little Loch (1936). In 1966 it was recorded from Black Loch (BWR) and can still be found there (NS4951, 1997). Today it has only been found at a few other deep-water sites: Walls Loch (NS4158, 1993), the Black Cart Water, Howwood (NS3960, 1994, CD), Drum Estate (NS3971, 1997), Knowes (NS3756, 2003) and Gateside (NS4858, 1992, DM); the SNH Loch Survey (1996) noted it from Brother and Little lochs (NS5052) and Castle Semple Loch (NS3659).

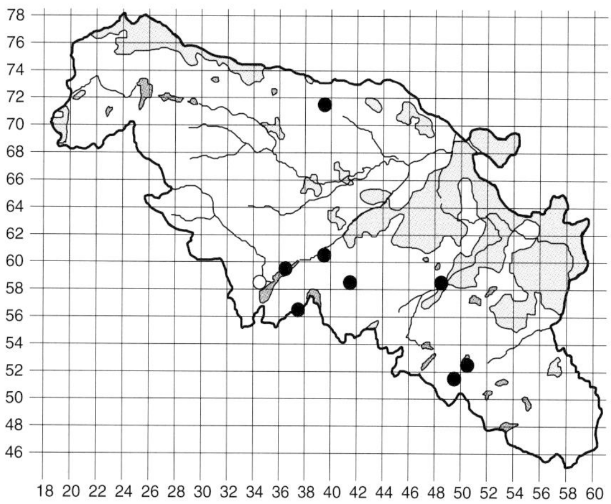

Schoenoplectus lacustris

Schoenoplectus tabernaemontani (C.C. Gmel.) Palla
Grey Club-rush

Britain: frequent In the S, often coastal.
Former native; 0/3; extinct.
1st Renfrew, 1865, RH.

Hennedy noted Grey Club-rush as being 'very abundant' and in 1891 described it as 'still plentiful on the south side of the Clyde from Newshot Island to Erskine'; it was collected (GL) from both localities in 1907 and 1892 respectively. The only more recent record is from Blythswood (1968, JP), but although there is still suitable habitat in the area it has not been seen during modern surveys, indicating a dramatic decline.

Eleocharis palustris (L.) Roem. & Schult.
Common Spike-rush

Britain: common throughout.
Native; 132/159; common in wetlands.
1st Cart, 1865, RH.

Common Spike-rush was considered a common plant in the 19th century but there are few named locations. Today it is widespread, mainly rural, but extending into the uplands where the water is not too acidic. It is commonly found forming swamp stands at the margins of open waterbodies and in the wetter parts of marshes, and also along watercourses including ditches and some brackish stretches, as along the Black Cart Water (NS4967, 2005).

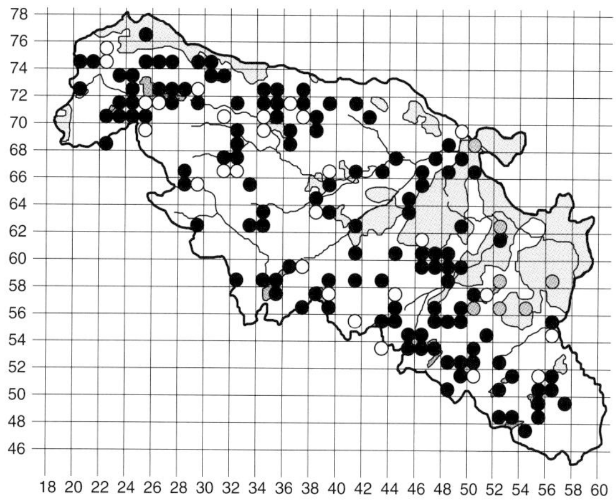

Eleocharis palustris

Eleocharis multicaulis (Sm.) Desv.
Many-stalked Spike-rush

Britain: frequent in the N and W.
Native; 1/4; very rare.
1st Renfrewshire, 1872, TB.

A specimen from Auchenbothie (NS3471), collected in 1980 by DM (PSY), was determined as this spike-rush by AJS; Many-stalked Spike-rush was also recorded from an excursion to Castle Semple in 1887 (NHSG). The sole modern record was made by the SNH Loch Survey from a marsh by the edge of Caplaw Dam (NS4358, 1996). It is likely that this is a rare plant in the area but it may well have been overlooked.

Eleocharis quinqueflora (Hartmann) O. Schwarz
Few-flowered Spike-rush

Britain: frequent in the N and W.
Native; 17/23; scarce.
1st Lochwinnoch, 1845, NSA.

Hennedy (1891) noted 'Hills above Greenock' and Gourock but there are few other records for Few-flowered Spike-rush and it was only rarely noted during the 1970s (BWR): Dod Hill (NS4953, 1975), Daff Reservoir (NS2270, 1977) and Blackettywater (NS3067, 1978). It has been recorded several times during the modern period, usually from small base-rich flushes. Its distribution is widespread: it is well represented in the north-west but more scattered in the central area, e.g. Corsliehills Wood (NS3869, 1998), Barbeg (NS4071, 1998), Linthills (NS3359, 2001) and near Kaims Dam (NS3362, 2001). It is only known from a couple of places in the east: at Killoch Hill (NS4759, 1999) and by the Earn Water at Brownside (NS5151, 2005).

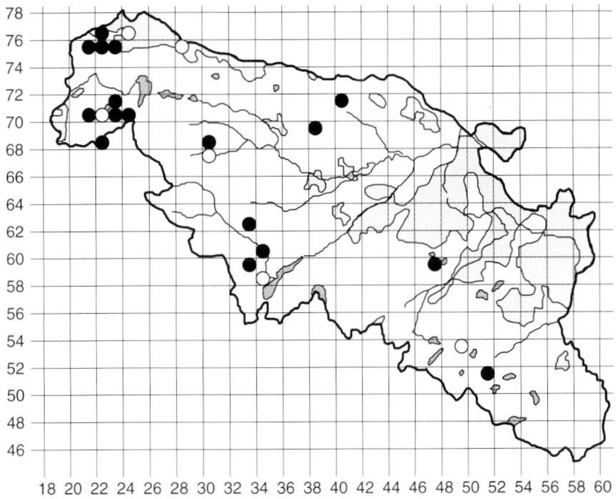

Eleocharis quinqueflora

Eleocharis acicularis (L.) Roem. & Schult.
Needle Spike-rush

Britain: scattered, local in C Scotland.
Native; 9/14; rare.
1st Lochwinnoch, 1834, MC.

This diminutive spike-rush was collected from Brother Loch in 1865 (GGO) and 1883 (GL), and was refound there in 1996 (NS5052, SNH-LS). Other old or recent past records are from Castle Semple Loch (1887, GL), Black Loch (1899, NHSG), Bennan Loch (1940, RM), Harelaw Dam (1951, RM), Glanderston Dam (1968, RM), Loch Libo (1979, AJS) and Barr Loch (1974, JM). Needle Spike-rush may have been overlooked in the modern period, often due to high water levels, but the SNH Loch Survey (1996) recorded it from Caplaw Dam (NS4358), Loch Libo (NS4355), Barr Loch (NS3557) and the north edge of Kilbirnie Loch (NS3355). It was also found at Kaim Dam (NS3462, 2003, IPG) and it can still be found at Balgray Reservoir (NS5057, 2000), as reported in CFOG, when the water is low.

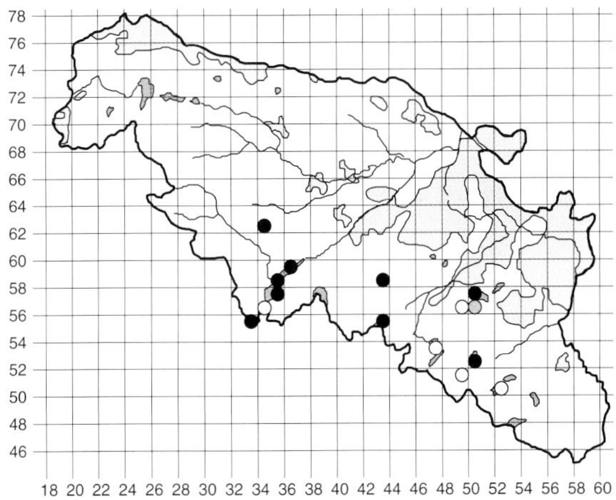

Eleocharis acicularis

Isolepis setacea (L.) R. Br.
Bristle Club-rush

Britain: widespread, more so in the W.
Native; 45/55; occasional, scattered.
1st Gourock, 1837, GL.

Hennedy described Bristle Club-rush as frequent but only repeated the record from Gourock; other old records include Bridge of Weir (1886, PSY) and Gourock (1872, GLAM), and RM (1925) noted Mac's Wood and Ravenscraig. Today it has a markedly widespread, scattered distribution. It is very likely overlooked, being a small plant and ephemeral, but it frequently appears in disturbed or gravelly poorly draining places.

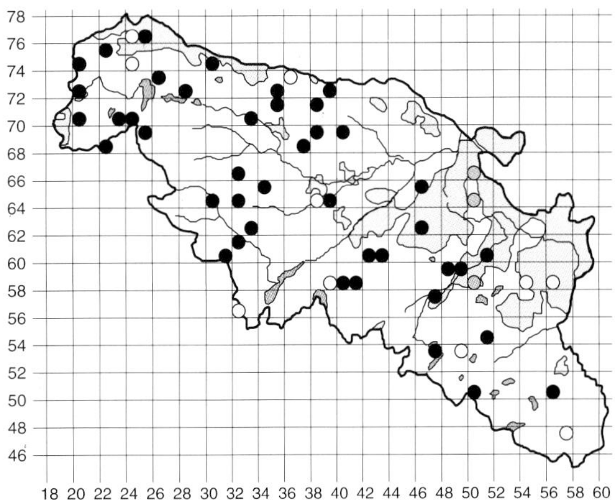

Isolepis setacea

Eleogiton fluitans (L.) Link
Floating Club-rush

Britain: frequent in the N and W.
Native; 2/4; very rare.
1st Gourock, 1865, RH.

Hennedy (1891) also noted 'moor near Kilmacolm, June 1889' but there are no other old records for Floating Club-rush and it was considered extinct, or overlooked, in the VC. However, in 1996 it was found in deep water in the North Loch at Drum Estate (NS3971, SNH-LS). A more recent find was from a small pool to the north of Craigmarloch (NS3472, 2006).

Blysmus compressus (L.) Panz. ex Link
Flat-sedge

Britain: local in S and N England, very rare in Scotland.
Former native; 0/1; extinct.
1st Below Greenock, 1865, RH.

Flat-sedge was described by Hennedy as frequent but there are no other records for this rare Scottish sedge, although AMcGS (1950) recorded it from '1 mile south of Skelmorlie', just into VC75.

Blysmus rufus (Huds.) Link
Saltmarsh Flat-sedge

Britain: coastal, northern.
Former native; 0/2; extinct.
1st Wemyss Bay, 1844, GL.

Hennedy (1891) noted Saltmarsh Flat-sedge as frequent but there have been no other records apart from one in the BSBI database from Levanne House near Gourock (NS2176, 1963, J Boyd).

Schoenus nigricans L.
Black Bog-rush

Britain: common in the N and W, local in the S.
Former native; 0/1; extinct.
1st Gourock, 1849, HLU.

Hennedy (1891) described Black Bog-rush as common on wet moors 'from Gourock all around the Firth'. There are no other records and it is doubtful that it was ever really that common in the VC.

Rhynchospora alba (L.) Vahl
White Beak-sedge

Britain: frequent in the N and W.
Former native; 0/3; extinct.
1st Gourock, 1865, RH.

Hennedy (1865) described White Beak-sedge as 'not common' and noted 'hills above Gourock'; there are other old records from Linwood Moss (1896, NHSG) and it was collected from 'Houston' Moss in the same year (PSY), possibly the same place. It has not been reported from any local bogs for more than 100 years, but it was included in the Renfrewshire list (TPNS 1915).

Carex paniculata L.
Greater Tussock-sedge

Britain: widespread but more local in Scotland.
Native; 11/14; scarce.
1st Wet meadow, Ravenscraig, 1880, NHSG.

This large, impressive sedge was collected from Loch Libo in 1892 (GL) but there are no further records until it was refound there in 1965 (NS4355, BWR); it is still present at Loch Libo. Other recent past records include the mire at Dargavel Burn (1971, BWR) and a swampy wood at Lochwinnoch (NS3660, 1979, AJS). Modern records are all from central wet mires, some in the open, including Walls Hill (NS4158, 2011), Barmufflock (NS3664, 1998), Hartfield Moss, Plymuir (NS4357, 1999), Barochan Marsh (NS4069, 1997), Dargavel Burn (NS3771, 1997), Reivoch (NS3957, 2010) and Corsehouse Reservoir (NS4850, 1997).

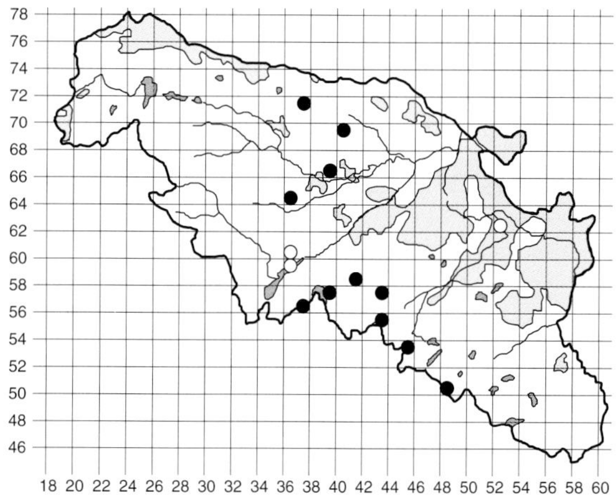

Carex paniculata

Carex diandra Schrank
Lesser Tussock-sedge

Britain: local, mainly in the N and W.
Native; 13/15; scarce.
1st Loch Libo, 1892, GL.

There are no relevant old literature localities for Lesser Tussock-sedge (as '*C. teretiuscula*'), which, perhaps surprisingly, appears to have been rare. Other older

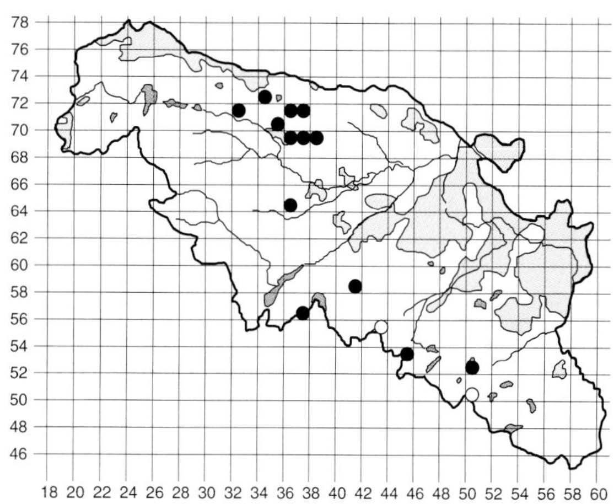

Carex diandra

records include Newton Mearns (1936, E), Little Loch (1936, GLAM), Barmufflock Dam (NS3664, 1961, BWR) and Dargavel Burn (1971, BWR). Today it is found at scattered species-rich fen mires, with a cluster of records from about Kilmacolm, including Dargavel (NS3771, 1997) and Glen Moss (NS3769, 1998), and slightly further west at Mathernock (NS3271, 2005) and Craigmarloch (NS3472, 2006); it was last recorded at Barmufflock (NS3664) in 1989 (SNH files). The southern mires are at Walls Hill (NS4158, 1993), Little Loch (NS5052, 1996), Roebank Burn, Knowes (NS3756, 2003) and Knockmade (NS4553, 1999).

Carex otrubae Podp.
False Fox-sedge

Britain: common in the SE but more coastal in the N.
Native; 7/8; rare, coastal.
1st Renfrewshire, 1883, TB.

False Fox-sedge was noted from the Clyde and described as frequent in the 19th century, but no VC76 localities were named. Today it is known from a few sites, all on the west coast, from Wemyss Bay (NS1869, 2012) to Ardgowan (NS1972, 2003, IPG), Lunderston (NS2073, 2008) and Gourock (NS2477, 2001).

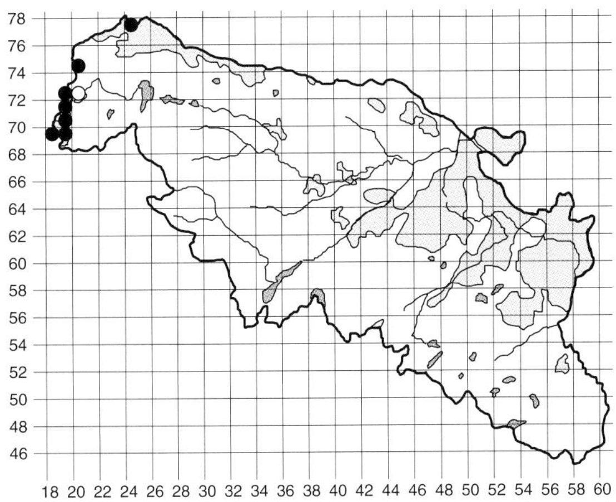

Carex otrubae

Carex vulpinoidea Michx.
American Fox-sedge

Britain: neophyte (N America); very rare casual.
Former accidental; 0/1; extinct casual.
1st Williamwood, 1980, AMcGS.

The original find was from waste ground by a car park (NS5658) and was considered to have been there for some time (CFOG), but the site was subsequently developed. There are no other records.

Carex spicata Huds.
Spiked Sedge

Britain: common in the S, rare in Scotland.
Native; 4/5; rare.
1st Williamwood, 1937, RM.

RM collected Spiked Sedge (1937, E) from a 'wet meadow at Clarkston', presumably the same place. It was also recorded during the CFOG period, when it was found at Ralston Golf Course (NS5063, 1987) and a meadow at Darnley (NS5258, 1988). The only other modern survey records are from grassland near Floors Burn, near Johnstone (NS4262, 1993) and rough grassland at the Royal Ordnance Factory, Georgetown (NS4268, 2005).

Carex muricata L. ssp. *pairae* (F.W. Schultz) Celak.
Carex muricata ssp. *lamprocarpa* auct. non (Wallr.) Celak.
Prickly Sedge

Britain: frequent in the S, local in the S and E of Scotland.
Native; 1/3; very rare.
1st Cathcart Castle, 1821, FS.

Hennedy (1865) also noted Inverkip. There is some doubt about which species is referred to by these early records (CFOG). Prickly Sedge was first found at Bishopton in 1980 (BWR), perhaps the same station as the record from the BSBI monitoring survey (1987), which corresponds with the 2003 record from Formakin Estate (NS4170, IPG).

Carex arenaria L.
Sand Sedge

Britain: coastal throughout.
Native; 2/3; very rare.
1st Renfrewshire, 1883, TB.

Hennedy (1891) described Sand Sedge as common on the coast but no VC76 sites are noted; it was collected from Inverkip in 1904 (GL). In 1975 it was found at Erskine Harbour (NS4672, AJS), but the site was considered lost to car parking. There was also a record from Lunderston Bay (1978, BWR), where it still occurs (NS2074, 1996), and in 2003 it was found at nearby Ardgowan Point (NS1972, IPG).

Carex disticha Huds.
Brown Sedge

Britain: frequent, rare in N Scotland and SW England.
Native; 10/15; rare.
1st Commore Dam, 1892, GL.

Brown Sedge was not mentioned in the Clyde area by authors before Hennedy (1891), so modern records appear to represent recent spread. RM collected specimens (1938, GLAM) from a marshy field south-west of Neilston and a quarry at Mearns, and also reported it from Braco, near Uplawmoor (1972); it was further recorded from Nether Carswell (1983, BWR) and in Pollok Country Park (NS5363, 1991, CFOG). Most modern records are from the south, where it has been found in marshy grasslands by the Levern Water at Commore Dam (NS4654, 1998), Carswell (NS4653, 1999), West Carswell (NS4552, 1999) and mire at Knockmade (NS4553, 1999), and further east by the Earn Water at North Moorhouse (NS5152, 2005) and Blackhouse (NS5353, 1991, LB). It was recently found in a moorland flush at Whitelee Windfarm (NS5348, 2011). The northern outliers are from the brackish banks of the Black Cart Water near Glasgow Airport (NS4868, 2005) and by Spango Burn (NS2374, 2006).

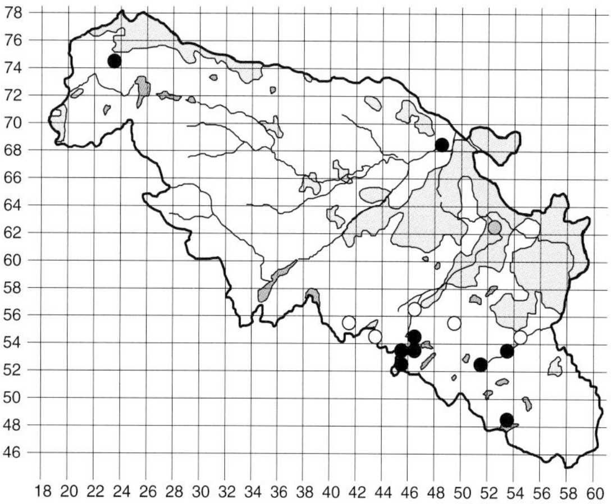

Carex disticha

Carex divisa Huds.
Divided Sedge

Britain: rare, mainly coastal in SE England.
Former accidental; 0/1; extinct.
1st Williamwood, 1927, RM.

This sedge was considered to be 'one of the most geographically remarkable plants in Glasgow's flora' (CFOG). RM collected it in 1940 (GL and GLAM), from marshy waste ground, and it was last recorded in 1962. It has not been seen since.

Carex remota L.
Remote Sedge

Britain: common in the S, rare in N and E Scotland.
Native; 60/65; frequent in wet woodlands.
1st Gourock, 1865, RH.

Other old records include Thornliebank (1891, GL) and Linthills (1912, GL). Today Remote Sedge is well represented in the extreme west and quite frequent in the central area but absent from much of the east except about the White Cart and Levern waters. It is typically found in shaded moist woodlands, often on flushed clayey soils.

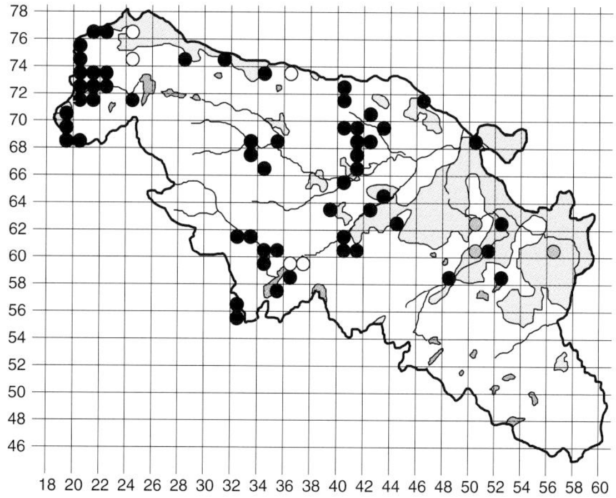

Carex remota

Carex leporina L.
Carex ovalis Gooden.
Oval Sedge

Britain: common throughout except in E England.
Native; 257/320; very common.
1st Renfrewshire, 1869, PPI.

There are few 19th-century named localities for this sedge which at the time was considered common. Today Oval Sedge is widespread and occurs throughout the area. It is found in damp, short usually neutral to slightly acidic grassland, and can also be found on waste ground, especially where compacted and clayey.

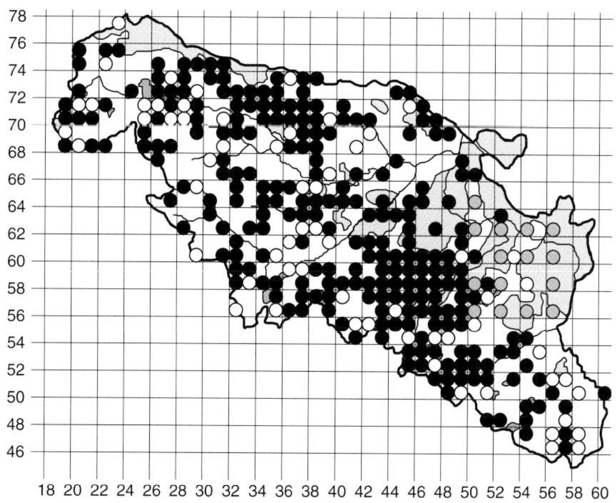

Carex leporina

Carex echinata Murray
Star Sedge

Britain: common in the N and W, rare in the SE.
Native; 222/251; common in rural uplands.
1st Gourock, 1837, GL.

Star Sedge was common in the 19th century and remains so today, although it is likely to have declined with draining or enriching of lowland marshes. It shows a strong rural upland pattern, and is rare in the lowlands; records from the latter include Southbar (NS4669, 1997), Paisley Moss (NS4665, 2000) and Jenny's Well (NS4962, 1992, DM). It is usually found in more acidic marshes, mires, flushes and rush pastures, often where associated with peatbogs.

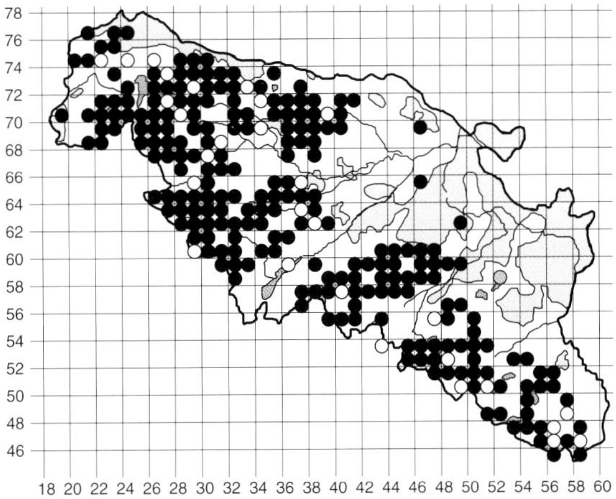

Carex echinata

Carex dioica L.
Dioecious Sedge

Britain: common in the N and W.
Native; 30/36; occasional in upland areas.
1st Gourock Hills and Inverkip, 1865, RH.

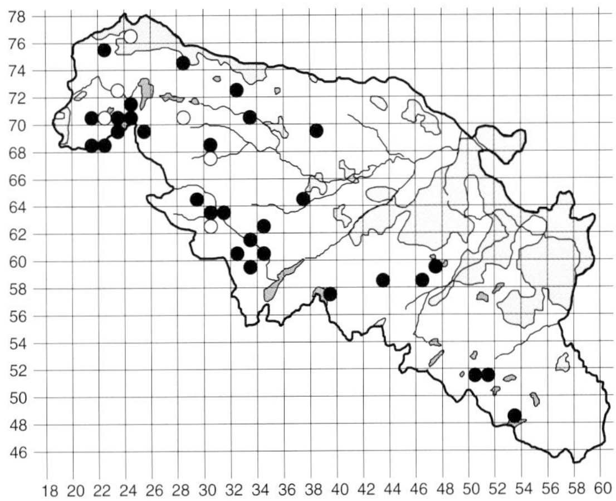

Carex dioica

Dioecious Sedge was considered 'not common' in the 19th century; it was collected from Kilmacolm in 1891 (GL). More recently it was recorded from six places in the 1960s–80s period, including south of Daff Reservoir (NS2270,1977, BWR) and Blackettywater (NS3076,1978, BWR). The modern distribution shows a strong western rural pattern, mainly to the fringes of the Renfrewshire Heights. It is associated with base-rich flushes, usually along burn sides with shallow peat over rocks. The seven sites to the south-east are from Reivoch (NS3957, 2010), Killoch Hill (NS4759, 1999), Witch Burn (NS4658, 1999), Caplaw Dam (NS4358, 2001), Earn Water, Brownside (NS5151, 2005), Brown Castle (NS5051, 2005) and Brown Hill (NS5348, 2011).

Carex canescens L.
Carex curta Gooden.
White Sedge

Britain: common in the N and W.
Native; 112/155; common in acidic mires.
1st Near Paisley, 1865, PSY.

Hennedy (1891) described White Sedge as 'Very common. In most of our bogs'. Today it is locally common and has a strong rural distribution and extends into the uplands. It is found in wet peaty places, especially ditches through bogs, but can also occur in wet acidic mires. The lowland sites include Barochan Moss (NS4268, 1994) and Paisley Moss (NS4656, 2003).

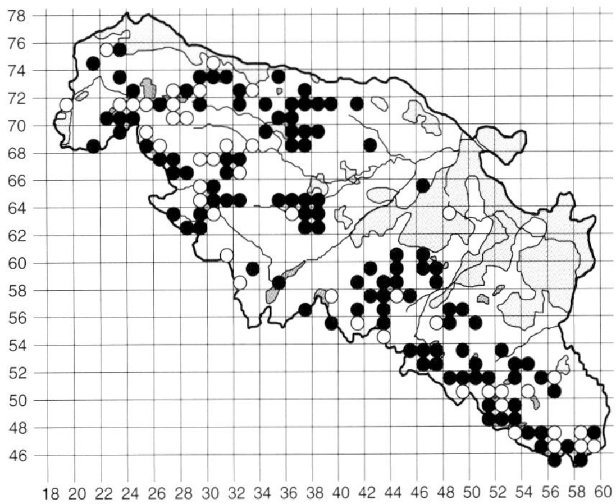

Carex canescens

Carex hirta L.
Hairy Sedge

Britain: common, rare in N Scotland.
Native; 72/89; frequent in lowlands.
1st Lochwinnoch, 1845, NSA.

Other old records included Paisley Canal (1858, GLAM) and Gourock (1876, FFWS). Today Hairy

Sedge is widespread but strongly lowland and often found along watercourses; the sole Renfrewshire Heights record is from Muirshiel Barytes Mine waste (NS2865, 2003). It is found in less acidic grasslands, somewhat damp or clayey, and in such areas is capable of competing with coarse grasses.

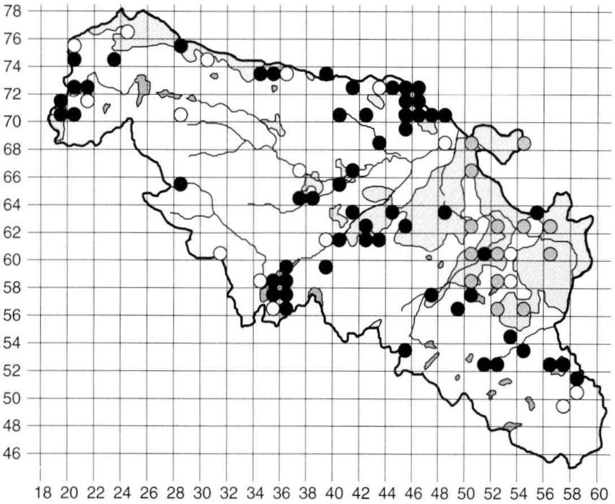

Carex hirta

Carex lasiocarpa **Ehrh.**
Slender Sedge

Britain: local in the S, frequent in NW Scotland.
Native; 2/3; very rare.
1st Castle Semple Loch, 1901, FFCA.

This rare sedge is only known from a couple of local sites. It is found in the species-rich mire at Walls Loch (NNS4158, 1993), where it was first recorded by AJS in the late 1970s, and the marginal ('lagg') fen by the raised bog at Hartfield Moss (NS4257, 1999).

Carex acutiformis **Ehrh.**
Lesser Pond-sedge

Britain: frequent in the S, rare in the W and N.
Native; 16/18; scarce.
1st Banks of Cart, 1865, RH.

Hennedy (1891) described Lesser Pond-sedge as common and specimens (GL) were collected from Hawkhead Farm (1892) and by Inverkip Road (1893). RM recorded or collected it on several occasions: Newton Mearns (1938, GLAM), Earn Water, Hazeldean (1940) and Loch Libo (1940, GLAM) – where it still occurs (NS4355, 1996); AJS reported it from marshes at Paisley (Paisley Moss, NS4665) in 1978, but it is not there now. Modern records are limited and mainly from the south, where it is found in wetlands and along watercourses, e.g. the Spateston Burn (NS4161, 1994, CD), Humbie Burn (NS5454, 1997), about Blackhouse (NS5352, 1992, LB), and Levern Water at Commore (NS4654, 1998), Kirkton Dam (NS4856, 1989) and Neilstonside (NS4655, 1998); it can still be found in wet woodland at Hawkhead (NS5062, 2002). The records from western sites are from Shovelboard (NS3869, 1989, PSY), Earns Hill (NS2275, 2006) and by Spango Burn (NS2475, 2011).

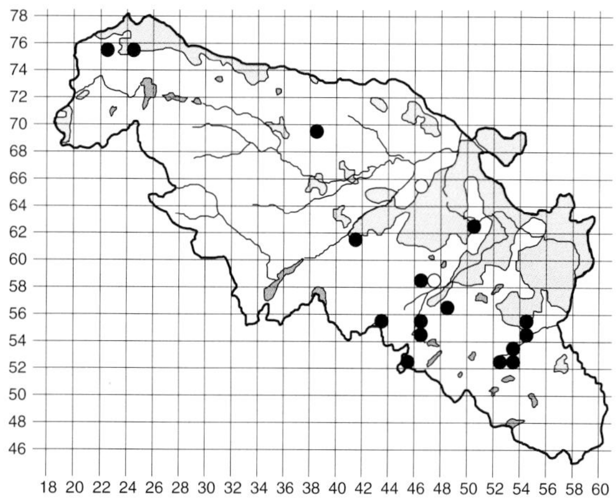

Carex acutiformis

Carex riparia **Curtis**
Great Pond-sedge

Britain: common in the SE, local in the N and W.
Former native; 0/4; extinct.
1st Cart, 1865, RH.

Great Pond-sedge was also recorded from Dunrod Castle (1880, NHSG), Neilston (1884, NHSG) and the old canal, Hawkhead (1904, PSY); an RM specimen (1965, E) from Commore Dam (NS4654) is Lesser Pond-sedge. There are no modern records from any of these places.

Carex rostrata **Stokes**
Bottle Sedge

Britain: common in the N and W.
Native; 228/247; common in wetlands.
1st Near Paisley, 1865, PSY.

Hennedy (1891) described Bottle Sedge as frequent and noted 'Gourock Hills'; other sites include Loch Libo (1887, NHSG). Today it is common in rural areas but absent from the urban lowlands; it can extend into the uplands, where it can grow in peaty pools, drains and swamps, at least where there is some mineral movement. It is one of the commonest dominant sedges of loch margins, swamps and mires, generally where nutrients are low.

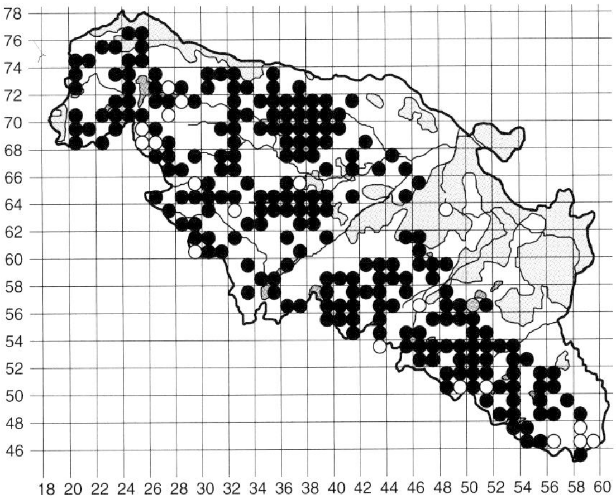

Carex rostrata

Carex vesicaria L.
Bladder-sedge

Britain: widespread but scattered, more frequent in N and W.
Native; 10/12; rare.
1st Busby, 1865, RH.

Although described as frequent by Hennedy (1891), there are no other old localities provided for Bladder-sedge. RM knew it from Brother Loch (1942) and the moor at Mearns (1945). It was collected from Castle Semple Loch in 1887 (GL), where it still occurs at several places, e.g. NS3558 and NS3658 (2010, MG). RM (1934) collected it from a boggy woodland near Lochwinnoch, which may relate to one of the modern records: willow scrub, Knockbartnock (NS3560, 1995) or Boghead (NS3560, 1996). A small population occurs at Paisley Moss (NS4665, 2003), where it was first recorded in 1981 (BWR). The other modern records are from Lawfield Dam (NS3769, 1989, TPNS), Ladymuir (NS3463, 1989, PSY), Corehouse Reservoir (NS4850, 1997), Kays Wood, Auchenbothie (NS3471, 1998) and Picketlaw Reservoir (NS5651, 1998). It has a widespread distribution and is perhaps overlooked, but must be considered a rare plant now.

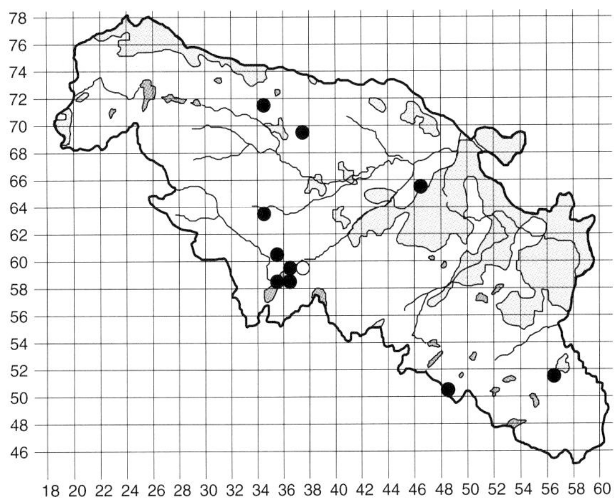

Carex vesicaria

Carex pendula Huds.
Pendulous Sedge

Britain: common in the S, local in lowland Scotland.
Native and hortal; 33/39; occasional.
1st Near Cloch, 1873, TGSFN.

There are a few old records for this large sedge from the north-west of the VC, including Gourock (1876, FFWS), Shielhill Glen (1888, NHSG) and the Inverkip–Ashton area (1904, NHSG). Today it shows a disjunct distribution with a strong north-west bias, where it occurs in the woodlands from Kelly Glen (NS1969, 1995) north to Cloch (NS2075, 1997); it can also still be found near Gourock, at Port Glasgow (*c.* NS3074), but here it is found as a colonist on waste ground. Eastern records are from mainly estate-type woods or plantations, where Pendulous Sedge has perhaps been introduced, e.g. Houston (NS4167, 1999), Cowdon Hall (NS4757, 1997), Linwoodmoss (NS4466, 1999) and Durrockstock (NS4662, 1999). It is quite well known in Glasgow (as reported in CFOG), with more recent records (2005) from Rouken Glen (NS5458), Pollok Park (NS5561), Darnley (NS5258) and Linn Park (NS5857). The southernmost locality is the refuse tip at Mearns Muir (NS5152, 2006).

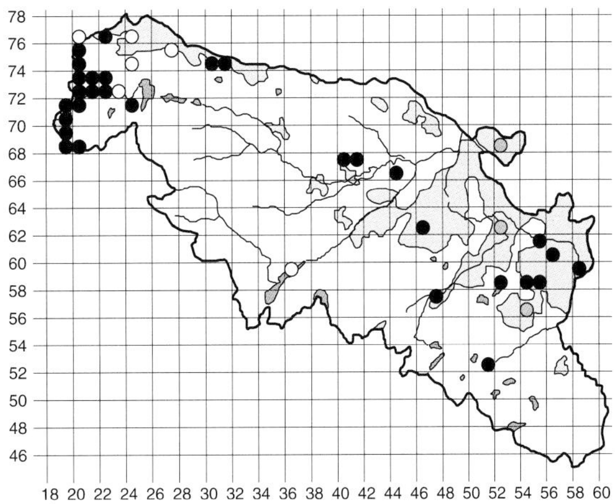

Carex pendula

Carex sylvatica Huds.
Wood-sedge

Britain: common in the S, local in N Scotland.
Native; 45/53; frequent in valley woodlands.
1st Langside and Gourock, 1865, RH.

Hennedy (1891) described Wood-sedge as frequent 'In moist woods'. Today it is well represented in the extreme west of the VC and more scattered elsewhere.

It is found in many of the wooded burn valleys feeding off the Renfrewshire Heights, but also occurs in woodlands to the south-east, e.g. Brock Burn, Levern Water and the White Cart Water. It is associated with woodlands on less acidic soils, usually along watercourses.

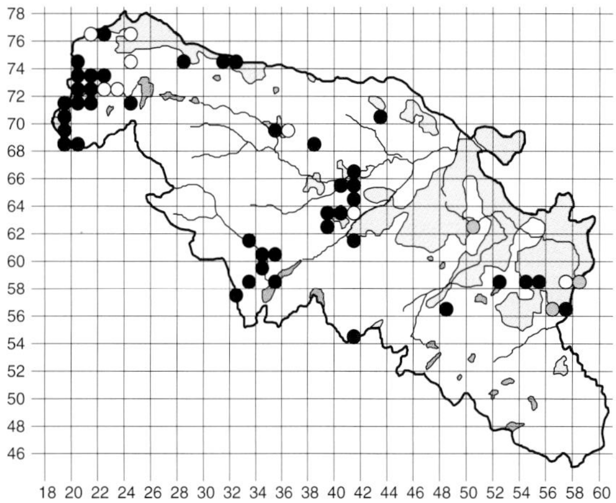

Carex sylvatica

Carex flacca Schreb.
Glaucous Sedge

Britain: common throughout.
Native; 166/201; common.
1st Near Barrhead, 1843, GLAM.

Glaucous Sedge is a common sedge, as noted in the 19th century by Hennedy (1891). It has a widespread distribution: it is mostly absent from the peaty uplands but occurs there on some burn sides, and can be found on waste ground in the lowlands. It typically occurs in short flushed grassland on clayey soils with some base enrichment, including some amenity managed parks and open spaces.

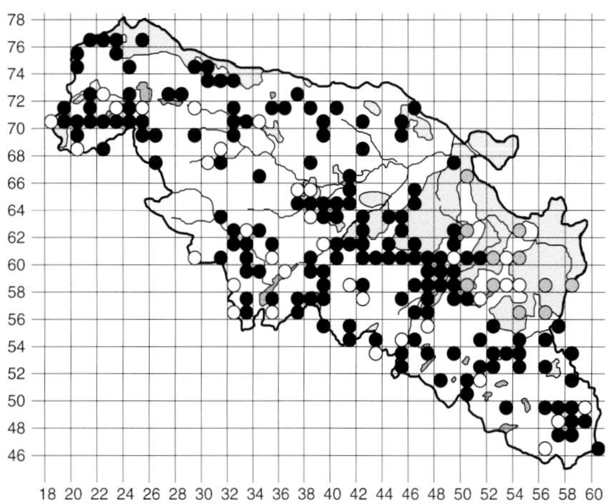

Carex flacca

Carex panicea L.
Carnation Sedge

Britain: common throughout except in E England.
Native; 259/290; very common in rural uplands.
1st Mearns, 1845, NSA.

The original record stated that Carnation Sedge was found in 'Higher places' and in general it was described as common 'in bogs and marshes'. Today it is widespread in rural areas and well represented in the uplands. It is found in marshy acidic grasslands and wet heaths, and in the drier parts of mires.

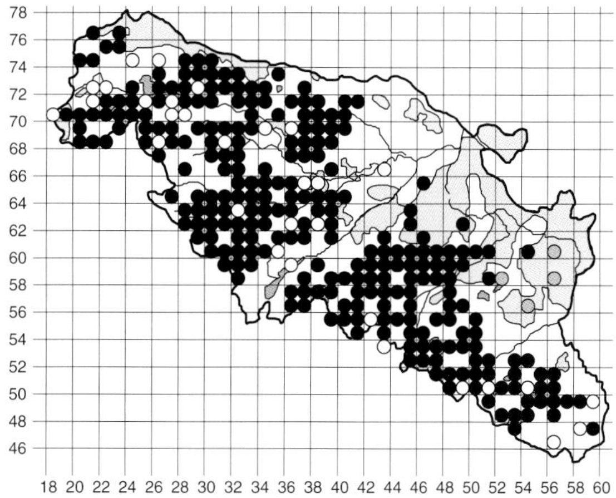

Carex panicea

Carex laevigata Sm.
Smooth-stalked Sedge

Britain: widespread, mainly western.
Native; 20/24; scarce.
1st Renfrewshire, 1883, TB.

This sedge was considered rare by Hennedy (1891) and not common by Lee (1933), and there are no other old localities in the VC. Recent past records include

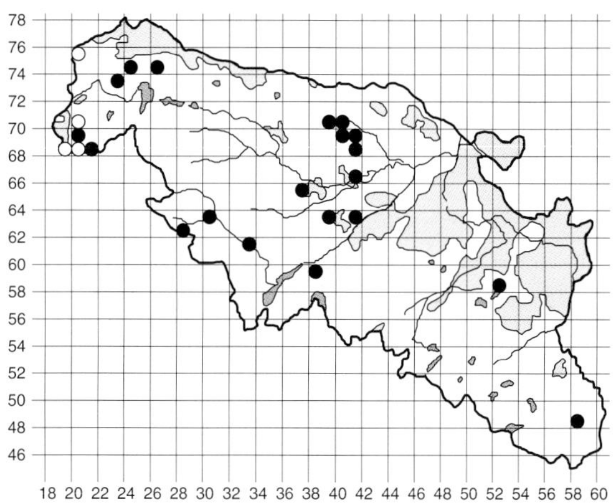

Carex laevigata

Kelly Glen (NS1968, 1959, BWR) and Cloch (NS2075, 1968, BWR); there are couple of modern records from near the former and it can still be found about the Greenock Cut area (N2474, 1965, RM). It is usually found in wet woodlands, often along burn sides, as along the River Calder (NS3361, 1997), but can occur in open marshy places, as at Barochan (NS4069, 1998). The south-eastern records are from Waulkmill Glen (NS5258, 2006) and from the steep valley side of Munzie Burn (NS5848, 2003).

Carex binervis Sm.
Green-ribbed Sedge

Britain: common in the N and W.
Native; 215/253; common.
1st Stanely Reservoir, 1865, PSY.

Green-ribbed Sedge was considered common in the 19th century, and it is still common today, although it has probably declined in the lowland agricultural areas. Its modern distribution shows a strong rural upland pattern, where it is a common feature of the more acidic grasslands and some wet or dry heaths.

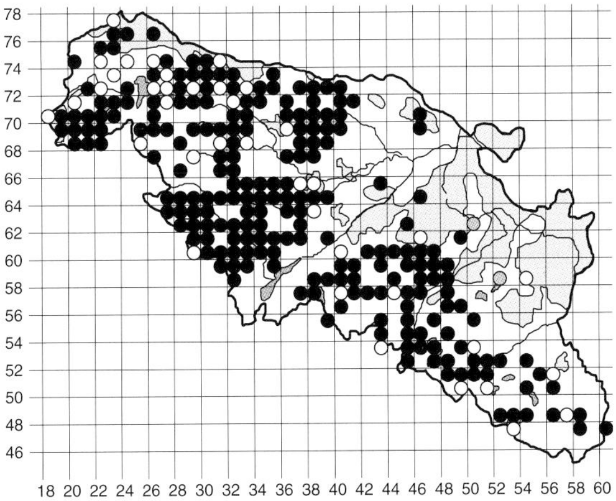

Carex binervis

Carex distans L.
Distant Sedge

Britain: widespread, coastal in the N.
Native or accidental; 1/3; very rare.
1st Langbank, 1965, BWR.

Distant Sedge was considered by Hennedy (1891) to be a very rare sedge of 'muddy marshes by the sea', but he gave no VC localities. There is a reference to it from Inverkip (1979, ERTC) in the VC76 card index, but no other details. The sole modern record is from the edge of a new path by Bottombow saltmarsh (NS4671, 1995), where it may have been introduced during construction.

Carex extensa Gooden.
Long-bracted Sedge

Britain: coastal, mainly in the W.
Native; 0/1; extinct or very rare and overlooked
1st Cardwell, 1979, AMcGS.

Hennedy (1865) knew this coastal sedge from Skelmorlie (just over the border, in VC75) but the first, and only, VC76 record was from saltmarsh turf, between Cloch and Inverkip (NS2074).

Carex hostiana DC.
Tawny Sedge

Britain: common in the N and W.
Native; 27/36; occasional, western.
1st Gourock, 1865, RH.

Tawny Sedge cannot now be called frequent as it was by Hennedy in 1891. Today it has a strongly rural western and upland-fringe distribution, but is perhaps overlooked. It is typically found in base-rich flushed grasslands. The easternmost site is from beside Witch Burn (NS4658, 1999), but it has not been seen in the south-east since 1960, when it was recorded from Carrot Burn (NS5748, BWR).

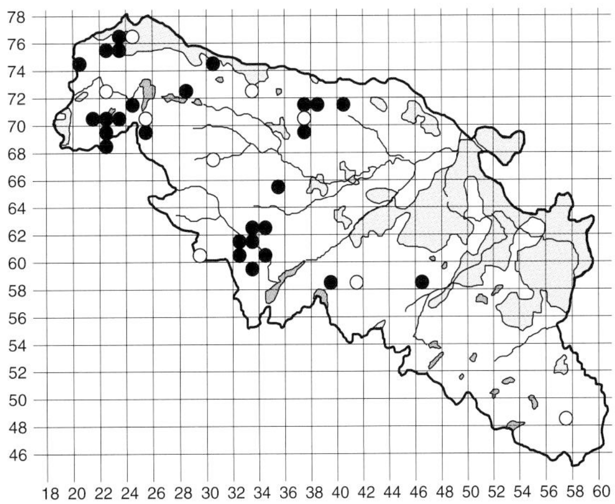

Carex hostiana

Carex hostiana × *C. demissa*

Britain: frequent hybrid, mainly in the N and W.
Native; 6/6; rare.
1st Calder, Muirshiel (NS3163), 1995, AMcGS.

This hybrid sedge is perhaps overlooked but has been found at several places, not always with both parents: Shepherd's Burn (NS2569, 1997), Witch Burn (NS4558, 1999), Cample Burn (NS3062, 2001), Gimblet Burn (NS2370, 2005) and Brown Castle (NS5051, 2005).

Carex lepidocarpa Tausch
Carex viridula Michx. ssp. *brachyrrhyncha* (Celak) B. Schmid
Long-stalked Yellow-sedge

Britain: frequent in the N, scarce and local in S England.
Native; 8/9; rare.
1st Little Loch, 1936, RM.

It is not clear which of the two yellow-sedges earlier authors referred to when they recorded '*C. flava*', which was considered frequent by Hennedy (1891). Long-stalked Yellow-sedge is undoubtedly much rarer, and presumably always has been, than Common Yellow-sedge and occurs in more distinctly basic species-rich mires. It was found in the mire at Dargavel Burn in 1971 (BWR) and still occurs there today (NS3771, 1997), but has not been seen again at Darnley (NS5258, 1991, CFOG). Other modern records are from Floors Burn (NS4262, 1993), the mire at Lawfield Dam (NS3769, 1997), base-rich flushes at Glenshilloch Burn (NS2269, 2001 and NS2268, 2007), flush at Everton (NS2170, 2004) and by the Gryfe Water, Cauldside (NS3270, 2005).

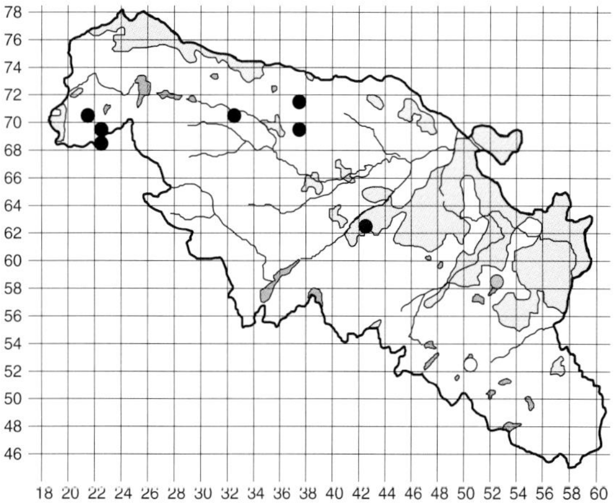

Carex lepidocarpa

Carex demissa Hornem.
Carex viridula Michx. ssp. *oedocarpa* (Andersson) B. Schmid
Common Yellow-sedge

Britain: common in the N and W, rare in SE England.
Native; 155/193; common.
1st Gourock, 1856, GLAM.

The specimen from Gourock and another collected from near Paisley (1865, PSY) confirm the early presence of Common Yellow-sedge; other old records include Loch Libo (1887, NHSG) and Gourock (1876, FFWS). Today, as its name suggests, Common Yellow-sedge is quite common and well distributed. It is found throughout the rural area, extending into the upland fringes, where it occurs in flushes and marshy pastures, generally reflecting some mineral enrichment.

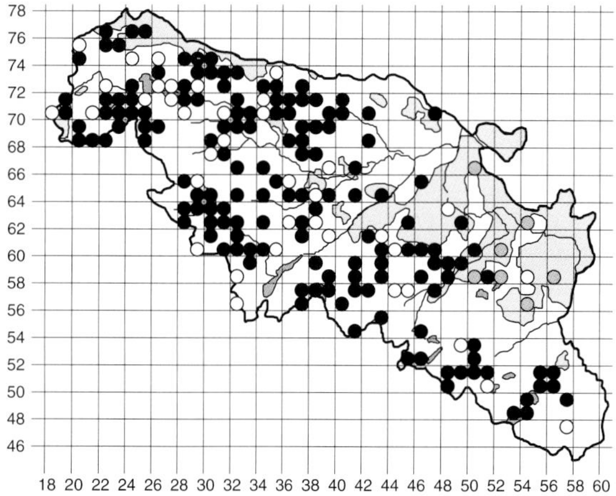

Carex demissa

Carex pallescens L.
Pale Sedge

Britain: common in the N and W and uplands elsewhere.
Native; 10/19; rare, possibly overlooked.
1st Near Blackland Mill (Paisley), 1865, PSY.

Pale Sedge was also reported from Mill Burn, Kilmacolm (1874, TGSFN) and also Castle Semple Loch (1887, NHSG). BWR found it at Lochwinnoch (NS3558, 1956), Knockmade (NS4353, 1973) and Knocknairshill (NS3074, 1983). It was recorded from Langside and Netherlee during CFOG surveys, but there are few other modern records, which include damp grassland by the burn at Shielhill Glen (NS2471, 2000), marsh at Cauldside (NS3270, 2005), a roadside near Gavilmoss (NS3257, 2005), Ladymuir (NS3463, 1989, TPNS), Green Water, Hillside (NS3069, 2006), Linthills (NS3460, 2008) and Muirend (NS3958, 2010).

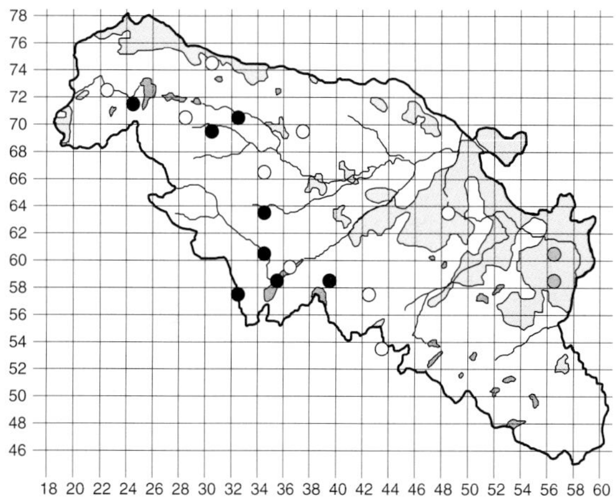

Carex pallescens

Carex caryophyllea Latourr.
Spring-sedge

Britain: widespread except in N Scotland and E England.
Native; 69/76; frequent.
1st Gourock, 1865, RH.

Spring-sedge was considered common in the 19th century but it is likely to have declined with agricultural improvement and urban spread. Its distribution is widespread but with a strong rural pattern; it avoids the more acidic peaty uplands. It is most typically found on dry unimproved pasture banks or ridges, usually indicating less acidic soils.

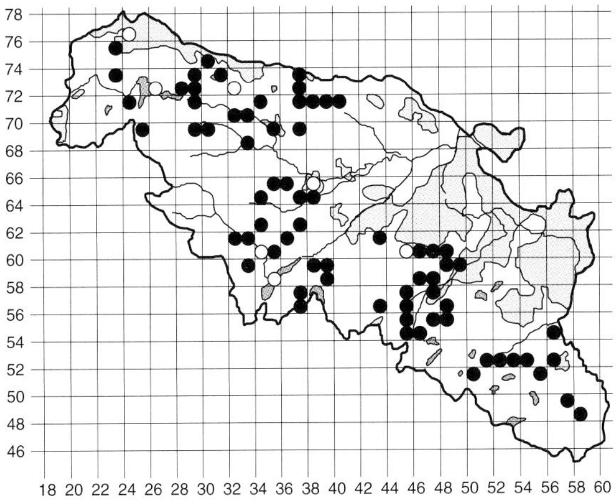

Carex caryophyllea

Carex pilulifera L.
Pill Sedge

Britain: widespread, commoner in the N and W.
Native; 131/159; common.
1st Gourock, 1865, RH.

Pill sedge was described as frequent from heathy pastures in the 19th century, which is still accurate

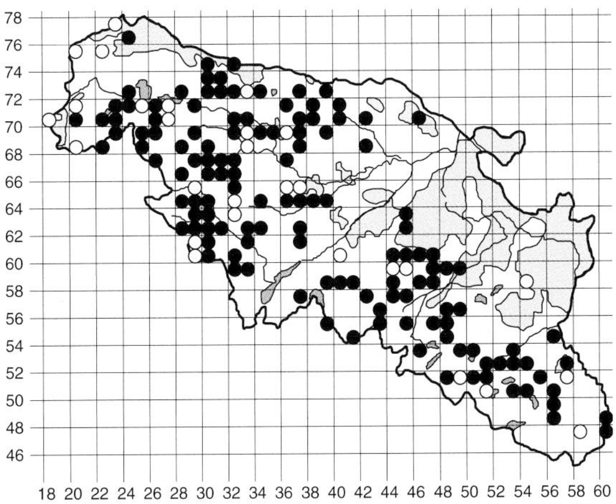

Carex pilulifera

today. It shows a strong rural and upland pattern, and is absent from enriched lowland sites and apparently scarce in the extreme north-west; it is also rare in the south-east moorland and parts of the Renfrewshire Heights where it avoids boggy soils. It is typically found in freer-draining acidic grasslands and heaths, as on embankments and ridges, and it frequently occurs under heather and bracken.

Carex limosa L.
Bog-sedge

Britain: frequent in the N and W.
Native; 6/9; rare.
1st Kilmacolm, 1893, GL.

The collector (RD Wilkie) reported his find as new to Renfrew (Wilkie 1894) and there are no other local 19th-century records for this rare sedge, although it is likely to have been present at more sites at the time. Lee (1933) repeated the record from Kilmacolm and RM added Little Loch in 1936, and in 1971 he collected a specimen from the mire at Dargavel Burn (NS3771). In 1961 (BWR) found it at Barmufflock (NS3664) and later DM collected it from Shovelboard (NS3869, 1989). Today Bog-sedge is still found at Little Loch (NS5052, 1996, SNH-LS), Glen Moss (NS3669, 1996, SNH-LS), Shovelboard (NS3869, 2011) and Dargavel (NS3771, 1997); it is also known from mires at Walls Hill (NS4158, 1993) and Corsliehill Wood (NS3869, 1998). At all sites it occurs in species-rich fen mire.

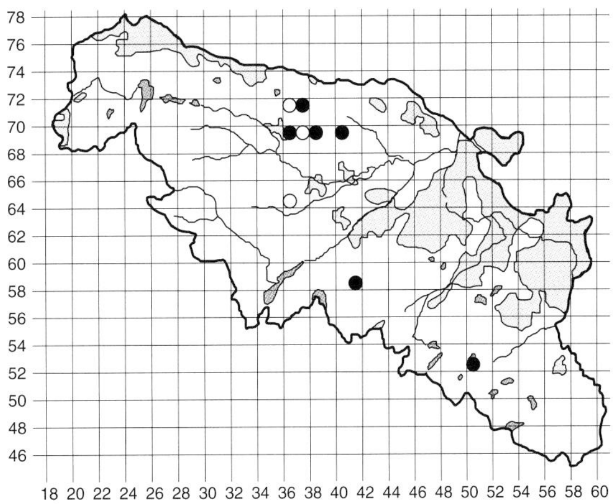

Carex limosa

Carex magellanica Lam.
Tall Bog-sedge

Britain: very local in the N and W.
Native; 1/3; very rare, endangered.
1st Loganswell, 1957, RM.

ERTC reported Tall Bog-sedge from 'moorland west of Eaglesham' (Conacher 1959) but accredited this find to VC75, adding that it was from 'close to the recently discovered Renfrewshire station', which Mackechnie (1958) had recently reported. A few years later RM collected it from nearby Craigenfaulds Moss (1969, GLAM) on the VC75 border and later (1977) reported it from Lochgoin Reservoir (NS5347). There have been no recent reports of this sedge from this location. However, a new station was recently discovered at Black Hill moss (NS4851, 2011); the site is within yards of the VC75 border, occurring on the 'lagg' transition of the small peat bog.

Carex aquatilis Wahlenb.
Water Sedge

Britain: northern, occasional.
Native; 53/65; frequent.
1st Shielhill Glen, 1881, NHSG.

There may be doubts about the identification of old literature records but specimens (GL) exist from Loch Libo (1892) and Lochwinnoch (1938), and Water Sedge still occurs at both (1996, SNH-LS); other old records include a farm pond at Malletsheugh (1933, RM) and Williamwood (1926, GLAM). In 1895 Professor T King considered it the 'commonest sedge' at Barr Meadow, near Lochwinnoch, where he noted it was known locally as 'black star' (King 1897). Today it is quite frequent in the south-central part of the VC, with a western outlier at the Compensation Reservoir (NS2472, 2007), close to the original station.

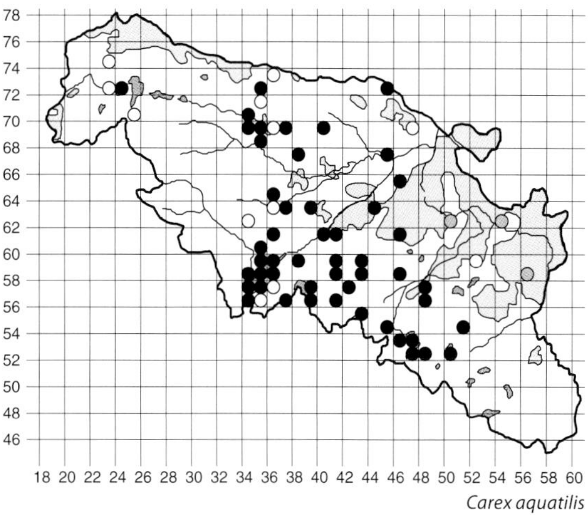

Carex × *hibernica* A. Benn.
(*C. aquatilis* × *C. nigra*)

Native; 1/2; very rare, probably overlooked.
1st Loch Libo, 1945, GLAM.

This hybrid sedge has not been detected during recent surveys of Loch Libo or other wetlands, but the two parents co-exist at many sites, so it is likely more widespread. The only modern record is from the fen by Roebank Burn at Knowes (NS3756, 2003).

Carex acuta L.
Slender Tufted-sedge

Britain: frequent in the S, absent north of C Scotland.
Native; 4/5; very rare.
1st Gourock, 1865, RH.

Slender Tufted-sedge was found at Paisley Moss in 1961 (BWR) and now forms large stands to the west side of the marsh (NS4665, 2000). It also occurs sparingly on the north bank of the Black Cart Water at Inchinnan (NS4767, 1999), the River Cart at Blythswood (NS4968, 2007), and in a wet depression by a track at the Royal Ordnance Factory, Georgetown (NS4467, 2005). An RM specimen from Loch Libo (1944, E) is an error for Water Sedge.

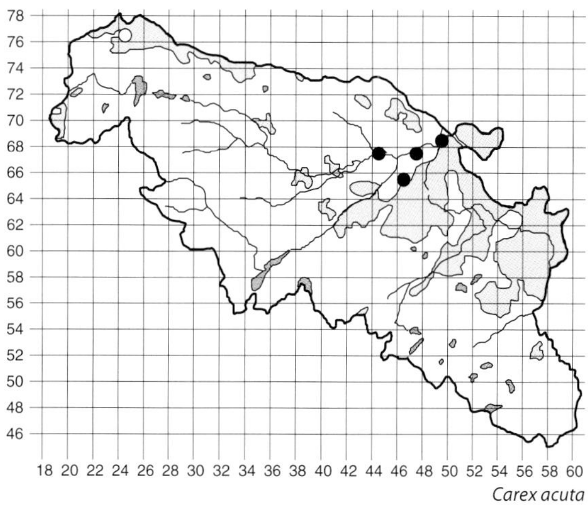

Carex nigra (L.) Reichard
Common Sedge

Britain: common throughout.
Native; 279/346; very common.
1st Near Paisley, 1865, PSY.

Common Sedge was considered 'common' by Hennedy (1891) but there are few named stations. Today it is widespread, occurring from the lowlands into the uplands, but rare in the agriculturally improved lowlands. It is found in wet grasslands and generally more acidic mires. It is thought that old records for '*C. stricta*' or '*C. elata*' (e.g. TPNS 1915) referred to tussocky wetland forms of Common Sedge.

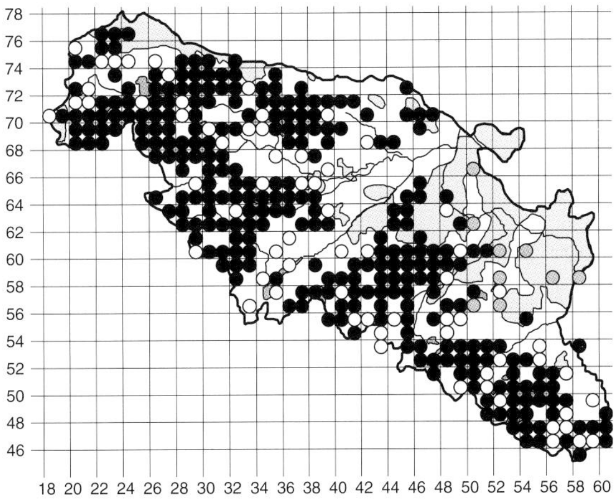

Carex nigra

Carex bigelowii Torr. ex Schwein.
Stiff Sedge

Britain: frequent in the uplands of the NW.
Native; 1/3; very rare.
1st Myres Hill (NS5646), 1975, BWR.

There are no other old records for Stiff Sedge and it has not been relocated at the original location during modern surveys in the hills above Eaglesham. BWR (1984) recorded it from moorland near Loch Thom (NS2571), but this location is at less than 250m so this is a doubtful record. Today the only known population is a small patch on the north side of the Hill of Stake summit ridge (NS2763, 1997).

Carex pauciflora Lightf.
Few-flowered Sedge

Britain: frequent in NW Scotland, very rare further S.
Native; 2/2; very rare.
1st Brownhill Moss, 2001.

Apart from a vague 'near Glasgow' record (1821, FS – repeated by RH in 1865), there are no other old records for Few-flowered Sedge. The modern finds are therefore noteworthy: several clumps in the peat bog at Brownhill (NS2369), with similar patches near the River Calder below Muirshiel (NS3162, 2005). It may well occur in other boggy parts of the Renfrewshire Heights or perhaps the south-east uplands, where it has been found just over the border, in VC77 (PM pers. comm.).

Carex pulicaris L.
Flea Sedge

Britain: common in the N and W.
Native; 69/91; frequent.
1st Near Blackland Mill (Paisley), 1865, PSY.

Other old records include Gourock (1865, RH), Paisley Canal (1876, FFWS) and Castle Semple Loch (1887, NHSG). Today Flea Sedge shows a strong rural pattern but is absent from several areas. It grows in marshy grasslands and mires, usually where there is some flushing or mineral enrichment.

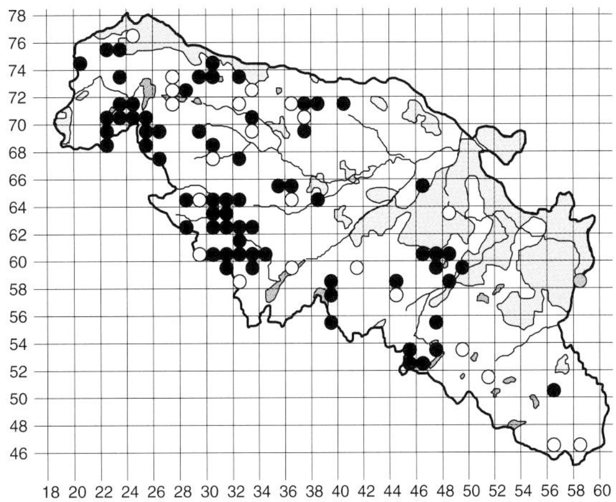

Carex pulicaris

POACEAE

Sasa palmata (Burb.) E.G. Camus
Broad-leaved Bamboo

Britain: neophyte (Japan); scattered, mainly in the S.
Hortal; 6/6; rare in old estates.
1st Drum Estate, 1997.

This tall vigorous bamboo forms dense stands at several, mainly central, estates: Drum Estate (NS3971), Rouken Glen (NS5458, 1997), Barmufflock (NS3664, 1998), Corsliehill House (NS3969, 1998), Glentyan House (NS3963, 1999) and Barochan Estate (NS3969, 2004). A few other bamboos are planted in old estates, e.g. Glentyan, but do not appear to have spread to any great extent.

Nardus stricta L.
Mat-grass

Britain: common in the N and W.
Native; 245/287; very common in uplands.
1st Lochwinnoch, 1845, NSA.

Mat-grass is a very common and often dominant plant of upland acid grasslands. It shows a strong rural distribution pattern; it is absent from lowland intensive farms and urban areas, but also scarce in the wetter boggy areas of the uplands.

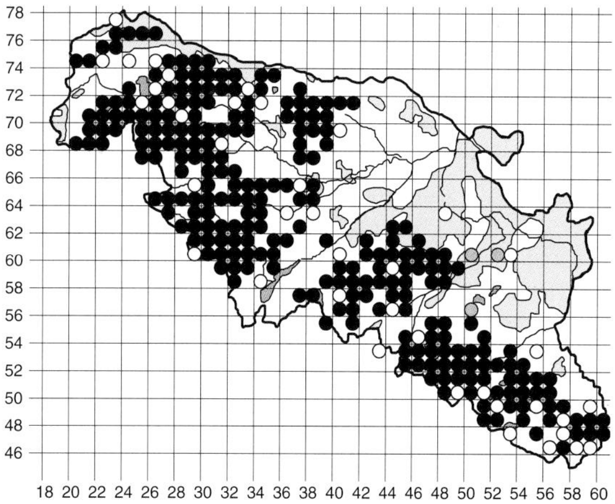

Nardus stricta

Anemanthele lessoniana (Steud.) Veldkamp
Pheasant's-tail

Britain: neophyte (New Zealand); rare in the S.
Hortal; 1/1; very rare.
1st Lochwinnoch, 2005, IPG.

Young seedlings, close to three planted clumps, were found in pavement cracks and flowerbeds in the centre of Lochwinnoch; plants appear to persist at the latter.

Milium effusum L.
Wood Millet

Britain: common, except in N Scotland.
Native; 24/30; occasional.
1st Woodland, Langside, 1854, GGO.

Wood Millet was considered frequent in the 19th century; old localities included Kelly Glen (1897,

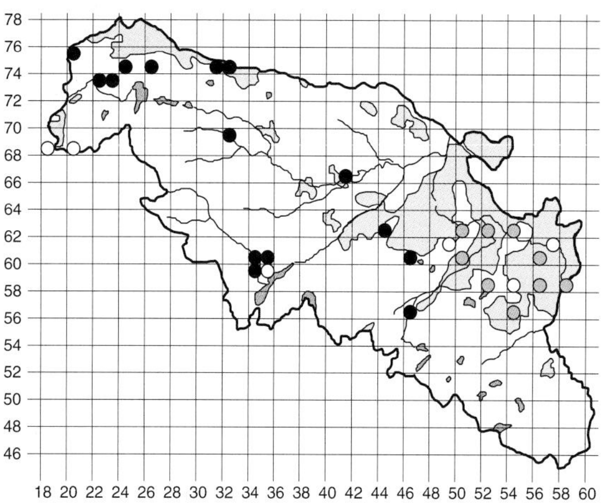

Milium effusum

NHSG), Crookston Castle (1889, AANS), Hawkhead (1895, PSY) and Ravenscraig (1923, RM). Today it is scattered throughout the area with a cluster of sites in the north-west about Spango (NS2373, 1998), River Calder (NS3460, 2006) and south-west Glasgow (CFOG). It is always associated with semi-natural or older plantation woodlands.

Schedonorus pratensis (Huds.) P. Beauv.
Festuca pratensis Huds.
Meadow Fescue

Britain: common except in N Scotland.
Native; 43/89; frequent in the SE, perhaps overlooked.
1st Paisley, 1865, PSY.

Meadow Fescue appears to have undergone a decline since the 19th century and was better recorded in the 1960s–80s, and during the CFOG period, than more recently. It is a grass of less-improved meadows and riverbanks and the decline likely reflects changes in agricultural practice. Today, although it is probably overlooked, most records are from the south and east, often along watercourses or at wayside grasslands, rather than traditional meadows.

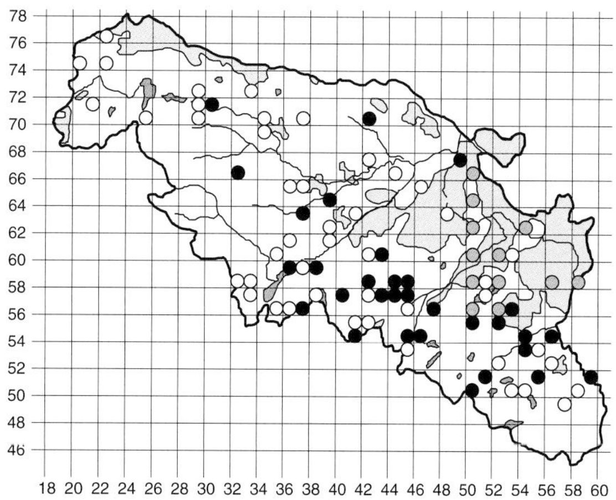

Schedonorus pratensis

Schedonorus arundinaceus (Schreb.) Dumort.
Festuca arundinacea Schreb.
Tall Fescue

Britain: common except in N Scotland.
Native; 73/84; frequent, lowland.
1st St Bride's Mill, Kilbarchan, 1845, NSA.

Tall Fescue can no longer be said to be very common, as Lee described it in 1933, but it is still widespread with a strong preference for the lowlands, and notably the north-west coast. The decline is due in part to the decrease in hayfields as fields are cut for silage. It can still be found on farms but usually at the less intensive margins, and also along riverbanks, at some waste-ground grasslands and can be common by the strandline at the coast.

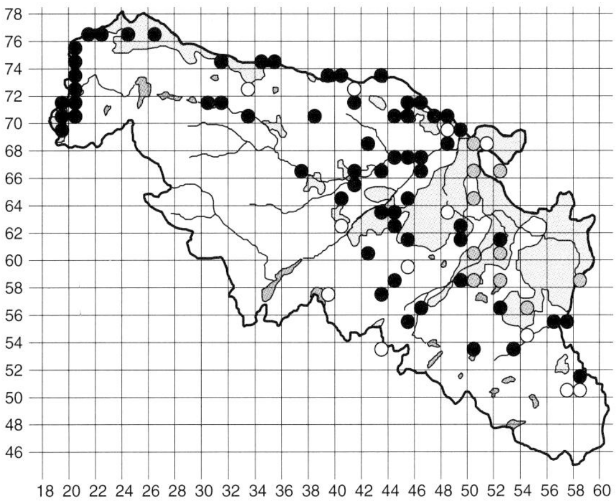

Schedonorus arundinaceus

Schedonorus giganteus (L.) Holub.
Festuca gigantea (L.) Vill.
Giant Fescue

Britain: common except in N Scotland.
Native; 93/113; frequent, lowland.
1st St Bride's Mill, Kilbarchan, 1845, NSA.

Giant Fescue was known 'about Glasgow' (1821, FS) but Gourock is the only local place mentioned by Hennedy (1891), although he did note it as frequent. Today it is widespread but almost exclusively confined to wooded watercourses.

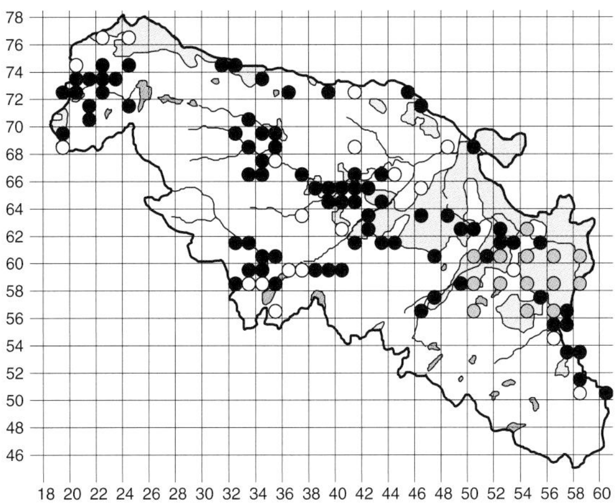

Schedonorus giganteus

× *Schedolium* × *loliaceum* (Huds.) Holub
(*Schedonorus pratensis* × *Lolium perenne*)
× *Festulolium loliaceum* (Huds.) P. Fourn.
Hybrid Fescue

Britain: frequent in the S, scattered in S and C Scotland.
Native; 2/2; very rare.
1st Commore Dam, 1998.

Hooker (FS 1821) described this hybrid grass as occasional ('about Glasgow') and Hennedy (1891) mentioned var. *loliacea* as being 'not common' but named no local sites. Today Hybrid Fescue has only been recorded from Commore Dam (NS4654) and by the Earn Water at Waterfoot (NS5654, 2003).

Lolium perenne L.
Perennial Rye-grass

Britain: common throughout.
Native and hortal; 282/378; very common in lowlands.
1st Near Hurlet, 1842, GLAM.

Hennedy (1891) considered Perennial Rye-grass as very common in meadows, pastures and waysides, and today it is still common and widespread and found in the same places, but is notably absent from the highest ground. It is now very commonly sown as short-term leys on farmland and as an amenity turf in parks, but can still be found in old agriculturally improved pastures or semi-improved neutral grasslands; it is also frequent on waste ground.

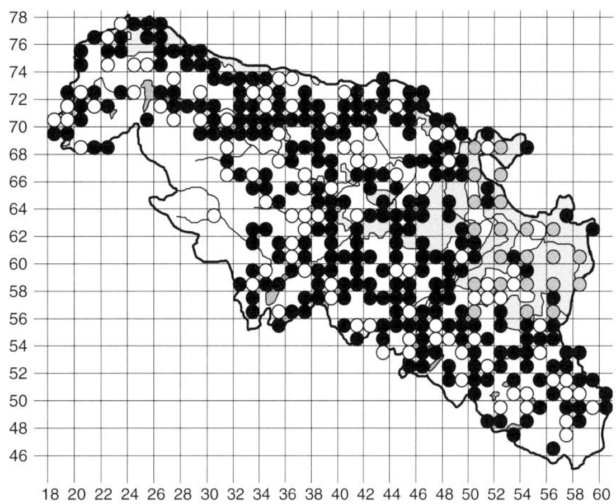

Lolium perenne

Lolium multiflorum Lam.
Italian Rye-grass

Britain: neophyte (S Europe); common, rare in NW Scotland.
Hortal; 10/18; rare.
1st Clarkston, 1926, GL.

Hennedy (1891) stated that Italian Rye-grass was never found except where cultivated. Later it was recorded from Ralston and Kilbarchan (1944, TPNS).

It was found on several occasions in the 1960s–80s period and during CFOG surveys. Probably overlooked, the four most recent records are from waste ground, Inchinnan (NS4768, 1996), an arable field, Craigmuir (NS4960, 1993), Barcraig Wood (NS3764, 1998) and Gryfe Wraes (NS3966, 1998). The hybrid with Perennial Rye-grass (*L.* × *boucheanum* Kunth) was reported from allotments, Merrylee, in 1942 by RM.

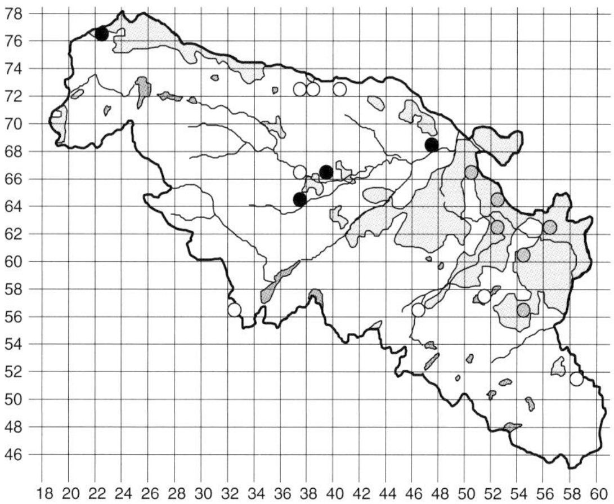

Lolium multiflorum

Lolium temulentum L.
Darnel

Britain: archaeophyte (Mediterranean); scattered in the S.
Former hortal; 0/2; extinct.
1st Renfrewshire, 1895, NHSG.

The first record is accredited to R and T Wilkie, who collected Darnel from Giffnock (GL – undated but likely *c.* 1892); it was also collected from Glenkilloch in 1904 (PSY), which appears to be the last record.

Festuca rubra L.
Red Fescue

Britain: common throughout.
Native; 404/465; very common.
1st Renfrewshire, 1872, TB.

Past distribution is unclear due to taxonomic confusion in the 19th century, but today it is obvious that this fescue is widespread and very common, and presumably always has been. Red Fescue occurs in various, usually open, habitats: it is common, often dominant, in lowland grasslands, where not heavily improved, and even on waste ground, usually where freely draining, and it can persist in the shade of secondary scrub. However, it can also be found in upland grasslands and in some mires; it is also present along the coastal shores and at relict saltmarshes. Most recording has been at the species level but a few of the tentative subspecies records have been checked by Professor Clive Stace.

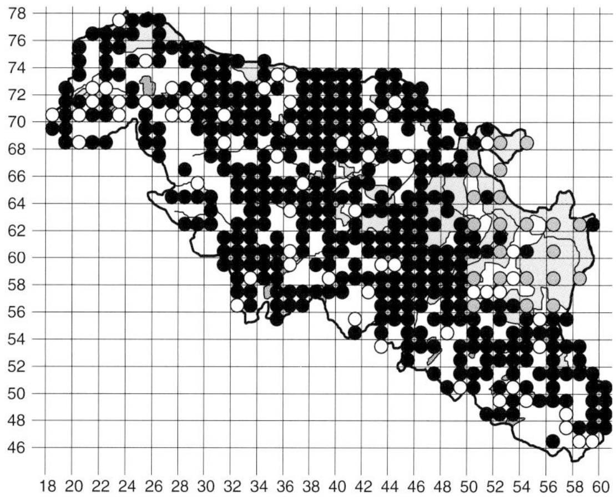

Festuca rubra

Festuca rubra ssp. *juncea* (Hack.) K. Richt.

There are two confirmed records for this subspecies from Misty Law (NS2862, 1997) and Wardlaw Burn, Gleniffer Braes (NS4660, 1995). There have also been a several other tentative records from upland mires, e.g. at Long Wood (NS5452, 1998) and Witch Burn (NS4557, 1999). Coastal forms with markedly pruinose leaves (ssp. *pruinosa* (Hack.) Piper) are occasionally found about the tide line. One specimen has been determined as such, from the shore line at Lunderston Bay (NS2074, 1998), and other field records include Blythswood (NS5068, 2005) and Wemyss Bay (NS1869, 2012), indicating a likely wide coastal distribution.

Festuca rubra ssp. *litoralis* (G. Mey.) Auquier

This subspecies has been confirmed from saltmarshes at Lunderston Bay (NS2074, 1998) and Erskine Park

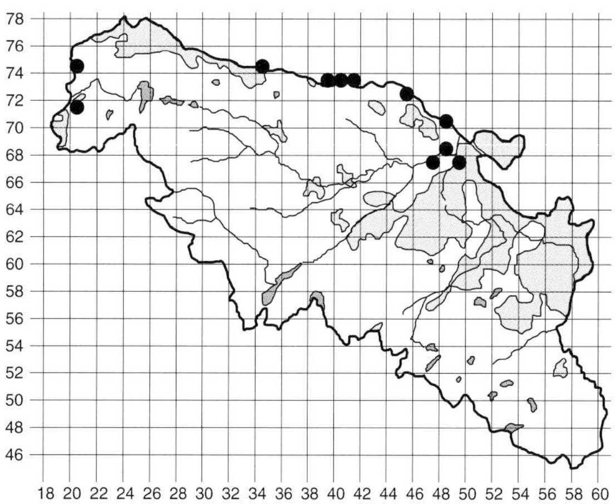

Festuca rubra ssp. *litoralis*

(NS4572, 1995); there are nine other saltmarsh records (unconfirmed) from the coast and also along the estuarine Black Cart and White Cart waters.

Festuca rubra* ssp. *commutata Gaudin
Chewing's Fescue

This subspecies has been recorded from 12 mostly urban tetrads, mainly in the east.

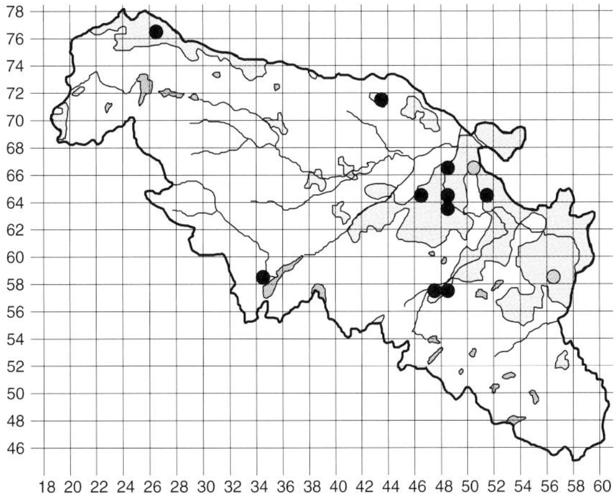

Festuca rubra ssp. *commutata*

Festuca rubra* ssp. *megastachys Gaudin

This tall neophyte was first recorded in 1977 from Williamwood (NS55U, CFOG) and has since been recorded from 28 mainly urban localities.

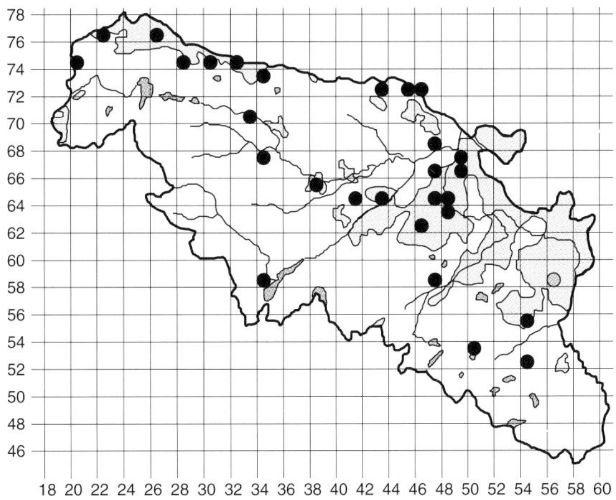

Festuca rubra ssp. *megastachys*

Festuca ovina L.
Sheep's-fescue

Britain: common throughout.
Native; 303/369; very common in upland fringes.
1st Renfrewshire, 1869, PPI.

Sheep's-fescue is one of the commoner grasses of upland unimproved grasslands. It is a feature of acidic grasslands but also found in heaths and at the margins of some mires; occasionally it can be found on lowland waste ground. There are 18 widespread records for hairy forms – ssp. *hirtula* (Hack. ex Travis) M.J. Wilk.

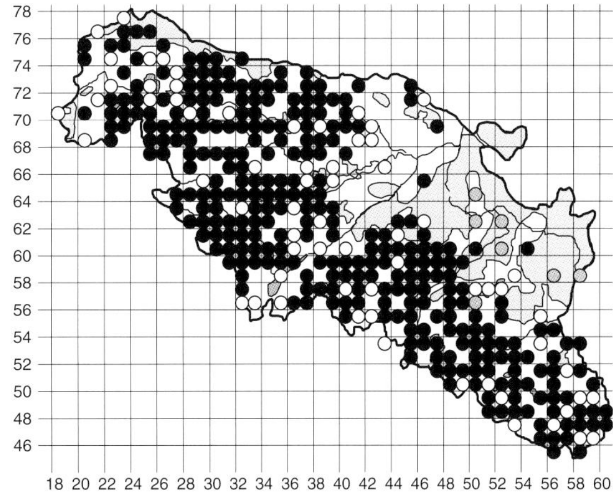

Festuca ovina

Festuca vivipara (L.) Sm.
Viviparous Sheep's-fescue

Britain: common in NW Scotland, and uplands further S.
Native; 22/25; occasional in uplands.
1st Calder Dam, 1976, BWR.

Earlier authors make only passing reference to 'var. *vivipara*' but no VC76 localities are named, although it must have been frequent in the uplands. BWR recorded Viviparous Sheep's-fescue from eight upland sites in the 1960s–80s period, including about Darndaff (NS2772, 1977) and Greenock Cut (NS2674, 1965), where it has not be seen since. Today it is quite well represented in the Renfrewshire Heights but

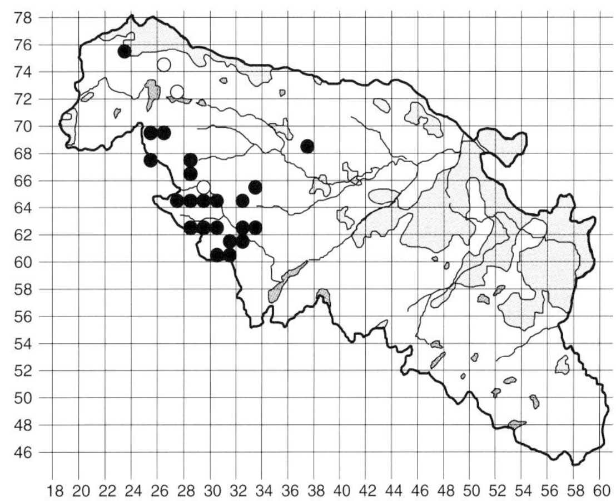

Festuca vivipara

there are no records from the south-eastern uplands. It is also known from a couple of isolated sites: a rocky outcrop at North Barlogan (NS3768, 1994, CD) and Earns Hill (NS2375, 1998).

Festuca filiformis Pourr.
Fine-leaved Sheep's-fescue

Britain: widespread, but unclear because of identification problems.
Native; 45/48; occasional.
1st Brother Loch, 1937, RM.

Fine-leaved Sheep's-fescue has been recorded from several locations across the VC and is presumably widespread, although there may have been confusion with stunted forms of Sheep's-fescue. It is usually encountered on the shallow soils associated with rocky ridges or ledges.

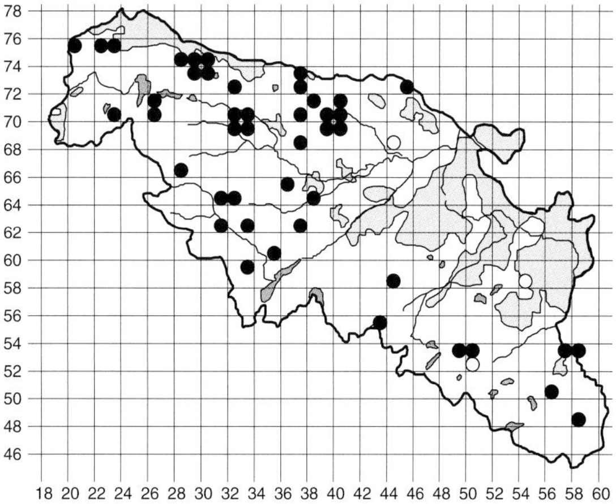

Festuca filiformis

Vulpia bromoides (L.) Gray
Squirreltail Fescue

Britain: common in the S, scarce in Scotland.
Native; 4/9; rare.
1st Renfrewshire, 1869, PPI.

Squirreltail Fescue was considered not common by Hennedy in 1891 and only Gourock was named; other old records include Paisley (1870, PSY), Wemyss Bay (1887, NHSG), allotments, Nether Auldhouse (1945, RM) and lanes, Bogside (1978, BWR). The modern records are from sandy coastal soils at Erskine (NS4572, 1996) and from rocky shallow soils supporting species-rich grassland on the summit plateau of Barscube Hill (NS3871, 1997), Knockmountain (NS3671, 2007) and Linthills (NS3359, 2001).

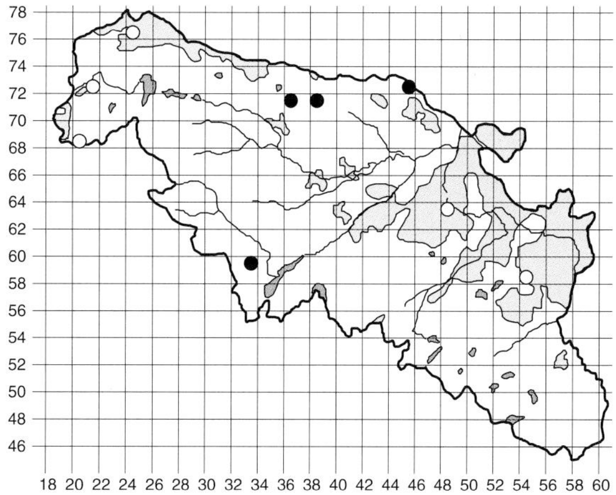

Vulpia bromoides

Vulpia myuros (L.) C.C. Gmel.
Rat's-tail Fescue

Britain: archaeophyte (Eurasia); frequent in the S.
Accidental; 10/11; rare casual.
1st Yoker Ferry, 1985, AMcGS (CFOG).

There are only a few records of this small grass of open waste-ground places. There is a cluster of sites about the streets of Greenock and it is also known from Linwood (NS4463, 1995), Gilmour St, Paisley (NS4864, 2001), Yoker (NS5169, 2011) and Polmadie (NS5962, 2011).

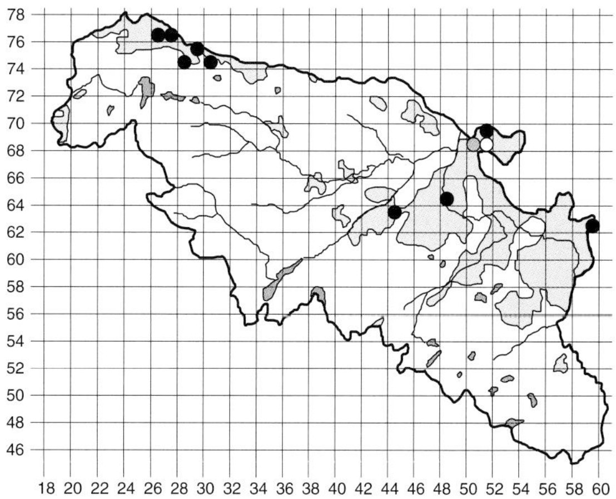

Vulpia myuros

Cynosurus cristatus L.
Crested Dog's-tail

Britain: common throughout.
Native; 328/397; very common.
1st Renfrewshire, 1869, PPI.

Crested Dog's-tail is a widespread grass that only becomes rare or absent from upland peaty soils or on intensive lowland farmland. It occurs in a range of grasslands, often indicative of semi-improvement, especially in upland pastures; it can also be found on waste ground, often where clayey and compacted.

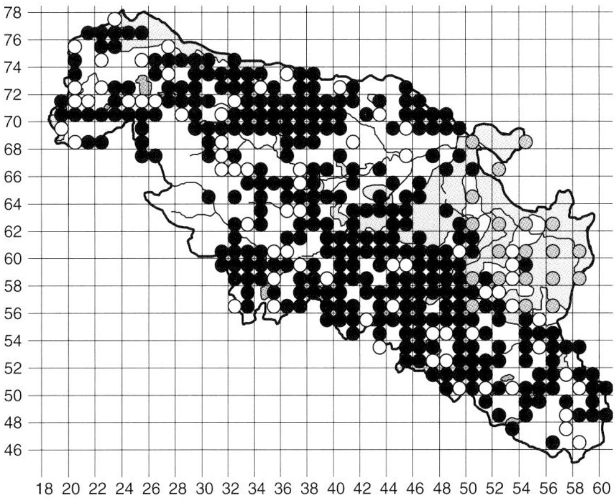

Cynosurus cristatus

Cynosurus echinatus L.
Rough Dog's-tail

Britain: neophyte (S Europe); scarce, mainly in the extreme S.
Former accidental; 0/1; extinct casual.
1st Crossmyloof, 1892, NHSG.

Puccinellia maritima (Huds.) Parl.
Common Saltmarsh-grass

Britain: common, coastal.
Native; 23/25; frequent on the coast.
1st Renfrewshire, 1834, MC.

Hennedy (1865) described this grass as common around the coast. Today Common Saltmarsh-grass is more scattered and isolated but can still be frequently encountered on tidal mud and rocks, although often in small patches. It extends up the brackish White Cart and Black Cart waters to Porterfield (NS4967, 2001) and Inchinnan (NS4868, 2005) respectively. It was also noted as growing on abandoned boat ramps at Clydeport (NS3174, 2002).

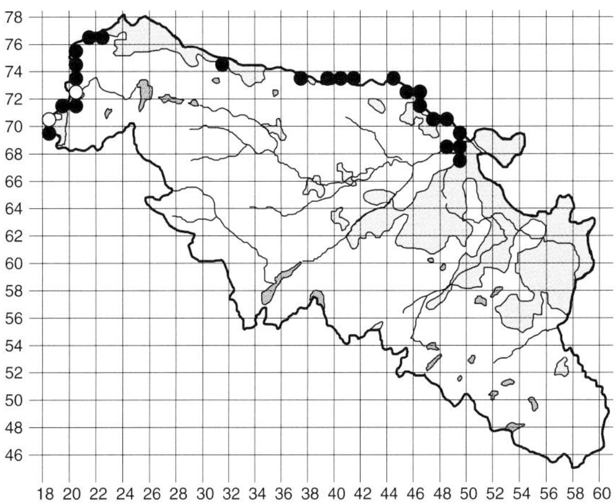

Puccinellia maritima

Puccinellia distans (Jacq.) Parl.
Reflexed Saltmarsh-grass

Britain: coastal, mainly in the E, increasing inland as alien.
Native or accidental; 5/6; rare.
1st Crossmyloof, 1894, GL.

This coastal grass has been recorded from inland roadsides in the Glasgow area in recent years (CFOG). Two of the more recent records are from such places: roadside, Dod Hill (NS4953, 1998) and roadside, Shielhill (NS5148, 2001). There are also a couple of coastal records from the Erskine area: saltmarsh at Bottombow (NS4671, 1997) and by the slipway (NS4672, 1998, AMcGS), but two others from this area, from a roadside at Parks Mains (NS4669, 1997) and by a seeded new path edge, (NS4671, 1998), could be introductions or represent local spread.

Briza media L.
Quaking-grass

Britain: common in the S, rare in N and W Scotland.
Native; 1/3; very rare.
1st Lochwinnoch, 1845, NSA.

Quaking-grass was collected from Ferguslie in 1903 (PSY) and MY noted it in 1865 but gave no location. This distinctive lime-loving grass had no other records until A Hamilton, a CMRP ranger, recorded it at the edge of Shielhill Glen (NS2471, 1996).

Poa annua L.
Annual Meadow-grass

Britain: common throughout.
Native; 348/455; very common.
1st Renfrewshire, 1869, PPI.

Annual Meadow-grass is found throughout the area

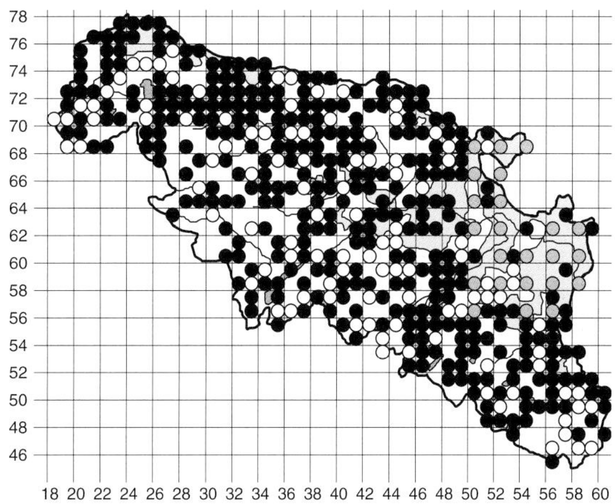

Poa annua

at sites including urban waste ground, streets, lawns and parks, and it can occur in upland grasslands, usually where sheep gather; it also grows in farmland grasslands, usually where more disturbed, and along waysides and to the open margins of rivers and lochs or reservoirs.

Poa trivialis L.
Rough Meadow-grass

Britain: common throughout.
Native; 368/431; very common.
1st Paisley, 1865, PSY.

Rough Meadow-grass is a very common and widespread grass growing in many habitats usually where not very acidic. It occurs in most neutral grasslands including marshy pastures or improved leys. It is also found in woodlands, along open or shaded riverbanks and commonly at waste-ground sites.

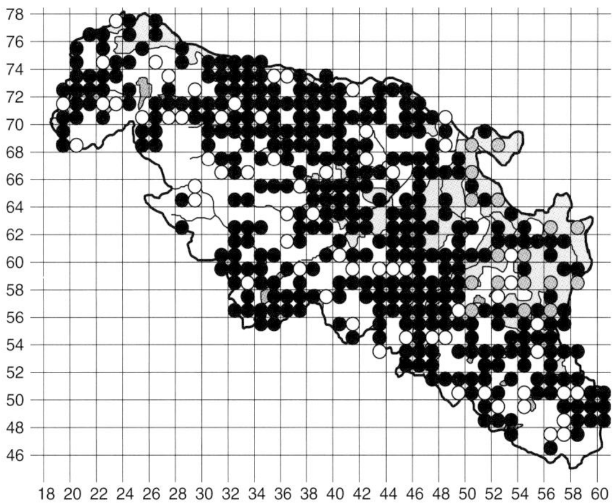

Poa trivialis

Poa pratensis s.l.

Britain: common throughout.
Native and alien; 312/409; very common.
1st Renfrewshire, 1869, PPI.

The map shows combined records for the following two meadow-grasses, and includes many 1960s–80s records, and some from CFOG, recorded as *Poa pratensis* s.l. However, it shows the former and current widespread occurrence of this grass, which is a feature of less acidic pastures and meadows but is also found on waste ground and at waysides and hedge banks.

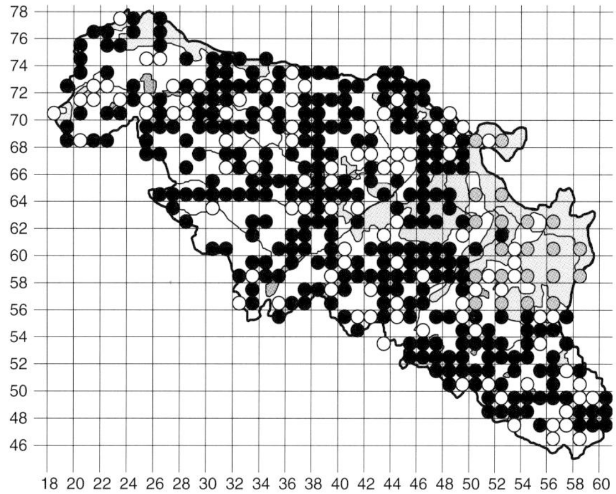

Poa pratensis s.l.

Poa humilis Ehrh. ex Hoffm.
Spreading Meadow-grass

Britain: common, especially in the N and W.
Native; 241/244; common in rural areas.
1st Daff Reservoir, 1973, BWR.

Spreading Meadow-grass is more frequently recorded than Smooth Meadow-grass, although distinguishing the two can be difficult. The plants referred to here tend to be short-growing pasture plants most typically found in less acidic bent-fescue grasslands in rural areas to the upland fringes, but many records also refer to wayside plants. It also occurs on sandy soil to the Clyde edge at Erskine (NS4572, 1996) and on waste ground as at Port Glasgow (NS3174, 2001).

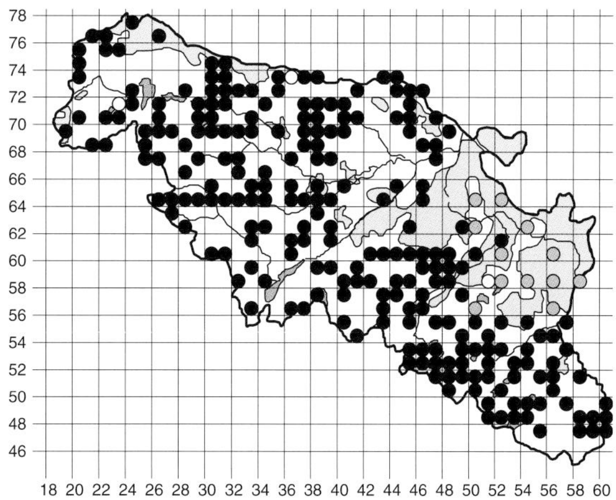

Poa humilis

Poa pratensis L.
Smooth Meadow-grass

Britain: common and widespread, except in the NW.
Native and alien; 82/82; common in lowlands.

Records for Smooth Meadow-grass are widespread but

distinctly lowland. Most records are for taller-growing grasses, often on waysides, hedge banks, riverbanks or waste ground, but distinguishing this species from Spreading Meadow-grass is not always possible.

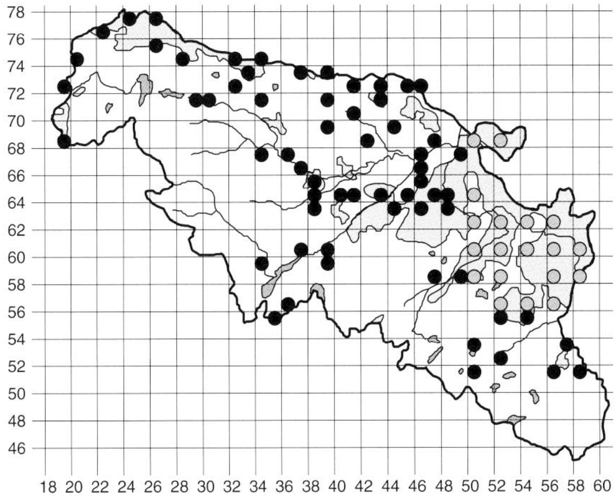

Poa pratensis

Poa angustifolia L.
Narrow-leaved Meadow-grass

Britain: frequent in the SE.
Native or accidental; 1/4; very rare or overlooked.
1st Garden weed, Pollokshaws, 1962, GLAM.

The recognition and status of this species was discussed in CFOG where it was noted that this grass is locally rare, although possibly overlooked. The most recent record for Narrow-leaved Meadow-grass is that of AJS from the M8 at Hillington (1988, CFOG). There are also anonymous file notes for two records from a railway survey: from near Quarriers, Kilmacolm (NS3766, 1980) and Dunrod (NS2272, 1980), but there are no further details.

Poa chaixii Vill.
Broad-leaved Meadow-grass

Britain: neophyte (Europe); scattered, more in the N.
Hortal; 9/11; rare.
1st Lochwinnoch, 1958, RM.

Broad-leaved Meadow-grass was also recorded from White House, Kilbarchan, in 1976 (RM) and from five relevant localities during CFOG surveys. More recently it was found in the original Lochwinnoch area (NS3558) in 2003 by IPG and it is also known from nearby Lochside House (NS3658, 1994, CD). Two other adjacent records are from the woodland in Kilmacolm (NS3569, 1997) and from nearby Duchal House estate woodlands (NS3568, 2000).

Poa compressa L.
Flattened Meadow-grass

Britain: common in the S.
Native or accidental; 12/16; scarce.
1st Crossmyloof, 1892, NHSG.

This grass was considered rare in the 19th century, although it was known 'about Glasgow' in 1821 (FS); by 1930 RG noted Flattened Meadow-grass as being 'frequent and well established on dry waste ground round Glasgow'. RM reported it from a quarry at Mearns in 1950 and collected it from waste ground in Clarkston in 1945 (GLAM); it was reported from Bishopton Station in 1975 (BWR) and there are a few relevant records in CFOG. Other modern records include waste ground, such as at Boylestone Quarry (NS4959, 1999), and disused railways at Millarston (NS4563, 1993) and Kilbarchan (NS4264, 2003), with outliers at Lochwinnoch Station (NS3557, 2005, IPG) and waste ground, Port Glasgow (NS3075, 2012).

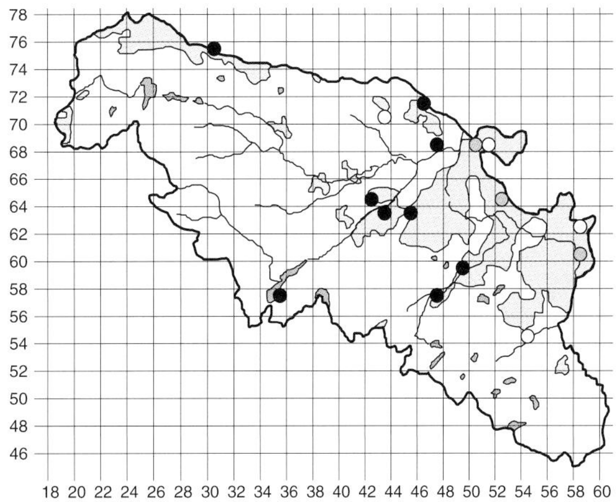

Poa compressa

Poa palustris L.
Swamp Meadow-grass

Britain: neophyte (Europe); scattered.
Accidental; 0/1; very rare.
1st Erskine Bridge, 1979, AMcGS (E).

Swamp Meadow-grass may be overlooked, but the record from waste ground by the Erskine Bridge (NS4572) is the only one.

Poa nemoralis L.
Wood Meadow-grass

Britain: common except in NW Scotland.
Native; 57/74; frequent.
1st Kelly Glen, 1872, TB.

Wood Meadow-grass was considered common by Hennedy (1891). The map shows a widespread distribution, although there are few modern records from the north-west, where it may have been overlooked, and it is not found in the more acidic uplands. It is most likely to be encountered in shaded woodlands, often along watercourses, but can sometimes be found on waste ground as at Port Glasgow (NS3174, 2001) and Underheugh (NS2075, 2012).

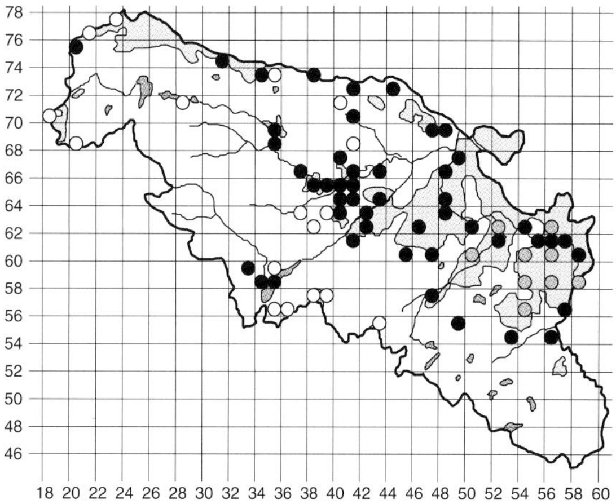

Poa nemoralis

Dactylis glomerata L.
Cock's-foot

Britain: common throughout except in the uplands of N Scotland.
Native; 341/421; very common.
1st Paisley, 1869, RPB.

Cock's-foot can still be described as very common, as it was by Hennedy in 1891. It is widespread but notably absent from the uplands of the Renfrewshire Heights, but less so from the south-east uplands. It is found in rough rank grasslands of abandoned pastures and also on long-neglected waste ground; it is also common along waysides and riverbanks and can often be found in woodlands, although seldom in any abundance.

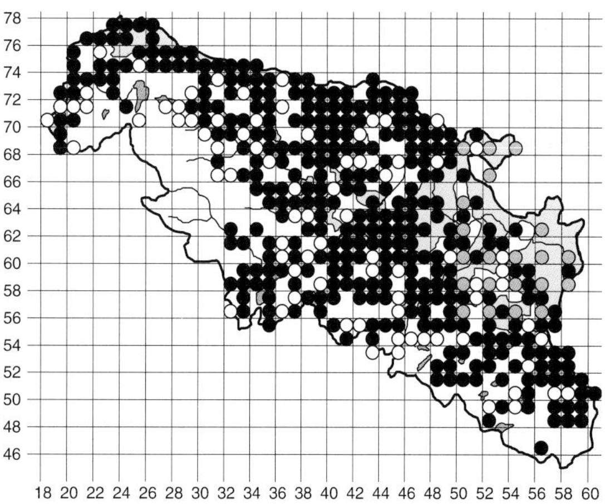

Dactylis glomerata

Dactylis polygama Horv.
Slender Cock's-foot

Britain: neophyte (Europe); very rare.
Former hortal or accidental; 0/1; extinct but possibly overlooked.
1st Rouken Glen, 1960, GLAM.

The specimen collected by RM is annotated as '1st Scottish Record', but it has not been seen in the VC since, or indeed anywhere else in Scotland.

Catabrosa aquatica (L.) P. Beauv.
Whorl-grass

Britain: occasional, more coastal in Scotland.
Native; 4/6; rare.
1st Dennistoun, Kilmacolm, 1890, NHSG.

Whorl-grass was considered rare in the 19th century, although Professor T King described var. *littoralis* as 'common on sandy shores of the Firth' (King 1892). In 1975 it was recorded from Dod Hill (BWR). Today, though perhaps overlooked, Whorl-grass is still rare, known from three inland sites in the south-east: a small flushed drain, West Carswell (NS4953, 1999), floating swampy vegetation at Kirkton Dam (NS4856, 1999) and a vegetated drain, Langton (NS5054, 2004). The only coastal record is from Lunderston Bay (NS2073, 1989, AMcGS), presumably indicating a dramatic decline.

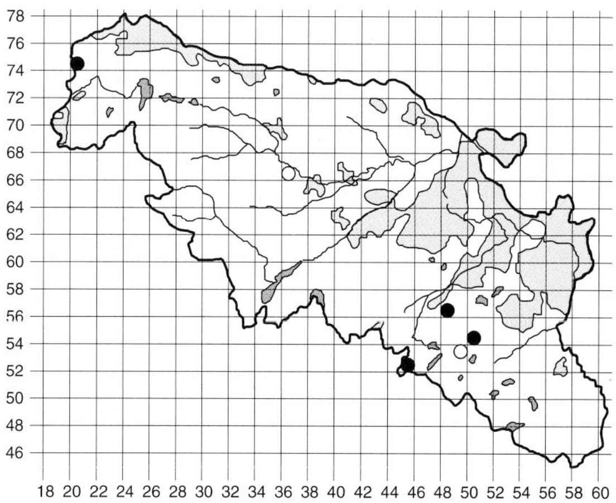

Catabrosa aquatica

Catapodium marinum (L.) C.E. Hubb.
Sea Fern-grass

Britain: frequent on coasts, rarer in N Scotland.
Native or accidental; 1/2; very rare.
1st One mile south of Gourock, 1968, JP.

A small population of Sea Fern-grass on a roadside just east of Glasgow Airport (NS4866, 2006) was a surprising inland find for this coastal grass.

Parapholis strigosa (Dumort.) C.E. Hubb.
Hard-grass

Britain: coastal, only in the S in Scotland.
Former native; 0/3; extinct.
1st Gourock, 1878, GL.

A specimen was also collected in 1880 by P Ewing from a 'marsh at sea level, Gourock' and in 1880 it was noted west of Greenock (NHSG), presumably the same place. In 1882 (NHSG), a record 'of some years previously' by T Scott reported Hard Grass as 'growing in abundance in salt marshes on the shore above Gourock'. The records are repeated in later literature but there have been no modern sightings.

Avenula pubescens (Huds.) Dumort.
Helictotrichon pubescens (Huds.) Pilg.
Downy Oat-grass

Britain: frequent but rare in SW England and N and E Scotland.
Native; 3/6; very rare.
1st Cathcart, 1821, FS.

Downy Oat-grass was reported from Shielhill Glen in 1881 and again in 1888 (NHSG), but the next record was not made until 1977 when it was found along the Earn Water near Windhill (NS5654, BWR) and also nearby at Burnhouse (NS5554, 1983, BWR). Today it can still be seen on rocks by a small waterfall at the latter site (2005). Another record is from a similar location along the Gryfe Water at Cauldside (NS3270, 2005) and it was recently found on a coastal rocky outcrop at Cloch (NS2071, 2006).

Arrhenatherum elatius (L.) P. Beauv. ex J. & C. Presl
False Oat-grass

Britain: common throughout.
Native; 292/347; very common.
1st Lochwinnoch, 1845, NSA.

False Oat-grass is a very common tall grass found throughout the lowlands, but which becomes rare or absent from poorer upland soils. It can be found in woodlands but is most readily encountered along hedge banks and as a dominant of drier, coarse or rank grasslands in urban situations or on neglected farmland. A couple of records refer to Onion Couch (var. *bulbosum* (Willd.) St-Amans) which is likely to be common in the lowlands.

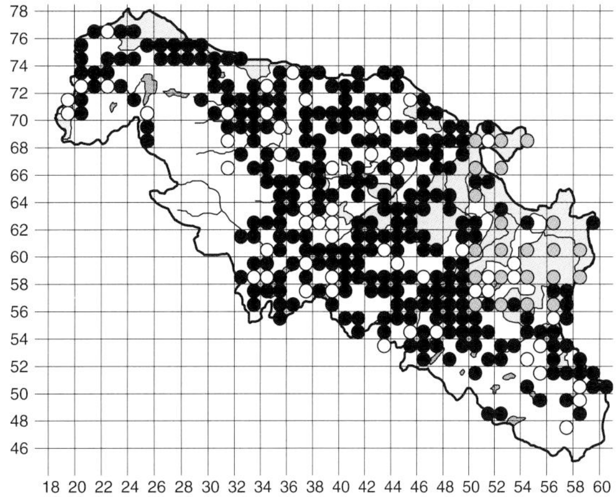

Arrhenatherum elatius

Avena strigosa Schreb.
Bristle Oat

Britain: neophyte (Spain); a few very scattered, lowland records.
Former hortal or accidental; 0/1; extinct casual.
1st Cathcart, 1865, RH.

Avena fatua L.
Wild-oat

Britain: archaeophyte (SE Europe); common in the SE, occasional in lowland Scotland.
Accidental; 3/5; very rare.
1st Cornfield, Whinhill, 1880, NHSG.

Wild-oat was considered rare by Hennedy (1891) and there are no other old records for it except for the collection from wasteland, Paisley (1910, PSY). In the modern period it has been seen on three occasions: by the Black Cart Water, Inchinnan (NS4767, 1999), on waste ground, Barrangary Tip (NS4469, 2001) and in a barley field, Knowes (NS4366, 2002).

Avena sterilis L. ssp. *ludoviciana* (Durieu) Gillet & Magne
Winter Wild-oat

Britain: neophyte (S Europe); frequent in SE England, very rare in the N.
Former accidental; 0/1; extinct casual.
1st Near Inchinnan, 1970, JP.

Avena sativa L.
Oat

Britain: neophyte (W Mediterranean); scattered in lowlands.
Hortal; 11/13; scarce.
1st Barrhead, 1960, BWR.

Not mentioned by earlier authors, Oat was also

recorded from Inverkip (1960, BWR) and it was observed a few times during CFOG surveys. Today it is known from eight sites, mainly roadsides or waste ground: Kirkton (NS3868, 1996), Greenock (NS2875, 1998), Ferguslie Park (NS4664, 2000), Barrangary Tip (NS4469, 2001), Jordanhill (NS5468, 2002, AR), Lunderston Bay (NS2074, 2008) and barley fields at Knowes (NS4366, 2002) and East Fulwood (NS4567, 2007).

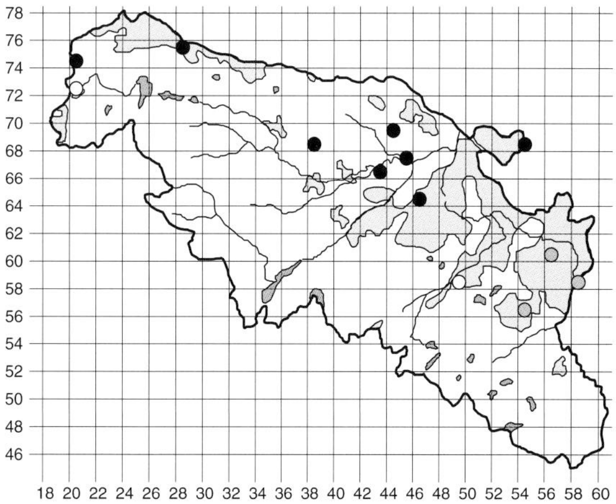

Avena sativa

Trisetum flavescens (L.) P. Beauv.
Yellow Oat-grass

Britain: common except in N and W Scotland.
Native or accidental; 1/7; very rare.
1st Renfrewshire, TB, 1872.

There are several old records for this grass but only one modern find: Formakin Estate (NS4170, 2003, IPG); perhaps it has been overlooked but it is likely to be very rare as a native. Old localities include Johnstone (1876, FFWS), Rouken Glen (1885, NHSG and 1961, RM), Barfod (1895, NHSG), Mearns (1917, GL), Braidbar and Giffnock (1946, RM) and Auchenames (NS3962, 1976, BWR).

Deschampsia cespitosa (L.) P. Beauv.
Tufted Hair-grass

Britain: common throughout.
Native; 458/506; very common.
1st Langside, 1854, GGO.

Tufted Hair-grass is a very common and widespread grass of damp places only becoming rare or absent in the more acidic peaty uplands or strongly urban areas, although it is often found on waste ground. It can dominate large areas of damp grassland, notably in the drier fringes of rush pastures and wetlands, and can also be common in wet woodlands.

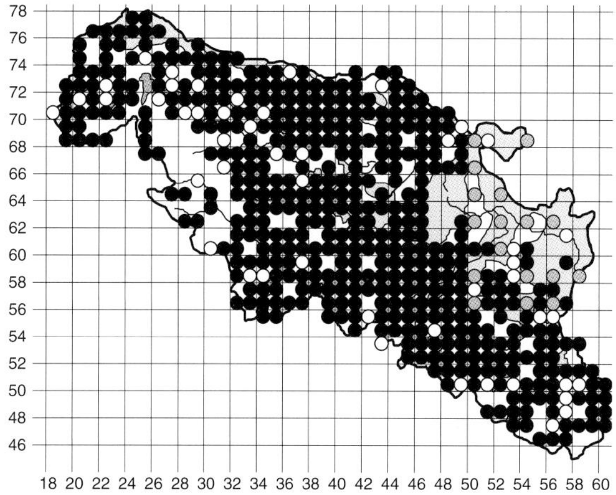

Deschampsia cespitosa

Deschampsia flexuosa (L.) Trin.
Wavy Hair-grass

Britain: common throughout except in SE England.
Native; 352/399; common.
1st Lochwinnoch, 1845, NSA.

A common and widespread grass of acidic soils found throughout the area, extending into the peaty uplands but rare in some urban areas and in lowland improved farmland. Wavy Hair-grass is typically found in more acidic grasslands, usually in the uplands, but it also occurs in the lowlands on sandy soils or peat, and often on north-facing rock outcrops; it is also common in upland bogs or mires and can dominate the ground flora in shaded acidic oak, beech or birch woodlands.

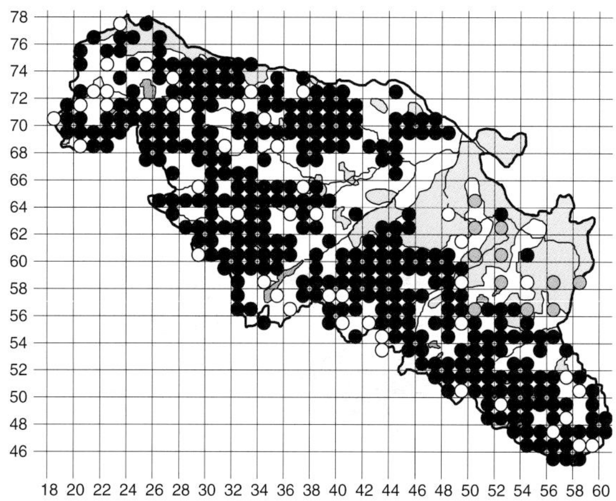

Deschampsia flexuosa

Holcus lanatus L.
Yorkshire-fog

Britain: common throughout.
Native; 457/509; very common.
1st Shawlands, 1857, GGO.

Yorkshire-fog is one of the commonest grasses, found throughout the area, which only becomes rare in the peaty uplands. It is most often encountered in poorly draining pastures, often where rushy, but can also be found in most other grasslands and on waste ground and scrub woodlands.

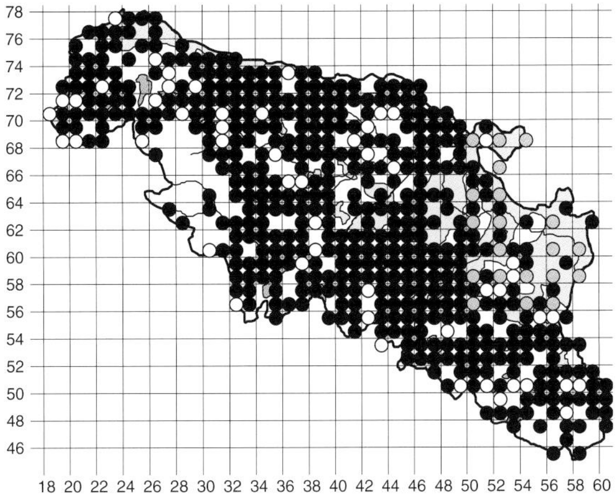

Holcus lanatus

Holcus mollis L.
Creeping Soft-grass

Britain: common throughout.
Native; 396/463; very common.
1st Langside, 1856, GGO.

Like the previous species, Creeping Soft-grass is widespread and very common, only becoming rare in the peaty uplands. It can be abundant in oak-type woodlands, where it can form extensive carpets; it also occurs in grasslands, often where damp or flushed, and can invade acid grassland pastures, often becoming abundant when not grazed, e.g. at many golf-course roughs.

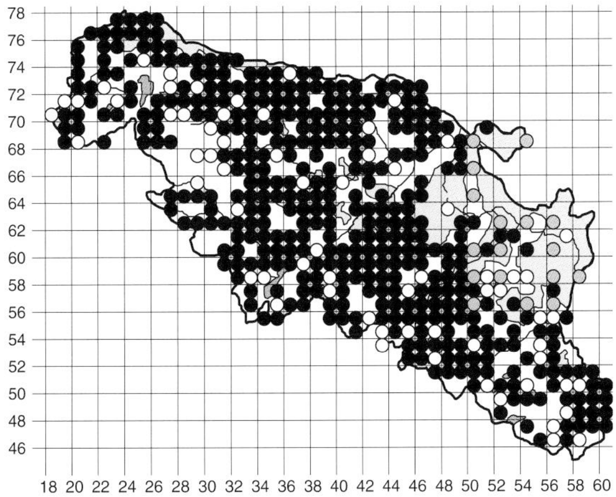

Holcus mollis

Aira caryophyllea L.
Silver Hair-grass

Britain: widespread.
Native; 21/28; occasional.
1st Paisley Canal bank, 1865, RH.

Although considered frequent in the 19th century there are only a couple of other named old locations for this delicate grass: Barochan (1882, PSY) and Loch Libo (1887, NHSG). Silver Hair-grass is widespread and perhaps overlooked. It is found on waste ground, as at Pilmuir Quarry (NS5154, 2000), Ferguslie Park (NS4664, 2000) and Ladyburn (NS3075, 1998), and also along the disused railway (now a cycleway) between Scart (NS3667, 2001) and Auchenbothie (NS3271, 2005). It is also found in more 'natural' settings as along the Levern Water at Harelaw (NS4654, 1998) and Glenbrae (NS2872, 1999). AMcGS recorded ssp. *multiculmis* (Dumort.) Bonnier & Layens from a garden in Jordanhill (NS5368, 1983).

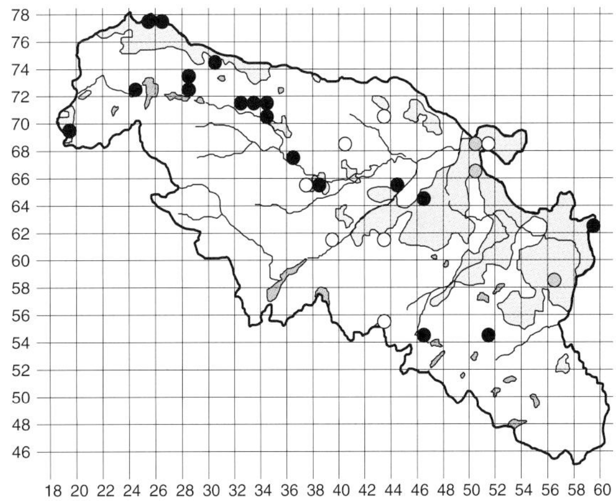

Aira caryophyllea

Aira praecox L.
Early Hair-grass

Britain: widespread.
Native; 130/195; common.
1st Lochwinnoch, 1845, NSA.

Other old records include Gourock (1876, FFWS) and Stanely Reservoir (1882, PSY) and RM recorded Early Hair-grass from near Ravenscraig in 1923. The map shows a widespread distribution with records from urban waste ground, such as at disused railways. It is readily found in upland pastures, most typically on the open, shallow soils of rocky ridges or outcrops.

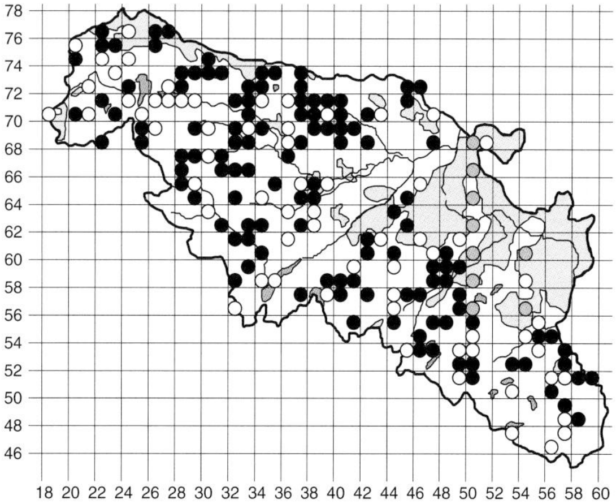

Aira praecox

Hierochloe odorata (L.) P. Beauv.
Holy-grass

Britain: very rare, a few sites in N and S Scotland.
Native; 1/1; very rare.
1st Near Blythswood, 1931, GL.

This nationally rare grass has been recorded and collected several times since Lee's first gathering. Today Holy-grass is known from at least three populations on the east bank of the River Cart at Blythswood (NS4968); it also occurs to the west side on a small island (Tait 2000). It is generally found at the high watermark, where it appears to be able to compete with the vigorous Couch-grass. The location is near the site of an old church, so there may be a link to the traditional use of Holy-grass in church festivals. However, in a Botanical Exchange Club report of 1931 (see Tait 2000), Lee suggested that the plant had not long been at the site and speculated that it arrived with ship's ballast, a perhaps more likely source.

Anthoxanthum odoratum L.
Sweet Vernal-grass

Britain: common throughout.
Native; 424/495; very common.
1st Renfrewshire, 1869, PPI.

Sweet Vernal-grass is one of the commonest grasses in the area, although it is seldom found in any abundance. However, it can be found in most unimproved grasslands, whether lowland or upland; it can also be found at waste-ground sites and to the drier margins of marshes.

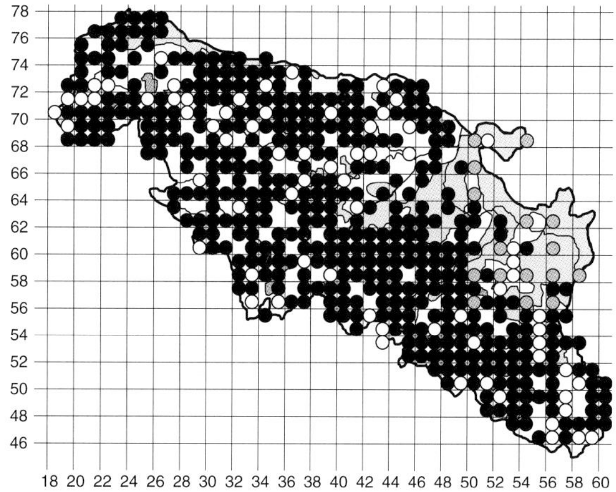

Anthoxanthum odoratum

Anthoxanthum aristatum Boiss.
Annual Vernal-grass

Britain: neophyte (S Europe); rare casual, southern.
Former accidental; 0/1; extinct.
1st Giffnock, 1892, NHSG.

Phalaris arundinacea L.
Reed Canary-grass

Britain: common except in NW Scotland.
Native; 255/283; very common in wet places.
1st Renfrewshire, 1869, PPI.

This tall grass was very common in the 19th century and this remains the case today at wet places in the lowlands. Reed Canary-grass is common along lowland rivers, although it often only forms a narrow fringe at the lower embankments, and can dominate summer-dry 'swamps', often to the exclusion of other species. It is absent from upland peaty places and is fairly scattered in the north-west, reflecting a lack of suitable soils. The variegated hortal (var.

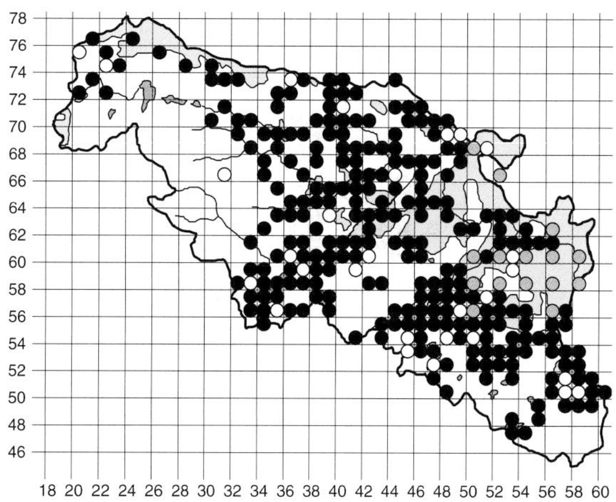

Phalaris arundinacea

'*picta*') has been recorded from Teucheen Wood (NS4869, 1996), Gryfe Wraes (NS3966, 1998), Inverkip (NS2070, 1999) and the White Cart Water at Paisley (NS4863, 2000).

Phalaris canariensis L.
Canary-grass

Britain: neophyte (NW Africa); frequent in the S.
Accidental; 2/11; very rare casual.
1st Renfrewshire, 1869, PPI.

There are a few old records for this grass from places such as Cathcart (1884, GL), Eastwood Cemetery (1908, GL), Inchinnan (1907, GL) and wasteland at Paisley (1910, PSY). More recently Canary-grass was recorded from Greenhags Coup (NS5152, 1966, BWR) and Paisley (NS4863, 1975, AJS). It was found in Yoker (NS5068, 1986, CFOG), but the only other modern record is from Main Street, Lochwinnoch (NS3558), where it was found by IPG in 2003.

Phalaris paradoxa L.
Awned Canary-grass

Britain: neophyte (S Europe); rare in SE England.
Former accidental; 0/1; extinct casual.
1st Crossmyloof, 1892, NHSG.

Agrostis capillaris L.
Common Bent

Britain: common throughout.
Native; 428/457; very common.
1st Langside, 1854, GGO.

Common Bent is one of the commonest plants in the area, being a dominant of the extensive bent-fescue pasture in rural areas; it also occurs on lowland urban waste ground but can extend to the highest hills where not too boggy or acidic. It can occur in some grazed woodlands where not too heavily shaded. It is a variable grass and some records will be of seeded cultivars.

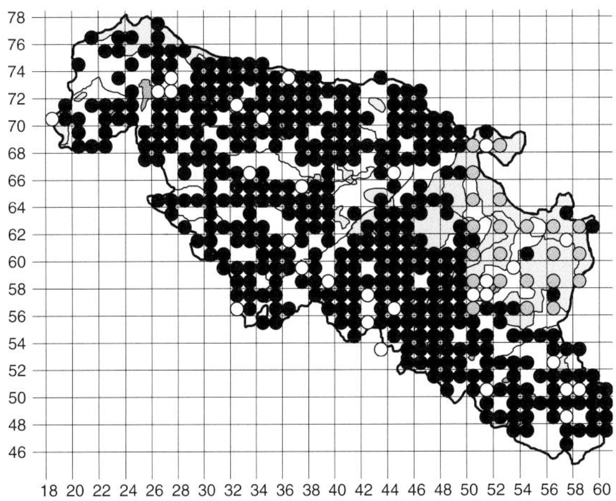

Agrostis capillaris

Agrostis gigantea Roth
Black Bent

Britain: archaeophyte (Eurasia); common in the S, lowland in Scotland.
Accidental; 20/28; occasional.
1st Nether Auldhouse, 1943, RM.

There are no old records for this tall bent, but it was noted eight times between 1962 and 1984 (BWR), and it may well have been overlooked for large cultivars of Common Bent. The modern records for Black Bent are from waste ground, rough grasslands and occasionally arable field margins.

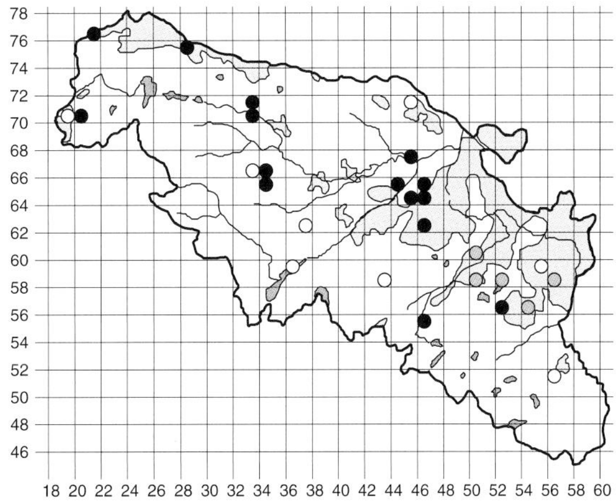

Agrostis gigantea

Agrostis castellana Boiss. & Reut.
Highland Bent

Britain: neophyte (S Europe); occasional in the S.
Hortal; 0/1; extinct, but probably overlooked.
1st Earn Water near Waterfoot, 1977, BWR.

There is only one record for this species, which is often used in seed mixes. It is presumably overlooked.

Agrostis stolonifera L.
Creeping Bent

Britain: common throughout.
Native; 405/416; very common.
1st Renfrewshire, 1883, TB.

Creeping Bent occurs throughout the area in wet places but also on urban waste ground, roadsides and drier grasslands. It is most typically found along watercourses, at the margins of open waterbodies and in marshes, often forming luxurious swamp mats where the waters are more enriched; it can also be found along the coast at the onshore margins of saltmarshes and about the strandline.

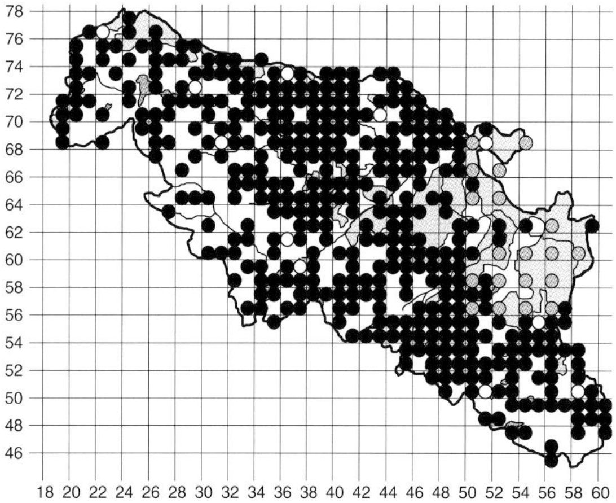

Agrostis stolonifera

Agrostis canina s.l.
Velvet/Brown Bent

Britain: common in uplands and acidic areas.
Native; 279/316; very common in upland areas.
1st Renfrewshire, 1869, PPI.

Early recorders up until the late 1960s did not distinguish Velvet and Brown Bent, although no doubt both were very common. The map shows all records, indicating a widespread rural distribution extending into the uplands.

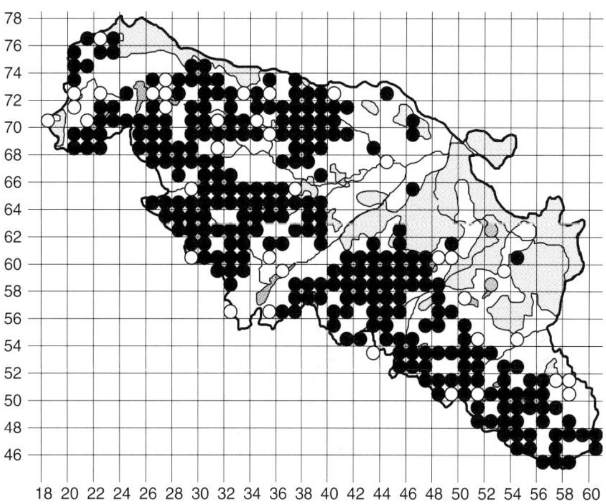

Agrostis canina s.l.

Agrostis canina L.
Velvet Bent

Britain: common throughout in wetter acidic soils.
Native; 239/245; common.
1st Garvock (NS2771), 1968, BWR.

Velvet Bent is a common grass of mires and wetter acidic grasslands, and some wet heaths. The map shows a widespread rural distribution extending into the uplands, with a few sites in urban-fringe areas.

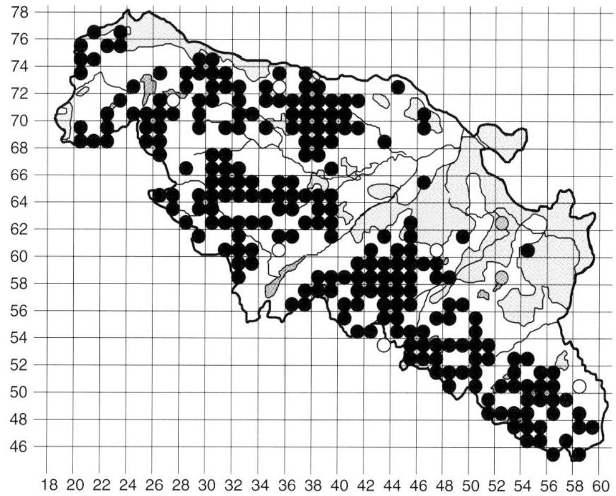

Agrostis canina

Agrostis vinealis Schreb.
Brown Bent

Britain: widespread, upland.
Native; 160/160; common in uplands.
1st Peesweep, 1993.

Brown Bent is a common grass of upland acidic grasslands and heaths, usually found on less marshy ground than the similar Velvet Bent. The map shows a widespread rural distribution, readily marking out the higher ground.

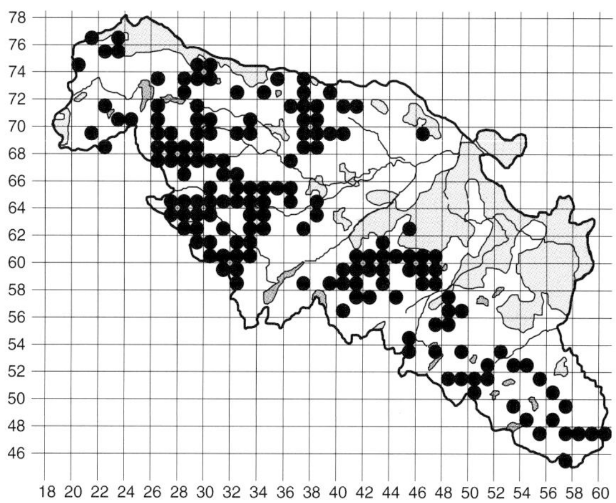

Agrostis vinealis

Agrostis scabra Willd.
Rough Bent

Britain: neophyte (N America); rare casual.
Accidental; 3/4; very rare.
1st Govan Docks, 1977, AJS.

The first record for Rough Bent was from the former railway yard at Govan Docks (NS5764), just inside the VC border. AJS also recorded it from the disused railway at Elderslie (NS4463, 2005) and more recently it was found by a path in Yoker (NS5169, 2011) and on waste ground, Battery Park (NS2577, 2012).

Calamagrostis epigejos (L.) Roth
Wood Small-reed

Britain: common in the SE, scattered in Scotland.
Native; 0/1; extinct.
1st Wemyss Bay, 1981, B Simpson.

This unexpected find, at a wood margin near the caravan site at Kelly Mains (NS1968), is the only one for the VC. This distinctive grass is not mentioned by any earlier recorders or authors

Ammophila arenaria (L.) Link
Marram

Britain: common on coasts throughout.
Native; 0/2; extinct or overlooked.
1st Inverkip, *c.* 1830, GL.

An undated specimen collected by Professor Arnott is the only named 19th-century locality of Marram; Lee (1933) does not record it for the VC. It was recorded in 1983 from Lunderston Bay (NS2074, BWR) but has not been reported since.

Apera spica-venti (L.) P. Beauv.
Loose Silky-bent

Britain: archaeophyte (Europe); occasional in the SE.
Former accidental; 0/2; extinct casual.
1st Crossmyloof, 1892, NHSG.

Loose Silky-bent was also reported from Scotstoun (Wishart 1912).

Polypogon monspeliensis (L.) Desf.
Annual Beard-grass

Britain: native in S, scattered alien elsewhere.
Former accidental; 0/1; extinct casual.
1st Crossmyloof, 1892, NHSG.

Alopecurus pratensis L.
Meadow Foxtail

Britain: common throughout except in NW Scotland.
Native; 205/278; common.
1st Renfrewshire, 1869, PPI.

There are no named old localities for this grass, which was considered very common by Hennedy (1891). Today Meadow Foxtail is widespread although there are few records in the west, perhaps due to the more acidic or shallower soils about the Renfrewshire Heights; it is also absent from similar upland areas in the south-east. It is not as common in farmland meadows as in previous years, due to improvements and silage crops, although it can still be found in marginal grasslands, often along burns and larger rivers; it is also not uncommon on waste ground and wayside grasslands.

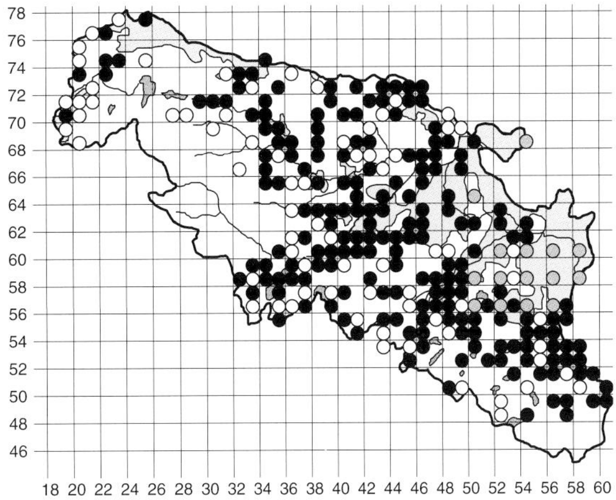

Alopecurus pratensis

Alopecurus geniculatus L.
Marsh Foxtail

Britain: common throughout except in NW Scotland.
Native; 250/333; common.
1st Renfrewshire, 1883, TB.

Marsh Foxtail is a common grass of wet grasslands occurring throughout the area apart from the more acidic uplands, although it can be found at the latter associated with waterbody margins. It is found along

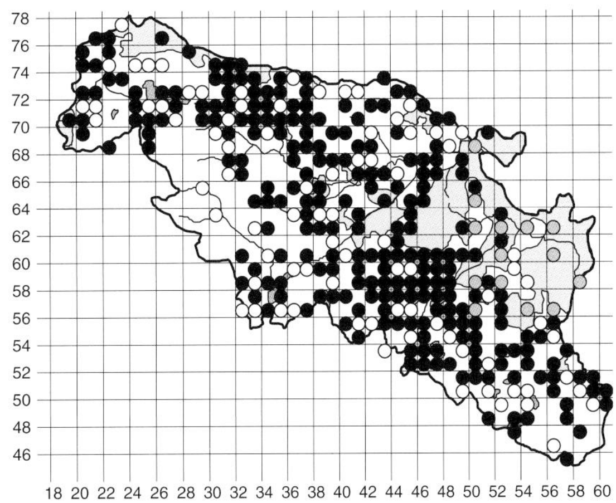

Alopecurus geniculatus

watercourses but is most readily associated with inundation ground, either in pastures or by the margins of waterbodies; it can also be common as an arable weed and is often recorded on waste ground.

Alopecurus myosuroides Huds.
Black-grass

Britain: archaeophyte; common in SE England, rare in Scotland.
Former accidental; 0/2; extinct.
1st Giffnock, 1892, GL.

A more recent record for Black-grass is from Pollokshields (1975, JP), but there are no modern records.

Phleum pratense L.
Timothy

Britain: common throughout except in NW Scotland.
Native and hortal; 118/187; common.
1st Paisley, 1869, RPB.

Hennedy (1891) considered Timothy a common grass of 'meadows and pastures'. Today it is widespread, occurring in the lowlands on neutral or enriched soils and frequently on waste ground or waysides; it is often sown for silage or hay, but seems to have declined at some old rural locations.

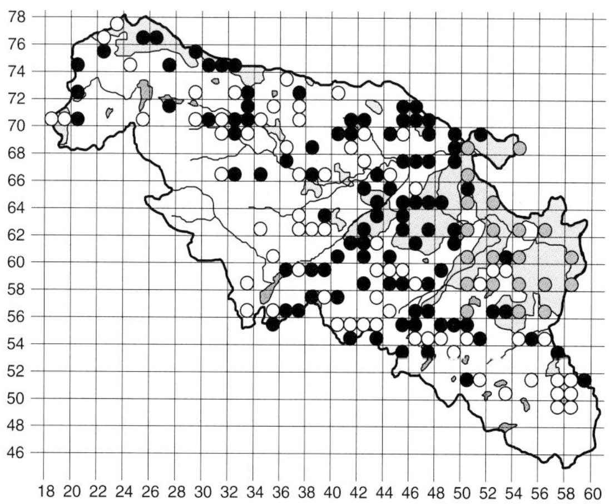

Phleum pratense

Phleum bertolonii DC.
Smaller Cat's-tail

Britain: common in the S and E.
Native and hortal; 11/19; scarce.
1st Polnoon, 1958, BWR.

Smaller Cat's-tail was not distinguished by earlier authors and some records for the previous may include this species, so the map is only tentative. It grows in similar places to Timothy and may also be sown.

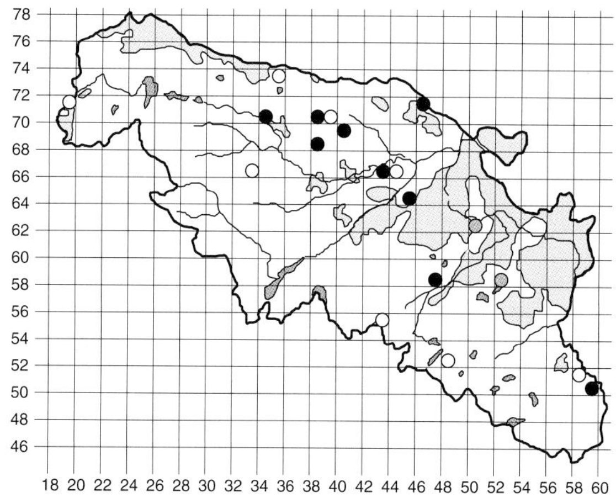

Phleum bertolonii

Phleum subulatum (Savi) Asch. & Graebn.

Britain: neophyte (S Europe); casual.
Former accidental; 0/1; extinct casual.
1st Crossmyloof, 1894, GL.

Glyceria maxima (Hartm.) Holmb.
Reed Sweet-grass

Britain: common in S and E England and C Scotland.
Native or accidental; 37/37; locally frequent.
1st Cart near bridge beyond Shawlands, 1865, RH.

RH was the first person to mention this tall grass, which is considered possibly a relatively recent arrival to the area (CFOG). However, it was described as frequent by 1891 (Hennedy), although later Lee (1933) stated 'not common'. Remarkably, there are no field records for Reed Sweet-grass during the 1960s–80s period (BWR), but it was known from Castle Semple in 1955 (ANG excursion) and from Barr Loch in 1969 (AMcGS). Today it is only found in the lowland south and east and is strongly associated with rivers, where

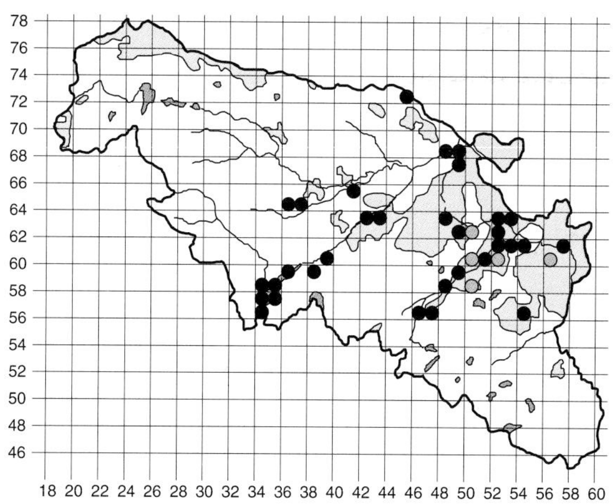

Glyceria maxima

nutrient enriched, notably the Black Cart, Levern and White Cart waters; it is also known from Barr Loch (NS3456 and NS3457, 1995), Barnbeth Loch (NS3664, 1997) and Erskine Park (NS4572, 1995).

Glyceria fluitans (L.) R. Br.
Floating Sweet-grass

Britain: common throughout.
Native; 270/315; very common in wet places.
1st Cartside, Paisley, 1865, PSY.

Floating Sweet-grass was considered common in the 19th century and remains so today. It is widespread with many records in rural areas but it becomes scarce in the more acidic peaty uplands. It is always found in wet places, such as along watercourses and at loch margins, and can often form luxuriant swampy mats in wetter parts of inundation grasslands, marshes and swamps.

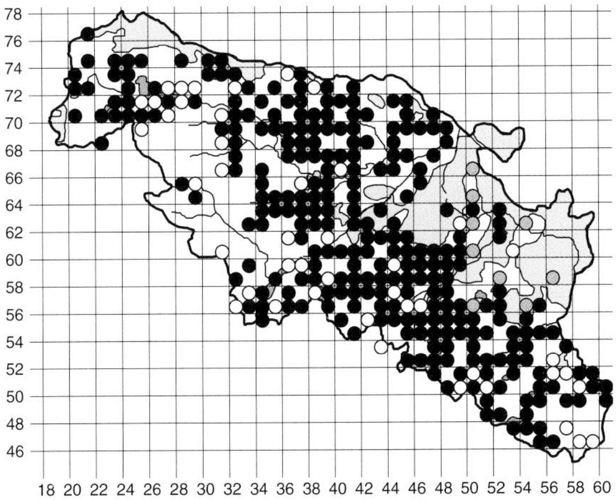

Glyceria fluitans

Glyceria × pedicillata F. Towns.
(*G. fluitans × G. notata*)
Hybrid Sweet-grass

Britain: scattered, mainly in the S.
Native; 0/2; extinct or overlooked.
1st Johnstone, 1883, GL.

There is only one other record (Blythswood, 1969, JP) for the hybrid which, like one its parents, may be overlooked.

Glyceria declinata Bréb.
Small Sweet-grass

Britain: common, western, rare in N Scotland.
Native; 142/154; common.
1st Bennan Loch, 1956, RM.

Small Sweet-grass was not distinguished by earlier authors, even by Lee as late as 1953 (GN), and there are no old specimens. It was recorded on several occasions during the 1960s–80s period (BWR) and other records include Glanderston Dam (1965, RM), Snypes Dam (1973, GLAM), Knapps Loch (NS3668, 1976, AJS) and Williamwood (NS5658, 1978, PM). It has a widespread distribution and occurs mostly in rural areas, usually in ditches, wetter flushes or slow-flowing burns, and sometimes on wet waste ground.

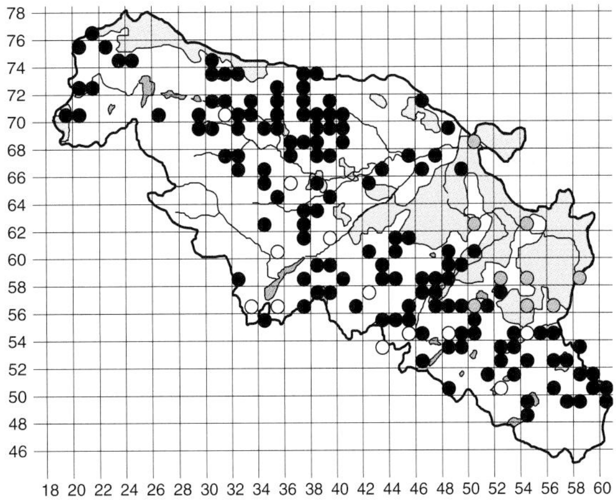

Glyceria declinata

Glyceria notata Chevall.
Plicate Sweet-grass

Britain: common in the S, rare or absent in N and W Scotland.
Native; 5/6; rare, probably overlooked
1st Renfrewshire, 1901, ASNH.

Plicate Sweet-grass was not distinguished by earlier authors and has likely been overlooked during more recent recording. Today it is only known from five sites: Darnley (NS55J, 1991, CFOG), a marshy field, Kirkton Burn (NS4857, 1998), Holehouse (NS4756, 1997), Formakin Estate (NS4170, 2003, IPG) and by the White Cart Water near Temples (NS6050, 2011).

Melica nutans L.
Mountain Melick

Britain: northern, in uplands.
Former native; 0/2; extinct.
1st Plantation opposite Busby Mill, 1865, RH.

In 1970 JP recorded Mountain Melick 'towards Muirshiel' (most likely in the River Calder valley), but this is the only recent report and, although a suitable location, should be viewed with some caution.

Melica uniflora Retz.
Wood Melick

Britain: common except in N Scotland.
Native; 36/42; frequent in wooded glens.
1st Harelaw Burn, 1858, PSY.

Wood Melick has not been refound at the original

location but it has been at other old localities such as Green Farm, Kilmacolm (1893, GL), Kelly Glen (1887, NHSG), Bardrain Glen (1942, TPNS) and Calder Glen (1909, GL). The map shows a widespread distribution. It favours steep banks or cliffs in shaded wooded glens and is well represented in the steep glens surrounding the Renfrewshire Heights; other localities include glens about Gleniffer Braes, e.g. Glenpatrick (NS44461, 1993), Wester Gavin (NS3759) and Linister Burn (NS3859, 2004), and it is also found along the White Cart Water as at Busby (NS5756, 2001), but not upstream where more open.

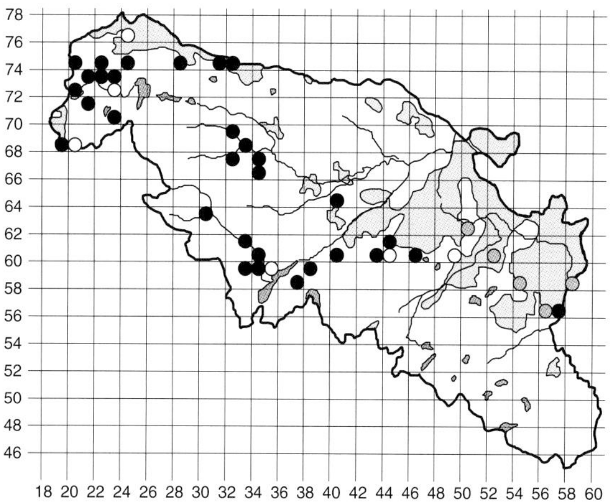

Melica uniflora

Bromus arvensis L.
Field Brome

Britain: neophyte (Europe); scattered in the S, declining.
Former accidental; 0/2; extinct casual.
1st About Cathcart, 1821, FS.

Hennedy considered this grass very rare and mentioned no local sites; it must be considered long extinct. It was collected from Crossmyloof in 1894 (GL).

Bromus commutatus Schrad.
Meadow Brome

Britain: southern, alien in Scotland.
Former accidental; 0/3; extinct.
1st 'New Kilmarnock road beyond Shawlands', 1865, RH.

Probably always rare in the VC, Meadow Brome was recorded from Castle Semple Loch (1877, NHSG) and was collected in Cathcart in 1894 (GL), but there are no other records, old or new.

Bromus racemosus L.
Smooth Brome

Britain: frequent in S England, rare alien in Scotland.
Former accidental; 0/2; extinct.
1st Cathcart and Gourock, 1876, FFWS.

As for the preceding species, there are very few old records for Smooth Brome, and none from modern times.

Bromus hordeaceus L.
Soft-brome

Britain: common except in N Scotland.
Native; 58/131; frequent.
1st Renfrewshire, 1869, PPI.

There are only a few old named localities for this grass of fields and waysides, considered very common by Hennedy (1891). The map shows a widespread, largely lowland distribution and Soft-brome can still be found by the edge of fields and along tracks and roadsides, or occasionally on waste ground. It was well recorded during the 1960s–80s period but there are many gaps in modern records, e.g. about Inverkip and Lochwinnoch, which may represent a decline but probably several populations will have been missed. Lesser Soft Brome (*B.* × *pseudothominei* P.M. Smith) was collected by RM from allotment waste ground at Merrylee (1942, E) and also recorded (BWR) from Witch Burn, Lochliboside (NS4557, 1972) and White House, Kilbarchan (NS4163, 1976).

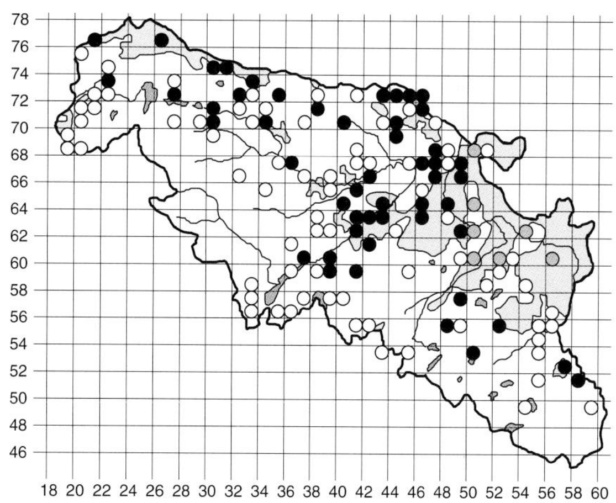

Bromus hordeaceus

Bromus lepidus Holmb.
Slender Soft-brome

Britain: neophyte (origin uncertain); occasional in S.
Former accidental; 0/2; extinct or overlooked.
1st Nether Auldhouse, 1944, RM.

The first record was from an allotment, and in 1969 RM found this brome at Auchendores Reservoir

(GLAM). Probably overlooked, there are no modern records for Slender Soft-brome.

Bromus secalinus L.
Rye Brome

Britain: archaeophyte (Europe); occasional in S.
Former accidental; 0/4; extinct.
1st Braidbar, Pollokshaws, 1864, NHSG.

Rye Brome was also recorded from Pollokshields (1865, RH), Gourock and Thornliebank (1891, RH) and collected from Crosssmyloof in 1894 (GL). RG (1930) thought it a fairly frequent casual and noted var. *submuticus* Reichb. from Newlands in 1922.

Bromus lanceolatus Roth
Large-headed Brome

Britain: neophyte (Eurasian); very rare.
Former accidental; 0/2; extinct casual.
1st Crossmyloof, 1892, NHSG.

RG (1930) also listed Paisley.

Anisantha rigida (Roth) Hyl.
Ripgut Brome

Britain: neophyte (S Europe); scarce, mainly in the SE.
Former accidental; 0/1; extinct casual.
1st Crossmyloof, 1892, NHSG.

Anisantha sterilis (L.) Nevski
Barren Brome

Britain: archaeophyte; common in England, mainly in E in Scotland.
Accidental; 9/13; rare.
1st Cathcart, 1851, GGO.

Barren Brome was considered frequent on waste ground and hedges by Hennedy (1891). It was recorded from Gourock (1876, FFWS) and collected from Paisley (1907, GL); more recently RM recorded it near Clarkston (1926) but BWR only reported it from Loanhead (NS4267, 1978). There are a couple of CFOG records and there are seven more recent ones: Gryfeside (NS3370, 1998), Conyston Plantation (NS4372, 2001), Crossmyloof (NS5762, 2009, PM), north Ranfurly (NS3864, 2010), Polmadie (NS5962, 2011), and two from Jordanhill (NS5368, 1995, AMcGS and NS5468, 2002, AR).

Anisantha tectorum (L.) Nevski
Drooping Brome

Britain: neophyte (Europe); rare, southern.
Accidental; 1/2; very rare.
1st Crossmyloof, 1892, NHSG.

In the modern period a solitary plant was found on waste ground in Greenock (NS2776, 2000).

Bromopsis ramosa (Huds.) Holub
Hairy-brome

Britain: common except in N Scotland.
Native; 62/70; frequent in woodlands.
1st Renfrewshire, 1869, PPI.

Hairy-brome is often encountered in lowland woodlands, on richer soils, nearly always where associated with a watercourse; it extends a little way into the Renfrewshire Heights along the surrounding wooded valleys, e.g. Mill Burn (NS3466, 1999), River Calder (NS3560, 1996) and Shielhill Glen (NS2471, 2000). It can also be occasionally found at wooded disused railways as at Langslie (NS3356, 2001) and occasionally appears on waste ground as at Port Glasgow (NS3274, 1998).

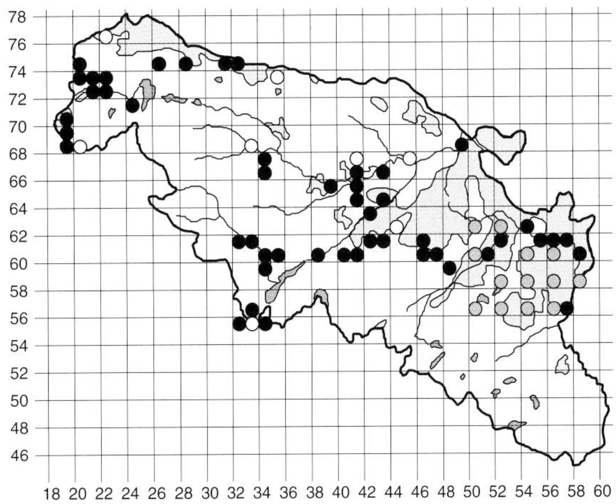

Bromopsis ramosa

Bromopsis erecta (Huds.) Fourr.
Upright Brome

Britain: southern, rare alien in Scotland.
Former accidental; 0/1; extinct casual.
1st Crossmyloof, 1894, GL.

Bromopsis inermis (Leyss.) Holub
Hungarian Brome

Britain: neophyte (Europe); scattered, mainly in the S.
Accidental; 2/2; very rare.
1st Bridge of Weir, 1998.

The original record for Hungarian Brome is from the marginal grassland of the disused railway (cycleway) at Bridge of Weir (NS3865). It was later found along the entrance track margin of the tip at Mearns Muir (NS5152, 2006).

Brachypodium sylvaticum (Huds.) P. Beauv.
False Brome

Britain: common, except in N Scotland.
Native; 40/46; frequent in valley woodlands.
1st Near Port Glasgow, 1841, GL.

Hennedy (1891) described False Brome as frequent, in woods and hedges, and named Gourock; Cathcart was named in 1876 (FFWS). Today it is found almost exclusively in steep-sided wooded valleys. It was found on waste ground at Port Glasgow (NS3174, 2002) – a site crossed by the lower reaches of the Devol Burn, where it is known to occur.

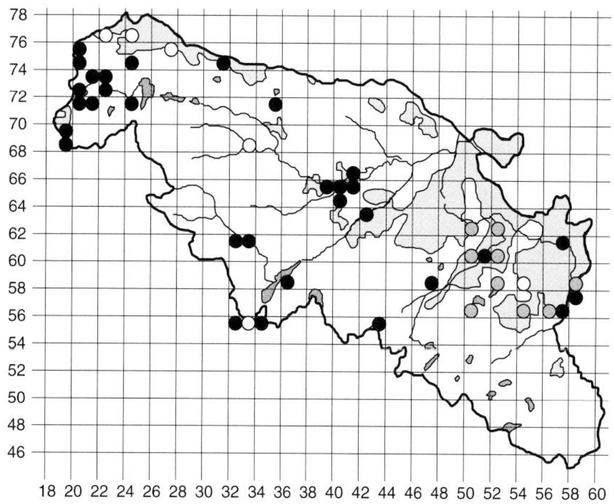

Brachypodium sylvaticum

Elymus caninus (L.) L.
Bearded Couch

Britain: common and widespread except in N Scotland.
Native; 53/57; frequent in valley woods.
1st Renfrewshire, 1872, TB.

There are few named old localities for this woodland grass but it was considered frequent by Hennedy

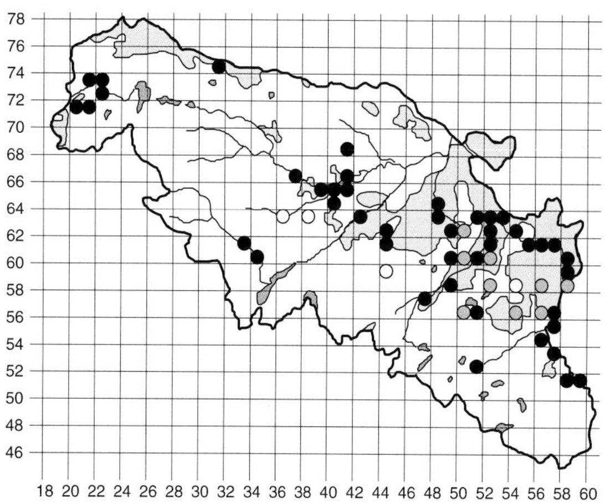

Elymus caninus

(1891). The map shows a scattered distribution and exclusively picks out wooded riverbanks, where Bearded Couch usually occurs on steep embankments. It is well represented on the White Cart Water and tributaries throughout the Glasgow conurbation, and also occurs along the Gryfe Water about Bridge of Weir, by the River Calder above Lochwinnoch and along several watercourses about Inverkip.

Elytrigia repens (L.) Desv. ex Nevski
Common Couch

Britain: common throughout except in N Scotland.
Native; 225/282; common in lowlands.
1st Renfrewshire 'weed', 1812, JW.

Common Couch shows a strong lowland distribution pattern, reflecting its preference for richer soils. It can be a pernicious weed of neglected arable land (as listed in the original record), abandoned pastures or old gardens, and can eventually dominate some waste-ground grasslands, particularly if poorly draining; it is also found along riverbanks. Common Couch is also a feature of the strandline along the west coast. Several awned specimens are referable to f. *aristata* (Schum.) Beetle, as along the Gryfe Water at Cauldside (NS3270, 2005).

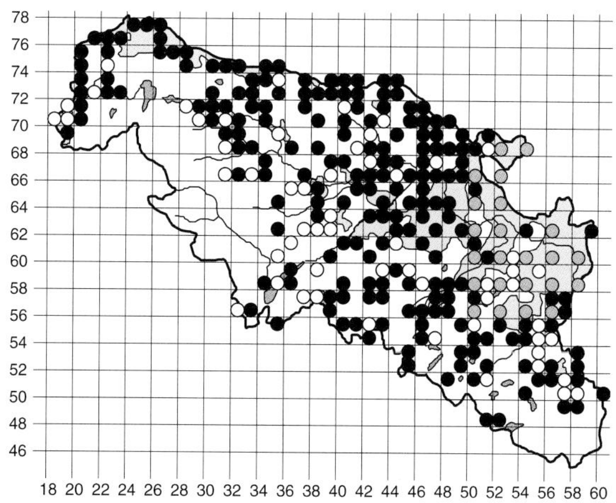

Elytrigia repens

Elytrigia juncea (L.) Nevski
Sand Couch

Britain: coastal throughout.
Native; 0/1; extinct or overlooked.
1st 'Levanne', Gourock, 1984, BWR.

Although described as frequent on sandy shores by Hennedy (1891), there are no VC76 records prior to the find at 'Levanne' (NS2176). It has not been reported from any other coastal surveys.

Leymus arenarius (L.) Hochst.
Lyme-grass

Britain: coastal, rarer in the S.
Native or hortal; 7/9; local on the coast.
1st Inverkip (NS1971), 1964, BWR.

There are only a few named old literature records for the region for this coastal grass, but none from the VC. Today it has been recorded on coastal sands from Wemyss Bay (NS1968, 2011) north to McInroys Point (NS2176, 1998) and also in the east at Langbank (NS3973, 1998) and Erskine Harbour (NS4672, 1996). It is not known whether the records represent natural dispersal or stabilizing planting.

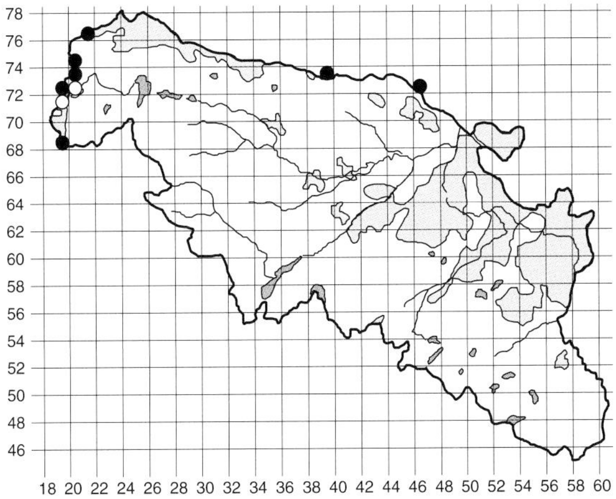

Leymus arenarius

Hordeum distichon s.l.
Cultivated Barleys

Britain: neophytes (SW Asia); frequent in the S.
Hortal; 10/11; rare casuals.
1st Hatton (NS4172), 1974, BWR.

Two-rowed Barley (*H. distichon* L.) is the most frequently recorded barley, found mainly as a casual at roadsides or occasionally river margins, as along the White Cart Water (NS4866, 2000) and Gryfe (NS4366, 2001). The original record may refer to Six-rowed Barley (*H. vulgare* L.), which was recorded twice locally during surveying for CFOG: from Hurlet (NS56A, 1986) and the banks of the Levern Water (NS55E, 1988).

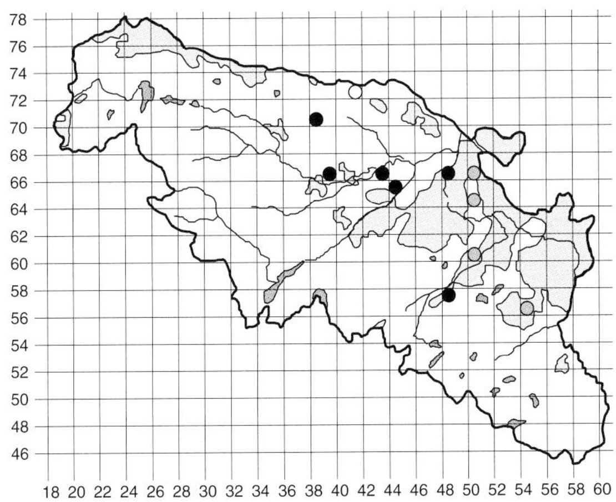

Hordeum distichon s.l.

Hordeum murinum L.
Wall Barley

Britain: archaeophyte (Eurasia); common in S and E England and SE Scotland.
Former accidental; 0/2; extinct casual.
1st Jenny's Well, Cart, 1900, PSY.

A further report was from Giffnock (1921, TPNS).

Hordeum jubatum L.
Foxtail Barley

Britain: neophyte (N America); scattered through lowlands.
Hortal or accidental; 7/10; rare.
1st Crossmyloof, 1892, NHSG.

Lee collected Foxtail Barley from Anniesland in 1918 (GL), on the edge of the VC, and more recently it was reported from the disused railway at Glenburn (NS4860, 1989, DM) and recorded during the CFOG period from Newlands (NS56Q, 1986), Barrhead (NS46V, 1989) and Whitecraigs (NS55N, 1987). It has also been found by a newly made path at Bottombow, Erskine (NS4571, 1997), and from a couple of places along the A737 roadside near Kilbarchan (NS4062 and NS4162, 1998), reflecting its salt tolerance.

Hordeum secalinum Schreb.
Meadow Barley

Britain: strongly south-eastern.
Former accidental; 0/1; extinct casual.
1st Paisley, 1930, RG.

Secale cereale L.
Rye

Britain: neophyte (cultivated); occasional in the S.
Hortal or accidental; 1/1; very rare.
1st Ardgowan, 2003, IPG.

The sole record is for a weed in an arable field (NS2073).

Triticum aestivum L.
Bread Wheat

Britain: neophyte (SW Asia); frequent, mainly in the S.
Hortal; 7/7; rare casual.
1st East Arkleston (NS56C), 1986, CFOG.

Bread Wheat was also recorded from by Newburgh Road in Glasgow (NS56Q, 1988, CFOG) and there are a handful of more recent records for casuals, usually from roadsides or tips: Gryfe Wraes (NS3966, 1998), Linwood (NS4465, 1998), near Pomillan (NS3466, 1999), Barrangary Tip (NS4469, 2001) and a barley field, East Fulwood (NS4567, 2007).

Danthonia decumbens (L.) DC.
Heath-grass

Britain: common, scarcer in SE England.
Native; 105/129; common in upland fringes.
1st Gourock, 1865, RH.

Hennedy (1891) described this grass as frequent on dry heaths and pastures, but there are no other old localities. Today Heath-grass is widespread but strongly rural and upland in its distribution. It is found on the drier shallower soils of upland pastures or heaths, most often encountered on rocky outcrops or ridges.

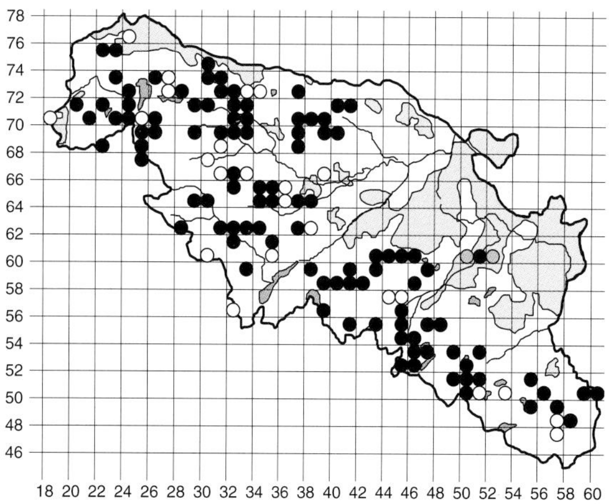

Danthonia decumbens

Cortaderia **spp.**
Pampas-grasses

Britain: neophytes (New Zealand/S America); scattered, mainly in the S and W.
Hortals; 14/14; scarce, scattered.
1st Bridgend, Lochwinnoch, 1995.

Records for Pampas-grasses are scattered and from urban-fringe areas. It appears to persist, and at the original location there are several individuals. The original locality, by the River Calder, supports Early Pampas-grass (*C. richardii* (Endl.) Zotov) (NS3459, determined by IPG) and this species has also been recorded from Jordanhill (NS5468, 2002, AR), from several locations on the north-west coast and from inland by a track near Laxlie Hill (NS2270, 2010, PRG). A few of the other records may include the South American neophyte Pampas-grass (*C. selloana* (Schult. & Schult. f.) Asch. & Graebn.).

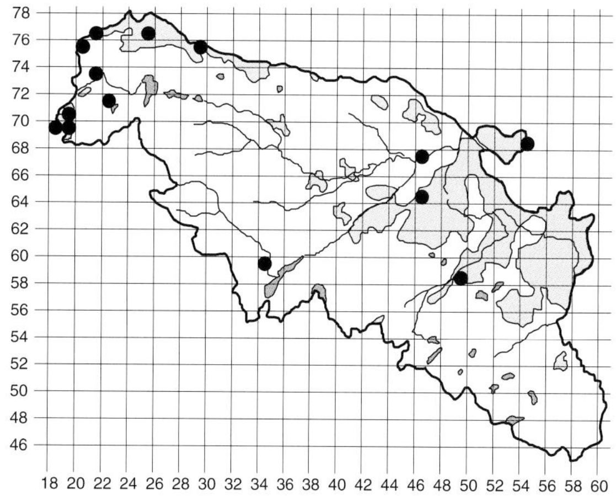

Cortaderia spp.

Molinia caerulea (L.) Moench
Purple Moor-grass

Britain: common in the N and W.
Native; 193/223; common.
1st Near Blackland Mill, Paisley, 1865, PSY.

Hennedy (1891) wrote that Purple Moor-grass was 'Common. On all our moors' and this is still an accurate description. The map shows a strong rural pattern, and it is mostly absent from urban areas and lowland improved agricultural land; an absence of records on the map clearly marks out the lowland 'Paisley Ruck' along the Black Cart Water. It is notably absent from large areas of the heather-dominated blanket bog in the higher ground of the Renfrewshire Heights. It is typically found in acidic peaty grasslands and wet heaths or in bogs and mires.

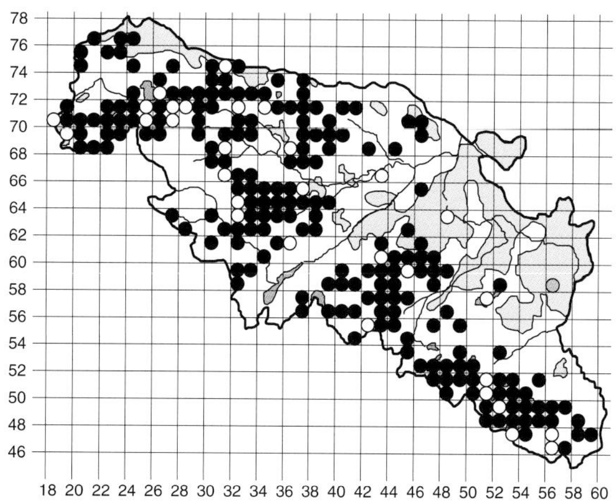

Molinia caerulea

Phragmites australis (Cav.) Trin. ex Steud.
Common Reed

Britain: widespread, rare in uplands.
Native; 29/35; occasional.
1st Kilbarchan and Lochwinnoch, 1845, NSA.

Hennedy (1891) considered this tall grass frequent, but noted it in abundance on the Clyde 'from Newshot Island to Erskine'; other old records include Loch Libo (1887, NHSG) and Williamwood (1928, GLAM). Today Common Reed can still be found in the brackish waters of the Clyde about Erskine, including Langbank in the west (NS3773, 1998), and extends up the Black Cart Water at Inchinnan (NS4767, 1999), the Gryfe Water at Selvieland (NS4567, 2001) and the Dargavel Burn at Georgetown (NS4467, 2005). Other 'inland' swamp sites are few and scattered but include Barr Loch (NS3557, 1996), Loch Libo (NS4355, 1996), a pond at Kelly Mains (NS2068, 1995) and Paisley Moss (NS4656, 1999).

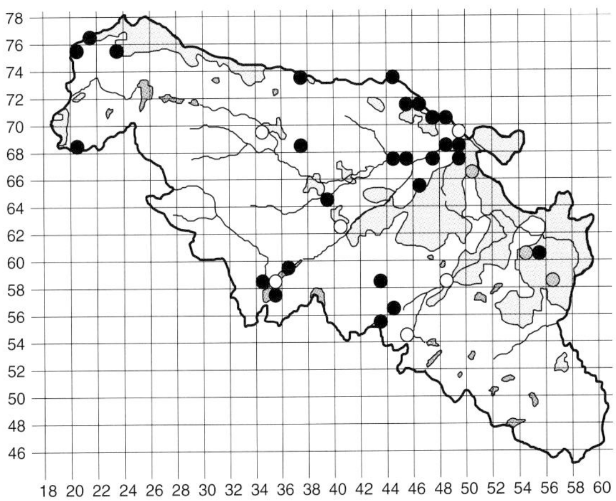

Phragmites australis

Eragrostis minor Host
Small Love-grass

Britain: neophyte (S Europe); very rare casual.
Former accidental; 0/1; extinct casual.
1st Crossmyloof, 1892, NHSG.

Panicum miliaceum L.
Common Millet

Britain: neophyte (Asia); scattered in the S.
Former hortal or accidental; 0/1; extinct casual.
1st Finlaystone Estate (NS3673), 1976, AJS.

Echinochloa crus-gallii (L.) P. Beauv.
Cockspur

Britain: neophyte (tropics); scattered in S England.
Former accidental; 0/2; extinct casual.
1st Giffnock, 1892, GL.

This grass was also collected from Newlands in 1921 (GL).

Setaria pumila (Poir.) Roem. & Schult.
Yellow Bristle-grass

Britain: neophyte (Eurasia); scattered in SE England.
Former accidental; 0/2; extinct casual.
1st Near Paisley, 1865, PSY.

Like the previous species, Yellow Bristle-grass was also collected from Newlands (1921, GL).

Setaria viridis (L.) P. Beauv.
Green Bristle-grass

Britain: neophyte (Eurasia); scattered in SE England.
Former accidental; 0/5; extinct casual.
1st Near Renfrew, 1873, TGSFN.

Other old records are from rubbish, Greenock (1880, NHSG), wasteland near Paisley (1910, PSY), for a garden casual, Bishopton (1911, GL) and from Newlands (1921, GL).

Setaria italica (L.) P. Beauv.
Foxtail Bristle-grass

Britain: neophyte (cultivated, China?); rare, mainly in the S.
Former accidental; 0/1; extinct casual.
1st Disused railway, Kilmacolm, 1984, BWR.

Sorghum bicolor (L.) Moench
Great Millet

Britain: neophyte (Africa); very rare casual.
Former accidental; 0/1; extinct casual.
1st Rubbish tip, Barr Loch, 1969, AMcGS.

There is uncertainty about the specific identity of this casual record, but it was thought to be Great Millet.

PART 3

Analysis

3.1 Distribution Patterns

There are several well-known reasons why a species shows a particular pattern of distribution, and the recent plant atlas *New Atlas of the British and Irish Flora* (Preston et al. 2002) readily demonstrates nationwide distributions. However, when an area is viewed at a 1km-square local scale several patterns can be seen which are not apparent from the 10km-square distribution maps used at the national level. At the local scale, Renfrewshire shows several patterns that reflect factors such as higher altitude (that is above 300m) or, in low-lying areas, intensive agricultural practices or the disturbance associated with conurbations; additionally, there is the coastline with its maritime influence, and the east–west orientation of the vice-county, with greater rainfall in the west. Other well-defined local patterns reflect a species association with open water or riparian habitats.

Ubiquitous pattern

There are a number of species which are widespread and cosmopolitan, showing a presence virtually throughout the vice-county (fig. 10). These are sufficiently widespread to be said to show a ubiquitous pattern, even though because of the highest reaches of the Renfrewshire Heights, notably the blanket peat and to a lesser extent the south-eastern hills, no one species has been found completely throughout the vice-county. The most commonly recorded species (shown in table 15, p. 357) are mostly perennials that are all tolerant of some urban disturbance or nutrient enrichment, occur along waysides and can extend into the uplands.

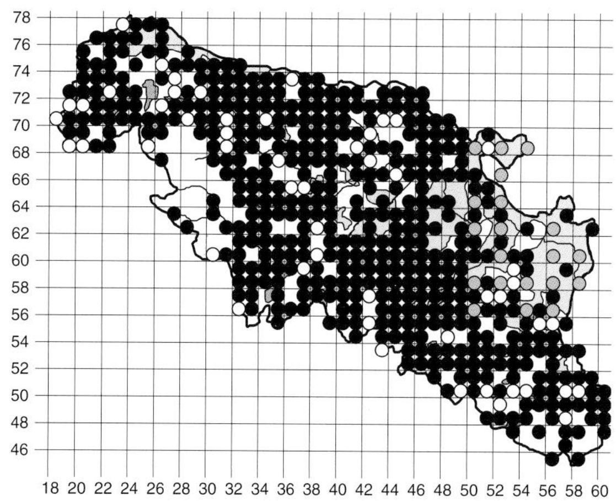

Figure 10. Ubiquitous pattern shown by Yorkshire-fog.

Lowlands

A number of species show a lowland pattern (fig. 11), shunning higher ground (above 300m); a good example is Great Willowherb. Many other species fit into this lowland category: a few further examples are Hedge Woundwort, Broom, Elder, Woody

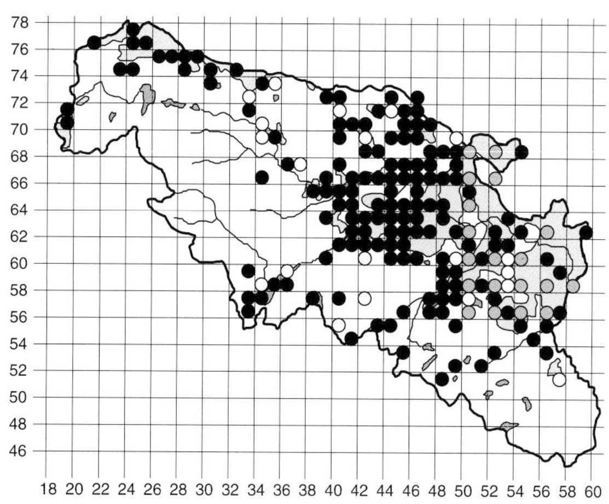

Figure 11. Lowland pattern shown by Great Willowherb.

Nightshade, Cow Parsley and Goat Willow. It is likely that such lowland patterns cannot be solely ascribed to climate and other factors such as grazing and nutrient levels may be important.

Uplands

The species indicative of the highest ground (above 300m) are only found in small numbers, typically in the Renfrewshire Heights but also to a lesser extent in the south-east uplands (fig. 12); strict 'uplanders' include Cowberry, Stiff Sedge, Mossy Saxifrage, Lesser Twayblade, Alpine Clubmoss, Cloudberry and Hairy Stonecrop.

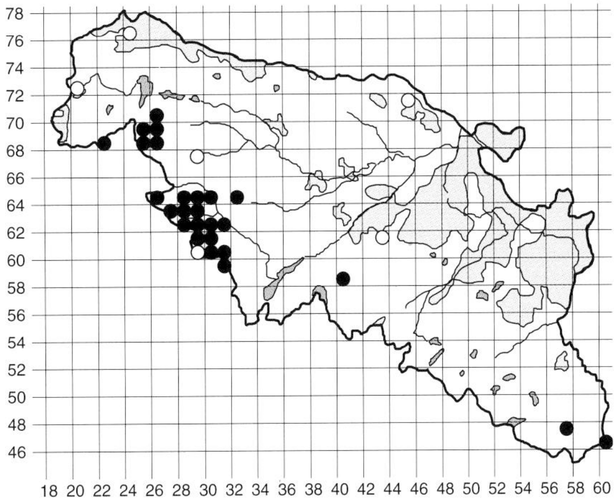

Figure 12. Upland pattern shown by Cowberry.

Peaty uplands

A number of widespread species are found in the Renfrewshire Heights and moorland about Eaglesham, and various isolated 'islands' of high ground such as about Howcraigs Hill or even Barscube Hill (fig. 13). They tend to be species tolerant of acidic and/or peaty soils, and can be quite widespread, and some can even extend into lowlands at relict raised bogs. Deergrass is a good example, and others include Brown Bent, Mat-grass, Hare's-tail Cottongrass, Crowberry, Green-ribbed Sedge, Star Sedge and Cross-leaved Heath.

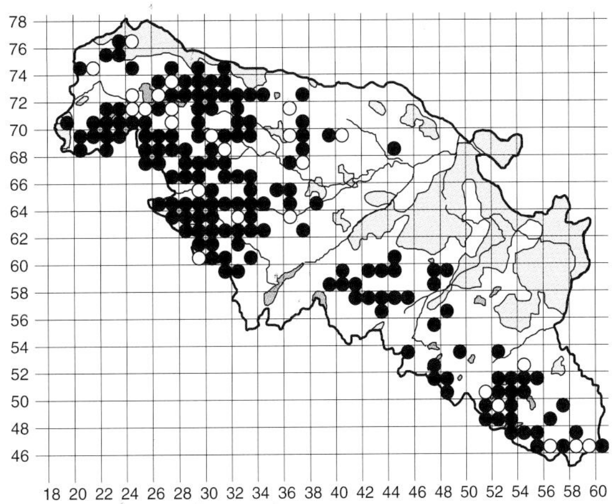

Figure 13. Peaty-uplands pattern shown by Deergrass.

The 'squeezed' pattern

A more subtle distribution zone appears to be the 'squeezed' pattern of lowland species shunning the higher, often peaty, ground, but also retreating from the lowland urbanization and agricultural intensification (fig. 14); species

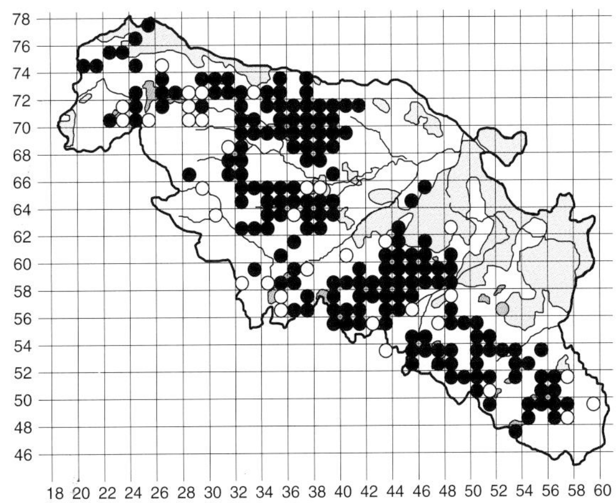

Figure 14. 'Squeezed' pattern shown by Marsh Cinquefoil.

occupying this area may well be nationally widespread but many are declining as a result of habitat deterioration. They tend to be species of low nutrient soils or water, perhaps with some base or mineral enrichment. Marsh Cinquefoil provides a good example, and others include Bugle, Water Avens, Marsh Hawk's-beard, Spring-sedge, Mountain Pansy, Primrose and Wood Sage.

Western

A western pattern would, without other influencing factors, simply reflect the higher rainfall and humidity of the west coast, but the undulating topography tends to distort such a smooth gradation (fig. 15). A prime example of a species with a distribution influenced by precipitation is English Stonecrop, which grows on rock exposures in the west but seemingly avoids similar locations further east; Whorled Caraway is perhaps an equally fine example. Other western species include Butterwort, Lesser Clubmoss, Bell Heather, Dioecious Sedge and notably several ferns: Beech Fern, Wilson's Filmy-fern and Intermediate Polypody. A further western group comprises alien urban plants, though their pattern is also influenced by chance factors of colonization, but they do tend to do well in the wetter and milder west – examples are Fuchsia, Montbretias (*Crocosmia* spp.) and Tutsan.

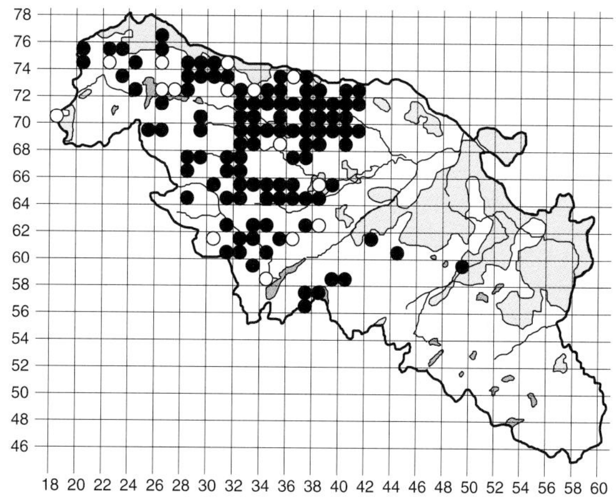

Figure 15. Western pattern shown by English Stonecrop.

Eastern

There are also a number of species which show a marked eastern preference. Again, the topography of the Renfrewshire Heights, and the lack of suitable lowlands on the narrow and urbanized north-western coastal plain, are likely important factors influencing this pattern (fig. 16). The species are mixed types, some preferring enriched soils and not all reflecting the drier soil conditions of reduced rainfall further east; examples include Zig-zag Clover, Bitter-vetch and Meadow Crane's-bill. A number are wetland species, but these presumably enjoy the more mesotrophic or eutrophic conditions of waterbodies or richer alluvial soils along watercourses, for example Water Sedge, Reed Sweet-grass, Branched Bur-reed, Water-plantain and Crack, White and Bay Willow.

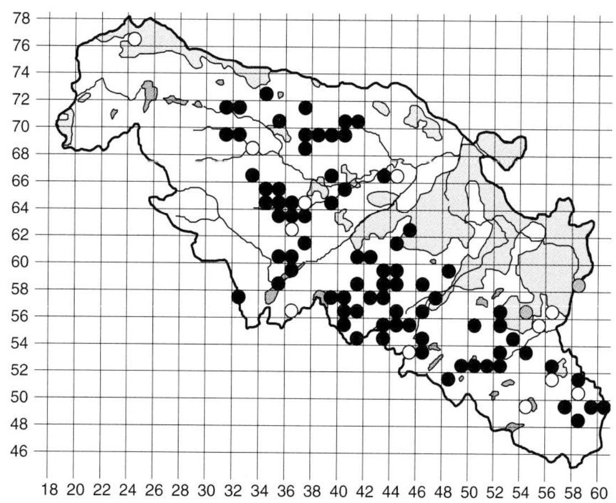

Figure 16. Eastern pattern shown by Bay Willow.

Coastal

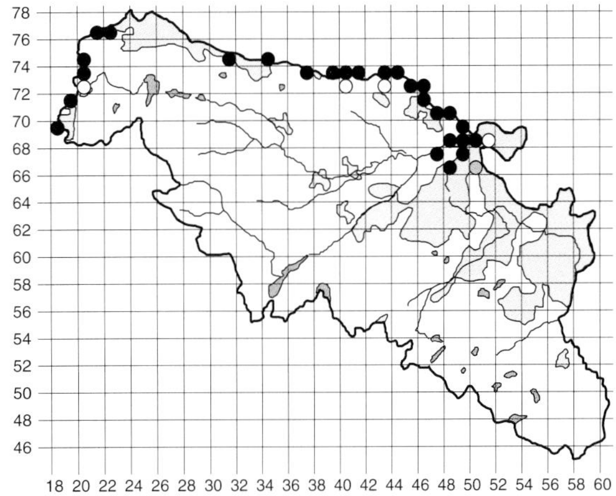

Figure 17. Coastal pattern shown by Sea Aster.

This pattern is an obvious one (fig. 17) with several maritime or salt-tolerant species extending from the western rocky shores to the estuarine east and some way up the River Cart and its major tributaries the White and Black Cart waters, although virtually all display a gap at the Gourock–Port Glasgow conurbation; maritime species include Sea Aster, Sea Plantain, Common Saltmarsh-grass, Sea Milkwort, Sea Arrow-grass, Sea Radish and Thrift; the latter shows a preference for the less estuarine west.

Urban

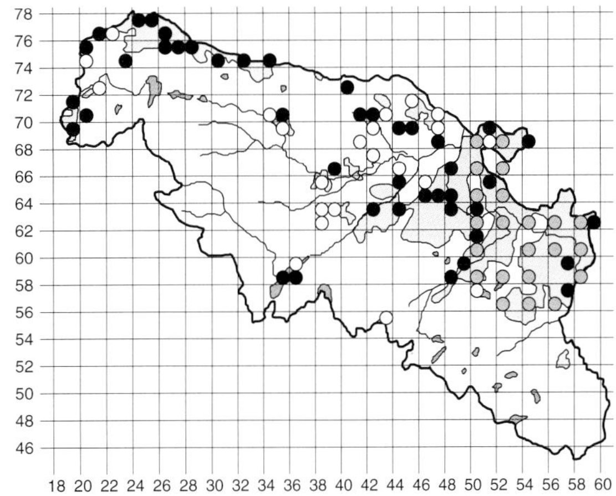

Figure 18. Urban pattern shown by Smooth Sowthistle.

In any populated area human disturbance and alien introductions combine to produce distinctive urban patterns (fig. 18). Urban areas spread from the Greater Glasgow and Paisley conurbations and along the coast from Port Glasgow to Gourock, with smaller outliers at places such as Wemyss Bay, Kilmacolm and Lochwinnoch. Urbanophile species – not all alien – include Confused Michaelmas-daisy, Welsh Poppy, Smooth Sowthistle, Garlic Mustard, Oxford Ragwort, Butterfly-bush, Spanish and hybrid Bluebell and Alsike Clover.

Disjunct

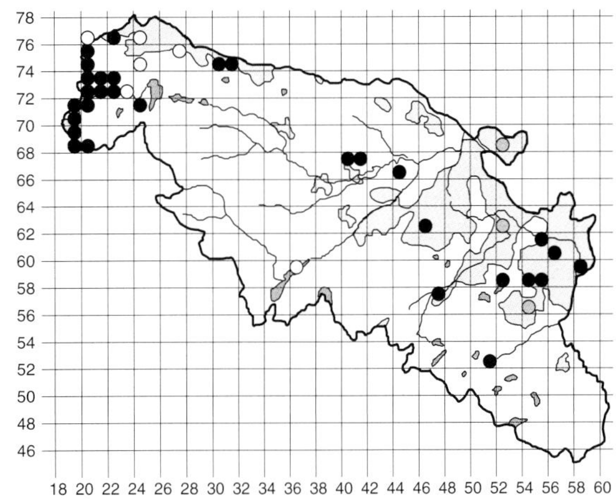

Figure 19. Disjunct pattern shown by Pendulous Sedge.

We have already seen that coastal species show a disjunct pattern due to the urban developments about Greenock. Similarly, other patterns are to some extent disjunct due to topography (fig. 19). A few native species show local centres of distribution, such as Lesser Water-parsnip and Burnet-saxifrage. The reasons for such gaps are unclear, but localized extinction due to land-use change and topography may be a factor. Pendulous Sedge shows a disjunction, perhaps reflecting native and introduced populations.

Spreading

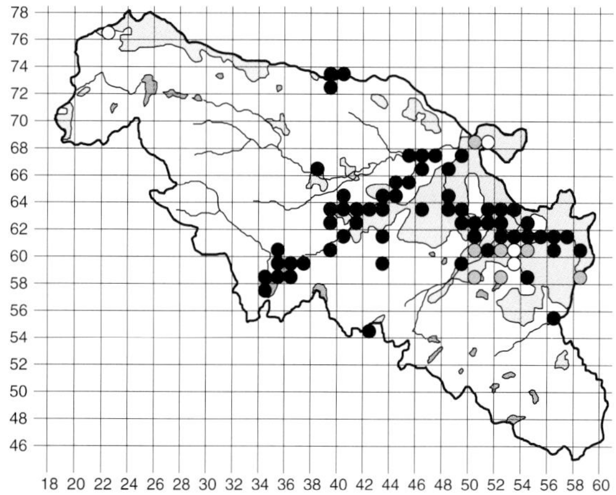

Figure 20. Spreading pattern shown by Indian Balsam.

Chance factors of colonization are responsible for many alien (neophyte) distribution patterns (fig. 20); good examples are the relatively recent aliens Butterfly-bush and Oxford Ragwort, which possibly show disjunctions due to multiple introductions or 'stepping stone' spread, and are in a state of rapid change. Indian Balsam may have had multiple introduction points but populations are merging and spreading along watercourses and the coast at various localities. In a similar way, Danish Scurvygrass is spreading rapidly along our main roads; this will perhaps soon to be followed by a few other salt-tolerant species such as Sea-spurrey.

Habitats

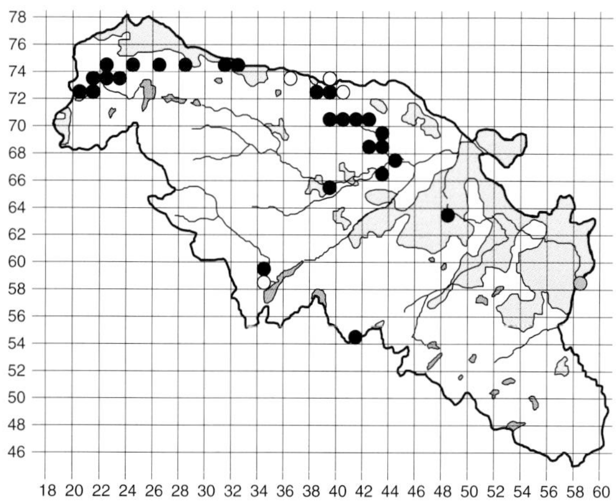

Figure 21. Habitat pattern shown by Wood Stitchwort.

Additional patterns reflect habitat features such as open water (fig. 21); some of the more widespread species include Alternate Water-milfoil, Lesser Marshwort, Canadian Pondweed and Broad-leaved Pondweed. Although some of these may be found along watercourses, the more reliable riparian patterns are associated with marginal species such as Butterbur, Winter-cress and the alien Indian Balsam; Wood Stitchwort is another species showing a riparian pattern, graphically representing the arc of wooded riverbanks draining from the Renfrewshire Heights.

Other more diffuse factors reflect the distribution of more specialist habitats where a particular species may occur; examples include the distribution of raised bogs, fen mires, dry heaths, ancient woodlands and unimproved grasslands, but these do not readily provide distinctive geographic distribution patterns at the vice-county scale.

3.2 Floristic Composition

The combined 143,000 records held within the botanical databases (see p. 47) provide a useful dataset from which to analyse the species – native and alien – growing wild in Renfrewshire, allowing study of their frequencies, the dates of first records and the gradual changes in the accumulated knowledge about the local flora.

This flora contains accounts of 1533 'species'. The total includes: 98 microspecies – for example brambles, hawkweeds and dandelions – of which 26 are alien; 124 hybrids – native and alien, or mixed; and 81 infraspecific taxa, mainly subspecies but with 17 varieties. Of the grand total, 788 species are native and 745 are alien, as shown in table 10.

Table 10. Summary of Renfrewshire flora totals showing aliens and natives.

	Extant	Extinct	Totals
Native	678	110	**788**
Alien	514	231	**745**
Totals	**1192**	**341**	**1533**

There are 1192 species that are extant (with records from post-1986 or from *The Changing Flora of Glasgow*), of which 514 (43 per cent) are aliens. Of the 341 species now considered extinct (with no post-1986 record), 110 are considered native and 231 alien (mainly casuals).

The 745 aliens make up a significant element (48 per cent) of the total Renfrewshire flora. Table 11 shows a breakdown of the categories of aliens. Using the definitions from the *New Atlas of British and Irish Flora* (Preston et al. 2002; see also p. 51), archaeophytes and neophytes account for 101 and 510 species respectively; these are species that are considered alien to the UK. A large number of species (134) are native in the UK but considered alien in the Vice-county of Renfrewshire as they have arrived as a result of human activity; it should be noted that this does not imply that such plants are actually of UK provenance, as some, or indeed many, may have arrived here from European sources or further afield, as was speculated by Grierson many years ago (1930). A small number of UK-native species are questionably native in the vice-county, but have been considered as native in the analysis. Some aliens have been classed as microspecies, for example dandelions and hawkweeds, but these are assumed to be accidental neophytes. A further category is alien hybrids, but really this group can be viewed as neophyte 'species', that is hortals introduced as hybrids, but the total does include eight apparently spontaneous hybrids (crosses between neophytes and a native species) occurring in the vice-county.

Table 11. Alien taxa in the Renfrewshire flora.

Aliens	Extant	Extinct	Totals
Archaeophyte	67	34	**101**
Neophyte	372	138	**510**
UK native	75	59	**134**
Totals	**514**	**231**	**745**

Somewhat more cautiously, aliens have been assigned to categories based on their presumed method of arrival in the vice-county, the two main categories being hortal and accidental (tables 12 and 13). The table shows 459 hortals and 286 accidentals, the last having a greater ratio of extinct records (52 per cent), mainly reflecting historical casuals.

Table 12. Alien taxa and main arrival methods.

Method of arrival	Extant	Extinct	Totals
Hortals	379	80	**459**
Accidentals	135	151	**286**
Totals	**514**	**231**	**745**

Table 13. Main categories of alien taxa.

	Extant	Extinct	Totals
Archaeophyte accidental	48	25	**73**
Neophyte accidental	66	85	**151**
UK native accidental	21	41	**62**
Archaeophyte hortal	22	9	**31**
Neophyte hortal	302	52	**354**
UK native hortal	55	19	**74**
Totals	**514**	**231**	**745**

Of the aliens with modern records, hortals (379) make up the largest category, and of these the majority are thought to be neophytes (302), with a few archaeophytes (22, mainly trees and medicinal plants) and the remainder (55) being 'UK natives'. Just under half of the 135 modern accidentals are neophytes, including microspecies and hybrids, with the remainder being archaeophytes (48) and UK natives (21).

Species frequencies

Table 14 shows the distribution of frequency classes for the species; the table covers all species, including those now extinct, and the second column shows the total number of 1km squares from which there are records (not the number of individual records).

It can readily be seen that there is wide range of frequencies represented. Well over half of the species (871) are found in 10 or fewer squares and over half of these are the 493 species which have only one or two records. This latter group represents many old extinct, often casual species (290 extinct species have only ever been recorded on one or two occasions), but also include some recent neophytes and very rare or now extinct native species.

Table 14. Species frequencies (number of 1km squares from which there are records).

Records	Number	Records	Number	Records	Number
1	329	1–10	871	1–100	1282
2	164	11–20	147	101–200	90
3	101	21–30	73	201–300	65
4	69	31–40	45	301–400	60
5	57	41–50	40	401–500	31
6	33	51–60	18	501–600	5
7	34	61–70	32		
8	24	71–80	19		
9	23	81–90	21		
10	37	91–100	16	**Total**	**1533**

At the other extreme, 251 species have been recorded from more than 100 1km squares and so can be classed as common. The 20 commonest or most widespread species, based on numbers of 1km-square records, are listed in table 15.

With one exception, the list favours perennials and interestingly the top four species are typically all plants of poorly draining pastures; most of the others are commonly found in open grassland habitats. The list includes one fern, Broad Buckler-fern, and one tree, Hawthorn. There are three tall herbs: Common Nettle, Creeping Thistle and Rosebay Willowherb, the sole alien.

Table 15. The 20 commonest Renfrewshire plants.

Name		No. of 1km-sq. records	Name		No. of 1km-sq. records
1	Creeping Buttercup	531	11	Ribwort Plantain	467
2	Soft-rush	525	12	Red Fescue	465
3	Yorkshire-fog	509	13	Creeping Soft-grass	463
4	Tufted Hair-grass	506	14	Rosebay Willowherb	460
5	White Clover	500	15	Common Mouse-ear	459
6	Common Nettle	498	16	Common Bent	457
7	Broad Buckler-fern	497	17	Annual Meadow-grass	455
8	Common Sorrel	495	18	Creeping Thistle	452
9	Sweet Vernal-grass	495	19	Hawthorn	449
10	Meadow Buttercup	469	20	Tormentil	444

Rates of change

The species accounts in Part 2 include dates of first records, which if used with caution provide a baseline for measuring the changing knowledge about the flora and, less reliably, the actual composition at a particular time. For many species, particularly neophytes, through literature and herbarium sheets the first record can be assigned with some certainty; but for many natives and archaeophytes the first record is more speculative. The first record indicates the earliest known recorded mention of the species in the vice-county (see Chapter 1.8). This ignores old archaeobotanical reports (see Chapter 1.6) and is particularly misleading for many common natives, which the earliest historical recorders did not think to mention from specific sites and thus provided no precise first locality; authors such as Hennedy (1865) described many such species as common, or similar, but frustratingly without naming a locality within the vice-county.

Table 16 and figure 22 show the first records by various date classes which have been amalgamated to reflect recording periods (see Chapter 1.8). Table 17 and figure 23 show the same records but accumulated to show the increasing totals. It should be noted that the species totals take no account of whether the taxa, native or alien, is still extant within the vice-county. The figures are also skewed in favour of more recent firsts due to the inclusion of infraspecific taxa and microspecies, many of which would have been present in the nineteenth century. It should be noted that the first record data includes eight native aggregate species, for example *Galeopsis tetrahit* s.l., where it is not possible to clarify which taxon the first record referred to, so the 'grand total' number of 'species' is inflated from 1533 to 1541.

The earliest group (those from before 1846) includes literature records – mainly the floras of Thomas Hopkirk and Sir William Jackson Hooker, the Montgomery Catalogue list, some earlier herbarium sheets, and notably a mention in *The Statistical Account of Scotland 1791–1799* or the edition of 1845. The next date class (1846–72) represents a more reliable estimate, reflecting the greater scientific recording of that time. It includes the first edition of Roger Hennedy's *The Clydesdale Flora* (1865) but because of the lack of local substantive records for many common species, the Renfrewshire lists of *c.* 1869 – of Morris Young and *The Flora of Renfrewshire (Paisley & Neighbourhood)* – have been included in this date class, and also the 1872 'Hennedy Catalogue' utilized in Topographical Botany (Baker and Newbold 1883). The period up to 1901 is covered by Hennedy's subsequent editions (notably that of 1891 – revised by Professor T King), the first natural history societies' publications supported by further herbarium collecting and Peter Ewing's Clyde area 'Phanerogram' list. This was a period of much more active recording and wider interest in local natural history, and the consequent steady rise in species numbers is not unexpected.

The 1902–50 period was a time of reduced recording, reflected in the drop in the number of new records, but the efforts of Daniel Ferguson and Robert Grierson, both great casual hunters, and also Lee's *Flora of the Clyde Area* (1933) and the early work of Robert Mackechnie help to keep the totals rising, albeit at a reduced gradient. The 1951–83 period records are mostly obtained from the recording activities of Basil Ribbons (BWR) and associates up until the time of the start of the dedicated recording for *The Changing Flora of Glasgow*. They also include the early activities of Allan Stirling (AMcGS) and Dr Alan Silverside (AJS) – who both reported a number of new native and alien microspecies – which,

Table 16. First record totals by date class.

	Before 1846	1846–72	1873–1901	1902–50	1951–83	After 1983
Alien	76	80	145	105	131	208
Native	255	265	87	39	79	71
Totals	**331**	**345**	**232**	**144**	**210**	**279**

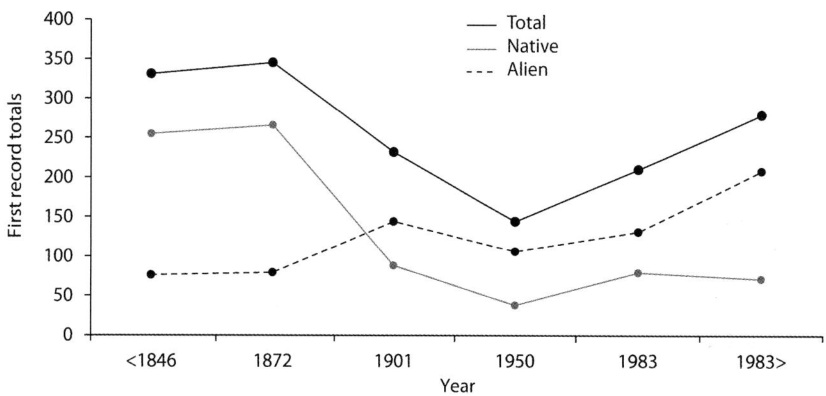

Figure 22. First record totals by date class.

along with more general critical recording, explains the sharp rise in new native firsts. The final category has been used to incorporate the relevant records from *The Changing Flora of Glasgow* (dated from 1984, when most detailed recording occurred, until the early 1990s) and also other field recording post-1986, mostly 1993 onwards.

The totals from before 1846 are clearly an underestimate, but by the time of Hennedy and his contemporaries, the picture is more realistic: by 1872 there were 676 recorded species, 520 of which were considered native. At the beginning of the twentieth century 908 species had been recorded, and of this total 607 were native and 301 alien. The first half of the twentieth century was a low point for new records, particularly of native taxa, which is not surprising given the preceding 100 years of discovery and documentation. New native discoveries remained at a low level throughout the rest of the century, but there was a slight upsurge, mainly reflecting more expert recording of microspecies, hybrids and infraspecific taxa. Thanks to the efforts of Ferguson and Grierson, the alien total does not show as dramatic a dip as that for natives between 1902 and 1950, and since that time the increased recording of aliens has helped to reverse the expected trend of a continued decline of new first records.

The graph of accumulated totals (fig. 23) shows a steady growth of the flora's total number of species. It is quite apparent that the sustained growth in the overall total is due to increased aliens, and the number of aliens is gradually catching up with the number of natives. This is to be expected as nearly all the natives to be found will have been recorded by now and the grand total can only be expected to continue to rise by more alien arrivals or more detailed taxonomic scrutiny of critical microspecies, hybrids and infraspecific taxa – unless climate change has the dramatic impact that some people predict, of allowing southern UK natives to gradually spread north.

Table 17. Accumulated first record totals by date class. (The 1541 grand total includes the eight native aggregate taxa as noted on p. 358).

	Before 1846	1846–72	1873–1901	1902–50	1951–83	After 1983
Alien	76	156	301	406	537	745
Native	255	520	607	646	725	796
Totals	**331**	**676**	**908**	**1052**	**1262**	**1541**

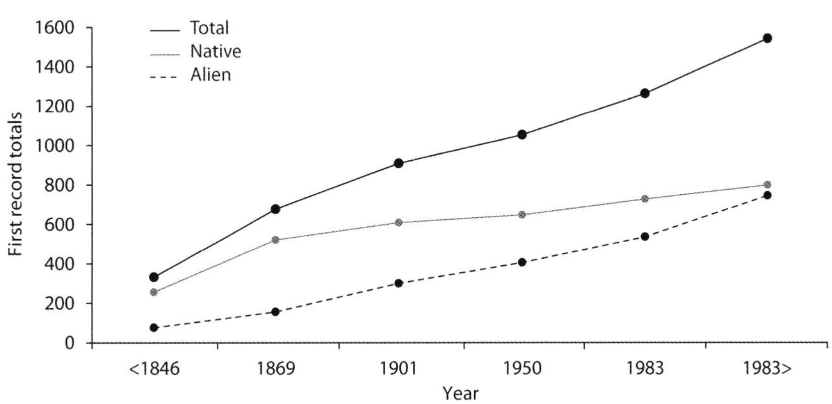

Figure 23. Increasing flora totals, from accumulated first record totals by date class.

Comparison with the 'Renfrewshire Plants' list of 1915

There were no complete catalogues or inventories of the plants found in Renfrewshire compiled in the nineteenth century, making it a difficult task to assess the composition of the flora at one particular time. However, the checklist 'Renfrewshire Plants' published in the *Transactions of the Paisley Naturalists' Society* in 1915 (TPNS 1915) does provide a valuable reference point for making an assessment of the changes over the last 100 years.

The list names 738 'species', both native and introduced (alien); the latter were included if considered established or, if a casual, found near the time of publication, but there is no indication in the text of which species are considered introduced. The species listed include a few infraspecific taxa not recognized now, so the total for comparison has been reduced to 731 species. 'Renfrewshire Plants' additionally listed a further 24 plants as 'Excluded', defined as 'not now found', producing an overall total of 755 for comparison. Another very useful feature of the list is the annotation of local rarities (205 species). Of the plants listed it is now considered that 69 were neophyte aliens, 49 being classed as rare at the time.

This flora's grand total of 1533 species is just over twice the total number of plants listed in 1915 (755). The comparison is distorted as this flora includes all plants recorded through the ages, and thus including many casual species and also critical taxa omitted or not recognized by the Paisley Naturalists' Society. If just the extant totals are compared, that is 1192 and 731, then there has been an increase of 461 species in the total Renfrewshire flora.

Table 18. Summary of species listed in 'Renfrewshire Plants' (TPNS 1915). Alien designations follow those used in this flora (see p. 51).

Species	Common	Rare	Excluded	Totals
Native	451	83	10	**544**
Alien	75	122	14	**211**
Totals	**526**	**205**	**24**	**755**

There are 626 species included in the 1915 list that are still present in the modern flora, but that leaves 566 now present that were unknown to the Paisley Naturalists; aliens account for 368 of this total – more than 300 being neophytes – with natives numbering 198. However, a substantial number, more than 200 taxa (alien and native), can be accounted for by taxonomic changes and the recognition of microspecies, hybrids, infraspecific taxa and other 'cryptic' taxa not readily recognized in 1915. The research for the current flora has benefited from a greater access to literature and herbarium records than would have been available to botanists working nearly 100 years ago. Also, in more recent times more recorders have been active, aided by improved transport and accessibility for fieldwork, with perhaps a sharper focus for new adventives.

Out of the 24 'Excluded' species 14 are aliens, although six of these are native in the UK. Interestingly eight species now have modern records, the most frequent being Teasel (with 14), Musk Mallow (15), Welted Thistle (7) and Astrantia (8). Several of the excluded species must be viewed with caution: where is the original evidence for species such as Great Water Dock, Slender Yellow Trefoil and 'Short Crowned-spiked Sedge' (*Carex vaginata* Tausch.)? Are these recording errors reflecting nomenclatural misunderstandings?

The 205 species formerly classified as rare represent an additional comparative dataset. There are 71 species that have not been recorded recently, of which 23 are former natives, including Sea Spleenwort, Annual Sea-blite, Herb-Paris, Common Cudweed, Bog Pimpernel and Marsh Stitchwort. Many former rarities (87) have remained as such, that is with ten or fewer records. The remaining 47 former rarities now have more than 10 modern records, 36 of which have 20 or more records, and 11 exceed 50 records each. Rosebay Willowherb stands proud with 386 records today – a remarkable increase. Not surprisingly, many of these are aliens such as Daffodil (110), Welsh Poppy (101), Red Currant (74), Londonpride (57), Ivy-leaved Toadflax (56), Black Currant (54) and Slender Rush (53). Formerly rare natives which cannot now be considered as such include Intermediate Enchanter's Nightshade, Water Sedge, Narrow Buckler-fern, Unbranched Bur-reed, Common Reed, Sea Radish, Water-purslane, Perfoliate Pondweed and Lesser Marshwort; is such an increase purely down to increased recording?

Of the 526 species not considered rare by the Paisley Naturalists, there have been 28 extinctions. Of these, 21 are native and many are from coastal or wetland places, for example Naval Pennywort, Brookweed, Eelgrass, Sea Rush, Flat-sedge, Saltmarsh Flat-sedge, Greater Pond-sedge, White-beaked Sedge and Black Bog-rush; others are from 'drier' land, for example Allseed, Wild Basil, Heath Cudweed, Small-white Orchid and Mountain Melick. Additionally, 90 species now have ten or fewer modern records, and so are now considered rare, of which 73 are native.

There are some 214 species missing from the 1915 list, which subsequent literature and herbarium research have shown to have been first recorded before 1915, making a total of 945 'known' species at the time, not 731. The majority of these are casuals or critical taxa, but the total includes trees such as Sessile Oak, Norway Maple, European Larch, Horse-chestnut, Laburnum, Apple, Plum, Cherry Laurel and Tea-leaved Willow, and a range of herbs including False Fox-sedge, Water Whorl-grass, Eastern Rocket, Northern Dock, Slender Pearlwort and, most surprisingly, Giant Fescue and Spring-sedge; the latter two, at least, must have been oversights.

Several absences may be accounted for by greater taxonomic scrutiny in more recent times, although several of the following were recognized taxa at the time. Examples include Common Spotted-orchid, Heath Milkwort, Alternate Water-milfoil, Sea Mayweed, Sea Rocket, Velvet Bent, Dwarf Eelgrass and several pondweeds and willows.

Other natives that are now rare may have just been overlooked, such as Distant Sedge, Common Stork's-bill, Broad-leaved Cottongrass, Holy-grass, Few-flowered Sedge, Lesser Meadow-rue, Coralroot Orchid, Cloudberry, Viviparous Fescue, waterworts and bladderworts.

Other presumed locally native species that were not noted but are now found, some frequently, may represent more recent arrivals, for example Purple Ramping-fumitory, Northern Dock, Spiked Sedge, Mudwort, Small Sweet-grass, Slender Pearlwort, Eight-stamened Waterwort, Fennel Pondweed, Hedge Bedstraw, False Fox-sedge and Wild Onion. If not actually introduced, could any of these increases be due to natural spread, perhaps influenced by a warming climate?

A number of species not present or not recorded in 1915 are now common features of the modern flora; missing species with more than 100 records today include Japanese Knotweed (205 records), American Willowherb (202), Pineappleweed (189), Rhododendron (132), Snowberry (124) and Dotted Loosestrife (119). Not surprisingly these are all established neophytes, as are many other newcomers discussed further in the following chapter.

3.3 The Changing Flora

Once the raw numeric analysis of changes has been completed, it is tempting to search for the reasons behind the apparent changes. Chapter 1.6, about the changing environment of Renfrewshire, provides an insight into how the landscape may have appeared after the Ice Age up until the turn of the nineteenth century, when documenting first began. Virtually all of the later changes are the result of large-scale land-use activities brought about by the growing human population. Even in the nineteenth century the rural landscape would have been highly modified; writing about the highly fertile lowland soils, Ferguson (1915) noted that 'it is extremely difficult, if not impossible, to form any adequate conception of the natural vegetation of these districts owing to market gardening and agricultural operations'.

Some changes, such as greater taxonomic knowledge and levels of recording, have already been highlighted. From the perspectives of nature conservation and land management, much interest lies in the extinction or decline of native species, but also in the species that are now increasing, notably neophytes.

Extinction for some native or casual species is reasonably straightforward to establish, as many have no subsequent record since the original nineteenth-century literature or herbarium record. In total 341 species are considered extinct today, 110 of these native. However, the term 'extinct' must always be used cautiously as unless the target species had a well-documented population there will always remain a doubt as some individuals may have been overlooked recently, and casuals by their very nature will keep reappearing for short stays. A few critical taxa first recorded in the 1970s or early 1980s have had to be classed as extinct as there have been no subsequent expertly determined records, but they are very likely still present.

To note a species in decline is perhaps more speculative as detailed old records of distribution are lacking. Hennedy's or Lee's comments on distribution or frequency can be helpful, through mention of Renfrewshire place names in the text or statements such 'Hills above Gourock' or 'All around the firth', which would indicate a former widespread frequency. A declining designation is most reliable for currently rare species where there is some indication of previous higher numbers, but for many now uncommon species an apparent lack of decline may reflect the increased level of more systematic modern recording contrasted with that of previous years, thus masking a real decline, which is a concern when interpreting the data for species conservation reasons. 'Renfrewshire Plants' (TPNS 1915) notes 205 species as being rare out of the 755 species listed (see p. 360); other authors provide further useful anecdotal or observational evidence (Anon 1869; Wood 1993; Paterson 1893).

The increasing category can be estimated with some confidence because of the level of modern recording and inference from the information – or lack of – supplied by earlier authors.

Losses and declines

Chapter 1.7 about habitats summarizes the commoner or indicative species comprising most of the local vegetation today, but the following examines some of the now rare, extinct or overlooked species of particular habitats. Many of these species – mostly native but some archaeophytes – were discussed at some length in *The Changing Flora of Glasgow* (Dickson et al. 2000) and many of the species listed there are relevant to the current flora.

Grasslands and heaths

Unimproved meadows or pastures are now very rare, the loss being mostly accredited to the pressure of urban land use in the lowlands or, more significantly, agricultural changes in both the lowlands and elsewhere; agricultural improvement is increasingly evident on remote or higher ground. Today any lowland 'species rich' meadow, as opposed to pasture, is more likely to be a secondary urban type on waste ground, rather than a relict of traditional farming. On farmland in rural fringes, pockets of unimproved pastures can still be found, but they are virtually restricted to marginal steep slopes or rocky ridges; even here they are invariably heavily grazed but have at least escaped the plough or the excesses of fertilizer applications. Some unimproved relicts persist in the margins of golf courses or other urban-fringe open spaces, but here they lack grazing and none receives a sympathetic 'cut and lift' management regime.

Although still presenting an open countryside landscape much of the lowland and upland fringe grasslands support well-improved or at best species-poor pasture; such swards, often bright green, can even be seen on many hillside slopes, extending to the higher plateau lands. On the higher ground, above 200–250m, the grasslands have mostly escaped heavy improvement treatments but are still negatively influenced by heavy grazing pressure, resulting in extensive areas of somewhat uniform acidic grassland, punctuated by rushy pastures or mires in depressions, or relict heaths and scrub on raised outcrops and ridges. Dry heathland is restricted now to the shallow soils of upland areas or some drained peats, but a few lowland relicts can persist, notably at some golf-course roughs. Wet heaths away from the highest ground are similarly restricted to local shallow peaty areas, where they often grade into deeper and wetter mire depressions.

In view of the above, it is not surprising that there is quite an extensive list of species which are now rare in grasslands and

Table 19. Extinct, very rare or declining species of grasslands and heaths. Extinct native species are marked with *.

Adder's-tongue	Heath Fragrant-orchid
Agrimony*	Heath Groundsel
Allseed*	Heath Pearlwort
Annual Knawel	Kidney Vetch*
Bird's-foot	Lesser Butterfly-orchid
Bulbous Buttercup	Lesser Meadow-rue
Burnet-saxifrage	Little Mouse-ear
Chaffweed*	Meadow Saxifrage
Common Centaury	Moonwort
Common Cow-wheat	Mountain Everlasting
Common Milkwort	Pale Sedge
Common Restharrow	Quaking-grass
Common Stork's-bill	Rough Hawkbit
Downy Oat-grass	Sand Spurrey
Early-purple Orchid	Sheep's-bit
Field Gentian	Shepherd's Cress*
Field Madder	Small-white Orchid*
Field Mouse-ear	Squirreltail Fescue
Field Scabious*	Trailing St John's-wort
Frog Orchid	Wild Carrot*
Heath Cudweed*	Wood Bitter-vetch*
Heath Dog-violet*	

heathlands, many of which are known or presumed to have declined in frequency. However, the number of actual extinct native species appears to be low. Table 19 lists many of these species; the grasslands include short open types as well as rank or poorly draining types. The species listed refer to presumed native occurrences only.

Uplands

It is tempting to consider that the vegetation of the highest upland ground has remained constant for many centuries, indeed some of the pollen diagrams from Walls Hill (p. 19) could well describe today's surface vegetation locally. Ferguson (1915) described natural and semi-natural grass of the upland shallow volcanic soils as being 'well adapted for grazing purposes', noting the wet grassy moors on clayey soils (where Mat-grass and Purple Moor-grass were the 'outstanding plants') and deep wet peat which was characterized by '*Eriophoron [Eriophorum], Sphagnum, Carex* and *Scirpus*, forming cotton-grass moors, truly waste land from the agriculturist's point of view.' He also wrote of heather moors, dominated by Heather: 'the grouse moors of the sportsman.' He considered the flora of the 'Renfrewshire plateaux' as 'largely a natural one, little influenced by the operations of man – with the exception of the "firing" of the heather and the opening up of "sheep drains"'. Although the 'naturalness' of such activities can be questioned, arguably these conditions still prevail on much of the higher ground of the Renfrewshire Heights, but undoubtedly burning, draining and grazing have modified the floristic diversity here, and markedly so towards its fringes and at other upland zones further east. There can be a marked contrast between the more heather-dominated ground within parts of Clyde Muirshiel Regional Park and an increased graminoid cover elsewhere in the Renfrewshire Heights, often clearly demarcated by a boundary wall; often the species diversity may be higher at the latter, but the sward is usually grazed short.

The modern-day scarcity of species listed in table 20 may well be down to the continued application of the above management activities, but there do not appear to be any extinct upland species; indeed, quite a few were omitted from earlier species lists. It is perhaps on the highest ground that the predicted climate change may have an impact – will all of these species be able to survive on the limited high ground available?

Table 20. Rare or declining native upland species.

Alpine Clubmoss	Mossy Saxifrage
Cloudberry	Parsley Fern
Common Juniper	Stag's-horn Clubmoss
Few-flowered Sedge	Starry Saxifrage
Hairy Stonecrop	Stiff Sedge
Lesser Twayblade	

Coastal

The coastal developments between Gourock and Port Glasgow have obliterated any natural coastal vegetation which may have been familiar to nineteenth-century botanists. Elsewhere the coastal strip has been squeezed by developments and infilling due to land reclamation. Such encroachment and changed land use are likely to have been the main causes of species loss. Today it is only south from Cloch Point down to Wemyss Bay where relict maritime vegetation occurs, but even here it is restricted to a narrow strip, except perhaps about Lunderston Bay. East from Port Glasgow as far as Inchinnan, the Clyde and Cart banks are highly artificial, but there is a narrow fringe of estuarine vegetation on the extensive mudflats, including a small patch of Dwarf Eelgrass at Langbank; only a few patches of relict saltmarsh occur, but about Newshot Island and the Cart confluence there are larger coastal swamp stands of Common Reed and Sea Clubrush.

Table 21 lists a remarkably large number of coastal species that have not been recorded for a long time. Many of these only ever had one or two records but a number were formerly considered more widespread. A few other coastal species are only known from a few places, so can be considered as vulnerable or threatened.

Table 21. Extinct, rare or declining coastal plants. Extinct native species are marked with *.

Annual Sea-blite*	Marram*
Beaked Tasselweed	Navelwort*
Brackish Water-crowfoot*	Oysterplant*
Buck's-horn Plantain	Parsley Water-dropwort
Danish Scurvygrass*	Ray's Knotgrass
Dwarf Eelgrass	Reflexed Saltmarsh-grass
Eelgrass*	Saltmarsh Flat-sedge*
False Fox-sedge	Sand Couch*
Frog Rush	Sand Sedge
Frosted Orache*	Scots Lovage*
Glasswort*	Sea Beet*
Greater Sea-spurrey*	Sea Campion

Woodlands

We have already seen that the woodland resource was seriously depleted 200 years ago, but if anything, coverage is greater now than it has been for a long time. Old woodlands though remain very rare, but arguably in recent times few have been totally lost. However, on a qualitative level the impact of canopy planting, introduced exotics, stock grazing and encroaching developments or associate leisure infrastructure continue to negatively impact on woodlands. Most of the true ancient woodland relicts have survived on the steeper valley sides and occasionally in estate woodlands, but here are highly modified.

Extinct species which are solely dependent on the woodland habitat are very few. Herb-Paris is an example: it was formerly found at three places even though at these former haunts the woodlands are still in a relatively good condition. Several species included in table 22 tend to be marginal species often found on rocky ledges and some are from wet or boggy woodlands. More woodland species fit into the rare or presumed declining category, although few of these were unlikely to have been widespread in previous centuries.

Table 22. Extinct, rare or declining native woodland plants. Extinct native species are marked with *.

Aspen	Mountain Melick*
Bird's-nest Orchid	Rock Whitebeam*
Black Spleenwort	Serrated Wintergreen*
Chickweed-wintergreen	Shady Horsetail*
Common Wintergreen	Soft Shield-fern
Coralroot Orchid	Stone Bramble
Globeflower	Toothwort
Goldilocks Buttercup	Tunbridge Filmy-fern*
Great Horsetail	Wild Basil*
Green Spleenwort	Wilson's Filmy-fern
Hay-scented Buckler-fern	Wood Crane's-bill
Herb-Paris*	Wood Small-reed*

Arable farmland

The nationwide decline of arable weeds is well documented and even at the time of Hennedy's first edition (1865) the demise of once familiar weeds was being noticed. Changes that came with mechanization in the nineteenth century – notably weed seed decontamination – are thought to account for the decline and loss of many arable species long before the arrival of modern-day agricultural intensification; modern use of herbicides and sowing cycles mean there is very little opportunity left for many once common arable weeds. Additionally, today there is little arable farming remaining, with many fields converted to improved pasture or lost to lowland developments.

Table 23 shows a number of species that are long gone or now very rare. Many of these appear to have been lost from farmland but some have records from waste ground or gardens – all places where other populations may well lurk unreported; additionally, a few, often colourful, species appear in wildflower seed mixes (deliberate or as contaminants).

Table 23. Extinct, rare or declining species of arable farmland. Extinct natives (or archaeophytes) are marked with *.

Barberry	Field Gromwell*
Barren Brome	Field Pansy
Black Bindweed	Field Woundwort*
Bristle Oat	Green Field-speedwell
Bugloss	Henbit Dead-nettle
Common Cornsalad	Northern Dead-nettle*
Common Cudweed*	Petty Spurge
Common Poppy	Prickly Poppy*
Corn Buttercup*	Rye Brome*
Corn Chamomile*	Scarlet Pimpernel
Corn Marigold	Shepherd's-needle*
Corn Mint	Small Cudweed*
Corncockle*	Small Nettle
Cornflower*	Stinking Chamomile*
Cut-leaved Dead-nettle	Tall Ramping-fumitory
Darnel*	Wild Pansy

Wetlands

The main factors that have impacted on the flora of poorly draining wetland localities are direct loss of the habitat – due to total drainage, infilling or partial drainage and nutrient enrichment – and subsequent grazing and other agricultural treatments. There is quite an extensive list of wetland species with no modern records, some only recorded once or twice in the nineteenth century and quite a few with only a few known modern stations; these wetland species have been divided into three wetland habitat categories, but several species overlap.

A large number of open water bodies are still to be found in Renfrewshire, perhaps more so than 200 years ago, but many are artificial or highly modified in some manner. A concern of recent years has been the trend of breaching old dam walls in response to demanding inspection regulations. However, a number still support a diverse aquatic macrophyte flora and marginal emergent vegetation, but a few species have been lost or have become quite rare; a small number of the species listed in

Table 24. Extinct, rare or declining marsh, ditch and swamp species. Extinct natives species are marked with *.

Bladder-sedge	Marsh Stitchwort*
Clustered Dock	Nodding Bur-marigold
Common Club-rush	Skullcap
Cowbane	Slender Tufted-sedge
Dark-leaved Willow	Small Water-pepper
Dwarf Elder*	Tea-leaved Willow
Fool's-water-cress	Tubular Water-dropwort*
Gipsywort	Tufted Loosestrife
Great Pond-sedge*	Water-cress
Greater Spearwort	Whorl-grass
Green Figwort	Yellow Loosestrife
Lesser Water-parsnip	

Table 25. Extinct, rare or declining open water species. Extinct native species are marked with *.

Awlwort*	Pillwort*
Common Water-crowfoot	Pond Water-crowfoot
Eight-stamened Waterwort	Red Pondweed
Floating Bur-reed	Rigid Hornwort*
Floating Club-rush	Shining Pondweed*
Horned Pondweed*	Six-stamened Waterwort
Least Bur-reed	Spiked Water-milfoil
Lesser Pondweed	Stream Water-crowfoot
Lesser Water-plantain*	Thread Rush
Mudwort	Thread-leaved Water-crowfoot*
Narrow-leaved Water-plantain*	Various-leaved Pondweed

table 25 may reflect previous taxonomic or nomenclatural uncertainty.

There has also been a serious loss of plants associated with deep peat: mires, fens, flushes and bogs (table 26). Several of these were documented a long time ago, occurring with the loss and reclamation of Paisley Moss and other mosses along the Black Cart Water floodplain; similar drainage will have also taken its toll in the uplands.

Table 26. Extinct, rare or declining peaty flush, fen, mire or bog species. Extinct species are marked with *.

Blunt-flowered Rush	Few-flowered Spike-rush
Bog Orchid*	Flat-sedge*
Bog Pimpernel*	Great Sundew*
Bog-rosemary*	Lesser Tussock-sedge
Bog-sedge	Long-stalked Yellow Sedge
Broad-leaved Cottongrass	Many-stalked Spike-rush
Brown Sedge	Oblong-leaved Sundew*
Crested Buckler-fern*	Slender Sedge
Early Marsh-orchid	Tall Bog-sedge*
Fen Bedstraw	Tawny Sedge

Gains and increases

As demonstrated for Renfrewshire, through the floristic analysis in Chapter 3.2, the flora of any vice-county in Britain, particularly one with a high urban exposure, will have changed dramatically over the last 200 years. We have already seen that many now common aliens would have been unfamiliar to early botanists, or were at least thought of as insignificant or irrelevant to floristic studies.

During recent research on the British flora, highlighted in the *New Atlas of the British and Irish Flora* (Preston et al. 2002), many familiar plants of the British landscape have been reclassified as archaeophytes. Some of these plants have long been considered a part of our native flora, and there is little reason to think any differently in light of such pronouncements. The arrival and establishment of many neophytes in more recent times has resulted in many more aliens, or non-natives, becoming familiar and a large number can now be considered a part of our flora; the notorious 'invasive non-native species' (INNS), by virtue of their innate invasive abilities, from whatever cause, are at the forefront of this integration.

Of the 745 aliens recorded, most are not increasing or even constant features. A very large number can be classed as casual species which have only ever appeared once or twice in the vice-county and their detection has been dependent on the vigilance of keen naturalists and chance encounters; many of these will keep reappearing, some may still be around and overlooked recently, but how many others have been missed entirely? A smaller number of long-established aliens, introduced for their historical medicinal, ornamental or culinary value, have now become a rare sight or, indeed, extinct. Their demise can be caused by a number of factors, but fundamentally reflects their modern-day lack of use and changes in gardening preferences, and thus the reduced chance of new introductions.

A good number of the more familiar aliens are neophytes, which are most likely to be encountered at habitats near to urban locations, and most notably on waste ground, along waysides and at the margins of larger watercourses, themselves highly artificial and enriched by nutrients. The commonest 50 aliens now recorded in the vice-county are shown in table 27; these are species occurring in 60 or more 1km grid squares. Of these, 29 were listed in 'Renfrewshire Plants' (TPNS 1915) (see p. 360), six of them considered

Table 27. The 50 commonest alien plants (in modern 1km-square records).

Records	Name	Status	1st record	TPNS 1915	Habitat
386	Rosebay Willowherb	Neophyte	1845	Rare	Open ground
352	Sycamore	Neophyte	1777	Recorded	Woodland
296	Beech	UK native	1777	Recorded	Woodland
284	Ground-elder	Archaeophyte	1869	Recorded	Woodland
205	Japanese Knotweed	Neophyte	1926		Riverbanks
202	American Willowherb	Neophyte	1969		Open ground
189	Pineappleweed	Neophyte	1910		Open ground
158	Pink Purslane	Neophyte	1863	Recorded	Woodland
142	Field Forget-me-not	Archaeophyte	1860	Recorded	Open ground
132	Rhododendron	Neophyte	1959		Woodland
125	Shepherd's-purse	Archaeophyte	1869	Recorded	Open ground
124	Snowberry	Neophyte	1961		Woodland
119	Dotted Loosestrife	Neophyte	1969		Open ground
115	Scentless Mayweed	Archaeophyte	1845	Recorded	Open ground
111	Monkeyflower	Neophyte	1879	Recorded	Riverbanks
108	Scots Pine	UK native	1834	Recorded	Woodland
104	Horse-chestnut	Neophyte	1845		Woodland
101	Welsh Poppy	UK native	1890	Rare	Woodland
99	Lime	UK native	1842	Recorded	Woodland
98	Gooseberry	Neophyte	1865	Recorded	Woodland
95	Irish Ivy	UK native	1986		Woodland
87	Montbretia	Neophyte	1971		Open ground
85	Equal-leaved Knotgrass	Archaeophyte	1985		Open ground
84	Hybrid Bluebell	Neophyte	1974		Woodland
80	Crack-willow	Archaeophyte	1887	Recorded	Riverbanks
78	Confused Michaelmas-daisy	Neophyte	1916		Open ground
78	Daffodil	Neophyte	1871		Woodland
78	Mugwort	Archaeophyte	1845	Recorded	Open ground
77	Norway Maple	Neophyte	1842		Woodland
77	Pick-a-back-plant	Neophyte	1969		Woodland
75	New Zealand Willowherb	Neophyte	1928		Upland
75	Sweet Cicely	Neophyte	1834	Recorded	Open ground
74	Red Currant	UK native	1912	Rare	Woodland
71	Osier	Archaeophyte	1865	Recorded	Riverbanks
69	Corn Spurrey	Archaeophyte	1869	Recorded	Open ground
65	Indian Balsam	Neophyte	1937		Riverbanks
62	Russian Comfrey	Neophyte	1915		Riverbanks
62	Snowdrop	Neophyte	1834	Recorded	Woodland
61	Spanish Bluebell	Neophyte	1974		Woodland
61	Yellow Archangel cultivar	Neophyte	1984		Woodland
60	Wild Privet	UK native	1845	Recorded	Woodland
59	Canadian Waterweed	Neophyte	1890	Recorded	Open water
57	Alsike Clover	Neophyte	1899	Recorded	Open ground
57	Dame's Violet	Neophyte	1895	Recorded	Riverbanks
57	Londonpride	Neophyte	1813	Rare	Open ground
56	Ivy-leaved Toadflax	Neophyte	1869	Rare	Stonework
55	Brideworts	Neophyte	1834		Open ground
54	Black Currant	Neophyte	1834	Rare	Woodland
53	Slender Rush	Neophyte	1863	Rare	Open ground
52	Cut-leaved Crane's-bill	Archaeophyte	1845	Recorded	Open ground

rare. Ten are archaeophytes, 32 neophytes and the remainder native elsewhere in the UK. There were 27 already recorded by 1872, but 15 were first recorded after 1915; of the latter, Japanese Knotweed and American Willowherb are the commonest. If a main habitat type is tentatively assigned to each species, then the vast majority are from woodland or open places – waste ground and waysides, with 23 and 17 respectively – and seven are often found primarily along riverbanks.

At most traditionally managed semi-natural habitats, which are quite rare, the impact of newcomers appears to be low. Woodland is probably the habitat on which there has been the most impact. Woodlands have been influenced by planting and invasion of non-native species for several hundred years; this relates to the canopy, shrub layer and ground flora. Heavy shade species such as Beech and conifers, but also some shrubs, notably Rhododendron, can drastically change the woodland's floristic composition. The floras of many mature lowland estate woodlands and also several more 'natural' riverbank woodlands include many alien species, as noted in Chapter 1.7, and a fair number of these are included in the commonest aliens listed above.

Arable land has long been noted for its suite of archaeophytes, but this is not the case for grazed pasture grasslands, where it is difficult to think of a neophyte that has become common; this is not the case in secondary urban grasslands. However, the now widespread improved forms of seeded grass leys are essentially wholly comprised of alien cultivated stock, even if nominally 'native'.

In rivers and open waterbodies Canadian Waterweed and increasingly Nuttall's Waterweed are widespread, but other aquatic aliens are rarely found at open waterbodies; New Zealand Pygmyweed has recently appeared and will need to be watched. Along smaller watercourses, Monkeyflower and various mints may be found at the margins, but perhaps the biggest change is the abundance of Hybrid Water-cress – although this is not strictly a neophyte it is most likely 'introduced'. At urban marshy ground Confused Michaelmas-daisy can be well established and American Willowherb is fast becoming widespread. In general there are few aliens at marshy sites, and peatlands, including bogs, mires, fens and flushes, tend to have retained largely native floras.

An excellent place to see neophytes, although not strictly wetland species, is along riverbanks. This is where a number of the notorious 'invasive non-native species' (INNS) can be found, often in abundance, including the 'big three' of Indian Balsam, Giant Hogweed and Japanese Knotweed. These species are most likely responding to disturbance, including loss of shade, and the enriched water that inundates the banks, giving the invasive species a competitive edge over 'traditional natives'; other riverbank aliens include Few-flowered Garlic, Dame's-violet and Russian Comfrey, which along with a number of native tall growing herbs and grasses present quite a different flora to that of times gone by. The widespread planting and natural spread of several alien willow and osier species, and hybrids, also contribute to a changed riverbank landscape.

A list of the alien species readily encountered on waysides and waste ground, notably near urban areas, would be long. The commoner ones are listed in table 27, and most are found on open ground; other prominent species frequently seen include Perennial Cornflower, Soft Lady's-mantle, Purple Crane's-bill, Canadian Goldenrod, Fox-and-cubs, Garden Lupin, Oxford

Ragwort, Butterfly-bush and various types of mints, hawkweeds, cotoneasters and dandelions.

A few aliens are now characteristic of more unusual habitats and could be viewed as specialists exploiting new niches; examples include Ivy-leaved Toadflax (old stonework), Slender Rush (paths and trampled areas), Slender Speedwell (amenity cut grass), Danish Scurvy-grass as an introduced 'native' (salted main roads) and New Zealand Willowherb (upland rocks and streamsides).

It should be noted that not all newcomers, or increasing taxa, are aliens. A few increasing 'native' species were noted above (p. 361) in the comparison with the list 'Renfrewshire Plants' (TPNS 1915). It is not clear whether these increases are due to recent natural arrivals, past oversight or just a greater modern taxonomic scrutiny. However, a few reasonably distinct taxa are now quite frequent, as listed in table 28. As noted previously, Spring-sedge and Giant Fescue must be oversights but some taxa may represent actual modern increases, or indeed first arrival, as indicated by the late first record.

Table 28. Native taxa with an apparent recent arrival date or showing an increase in frequency.

No. of 1km-sq. records	Name	1st record	TPNS 1915 record
142	Small Sweet-grass	1956	
82	Northern Dock	1876	
53	Water Sedge	1881	Rare
49	Narrow Buckler-fern	1876	Rare
41	Spear-leaved Orache	1883	Rare
41	Unbranched Bur-reed	1850	Rare
36	Intermediate Water-starwort	1860	
34	Slender Pearlwort	1883	
33	Lesser Marshwort	1845	Rare
33	Pendulous Sedge	1873	Rare
31	Perfoliate Pondweed	1843	Rare
30	Common Reed	1845	Rare
29	Purple Willow	1865	Rare
28	Creeping Yellow-cress	1845	Rare
27	Tansy	1845	Rare
24	Water Purslane	1858	Rare
23	Blunt-leaved Pondweed	1938	
22	Viviparous Sheep's-fescue	1976	
19	Sea Radish	1865	Rare
11	Wild Onion	1957	

Summary of changes

There are several factors that have brought about the numerous examples of floristic changes highlighted earlier in this chapter. The two main reasons for species change are the forces of agriculture and urbanization; the former is very widespread, and increasingly pronounced in its impact; the latter is more local and lowland, but tending to be absolute. As yet there is no evidence of changes induced by climate change, though a more careful study of upland plant populations may be illuminating. There is also no evidence of factors such as the introduction of aliens being the sole cause of local extinctions.

In the wider rural area changes in agricultural practices are the key agent of change. Historically, there has been the conversion of wetlands, heaths, scrub or woodlands to rough grazing or arable fields. This has led to declines in the extent of these more natural habitats and often an associated loss in quality and diversity. However, originally at least, such changes enabled the spread of many open grassland species and the creation of a niche for a host of arable weeds, both natives and archaeophytes.

Such past agricultural changes are well illustrated by a couple of quotes from the nineteenth century. In the description of Kilbarchan in *The New Statistical Account of*

Scotland (1845) it was stated that 'The strong impetus given to agriculture, from obvious causes, during half a century bypast, in so narrow a district as Renfrewshire, teeming as it does with a rapidly growing population, has greatly narrowed the field of the botanist's researches; and the woods are, with few exceptions, of recent origin.' A few years later, writing in the *Paisley & Renfrewshire Gazette*, the anonymous author (Anon. 1869) lamented the loss of favourite places to manufacturing, but added that 'it is no less the case that the advancement of agriculture has encroached on the verdurous borders and tree fringed seclusions of many a wild glen, and swept away the stripes of unenclosed land that used to be left as a margin by the riverside or to rim the inland lochs.'

In recent times the use of herbicides and fertilizers, high stocking densities and expansion of improved, permanent pastures or seeded leys has resulted in even more dramatic changes to the farmed landscape; such modern changes are invariably negative for the local flora, with monocultures and severely depleted species complements the end result. Much of the interesting farmland habitat remaining tends to be restricted to steeper slopes and ridges, or poorly draining features.

Urbanization has also accounted for changes, chiefly in the lowlands. Much of this land would have undergone early historical modification due to farming, but the subsequent rural landscape is now long lost, replaced by industry, housing, parks, roads, quarries, dumps and associated infrastructure. However, some compensation for the plant hunter has been provided by the resultant urban waste ground and associated marginal land, which have proved attractive refuges to a range of casual arrivals or garden outcasts and escapes. In *The Changing Flora of Glasgow* (Dickson et al. 2000) it is well documented how large areas of waste ground created by industrial decline along the Clyde in the 1980s provided happy hunting grounds for botanists. In more recent times urban regeneration and amenity landscape management has rapidly diminished even this resource. Much of the open urban 'greenspace', including parkland, is routinely mowed, resulting in fairly monotonous, limited diversity turf, which is seldom allowed to flower.

In the twentieth century railways have come to be sanctuaries for wildlife, in contrast with adjacent intensively managed land – a reversal of the situation in the nineteenth century, when railways were a development in the traditional agricultural landscape. In the latter part of the last century the closure of railway lines resulted in an increase in floristic diversity and the development of compact vegetation succession zones: from the central open trackbed ephemerals, through short and long grasslands to low scrub at the margins. Over time scrub has become denser and dominant, and the recent conversion of many disused railways to cycleways, with the heavy – perhaps excessive – use of asphalt, has virtually obliterated the former trackbed specialists from many local sites.

The road network is not traditionally looked upon as being an important habitat, but it does receive attention from recorders due to its accessibility. With the demise of hedgebank floras in the wider agricultural landscape, due to total removal or heavy grazing, those remaining on roadsides are usually now the most diverse, even though they seldom receive sympathetic management, being either uncut or repeatedly mown. In more rural areas old tracks and roadsides tend to be important refuges for species now lost from the adjacent productive fields – rocky knolls and associated road bends being the best places.

In urban areas roadside verges tend to comprise frequently mown amenity grasslands or neglected rank versions, both of limited diversity. A few stretches of newly constructed roads have received wildflower seeding, but not necessarily subsequent sympathetic meadow management. Although the actual carriageway is not amenable to plant colonization, recent gritting with salt and sand in winter has encouraged a few halophytes – notably Danish Scurvy-grass – to exploit a new niche created to the margins or central reservations of treated roads.

Most open water today exists as dams and reservoirs, of which there are a greater number than would have been the case 200 years ago. Initially at dammed locations there would have been a loss of any pre-existing mire and swamp communities, but with some compensation in the form of increased open water and an exaggerated 'draw-down' marginal zone. In recent times breaching of unused waterbodies, brought about mainly by inspection regulations and liability costs, has created a trend of loss of open water, but for some a return to marshy conditions; open waters recently lost include those at Whittliemuir (Midton Loch), Craighall and Leperstone; older losses – most now with well-established, often high-quality marshy vegetation – include the former dams at Dyke, Picketlaw, Calder (Muirshiel), Kirktonmill, Glen Moss, Barmufflock, Matherneuk and Carswell.

Even though today water quality has dramatically improved, nutrient levels will be much higher than of old, as is indicated by the often luxurious tall herb stands, many dominated by notorious invasive alien species, that occur in the inundated margins of large lowland watercourses. Today, even though some upland burns may appear to have water which is relatively 'unpolluted', they are nevertheless increasingly enriched by agricultural run-off and affected by canalization and drainage from the surrounding fields.

Evidence of changes to the floristic diversity of watercourses is limited but given various accounts of pollution and bank engineering works over the past 200 years it is not difficult to surmise that changes will have taken place with many losses. A couple of older quotations provide some historical insight: Crawford and Robertson (1818) wrote that in the 'White Cart above the town of Paisley, there are found pearls so fine and big', which they compared with oriental pearls, and Wood (1893), who started his excursion along the Capelrig Burn at Thornliebank, noted, 'We are here above the works, and the stream is unpolluted'. However, he wrote of the Aurs Burn, '... what a contrast this little stream presents above and below Barrhead. Below, it is foul, foetid, livid with dye, fermenting with filth ... above, it is pure, clear, sparkling as ether as a stream ought to be...', adding, 'Above, the spirit of nature holds sway; below, the burn is deformed by the spirit of the age – commerce, money.'

PART 4

Conservation

4.1 The Need for Botanical Conservation in a Changing Renfrewshire

The discussions and tables in the preceding two chapters, along with the historical background in Part 1.6, have clearly demonstrated the dynamic and changing nature of the local flora, and also introduced many of the factors influencing it. The current situation is one of loss or decline of native species, and their traditional habitats, and the rise of a number of new aliens that are fast becoming established features in urban and to a lesser extent rural locations. There is no reason to believe that such trends will not continue in future years, but is this outlook inevitable or indeed desirable? Human activity is at the heart of the majority of changes – perhaps all if it is considered to be the prime mover of climate induced impacts – so arguably we can greatly influence the shape and character of the botanical inheritance of future generations.

Being a largely rural vice-county, the impact of farming in Renfrewshire has been repeatedly mentioned, but deserves re-emphasizing here as it remains the key pressure on the relict natural habitats, and is largely outwith the control of regulatory planning. At the vast majority of farms the monotonous bright-green sward of improved pastures or sown leys has become the standard grassland throughout the lowlands, except on the steepest or poorly draining ground, and local marshy features continue to be drained or occasionally infilled. Agri-environment schemes offer some potential for redressing the incessant loss in recent times, but it is hard to point to a site which has received sensitive management to promote typical, yet locally uncommon, species based on a systematic assessment. Species-rich grasslands, in particular, are very poorly funded under current agri-environment schemes (compared to more well-funded options such as prescriptions for birds and beetle banks and so on which have less of an impact on farmers' incomes), and recipient sites are rarely selected on the strength of their existing botanical complement. Until this situation is addressed, species-rich grasslands species in particular will continue to decline or disappear. Farmers, who created many of these habitats in the first place, are best placed to conserve and sympathetically manage this dwindling resource, but need support and encouragement.

The impact of development and urbanization has also been repeatedly noted, and further comment on legislation follows later in this chapter. However, even when potentially negative developments are strongly regulated, pressure on sites is coming increasingly from wider environmental improvement efforts. Current concern about climate change is resulting in pressure for land to be used for forestry and wind farms; unfortunately, for many of these sites 'unproductive' marginal land is targeted. Recently proposed or implemented 'Community Woodlands' can and will result in the direct and indirect – for example though changes to drainage and grazing – loss of wet and dry heath, mires, flushes and relatively less improved grasslands; these schemes are grant-aided by public funds and generally claim to promote biodiversity, which may only be partially true. The Forestry Commission (Scotland) is aware of some of these issues and does issue guidance but even where mosaics are retained, implementation of appropriate management is hard to achieve. New native woodlands are in general to be welcomed and – given sympathetic design, recognition of and integration with existing woodland features, and respect for existing open habitats – these efforts can enrich local biodiversity, as well as complement the limited relict woodland resource. Either way the botanical or vegetation benefits of new woodlands will take a number of years to become apparent, even with careful planning.

At the time of writing this flora, there have been a number of wind farms erected and several more proposed within the vice-county, including one of Britain's largest on-shore ones, at Whitelee above Eaglesham; some of these applications have been rejected but not on botanical grounds. Aside from arguments about landscape and avifauna, the impact of wind farms on plant diversity is debatable. As they tend to be sited at remote upland locations the biggest threat is to the hydrology of peatland habitats, for example bogs, mires, flushes and fens, particularly from the complex network of accompanying service roads and additionally from the changes in land management that may subsequently ensue. As many upland sites have been highly modified by past drainage and intensive stock grazing, with cautious siting of turbines and considered infrastructure design the impact would be much reduced and associated conservation measures could provide an opportunity to restore some habitats; the reinstatement of sympathetic management is fundamental to effective mitigation. At present there is no documented loss of a particular local rarity, but equally there appears to be little sympathetic habitat-restoration work being implemented.

One direct consequence of climate change is likely to be further pressure on upland vegetation. Although Renfrewshire has a very small subalpine flora there is little room for movement for several species because of the limited altitude of the vice-county and extensive peaty soils. Given that much of the Renfrewshire Heights are publicly owned uplands, there are opportunities for sympathetic management. However, currently much of the Clyde Muirshiel Regional Park is still largely 'traditionally' managed in the interests of game shooting and upland farming, with consequent burning of heather moorland, locally heavy grazing and drainage. There exists considerable scope for greater positive management, at local sites and in the wider landscape, to benefit the large areas of relatively unimproved pastures and peatlands, and to promote sympathetic heather management and introduce woodland restoration.

A potentially positive response to climate change is the aim to ameliorate the impact of past drainage through flood alleviation schemes, including wetland creations and some woodland planting. However, such water retention initiatives are taking place against a backdrop of the breaching of a number of former dams or reservoirs; other such waterbodies are retained with high water levels, such as Balgray (NS5157), denying the seasonal drawdown zone so important for the Scottish rarity Mudwort and a number of locally rare bryophytes. In addition to the actual loss of open water, the perhaps more widespread hydrological impact is seen through water enrichment and pollution. Scottish Water and the Scottish Environment Protection Agency have worked hard to reduce the impact of human populations, and farming, on water quality, but the enriched water is likely to have been one of the key factors in past floristic changes, notably through the encouragement of alien species.

The issue of alien species – now more usually referred to as 'non-natives' – and in particular invasive non-native species (INNS), is increasingly high on the conservation agenda. Current UK and Scottish Government guidelines on 'non-natives', although widely interpreted as being aimed at control, recognize the difficulty of removal of long established aliens. A strong and welcome emphasis is on prevention and early intervention. Although there can be little argument that a number of species, though not always the more familiar larger ones, are a problem to the quality and diversity of a particular habitat (Rhododendron and native woodland, perhaps, being the most significant locally), the real impact of an alien at a particular location is not always clear cut. There is no evidence of a native species becoming extinct solely due to the arrival of an alien species. Additionally there is very little recognition of the fact that at many locations the substrate or wider habitat has been fundamentally changed by human activity; the presence of aliens is more likely a symptom of this situation and, notably at urban locations, can be seen as integral to the modified local ecology. The floras of all our lowland watercourses, notably the riparian banks, are prime examples. For many well-established aliens removal is unlikely to be achievable and local-scale attempts, in particular chemical spraying, could actually be more damaging to the native flora. Even if the removal of aliens was practical, the return to a particular preconceived 'natural' state is by no means guaranteed. One consequence of these actions is that limited conservation resources are being drained towards well-meant but often damaging and ill-advised control treatments. How much more beneficial to native species would it be if such resources were spent on site protection, sympathetic management and habitat restoration?

Alien introductions and habitat disturbance still occur, and are unlikely to diminish given current activities, attitudes and limited resources for prevention. Fly-tipping, ranging from the dumping of garden waste by individual householders to the actions of commercial operators, still occurs, and larger-scale tipping down riverbanks or at woodland margins is often overlooked or, even when not, is not acted upon sufficiently to bring about full, if any, restoration. A small-scale, but widespread example of disturbance occurs due to footpaths and access issues. Although increasing responsible access to wilder areas is welcomed, footpaths are often directed towards restricted remnants of semi-natural vegetation, with the result of habitat loss along the route; this is compounded by the seemingly accepted practice of spreading excess soil to the sides of footpaths during construction or routine maintenance, which invariably encourages more nutrient-demanding ruderals (natives and aliens) at the expense of the existing, now smothered, original vegetation.

Arguably another source of alien introductions is brought about by the increasing use of wildflower seeds or stock at various habitat-creation or restoration schemes, including larger woodland projects and agri-environment grant-aided work. These invariably contain species not native to the vice-county (although usually to the UK) and even for native species, not of local provenance and thus of differing genetic stock. Again, well-meant actions but ones that could be better employed if they were more sympathetic to local habitats and carried out utilizing local species. Perhaps the most dangerous consequence is that it engenders an 'it's replaceable mentality' in decision makers: such as 'destroy a relict area and then recreate one elsewhere with a packet of seeds – just add water!'. This would rightly never be allowed in relation to native species of birds or mammals, but it is generally seen as a positive and commonplace action for plants.

Knowledge, communication and informed decision-making are essential if further disturbance and damage are to be avoided, and for the limited resources available to conservation to be employed efficiently and effectively. Some losses could be avoided if better communication occurred. A cogent example is the devastation wrought on the last remaining significant patch of Bog-myrtle in the vice-county, at Gall Moss (NS2672), during water pipeline work (see Plate 24); a little consultation could have avoided such extensive, and excessive, damage.

The deliberate digging-up or other removal of locally rare species is unlikely to be a current cause for concern, in contrast with the situation in the nineteenth century, as is reflected in Part 2, particularly for a remarkable number of ferns; however, although the Wildlife and Countryside Act (1981) appears to confer protection from such action, it only really gives protection to landowners' property – preventing removal without permission.

What is the most effective way to communicate to landowners or managers the presence of local rarities on their land? Would such information be welcomed or viewed as a restriction or impediment to otherwise legitimate activities? Although most responsible developers, landowners and managers will seek ecological advice during regulated developments, which excludes a number of land-use changing activities, the information is not always readily to hand, but even where it is, assessments and decisions are often focussed on national, statutory issues not on the intrinsic value of local sites or species. Pulling all the information together into a readily accessible format is a difficult task. Local record centres, in coordination with council-based survey information, and many local naturalist organizations, societies and individuals already possess much of the information needed, but the situation is far from perfect.

Although local authorities have made great progress in increasing their knowledge of sites and species through recent surveys, most lack experienced personnel or general staff resources to adequately appraise the varied activities that affect semi-natural habitats. Staff at Scottish Natural Heritage are stretched just keeping on top of issues relating to national (statutory) sites and priorities, so have little time to respond to local site or species issues. Commercial ecologists' and consultants' efforts are similarly focussed on the presence of 'priority' species, at the expense of the wider local ecology, overlooking the fundamental point of protecting the habitats that these species depend upon. This short-sighted approach results in unsatisfactory appraisals that can be readily interpreted as a green light to developers and planning regulators. Opposition from knowledgeable naturalists and interested members of the local community, although not always readily heard regarding remote areas, can be the last hope of ensuring nature conservation interests are fully assessed.

It is undoubted that there are a number of pressures – from various sources, as summarized above – that will continue to impact on the wild flora of Renfrewshire. However, despite the loss of many populations and quite a few species, there are many that have shown a remarkable resilience in the face of a number of threats. Today there is a much greater interest shown by decision makers and members of the wider public in nature conservation and biodiversity, which can help to ameliorate or even resist the negative impact of land-use changes. A range of legislative measures, at both local and national levels, community consultations, non-governmental organization's guidelines and campaigns by conservation charities and passionate individuals all help to a give a much louder voice to concerns about nature than has been heard in preceding years. The various mechanisms of species and habitat conservation are considered in the following chapter.

4.2 Legislative Background

The more recent surveying for this flora has taken place at a time of increasing public awareness of nature conservation and environmental issues and the recognition by local authorities and central government, and other public bodies, of the growing need to protect the local natural heritage. This interest is manifest in the range of site- and species-protection designations delivered by various legislative means. The existing national system of designated Sites of Special Scientific Interest (SSSIs) and National Nature Reserves (NNRs) has been augmented by recent designations such as Natura 2000 areas. Species protection has been brought in through acts such as the Wildlife and Countryside Act (1981) and more recent revisions (both UK-wide and independently for Scotland).

In the years following the Rio Earth Summit in 1992 and the UK Government's Biodiversity Action Plan (BAP) in 1995, followed by the Scottish Biodiversity Strategy (1996), there has been an even greater drive to protect and promote 'biodiversity' at both local and international levels. Local authorities are delivering local biodiversity action plans (LBAPs) and are expected to exercise their 'biodiversity duty' according to the 'Nature Conservation Act (Scotland) 2004'.

So how is this potential backdrop of legislation, driven by public desire and some political will, manifest in the conservation of the botanical heritage of Renfrewshire?

Species protection

There are various levels of species-conservation designations that have been applied to plants found growing in the vice-county. A recent national list has collated a number of these designations for qualifying species (see table 29 and its legend). There are a total of 125 species which have been recorded from within the vice-county that appear on this list. Of these 61 are thought to be now extinct in the vice-county and of the total 73 are considered definitely alien. Removal of the most obvious neophytes, and all extinct aliens, reduces the list to the 66 species shown in table 29.

There are a number of quite frequent species included in the table, which may cause some surprise; several of these are declining archaeophytes of arable fields included in the Scottish Biodiversity List or the UK Red Data List (Corn Spurrey and Charlock are perhaps the most notable). Some locally frequent native species indicate the importance of local populations at the national scale (Near Threatened), for example Spignel, Greater Butterfly-orchid and to a lesser extent Lesser Tussock-sedge. A few extant but locally rare natives are also present on the Vulnerable criterion, for example Frog Orchid, Coralroot Orchid, Field Gentian, Small Water-pepper, Mossy Saxifrage, Lesser Butterfly-orchid and Dwarf Eelgrass. Annual Knawel is listed as Endangered; it is certainly endangered locally. A few other local natives in the higher categories have not been seen recently, for example Crested Buckler-fern (Critically Endangered) and Heath Cudweed (Endangered).

There are 18 species in table 29 that are already considered extinct in the vice-county. Many of the other species are rare natives or of questionable status, and have already been highlighted in the losses and declines section (see p. 363); these national listings give additional weight to their local conservation status. However, in this section there are also quite a number of locally rare native species listed by habitats which do not appear on any of the national lists. These plants deserve equal conservation measures at the local level, especially as the majority are indicators of relict-quality habitat and are not easy to replace; mitigation measures can seldom adequately compensate for the deterioration of natural quality that further losses of already seriously depleted populations would represent.

Table 29. Species with national conservation designations recorded from the Vice-county of Renfrewshire (natives and archaeophytes). Species marked * are native but considered likely to be extinct; italicized species are not considered native in the vice-county (majority archaeophytes); species marked # are of questionable former status.

Designation Codes:
RL: CR/EN/VU/NT: Red Data List with 2001 IUCN guidelines (Cheffings and Farrell 2005): Critically Endangered, Endangered, Vulnerable and Near Threatened
NR/NS: Nationally Rare and Nationally Scarce species (not based on IUCN criteria)
SBL: Scottish Biodiversity List

No. of 1km squares	Name	Designation	No. of 1km squares	Name	Designation
0	Allseed*	RL: NT	0	Isle of Man Cabbage*	NS
2	Annual Knawel	RL: EN; SBL	1	Ivy-leaved Bellflower*	RL: NT; SBL
1	Bird's-nest Orchid	RL: NT	24	*Large-flowered Hemp-nettle*	RL: VU; SBL
8	*Black Bindweed*	SBL	2	Lesser Butterfly-orchid	RL: VU; SBL
1	*Caraway*	RL: EN; NS; SBL	13	Lesser Tussock-sedge	RL: NT
0	Chaffweed*	RL: NT	0	Lesser Water-plantain*	RL: NT
46	*Charlock*	SBL	1	Long-stalked Orache	NS
0	Common Cudweed*	RL: NT; SBL	0	Marsh Stitchwort*	RL: VU
7	Common Juniper	SBL	5	*Masterwort*	RL: NT; NS
5	Coralroot Orchid	RL: VU; NS	8	Mossy Saxifrage	RL: VU; SBL
1	*Cornflower*	SBL	2	Mudwort	NS
10	*Corn Marigold*	RL: VU	2	Northern Yellow-cress	NS
4	Corn Mint	SBL	0	Oysterplant*	RL: NT
69	*Corn Spurrey*	RL: VU	6	Purple Ramping-fumitory	NS; SBL
3	Cowbane	NS	2	Round-fruited Rush#	RL: NT; SBL
0	Crested Buckler-fern*	RL: CR	0	Sea Holly*	SBL
1	Dewberry#	SBL	0	Shepherd's Cress*	RL: NT; SBL
1	Dwarf Eelgrass	RL: VU; NS	5	Small Water-pepper	RL: VU
0	Eelgrass*	RL: NT	0	Small-white Orchid*	RL: VU
6	Eight-stamened Waterwort	NS	1	*Smooth Rupturewort*	NR
2	Field Gentian	RL: VU	57	Spignel	RL: NT; NS
2	Field Madder	SBL	12	*Sun Spurge*	SBL
1	*Field Pepperwort*	SBL	1	Tall Bog-sedge	NS
0	Flat-sedge*	RL: VU	6	Thread Rush	NS
4	Frog Orchid	RL: VU	0	Tubular Water-dropwort*#	RL: VU; SBL
1	*Good-King-Henry*	RL: VU; SBL	9	Tufted Loosestrife	NS
88	Greater Butterfly-orchid	RL: NT; SBL	2	*Wall Whitlowgrass*	NS
2	*Greater Celandine*	SBL	1	Water x Common Sedge	RL: VU
0	Great Sundew*	RL: NT	3	*White Mustard*	SBL
6	Hairy Stonecrop	RL: NT; NS	6	*Wild Pansy*	RL: NT; SBL
0	Heath Cudweed*	RL: EN; SBL	9	Wilson's Filmy-fern	RL: NT
0	Heath Dog-violet*	RL: NT	0	Wood Bitter-vetch*	RL: NT; NS; SBL
1	Holy-grass	NR; SBL	3	Yellow Bartsia#	SBL

For all of the species listed above, and those mentioned in Part 3.3, further detailed survey work is needed to establish baseline population sizes, and to hopefully find new nearby populations. This will enable a realistic assessment of a species' local rarity, assess its viability and assist in monitoring future changes. It is difficult to see how many of the locally rare and isolated populations, most of which seem very small, are not suffering from inbreeding depression and most will be prone to chance detrimental disturbance events, which over time will seriously bring into question a local population's long-term survival.

Site conservation

During the last 25 years a number of systematic habitat surveys, supplemented by data from earlier surveys, for example notably through Paisley Museum and the forerunner agencies of Scottish Natural Heritage (SNH), plus other existing floristic and fauna records, have provided the raw data for the production of lists of 'Sites of Importance for Nature Conservation' (SINCs), as designated for the four local authority areas represented in the Vice-county of Renfrewshire. All of the land covered by this flora has been 'habitat-mapped' to some degree between 1987 and 1999. Nature conservation policies have been produced by the individual local authorities in recent years, resulting in a much greater knowledge and protection of the natural heritage than was the case 25 years ago; all four local authorities have produced designation maps and associated Local/City Plans which list the sites or areas of local natural heritage interest, supported by various policies. Details about designated local sites are listed in the relevant local authorities' Local Plans and can be seen on their respective web pages; additional location information can be obtained from local habitat survey site reports (mainly maps and target notes), some of which are held by Glasgow Museums' Biological Records Centre; SNH web pages provide information about Local Nature Reserves (LNRs), SSSIs and other national designations.

It should be remembered that a SINC designation only confers limited protection from land development, and has no impact on agricultural activities and can be easily overlooked during 'environmental' developments such as forestry or agricultural grant schemes, as discussed above. All too often an area's botanical (and faunal) interest can be ignored or dismissed due to the lack of national or European protected species or habitats. If the diminishing local natural heritage is to be protected, the SINC system must be given a higher profile at both national and local levels when assessing the inherent value of land.

This situation is exacerbated by the limited, unsystematic coverage of the SSSIs in the vice-county. In total there are 19 SSSIs (including five geological and three ornithological, although most have associated habitat interest); the remainder are wetlands, including quite a few good-quality mires, and two predominantly grassland sites and a couple of woodlands. The vast majority of high-quality, relict semi-natural vegetation receives no recognition at this national level. A system of designations, at least recognized at the local-authority level but supported by national policies, of high-quality relict examples of the main local habitat or vegetation types needs to be developed so that coming generations are able to appreciate some of the fine examples that have survived thus far.

The SINCs are focused chiefly on the protection of the remaining semi-natural habitats such as woodlands, grasslands, heaths, mires and other wetlands, but rarely on the more urban, secondary waste-ground sites. The latter, which may hold high floristic diversity, ranging from waste-ground ephemerals to relicts of former farmland, have received increased attention of late for development and the desire to 'tidy up' marginal land; many gap sites and relict habitat mosaics about Paisley and Erskine, and the Inverclyde coast, have been lost in recent years. Although this flora makes the case for the fundamental need for the protection of the remaining natural heritage resource and its constituent floristic communities, there should also be strong emphasis on the wider role of the flora not just at wildlife 'sites', but throughout the vice-county, on waste ground, spoil heaps, buildings, street corners, roadsides, railways, gardens, parks – in fact, just about everywhere that a bit of 'wild' space is left to flourish.

4.3 Plants and Places

There are too many sites supporting habitats or vegetation of interest to meaningfully list here; information about designated sites can be found at sources noted on p. 380. However, it would be remiss not to attempt to bring together a few places mentioned in the text and some of the more notable species that have been highlighted.

The following table, therefore, presents a selection of sites or places that feature particular botanical hotspots chosen for their floristic highlights and not necessarily overall diversity or vegetation quality, although these are usually correlated. There is no space to itemize all the brightly coloured species-rich grasslands, the hidden colourful spring floras seen at many valley woodlands, large and small, the still diverse, but cramped remnants of coastal vegetation, the scattered, secluded mires packed with sedges and bryophytes and the extensive peaty moorlands and subalpine flora of the remoter uplands, or even the locally distinctive Spignel banks and Whorled Caraway flushes. Also missing are the dynamic, varied and usually colourful waste ground, riverbank and wayside floras, supporting native and aliens in complex mosaics.

Some of the sites listed in table 30 are quite small but others represent large areas comprising a few amalgamated sites. This is not intended as a comprehensive list, or indeed to indicate the 'best' sites, but it does include many places worthy of special conservation and certainly well worth a visit to explore the botanical features. Many are already designated sites (SSSIs, LNRs, SINCs or Wildlife Corridors), or include parts of such, but some receive no protection or recognition.

Table 30. Renfrewshire sites with particular floristic interest.

Place name	Grid Reference (approximate)	Key species * indicates extinct; possible aliens are italicized
Kelly Glen and Reservoir	NS2168	Aspen, Black Spleenwort, Floating Bur-reed, Grass-of-Parnassus, Wilson's Filmy-fern, Wood Small-reed
Daff Reservoir/Glenshilloch	NS2269	Bog-myrtle, Broad-leaved Cottongrass, Grass-of-Parnassus, Knotted Pearlwort, Stag's-horn Clubmoss, Yellow Saxifrage
Everton/Leapmoor	NS2170	Grass-of-Parnassus, Hay-scented Buckler-fern, Long-stalked Yellow-sedge
Brownhill Moss/Crawhin Reservoir	NS2369	Few-flowered Sedge, Floating Bur-reed
Lunderston Bay/Cloch	NS2074	Bulbous Buttercup, *Elecampane*, Hairy Oat-grass, Hemp Agrimony, Long-bracted Sedge*, Ray's Knotgrass, Sand Sedge, Soft Shield-fern
Burneven/Earn hills	NS2275	Globeflower, Grass-of-Parnassus, Heath Pearlwort, Stone Bramble
Shielhill Glen	NS2372	Early-purple Orchid, Giant Horsetail, Grass-of-Parnassus, Heath Fragrant-orchid, Quaking-grass
Loch Thom/Gryfe reservoirs	NS2672	Bog-myrtle, Hairy Stonecrop, Thread Rush
North Rotten Burn	NS2568	Aspen, Green Spleenwort, Hairy Stonecrop, Mossy Saxifrage, Northern Bedstraw, Starry Saxifrage, Wood Crane's-bill
Queenside Muir/Hill of Stake	NS2863	Alpine Clubmoss, Aspen, Cloudberry, Lesser Twayblade, Stag's-horn Clubmoss, Stiff Sedge

Place name	Grid Reference (approximate)	Key species * indicates extinct; possible aliens are italicized
Raith Burn	NS2962	Common Juniper, Few-flowered Sedge, Mossy Saxifrage, Parsley Fern, Smooth-stalked Sedge
Mistylaw/Maich Water	NS3061	Alpine Clubmoss, Aspen, Hairy Stonecrop
Calder Glen	NS3460	Early-purple Orchid, Globeflower, Mountain Everlasting, Wilson's Filmy-fern
Castle Semple Loch	NS3658	Bird's-nest Orchid, Bladder-sedge, Cowbane, Eight-stamened Waterwort, Northern Yellow-cress, Tea-leaved Willow, Tufted Loosestrife
Roebank Burn, Knowes	NS3756	Early-purple Orchid, Lesser Tussock-sedge, Skullcap
Barcraigs Reservoir	NS3857	Eight-stamened Waterwort, Mudwort, Small Water-pepper, Thread Rush
Knocknairshill/Devol Glen	NS3074	Common Cow-wheat, Green Spleenwort, Hairy Stonecrop, Lesser Butterfly-orchid
Cauldside/Strathgryfe	NS3270	Burnet-saxifrage, Common Milkwort, Hairy Oat-grass, Heath Fragrant-orchid
Craigmarloch	NS3472	Burnet-saxifrage, Floating Club-rush, Heath Fragrant-orchid, Lesser Tussock-sedge, Small-white Orchid*
Knockmountain	NS3671	Common Wintergreen, Lesser Tussock-sedge, Squirreltail Fescue
Dargavel Burn	NS3771	Bog-sedge, Early Marsh-orchid, Greater Tussock-sedge, Lesser Butterfly-orchid*, Lesser Tussock-sedge, Long-stalked Yellow-sedge
Barscube Hill	NS3871	Adder's-tongue, Common Centaury, Moonwort, Nodding Bur-marigold, Tufted Loosestrife
Drum Estate	NS3971	Common Water-crowfoot, Floating Club-rush, Least Bur-reed
Gryfe Water, Kilmacolm	NS3569	Bulbous Buttercup, Burnet-saxifrage, Globeflower, Stream Water-crowfoot
Glen Moss	NS3669	Bladderworts, Bog-sedge, Coralroot Orchid, Least Bur-reed, Lesser Tussock-sedge, Nodding Bur-marigold, Tufted Loosestrife
Lawfield Dam	NS3769	Bladder-sedge, Lesser Tussock-sedge, Long-stalked Yellow-sedge, Tufted Loosestrife
Shovelboard	NS3869	Bog-sedge, Coralroot Orchid, Dark-leaved Willow, Lesser Pond-sedge
Barmufflock	NS3664	Bog-sedge, Creeping Willow, Lesser Tussock-sedge, *Melancholy Thistle*
Whinnerston/Locher Water	NS3864	Adder's-tongue, Globeflower, Heath Fragrant-orchid, Stream Water-crowfoot
Gryfe Water, Bridge of Weir	NS3965	Barberry, Goldilocks Buttercup, Northern Bedstraw, Wood Crane's-bill
Formakin/Park Glen	NS4070	Black Spleenwort, Prickly Sedge, Stream Water-crowfoot, Yellow Oat-grass
Corsliehill/Barochan	NS4069	Annual Knawel, Coralroot Orchid, Smooth-stalked Sedge
Bardrain Glen	NS4360	Chickweed-wintergreen, Common Water-crowfoot, Common Wintergreen, Herb-Paris, Serrated Wintergreen*, Shady Horsetail*
Walls Hill	NS4158	Bog-sedge, Cowberry, Fen Bedstraw, Greater Tussock-sedge, Lesser Tussock-sedge, Slender Sedge, White Water-lily
Caplaw Dam	NS4358	Cowbane, Greater Spearwort, Many-stalked Spike-rush
Old Patrick Water, Plymuir	NS4357	Fen Bedstraw, Greater Tussock-sedge, Slender Sedge
Loch Libo	NS4355	Cowbane, Crested Buckler-fern*, Fool's-water-cress, Greater Tussock-sedge, Lesser Pond-sedge, Six-stamened Waterwort

Place name	Grid Reference (approximate)	Key species * indicates extinct; possible aliens are italicized
Sergeantlaw Moss	NS4459	Coralroot Orchid, *Royal Fern*, Tea-leaved Willow
Witch Burn	NS4658	Dioecious Sedge, Lesser Pond-sedge, Mountain Everlasting
Longhaugh/Erskine Park	NS4473	Beaked Tasselweed, Dwarf Eelgrass, Field Mouse-ear, oraches, Parsley Water-dropwort, Sea Pearlwort
Erskine Harbour	NS4672	Bird's-foot, Common Stork's-bill, dandelions (Section *Erythrosperma*), Little Mouse-ear, Sand Sedge, Squirreltail Fescue
River Cart, Inchinnan area	NS4968	English Scurvygrass, Holy-grass, Purple Loosestrife, Slender Tufted-sedge
Paisley Moss	NS4665	Adder's-tongue, Bladder-sedge, Slender Tufted-sedge
Gleniffer Braes	NS4660	Alternate-leaved Golden-saxifrage, Moonwort, Spignel
Harelaw Reservoir	NS4859	Globeflower
Balgray Reservoir	NS5157	Mudwort, Needle Spike-rush, Trifid Bur-marigold, Various-leaved Pondweed, *Yellow Bartsia*
Commore Dam/Carsewell	NS4654	Brown Sedge, Common Milkwort, Field Gentian, Frog Orchid, Lesser Pond-sedge, Mountain Everlasting
Snypes/Muirhead	NS4859	Burnet-saxifrage, Field Gentian, Frog Orchid, Heath Groundsel, Mountain Everlasting, Spignel
Black Hill Moss	NS4851	Tall Bog-sedge
Little/Brother lochs	NS5052	Bog-sedge, Common Club-rush, Creeping Willow, Lesser Tussock-sedge, Needle Spike-rush, Small Water-pepper, Various-leaved Pondweed, White Water-lily
Dunwan Burn	NS5749	Common Butterwort, Fen Bedstraw, Hairy Stonecrop*, Tea-leaved Willow
Picketlaw	NS5651	Bladder-sedge, Dark-leaved Willow, Knotted Pearlwort, Red Pondweed
Ardoch/Carrot	NS5848	Alpine Clubmoss, Cowberry, Early-purple Orchid, Smooth-stalked Sedge, Stag's-horn Clubmoss
Earn Water	NS5554	Bulbous Buttercup, Burnet-saxifrage, Common Milkwort, Hairy Oat-grass, Hybrid Fescue, Lesser Pond-sedge, Maidenhair Spleenworts
White Cart Water, Busby–Eaglesham	NS5755	Black Spleenwort, Bulbous Buttercup, Burnet-saxifrage, Early-purple Orchid
White Cart Water, Glasgow	NS5859	Alternate-leaved Golden-saxifrage, Meadow Saxifrage, Toothwort

Afterword

The information within this flora provides a useful baseline for further, more directed surveying and also for more detailed research studies, and it would be a pleasing outcome if it helped inform decision makers about issues related to land management, built developments, environmental improvements and the general conservation of the natural heritage. The many interesting species recorded in this flora deserve to be given full consideration during such deliberations, so that their value as wild plants in their natural setting can be fully assessed and appreciated, and the species/plants secured for future generations. It would also be a satisfying legacy if the book stimulated interested residents and local naturalists to take a closer look at the flora of their local patch, and take a wider interest in the conservation of species and habitats.

Without the passion and enthusiasm of informed local people, supported by good scientific rigour, botanical conservation would struggle to achieve the attention that it merits. It is interesting to read Ferguson's concluding remarks to his work (1915) where he laments the alienation of the amateur worker and the decline in botanical fieldwork by the 'man of science': 'That he [the 'man of science'] will succeed in again interesting the amateur botanist in his work and its results must be the wish of all who recognise the value of going direct to Nature for the solution of Nature's mysteries.' Although, with a few exceptions, the amateur's efforts subsequently remained rare, since the late 1950s enthusiastic and knowledgeable field botanists have managed to revive an interest and passion for the botany of the vice-county. The efforts of the last 50 years or so have allowed a more accurate, but ever changing, picture of the flora to be realized. The many records and interesting finds contained within these pages are intended to provide a flavour and reliable account of the diversity and richness of the flora that can still today be found in the old County of Renfrewshire.

It is perhaps fitting to leave the final sentiment to one of the more eloquent and passionate admirers of the local botanical heritage, John Wood, who in 1893 wrote, 'But through it all I claim to have proved that so far as Renfrewshire is concerned every nook and corner of it is worth a more thorough search from botanists than it has yet received.' It is to be hoped that this publication has demonstrated the rewards of such an endeavour, but it is not a closed book: there is still much to be discovered, studied and enjoyed.

Gazetteer of Renfrewshire Place Names

The following list provides 1km-square grid references for many places mentioned in the text; most modern records referred to in the Catalogue of Species (Part 2) include a grid reference, but older records may not. The grid squares shown for many larger sites or linear features relate to a central or key location; most of the larger urban centres and watercourses are shown in figure 1 (p. 4).

Ardgowan Estate	NS2073
Ardoch Burn	NS5849
Auchenbothie	NS3471
Auchendores Reservoir	NS3572
Auldouse Burn, Glasgow	NS5560
Aurs Burn	NS5058
Balgray Reservoir	NS5157
Ballageich Hill	NS5350
Barcraigs Reservoir	NS3957
Bardrain Glen	NS4360
Barhill	NS4163
Barmufflock	NS3664
Barnbrock (CMRP)	NS3563
Barochan Estate	NS4069
Barochan Moss	NS4268
Barrhead	NS4959
Barr Loch	NS3557
Barscube Hill	NS3871
Bennan Loch	NS5250
Bent	NS4359
Bishopton	NS4371
Blackbyres	NS5060
Black Hill	NS4851
Black Loch	NS4951
Blackstone Mains	NS4666
Blackstoun	NS4566
Blythswood Estate	NS5068
Boden Boo	NS4571
Bottombow	NS4671
Boylestone Quarry	NS4959
Braidbar	NS5659
Bridge of Weir	NS3865
Broadfield Hill	NS4059
Brother Loch	NS5052
Brown Hill moss	NS2369
Brownside Braes	NS4860
Burnbank Water	NS3068
Burneven	NS2175
Busby	NS5756
Calder dam (old)	NS2965
Calder Glen	NS3460
Caldwell Estate	NS4154
Caplaw Dam	NS4358
Carrot Burn	NS5747
Carruth Bridge	NS3565
Carswell	NS4653
Castle Semple Loch	NS3659
Cathcart	NS5860
Cauldside	NS3270
Chlochodrick	NS3761
Cloch	NS2075
Clydeport, Port Glasgow	NS3074
Commore Dam	NS4654
Corkindale Law	NS4456
Corlick Hill	NS2972
Corse Hill	NS5946
Corsehouse Reservoir	NS4850
Coves Reservoir	NS2476
Cowdon Burn	NS4555
Cowdon Hall	NS4657
Craighall Dam	NS4755
Craigmarloch Wood	NS3471
Craig Minnan	NS3264
Crawhin Reservoir	NS2470
Creuch Hill	NS2668
Crookston Castle	NS5262
Crossmyloof	NS5762
Daff Burn	NS2171
Daff Reservoir	NS2270
Dargavel Burn SSSI	NS3771
Dargavel House	NS4369
Darndaff Moor	NS2673
Darnley [Waulkmill] Glen	NS5258
Denniston	NS3667
Devol Glen	NS3174
Dickman's Glen	NS5847
Dod Hill	NS4953
Drumduff Hill	NS5846
Drum Estate	NS3971
Duchal Estate	NS3568
Duchal Moor	NS2767
Duncarnock Hill	NS5055
Dunrod Hill	NS2472
Dunwan Dam	NS5549
Eaglesham	NS5751
Earn Hill	NS2375
East Girt Hill	NS2862
Eastwood	NS5558
Elderslie	NS4462
Erskine	NS4571
Erskine Harbour	NS4672
Erskine Park	NS4572
Everton valley	NS2170
Fereneze Hills	NS4859
Ferguslie Park	NS4664
Ferryhill Plantation	NS4072
Finlaystone Estate	NS3673
Flow Moss	NS5547
Forkings of Raith	NS2862
Formakin Estate	NS4171
Foxbar	NS4561
Gall Moss	NS2672
Garvock forest	NS2771
Gavin Braes	NS3859
Georgetown	NS4367
Giffnock	NS5658
Glanderston Dam	NS4956
Glasgow Airport	NS4766
Gleddoch Estate	NS3872
Gleniffer Braes	NS4460
Glen Moss	NS3769
Glentyan House	NS3963
GMRC, Nitshill	NS5160
Gourock	NS2477
Greenock	NS2875
Greenock Cut	NS2474
Greenside	NS3661
Gryfe Reservoirs	NS2771
Hannah Law	NS3061
Hardridge Hill	NS3167
Harelaw Dam	NS4753
Harelaw Reservoir	NS3173
Harelaw Reservoir (Barrhead)	NS4859
Harestane Burn	NS2568
Hartfield Moss	NS4157
Hawkhead	NS5062

Hill of Stake	NS2763
Houston	NS4066
Houston Burn	NS3867
Houston Estate	NS4167
Houstonhead Dam	NS3966
Howcraigs Hill	NS4555
Howwood	NS3960
Hyndal Hill	NS2866
Inchinnan	NS4769
Inverkip	NS2071
Jenny's Well	NS4962
Jock's Craig	NS3269
Johnstone	NS4363
Jordanhill	NS5468
Kaim Dam	NS3462
Kelly Cut	NS2371
Kelly Reservoir	NS2268
Kelly Water/Glen	NS2068
Kilbarchan	NS4063
Killoch Hill	NS4758
Kilmacolm	NS3569
Kip Water	NS2273
Kirktonmill Dam	NS4856
Knapps Loch	NS3668
Knockindon Burn	NS4659
Knockmountain	NS3671
Knocknairs Hill	NS3074
Knowes	NS3756
Langbank	NS3773
Langside Wood	NS5761
Larkfield	NS2475
Lawfield Dam	NS3769
Leapmoor Forest	NS2270
Leperstone Reservoir	NS3571
Levan Burn	NS2176
Linn Park	NS5858
Linwood Moss	NS4465
Little Loch	NS5052
Lochcraig Reservoir	NS5351
Locher Water	NS3763
Lochgoin Reservoir	NS5347
Loch Libo	NS4355
Lochliboside Hills	NS4557
Lochside	NS3658
Loch Thom	NS2572
Lochwinnoch	NS3558
Long Loch	NS4752
Lunderston Bay	NS2073
Lurg Moor	NS2973
Maich Water	NS3258
Malletsheugh	NS5255
Marshall Moor	NS3762
Mathernock	NS3271
Mearns	NS5455
Melowther Hill	NS5648
Midton Loch	NS4158
Mill Burn	NS3266
Misty Law	NS2961
Mosspark	NS5463
Moyne Moor	NS4752
Muirshiel	NS3163
Neilston	NS4757
Neilston Pad	NS4755
Newlands	NS5760
Newshot Island	NS4870
Newton Mearns	NS5355
North Rotten Burn	NS2667
Old Partrick Water	NS4461
Paisley Abbey	NS4863
Paisley Moss	NS4665
Park Glen	NS3970
Picketlaw	NS5650
Pilmuir Reservoir	NS5154
Pollok Park	NS5461
Pollokshaws	NS5561
Polnoon Water	NS5851
Port Glasgow	NS3274
Queenside Hill	NS2963
Queenside Loch	NS2964
Raith Burn	NS3063
Ranfurly Golf Course	NS3664
Ravenscraig Wood	NS2575
Renfrew	NS5067
Roebank Burn	NS3455
Rouken Glen	NS5458
Scart Wood	NS3867
Scotstoun	NS5268
Shawlands	NS5661
Shielhill Glen	NS2372
Shovelboard mire	NS3869
Snypes Dam	NS4855
Spango	NS2374
St Brydes	NS3860
Stanely Reservoir	NS4661
Strathgryfe	NS3370
Thornleybank Hill forest	NS3363
Thornley Dam	NS4860
Thornliebank	NS5459
Uplawmoor	NS4355
Walls Hill/Loch	NS4158
Walton Dam	NS4955
Waulkmill Glen (Darnley)	NS5258
Waulkmill Reservoir	NS5257
Wemyss Bay	NS1968
Whinnerston	NS3864
Whitelee Windfarm	NS5348
White Loch	NS4852
Whitemoss Dam	NS4171
Whittliemuir Moss	NS4259
Williamwood	NS5658
Woodhall	NS3473

Bibliography

Anon. (1863) List of localities of ferns around Glasgow and its watering places, *Manuscript Magazine of the Glasgow Naturalists Society*, 4, 158.

Anon. (1869) Rambles with the Paisley Botanists, *Paisley & Renfrewshire Gazette*, 13 February 1869.

Anon. (1869a) Session 1859–60 specimens exhibited, *Proceedings of the Natural History Society of Glasgow*, 1, 6.

Anon. (1945) Botanical observations, 1942, *Transactions of the Paisley Naturalists' Society*, 5, 8.

Anon. (1981) *The Climate of Great Britain: Climatological Memorandum 124 Glasgow and the Clyde Valley*, Edinburgh, Meteorological Office.

Anon. (1989) *The Climate of Scotland: Some Facts and Figures*, HMSO.

Baker, JG and Newbold, WW (1883) *Topographical Botany*, 2nd edn, Kew.

Bennett, A (1902) Records for Scottish plants for 1901, additional to Watson's 'Topographical Botany' 2nd edition (1883), *The Annals of Scottish Natural History*, 41, 32, Edinburgh.

Blackwood, J (1976) *Pseudorchis albida*, *Glasgow Naturalist*, 19, pt 4, 344–345.

Bos, JAA, Dickson, JH, Coope, GR and Jardine, WG (2004) Flora, fauna and climate of Scotland during the Weichselian Middle Pleniglacial – palynological, macrofossil and coleopteran investigations, *Palaeogeography, Palaeoclimatology, Palaeoecology*, 204, 1–2, 65–100.

Bown, CJ, Shipley, BM and Bibby, JS (1982) *Soil and Land Capability for Agriculture, South-west Scotland*, Soil Survey of Scotland, Aberdeen, Macaulay Institute for Soil Research.

Boyd, DA (1887) Notes on some of the plants of the Clyde District, *Proceedings and Transactions of the Natural History Society of Glasgow*, 1 ns, 151–6.

Boyd, J (1908) The last of the Pollok Wych Elms, *Annals of the Andersonian Naturalists' Society*, 3, 127.

Boyd, WE (1986) Vegetation history at Linwood Moss, Renfrewshire, Central Scotland, *Journal of Biogeography*, 13, 207–223.

British Geological Survey (1985) *British Regional Geology: Midland Valley of Scotland*, British Geological Survey, Keyworth.

British Geological Survey (1989) *Greenock. Scotland Sheet S030W and parts of S029E. Drift Geology, 1:50,000*, British Geological Survey, Keyworth.

British Geological Survey (1990) *Greenock. Scotland Sheet S030W and parts of S029E. Solid Geology, 1:50,000*, British Geological Survey, Keyworth.

British Geological Survey (1993) *Glasgow. Scotland Sheet S030E. Solid Geology, 1:50,000*, British Geological Survey, Keyworth.

British Geological Survey (1994) *Glasgow. Scotland Sheet S030E. Drift Geology, 1:50,000*, British Geological Survey, Keyworth.

Cheffings, C and Farrell, L (eds), with Dines, TD, Jones, RA, Leach, SJ, McKean, DR, Pearman, DA, Preston, CD, Rumsey, FJ, Taylor, I (2005) *The Vascular Plant Red Data List for Great Britain*, Species Status 7: 1–116, Peterborough, Joint Nature Conservation Committee.

Clement, EJ and Foster, MC (1994) *Alien Plants of the British Isles*, London, BSBI.

Conacher, ERT (1959) Some recent records of flowering plants from the Clyde area, *Glasgow Naturalist*, 18, pt 2, 82.

(1966) Annual Mercury (*Mercurialis annua*) in Renfrewshire, *Glasgow Naturalist* 18, pt 8, 451.

(1974) *Ceterach officinarum*, *Glasgow Naturalist*, 19, pt 2, 134.

Crawford, G and Robertson, G (1818) *General Description of the Shire of Renfrewshire*, Paisley, J Neilson.

Dickson, C (1996) Food, medicinal and other plants from the 15th century drains of Paisley Abbey, Scotland, *Vegetation History and Archaeobotany*, 5, 25–31.

Dickson, C and Dickson, JH (2000) *Plants and People in Ancient Scotland*, Stroud, Tempus.

Dickson, JH, Macpherson, P and Watson, K (2000) *The Changing Flora of Glasgow*, Edinburgh University Press.

Dudman, AA and Richards, AJ (1997) *Dandelions of Great Britain and Ireland*, London, BSBI.

Edees, ES and Newton, A (1988) *Brambles of the British Isles*, London, The Ray Society.

Elliot, GFS, Laurie, M and Murdoch, JB (eds) (1901) *Fauna, Flora and Geology of the Clyde Area*, Glasgow, Local Committee for the Meeting of the British Association.

Ewing, P (1890) A contribution to the topographical botany of the West of Scotland, *Proceedings and Transactions of the Natural History Society of Glasgow*, 2, ns, 309–321.

(1892) Second contribution to the topographical botany of the West of Scotland, *Proceedings and Transactions of the Natural History Society of Glasgow*, 3, ns, 159–160.

(1892) Third contribution to the topographical botany of the West of Scotland, *Proceedings and Transactions of the Natural History Society of Glasgow*, 3, ns, 161–165.

(1897) Contribution to the topographical botany of the West of Scotland, *Proceedings and Transactions of the Natural History Society of Glasgow*, 4, ns, 199–214.

(1899) *The Glasgow Catalogue of Native and Established Plants*, 2nd edn.

(1901) Phanerogams, in Elliot et al. 1901.

Forestry Commission (2001) *Consensus of Woodlands*, Edinburgh, Forestry Commission.

Ferguson, D (1911) Notes on alien plants found near Paisley, *Glasgow Naturalist*, 3, 28–31.

(1915) Renfrewshire plants, *Transactions of the Paisley Naturalists' Society*, 2, 1–11 [the introduction to the list of plants – see TPNS 1915].

Gilbert, OL (1989) *The Ecology of Urban Habitats*, London, Chapman and Hall.

Grierson, R (1930) Clyde Casuals 1916–1928, *Glasgow Naturalist*, 9, 1931, 5–51.

Groome, FH (ed.) (1882–5) *Ordnance Gazetteer of Scotland: A Survey of Scottish Topography, Statistical, Biographical and Historical*, Edinburgh, TC Jack.

Hall, CA (1915) Introductory notes to Renfrewshire plants, *Transactions of the Paisley Naturalists' Society*, 2, v–xvi.

Hall, IHS (1998) *Glasgow district. Memoir for sheet S30E*, British Geological Survey, Keyworth.

Hennedy, R (1865) *The Clydesdale Flora: A Description of the Flowering Plants and Ferns of the Clyde District*, Glasgow, D Robertson.

(1891) *The Clydesdale Flora: A Description of the Flowering Plants and Ferns of the Clyde District*, 5th edn, revised by T King, Glasgow, H Hopkins.

Henry, T (*c.* 1869) *The Flora of Renfrewshire (Paisley & Neighbourhood)*, Paisley Philosophical Institution (Botanical Section).

Hooker, WJ (1821) *Flora Scotica*, London, Archibald.

Hopkirk, T (1813) *Flora Glottiana: A Catalogue of the Indigenous Plants on the Banks of the River Clyde, and in Neighbourhood of the City of Glasgow*, Glasgow, J Smith & Son.

Houston, RS (ed.) (1934) *Silene* x *hampeana* 1913–14 report, *Transactions of the Paisley Naturalists' Society*, 3, 1914–32, 5.

Jones, G (1980) The herbarium of the Glasgow Museum and Art Gallery, *Glasgow Naturalist*, 20, pt 1, 51.

Kent, DH and Allen, DE (1984) *British and Irish Herbaria*, London, BSBI.

King, T (1892) Proceedings: 39th AGM – 28th October 1890, *Proceedings and Transactions of the Natural History Society of Glasgow*, 3, ns, 1889–92, lx.

(1895) Proceedings of the Society: summer session 7th August 1894, *Proceedings and Transactions of the Natural History Society of Glasgow*, 4, ns, 1892–6, 283.

Lee, JR (1933) *The Flora of the Clyde Area*, Glasgow, J Smith & Son.

(1953) Additions to *The Flora of the Clyde Area*, *Glasgow Naturalist*, 17, 65–82.

Lennie, J (1967) *Sanguisorba canadensis* at Castle Semple Loch, *Glasgow Naturalist*, 18, pt 9, 522.

Lloyd, B (1964) The herbarium of the Royal College of Science & Technology, Glasgow, *Glasgow Naturalist*, 18, pt 7, 363.

McCosh, DJ & Rich, TCG (2011) *Atlas of British and Irish Hawkweeds (Pilosella Hill and Hieracium L.)*, London, Botanical Society of the British Isles.

MacDonald, H (1854) *Rambles Round Glasgow*, Glasgow, J Hedderwick.

(1856) *Rambles Round Glasgow*, 2nd edn, Glasgow, T Murray and sons.

McKay, R (1876) Vegetabilia, in *Notes on the Fauna and Flora of the West of Scotland*, 54–84, Glasgow, Blackie & Son.

McKay, R and Horn, G (1873) Botanical Report, *Reports and Transactions of the Glasgow Society of Field Naturalists*, 1, 1872–3, 13.

McKim, J (ed.) (1942) 1937–1938 meeting reports, *Transactions of the Paisley Naturalists' Society*, 4, 1933–42, 55.

Mackechnie, R (1958) *Carex paupercula* Michx. in Renfrewshire, *Glasgow Naturalist*, 18, pt 1, 28.

(1961) Noteworthy plants, 1959, *Glasgow Naturalist*, 18, pt 3, 147.

(1964) New localities of Renfrewshire plants, *Glasgow Naturalist*, 18, pt 7, 384.

(1966) *Hieracium chloranthum* Pugsley in Renfrewshire, *Glasgow Naturalist*, 18, pt 8, 453.

(1971) *Polystichum setiferum* in the Clyde area – compiler's note, *Glasgow Naturalist*, 18, pt 10, 581.

(1974) *Epilobium adenocaulon*, *Glasgow Naturalist*, 18, pt 2, 136.

(1975) *Lathraea squamaria*, *Glasgow Naturalist*, 19, pt 3, 204.

Macpherson, AC and Macpherson, P (1978) Thyme Broomrape, *Glasgow Naturalist*, 19, pt 3, 427.

Macpherson, P and Lindsey, ELS (1993) Cotoneasters update, *Glasgow Naturalist*, 22, pt 3, 239–42.

(1996) Cotoneasters continued, *Glasgow Naturalist*, 23, pt 1, 11–13.

Macpherson, P and Watson, K (1996) Strathclyde University herbarium – computerised database, *Glasgow Naturalist*, 23, pt 1, 7–8.

Macpherson, P, Dickson, JH, Ellis, RG, Kent, DH and Stace, C (1996) Plant status and nomenclature, *BSBI News*, 72, 13–16.

Manuscript Magazine of the Glasgow Naturalists' Society, IV (1863).

Mitchell, BD and Jarvis, RA (1956) *The Soils of the Country around Kilmarnock (Memoirs of the Soil Survey of Great Britain)*, Scotland, Edinburgh, HMSO.

Mitchell, J (1976) *Limosella aquatica*, *Glasgow Naturalist*, 19, pt 4, 341,

Montgomery, J (1834) LVIII Renfrewshire: a catalogue of plants observed in Renfrewshire (communicated by Mr Joseph Hooker), in Watson 1837, 417.

Moore, JN (2000) *The Early Cartography of Renfrewshire to 1864*, https://dspace.gla.ac.uk/bitstream/1905/55/1/carto.html.

Murray, J (1898) *Kilmacolm: A Parish History*, Paisley, Alexander Gardner.

The New Statistical Account of Scotland, 15 vols, Edinburgh, William Blackwood & Sons, 1845.

Newton, A and Randall, RD (2005) *Atlas of British and Irish Brambles*, BSBI Publications.

Paterson, J (1893) Records of excursions in Renfrewshire with additional matter, *Annals of the Andersonian Naturalists' Society*, 1, 8–45.

Patterson, IB (1990) *Greenock district. Memoir for sheet S30W and part S29E*, British Geological Survey, Keyworth.

Patton, D (1954) The British Herbarium of the Botanical Department of Glasgow University, *Glasgow Naturalist*, 17, pt 3.

Preston, CD, Pearman, DA and Dines, TD (2002) *New Atlas of the British and Irish Flora*, Oxford University Press.

Ramsay, S (1996) Human impact on the vegetation around Walls Hill in Prehistoric Renfrewshire, in *Renfrewshire Local History Forum*, ed. D Alexander, 59–63.

Ramsay, S and Dickson, JH (1997) Vegetational history of Central Scotland, *Botanical Journal of Scotland*, 49, pt 2, 141–50.

Ribbons, BW (1964) Salad Burnet (*Poterium sanguisorba*) in Renfrewshire, *Glasgow Naturalist*, 18, 381.

(1971) Moonwort in Renfrewshire, *Glasgow Naturalist*, 18, pt 10, 581.

(1976) *Ledum* in Britain, *Glasgow Naturalist*, 19, pt 4, 219.

Rodwell, JS (1991–2000), *British Plant Communities*, Vols 1–5, Cambridge University Press.

Rogers, WM (1902) Some Clydesdale and SW Ayrshire plants, *Journal of Botany*, 40, 54–59.

Ross, Rev. W (1883) *Busby and its Neighbourhood*, botany appendix 121.

Scott, T (1887) Notes on *Silene maritima* Linn., *Transactions and Proceedings of the Natural History Society of Glasgow*, 1 ns, 1883–6.

Sell, PD (1986) The genus *Cicerbita* in the British Isles, *Watsonia*, 16, pt 2, 121–9.

Shanks, A (1915) The occurrence of *Claytonia sibirica* L. in the Clyde area, *Glasgow Naturalist*, 7, 101.

Smith, D (1915) Session 1902–1903: notes on the work of the society, *Transactions of the Paisley Naturalists' Society*, 2, 113.

Somerville, A (1907) Proceedings of the society: session 1902–1903, *Transactions of the Natural History Society of Glasgow (including the Proceedings of the Society)*, 7, ns, 1902–5, 104.

Stace, CA (1997) *New Flora of the British Isles*, 2nd edn, Cambridge University Press.

(2010) *New Flora of the British Isles*, 3rd edn, Cambridge University Press.

The Statistical Account of Scotland (1791–1799), ed. Sir John Sinclair, 20 vols, Wakefield, EP Publishing, 1973–83.

Stewart, AM (1930) AGM October 1944: chronicles of the Stickleback Club, *Transactions of the Paisley Naturalists' Society*, 5, 1942–5, 30.

Stirling, A McG (1980) Adventive plants in Paisley, *Glasgow Naturalist*, 20, pt 1, 88.

Tait, TN (2000) The history, habitat and present status of holy grass (*Hierochloe odorata* (L.) P. Beauv.) at Blythswood, Renfrew District, VC 76, *Glasgow Naturalist*, 23, pt 5, 17–20.

Thomson, WH (1944) AGM October 1944: some notes on Bardrain Glen, *Transactions of the Paisley Naturalists' Society*, 5, 1942–5, 30.

TPNS (1915) Renfrewshire plants, *Transactions of the Paisley Naturalists' Society*, 2, 12–40 [the introduction to this list was written by Daniel Ferguson – see Ferguson 1915].

Trail, JWH (1898) Topographical Botany of Scotland, *The Annals of Scottish Natural History*, 25, 46.

(1902) Scottish Rubi, *The Annals of Scottish Natural History*, 43, 175.

Watson, HC (1837) *The New Botanist's Guide to the Localities of the Rarer Plants of Britain. Volume II: Scotland and Adjacent Isles*, London, Longman.

(1873) *Topographical Botany*, London, Thames Ditton.

Weddle, R (2008) Morris Young's 'Flora of Renfrewshire' (vc76), *Glasgow Naturalist*, 25, pt 1, 29.

Wilkie, R (1894) Proceedings of the Society: Summmer Session 23rd May 1893, *Proceedings and Transactions of the Natural History Society of Glasgow*, 4, n.s., 146.

Wilson, J (1812) *General View of the Agriculture of Renfrewshire*, Bristol, Cedric Chivers, 1992 for Renfrew District Council.

Wishart, RS (1912) Proceedings: exhibited *Apera spica-venti* 29/10/1912, *Glasgow Naturalist*, 5, 1913, 39.

Wood, J (1893) Rarer flowers of East Renfrewshire, *Annals of the Andersonian Naturalists' Society*, 1, 46–54.

Index